HUMAN EVOLUTIONARY GENETICS

HUMAN EVOLUTIONARY GENETICS

Origins, Peoples & Disease

Mark A. Jobling

Department of Genetics, University of Leicester, Leicester, UK

Matthew Hurles

McDonald Institute for Archaeological Research, University of Cambridge, Cambridge, UK
Current address: The Wellcome Trust Sanger Institute, Wellcome Trust Genome Campus, Cambridge, UK

Chris Tyler-Smith

Department of Biochemisry, University of Oxford, Oxford, UK
Current address: The Wellcome Trust Sanger Institute, Wellcome Trust Genome Campus, Cambridge, UK

GS Garland Science
Taylor & Francis Group

Vice President: Denise Schanck
Managing Editor: Nigel Farrar
Editorial Assistant(s): Eleanor Hooker
Production Editor: Andrew Watts
Illustration: Chartwell Illustrators, Croydon, UK
New Media Editor: Michael Morales
Copyeditor: Moira Vekony
Typesetter: J&L Composition, Filey, UK
Printer: Ajanta Offset, New Delhi, India

ISBN 08153 41857

Library of Congress Cataloging-in-Publication Data

Jobling, Mark.

Human evolutionary genetics / Mark Jobling, Matthew Hurles, Chris Tyler-Smith.

        p. ; cm.
Includes bibliographical references and index.
ISBN 0-8153-4185-7 (pbk. : alk. paper)
1. Human genetics—Variation. 2. Human evolution. 3. Human molecular genetics.
[DNLM: 1. Evolution, Molecular. 2. Genome, Human. 3. Genetics, Medical. 4. Variation
(Genetics) GN 289 J62h 2004] I. Hurles, Matthew. II. Tyler-Smith, Chris. III. Title.
QH431.J53 2004
599.93'5--dc22
        2003018294

Published by Garland Science, a member of the Taylor & Francis Group, 29 West 35th Street, New York, NY 10001–2299, and 4 Park Square, Milton Park, Abingdon OX14 4RN

Printed in India

15 14 13 12 11 10 9 8 7 6 5 4 3 2 1

Chapter list

Contents

Preface

We wrote this book primarily because there were none available which covered the areas that interested us in the field of human evolutionary genetics. Students would ask us to recommend a textbook, and we would apologetically have to refer them to a miscellaneous collection of review articles focused on individual topics. There was, of course, Cavalli-Sforza *et al.*'s monumental and ground-breaking *The History and Geography of Human Genes*. However, because of the time at which it was written, this book could not encompass the many recent developments in human molecular genetics that have led to an explosion of studies of human genetic diversity. Nobody that we knew of was planning a modern synthesis. When we decided to go ahead, wrote a synopsis and sent it out for review, we realized why nobody had done it already. First, it is a wide subject, touching on many disciplines; second, there is a large and all-too-rapidly evolving literature; and third, the field is often highly contentious, making the setting down of an unbiased account of the state of knowledge difficult or impossible. One of our US reviewers used a baseball analogy about our proposal: *'someone has decided to "step up to the plate"'*.

To non-baseball watchers such as ourselves, the only phrase that we automatically associate with that one is: *'three strikes and you're out'*. The strikes that we anticipate come from three areas that we have had to trespass upon, but really know little about. These are archaeology/paleontology (we visit museums like everyone else), anthropology (we have worked on human populations from around the world, but rarely meet our DNA donors) and linguistics (we are reasonably fluent in English and can order beer in several Indo-European languages). Nonetheless we hope that archaeologists, paleontologists, anthropologists, and even historical linguists will find the book useful, and are not too upset by bias or oversimplification on our part.

Unlike baseball, there will certainly be a fourth strike – from fellow geneticists. Apart from the fundamentals, almost everything we say about genetics in this book will be vehemently contested by somebody; while we await the criticism with some trepidation, it is true to say that this controversy is one of the aspects of human evolutionary genetics that make it an interesting field in which to work.

What are our aims?

This book provides an extensive introduction to the analysis of human genomic variation from an evolutionary perspective. The fundamental concepts underpinning anthropological, medical and forensic applications of human diversity studies are all explored with illustrative examples.

This book is *not* intended to provide a comprehensive anthropological account of current evidence as to how humans have expanded and colonized the world from the origin of modern humans to the present day. The literature on this subject is already vast and there is no consensus on many issues. Nor is it intended to provide extensive accounts of medical or forensic genetics; more detail on medical genetic aspects can be found in Strachan and Read (2003). Instead, we focus on the principles behind these fields, and provide examples to illustrate them. We have chosen not to adopt an overly historical approach to human evolutionary genetics; however, *Figures 1.6* and *1.7* in Chapter 1 show time-lines of important events in the development of the field, and in the evolutionary history of humans.

In writing this book we have tried to aim at interested students and researchers from different disciplines. We therefore explain specialist terms where they are first used, and include a glossary at the back of the book that defines all terms in the text that are in bold type, as well as an extensive list of supporting references in each chapter. Using the internet is an essential part of human evolutionary genetics, and each chapter also includes the addresses of key websites.

How to use this book

The book is divided into sections, allowing it to be read by students and researchers from a broad range of backgrounds. In the introductory section **'Why study human evolutionary genetics?'** (Chapter 1) we explore the reasons why this subject is interesting and useful, and raise some general caveats about what questions we can meaningfully ask, and what we can and cannot know about the past. The sections **'How do we study genome diversity?'** (Chapters 2–4) and **'How do we interpret genetic variation?'** (Chapters 5–6) provide the necessary tools with which to understand the rest of the book. The first of these sections surveys the structure of the genome, different sources of genomic variation and the methods for assaying diversity experimentally. The second introduces the evolutionary concepts and analytical tools with which to interpret this diversity. The section **'Where and when did humans originate?'** (Chapters 7–8) considers our origins at two levels: first, our links to our closest living non-human relatives, the great apes; and second, the more recent African origin of our own species. '**How did humans colonize the world?**' (Chapters 9–12) discusses the evidence for early human movements out of Africa, and the subsequent processes of migration and mixing that have shaped human genetic diversity as it is found today. Finally, '**How is an evolutionary perspective useful?**' (Chapters 13–15) demonstrates the wider applications of analyzing human genetic diversity for our understanding of phenotypic variation, the genetics of disease and the identification of individuals. Extensive cross-referencing between these sections facilitates different routes through the book for readers with divergent interests and varying amounts of background knowledge.

An important feature of the book is the use of 'Opinion Boxes' – short contributions by guest authors who are experts in different aspects of this diverse subject area. These help to give a flavor of scientific enquiry as an ongoing process, rather than a linear accumulation of facts, and encourage the reader to regard the published literature with a more critical eye.

Additional resources have been incorporated to permit interested readers to explore the topics in greater depth. Electronic references to internet sites are given throughout the book, both for additional information and for useful software and databases. The constant turnover of websites on the internet will mean that some will go out of service; however, enough information has been provided to allow the reader to use a search engine to find the new location of any relocated pages. Teachers may be interested to know that most of the figures in this book are freely available from the Garland Science website (www.Garlandscience.com) for use in teaching materials.

The authors would appreciate feedback on the experiences of readers with different interests. Although it will not be possible to reply to all comments, constructive criticism will undoubtedly improve future editions. Feedback can be sent to: science@garland.com

Acknowledgments

We have many people to thank for contributions to this book. More than 30 researchers found time to write Opinion Boxes; we are very grateful to: Bill Amos, Ulfur Arnason and Axel Janke, Michael Ashburner, Philip Awadalla, David Balding, Guido Barbujani, Jim Bindon, Neil Bradman and colleagues, Alan Cooper, Tim Crow, Alan Fix, Rob Foley and Marta Mirazón Lahr, Hank Greely, Evelyne Heyer, Luba Kalaydjieva, Heikki Lehväslaiho, Randolph Nesse, Magnus Nordborg, Thomas Parsons, Leena Peltonen and Antti Sajantila, Hendrik Poinar, Carole Ober, Stephen Oppenheimer, Colin Renfrew, Martin Richards, Merritt Ruhlen, Richard Sturm, Naoyuki Takahata, Alan Templeton and Milford Wolpoff. We must stress that any opinions outside these Boxes are our own, and not necessarily endorsed by the Opinion Box contributors.

Other people kindly provided data or advice, or commented on Figures; thanks to: DiAnne Bradford, Peter Forster, Ed Hollox, Marta Mirazón Lahr, Vincent Macaulay, Celia May, Colin Renfrew, David Richards, Antonio Salas and Peter Underhill. We also thank the companies who kindly sponsored color illustrations.

Most of the chapters were sent for external specialist review, and we are grateful for the time and trouble that these reviewers took in improving our manuscript. We thank Guido Barbujani (University of Ferrara), Graeme Barker (University of Leicester), Peter de Knijff (Leiden University), Anna di Rienzo (University of Chicago), Pascal Gagneux (University of California at San Diego), David B. Goldstein (University College London), Alec Jeffreys (University of Leicester), Fabrício Santos (Federal University of Minas Gerais), Mark Stoneking (Max Planck Institute for Evolutionary Anthropology), Chris Stringer (Natural History Museum, London), Sarah Tishkoff (University of Maryland) and an anonymous reviewer of Chapter 5. Our most sincere gratitude, however, must be reserved for John Mitchell of La Trobe University, Melbourne, who has heroically read the entire book and offered constructive and helpful criticism. Of course, despite the efforts of reviewers, errors and omissions will remain, and we take full responsibility for these.

We thank Nigel Farrar and Jonathan Ray at BIOS Scientific Publishers/Garland Science for their enthusiasm for this project and their practical help. Eleanor Hooker's encouragement and efficient administration of the review and submission process has been greatly appreciated, and we also thank Andrew Watts for taking us through the production phase.

We are also grateful to the funding bodies and institutions that have allowed us to maintain our research interest in human evolutionary genetics. In particular, MAJ thanks the Wellcome Trust for funding through a Senior Fellowship, and the University of Leicester for providing a supportive working environment; MEH thanks the McDonald Institute for Archaeological Research for its support and encouragement; and CTS thanks Oxford University and The Wellcome Trust.

This project has been interesting and educational, but also onerous. Much of this burden has fallen on our families and colleagues, as well as ourselves. MAJ thanks Nicky, Billy and Isobel for cheerfully putting up with disruption to normal family life, and Sue Adams, Elena Bosch, Turi King, Andy Lee, Zoë Rosser and Morag Shanks in the lab for tolerating distracted supervision. MEH thanks Lizzie for her unerring support and enthusiasm, and Peter Forster for so willingly putting up with a preoccupied colleague in the lab. CTS's thanks for their comments and forbearance go to all members of his lab, particularly Silvia Paracchini and Tatiana Zerjal.

Abbreviations

α-MSH	α-melanocyte stimulating hormone
μg	microgram
^{14}C	Carbon-14
A	adenine
AD	Alzheimer disease
ADH	alcohol dehydrogenase
ADHD	attention-deficit hyperactivity disorder
aDNA	ancient DNA
AFLP	amplified fragment length polymorphism
AI	artificial insemination
AIDS	acquired immune deficiency syndrome
AIM	ancestry informative marker
AIS	androgen insensitivity syndrome
ALDH I	aldehyde dehydrogenase I
AMH	anatomically modern human
AMOVA	analysis of molecular variance
ARMS	amplification refractory mutation system
ART	assisted reproductive technology
ASD	average squared distance
ASP	affected sib pairs
BAC	bacterial artificial chromosome
bp	base pairs
BSE	bovine spongiform encephalopathy
C	cytosine
cAMP	cyclic AMP
cc	cubic centimeters
CD/CV	common disease/common variant
CEPH	Centre d'Etude du Polymorphisme Humain
CF	cystic fibrosis
CI	cephalic index (or cranial index)
CJD	Creutzfeldt-Jakob disease
cM	centiMorgan
CODIS	Combined DNA Index System
cpDNA	chloroplast DNA
CpG	C followed 3′ by G, where 'p' indicates a phosphate group
CRS	Cambridge Reference Sequence
CYP	cytochromes P450
DARC	Duffy antigen receptor for chemokines
DHPLC	denaturing high performance liquid chromatography
DI	donor insemination
DMD	Duchenne muscular dystrophy
DNA	deoxyribonucleic acid
DOP-PCR	degenerate oligonucleotide-primed PCR
DSB	double-strand break
DTD	diastrophic dysplasia
DZ	dizygotic
EHH	extended haplotype homozygosity
EM	expectation-maximization (algorithm)

ESR	electron spin resonance
FISH	fluorescence *in situ* hybridization
G	guanine
G6PD	glucose-6-phosphate dehydrogenase
GRR	genotypic relative risk
Hb^S	hemoglobin S
HD	Huntington disease
HERV	human endogenous retrovirus
HGDP	human genome diversity project
HIV	human immunodeficiency virus
HKA	Hudson–Kreitman-Aguadé
HLA	human leukocyte antigen
HR	homologous recombination
HTU	hypothetical taxonomic unit
HVR	hypervariable region
HVS I	hypervariable segment I
HVS II	hypervariable segment II
IBD	isolation by distance, or identical by descent
ICSI	intracytoplasmic sperm injection
ID	immunological distance/index of dissimilarity
IGF2	insulin-like growth factor 2
IR	ionizing radiation
IVF	*in vitro* fertilization
JC	Jukes–Cantor
kb	kilobases/kilobase-pairs
kg	kilogram
KY	thousand years
KYA	thousand years ago
LBK	*linienbandkeramik*
LCR	low copy repeat
LD	linkage disequilibrium
LEA	likelihood estimation of admixture
LGM	Last Glacial Maximum
LINE	long interspersed nucleotide element
LOD	log(arithm) of the odds
LP	Lower Paleolithic
LSA	Later Stone Age
LTR	long terminal repeat
m	meter
MALD	mapping by admixture linkage disequilibrium
MALDI	matrix-assisted laser desorption-ionization
Mb	megabases / megabase-pairs
MC1R	melanocortin-1 receptor
MDA	multiple displacement amplification
MDS	multidimensional scaling
MED	minimum-erythremal dose
MEPS	minimum efficient processing segment
MHC	major histocompatibility complex
ML	maximum likelihood
MP	maximum parsimony/Middle Paleolithic

MRCA	most recent common ancestor	QTL	quantitative trait locus
mRNA	messenger RNA	RAFI	Rural Advancement Foundation International
MS	mass spectrometry	RAPD	randomly amplified polymorphic DNA
MSA	Middle Stone Age	rDNA	ribosomal DNA
mtDNA	mitochondrial DNA	REV	general reversible model
MVR-PCR	minisatellite variant repeat PCR	RFLP	restriction fragment length polymorphism
MY	million years	RNA	ribonucleic acid
MYA	million years ago	SCAR	sequence characterized amplified region
MZ	monozygotic	SCID	severe combined immunodeficiency
NAHR	non-allelic homologous recombination	SGM	Second Generation Multiplex
NCA	nested cladistic analysis	SINE	short interspersed nucleotide element
NCBI	National Center for Biotechnology Information	SIV	simian immunodeficiency virus
		SLP	single-locus probe
NDNADB	National DNA Database	SMM	stepwise mutation model
ng	nanogram	SNP	single nucleotide polymorphism
NIDDM	non-insulin dependent diabetes mellitus	SRY	sex-determining region, Y
NJ	neighbor joining	SSCP	single-strand conformation polymorphism
nm	nanometer	SSR	simple sequence repeat
NMDA	N-methyl-D-aspartate	STR	short tandem repeat
NOR	nucleolar organizer region	T	thymine
NR	neurotransmitter receptor	Taq	*Thermus aquaticus*
NTD	neural tube defect	TDT	transmission disequilibrium test
numt	**nu**clear **mt**DNA insertion	TL	thermoluminescence
OCA	oculocutaneous albinism	TMRCA	time to most recent common ancestor
OLA	oligonucleotide ligation assay	TOF	time of flight
OTU	operational taxonomic unit	tRNA	transfer RNA
OWM	old world monkeys	TSE	transmissible spongiform encephalopathy
PAR1	pseudoautosomal region 1	U	uracil
PAR2	pseudoautosomal region 2	UEP	unique event polymorphism
PC	principal component	UP	Upper Paleolithic
PCA	principal components analysis	UPGMA	unweighted pair-group method with arithmetic mean
PCR	polymerase chain reaction		
PFGE	pulsed-field gel electrophoresis	UVR	ultraviolet radiation
pg	picogram	VNTR	variable number of tandem repeats
PKU	phenylketonuria	WGA	whole genome amplification
PPARγ	proliferator-activated receptor-γ	WLS	weighted least-squares
PSA	population-specific allele	YA	years ago
PSV	paralogous sequence variant	YAC	yeast artificial chromosome

SECTION ONE

Introduction

Human evolutionary genetics is the study of how one human genome differs from another, and the implications of this for our understanding of our species in the past and in the present. Differences between genomes form the basis of anthropological, medical and forensic genetics. All of these fields are experiencing massive advances as a result of the Human Genome Project. But before we leap into the fascinating world of DNA sequences, prehistoric migrations and pathogenic mutations, we must first appreciate the full range of information at our fingertips, and explore our own motivations and subconscious biases. It is much easier to examine our own prejudices in the cold light of prior reflection than in the midst of a heated debate.

CHAPTER 1   *Why study evolutionary genetics?*

In the opening chapter, we examine the range of records of the human past that are available to us. These complementary records tell us about different aspects of the past, and are informative over different time-scales. We also ask what questions can and cannot be answered using genetic data, and explore the value of an evolutionary perspective to disciplines that may not be traditionally regarded as being concerned with the past. Finally, we outline the structure of the book, which first assembles the tools and understanding necessary for investigating genetic diversity before exploring the past 7 million years of human evolution, and subsequently applying an evolutionary perspective to the genetic dissection of phenotypic traits, genetic diseases and individual identity.

CHAPTER ONE

Why study human evolutionary genetics?

CHAPTER CONTENTS

BOXES

1.1 A genetic record of the past

We are all curious about our origins. This common interest manifests itself in different ways, at both the personal level (local historical research and genealogical investigation) and at the level of society (a fascination with archaeology and history, and public funding for the investigation of our past). This curiosity is not new: some of the oldest historical texts from different societies detail origin myths, as do oral histories passed down over generations. Distinguishing fact from fiction in these stories is difficult and differences of opinion can be a source of political conflict.

1.1.1 Records of the past

Any scientific investigation of the past should start with a consideration of the different types of evidence available, since this provides the basis for testing hypotheses and refining models. While there was only one past, it is revealed to us in many ways (see *Figure 1.1*):

▶ The historical record comprises written texts, the oldest of which are from Mesopotamia and date from as far back as 4000 years ago ('The Epic of Gilgamesh'). Writing itself goes back another 1000 years but appears to have been mostly associated with accounting practices. These early texts were written in cuneiform (wedge-shaped) symbols indented into clay tablets using reeds, and it is only later that papyrus and modern alphabets were invented. It must be remembered that very few ancient languages were written down, and some of those that were recorded are presently indecipherable (e.g. the Cretan script known as 'Linear A'). There are also oral histories, and folklore handed down through the centuries. It is difficult to judge the factual content of these inter-generational 'Chinese whispers', let alone their time-depth. The boundaries between oral and written histories are often blurred; some histories that were initially oral have achieved special prominence through being written down in antiquity.

▶ **Spoken languages** retain evidence of their origins over thousands of years (see *Box 1.1*). The discipline of historical linguistics seeks to trace the ancestry of all ~6500 languages currently spoken in the world. Many of these languages can be traced to a number of ancestral languages known as proto-languages. For example, English, French, German, Russian and Sanskrit all belong to the Indo-European language family and share a common ancestral language known as proto-Indo-European. Although the origins of spoken languages can be traced back further than historical records, a common ancestry for all human languages cannot be identified, and some historical linguists have suggested that languages do not retain evidence of their origins over more than ~6000 years.

▶ The **archaeological record** consists of physical objects that have been shaped by human contact. These include not only tools, ornaments, and pottery, but also soils, waste deposits, houses and landscapes. The earliest recognizable stone tools date from about 2.5 million years ago (MYA; see Section 8.3). Humans are not the only animals to make tools, but tools produced by earlier human ancestors or nonhuman species are difficult to distinguish from naturally occurring objects.

▶ The **paleontological record** comprises the fossilized remains of living organisms or their traces, like preserved footprints. The earliest microfossils are suggested to date from 3500 MYA. These would be roughly three-quarters of the age of the planet itself, suggesting that life on Earth started almost as soon as conditions were suitable. Paleoanthropologists focus on the remains of humans and their ancestors. The earliest fossils that appear to be more closely related to humans than to any other living species are dated from 5–7 MYA (Section 8.2).

▶ **Paleoclimatology** seeks information on past climates and aims to reconstruct paleoenvironments. Data come from physical remains, both organic and geological in origin. An example of geological evidence comes from the cores of ice taken from ice sheets, which at different depths exhibit varying isotope ratios of certain elements that can be related to temperature changes over the past million years. Similarly, cores of lake sediments reveal the predominance of pollen from different plants at different times, which provides evidence about the biotic environment over shorter

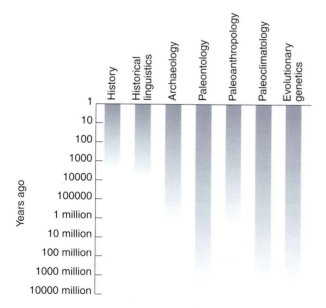

Figure 1.1: The informative time-depth of different records of the past.

BOX 1.1 Opinion: Linguistic evidence for human origins

It has been known for more than two centuries that the comparison of extant or historically-attested languages can lead to the identification of extinct languages that existed before the invention of writing – that is, in prehistoric times. In 1786 an English jurist serving in India, Sir William Jones, noted that similarities in the words of Latin, Greek, and Sanskrit were so striking that they could only have arisen 'from some common source, which perhaps no longer exists.' What struck Jones was that both the roots and endings in verbs in these languages were almost identical: Sanskrit bhár-anti 'they carry,' Greek phér-onti 'they carry,' Latin fer-unt 'they carry', and he recognized that it would be absurd to suppose that such similarities had arisen by chance. Jones thus identified a language family that later came to be known as Indo-European, and which includes most of the languages of Europe (except for Basque, Hungarian, Finnish, and Estonian) as well as languages ranging as far east as northern India (Kurdish, Farsi, Pashto, Hindi, Bengali). But even more importantly Jones offered an evolutionary explanation for linguistic diversity and he did so some 73 years before Darwin's evolutionary explanation via natural selection for biological diversity.

What is surprising is that two centuries after Jones' fundamental discovery, most historical linguists still maintain that Indo-European represents the limit of comparative linguistics. According to this widely held view language change is so rapid and unrelenting that after around 6000 years – conveniently the supposed age of Indo-European – all trace of genetic affinity is lost so that even if Indo-European did once have some relatives, there simply could be no trace of this relationship still surviving today. What is even more surprising, however, is that this view was known to be incorrect a century ago. At the beginning of the twentieth century Holger Pedersen, Alfredo Trombetti, and others pointed out numerous traits shared by Indo-European and other language families, among the most prominent of which were an M/T 'I/you' pronominal pattern and a dual/plural opposition expressed by K/T suffixes.

This larger family, of which Indo-European was just one branch, was initially called Nostratic by Pedersen. After a period of quiescence in the first half of the twentieth century, work on Nostratic was revived in Moscow during the 1960s by Vladislav Illich-Svitych and Aron Dolgopolsky. A family similar to Nostratic has recently been proposed by Joseph Greenberg, who calls the family Eurasiatic (see *Figure*). The Eurasiatic family includes, in addition to Indo-European, the Uralic (Hungarian, Finnish), Altaic (Turkic, Mongolian, Tungus), Korean, Japanese, Ainu, Gilyak, Chukchi-Kamchatkan, and Eskimo-Aleut families.

In addition to Eurasiatic, other large families were discovered in the second half of the twentieth century, primarily by Greenberg, whose taxonomic work forms the basis of our

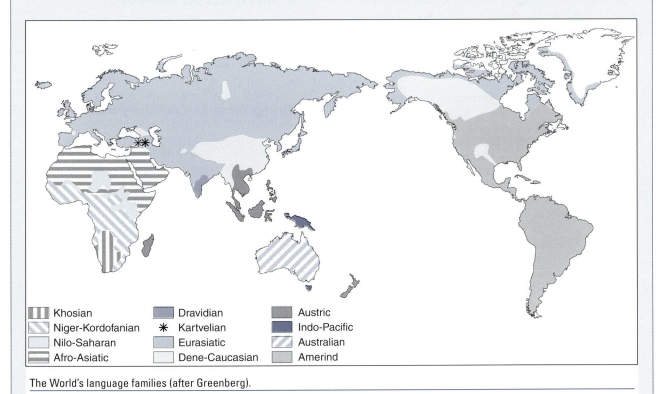

▥ Khosian	▦ Dravidian	▨ Austric
▧ Niger-Kordofanian	✳ Kartvelian	▮ Indo-Pacific
▢ Nilo-Saharan	▥ Eurasiatic	▨ Australian
▤ Afro-Asiatic	▢ Dene-Caucasian	▨ Amerind

The World's language families (after Greenberg).

knowledge of the relationships among the languages of Africa, New Guinea, and the Americas. The *Figure* shows the distribution of the 12 families into which Greenberg classified all of the world's roughly 6000 languages. Each of these families is defined by a specific set of traits. Whereas Eurasiatic is characterized by the M/T 'I/you' pronominal pattern, the Amerind family is characterized by N/M, and other families have pronominal patterns different from either of these.

Perhaps the most surprising finding of all is that even among these dozen families, most of which are quite ancient, there are still roots that are widely shared and which imply a recent common origin for all extant languages. Substantial evidence for monogenesis was already given by Trombetti a century ago, and recently John Bengtson and I have enlarged and refined Trombetti's evidence based on much better resources now available (Ruhlen, 1991, 1994). Examples of roots whose distribution ranges from Africa to the Americas are TIK 'finger, 1'; PAL '2'; AKWA 'water'; KAPA 'cover(ing), skin, bark'; and PUT 'vulva.'

While this emerging linguistic picture is still considered anathema to most mainstream historical linguists, one cannot fail to see that it fits well with the archaeological and genetic evidence for human origins, according to which a behaviorally modern group of humans left Africa around 50 KYA and in a short time populated the entire world, replacing all of the earlier inhabitants, such as the Neanderthals. Traces of this rapid expansion out of Africa – and then throughout the world – can be seen in the archaeological record, in the genes of modern humans, and in modern languages. Furthermore, not only is the linguistic evidence consonant with the archaeological and genetic evidence, but several scholars, including Richard Klein and Jared Diamond, have proposed that it was in fact the emergence of fully-modern language around 50 KYA that was directly responsible for this expansion out of Africa and the subsequent disappearance of all earlier humans or human-like species. It would appear that human language – and its correlate human culture – was a tool with which earlier humans, such as Neanderthals, could not successfully compete.

Merritt Ruhlen

Department of Anthropological Sciences, Stanford University, USA

time-scales [typically up to 100 KYA (thousand years ago)]. **Geological records** in rocks provide information on older climates as far back as 3800 MYA. We will see that climate has varied greatly over the last few million years, and has had a major influence on human populations.

▶ In principle, the **genetic record** of life on Earth contained in the genomes of living species could provide evidence on evolutionary processes and relationships all the way back to the last universal common ancestor of all extant species. To find out about this lucky individual we have to compare and contrast the most distantly related branches on the tree of life. Comparisons among much more closely related individuals, perhaps from the same species, should provide evidence on much more recent evolutionary processes. Our ability to read this genetic record is a relatively recent development, although our confidence that information on our past exists within our heritable material is somewhat older. Genetic evidence comes from two main sources:

▶ the genomes of living individuals that must have been passed down from ancestors;

▶ ancient DNA from well-preserved organic remains, which may or may not have been passed down to living descendants.

1.1.2 All records are selective

It is important to bear in mind that none of these records represents an unbiased picture of the past. We do not have a time machine and therefore must rely upon evidence that has survived to the present. This process of survival is selective. In the archaeological record we find many stone tools but few wooden ones: arrowheads but not shafts. In the paleontological record we find plenty of skeletal fossils, but soft tissues leave traces only very rarely. In the historical record we may not encounter those texts that displeased contemporaneous or subsequent heads of state, either because they were destroyed, or not written in the first place. Similarly, in the genetic record, survival is quite literally selective. Natural selection and other processes have shaped, and continue to shape, our genome in different ways. Even ancient DNA evidence, although not influenced by subsequent natural selection, is biased: for technical reasons it can tell us far more about the genetic diversity of our female ancestors than it can about our male forebears. Since the survival of ancient DNA is influenced by physical and chemical conditions, samples will be more plentiful from some regions of the world than from others.

1.1.3 The palimpsest metaphor

In times past, when writing materials were in short supply a scribe would often reuse a manuscript rather than obtain a new parchment. The manuscript would be turned through

90°, and overwritten. These overwritten manuscripts bearing the imprint of more than one text are known as **palimpsests**. The genetic record is a complex palimpsest. Variation among modern individuals is shaped by cumulative past processes. Extracting information on any one past period or event requires careful interpretation to isolate it from previous and subsequent processes. In addition, natural selection is ever present, potentially influencing any variation that affects phenotypic fitness. As a result, researchers have sought loci within our genomes that are selectively neutral, in the hope that these might provide a more representative record of the past. Studies of ancient DNA only partly circumvent the palimpsest problem, since organic remains themselves have complex histories, and there are formidable technical difficulties in extracting information (Section 4.11).

Different strata of the past are accessible through the analysis of genetic diversity. Moving from the most ancient to the most recent, we encounter:

▶ our phylogenetic relationship to other species (Chapter 7);

▶ the origin of our species (Chapter 8);

▶ prehistorical migrations (Chapters 9, 10 and 11);

▶ historical migrations (Chapter 12);

▶ genealogical studies (Chapter 15);

▶ paternity testing (Chapter 15);

▶ individual identification (Chapter 15).

1.1.4 Comparisons between different records of the past

No single record of the past is more important than any other, but each records different features of the past. Therefore the record chosen for investigation in a given study is largely determined by the specific questions under consideration. If we want to address a cultural question, such as 'What was life like for humans 10 000 years ago?', archaeology will give us more insight than will linguistics or genetics. In contrast, if we want to answer a biological question, such as 'Which nonhuman species is most closely related to us?', archaeology and linguistics will be of little use.

To compile a fuller picture of the past we often seek to combine information from multiple records into a single 'synthesis'. This requires that we consider a little more closely the similarities and unique characteristics of different records. For example, while we can be sure that the speakers of modern languages have ancestors, we cannot be sure that the makers of a certain style of pottery left any descendants. When similar information is being sought from several records of the past, it becomes particularly critical to appreciate the differences between them. For example, if we are interested in migrations we must appreciate that artifacts can move through trade, without the concomitant movement of genes. Similarly, not all gene flow need be accompanied by the movement of languages.

These different records are independent reflections of a single past – but they need not all tell us the same thing.

Rather, conflicting signals allow us to reconstruct subtle and nuanced views of a past that must have been just as complex as the present. Nevertheless, readers should maintain a healthy skepticism of interdisciplinary syntheses (*Figure 1.2*). Each individual discipline that interprets a record of the past contains competing hypotheses and many issues upon which there is no consensus. This could enable researchers in one field to 'cherry-pick' hypotheses from other disciplines that agree best with their own thinking. Stringing together an initial contentious hypothesis from field A with equally contentious theories from fields B and C, may make for a more complete and interesting narrative, but does not make the original hypothesis any less contentious. As a result, the field of human evolutionary studies is especially prone to heated debates, and the reader should realize that almost everything we say in this book will be contested by someone. We recognize the importance of these ongoing debates and try to give readers a flavor of the diversity of opinions by incorporating 'Opinion Boxes' written by active researchers who have new, interesting or challenging theories, or who are particularly well placed to comment on an area of controversy.

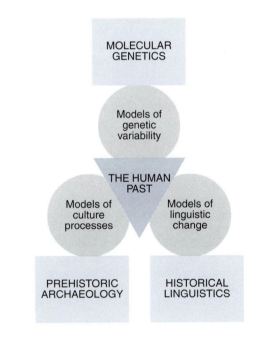

Figure 1.2: Reasonable optimism? Renfrew's 'New Synthesis'.
Redrawn from Renfrew (1999).

1.1.5 The importance of chronology

If we are to provide a wider context for events visible in a single record of the past, we need to have some method to relate this event to other events in other records. For example, we might wonder whether a period of population growth evident in the genetic record is matched by specific innovations apparent in the archaeological record that may indicate why a larger population could be supported.

The natural way to achieve this cross-referencing between multiple records is to relate them all to chronological time, by dating events and processes visible in the different records. Consequently, methods for dating are of prime importance in the analysis of all the records of the past outlined earlier in this chapter. Dating methods are typically dependent on cumulative processes (often known as 'clocks' for obvious reasons) that occur, on average, at known uniform rates and can be accurately quantified. However, dating methods in different disciplines are not equally reliable:

▸ The physicochemical methodologies used for dating in paleoclimatology, paleontology and archaeology frequently depend upon the decay of radioisotopes. This process is stochastic but on average highly uniform and accurately quantifiable, and as a consequence can produce very accurate dates, although these sometimes need to be calibrated by adjusting for past fluctuations in radioisotope levels (see *Box 9.4*).

▸ Dating in linguistics is difficult and controversial, as rates of language change are highly variable. As a consequence, linguistic processes are generally not reliably dated.

▸ Dating the genetic record can utilize a number of different 'molecular clocks' (see Section 6.6), but can also be controversial. First, these molecular clocks are difficult to calibrate accurately, and second, statistical confidence in these dating estimates is determined at least in part by the history of population size and subdivision (demography), about which we have little information.

Having used chronology to identify a *correlation* between events and processes observed in different records of the past, there is a temptation to ascribe *causal* relationships. A good example in the scenario given above would be that technological innovations revealed through archaeology allowed the population growth that is apparent within the genetic record. However, we must realize that it may not be possible to prove these kinds of causal relationships with the same degree of certainty available to other branches of science. The study of evolution is in many ways a historical discipline; being limited to a single past prevents us from demonstrating causal relationships using the principle of reproducibility.

1.2 Complications: potential misuse, 'knowability' and meaningless questions

1.2.1 Our collective ethical responsibility

It is human nature to use knowledge of the past to guide actions in the present. This places an ethical responsibility upon those who seek to explore the past to do all they can to ensure that their work is not misused. This is not just a theoretical possibility: history is rife with the misuse of anthropological research to justify regimes that have cost the lives and livelihoods of many innocent people. So much damage has been caused that some have even questioned

whether work to reconstruct our evolutionary past should be undertaken at all. We believe that the potential intellectual and medical benefits of this work outweigh the potential dangers, but only when researchers take responsibility for the accurate popularization and public dissemination of this research, including active opposition to misinterpretation. Having said that, we must acknowledge that much of this work (including this book) is published by an unrepresentative subset of our species, namely men in developed countries, and that this cultural framework undoubtedly has an influence on interpretation.

As scientists, we should recognize that our work rightly depends upon the approval of wider society; indeed, most funding for evolutionary studies around the world comes from the public purse. Irrespective of the source of funding, public concerns about the wider implications of our work must be addressed. Although the most notorious historical misuse of anthropology has been the justification of genocide, public anxiety is currently more focused on issues of ownership, commercialization and privacy. When work is being conducted in the public interest, the *perception* of misuse can be as important as the *reality* of misuse. Steps must be actively taken to ensure both that such misuse does not occur, and that misuse is seen not to occur. It is for these reasons that research projects on human subjects, whether they are medical patients or volunteers contributing a few cheek cells, should be scrutinized and approved by ethical committees prior to their initiation.

1.2.2 Beware of seeking unknowable answers

What we *would like to* know about the past, and what we *are able to* know about the past are two separate things. Some of the answers we would like to find are quite simply unknowable. It is useful to explore the concept of 'knowability' so that we can understand where these boundaries lie. It would be futile to expend time, energy and money on questions that cannot be resolved. Clearly, the boundaries of knowability are not static: technological innovation continues to 'push the envelope'. For example, before the advent of methods for sequencing ancient DNA, the genetic divergence between Neanderthals and modern humans was unknowable. Now we have a measure of this divergence, for one locus at least (Section 8.7). Nevertheless, despite recent advances that have brought this issue into the realm of the knowable, many others, such as whether any two extinct ape species were interfertile, remain unknowable.

1.2.3 Beware of asking meaningless questions

Intuition is a fertile source of the questions we ask of ourselves and of our surroundings, but it is not infallible. Not all questions have answers: some are nonsensical even if grammatically correct, for example (to paraphrase Noam Chomsky):

Do colorless green dreams sleep furiously?
Other, intuitively attractive, questions appear to make sense even if they do not, for example:

What was the ancestral biological homeland of population X?

The intuitive nature of such questions means that they are often posed, despite their meaninglessness. They often arise out of a conflict between our commonsense view of the world, which is conditioned to some degree by the way our minds work and the society in which we live, and the way the world actually is.

Taking the example above, to conceive of a single origin of an entire population is to misunderstand the nature of how genetic diversity accumulates in a sexually reproducing, outbreeding species such as humans. The term 'population' is itself problematic: are humans really divided into separate populations, and if they are, how do we define them (*Box 1.2*)? We will defer further discussion of this question until Section 9.2.4, and here assume that it is possible to talk about populations. Even so, a population could have a single biological homeland only if all genetic diversity within that population could be traced to a single time and place.

The processes of recombination and independent chromosomal assortment divide the genome into segments that have independent genealogical histories: while the ancestry of a single segment of the genome converges on a single common ancestor, each segment has a *separate* common ancestor. These common ancestors are almost always far older than the population itself, and it is highly unlikely that they were ever present in the same place at the same time. The boundaries of these genomic segments are not determined by gene function, but by recombination events, and so segments may contain no genes, one gene or several genes. Later in the book, we will encounter mitochondrial DNA and the Y chromosome, segments of DNA that have received enormous attention. Despite their popularity, these particular segments do not represent an individual's ancestry any more completely than any other single segment.

Alternatively, if the imagined population homeland is not a place of ultimate common ancestry, maybe it could be a more proximate location through which all alleles currently residing within the modern population came at the same time? As we shall see in this book, human populations are fluid, outbreeding entities, giving and receiving genes from neighboring populations all the time. Alleles enter the population at different times and from different places.

The three authors of this book are all British, each with more than 30 000 genes in their genomes. Some of these genes will have journeyed to Britain when it was still joined

---

BOX 1.2 Caucasians, Caucasoids, European–Americans, Whites? The confusing classification of human social groups

Population geneticists, forensic geneticists, anthropologists and archaeologists need labels to refer to social groups of human beings. These labels differ between fields, and even within fields. In human genetics papers describing DNA diversity in a group of people living in the USA whose ancestors came from Europe, for example, you may find them referred to as:

▶ a **US population** – because of population admixture in the last 500 years, the people of the USA are an extremely heterogeneous group, of which those of European descent are only a part;

▶ **Caucasians**, or **Caucasoids** – this is not meant to imply an origin in the Caucasus mountains, but refers to 'beautiful people' in a racial classification scheme of the German anatomist Blumenbach (1752–1840). The skull that was claimed to best represent the characteristics of this group came from the Caucasus. The other classifications in this scheme were Mongoloid, Malay, Ethiopian and American (referring to Native Americans). Issues of racial classification are discussed further in Chapter 9;

▶ **European–Americans** – usually in contradistinction to African–, Hispanic–, Japanese–, Native– and other Americans. There is heterogeneity within this grouping – people who would classify themselves as European–American have ancestors from many different parts of Europe, such as Ireland, Italy, Poland, Russia and Turkey;

▶ **Whites** – this classification is favored by some journals over 'Caucasians', which might quite reasonably be reserved for people who really *do* come from the Caucasus. However, it seems odd to use 'Whites', when authors of a manuscript may have no idea what skin color the donors of their DNA samples had.

Often these racial or ethnic labels disguise a great deal of biological heterogeneity, and the identification of DNA donors as members of social groups is not self-evident. Confusion is possible when a paper has the title: 'Strong Amerind/White sex bias and a possible Sephardic contribution among the founders of a population in northwest Colombia' (Carvajal-Carmona *et al.*, 2000). It includes labels based on indigenous continental affiliation, skin color, membership of a group defined on religious–historical grounds, and current small-scale geography. A DNA donor may belong to a large number of categories simultaneously (reviewed by Foster and Sharp, 2002). In addition, when populations are compared in genetic studies, the level of classification in different samples may be unequal. The Hadza of Tanzania, with a population size of only 1000, have in some studies occupied the same analytical status as the South Chinese, whose population size is 600 000 times greater (reviewed by MacEachern, 2000).

In general, the most suitable default method for classification is to use geographical information, rather than national, cultural or phenotypic labels.

to the European continent as the ice-sheets of the last Ice Age receded. Other genes will have been introduced by migrants as the sea levels rose, making what is now the bed of the North Sea uninhabitable. Still more genes may have been contributed by the first farmers, the Romans, the Angles, the Saxons, the Vikings and the Normans, as well as other travelers less well-documented in historical or archaeological records. What hope is there for a single biological homeland for the British?

Let us now consider a second intuitively attractive yet meaningless question:

Where did my ancestors live, a thousand years ago?
As a result of sexual reproduction, individuals have ever-increasing numbers of ancestors as we look back further over the generations (2, 4, 8, 16, 32, 64, 128 and so on). It is worth considering a few consequences of this seemingly trivial calculation (Ohno, 1996). After 30 generations (< 900 years), each of us has more than a billion potential ancestors, and after 40 generations (< 1200 years) more than a thousand billion (*Figure 1.3*). Since the current world population is less than seven billion, and past populations were much smaller in size, these large numbers of ancestors never existed. This is because these potential ancestors were not all different individuals: the same person appears in different places in the family tree. There is another consequence of this reasoning that is worth spelling out. In *Figure 1.3*, the curves representing potential ancestors *of a single individual* and world population growth cross within the past 900 years. Before this time, everyone in the world was potentially an ancestor. Population substructure, the nonrandom breeding of individuals from different places, complicates this over-simplified calculation, but the essential point remains: the best answer to the question 'Where did my ancestors live?' is 'Everywhere'.

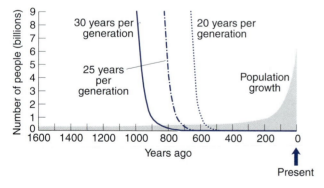

Figure 1.3: Past population growth plotted against the accumulation of potential ancestors.

The filled segment shows global population growth over the past 1600 years. The lines correspond to the potential numbers of ancestors of a single individual expected at a given point in time. The potential number of ancestors depends on the length of generation time that is assumed.

So why are the above questions about single origins of a population, or set of ancestors so intuitively attractive? Our conception of ourselves as individuals, of having a single identity is well founded: each of us has a sense of self, not a sense of 2, 4, 8, 16, 32, 64, 128 or a billion selves. We have minds that like to impose order on chaos, to classify continua; the spectrum of visible light is continuous yet we have fixed notions of individual colors. We live in societies that classify people into groups despite our rejection of overtly racial thinking. These are some of the factors that lead us to pose such questions. We must be aware of these tendencies, and question our intuitions accordingly.

1.2.4 The fallacy of the contemporary ancestor

We want to find out about our ancestors and to collate information from every source that we can. However, this insatiable thirst for knowledge can lead us to adopt unreliable sources of information. Unfortunately, we cannot dig up a living, breathing ancestor, or use a time machine to study the past directly.

All living organisms are 'cousins' that share a common ancestor some time in the past. Each pair of contemporary organisms is equally derived in terms of time from this common ancestor (see *Figure 1.4*), yet the concept of a 'living fossil' or a 'contemporary ancestor' is pervasive. This concept insinuates itself implicitly or explicitly into all levels of the tree of life, as illustrated by misguided statements such as these:

The first living organisms were like bacteria.

The coelacanth is a 'living fossil' of the first four-limbed vertebrate.

Humans evolved from chimpanzees.

Modern hunter–gatherers resemble humans before the advent of agriculture.

The Basques are a Paleolithic relict population.

At the inter-specific level, this concept derives in part from an anthropocentric view of nature: that humans have transcended the rest of the natural world, and that we have progressed beyond the bounds that limit other species (see Sections 13.5–13.7). Potential dangers lie in the assumption that in the absence of evidence about the common ancestor of humans and any other species, the nonhuman species is a closer approximation to this common ancestor.

A misconception of inevitable progress also dominates popular thinking about diversity within humans. Ancient populations are often assumed to be less like modern Western societies and more like 'less-developed' indigenous peoples. Yet it is well accepted within anthropology that ethnography (the study of living peoples and cultures) can be a poor guide to the past.

If there is any basis to this common misconception, it must lie in some measurement of evolutionary derivation other than time. In principle evolutionary derivation may be measured in terms of:

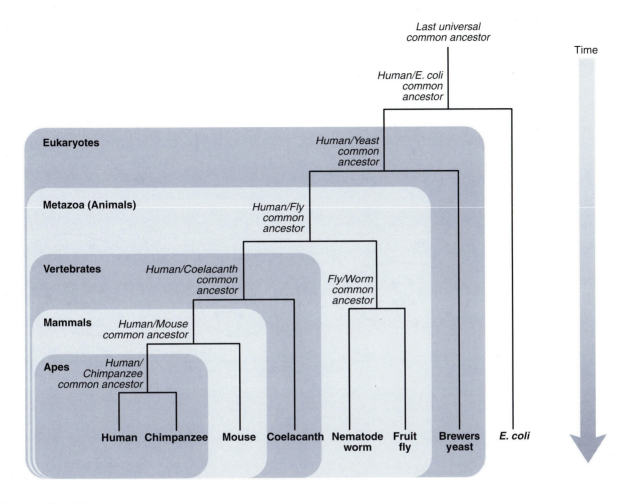

Figure 1.4: 'Cousins', not contemporary ancestors.

A phylogenetic tree relates different branches of modern species, showing that they are all equally derived from their common ancestors in terms of time. The common ancestor of humans and *Escherichia coli* may also be the last universal common ancestor.

▶ time;

▶ phenotypic change;

▶ ecological change;

▶ genetic change.

It could be argued that many of our 'cousins', whether human or not, have maintained modes of life more similar to those of our common ancestor, and that this has led them to undergo less phenotypic change. However, it must be remembered that all forms of life are adapted to their present (or at least very recent) ecological niches. Selection is ongoing, and as a consequence important adaptive changes have taken place in all organisms, which may reveal themselves at the morphological or physiological level.

Mutational change is blind to ecological circumstance; neutral mutations that do not affect the fitness of the individual occur in all organisms. If the annual mutation rate is equal between two species, each will be equally derived from the common ancestor with respect to neutral mutations. However, annual mutation rates can be shown to have varied between lineages, such that humans have accumulated fewer neutral mutations since their common ancestor with Old World Monkeys, and Rodents. In these cases, it is the humans who are less derived.

1.2.5 Interpretation, interpretation, interpretation

In many fields, as time passes, opinion upon how data should be interpreted changes. Indeed, there are often differences in opinion about data interpretation at any one time. This is particularly true of genetic data on human diversity. Debates described in Chapters 8 and 10, on the origins of modern humans and the genetic impact of the spread of agriculture in Europe, illustrate this. Particular methods of analysis, with different underlying paradigms, can be adopted by opposing 'camps' within a particular field, and reconciliation becomes difficult. Some methods for analyzing diversity data seem

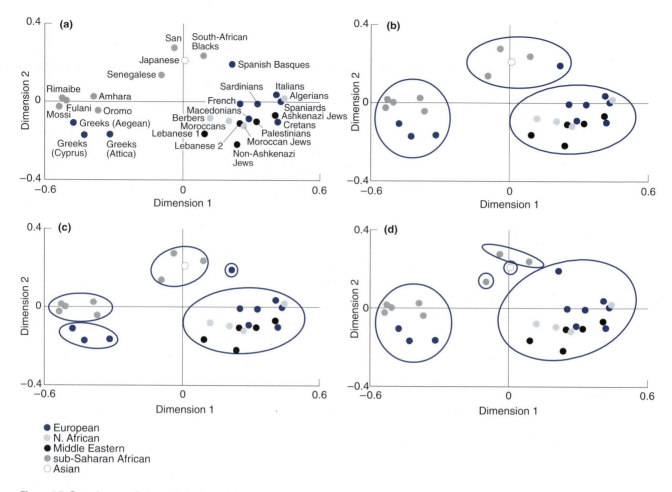

Figure 1.5: Grouping populations – take your pick.

Relationships between populations based on DNA sequence diversity data at the *HLA-DRB1* locus, displayed as a correspondence analysis plot (similar to principal components analysis; see Chapter 6) in which clustered populations are genetically similar. (a) Populations, with names indicated; (b, c, d) Three alternative groupings of the populations (there are others). The grouping chosen by Arnaiz-Villena *et al.* (2001) is (d) (adduced as support for a sub-Saharan origin for the Greeks) but is essentially arbitrary. Why is it preferred to alternative groupings shown in (b) and (c)? If the population origins were unknown when the groupings were made, would it affect the outcome? Note that this locus is generally regarded as being under strong selection. Adapted from Arnaiz-Villena *et al.* (2001).

particularly open to different interpretations. As an example, *Figure 1.5* illustrates the arbitrariness of different possible population groupings based upon DNA sequence diversity at an HLA locus. Often an objective way to choose between different interpretations is not obvious (though objective methods are discussed later in this book), and in its absence, simple assertion often fills the vacuum.

1.3 Understanding phenotypes and disease

What use is an evolutionary perspective on human genetic variation beyond the reconstruction of the past for its own sake?

1.3.1 The importance of a shared evolutionary history

The great twentieth century evolutionary biologist Theodosius Dobzhansky wrote that:

'Nothing in biology makes sense except in the light of evolution.'

All the size, shapes, chemistries and genes of organisms alive today derive from ancestors that can be traced back over billions of years. All of these features have been shaped by the environmental challenges faced by these organisms and their ancestors. If it were not the case that humans share a common ancestor with every other species on the planet, there would be no value in performing any form of comparative analyses. There would be nothing that the *Escherichia coli* bacterium, brewers yeast, fruit fly, nematode worm, zebrafish, mouse or chimpanzee could tell us about ourselves. It is our shared evolutionary heritage with these species that makes them such powerful 'model organisms'.

To take just one example, sequencing the mouse genome allows us to identify more genes in the human genome than does sequencing the human genome alone. By identifying segments of DNA that are more similar between the two species than could be expected by chance, we can identify regions whose evolution has been constrained by the need to perform a specific function. In other words, we can identify a gene not because it *looks* like a gene, nor because an organism *treats* it like a gene (i.e., makes a product from it), but because it *evolves* like a gene. This approach, sometimes called 'phylogenetic footprinting' or 'shadowing' can be extended by examining genomes from a range of primates (Boffelli *et al.*, 2003).

1.3.2 Understanding the present

If we were to take a perspective to the biology of modern humans that neglects evolutionary history, what might we predict about the genetic diversity of our species, the significance of our phenotypic differences and the prevalence of disease-causing mutations?

First, we would be struck by the huge numbers of humans, especially when compared to other animals of similar body size. We might reasonably think that this should be mirrored by a correspondingly greater genetic diversity. Second, we might be struck by the clustered distribution of phenotypic diversity among modern human groups and might expect this to be matched by a similar structuring of genetic diversity. Third, we might expect that disease-causing mutations would be specific to different continental groups, in a similar manner to some of their easily observable 'normal' phenotypes. As we shall see in this book, all of these conclusions would be wrong.

To understand why this is so, we must comprehend that the past is not simply something that happened, and is packaged up and studied for its own sake, but is more properly considered as the source of the present. The present should only be seen as another small step in shaping this past. If we are to improve our present circumstances, we must take account of how that present has come to be. An evolutionary perspective does not just answer the question 'What happened in the past?', but also the question 'Why is the present like it is?'

Once we understand that the obvious differences between peoples' appearances can be unreliable indicators of biological origins, we start to appreciate the other factors that have shaped and continue to shape human biology. The interaction of humans and their environments comes to the fore, as does an understanding of human adaptability in the face of huge variability in inhabited environments.

1.3.3 Improving the future

An evolutionary perspective on human genetic variation also allows us to make predictions about the future, both of biological research and of our species. We are able to pose many more questions within biology than we are presently able to answer. An evolutionary perspective tells us how we might go about answering these questions, and about what kinds of answers we might expect.

Phenotypic traits of humans, be they skin color, height or disorders such as diabetes, are controlled by a combination of inherited and environmental factors, and stochastic developmental and molecular processes. The easiest traits to dissect genetically are those determined in large part by single genes – so-called **Mendelian** traits. However, many of the phenotypic traits of most interest to both anthropologists and physicians are not so simple. These complex traits are governed by interactions between chance events, multiple genes and the environment, and disentangling these interactions will help to relieve the considerable burden of complex diseases on individuals and economies.

Genes involved in complex traits can be identified in one of two ways. In a physiological or biochemical approach, we identify the gene product as a causal factor, based on an understanding of the physiological and molecular basis of the trait. In the alternative positional approach, we locate the position of the gene within the genome by identifying a chromosomal segment that is consistently co-inherited with the trait. The physiological approach has had limited success, and as we shall see in this book, an evolutionary perspective greatly enhances our ability to adopt the positional approach to gene identification.

A knowledge of our past allows us to predict something about the numbers and frequencies of gene variants expected to be involved in a given trait and to choose the best strategy of finding them: what populations to choose, and which segments of the genome to concentrate on. Not only that, but an evolutionary perspective helps to understand and predict which individuals will respond best to each therapy and how best to focus limited screening resources. Finally, genes of medical relevance are often sites of past selection, and evolutionary analyses of human diversity can offer a shortcut to identifying these important regions of the genome.

1.4 Chronological considerations

We have said that 'the past is the source of the present', and this is true in the academic field of human evolutionary genetics as much as in real life. This exciting subject owes its current status to developments and debates over the last 150 years in genetics, paleontology, archaeology, anthropology and linguistics. In this book we have avoided cataloging this history, instead taking a twenty-first century perspective, but we discuss key developments where they are relevant, and provide a time-line in *Figure 1.6*. In describing the origins of our species and its spread across the world it would be inappropriate not to take a chronological course, and this we do in Chapters 7 to 12; *Figure 1.7* summarizes many of the important events in human evolution in a time-line.

Data on human genetic variation are accumulating at an unprecedented rate. The only practical way to keep abreast of this tide of discovery is to use the internet, both to survey the latest publications on a topic through web databases such as PubMed (http://www.ncbi.nlm.nih.gov/entrez/query.fcgi?db=PubMed), and to access the diversity data as they appear through many websites, some of which are listed in *Box 1.3*.

1786	Recognition of language families
1856	Discovery of Neanderthal type specimen
1859	Publication of Darwin's 'The Origin of Species'
1866	Publication of Mendel's 'Experiments in Plant Hybrids'
1871	Publication of Darwin's 'The Descent of Man'
1900	First genetic polymorphism –- ABO blood group (Landsteiner)
1908	Hardy-Weinberg principle formulated
1918	Fisher reconciles Darwin's natural selection and Mendel's mechanism of inheritance
1925	*Australopithecus* fossil described from South Africa
1930 –32	Fisher, Haldane and Wright publish the foundations of modern population genetics
1944	DNA shown to be heritable material
1949	Radiocarbon dating introduced
1953	Double-helical structure of DNA described
1956	Human chromosome number described
1957	First molecular basis for a disease elucidated (sickle cell anemia)
1959	Y chromosome shown to be sex-determining
1966	Genetic code deciphered
1967	5 MY timescale for hominid evolution published
1968	Neutral theory of molecular evolution (Kimura)
1969	Internet first successfully tested
1977	Publication of DNA sequencing methods
1978	First human *in vitro* fertilization

1978	First human RFLPs described
1980	First genome (bacteriophage φX174) sequenced
1981	Human mtDNA genome sequenced
1984	DNA-DNA hybridization shows human-chimp common ancestry DNA fingerprinting (minisatellites) discovered
1985	First human ancient DNA results published Invention of PCR First Y-chromosomal polymorphism
1987	Development of laser-induced fluorescent detection of DNA
1988	Launch of Human Genome Project
1989	First human microsatellites described
1990	Development of capillary electrophoresis for sequencing
1991	Human Genome Diversity Project proposed
1996	First mammal cloned from adult cell (Dolly)
1997	First Neanderthal mtDNA sequence
1998	SNP consortium launched
1999	First human chromosome sequenced (Ch 22)
2001	Release of draft human genome sequence
2002	Release of draft mouse genome sequence Launch of HapMap project

Figure 1.6: Time-line of important developments in the field of human evolutionary genetics.

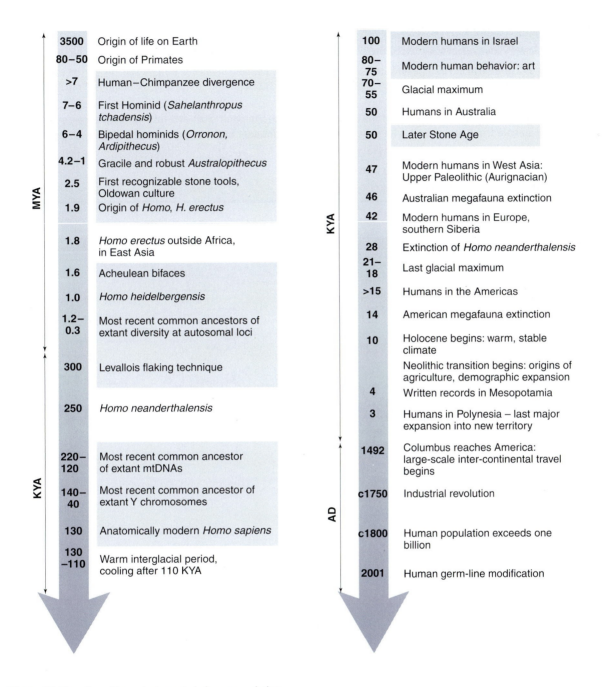

Figure 1.7: Time-line of important events in human evolution.

All times are approximate, and many are disputed. Blue shading indicates events believed to have occurred in Africa, and illustrates the importance of Africa in our ancestry. Some crucial events, like the origin of language, cannot be dated.

BOX 1.3 Online databases of genetic diversity

The internet resources available to the evolutionary geneticist are vastly greater now than even 5 years ago, but can be expected to develop significantly over the next 5 years. These improvements are just one strand of the present extremely rapid rate of change in our genetic knowledge.

At the time of writing, a researcher (or member of the public) can browse through draft and finished sequences of human chromosomes using standard web browsers like Netscape Navigator and Internet Explorer at a number of sites:

▸ UCSC: http://genome.ucsc.edu/cgi-bin/hgGateway?org=Human

▸ NCBI: http://www.ncbi.nlm.nih.gov/mapview/map_search.cgi

▸ Ensembl: http://www.ensembl.org/Homo_sapiens/)

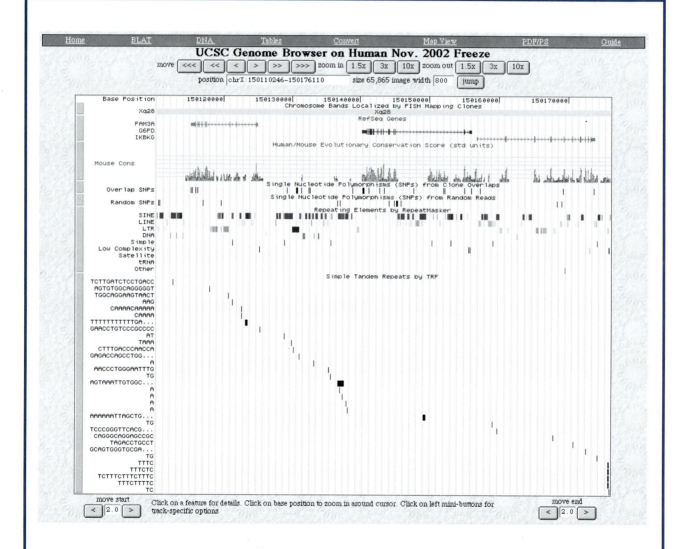

Browsing the genome.

Screenshot of a genome browser (UCSC) illustrating SNP and tandem repeat variation around the *G6PD* gene on the tip of the long arm of the X chromosome. Underneath the chromosomal position of a ~60-kb interval containing the *G6PD* gene are tracks detailing different features of this sequence including (from top to bottom): the three genes found in the interval, regions of high conservation between human and mouse, the position of single nucleotide polymorphism found by two different methods, dispersed repeats of different types and tandem repeats with different repeat unit sequences.

These 'genome browsers' integrate information of many different aspects of DNA sequence contained in linked databases, for example, the positions of genes and repeated sequences, sites of variation within humans and homology to other genomes (see *Figure*). This publicly accessible information represents the culmination of millions of dollars worth of research to sequence the human genome and uncover variation within it.

These linked databases allow researchers to identify sites of variation that they can then study in their own laboratory using population samples of their choosing. However, at present there are no databases summarizing all that is known about the frequencies of different polymorphisms in different human populations. The best alternatives are databases that contain subsets of this information, for example:

▶ ALFRED (Allele frequency database),

http://alfred.med.yale.edu/alfred/index.asp

▶ dbSNP (database of Single Nucleotide Polymorphisms),

http://www.ncbi.nlm.nih.gov/SNP/

▶ HGVbase (Human Genome Variation database)

http://hgvbase.cgb.ki.se/

This means that whereas the evolutionary geneticist interested in inter-specific relationships can download much of the data required for novel analyses from databases such as Hovergen (database of HOmologous VERtebrate GENes http://pbil.univ-lyon1.fr/databases/hovergen.html), human evolutionary geneticists presently generate much of their own data themselves.

Some human genetic variation causes genetic diseases. These are described in a constantly updated database, Online Mendelian Inheritance in Man (http://www.ncbi.nlm.nih.gov/Omim/), and we use OMIM numbers throughout this book when we refer to particular diseases, e.g. cystic fibrosis (OMIM 219700).

What does the future hold? Two current projects stand out as being particularly important future advances for human evolutionary genetics. First, the chimpanzee genome project, which is expected to produce 'rough draft' quality sequence by the time this book is published and reliably sequence the genome within the next few years. Second, the HapMap project, which seeks to characterize the haplotype structure of human single nucleotide polymorphism in four continental populations by the end of 2005. As well as providing quantum leaps of data for further investigation, both of these projects will help to identify sites of recent selection in the human genome and to illuminate many of the genetic changes that make us human. Beyond these large-scale projects there will undoubtedly be numerous unforeseen advances in our understanding of human evolution, especially now that the tools available for such work are readily available.

Further reading

Strachan T, Read AP (2003) *Human Molecular Genetics 3.* Garland Publishing Inc., New York.

References

Arnaiz-Villena A, Elaiwa N, Silvera C et al. (2001) The origin of Palestinians and their genetic relatedness with other Mediterranean populations. *Hum. Immunol.* **62**, 889–900. (This paper was subsequently retracted)

Boffelli D, McAuliffe J, Ovcharenko D et al. (2003) Phylogenetic shadowing of primate sequences to find functional regions in the human genome. *Science* **299**, 1391–1394.

Carvajal-Carmona LG, Soto ID, Pineda N et al. (2000) Strong Amerind/white sex bias and a possible Sephardic contribution among the founders of a population in northwest Colombia. *Am. J. Hum. Genet.* **67**, 1287–1295.

Foster MW, Sharp RR (2002) Race, ethnicity, and genomics: social classifications as proxies of biological heterogeneity. *Genome Res.* **12**, 844–850.

MacEachern S (2000) Genes, tribes, and African history. *Curr. Anthropol.* **41**, 357–384.

Ohno S (1996) The Malthusian parameter of ascents: what prevents the exponential increase of one's ancestors? *Proc. Natl Acad. Sci. USA* **93**, 15276–15278.

Renfrew C (1999) Reflections on the archaeology of linguistic diversity. In: *The Human Inheritance: Genes, Language, and Evolution* (ed. Sykes B). Oxford University Press, Oxford, pp. 1–32.

Ruhlen M (1991) *A Guide to the World's Languages, Vol. 1: Classification.* Stanford University Press, Stanford.

Ruhlen M (1994) *On the Origins of Languages: Studies in Linguistic Taxonomy.* Stanford University Press, Stanford.

SECTION TWO

How do we study genome diversity?

The success of the Human Genome Project has given us an unprecedented understanding of the structure and organization of our genome. In turn this has led to profound insights into its function and evolution, and the nature and dynamics of genome variation. This variation is the raw material of human evolutionary genetics.

CHAPTER 2   *Structure, function and inheritance of the human genome*
This chapter introduces the structure and function of DNA, genes and the genome, describing the packaging of DNA into chromosomes, and the means by which information encoded in genes is expressed as proteins. The different inheritance patterns of different segments of the genome, and the process of genetic recombination, both key ideas in understanding the patterns of genome variation among human populations, are explored.

CHAPTER 3   *The diversity of the human genome*
Next we ask how the sequence and structure of the genome varies between different genome copies. This variation is over a wide range of different scales, from single nucleotide polymorphisms through insertion polymorphisms of dispersed repeat sequences and variation in the numbers of tandemly repeated sequences, to large-scale changes in the structures of chromosomes. Each of these different types of polymorphism has characteristic mutation rates and processes. We explore the structures of haplotypes – combinations of different polymorphisms along a single segment of the genome, which are powerful tools to investigate our evolutionary past.

CHAPTER 4   *Discovering and assaying genome diversity*
This chapter describes the methods for discovering and assaying this genome variation. The polymerase chain reaction, together with the availability of the human genome sequence, have revolutionized polymorphism discovery. We discuss the methods available for typing the different classes of polymorphism in population samples, and the specialized, technically difficult, but potentially highly informative field of 'ancient DNA' studies.

CHAPTER TWO

Structure, function and inheritance of the human genome

Understanding the subject of human evolutionary genetics relies on some basic biological knowledge that this chapter summarizes:

▶ the structure of DNA;

▶ how information in DNA is used to make RNA and proteins (gene expression);

▶ the organization of the human genome;

▶ the structure of chromosomes;

▶ how information in DNA is passed on from one cell to daughter cells, and from one generation to the next (inheritance).

The detail of the structure of our genetic material, and the biochemistry and molecular biology of the processes by which it is packaged, replicated and expressed is beyond the scope of this book, and is covered elegantly and extensively in sources given in the bibliography to this chapter. This chapter presents some of these fundamentals in a selected and focused way, with an eye to the topics that we will address later on.

Figure 2.1 gives an overview of the human genome, showing its size and how it is partitioned among the **nuclear** chromosomes and **cytoplasmic** mitochondrial DNA (mtDNA). *Figure 2.2* gives an overview of how different segments of our genetic material follow the path from one generation to the next, via the **germ-line** (eggs and sperm).

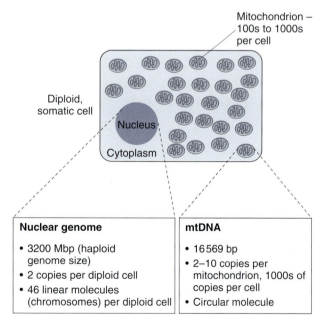

Figure 2.1: Overview of the human genome.

2.1 Structure of DNA

All organisms, with the exception of a few kinds of viruses, use **deoxyribonucleic acid (DNA)** as their genetic material. DNA is an extraordinary informational macromolecule that plays two central biological roles:

▶ it carries the instructions for making the components of a cell (mostly proteins, which themselves can manufacture further components);

▶ it provides a means for this set of instructions to be passed to the daughter cells when a cell divides.

DNA is a polymer, and its monomeric subunits are molecules called **nucleotides**. There are four varieties of nucleotide, which differ in portions known as **bases**. The bases are **adenine, guanine, cytosine** and **thymine**, abbreviated as A, G, C and T, and it is the sequence of these parts of nucleotide molecules that carries the genetic information. Adenine and guanine, double-ringed molecules, are **purines**, and cytosine and thymine, single-ringed, are **pyrimidines**. Each base is joined to a sugar molecule, **deoxyribose**, and each deoxyribose has a phosphate group attached to it; the sugar and phosphate play only a structural role in DNA, and themselves carry no information. The structures of these nucleotides are shown in *Figure 2.3*. Some nucleotides can be chemically modified after their incorporation into DNA: an important example is the **methylation** of cytosine, to give 5-methylcytosine. Methylation plays a role in gene regulation, and its consequences for mutation are discussed in Chapter 3 (Section 3.2.5).

The phosphate of one nucleotide is joined to the sugar of another (*Figure 2.4a*), and so on, and this forms the '**sugar-phosphate backbone**' of DNA. Because the deoxyribose group itself is asymmetrical, it provides a **polarity** to the backbone. The carbon atoms making up the deoxyribose molecule are given numbers from 1′ (pronounced 'one prime') to 5′. Phosphate groups attach to the 3′ and 5′ carbon atoms of the ring, and this provides a way to refer to the different ends of a DNA molecule: the **3′ end** has a free hydroxyl (-OH) group (unattached to another nucleotide) on the 3′ carbon, and the **5′ end** has a free hydroxyl group on the 5′ carbon. The polarity of a nucleic acid molecule [whether DNA or RNA (ribonucleic acid)] is important because fundamental processes like DNA **replication**, the **transcription** of DNA into RNA, and the **translation** of RNA into protein will proceed in one direction only along its length. By convention, the sequence of bases in a DNA or RNA molecule is always written left to right in the direction 5′ to 3′.

The structure described above is that of **single-stranded DNA**. Within a cell, DNA exists for most of the time in the **double-stranded** form, as two long strands, spiraling round each other in a double helix (*Figure 2.4b*). The bases of one strand project into the core of the helix, and here they pair with the bases of the other, **complementary**, strand. **Base-pairing** (*Figure 2.5*), where an A pairs strictly with a T, and a C with a G, underlies the two fundamental roles of DNA – expression of genetic information via RNA, and replication of genetic information prior to cell division. A single strand of DNA can act as a **template** either for the enzymatic synthesis of a complementary strand of RNA

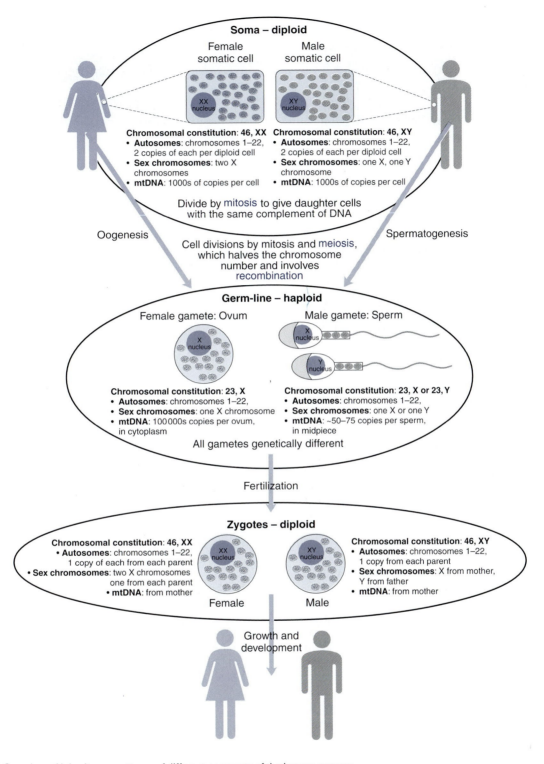

Figure 2.2: Overview of inheritance patterns of different segments of the human genome.

(transcription), using **ribonucleotides** as building blocks (see Section 2.2), or of a complementary DNA strand (replication), when the building blocks are deoxyribonucleotides. The lengths of double-stranded DNA molecules are described in units of base-pairs (bp), and for longer molecules kilobase-pairs (usually abbreviated to **kilobases, kb**) or megabase-pairs (**megabases**, Mb).

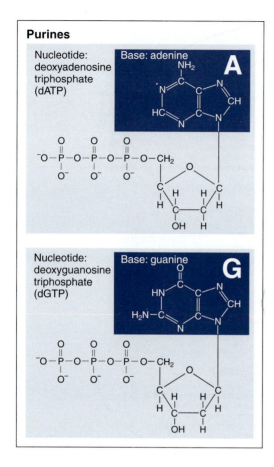
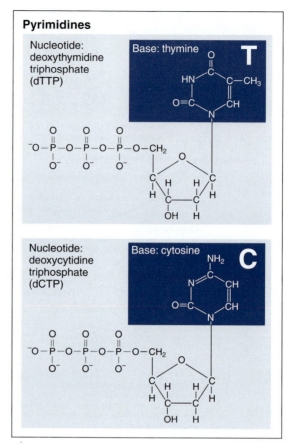

Figure 2.3: Chemical structures of the nucleotide components of DNA.

These are the standard four nucleotides; some bases in DNA are modified by the addition of other chemical groups. When nucleotides are incorporated into DNA, two of the phosphate groups are lost.

The chemical bonds that tie the atoms together within a single strand of DNA are strong and require considerable energy to break them – these are **covalent bonds**. The **hydrogen bonds** which exist between the bases of one strand and another within a double helix of DNA are much weaker, and can be broken *in vitro* by relatively gentle processes such as brief heating to 95°C or exposure to alkaline pH, or by active processes within cells. This separation of the strands of a double helix is called **denaturation**, or '**melting**', and must occur prior to either transcription or replication. There are three hydrogen bonds between a G and a C, and two between an A and a T; more energy is needed to denature GC-rich DNA than an equivalent length of AT-rich DNA.

In February 2001, a milestone was reached in the attempt to sequence the entire human genome: substantially complete draft versions of the sequence were announced (*Box 2.1*) (Lander *et al.*, 2001; Venter *et al.*, 2001). Note that this was a triumph not for the science of genetics, which is concerned with inheritance and variation, but for the newer and biochemically driven science of **genomics** – the mapping, sequencing and analysis of genomes. *Box 2.2*

discusses the ethical issues behind the competing efforts to sequence the human genome.

2.2 Genes

As we shall see later, much human DNA appears to have no specific function. However, some segments of DNA contain the instructions for the synthesis of proteins, or in some cases RNA molecules that function in their own right. These segments are called **genes**.

Protein production from a gene does not proceed directly from the DNA template. DNA is first transcribed to make an intermediate RNA molecule, known as **messenger RNA** (mRNA), and this then acts as the template for protein production. Production of many copies of mRNA from a single gene amplifies the number of copies of the corresponding protein that can be made, and also provides many opportunities for regulatory processes to act to influence the final amount and properties of active protein. RNA differs from DNA not only in the kind of sugar molecule it contains (ribose, rather than deoxyribose), but also in one of its bases. RNA contains the pyrimidine base

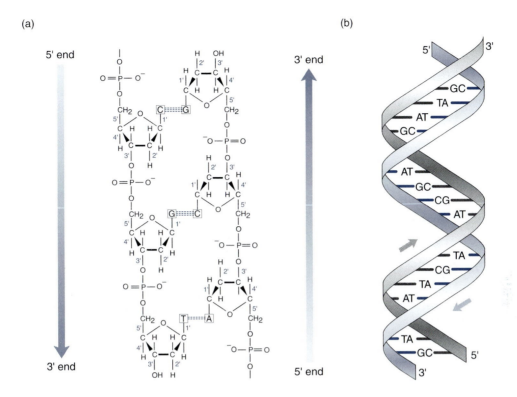

Figure 2.4: Double-stranded helical structure of DNA.

(a) The two strands of DNA are anti-parallel since the linking of the 3′ carbon atom of one base to the 5′ carbon of the next is in opposite directions on the two strands. Since sequences are written 5′ to 3′, the sequence of this three-base DNA molecule on the left-hand strand is CGT, while on the right-hand strand it is ACG.

(b) The two strands are wound round each other in a double helix. Arrows indicate the direction of each strand, 5′ to 3′.

BOX 2.1 The draft sequence gives an overview of the human genome

The near-completion of the human genome sequence is an enormous achievement, and provides us with a rich resource for:

▶ understanding the evolution of human genes and the human genome;

▶ comparative studies of other mammalian genomes;

▶ understanding chromosome structure and function;

▶ finding genes;

▶ identifying sequence variants for use in evolutionary and disease studies.

Humans, like almost all animals, are **diploid** – that is, we have two copies of our genome in each of our somatic cells, the cells that make up our tissues. The current estimate for the size of the human **haploid** genome (i.e., one copy of the genome) is about 3200 Mb and the two major publications on the draft sequence included 2693 Mb (Lander *et al.*, 2001) and 2907 Mb (Venter *et al.*, 2001) respectively. These two efforts differed from each other in methodology and philosophy, and these differences have led to much discussion and some controversy (Waterston *et al.*, 2002). The publicly funded effort (Lander *et al.*, 2001) is committed to immediate public release of data on the Internet, and this has greatly facilitated research, including that of the rival privately funded sequencing effort (see *Box 2.2*). The experience of freely downloading the entire draft sequence is as impressive for its demonstration of the freedom of information as it is of the grandeur of the sequencing project.

The draft sequence allows the examination of the large-scale properties of the genome. The nucleotide composition of a sequence is usually given as 'GC-content', which expresses the percentage of base pairs that are G–C, as opposed to A–T. Globally, this percentage is 41% for the human genome, but there are multi-megabase sections of the genome that depart substantially from this figure, ranging from 36% to 50% average GC-content. Domains of high GC-content correspond to the sections of the genome that are particularly rich in genes, rich in SINEs such as *Alu* elements, and are areas that do not stain darkly with the dye Giemsa after treatment to produce G-bands (light G-bands; see Section 2.4).

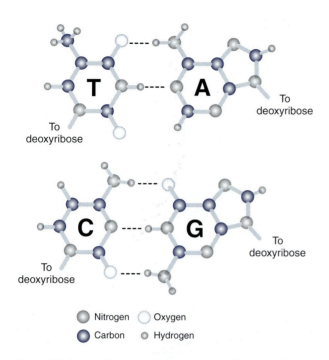

To deoxyribose

To deoxyribose

To deoxyribose

To deoxyribose

To deoxyribose

○ Nitrogen ○ Oxygen
● Carbon ○ Hydrogen

Figure 2.5: Base-pairing between thymine and adenine, and between cytosine and guanine.

Hydrogen bonds are indicated as dashed lines.

uracil (U) in place of thymine (T). Like thymine, uracil pairs with adenine (A).

While DNA and RNA are composed of four kinds of nucleotides, proteins (sometimes called **polypeptides**) are polymers composed of 20 kinds of **amino acids**, and the complex process by which information encoded in one kind of polymer is converted into the other is called **translation**. The nucleotide code, known as the **genetic code**, is illustrated in *Figure 2.6*. In this near-universal code, a set of

three adjacent nucleotides (a **codon**) specifies one of the amino acids, or alternatively represents an instruction to stop the process of translation (**stop codon**). The process of translating three bases into one amino acid involves one of a set of small intermediate RNA molecules (**transfer RNA, tRNA**) and is carried out in the cytoplasm by a large RNA–protein complex called the **ribosome** (*Figure 2.7*). Since there are 64 possible codons, there is more than enough capacity in the genetic code to specify the 20 amino acids and a stop signal. The code is therefore **redundant**: at one extreme each of the amino acids leucine, serine and arginine has six different corresponding codons, while at the other tryptophan and methionine each has only one. The pattern of redundancy in the code, and the relationship between the physicochemical properties of the amino acids and their codons, is nonrandom, and has important implications for the effects of mutations on protein function which are discussed in Chapter 3 (Section 3.2.6). The triplet codons within a gene are arranged adjacent to each other, without intervening 'punctuation'. The first codon is universally AUG, which both encodes a methionine and is part of the initiation signal; the last is one of the three possible stop codons (UAA, UAG, or UGA).

The structure of a typical human protein-coding gene is shown in *Figure 2.8*. There is much more to a gene than the DNA segment which actually encodes the protein: gene transcription proceeds from the 5′ to the 3′ end, and outside the coding region at the 5′ end lie sequences necessary for transcription to be initiated and regulated, including the **promoter** (see *Box 2.3*). At the 3′ end there are sequences which signal the termination of transcription, and the addition of a tail of adenosine nucleotides [**poly(A) tail**] to the end of the mRNA molecule, which is important for mRNA stability.

Almost all human genes are not a continuous run of adjacent codons, flanked by 5′ and 3′ untranslated regions, but are broken up into segments known as **exons**, each

AAA AAG	Lysine	Lys	K	CAA CAG	Glutamine	Gln	Q	GAA GAG	Glutamic acid	Glu	E
AAC AAU	Asparagine	Asn	N	CAC CAU	Histidine	His	H	GAC GAU	Aspartic acid	Asp	D
ACA ACG ACC ACU	Threonine	Thr	T	CCA CCG CCC CCU	Proline	Pro	P	GCA GCG GCC GCU	Alanine	Ala	A
AGA AGG	Arginine	Arg	R	CGA CGG CGC CGU	Arginine	Arg	R	GGA GGG GGC GGU	Glycine	Gly	G
AGC AGU	Serine	Ser	S								
AUA	Isoleucine	Ile	I	CUA CUG CUC CUU	Leucine	Leu	L	GUA GUG GUC GUU	Valine	Val	V
AUG	Methionine	Met	M								
AUC AUU	Isoleucine	Ile	I								

UAA UAG	STOP		
UAC UAU	Tyrosine	Tyr	Y
UCA UCG UCC UCU	Serine	Ser	S
UGA	STOP		
UGG	Tryptophan	Trp	W
UGC UGU	Cysteine	Cys	C
UUA UUG	Leucine	Leu	L
UUC UUU	Phenylalanine	Phe	F

Figure 2.6: The genetic code.

Three-base codons and their corresponding amino acids (including three-letter and single-letter abbreviations) or translation STOP signals are shown. This is the code used in the human nuclear genome; the code in mtDNA differs (see *Box 2.8*).

BOX 2.2 Opinion: The ethics of genome sequence publication

The availability of DNA sequence data has been the subject of a lively debate for over 20 years. As soon as the first sequences were published it became apparent that they could only be used if available in computer-readable form. In the early 1980s two organizations, the European Molecular Biology Laboratory (EMBL) and the National Institutes of Health (NIH) independently developed databases for nucleic acid sequence data; these were joined by the Japanese National Institute of Genetics a few years later and, in the mid-1980s, these organizations joined into an effective international collaboration to share data. Today (March 2003 release), this community database (EMBL-bank/GenBank/DDBJ) contains over 40 billion base pairs of sequence, from over 100 000 different organisms. All three databases implemented a policy that was both courageous and foresighted – to make all of their data freely available, without let, hindrance nor license, to all, whether they be companies, academics or John Smith from 1345 Main Street, Bloomsville, Ohio. By the late 1980s most reputable scientific journals were demanding deposition of sequence data in this database as a pre-condition for the publication of a scientific paper. This immediately gave rise to a conflict: commercial companies, and some academics, saw their DNA sequence data as intellectual property that might be turned into money. They were often reluctant to make their data freely available, since this might both compromise patent protection and reduce the value of their 'property'. Nowhere was this conflict so apparent as with the human genome sequence. Another reason for conflict is that in very large sequencing projects, such as that for the human genome, data are being produced from the sequencing centers in very large amounts months, if not years, before they are analyzed. Yet, these raw sequences are extraordinarily valuable for the scientific community. At the beginning of the Human Genome Project, in 1996, the major funding bodies, the Wellcome Trust in the UK and the NIH in the USA, convened a meeting in a chilly February in Bermuda to discuss the conditions of sequence data release. All present accepted what became known as the 'Bermuda Agreement' – all data from the Human Genome Project would be deposited in the public sequence databanks and, moreover, each sequencing center would release its own data every day. The decision by Craig Venter and colleagues, in May 1998, to compete with the public Human Genome Project by a commercially marketable human genome sequence, placed enormous strains on the public project. There was, on the one hand, a widespread revulsion that the sequence of 'the' human genome should be owned by a company; surely this must be information, like the calendar, which belongs to no one, but to all. There was also the realization that the public Human Genome Project and Celera Inc. had a very asymmetrical relationship: the former had no access to Celera's sequence data, the latter had free access to the former's data. Indeed, it is now clear that the Celera sequence could not have been 'completed' without incorporation of the public data. Finally, it was soon clear that Celera intended to publish an analysis of their sequence. Yet, to do so, public deposition of the sequence was a prerequisite for any respectable journal. The 'completion' of the human sequence, by both the Celera and public projects, was announced with high drama by Prime Minister Blair and President Clinton in June 2000. The agreement was for these to be published 'back-to-back' by the journal *Science*. The scientific community soon saw that this would be a major breach of the fundamental principle of publication being backed by the free availability of the underlying data. Despite widespread protest *Science* allowed Celera to publish (Venter *et al.*, 2001) whilst not making their data freely available. The public sequence of the human genome was, as a result, published not in *Science*, but in *Nature* (Lander *et al.*, 2001).

Why does this matter? There are two reasons. The first is ethical. It is that sequence data – of any organism – are fundamental to biology; they are as fundamental as the periodic table is to chemistry and physics and as Euclid's axioms are to mathematics. They rightly belong to all of us; they are not commodities to be sold in the market place like apples and oranges. The second reason is pragmatic. Modern science is enormously empowered by public databases. Data that are secret, or protected, are not part of science, which is best defined, in John Ziman's words, as 'public knowledge'. Databases are most useful if, within a given scientific domain, they are not fragmented. Much of modern genomics would simply be impossible if the universe of sequence data were split between hundreds of different databases. By allowing publication without sequence data deposition in the community sequence database *Science* has encouraged fragmentation of the universe of sequence data. This will be harmful to scientific progress. Finally, companies, such as Celera, which publish sequence data without public deposition are trying to have the best of two worlds – they crave the kudos that scientific publication brings, yet they protect their data from public eyes for the simple benefit of profit.

Michael Ashburner, Department of Genetics, University of Cambridge, UK

---

containing on average only 50 codons (and not necessarily an integral number of them). Between these lie noncoding segments known as **introns**. Transcription produces a **pre-mRNA** which contains both the exons and the introns, and a complex process known as **splicing** then removes the introns before mature mRNA can be transported from the nucleus into the **cytoplasm** and translated (*Figure 2.7*). This RNA molecule is stabilized by the addition of a cap consisting of a specialized nucleotide at the 5′ end, as well as the poly(A) tail at the 3′ end. As with every step in the process between DNA and protein, splicing is exploited as a powerful means to regulate the expression of genes (*Box 2.3*).

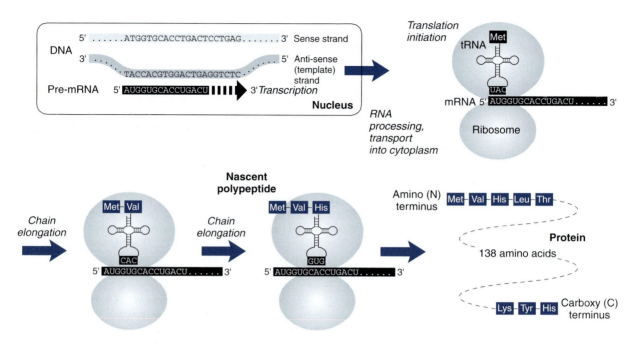

Figure 2.7: DNA makes RNA makes protein.

Transcription of a protein-coding gene (here the β-globin gene) occurs in the nucleus, and after RNA processing and transport into the cytoplasm the mRNA is translated in the ribosome. Each codon in the mRNA is recognized by the complementary anticodon in a transfer RNA (tRNA) molecule that bears the appropriate amino acid. As the ribosome moves in a 5′ to 3′ direction along the mRNA, amino acids are added to the growing chain by formation of peptide bonds between them. The 5′ end of the mRNA corresponds to the end of the protein bearing a free amino ($-NH_2$) group (N terminus), and the 3′ end to that bearing a free carboxy (-COOH) group (C terminus). Note that in this particular protein the N-terminal methionine is cleaved off after translation; many proteins undergo other important post-translational modifications.

While *Figure 2.8* shows a 'typical' gene, in reality human genes vary enormously in size and complexity (*Table 2.1*). The record-breaking dystrophin gene, mutations in which can cause the disease Duchenne muscular dystrophy (OMIM 310200), spans over 2.7 Mb of DNA, with 79 exons averaging only 180 bp in size marooned in an ocean of introns with an average size of 30 kb (and representing 99.4% of the gene). So large is this gene that the production of a

---

BOX 2.3 Regulating the expression of genes

While almost all human cells contain a complete complement of genes, only a subset of these genes is expressed in any one cell type at any one time. The cloning of Dolly the sheep shows that differentiation of a mammalian somatic cell (in this case, a mammary cell) does not involve the loss of genetic information. In other words, DNA that contains genes not expressed in a given cell type is not removed from the cell.

The quantitative, spatial and temporal regulation of the expression of genes is an enormously complex process involving interactions of a very large number of factors within the cell, and evolution has exploited every stage of the gene expression process as an opportunity for regulation; for example:

▸ different cell types often employ different transcription start sites, giving variability in the 5′ ends of transcripts, and thus the amino termini of proteins, which can give them different properties;

▸ proteins acting as positive and negative regulators bind 5′ to the transcription start site to influence the timing and rate of transcription, and, where the regulators are cell-type specific, the cell specificity;

▸ alternative splicing pathways, utilizing different exons from within a gene, can produce different forms of proteins with different functions, or even, under some conditions, nonfunctional proteins;

▸ different factors can act to influence mRNA stability and half-life, and thus the amount of protein produced;

▸ the properties of the final protein product are influenced by post-translational modifications such as specific cleavage, the addition of chemical groups such as phosphates or acetyl groups, cross-linking of chains via cysteine residues, and conjugation with sugar or lipid molecules.

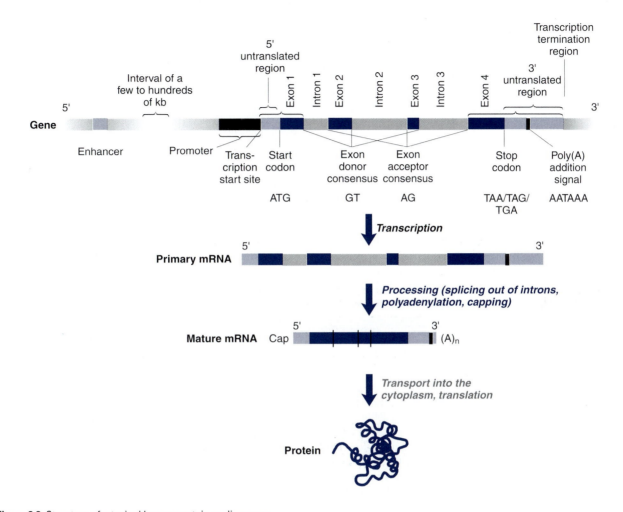

Figure 2.8: Structure of a typical human protein-coding gene.

Major structural and functional features of a typical gene are shown, together with the sequences of absolutely or very highly conserved elements. Primary and mature mRNA products from the gene are also shown; note that splicing can commence on incompletely transcribed mRNAs. The different elements of the gene are not to scale, and in particular the introns illustrated here are small relative to the exons (see *Table 2.1* for typical sizes of these elements).

TABLE 2.1: PROPERTIES OF HUMAN GENES.

	Mean values for 1804 genes[a]	Dystrophin gene[b]	*SRY* gene[c]
Exon length	145 bp	180 bp mean	612 bp
Exon number	8.8	79	1
Intron length	3365 bp	30 000 bp mean	–
5′ UTR length	300 bp	200 bp	140 bp
3′ UTR length	770 bp		133 bp
Coding sequence length	1340 bp (447 aa)	14 000 bp	612 bp (204 aa)
Genomic extent	27 kb	2700 kb	1 kb

[a]From Lander *et al.* (2001). [b]From Koenig *et al.* (1987) and Roberts *et al.* (1993). [c]From Behlke *et al.* (1993).

aa, Amino acid.

single complete transcript is estimated to take as long as 16 h (Tennyson *et al.*, 1995). The *SRY* gene, on the other hand, which is responsible for the initiation of testis development and thus male sex-determination, spans only a kilobase or so and has no introns at all.

A particularly controversial issue has been the number of genes in the human genome; the debate has been intensified by the publication of the draft sequence, and is discussed in *Box 2.4*.

2.3 Noncoding DNA

Despite the importance of genes in determining our **phenotypes**, about 98.5% of our DNA does not comprise coding sequences of genes, and about 70% of our DNA is not transcribed. The function of most of this nongenic material is not known, but some of it certainly does play essential roles in cells. The human genome is divided into discrete packages called chromosomes, and each chromosome has essential structural features that have corresponding essential noncoding DNA sequences (see Section 2.4). However, the vast majority of the DNA that is not involved in the processes of RNA and protein production has no known function. Indeed, most of it may have no function at all, although this view could simply reflect our lack of understanding; the experiment of removing all but the apparently 'essential' components of the genome and observing the consequences cannot be done.

The distinction is often made between 'single-copy' and 'repetitive' DNA, and genes tend to be placed in the former category. However, many genes have more than one copy, some have related copies which have arisen through gene duplication events (*Box 2.5*), and some lie in duplications of large chromosomal segments (**segmental duplications**) that cover many megabases and have their origins in recent primate evolution (Eichler, 2001). At the deepest level, some genes show sequence similarities to others which reflect ancient duplications of an entire ancestral genome which

---

BOX 2.4 Opinion: How many human genes?

A general assumption exists that higher vertebrates have greater phenotypic complexity, and therefore have more complex genomes, than all other organisms. This assumption was challenged when the International Human Genome Sequencing Consortium announced in February 2001 (Lander *et al.*, 2001) that there are only 32 000 genes in the human genome. This is only slightly more than the number found in the nematode worm (23 000 genes) or fruit fly (26 000) and fewer than in rice (45 000). Previous estimates had varied widely between 30 000 and 170 000 genes, with a popular round figure having been 100 000.

Despite the official announcement of 2001, the human genome sequence is not complete. New sequences are still being added into the chromosome maps and their relative locations are being adjusted. Gene prediction is done on top of this constantly changing base. Given the size of the human genome, and difficulties in defining the exact numbers and locations of exons, the preferred method for finding genes is to use computer programs that initially over-predict putative exons. Then, more sophisticated programs are used to prune away the spurious exons and combine the remaining ones into genes that are supported by independent experimental evidence.

Only 15 000 complete human genes have been experimentally identified and named by the HUGO Gene Nomenclature Committee. Roughly the same number are defined in the partly overlapping RefSeq reference sequence database. Evidence for other genes, which is sometimes only fragmentary, is present in various other databases. Incomplete information leads to fragmentation of real genes into several smaller annotated genes. Since the initial announcement, the overall gene fragmentation has rapidly decreased but the percentage of the genome represented in the sequence has not significantly increased. In consequence, the number of genes located by the Ensembl genome annotation system has decreased from the initial 29 700 to 22 800 in 1 year. This has, however, not changed the extrapolated total gene count.

Initially, a great worry was that there might be 'dark matter' genes whose structure current methods cannot predict. Also, gene prediction algorithms are generally unable to deal with open reading frames smaller than 100 nucleotides. Evidence for these unknown genes has not materialized, making their existence in large numbers less likely.

It is becoming clear that the gene count is not the right measure to explain human and higher vertebrate complexity. Many plants evolve by polyploidy, using their gene copy redundancy to start evolving in new directions, as exemplified by the high gene count in rice. The vertebrates seem to have a different preferred mechanism, which is rarely used in plants: alternative splicing. At least 40% of human genes have evidence of more than one transcript. So common are the different versions of genes that the **transcriptome**, the combined variability of all transcripts produced by the genome, might yet reach the elusive count of 100 000.

Heikki Lehväslaiho, European Bioinformatics Institute, Cambridge, UK

Electronic references:

Ensembl Genome browser:
http://www.ensembl.org/Homo_sapiens/

HUGO Gene Nomenclature Committee:
http://www.gene.ucl.ac.uk/nomenclature/

RefSeq database
http://www.ncbi.nlm.nih.gov/LocusLink/refseq.html

BOX 2.5 Evolving genes: Gene duplication and exon shuffling

How does a new gene arise? This is an important question, because new genes might underlie important new cellular functions and organismal phenotypes, and be important in the development of new species. Genes do not arise out of nonfunctional DNA by the accumulation of mutations: gene evolution proceeds by what François Jacob has called 'molecular tinkering', including the following:

▶ **Gene duplication**: duplication of a segment of DNA including an entire gene provides a second gene copy which can mutate, acquiring a specialized function (Prince and Pickett, 2002). A good example is the evolution of the two clusters of globin genes – five functional β-like globin genes on chromosome 11, and three functional α-like globin genes on chromosome 16. These two clusters evolved from initial duplication of a single common ancestral gene around 500 MYA, and then underwent further gene duplications. The duplicated genes have acquired specialized roles at different developmental stages from human embryonic life to adulthood, where requirements for efficient uptake and release of oxygen by hemoglobin (a tetramer of two β-like and two α-like globin molecules) differ. There are three possible outcomes of gene duplications: neofunctionalization (the adoption of a novel function by one of the copies); subfunctionalization (the division of the functions of the progenitor gene between the copies); and pseudogene formation, where one of the copies accumulates mutations and becomes nonfunctional.

▶ **Exon shuffling**: because of the segmental, exonic organization of genes, DNA rearrangements can bring together novel combinations of exons that can acquire novel functions. This process can occur intergenically, or intragenically, in which case it can amplify a particular exon to provide a structurally repetitive protein. Useful outcomes of shuffling are facilitated by the correspondence of many exons to discrete structural domains of proteins that can also possess specific functional roles, such as binding membranes, DNA, ions or other ligands, or a particular enzymatic activity such as proteolysis. Exon amplification and shuffling have been particularly important in the evolution of the immunoglobulin superfamily, which includes many functionally diverse proteins including cell adhesion molecules and receptors as well as antigen recognition and binding proteins.

---

occurred early in the evolution of the vertebrate lineage, several hundred million years ago (Wolfe, 2001).

Aside from the relatively low-copy-number repetition described above, the human genome contains a vast number of very highly repetitive sequences (*Table 2.2*). This highly repetitive component comprises dispersed repeat sequences with copy numbers in the hundreds to hundreds of thousands, comprising about 45% of the genome (Lander *et al.*, 2001). The extravagant production of this kind of DNA seems to be the result of 'ignorant' or 'selfish' processes within the genome that amplify sequences which then cannot readily be removed. Provided the presence of these sequences is not strongly deleterious, they will be tolerated and persist, although in some cases they may have evolved genic functions. The structures of an **L1 element** and an ***Alu* element**, members of the two commonest classes of dispersed repeats, respectively the **LINEs**

(long interspersed nuclear elements) and **SINEs** (short interspersed nuclear elements), are shown in *Figure 2.9*. *Box 2.6* explains some of the history of repeat nomenclature for the curious reader.

2.4 Human chromosomes and the human karyotype

The linear nature of DNA means that large genomes, such as our own, correspond to extremely long molecules. So great is the size of our haploid genome that, if it were a single DNA duplex and stretched out, it would be about 1 m long. Clearly, since each of our nucleated somatic cells contains two copies of this genome in its nucleus, a mere 10 μm or so in diameter, it must be efficiently packaged. Apart from the issue of simply condensing the DNA, there is the added

---

TABLE 2.2: CLASSES OF DISPERSED REPEATS IN THE HUMAN GENOME.

Class	Copy no. per haploid genome	Fraction of genome	Autonomous transposition or retrotransposition?	Length
LINEs	850 000	21%	Yes	Up to 6–8 kb
SINEs	1 500 000	13%	No	Up to 100–300 bp
Retrovirus-like elements	450 000	8%	Complete copies, yes	6–11 kb (1.5–3 kb)
DNA transposon copies	300 000	3%	Complete copies, yes	2–3 kb (80–3000 bp)

Values given in parentheses are lengths of incomplete elements, incapable of autonomous transposition (see Section 3.4). Adapted from Lander *et al.* (2001).

BOX 2.6 *Alu* and *Kpn* sequences – the history of names

Alu sequences were first isolated as a result of experiments to analyze the complexity of the human genome (i.e., the proportions of repetitive and single copy DNA it contains). DNA was sheared to about 2 kb in size, then denatured by boiling. It was then allowed to reanneal, and any nonreannealed segments removed using a single-strand-specific nuclease, S1. The remaining fraction contains highly repetitive sequences that have reannealed rapidly. When run out on an agarose gel, a prominent band of about 300 bp was seen, representing about 5% of the total starting DNA. To ask whether this 300-bp fraction represented a single family of repeats, or several different families, it was digested with different restriction enzymes (see Chapter 4). The vast majority of the fraction was specifically cleaved by the enzyme *Alu* I (recognizing the sequence AGCT) into fragments of about 120 bp and 170 bp, showing that this was indeed a single species – the *Alu* family (Houck *et al.*, 1979).

L1 sequences were also originally named after a restriction enzyme. In this case, digestion of genomic DNA from humans and other primates with the enzyme *Kpn*I (recognizing GGTACC) followed by agarose gel electrophoresis showed, as well as a high molecular weight smear, four prominent bands (of 1.2, 1.5, 1.8, and 1.9 kb), representing a highly repeated sequence family. These were cloned and shown to be interspersed and unrelated to known sequences (Shafit-Zagardo *et al.*, 1982). They were later confirmed to be part of a single large repeated unit that came to be called the L1 element. The name '*Kpn* family' can still be encountered in the literature.

---

complication that, despite this condensation, it must be able to be faithfully replicated and segregated at cell division, and that genes must be accessible for expression.

Packaging of DNA occurs on many levels (*Figure 2.10*): the DNA molecule is first coiled round nucleosomes – octamers of proteins called **histones**. This nucleosomal fiber is then itself coiled, and several levels of higher order packaging give the required degree of condensation.

The problem of managing such a large genome is also alleviated by subdivision: instead of existing as a single DNA molecule, the genetic material of the somatic cells of normal human individuals is divided into 46 separate molecules. These are the **chromosomes**.

Figure 2.11 shows the structural features of a typical chromosome. It is important to stress that this kind of representation, showing a condensed, compact structure with easily recognizable chromosomal arms, reflects the situation

for only a small proportion of the cell cycle, and that the banding pattern is produced artificially by staining methods. Except for a period during mitosis or meiosis, when cells are dividing, chromosomes are diffuse, and discrete structural features are difficult or impossible to recognize. Each chromosome has two essential elements to ensure its correct segregation and stability:

▶ a structure known as the **centromere**, and seen as a constriction in the chromosome at some stages of the cell cycle, is the point at which the **kinetochore** and **spindle fibers** attach when chromosomes are segregated into daughter cells at cell division;

▶ at each end of a chromosome lies a structure known as the **telomere**, which is essential to stop chromosomes fusing with each other, and which protects the chromosome against loss of essential sequences from its ends.

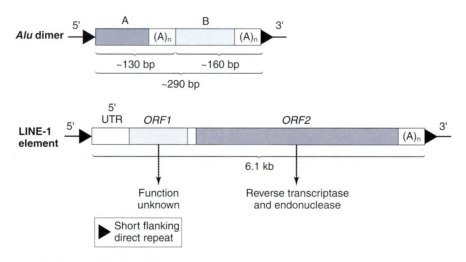

Figure 2.9: Structural features of full-length *Alu* and LINE-1 elements.

The two monomers within the *Alu* dimer differ in length because of a 32-bp insertion in the B monomer. Lengths are approximate because of variation in the length of the poly(A) tails [(A)ₙ]. Note that the internal poly(A) tract is significantly more diverged than the tract at the 3′ end. Full-length LINE-1 elements are rare: most are truncated at the 5′ end. ORF, open reading frame; UTR, untranslated region. Diagrams are not to scale.

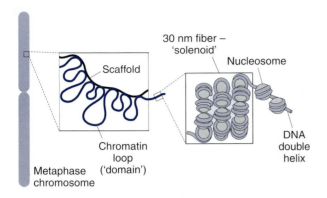

Figure 2.10: The packaging of DNA into chromatin and the chromosome.

The chromosomes of some organisms, such as bacteria and yeast, possess a third feature essential for stable inheritance: these are **replication origins**, which permit the initiation of DNA replication, and which are associated with specific DNA sequences. In human chromosomes, while replication is initiated at many sites, it appears that there is little or no sequence specificity involved. A notable exception is

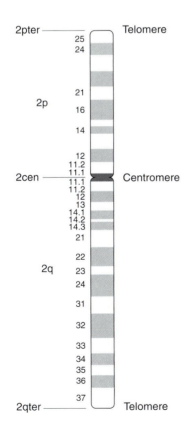

Figure 2.11: Structural features of a typical human chromosome.

This representation, of a G-banded chromosome 2, is known as an idiogram. The numbers of the dark and light bands are given to the left.

mitochondrial DNA, with its bacterial ancestry (see Section 2.7.2), which has two discrete origins of replication, one on each strand.

Both centromeres and telomeres are complexes of DNA and proteins, and in each case the DNA component includes a tandem array of repeated sequences. In the case of centromeres, the **alphoid repeat** unit, of 170 bp, is arranged into a higher order repeat unit of a few to several kilobases, varying between chromosomes. This is then itself repeated many times to produce an array which is variable in length between chromosomes and individuals, but can be as large as 5 Mb. Telomeres have a 6-bp repeat unit, TTAGGG, arranged into tandem arrays of tens of kilobases at the very ends of chromosomes. Just centromeric to these arrays are other repeated sequences that tend to vary between telomeres. These more complex repeated sequences are known as subtelomeric repeats.

Without the use of staining methods, visible differences between chromosomes are their lengths, and the position of the centromere (seen as a constriction in most cases), which allow them to be divided into groups. While some human chromosomes are **metacentric**, with the centromere dividing the chromosome into two clearly recognizable 'arms', some are **acrocentric**, where the centromere is very close to one end. The smaller of a chromosome's arms is designated the **p-arm** (for 'petit' – small), and the larger, conventionally drawn at the bottom, is the **q-arm** (for 'queue' – tail). The use of particular staining methods reveals specific banding patterns, allowing all 24 different chromosomes in the human karyotype to be distinguished from each other. The most widely used procedure is trypsin treatment followed by Giemsa staining, which produces a pattern known as **G-banding**. Current methods allow 850 or more G-bands to be identified in the human karyotype. The underlying chemical basis of G-banding is unclear, though data from the Human Genome Project show that dark-staining G-bands correlate with regions of the genome that have relatively low GC-content and are poor in genes. Some chromosomes contain regions which behave abnormally in chromosomal banding, and which may vary in size between individuals as a result of their high repeat content. These regions, termed **heterochromatin**, are often associated with centromeres and represent highly condensed, transcriptionally inert segments of the genome. The remainder of the genome is termed **euchromatin**.

The 46 chromosomes are divided into 23 pairs (*Figures 2.12* and *2.13*): one of each pair is received from each parent. Twenty-two of the pairs, the **autosomes**, are identical between the sexes, and are numbered from 1 (the largest) to 22 (the smallest). The remaining two chromosomes are known as the **sex chromosomes**, because they differ between the sexes. Females have two copies of the **X chromosome**, which, at 170 Mb, is about the size of chromosome 7. Males have one X chromosome, but in addition the smaller (~60 Mb) **Y chromosome**; the Y is sex-determining in mammals, through the action of a gene, *SRY* (sex-determining region, Y), which causes the bipotential gonad to become a testis, rather than an ovary,

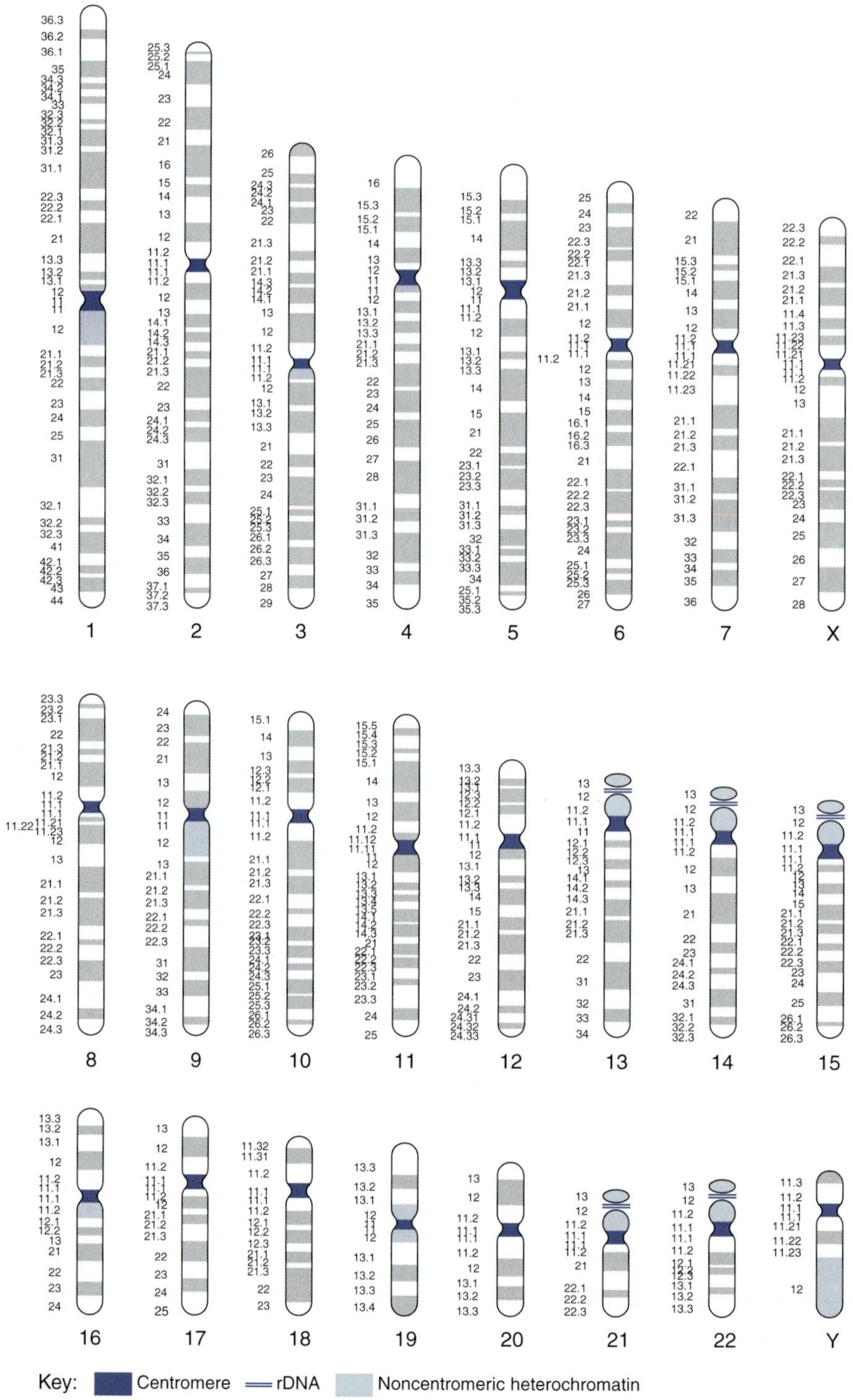

Figure 2.12: Human G-banded karyotype.

Knob structures on the short arms of chromosomes 13, 14, 15, 21, and 22 are arrays of ribosomal DNA repeats. Reproduced from Strachan and Read (2004) Garland Science.

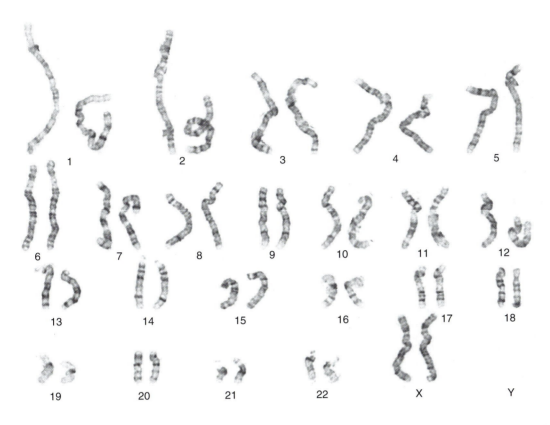

Figure 2.13: G-banded prometaphase chromosome karyogram of mitotic chromosomes from lymphocytes of a normal female. Compare with idealized banding patterns in *Figure 2.12*. Reproduced with permission from Strachan and Read (2004).

early in development. Subsequent steps in sex determination are mediated by hormones released by the gonads. The chromosomal constitution of an individual is known as their **karyotype**, and the karyotypes of normal females and males are denoted 46,XX and 46,XY respectively.

2.5 Mitosis and meiosis

When a cell divides, it passes on its genetic material to its daughter cells. In somatic tissues this process occurs exclusively through **mitosis**: the DNA replicates, so that the cell is temporarily tetraploid (has four copies of the genome). The chromosomes condense, and the two copies of each chromosome are associated at their centromeres – these identical associated copies are known as **sister chromatids**. After the nuclear envelope has dissolved, the chromosomes align at the metaphase plate, a region in the center of the cell. The associated centromeres of sister chromatids then separate, and the two chromatids of each chromosome move to opposite poles of the cell. After this, the nuclear envelope reforms around each set of segregated chromosomes as they decondense, the cytoplasm divides, and cell division is complete, resulting in diploid daughter cells which are genetically identical to the diploid parent.

Mitosis is a fundamental process; each of us starts life as a single cell, the fertilized egg, and develops and survives as a result of the enormous number of mitotic cell divisions (estimated to be about 10^{17}) necessary for growth, development and maintenance. Mitosis has great importance in disease, since errors in mitotic divisions can lead to cancers. However, in evolutionary terms the most important class of cell divisions is that which gives rise to the gametes, and it is these that enable the passage of genetic information to the next generation.

A **gamete** (egg or sperm) is haploid; it contains only one copy of the genome, as opposed to the usual two copies in our somatic, diploid cells. The production of gametes in the **germ-line** proceeds first by a series of mitotic cell divisions, leading in females from oogonia to primary oocytes, and in males from spermatogonia to primary spermatocytes (*Figure 2.14*). Following this, however, the cells enter a different kind of cell division process, specific to the germ-line, called **meiosis**. There are two fundamental distinctions between meiosis and mitosis:

▸ like mitosis, meiosis involves a single round of DNA replication; however, meiosis involves *two*, not one, subsequent cell divisions, resulting in a reduction of the amount of genetic material from two copies to one;

▸ while cells produced by mitosis are genetically identical to each other (and to the parental cell), the haploid cells (gametes) produced by meiosis are all genetically different.

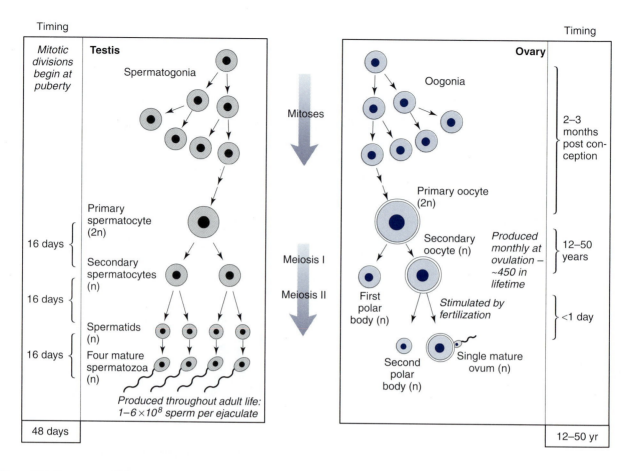

Figure 2.14: Gametogenesis in males and females.

Gametogenesis proceeds in males by mitotic divisions of diploid spermatogonia to give diploid primary spermatocytes. In females diploid oogonia divide to give primary oocytes. These cells can undergo meiosis. In females meiosis I occurs in a single primary oocyte at ovulation, and the fate of the secondary oocyte depends on whether or not it is fertilized. Unfertilized secondary oocytes are lost during menstruation, but contact of a sperm with the cell membrane triggers a rapid second meiosis. This entire process can take as much as 50 years. In contrast, spermatogenesis is rapid (48 days) and occurs throughout post-pubertal life.

The events of meiosis are summarized in *Figure 2.15*. The genetic differences between gametes arise from two distinct processes: clearly, the reduction from diploidy to haploidy necessitates the selection of either the paternal or the maternal copy of each chromosome to pass into the gamete. This **independent assortment** of chromosomes alone leads to differences between gametes – provided the choice is random, the possible number of different combinations of haploid chromosome subsets is very large: 2^{23} (8 388 608). However, there is a second, important level of modification in the passage of genetic material to the gamete: **recombination**. During meiosis paternal and maternal chromosomal homologs align and exchange segments through recombination, also known as **crossing-over**. This process is reciprocal, and there is no net loss of genetic information. Assortment and recombination, neither of which occurs during a normal mitotic cell division, ensure that any one gamete produced by a man or woman is genetically different from any other.

2.6 Recombination – the great reshuffler

Direct observation of molecular differences between homologous DNA sequences ('alleles' – see Chapter 3), or observation of differences indirectly manifested as phenotypes (such as diseases) arising from particular sequences, allows the inheritance of segments of DNA from parent to child to be followed in human pedigrees. Recombination events can be detected when they disrupt the coinheritance of molecular markers, or the coinheritance of a marker with a phenotype. The recognition of these recombination events allows an appreciation of the order of markers along DNA, and the counting of recombination events allows an estimation of recombinational, or **genetic distance** between markers; this is **genetic mapping**. Genetic distance is expressed in units of recombination, which are **centiMorgans**: one centiMorgan of recombination between a pair of markers is a genetic distance corresponding to a 1% recombination frequency between them. With the advanced knowledge we now have of the physical structure of the human genome (measured in

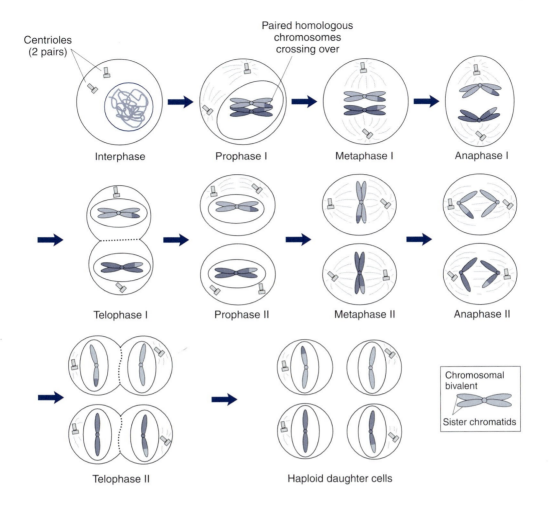

Figure 2.15: The stages of meiosis.

DNA is replicated in interphase, and chromosomes condense and thicken in prophase I – a single chromosome is shown. At this stage crossing over also occurs and the nuclear envelope breaks down. During metaphase I and anaphase I the chromosomal bivalents are pulled apart, and the diploid daughter cells have formed by prophase II. At metaphase II and anaphase II the sister chromatids are pulled apart by the spindle apparatus. Cell division is complete after telophase II, and each gamete contains one haploid copy of the chromosome. Two are recombinant, and two are nonrecombinant.

megabases, rather than centiMorgans), we can readily compare the **genetic map** with the **physical map** (*Figure 2.16*). As a genomewide average, one centiMorgan corresponds to one megabase of sequence (1 cM/Mb).

The distribution of recombination events varies between the sexes: females have more recombination events when they make eggs than males do when they make sperm: in other words, the genetic map of females is expanded on average about 1.65-fold with respect to that of males (*Figure 2.16*). Males have roughly 50 recombination events per meiosis, whereas females experience 80 such events. Also, the recombination rate across the genome within each sex is far from uniform. Recombination events are more frequent towards the telomeres of chromosomes (up to 3 cM/Mb), and less frequent towards their centromeres (< 0.1 cM/Mb). Regional variation is present on many scales: recombination rates are on average about twice as high (per Mb) on

the smallest chromosomes compared with the largest, but both recombination 'desert' and 'jungle' segments can be identified within chromosome arms.

These average values over large chromosomal regions could obscure substantial recombination rate heterogeneity at the sequence level. Direct molecular analysis of recombination indicates that, for some regions of the genome at least, events are concentrated in small (1–2-kb) segments of DNA known as recombination hotspots, separated by comparatively large regions of low recombinational activity (Jeffreys *et al.*, 2001). There could be as much as 1000-fold difference in recombination rates between hotspots and cold domains. This nonuniformity of recombination takes on great importance when we come to consider the issue of the distribution of sequence variants along chromosomes in different human populations, in Chapter 14, and is discussed further in Chapter 3 (Section 3.7). Furthermore, significant

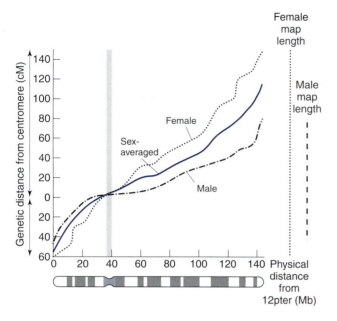

Figure 2.16: Genetic and physical distance compared.

Female, male, and sex-averaged genetic distance in cM plotted against physical distance in Mb for chromosome 12. Female recombination is elevated compared to male recombination. Recombination is generally elevated towards the telomeres, and depressed around the centromere. Adapted from Lander *et al.* (2001).

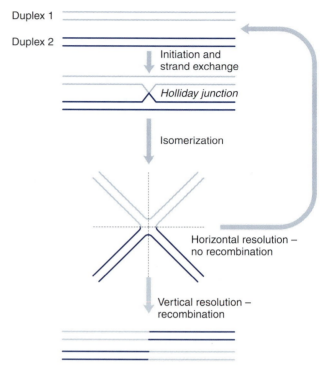

Figure 2.17: Simple molecular model of meiotic recombination.

variation in recombination rates in pedigree data can be detected between individual women, although not between individual men (Kong *et al.*, 2002).

While recombination undoubtedly plays an important role in generating new combinations of alleles, and thus genetic and phenotypic diversity, from one generation to the next, it also has importance for the behaviors of chromosomes in meiosis. Most chromosomes undergo on average just over one recombination event per chromosomal arm, and it seems that these events are necessary for the faithful segregation of chromosomes into daughter cells at division. Indeed, some chromosomal **aneuploidies** [departure from the correct number of chromosomes, due to **nondisjunction** (incorrect segregation)] can be shown to be accompanied by reduced recombination, or a failure to recombine (Hassold and Hunt, 2001).

A simple model of the mechanism of homologous recombination is illustrated in *Figure 2.17*. It involves a four-stranded intermediate structure called the Holliday junction (Holliday, 1964), and forms a basis not only for under-standing reciprocal exchange of genetic information through homologous recombination, but also nonreciprocal exchange through gene conversion (see Section 3.8).

2.7 Nonrecombining segments of the genome

While the great majority of our genome is biparentally inherited, and undergoes reshuffling each generation through recombination, there are two segments of our DNA that are atypical, being inherited from one parent only, and escaping recombination. These are the majority of the Y chromosome, and mitochondrial DNA (*Figure 2.18*).

2.7.1 The Y chromosome

Because of the Y chromosome's defining role in sex determination, it is male-specific, haploid, and passed from father to son. *Box 8.8* describes this locus in greater detail. Since it has no homolog with which to recombine, we expect the Y chromosome to avoid recombination. For the great majority (> 90%) of its length, known as the **nonrecombining portion of the Y** (sometimes **NRPY**, or **NRY**), this is so; *Box 2.7* addresses the question of whether there are circumstances under which recombination could occur in this region. Given the importance of recombination as a mechanism ensuring correct segregation of chromosomes (Section 2.6), it is not surprising to find that the Y does actually recombine with the X chromosome in male meiosis, albeit in specialized regions where sequence identity with the X is preserved (*Figure 2.19*). Sequences within these regions can be inherited from either parent, like sequences on autosomes, and the regions are therefore referred to as **pseudoautosomal**. One of these regions, pseudoautosomal region 2 (PAR2), lying at the tips of the long arms of the X and Y chromosomes is a recent evolutionary acquisition specific to humans, and of little importance in chromosomal segregation. However, PAR1, a 2.6-Mb region at the tips of the short arms, reflects the ancient origin of the mammalian

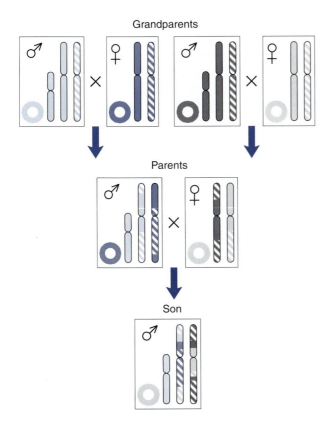

Figure 2.18: Inheritance of recombining and nonrecombining segments of the genome.

Schematic illustration of a three-generation pedigree. The son's Y chromosome (small chromosomal symbol) and mtDNA (circle) each descends from a single grandparental ancestor, his paternal grandfather and maternal grandmother, respectively. In contrast, his autosomes (large chromosomal symbols) descend from all of his grandparents, because of reshuffling of chromosomal segments in every generation through recombination.

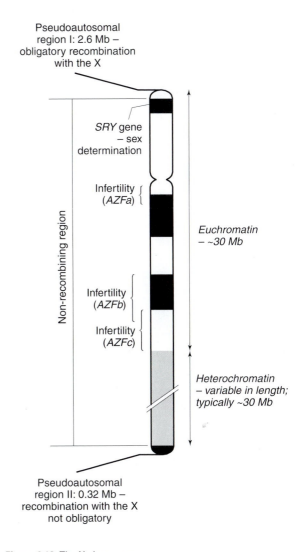

Figure 2.19: The Y chromosome.

Idiogram of a G-banded Y chromosome, showing position of intervals associated with specific phenotypes, and the sex-determining gene *SRY*. Correspondence of intervals with G-bands is schematic only.

sex chromosomes as a pair of homologous autosomes, and is the site of an obligate recombination event in male meiosis.

The recombinational behavior of both the X and the Y is unusual. The X chromosome behaves like the Y in male meiosis, recombining only in the pseudoautosomal regions, but in females it behaves like an autosome, with recombination able to occur along its entire length.

2.7.2 Mitochondrial DNA

The other constitutively nonrecombining region of our genetic material is mitochondrial DNA (mtDNA), a circular double-stranded DNA molecule about 16.5 kb in length (*Figure 2.20*; and see *Box 8.5* for a general description) whose entire sequence is known (Anderson *et al.*, 1981; Andrews *et al.*, 1999). This is contained not within the nucleus, but within mitochondria – cytoplasmic organelles in which the energy-generating process of oxidative phosphorylation takes place. Many features of mitochondria and mtDNA suggest

that these organelles originated as endosymbiotic bacteria (see *Box 2.8*).

The number of mitochondria in a cell varies with cell type: those requiring a lot of energy, such as nerve and muscle cells, contain thousands of mitochondria, each containing 2–10 copies of mtDNA, while other cell types may contain only a few hundred. Oocytes contain around 100 000 mitochondria, each containing a single mtDNA molecule, while sperm contain only about 50–75. Clearly, even if fertilization involved a complete mixing of paternal and maternal mtDNA molecules, the contribution of the father to the **zygote**'s pool of mtDNA would be relatively small. However, most evidence suggests that his contribution is actually zero, and that mtDNA is exclusively maternally inherited. Sperm mitochondria, required for motility, are

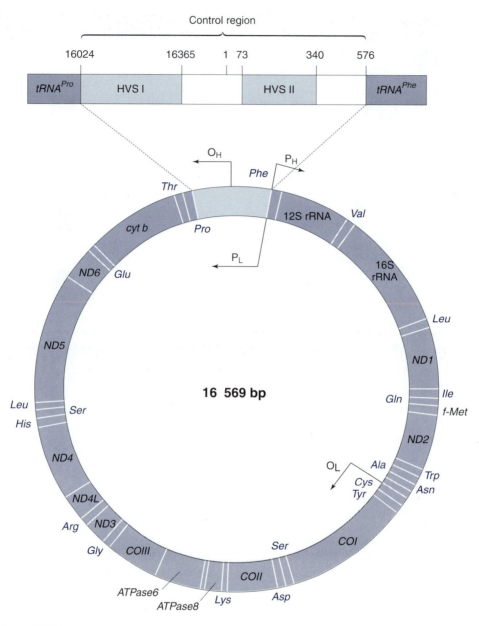

Figure 2.20: Mitochondrial DNA.

Human mtDNA is a circular double-stranded molecule with one strand (the heavy strand) relatively rich in G bases, and the other (the light strand) rich in C. tRNA genes are indicated by the three-letter name of the corresponding amino acid; remaining genes are protein coding, except for the two rRNA genes. Origins of replication of the light (O_L) and heavy (O_H) strands, and **promoters** for transcription of these two strands (P_L, P_H) are shown. The control region contains two **hypervariable segments** (also known as **hypervariable regions**, HVRs) that are commonly assayed for variability (HVS I and HVS II).

localized in the midpiece, between the sperm head and tail. After fertilization the midpiece and its paternal mitochondria can be seen within the zygote, and observed through several subsequent cell divisions. Note that some textbooks wrongly assert that the midpiece of the sperm does not enter the egg (Ankel–Simons and Cummins, 1996). The apparent absence of paternal mtDNA inheritance suggests that there may be an active system to eliminate paternal mitochondria; the one described case of paternal inheritance, involving a pathogenic

mtDNA mutation (Schwartz and Vissing, 2002), may reflect a defect in this elimination process.

If paternal mtDNA inheritance did occur, recombination between mtDNAs from different lineages could in principle happen – the mitochondrion contains the necessary molecular machinery to allow this. This fact, and some features of the pattern of polymorphic sites within the mtDNA molecule, have led to much debate over whether human mtDNA does indeed recombine (see *Box 2.9*).

BOX 2.7 Could the Y chromosome recombine?

Are there any circumstances where the Y chromosome could recombine with another Y chromosome? **Ectopic** recombination with the X and autosomes is well known but mostly leads to sterility because resulting Y chromosomes have lost genes necessary for spermatogenesis.

▶ While normal males have one X and one Y chromosome, about one in 1000 males has one X and two Y chromosomes, and in principle recombination could occur between the two Ys. However, both of these Y chromosomes come from the same father, and are therefore expected to be identical, which would limit the effects of recombination events. In fact, observations of meiosis in the fertile majority of XYY males suggest that YY bivalents (the term for paired, recombining chromosomes) do not occur.

▶ There are large segments of very similar (99%) sequences on the X and the Y chromosomes that reflect recent transpositions of material from the X to the Y. In principle it is possible for aberrant recombination to occur between blocks of such sequence, and to transfer segments of X DNA into a Y chromosome. Indeed, there is evidence that this has occurred in the sex chromosomes of the cat family (Slattery *et al.*, 2000). In humans, the maximum sequence divergence between different Y chromosomes (about 0.1%) is very much smaller than the sequence divergence between the most similar known blocks of nonpseudoautosomal homology (~1%), and we would therefore expect to recognize this kind of aberrant exchange readily.

▶ In rare cases an autosome can carry a segment of Y chromosome as a constitutive, apparently asymptomatic translocation. Recombination could occur between such a segment and a whole Y chromosome. This violation of the principle of no-recombination should be readily recognized within the robust framework of the Y phylogeny, or tree (see *Box 8.8*). As yet, there is no evidence for such a recombination event in human evolution.

BOX 2.8 The curious history of mitochondrial DNA

The theory originally proposed by Lynn Margulis in the 1960s (see Margulis, 1981) is now accepted – that mitochondria originated as **endosymbiotic** bacteria, taken up into proto-eukaryotic cells about 1.5 billion years ago, and provided energy generation in return for a safe environment. This prokaryotic past has left its traces in the many features of mitochondrial DNA which resemble those of modern bacteria:

▶ circular, rather than linear, genome;

▶ absence of **histones**;

▶ discrete origins of replication, unlike nuclear chromosomes;

▶ no introns in genes, no dispersed repeats, and very little intergenic DNA;

▶ polycistronic transcripts: transcription starts at only two **promoters**, one on each strand, and continues round the entire genome to produce a single RNA molecule from several genes;

▶ different genetic code – perhaps the most striking evidence of the exogenous origin of mtDNA. In mammals, five codons have different specificities in mtDNA compared to the nuclear genome:

Codon	Nuclear code	mtDNA code
UGA	STOP	Trp
AGA, AGG	Arg	STOP
AUA, AUU	Ile	Met

Since its origin, mtDNA has lost most of its genes, and hence its autonomy. The 37 genes it now carries all play roles in either the oxidative phosphorylation pathway, or mitochondrial protein synthesis. The other genes essential for mitochondrial function, including those encoding mtDNA polymerase, mtRNA polymerase and many structural and transport proteins, have been transferred to the nuclear genome.

As well as this ancient transfer of genes, segments of mtDNA have been inserted into the nuclear genome at various points in our more recent evolutionary history. Indeed, a survey of the draft sequence of the nuclear genome (Bensasson *et al.*, 2001) indicates that it contains over 400 kb of mtDNA sequences – 25 times as much as the mtDNA molecule itself. Some of these nuclear mtDNA insertions (**numts**) are common to all primates, while others are more recent, some even being polymorphic in human populations. As we shall see in Section 4.11.2, numts are a potential source of problems in mtDNA studies in ancient samples, but also have their uses in the rooting of phylogenetic trees constructed using mtDNA (Chapters 3, 6, 8).

BOX 2.9 Opinion: Does mtDNA recombine?

Human population geneticists normally accept without question that mitochondria are clonally inherited through the maternal line. The genetic evidence for this comes largely from pedigree analyses. However, cell biologists have debated whether paternal mitochondria have been transmitted through the germ-line for almost 40 years (reviewed by Ankel-Simons and Cummins, 1996). In 1999, a controversial claim by Eyre-Walker et al. (1999) caused a stir in the field of human population genetics. The claim was that human mitochondrial genomes exhibited too much homoplasy, more than could be accounted for by mutation alone, and therefore that:

▶ mitochondrial genomes recombine;

▶ mitochondria are not matrilinearly inherited.

These two points are not equivalent. First, mitochondria contain all of the enzymes necessary for recombination and repair. However, for recombination to be observed, it must occur between two different haplotypes within the same organelle. **Heteroplasmy** (multiple haplotypes in the same individual) is known in humans and is often discovered in relation to disease phenotypes. However, heteroplasmy has been found mostly in somatic tissue and not germ-line tissue. Second, a system is in place to degrade paternal mitochondria as they enter the ovum from the sperm (tail midpiece); therefore it is unlikely that one would observe paternally inherited mitochondria.

For recombination to occur, two double-strand breaks need to be created in order to maintain the integrity of the circular genome. Otherwise, two different sized molecules will be created, as has been suggested for the mitochondria of the nematode worm *Caenorhabditis* (Lunt and Hyman, 1997). Regardless of the mechanism of recombination, its occurrence at an effective rate could have substantial implications for population inferences based on mitochondrial data, including estimates of the effective population size, demography, and migration patterns. The question is, does it happen?

To test for recombination, Eyre-Walker et al. (1999) took two approaches. First, they counted the instances of parallel evolution (homoplasy) in a population dataset of complete genomes. However, recurrent mutation and recombination have similar effects on haplotype structure at segregating sites (see *Figure*). Under most models of molecular evolution at the population level, the infinite sites model is assumed (the genome is assumed infinitely large and therefore the probability of mutation at the same site is effectively nil). However, the mutation rate in mitochondria is known to be high and the assumptions of the infinite sites model are probably violated. Although it is possible that homoplasies might be a result of recombination, it is difficult to separate the two processes. A later study (Awadalla et al., 1999) demonstrated that the correlation of alleles among all pairs of segregating sites (linkage disequilibrium – LD) decays as sites become more distant from each other. The most parsimonious explanation is that the probability of recombination between pairs of sites increases with distance.

Both observations have been disputed on a number of grounds. It has been suggested that the pattern could be generated by regional variation in mutation (Worobey, 2001); this cannot explain the decay in LD (Innan and Nordborg,

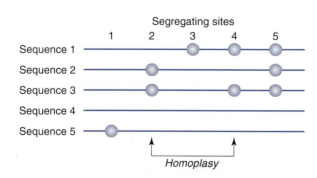

Homoplasy in a dataset of five sequences.

Recombination and recurrent mutation can have similar effects on variation by generating parallel patterns of evolution (homoplasy). The blue circles are mutations that have arisen on either the same or different genetic backgrounds. If recombination occurs between a pair of sites, then the two products of crossing over, plus the two parental types – in total, four different sequences – will be observed. Of the 10 possible pair-wise comparisons, only recombination between sites 2 and 4 can explain the presence of all five of the sequences represented here. An alternative explanation is recurrent mutation, and the probability of this increases when mutation rates are high, or are variable across regions. In this example, homoplasy is said to be present between sites 2 and 4.

2002). Furthermore, not all datasets exhibit the same decay (Eyre-Walker and Awadalla, 2001). Wiuf (2001) argued that the simple model of single crossing-over is not appropriate for circular genomes and that recombination is not possible if there is substantial substitution rate heterogeneity among sites. Finally, it was argued (McVean *et al.*, 2002) that under a model that assumes a finite-sites estimate of the mutation rate, the effective population rate of recombination (4Nr), was not significantly different from zero. This approach is attractive in that it models both the mutation and the recombination process in a sophisticated manner, and provides an estimate rather than merely assessing the scale of linkage disequilibrium. However, the approach is conservative and the test for significance may fail, even for data that are in known regions of recombination.

Where does the debate stand? Currently, the population genetic evidence is leaning away from recombination as an effective evolutionary force in human mtDNA. However, rare instances may be associated with disease and might indicate a breakdown in the process that degrades the paternal contribution. A pedigree was recently discovered in which heteroplasmy was inherited through the paternal line; this clearly shows that the mechanisms that degrade paternal mitochondria can leak (Schwartz and Vissing, 2002). Regardless, as people continue to investigate the process at the molecular level, and as observations in other taxa are made, it might be found that recombination does occur. Examples of recombination in other species clearly show that mitochondrial genomes are capable of exchange. However, the implications for our understanding of human evolution are not likely to be altered dramatically as it appears rare that these events occur in the human lineage.

Philip Awadalla, University of California, Davis, USA

Summary

▸ DNA is a macromolecule composed of four subunits, the bases A, G, C and T, whose order contains instructions for the production of cellular components, and for the transmission of identical instructions to daughter cells. DNA is double stranded, and its length is measured in base pairs.

▸ Most somatic cells are diploid (contain two copies of the genome), while germ cells (sperm and eggs) are haploid (one copy), containing about 3 200 000 000 base pairs of DNA; the order of these bases (the genome sequence) has been determined almost completely.

▸ Thirty percent of the genome comprises ~30 000 genes, containing instructions for producing proteins and RNA molecules which build cells. DNA is transcribed into RNA in the nucleus, and this nucleotide sequence is then translated in the cytoplasm into amino acid sequences to give proteins, via the genetic code. As well as coding sequences, genes contain regulatory sequences, and are usually broken up into coding segments (exons) separated by introns. Exons are stitched together after transcription by splicing.

▸ Most of our DNA is noncoding, and may have no function. About 45% is highly repetitive dispersed sequences including retrotransposons.

▸ The diploid genome is divided into 46 chromosomes: 22 pairs of autosomes are shared between men and women, and women have an additional two X chromosomes while men have an X and a Y chromosome, which determines sex. Chromosomes have structural features essential for stability and segregation at cell division, the telomeres and centromere.

▸ Somatic cell division (mitosis) ensures orderly segregation of chromosomes into daughter cells after DNA replication. Meiosis is a specialized cell division in the germ-line that reduces the number of chromosomes by half to give haploid gametes, and includes recombination – exchange of segments between pairs of homologous chromosomes. Women show more recombination than men in meiosis.

▸ Recombination is important because it affects how DNA sequences are inherited. Autosomes recombine in both men and women, X chromosomes mostly recombine only in women, and the paternally inherited Y chromosome doesn't recombine at all.

▸ Mitchondria contain their own circular genomes of 16.5 kb, which are maternally inherited and escape recombination because sperm fail to contribute paternal mitochondria to the zygote.

Further Reading

Strachan T, Read AP (2003) *Human Molecular Genetics 3*. Garland Publishing Inc. New York.

The double helix – 50 years (2003) *Nature* (special supplement). **421**, 395–453.

References

Anderson S, Bankier AT, Barrell GB, de Bruijn MHL, Coulson AR, Drouin J, Eperon IC, *et al.* (1981) Sequence and organisation of the human mitochondrial genome. *Nature* **290**, 457–465.

Andrews RM, Kubacka I, Chinnery PF, Lightowlers RN, Turnbull DM, Howell N (1999) Reanalysis and revision of the Cambridge reference sequence for human mitochondrial DNA. *Nature Genet.* **23**, 147.

Ankel-Simons F, Cummins JM (1996) Misconceptions about mitochondria and mammalian fertilization: implications for theories on human evolution. *Proc. Natl Acad. Sci. USA* **93**, 13859–13863.

Awadalla P, Eyre-Walker A, Smith JM (1999) Linkage disequilibrium and recombination in hominid mitochondrial DNA. *Science* **286**, 2524–2525.

Behlke MA, Bogan JS, Beer-Romero P, Page DC (1993) Evidence that the SRY protein is encoded by a single exon on the human Y-chromosome. *Genomics* **17**, 736–739.

Bensasson D, Zhang D-X, Hartl DL, Hewitt GM (2001) Mitochondrial pseudogenes: evolution's misplaced witnesses. *Trends Ecol. Evol.* **16**, 314–321.

Eichler EE (2001) Recent duplication, domain accretion and the dynamic mutation of the human genome. *Trends Genet.* **17**, 661–669.

Eyre-Walker A, Awadalla P (2001) Does human mtDNA recombine? *J. Mol. Evol.* **53**, 430–435.

Eyre-Walker A, Smith NH, Smith JM (1999) How clonal are human mitochondria? *Proc. Roy. Soc. Lond. B Biol. Sci.* **266**, 477–483.

Hassold T, Hunt P (2001) To err (meiotically) is human: The genesis of human aneuploidy. *Nature Rev. Genet.* **2**, 280–291.

Holliday RH (1964) A mechanism for gene conversion in fungi. *Gen. Res.* **5**, 282–304.

Houck CM, Rinehart FP, Schmid CW (1979) A ubiquitous family of repeated DNA sequences in the human genome. *J. Mol. Biol.* **132**, 289–306.

Innan H, Nordborg M (2002) Recombination or mutational hot spots in human mtDNA? *Mol. Biol. Evol.* **19**, 1122–1127.

Jeffreys AJ, Kauppi L, Neumann R (2001) Intensely punctate meiotic recombination in the class II region of the major histocompatibility complex. *Nature Genet.* **29**, 217–222.

Koenig M, Hoffman EP, Bertelson CJ, Monaco AP, Feener C, Kunkel LM (1987) Complete cloning of the Duchenne Muscular-Dystrophy (DMD) cDNA and preliminary genomic organization of the DMD gene in normal and affected individuals. *Cell* **50**, 509–517.

Kong A, Gudbjartsson DF, Sainz J, Jonsdottir GM, Gudjonsson SA, Richardsson B, Sigurdardottir S, et al. (2002) A high-resolution recombination map of the human genome. *Nature Genet.* **31**, 241–247.

Lander ES, Linton LM, Birren B, Nusbaum C, Zody MC, Baldwin J, Devon K, et al. (2001) Initial sequencing and analysis of the human genome. *Nature* **409**, 860–921.

Lunt DH, Hyman BC (1997) Animal mitochondrial DNA recombination. *Nature* **387**, 247.

Margulis L (1981) *Symbiosis in Cell Evolution: Life and its Environment on the Early Earth*. W. H. Freeman & Co., New York.

McVean G, Awadalla P, Fearnhead P (2002) A coalescent-based method for detecting and estimating recombination from gene sequences. *Genetics* **160**, 1231–1241.

Prince VE, Pickett FB (2002) Splitting pairs: the diverging fates of duplicated genes. *Nature Rev. Genet.* **3**, 827–837.

Roberts RG, Coffey AJ, Bobrow M, Bentley DR (1993) Exon structure of the human dystrophin gene. *Genomics* **16**, 536–538.

Schwartz M, Vissing J (2002) Paternal inheritance of mitochondrial DNA. *New Engl. J. Med.* **347**, 576–580.

Shafit-Zagardo B, Maio JJ, Brown FL (1982) Kpn I families of long, interspersed repetitive DNAs in human and other primate genomes. *Nucl. Acids Res.* **10**, 3175–3193.

Slattery JP, Sanner-Wachter L, O'Brien SJ (2000) Novel gene conversion between X-Y homologues located in the nonrecombining region of the Y chromosome in Felidae (Mammalia). *Proc. Natl Acad. Sci. USA* **97**, 5307–5312.

Tennyson CN, Klamut HJ, Worton RG (1995) The human dystrophin gene requires 16 hours to be transcribed and is cotranscriptionally spliced. *Nature Genet.* **9**, 184–190.

Venter JC, Adams MD, Myers EW, Li PW, Mural RJ, Sutton GG, Smith HO, et al. (2001) The sequence of the human genome. *Science* **291**, 1304–1351.

Waterston RH, Lander ES, Sulston JE (2002) On the sequencing of the human genome. *Proc. Natl Acad. Sci. USA* **99**, 3712–3716.

Wiuf C (2001) Recombination in human mitochondrial DNA? *Genetics* **159**, 749–756.

Wolfe KH (2001) Yesterday's polyploids and the mystery of diploidization. *Nat. Rev. Genet.* **2**, 333–341.

Worobey M (2001) A novel approach to detecting and measuring recombination: new insights into evolution in viruses, bacteria, and mitochondria. *Mol. Biol. Evol.* **18**, 1425–1434.

CHAPTER THREE

The diversity of the human genome

3.1 Introduction: Genetic variation and the phenotype

We are all different from each other, and much of this difference has a genetic basis: differences in **phenotype** caused by differences in **genotype**. Some of these differences are the stuff of common observation, and we know from our own experience that they 'run' in families – hair, eye, and skin color, stature, and some physical features. Others are the subject of university genetics department open days – color vision deficiencies, or the ability to roll the tongue, or to taste the chemical phenylthiocarbamide (before health and safety regulations prohibited this particular test). Other differences are more subtle but more important, and affect us medically – our blood group, should we require a transfusion, our HLA type, if we need an organ or bone-marrow transplant, and other factors which affect how we will respond to certain drugs, or how likely we are to contract infectious diseases such as malaria or AIDS, or disorders such as diabetes, cancer, asthma, schizophrenia or coronary heart disease. Some genetic differences are directly responsible for diseases, such as Huntington disease or cystic fibrosis, which we can pass on to our children. Some of these particular differences, such as those underlying Huntington disease (OMIM 143100), have a **dominant** effect, where only one copy of the gene is required to be mutated for an effect to occur. Others, such as cystic fibrosis (OMIM 219700), are **recessive** mutations, and both copies must be mutated for the phenotype to be seen (see *Box 3.1*).

The genetic basis of some of these phenotypic differences, such as whether or not a person suffers from cystic fibrosis, might seem straightforward and is quite well understood; however, even these 'simple' disorders have complex features. The genetic basis of other, often common, phenotypes (such as pigmentation, or schizophrenia predisposition), is certainly complex, and involves non-genetic factors. The path from genotype to phenotype is rarely simple (*Figure 3.1*):

▶ There are many observable differences between people that are not genetic, but due completely or in part to stochastic processes during development, or to environmental influences. Monozygotic (MZ) twins demonstrate this: they share all of their genes and usually many aspects of their environment, but have phenotypes that, though very similar in many respects, are certainly not identical. Megalomaniacs who might hope to create copies of themselves by cloning will be disappointed to find that they are no more similar to their clones than a pair of MZ twins separated at birth.

▶ While **monogenic** (also known as 'simple', or **Mendelian**) disorders show a relatively simple pattern of inheritance, there is often a more complex relationship between genotype and phenotype than the name suggests (reviewed by Badano and Katsanis, 2002): different mutant alleles in the same gene can have different effects, and alleles of other 'modifier' genes can influence phenotype. Such factors affect the cystic fibrosis phenotype, for example – whether or not the pancreas secretes enzymes normally and whether or not the *vas deferens* (the duct along which sperm pass) of affected males is occluded or open. There is therefore no sharp distinction between monogenic disorders and **complex**, or **multifactorial** disorders, such as asthma and diabetes, which are commonly said to involve the influence of more than one gene, as well as the environment (see Chapter 14).

▶ Although there are many differences between one copy of the human genome and another, the vast majority of these differences apparently play little or no role in the phenotype. Most of the genetic differences between individuals and populations used in human evolutionary genetic studies and discussed in this book are of this kind. They are often referred to as **neutral** mutations, since they are thought not to affect the evolutionary 'fitness' of someone who carries them, and hence their frequency is unaffected by natural selection. The neutral theory of molecular evolution is discussed in Section 5.6.3. In Chapters 13 and 14 we turn specifically to some mutations which are known, or thought, to have important effects upon our phenotypes. If we consider a severely deleterious allele, it is easy to think of a normal, or wild-type allele in contrast to it; however, in the context of the many mutations which are neutral or of unknown effect, it is difficult to conceive of a 'normal', or wild-type human genome.

This chapter is largely concerned with **mutation**, which is the ultimate source of all genetic diversity, and about the resulting kinds of differences at the DNA level which exist between people. The word used to describe a difference between two human DNA sequences varies between contexts and even between professions, and this can cause confusion (see *Box 3.2*). 'Mutation' is defined as any change in DNA sequence, and covers a very broad spectrum of events with different rates and different molecular mechanisms. It can range from the substitution of a single base in the genome, and small insertions and deletions of a few bases, through expansions or contractions in the number of tandemly repeated DNA motifs, insertions of transposable elements, insertions, deletions, duplications and inversions of megabase segments of DNA, to translocation of chromosomal segments and even changes in chromosomal number. *Figure 3.2* gives an overview of the scale and scope of these different mutational events, *Figure 3.3* shows the chromosomal distributions of the different classes of mutational event, and *Figure 3.4* summarizes the mutation rate ranges of the different classes.

Not all mutations are passed on to the next generation and contribute to evolutionary change; to do this they must:

▶ occur in the **germ-line** – the cell lineage culminating in the gametes (*Figure 3.5*, and see *Figure 2.14*). Because of this, germ-line mutation is what concerns us in evolutionary genetics; mutations in other cells of

---

BOX 3.1 Dominant and recessive mutations

A mutation occurs in one copy of a gene (one **allele**); however, we are diploid organisms, and the status of the other copy of the gene (allele) is important in determining whether a phenotype results from the mutation. Mutations that give a phenotype when only one allele is mutant (recognizable from following the inheritance of a disorder in affected pedigrees) are referred to as **dominant**. In some dominant disorders, such as achondroplasia (OMIM 100800), the commonest form of short-limb dwarfism, all patients carrying the mutation manifest the disorder fully; the dominant mutation is described as having full **penetrance**. Many dominant mutations, however, do not manifest fully in all people who carry them: this is referred to as reduced penetrance, and can complicate interpretation in family studies.

Most (> 90%) disease mutations require both copies of the gene to be mutant, and these are referred to as **recessive** – mutation in one copy alone is masked by the presence of a normal, **wild-type** allele. The mutations in the alleles of a patient manifesting a recessive disorder need not be identical at the sequence level, but need only have similar effects on the function of the gene. When alleles are identical, the patient is **homozygous** for the mutation; when they are different, the patient is **heterozygous**.

What is the molecular basis of recessiveness and dominance?

▶ Most mutations in a recessive disorder like cystic fibrosis result in either severe reduction or absence of gene product from a mutant allele, or production of an abnormal, nonfunctional product. A cystic fibrosis patient carries two mutant alleles, and so no functional gene product (the cystic fibrosis transmembrane conductance regulator, which functions as a chloride channel) is produced, leading to the phenotype. In a heterozygous **carrier**, who has one mutant and one normal allele, half as much gene product as normal is being produced, but this is enough for normal or near-normal function. This situation is referred to as **haplosufficiency** (literally, *a haploid dose is enough*). Many proteins are enzymes, and as catalysts of reactions it is not surprising that they will often function adequately in a haploid dose.

▶ The basis of dominance is much more varied than that of recessiveness. It can be **haploinsufficiency** – again, half as much product as normal is produced – but this half-dosage is *not* enough to ensure normal function. Dominance can also result from over-production of protein from a mutant allele, or expression in an inappropriate place or at the incorrect time. Some dominant mutations are in genes encoding structural proteins such as collagens: abnormal protein is produced from one allele, and this forms part of a structure which is therefore itself abnormal, even though the other allele is producing normal protein. Dominance can also result from a mutation that modifies protein function, or provides it with a new function that it did not previously possess. Such '**gain-of-function**' mutations cause achondroplasia: mutations in the fibroblast growth factor receptor 3 (*FGFR3*) gene lead to activation of the receptor at all times ('constitutive' activation), instead of activation in response to specific signals, disrupting normal bone maturation (reviewed by Webster and Donoghue, 1997). The basis of dominance in some disorders remains to be elucidated.

Disease-causing mutations on the X chromosome can behave differently in males and females, since males have only one X, while females have two. An X-linked recessive mutation (e.g., Duchenne Muscular Dystrophy; OMIM 310200) will manifest in males (who are **hemizygous**), but not in females, who are carriers. Dominant mutations will manifest in both sexes; an example is the disease hypophosphatemia (OMIM 307800), with a phenotype of vitamin-D-resistant rickets and short stature.

---

the body, the **somatic cells**, or **soma**, can have serious consequences such as cancers, but in evolutionary terms are not of direct importance;

▶ be nonlethal, and compatible with fertility. Some mutations could have lethal consequences to the bearer before birth or before they reach reproductive age, because of reduction or absence of expression of a particular gene, or its overexpression. These will not be passed on to the next generation. Similarly, mutations which do not affect survival, but lead to infertility, are also not passed on. Some large-scale mutations, in particular changes in chromosomal number (**aneuploidies**) are often lethal or severely deleterious. However, even in mild or neutral cases, such as males who possess an extra Y chromosome, they are not passed on to the next generation – they reflect a failure to segregate chromosomes correctly in a parental meiosis.

Most modern analysis of human genetic variation is at the level of DNA, and this book is primarily concerned with DNA sequence variation. However, before the recombinant DNA revolution of the 1970s and 1980s analysis was at the level of proteins; *Box 3.3* summarizes these classes of variation, sometimes known as **classical polymorphisms,** and the means used to analyze them.

3.2 Base substitutions (SNPs) and small indels in the nuclear genome

The simplest difference between two homologous DNA sequences is a **base substitution**, in which one base is exchanged for another (*Figure 3.6*). When a pyrimidine base in one sequence is exchanged for another pyrimidine (e.g., C for T), or a purine for another purine (e.g., A for G), the difference is called a **transition**. When a pyrimidine is exchanged for a purine, or vice versa, this is a **transversion**.

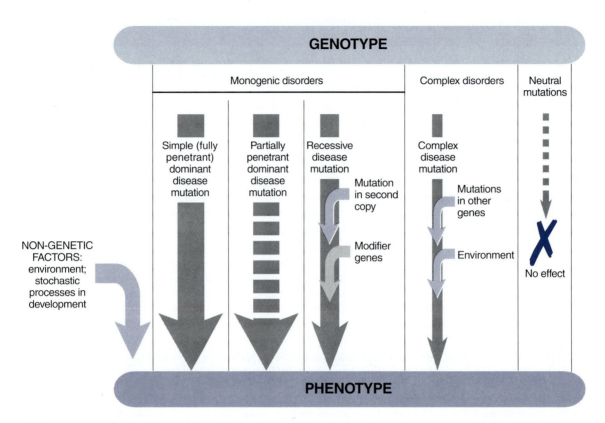

Figure 3.1: From genotype to phenotype.

The roles of genetic and nongenetic factors in generating the phenotype. Dominant mutations have a phenotypic effect even though a normal copy of the allele is present. This can be a full effect in all cases (full penetrance), or in only a subset of cases (reduced penetrance). Recessive mutations manifest a phenotypic effect only when no normal copy is present. Mutations involved in complex disorders require mutations in more than one gene and/or environmental factors to be manifested. Many mutations are neutral, having no known effect on phenotype. Much of human evolutionary genetics is concerned with analyzing and interpreting the patterns of diversity represented by these neutral mutations.

BOX 3.2 Polymorphism, variant, or mutation?

A simple base-substitutional difference between two human DNA sequences can be given a number of different labels, and unfortunately the rules for applying them are not always clear or consistent.

▶ A **polymorphism** has been rather arbitrarily defined as existing when at least two alleles are present in the population, and the minor allele is at a 1% or greater frequency (see e.g., Cummings, 2000).

▶ While all differences among sequences have mutation as their root cause, the word **mutation** is often reserved for use when pathogenic variation is being referred to, and is then used in contrast to 'polymorphism', which describes a sequence change in a gene which has no effect on function. This distinction in meaning is particularly prevalent among medical geneticists. However, there are clearly potential problems here, since it can be difficult or even impossible to know whether or not a sequence change is completely without a phenotypic effect (Cotton and Scriver, 1998). Also, in some populations disease-causing mutations are present at frequencies in excess of 1%, and therefore can under some definitions qualify as polymorphisms. Examples are a specific mutation in Europeans in the *CFTR* gene responsible for cystic fibrosis, or mutations in some African populations in the β-globin gene responsible for sickle cell anemia.

▶ An allele below the 1% frequency used to define a polymorphism is sometimes called a **variant**. Clearly, since allele frequencies often vary greatly between populations, one population's variant may be another's polymorphism. In addition, variant is sometimes used as a general term that includes both polymorphisms and mutations.

The term 'mutation' has negative connotations for the layperson (Condit *et al.*, 2002), and this fact has led to discussion about the choice of alternative words when scientists communicate about genetic diversity with the general public – 'variation', 'variant' or 'alteration' were favored.

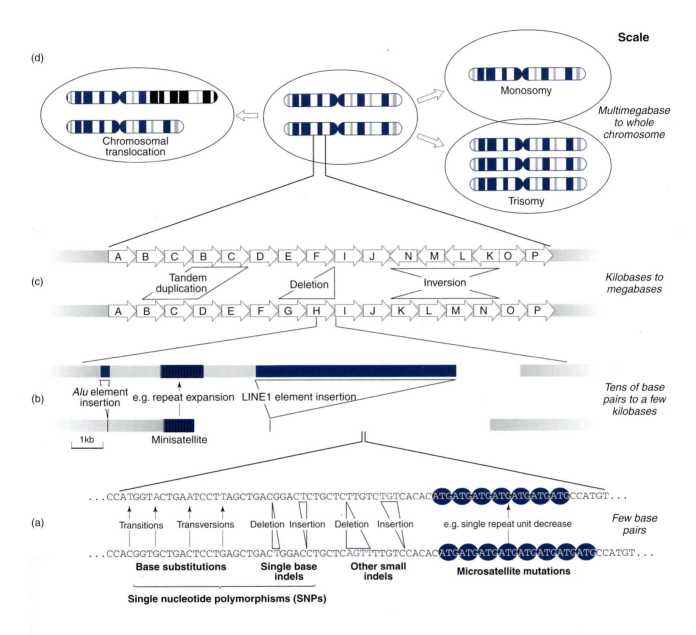

Figure 3.2: Overview of different classes and scales of mutation.

(a) Examples of mutations affecting a few base pairs, including examples of transition and transversion substitutions (see *Figure 3.6*), and a change in the number of ATG repeats at a trinucleotide microsatellite. (b) Examples of mutations affecting tens of base-pairs to a few kb, including insertion of a ~300-bp *Alu* element and a full length 6.1-kb LINE1 element, and the expansion by several repeat units of a minisatellite. (c) Examples of segmental mutations on the kilobase to megabase scale. Arrows indicate the order and orientation of adjacent segments of DNA. (d) Mutations on the multimegabase or whole chromosome scale. Most translocations are deleterious, but are important in interspecies differences. Most aneuploidies (departure from the 46-chromosome complement, shown here as monosomy or trisomy) are early lethals; some, e.g., trisomy 21 and aneuploidies of the sex chromosomes, are tolerated. No aneuploidy can be passed on to offspring.

These kinds of differences are examples of **single nucleotide polymorphisms**, or **SNPs** (pronounced 'snips'). The insertion or deletion (**indel**) of a single base is also included in the category of SNPs – perhaps unfortunately, since the mechanisms which underlie these indels, and the analytical treatment of them, differ from those for base substitutions.

Two fundamental processes give rise to base substitution mutations: the misincorporation of nucleotides during replication, and mutagenesis caused by the chemical modification of bases, or physical damage due, for example, to ultraviolet or ionizing radiation. These two processes are summarized in *Figure 3.7*.

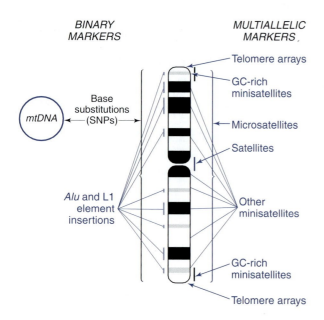

Figure 3.3: Overview of genomic distribution of different classes of polymorphic markers.

Base substitutions are widely distributed and frequent, and microsatellites and non-GC-rich minisatellites are also widespread. Telomere arrays are at the termini of chromosomes, and most GC-rich minisatellites are concentrated towards chromosome ends. All chromosomes carry alpha satellite sequences in their centromeric regions, although there are other satelllites on specific chromosomes at other positions. Although *Alu* elements are concentrated in GC-rich regions (light G-bands) and L1 elements are concentrated in AT-rich regions (dark G bands), recent insertions of both elements are concentrated in the latter regions (Lander *et al.* 2001). See text for more details on distributions of all classes of loci.

3.2.1 Base misincorporation during DNA replication

When a diploid cell divides, all 6400 Mb of its DNA must be replicated so that each daughter cell contains two copies of the haploid genome. DNA replication, the process that accomplishes this remarkable feat, proceeds with extremely high fidelity (reviewed by Kunkel and Bebebek, 2000). A new base is incorporated if it pairs with the existing base in the single-stranded template. However, the existence of the correct number of hydrogen bonds between bases is in itself insufficient to ensure that A pairs only with T, and G with C. The **DNA polymerase**, the enzyme responsible for DNA synthesis, also requires the correct overall geometry of the base pair before the bond will be made with the growing daughter strand. Occasional incorporation of the incorrect base does occur, possibly because of rare, transient chemical forms of bases that have different base–pairing properties and geometries. However, as well as the high fidelity involved in this incorporation step, the primary DNA replication machinery has a 'proof-reading' function: after incorporation of a new base at the 3′ end of the daughter strand, the

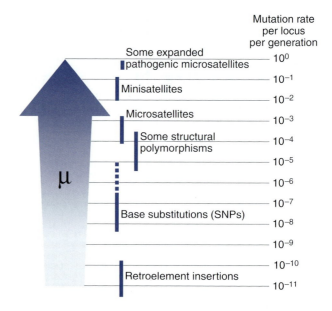

Figure 3.4: Overview of mutation rates (μ) of different classes of polymorphic markers.

The figure summarizes information given throughout this chapter. Mutation rates of micro- and minisatellites are very variable between loci. The rates given here reflect ascertainment bias, in that they are those of the mostly widely used and polymorphic loci. The average rates are probably much lower.

enzyme moves on so that the next base can be incorporated. The previously incorporated base is now re-examined by the enzyme, and if it is not observed to be part of a 'correct' base pair, an **exonuclease** activity of the enzyme excises it, and the enzyme tries again. This 'double checking' of each newly incorporated base greatly decreases the probability of misincorporating a base: replication errors occur with a frequency of about 10^{-9}–10^{-11} per nucleotide (Cooper *et al.*, 1995).

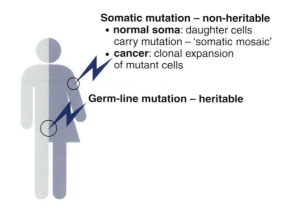

Figure 3.5: Somatic and germ-line mutations.

BOX 3.3 Classical polymorphisms

The ABO blood group system, discovered in 1900 (Landsteiner, 1900) is the first human genetic polymorphism to be defined. It was detected serologically: red blood cells carry **antigens** on their surfaces that can react with specific **antibodies** carried in the serum (the fluid component of the blood). When this reaction occurs, the cells clump together (agglutinate), which is easily visible and provides a simple assay. Two antigens, A and B, underlie the ABO system, and individuals carry in their serum antibodies to the antigen(s) that they do not possess. Mixing blood and serum samples reveals four classes of individuals: those carrying only the A antigen, those carrying only the B, those carrying both (AB), and those carrying neither (O; see *Figure*). Early studies showed differences in the frequencies of these blood groups in different populations (Hirszfeld and Hirszfeld, 1919), and later more blood groups (such as MN and RH) were defined and used in population studies. Each of these new blood group systems discovered was shown to behave independently of the ABO blood groups.

These and similar immunological methods were later used to detect specific variants of the immunoglobulins, proteins of the immune system found in serum or plasma, and also the extremely polymorphic **human leukocyte antigen** (**HLA**) system, analyzed since the late 1950s. HLA proteins are expressed on the surfaces of white blood cells, and are of great importance in transplant rejection or tolerance ('**histocompatibility**').

More markers became available with the introduction of **protein electrophoresis** to separate proteins in an electric field on the basis of their size and charge. This method was used to separate different hemoglobins, in particular that responsible for sickle cell anemia (Hb^S; OMIM 603903), and thus to demonstrate the first 'molecular disease' (Pauling *et al.*, 1949). Many different blood proteins were then shown by electophoresis to contain genetic differences, and these could be used in population studies. The amount of population data (in particular, sample sizes) accumulated using classical markers is very large, and probably has yet to be exceeded by DNA data. A review of these data is included in *The History and Geography of Human Genes*, by Cavalli-Sforza *et al.* (1994). Conclusions obtained with classical data tend to agree with those obtained with later DNA data when they can be compared – a good example is the excellent agreement on the apportionment of genetic diversity among different populations (see Section 9.3).

The molecular basis of many of these 'classical' polymorphisms at the DNA level is now known, and new PCR-based methods for their detection are available. For example, the gene underlying the ABO blood group system encodes a glycosyltransferase enzyme that adds a sugar residue to a carbohydrate structure known as the H antigen on the surface of red blood cells. The A allele codes for an enzyme which adds the sugar N-acetylgalactosamine, whereas the enzyme coded for by the B allele has two amino acid differences which alter its specificity so that it adds D-galactose, forming the A and B antigens respectively. The O allele has an inactivating mutation in the gene and therefore the H antigen remains unmodified (Yamamoto *et al.*, 1990).

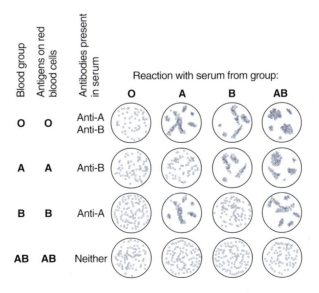

Defining the ABO blood group system by agglutination patterns.

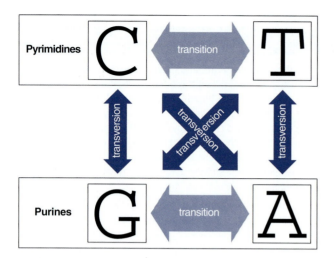

Figure 3.6: Transition and transversion mutations.

For each base there are two possible transversions, but only one transition. Nonetheless the mechanisms of mutation lead to transitions being more common than transversions (see Section 3.2.4).

3.2.2 Chemical and physical mutagenesis

The integrity of the genetic material, the sequence of DNA within a cell, is under constant assault by chemical and physical processes that alter bases or damage the physical structure of the DNA molecule (*Figure 3.7*). There are spontaneous, endogenous chemical processes going on in all cells that lead to base modification or loss: one example is the deamination (loss of an amine, $-NH_2$) of cytosine to produce uracil, which, unlike cytosine, pairs with adenine. It has been estimated that about 400 cytosines are deaminated daily in a human cell, and that between 10 000 and 1 000 000 bases per cell per day are damaged as a result of normal chemical reactions involving deamination, oxidation, methylation and depurination (loss of a purine base) (reviewed by Holmquist, 1998).

As well as this endogenous chemical damage, damage to the DNA molecule can be caused by chemical **mutagens**. Examples are:

▶ **base analogs**, with different base-pairing properties to the naturally occurring bases, which can be incorporated and subsequently lead to base substitution. An example is 5-bromouracil, a T analog that occasionally pairs with G;

▶ **base-modifying agents** which alter the base-pairing properties of bases within DNA. Hydroxylamine, for example, reacts with C to give a derivative which pairs with A, rather than with G;

▶ **intercalating agents** – flat molecules, such as acridine orange, which can insert between base pairs in the helix, distorting its structure and often leading to insertions or deletions of one or more bases;

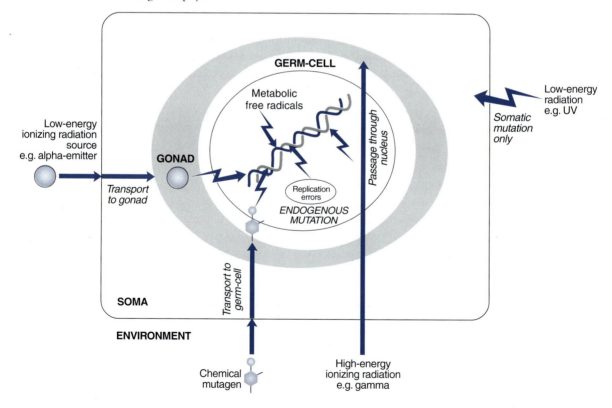

Figure 3.7: Chemical, physical and endogenous mutagenesis.

Overview of processes causing germ-line mutation. Ionizing radiation can either damage DNA directly, or cause mutations through the production of damaging ions.

▸ **cross-linking agents** such as mitomycin C, which cross-link different parts of a DNA helix, or different helices.

Physical agents, primarily forms of radiation, can also damage DNA. UV radiation is a low energy form of electromagnetic radiation that can induce chemical linking between adjacent thymidine nucleotides (producing thymidine dimers). **Ionizing radiation** (IR) is more energetic, and can break the DNA backbone, forming single- or double-strand breaks; alternatively, it can form reactive ions, known as free radicals, (also produced endogenously by oxidative metabolism) within cells which modify DNA leading, for example, to base substitutions. IR includes gamma radiation, which is a form of high energy electromagnetic radiation emitted by some radioisotopes, and X-rays, which are essentially the same as gamma rays, but generated by electron bombardment of heavy metal atoms. Forms of particulate radiation are also ionizing: neutrons, alpha particles and beta particles are emitted by radioisotopes, while cosmic radiation consists of a range of subatomic particles with varying energies, and originates outside the Earth.

Chemical and physical mutagens are important causes of or contributors to many cancers; however, their effects on germ-line mutation can be very different from their somatic effects. A good example is UV radiation, which is nonpenetrating and therefore has only external mutagenic effects, causing skin cancers – hence the worry about ozone layer depletion. Gamma radiation is highly penetrating and can cause germ-line and somatic mutations, while alpha and beta particles are more weakly penetrating, and for these to have an effect on the germ-line they must be emitted from an isotope which has been taken into the body and found its way to the gonads (*Figure 3.7*). The same consideration applies to chemical mutagens.

3.2.3 DNA repair

These physico-chemical assaults on DNA, if left unchecked, would lead to catastrophic levels of genome damage. Elaborate mechanisms have evolved to detect and repair DNA damage: the mutations that pass to the next generation are therefore the outcome of two processes – primary damage and secondary repair.

Many primary mutagenic events lead to a helix containing, on one strand, the 'correct' base, and on the other a mispaired base (when the helix is known as a **heteroduplex**), modified base or other lesion. If the mutation is to be passed on, the damaged helix must pass through DNA replication (*Figure 3.8*). Prior to replication, the damage can be repaired. Repair systems that carry this out include the **mismatch repair** and **nucleotide excision repair** systems, which have in common the removal of the lesion (for example a damaged base or thymidine dimer) and sometimes some surrounding bases, followed by resynthesis of the gap using the undamaged complementary single-strand as a template. As was described above, the primary DNA replication apparatus operates with very high fidelity. In contrast, many of the polymerases that carry out gap repair

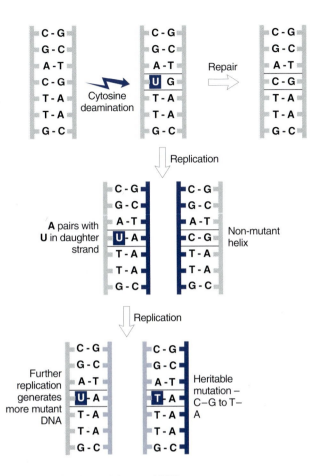

Figure 3.8: The fates of damaged DNA.

Damage to one strand of DNA, in this case cytosine deamination to yield uracil, can be repaired. Generation of a stable and heritable base-pair mutation requires two rounds of DNA replication.

synthesis have comparatively low fidelity. For example, most of the repair of damaged bases is carried out by the enzyme pol β, which excises and replaces only the damaged base. Pol β is the least accurate mammalian polymerase, with an average error rate (as determined *in vitro*) of about 10^{-3} (Osheroff *et al.*, 1999), so itself will lead to a large number of misincorporations during repair.

Double-stranded breaks (**DSBs**), which can be created by ionizing radiation, free radical action, and events during DNA replication, are particularly damaging lesions which can result in chromosomal aberrations, cell malfunction and cell death (reviewed by van Gent *et al.*, 2001). Complementary strand information cannot be used to repair such lesions, and instead a complex series of events can lead to repair of the DSB either through:

▸ homologous recombination (HR), in which the undamaged sister chromatid acts as template. This ensures accurate repair, but because it relies on the presence of the sister chromatid is dependent on the stage of the cell cycle at which damage occurs;

▶ nonhomologous end-joining, in which a protein complex recognizes the DNA ends, juxtaposes and ligates them. This sometimes involves removal or addition of bases at the DSB, and so is not as accurate as HR, and can itself be mutagenic.

In general, cell cycle progression is tightly coordinated with DNA repair and cell survival or death through a network of genome surveillance pathways (reviewed by Bartek and Lukas, 2001).

3.2.4 Estimating the base substitution mutation rate

Knowledge of the rates of the different classes of base substitution mutation is extremely important in evolutionary genetics. These rates can be incorporated into models of sequence evolution that underlie our attempts to interpret patterns of diversity within and between species (see Section 5.2). However, the mutation rate in nuclear DNA is so low that it cannot be investigated directly in pedigrees (though see *Box 3.6* on mtDNA), and an indirect approach is therefore required. Traditional methods for estimating mutation rates were based on observing the outcomes of specific kinds of mutations, namely genetic diseases (*Box 3.4*); these have their disadvantages, and methods based on a direct comparison of DNA sequences that are thought to be without function are preferable.

For neutral mutations, the rate of mutation can be shown to be equal to the evolutionary rate of change (Kimura, 1968; see Section 5.6.4). Therefore a direct comparison of nonfunctional DNA sequences between species whose divergence times are known can give an estimate of mutation rate. One example of this approach (Nachman and Crowell, 2000) compared ~16 kb of sequence from 18 **pseudogenes** (nonfunctional duplicates of genes) in chimpanzees and humans. A generation time of 20 years was assumed, together with a range of different published species divergence times

(see Section 7.4.4). Mutation rates deduced from this study are given in *Table 3.1*.

A number of important aspects of base substitutional mutation in humans emerge from these data, and from a larger scale study comparing 1 944 162 bp of noncontiguous chimpanzee sequences with human homologs (Ebersberger *et al.*, 2002):

▶ base substitutions are 10 times more frequent than indels, although this relative frequency varies substantially between loci, as it is heavily dependent on the sequence context;

▶ transitions are more than twice as frequent than transversions, despite the fact that for a given base there are two possible transversions to only one transition – if all mutations were equally probable we would expect a 2 : 1 ratio of transversions to transitions. This deviation from expectation might be explained by influences of efficiency of error detection and repair, sequence context, or perhaps by differences in misincorporation rates;

▶ rates of mutations (both transitions and transversions) at a particular dinucleotide, C followed 3′ by G (known as a **CpG**, where the 'p' indicates a phosphate group) are an order of magnitude higher than those at other dinucleotides. The reasons for this are discussed in the next section.

It is important to note that mutation rates of base substitutions are in general so low that a given mutation at a given position is unlikely to have recurred or reverted over the time-scale of the evolution of modern humans, such that independent occurrences are found at appreciable frequencies (see Chapter 8). Examples do, however, exist: one of the first Y-chromosomal SNPs to be discovered, SRY10831 (Whitfield *et al.*, 1995) is one (see *Box 8.8*): the original A to G transition occurred deep in the tree of Y

BOX 3.4 Mutation rates estimated from genetic diseases

The most straightforward method involves counting the number of individuals suffering from a dominant genetic disorder who have unaffected parents. This yields estimates of the mutation rate per locus, which range from 10^{-4} to 10^{-6} (reviewed by Vogel and Motulsky, 1997). In a specific example, achondroplasia (OMIM 100800), seven of a total of 242 257 newborns had the disorder (Gardner, 1977): since the mutation could have occurred in either parent, the rate of mutation is calculated as 7/(2 × 242 257), or 1.4×10^{-5} (with a large standard error, of $\pm 0.5 \times 10^{-5}$). Achondroplasia is an interesting example, because the mutation causing the disease turns out to be at exactly the same nucleotide in 97% of cases, leading to a glycine to arginine amino acid substitution. In this case, the rate of mutation to the disease phenotype is therefore almost equal to the base substitution rate, a rate which is highly atypical and, indeed, the highest rate known for any base in the nuclear genome (reviewed by Vajo *et al.*, 2000). For most dominant disorders the situation is much more complex:

▶ many different mutations in a gene can give rise to a particular phenotype;

▶ penetrance may be incomplete;

▶ mutations in other genes may give similar phenotypes (locus heterogeneity);

▶ **phenocopies** (imitations of phenotype through nongenetic effects) may exist.

Problems common to all dominant disorders are that many changes in the gene will not give a new phenotype at all, and that genes vary greatly in size, and therefore extrapolation from per-locus to per-base mutation rates is highly unreliable. Estimating mutation rates for most recessive disorders is almost impossible.

TABLE 3.1: ESTIMATES FOR MUTATION RATES OF BASE SUBSTITUTIONS AND SMALL INDELS.

Type of mutation	Rate per site per generation
All base substitutions	2.3×10^{-8}
Transition at non-CpG	1.2×10^{-8}
Transition at CpG	1.6×10^{-7}
Transversion at non-CpG	5.5×10^{-9}
Transversion at CpG	4.4×10^{-8}
Indel mutations	2.3×10^{-9}
All mutations	2.5×10^{-8}

Data from Nachman and Crowell (2000); assumes human–chimpanzee divergence time of 5 MYA, and a generation time of 20 years.

haplotypes, such that most modern Y chromosomes carry the G-allele; reversion from G to A then occurred defining a group of Y chromosomes, 'haplogroup R1a', which has spread rapidly and recently within Eurasia (Zerjal *et al.*, 1999). Note that even though the mutation rate is low (*Table 3.1*) the current population size ($\sim 6 \times 10^{9}$) is so large that the recurrence or reversion of any SNP in some individual(s) is now expected each generation. The frequency of such new mutations in the population will be extremely low, so in practice they are not encountered.

This low mutation rate means that in practice this class of mutation shows **identity by descent**, rather than **identity by state**: the presence of a given polymorphic base in two alleles usually implies that they have inherited the base from a common ancestor (*Figure 3.9a*). This stability also means that the direction of evolutionary change can usually be established (*Figure 3.9b*) by examining the homologous sequence in the DNA of great apes, who are our closest living nonhuman relatives (see Chapter 7). The low mutation rate also means that base substitutions belong to the class of **binary markers**, also known as **biallelic**, or **diallelic** markers (the term **unique event polymorphisms [UEPs]** is also sometimes used).

3.2.5 The CpG dinucleotide is a hot spot for mutation

About 75% of CpG dinucleotides in the human genome are the target of **DNA methylation**: a specific methyltransferase enzyme adds a methyl group to the 5-carbon of the cytosine ring to give 5-methylcytosine (*Figure 3.10a*). The complementary dinucleotide to CpG is also CpG – it is said to have **dyad symmetry** – and so methylation affects both strands at adjacent base pairs (*Figure 3.10b*). The function of this methylation is not entirely clear: it may play roles in the control of gene expression, the control of chromosome condensation, and even as a 'host defense' mechanism protecting against the damaging activities of transposable elements (reviewed by Robertson and Wolffe, 2000; see Section 3.4).

Spontaneous (or mutagen-induced) deamination of cytosine yields **uracil**, which is not a legitimate base in DNA, and is efficiently excised and replaced with cytosine by a repair system. In contrast, deamination of 5-methylcytosine yields thymine, and the T–G mispair can be repaired either as T–A or C–G: the repair machinery does not 'know' that the original base pair was C–G. The result of this difference is that methylated cytosine residues, and hence CpG dinucleotides, are hot spots for mutation. The tendency is for CpG to mutate to TpG or CpA (*Figure 3.10b*); in other words, the methylation of cytosine should lead to an increase of transition mutations. This is indeed observed (see *Table 3.1*); the accompanying increase of transversion mutations at CpG dinucleotides is not understood.

The highly mutable nature of methylated CpG dinucleotides means that the overall frequency of CpG in the genome is only about 20% of that expected on the basis of the frequency of C and G nucleotides. There are, however, regions of the genome, typically associated with the 5' ends of genes, which escape CpG methylation; this is probably due to the special role that these regions play in transcription initiation, perhaps requiring an open chromatin conformation. Absence of methylation means that deamination events in these regions are repaired more efficiently than elsewhere in the genome, and CpG dinucleotides in these '**CpG islands**', extending over hundreds of base pairs, and with a copy number of about 45 000, therefore exist at the expected frequency.

Evidence for hypermutability of CpG comes not only from inter-species comparisons (*Table 3.1*) but also from data on mutations in humans causing genetic disease. A specific example is the mutation underlying achondroplasia (OMIM 100800), discussed above (*Box 3.4*), in which the G to A transition at the hypermutable nucleotide 1138 of the *FGFR3* gene occurs at a CpG. Data based on an analysis of 7271 disease-causing base substitutions in 547 different genes (Krawczak *et al.*, 1998) show that 37% of transitions occur at CpG dinucleotides, which reflects a transition rate five times the base mutation rate [note that this figure for genes is not

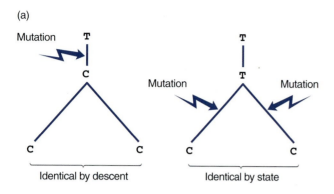

(a)

Mutation

Identical by descent   Identical by state

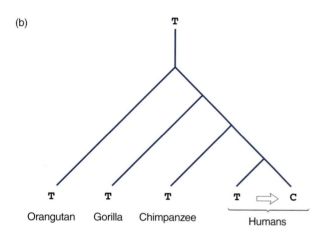

(b)

Orangutan   Gorilla   Chimpanzee   Humans

Figure 3.9: Identity by descent, and ancestral state determination of an SNP.

(a) When two alleles and their common ancestor share the same allelic state (here, a C-allele at SNP), they show identity by descent (left panel); when the common ancestor has a different allelic state (T-allele), they show identity by state (right panel). (b) Given the known phylogeny of our great ape relatives (see Chapter 7), the presence of a T at SNP site in great ape homologs of a human sequence showing a T/C polymorphism defines T as the ancestral state. The mutation underlying the polymorphism is a T to C transition. Note that polymorphism at the same base is on rare occasions expected to occur in the other primates. Analysis of more than one member of each species is useful.

in agreement with that for pseudogenes (*Table 3.1*), which may reflect differences in sequence composition].

3.2.6 Base substitutions and small indels within genes

Evolutionary genetic studies often focus on mutations that are assumed to be neutral – having no effect on phenotype. Many of these lie outside genes; however, because of the nature of the biological mechanisms which read information in DNA and express it as proteins, some sequence changes within genes are expected to have no or little effect on function. On the other hand, and as alluded to above, base substitutions within genes can be the cause of genetic disease,

and it is therefore important to understand the potential phenotypic effects of such changes in DNA sequences. Most mutations affecting phenotype do so adversely – there are more ways to impair the function of a gene than to enhance it. However, there is certainly an ascertainment bias here, since mutations causing diseases are noted and find their way into mutation databases. Mutations improving the performance of a protein or changing its function advantageously are not necessarily noted, but are nevertheless the stuff of positive selection and evolution. This discussion focuses on deleterious mutations.

Figure 3.11 shows some of the potential effects of base substitutions and small indels on a protein-coding gene; *Table 3.2* gives the relative frequencies of the different types of mutation causing genetic disease, as listed in the Human Gene Mutation Database.

Outside the open reading frame, mutations can affect expression of the gene by altering **promoters, enhancers,** or polyadenylation signals. Mutation of the consensus splicing signals at the ends of introns can affect splicing, leading, for example, to the retention of an intron in the transcript or the skipping of an exon, and consequent loss of function. While the GT/AG dinucleotides at the ends of introns (*Figure 3.11*) are highly conserved and necessary for correct splicing, other adjacent and even distant nucleotides can influence its efficiency or accuracy (reviewed by Cartegni *et al.*, 2002). Alternatively, a new splice site can be created in an intron. To complicate matters even further, a mutation within an intron can affect the regulation of an adjacent gene; this appears to be the case for the human lactase gene, whose downregulation in adulthood is prevented by a mutation 14 kb upstream, in an intron within an unrelated gene (see Section 13.4.3).

In the open reading frame itself, base substitutions can have a wide range of possible effects, from complete neutrality to complete abolition of protein production. The redundancy of the genetic code (see *Figure 2.6*) provides a buffer against deleterious effects of base substitution: many amino acids are encoded by more than one codon (six different codons encode serine, for example), and in general the third position is particularly insensitive to mutation. Substitutions that do not alter an amino acid are known as **silent-site,** or **synonymous** substitutions, and are often assumed to be selectively neutral. However, such substitutions could in principle alter mRNA stability, or efficiency of translation, and hence affect function. Cases are also known where they have a deleterious effect through altered splicing (reviewed by Cartegni *et al.*, 2002) – for example, a silent-site substitution in an exon of the phenylalanine hydroxylase gene gives rise to a phenylketonuria disease allele (OMIM 261600) by inducing skipping of the next exon (Chao *et al.*, 2001).

Base substitutions which lead to a change of amino acid are termed **nonsynonymous,** or **missense** mutations, and are denoted, for example as 'Arg702Trp', a substitution of tryptophan for arginine at amino acid 702 of a protein. Beyond straightforward redundancy, the genetic code

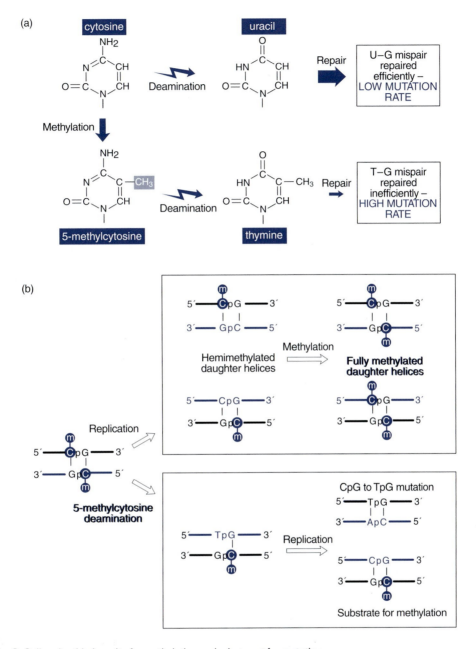

Figure 3.10: The CpG dinucleotide is a site for methylation and a hot spot for mutation.

(a) Spontaneous or mutagen-induced deamination of cytosine yields uracil, which is efficiently recognized and removed by uracil glycosidase. Deamination of 5-methylcytosine yields thymine, which is not efficiently repaired. (b) Most CpG dinucleotides are methylated at the 5-carbon of cytosine on both strands. Replication yields hemimethylated daughter helices that become fully methylated through the action of a specific methyltransferase. Unrepaired deamination followed by replication leads to a CpG to TpG mutation, or, if deamination affects the complementary strand, a CpG to CpA mutation.

appears to be organized to minimize the deleterious effects of base substitution (reviewed by Knight *et al.*, 1999): the assignment of amino acids among the 64 possible codons means that many missense mutations will tend to replace one amino acid with a different one which has similar chemical properties, for example hydrophobicity, or charge. Such a change is termed a **conservative** substitution. The effects of these substitutions depend on their position within the protein: conservative substitutions outside the active site of an enzyme, for example, might have little effect, while those within it might be deleterious. Some base substitutions, however, change one amino acid for another that has different properties; such **nonconservative** substitutions are more likely to be deleterious than are conservative changes.

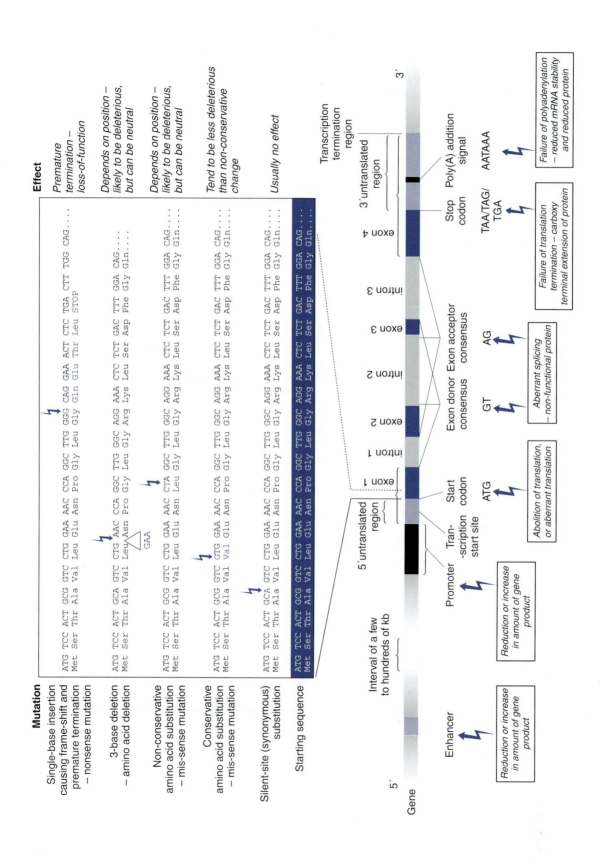

Figure 3.11: Small-scale mutations within a typical human protein-coding gene and their potential effects.

The different elements of the gene are not to scale, and in particular the introns illustrated here are small relative to the exons (see *Table 2.1* for typical sizes of these elements).

TABLE 3.2: SUMMARY OF STATISTICS FOR SMALL-SCALE PATHOGENIC MUTATIONS IN 1163 GENES.

Type of mutation	Number	Percentage
Base substitution	21 200	75% of total
Amino acid substitution (missense)	14 756	52% of base substitutions
Generation of stop codon (nonsense)	3198	11% of base substitutions
Abolition of stop codon	24	< 1% of base substitutions
Splice site mutation	2976[a]	11% of base substitutions
Regulatory	246	1% of base substitutions
Small deletion	5049	18% of total
Small insertion	1922	7% of total
Total	28 171	

Data taken from the Human Gene Mutation Database (02/09/2002 release). The database lists a total of 30 641 mutations, including 2470 gross lesions and other small indels not given here.

[a]Of which 1680 occur at the GT/AG consensus dinucleotides.

An example is the Hb[S] mutation, which changes a hydrophilic ('water-loving') glutamic acid on the molecule's surface for a hydrophobic ('water-hating') valine: the consequence is that the protein molecules associate together, rather than staying in solution, leading to changes in cellular behavior. A mutation changing an amino acid codon into a termination codon is termed a **nonsense** mutation, and is likely to be deleterious unless very close to the natural termination codon.

A small indel within an open reading frame is more likely to be deleterious than is a base substitution. Because the genetic code is based on triplets, any indel of a multiple of three bases will lead to an insertion or deletion of amino acid(s). Again, the effects of such changes will depend on their position in the protein. The best known example of a single amino acid deletion is the commonest cystic fibrosis mutation, ΔF508 (see Section 14.2.1), which deletes a phenylalanine from the CFTR protein, leading to defective intracellular processing. If the indel is not a multiple of three bases, it changes the reading frame of translation. Such **frameshift** mutations are highly deleterious, since the amino acid sequence 3′ of the mutation is completely different from that of the wild-type. Translation can proceed until a termination codon is reached, leading to prematurely terminated protein, or to abnormally extended protein, particularly if the mutation is close to the 3′ end of the coding region. Alternatively, the change of reading frame may lead to unstable mRNA or aberrant splicing which abolishes protein production. Among disease-causing mutations, at least, small deletions predominate over small insertions, and the great majority of events involve 1–5 bases (*Figure 3.12*).

3.2.7 SNP diversity and SNP distribution

Interest in the identification of SNPs is high, largely because of their potential use as **molecular markers** in **disease association studies** (*Box 3.5*). A number of large-scale re-sequencing (sequencing the same locus in several individuals) studies of particular loci have now been done, and provide a picture of SNP diversity in these regions (reviewed by Przeworski *et al.*, 2000). As well as this, efforts to sequence the entire genome have used libraries of clones [mostly **bacterial artificial chromosome** (**BAC**) clones] constructed from the DNAs of more than one individual. This means that overlapping sequences between different clones readily reveal SNPs: the draft and finished sequence of the publicly funded sequencing project, for example, represented a 7.5-fold coverage of the genome, and identified 971 000

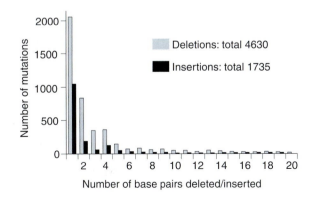

Figure 3.12: Incidence of small indels in human gene mutations.

Distribution of small indels in the Human Gene Mutation Database (21/06/2002 update) by size.

new candidate SNPs (International SNP Map Working Group, 2001). These and other SNPs detected in general comparative studies provide a relatively unbiased picture of SNP variation.

Overall, the average nucleotide diversity (π, representing the likelihood that a given nucleotide position differs across two randomly sampled sequences; see Section 6.2) in both genome-wide and locus-specific studies is about 8×10^{-4} (Przeworski *et al.*, 2000; International SNP Map Working Group, 2001; Venter *et al.*, 2001). This means that, on average, we expect to find one SNP about every 1250 bp.

The actual value of π varies significantly between chromosomes, from 5.19×10^{-4} for chromosome 22 to 8.79×10^{-4} for chromosome 15. Additionally, there is some suggestion that SNP density varies along chromosomes (Venter *et al.*, 2001), and explanations have been put forward based on variation in GC content or in the efficiency of DNA mismatch repair. However, as discussed in Chapter 6, both the genealogical history and mutation rate of a locus influence its sequence diversity. Both selection and population history have a part to play in shaping the genealogical history of a locus. For example, strong positive selection on a locus will cause it to have a recent genealogical history that means that we expect it to show lower diversity than loci not under positive selection. Indeed, it has recently been suggested that the genealogical history of a locus, rather than its mutation rate, is the primary determinant of this variation in sequence diversity (*Figure 3.13*; Reich *et al.*, 2002).

It is also worth noting that many of the SNPs in the databases have not yet been verified by independent detection in population samples, and that, therefore, a proportion of SNPs could be artifactual. Regions of the genome displaying apparent high SNP densities may represent misassignments (Bailey *et al.*, 2002; Estivill *et al.*, 2002) between sequences that represent not homologs, but instead highly similar ($> 97\%$ identical) **paralogs** (segmental duplications), which make up about 5% of the genome (see Section 3.5). A recent study has shown that the average apparent SNP density is elevated in duplicated regions, from 0.69 per kb to 1.33 per kb, suggesting that these 'SNPs' are actually **paralogous sequence variants** (**PSVs**). A specific example is on the short arm of the Y chromosome, which is annotated as having an extremely high SNP density (Venter *et al.*, 2001), despite being notorious for its low level of variation. In this case it seems probable that the putative SNPs really represent sequence differences between the highly similar (99% identity) blocks of paralogous sequence on Yp and Xq21. Having said that, frequent **gene conversion** events between some paralogs may result in their having higher sequence diversity than nonduplicated portions of the genome (see Section 3.5). In the context of these problems of interpretation and validation, discussions of the reasons underlying the existence of putative SNP 'jungles' and 'deserts' are perhaps premature.

3.2.8 The population within us: base substitutional mutation of mtDNA

Mutation in mtDNA warrants separate discussion because it has two unusual aspects – its rate with respect to nuclear DNA, and the manner in which mutations pass from one generation to the next.

Mutations occurring in mtDNA are largely base substitutions, and deletions that can involve much of the mtDNA molecule; deletions are the causes of maternally inherited mitochondrial diseases, while many base

---

BOX 3.5 Locating disease genes: linkage analysis and disease association studies

The search for human genome variation and the development of methods to analyze this variation in large numbers of samples have been driven by the desire to identify genes responsible for genetic disorders. Since the era of PCR the markers most often employed have been microsatellites and SNPs.

Linkage analysis is most readily applicable to **Mendelian** disorders (those that show the classic patterns of autosomal-dominant, autosomal-recessive, or sex-linked inheritance), and the principle behind it is to observe the cosegregation of a disease phenotype with molecular markers in pedigrees. Recombination events are likely to separate a marker allele from the causative gene if the two are far apart, and are less likely to separate them if they are close together. The results are commonly expressed as a **LOD score**, which is the logarithm (in base 10) of the odds of linkage – the ratio of the likelihood that loci are linked to the likelihood that they are not linked. For example, a LOD score of 3 represents odds of a thousand to one in favor of linkage, and is normally taken as the minimum value indicating linkage with a 5% chance of error. Pedigree studies can lead to the definition of a minimum recombinational interval which must contain the disease gene, an interval with a length expressed as a genetic distance, in **centiMorgans** (see Section 2.6). Linkage approaches can be applied to complex disorders, but are less powerful than association studies (see Section 14.3.1).

Disease association studies do not require pedigrees, and are used in attempts to elucidate the genetic basis of complex disorders. For example, through a study of the alleles of a set of molecular markers in a group of affected individuals and a group of control individuals (case–control study), a statement can be made about the relative likelihood of the affected group carrying a particular allele. Association between a marker allele and a disease susceptibility gene can mean that the marker is close to the gene (see Section 3.7), but can also arise from other causes, such as population admixture. Further examples of disease association study designs are discussed in Section 14.3.1.

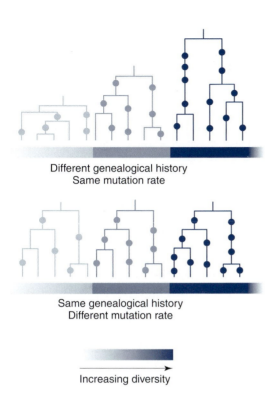

Different genealogical history
Same mutation rate

Same genealogical history
Different mutation rate

Increasing diversity

Figure 3.13: The effect of genealogical history and mutation rate on the sequence diversity of a locus.

substitutions appear to be neutral. Evolutionary genetic studies concentrate on these, but also consider other apparently neutral variations, such as change in length of a poly(C) tract in the control region, and the '9-bp deletion', a change in copy number of a 9-bp tandem repeat motif from two copies to one copy (rare mtDNAs even have three repeats). Early comparisons of human and other primate DNAs indicated that the base substitutional mutation rate in mtDNA is about 10 times higher than the average rate in nuclear DNA (Brown *et al.*, 1979). This high rate leads to a very large number of different sequences in human populations – one of the reasons for the popularity of mtDNA as an evolutionary tool. The observed mutation rate throughout the 16.5-kb molecule is nonuniform, with a relatively low rate in the coding regions, and a relatively high rate in the control region, which does not encode proteins or RNAs (see *Figure 2.20*). At some positions within the **hypervariable segments** (**HVS I** and **HVS II**, sometimes also known as **hypervariable regions**, or **HVR**s), of the control region, mutation rate is so high that mutations can often be observed in pedigrees, and comparisons between the pedigree rate and the rate calculated by other means have caused controversy (see *Box 3.6*).

A good estimate of mtDNA mutation rate is important for human evolutionary genetic studies. The widely employed estimates are indirect, based on calibrating the amount of divergence or diversity accumulated since a well-dated event, such as a phylogenetic split between humans and a closely

related primate, or a particular human settlement, such as that of New Guinea. Most interest has focused on the mutation rate of the control region, since this is the part of mtDNA most widely used in population studies. mtDNA mutation rates are often expressed as base substitutions per site per million years, rather than the 'per base per generation' expression that has been used elsewhere in this chapter. A range of rates covering an order of magnitude has been provided by these studies: 0.025–0.26 per site per million years (these figures include confidence intervals, and were compiled from the literature by Parsons *et al.*, 1997). This range, assuming a 20-year generation time, equates to approximately 5×10^{-7} to 5×10^{-6} per base per generation. Some mutation rates focus on an even smaller region, the more variable of the two **hypervariable segments** (**HVS I**). An average rate for a 275-bp section is given as one transition per 20 180 years (Richards *et al.*, 2000), which equates to 3.6×10^{-6} per base per generation, for a 20-year generation time. Outside the control region mutation rates have again been calculated by phylogenetic comparisons and assumptions about species divergence times: a rate from whole-mtDNA genome sequencing (Ingman *et al.*, 2000) is 3.4×10^{-7}.

There are a number of possible reasons for the high rate of mutation of mtDNA compared to nuclear DNA:

▶ because of its function in energy generation through oxidative phosphorylation, the mitochondrion contains a high concentration of mutagenic oxygen free radicals;

▶ mtDNA may have a higher turnover rate than nuclear DNA, requiring more replications per unit of time;

▶ because of the way it is replicated, mtDNA spends more of its time in the vulnerable single-stranded form than does nuclear DNA; indeed, the control region is sometimes called the 'D-loop' region to reflect its unusual structure;

▶ mtDNA is not packaged with histones, which may make it more vulnerable to mutation;

▶ repair systems in mitochondria are less effective than those in the nucleus (reviewed by Shadel and Clayton, 1997).

Mutation in the nuclear genome has been described in the text, and is easy to understand – mutations occur in the germ-line and are passed to diploid offspring via gametes in a single (haploid) dose. Mutations in mtDNA are different: as was explained in Chapter 2, each cell contains many mitochondria, each of which contains several mtDNA molecules; thus there is a large population of mtDNAs in any cell. A mtDNA mutation occurs in a *single* mtDNA molecule in a *single* mitochondrion within this population, yet can come to represent all the mtDNA in the soma of an individual in a subsequent generation. How does this occur?

Observations of mtDNA mutations, both neutral and disease-associated, indicate that the population of mitochondria passes through a bottleneck, at some point in oogenesis between the primordial germ cells and the primary oocyte

BOX 3.6 Opinion: A convenient number

It is a concept that is simple and elegant, and – if correct – extremely powerful for telling us just what we want to know: how long ago did a given set of DNA sequences (and the people they represent) diverge from one another? The concept is that of an accurately calibrated molecular clock, considered here as applied to mitochondrial DNA (mtDNA) sequences. In this concept, the 'clock' part comes from mutations that accumulate gradually over time as mtDNA lineages are propagated, and it is calibrated by knowing the diversity within a set of sequences that descended from a known point in time. In the case of mitochondrial DNA, the primary calibration comes from dividing the sequence divergence between humans and chimpanzees by the time since they diverged.

Sequences from the mtDNA control region, particularly hypervariable segment I (HVS I), have been predominant in molecular evolutionary studies of modern *Homo sapiens*. Vast mtDNA data have bolstered an 'out of Africa' hypothesis, and the molecular clock calibration rate – the mutation rate – has dated an African 'mitochondrial Eve' at ~150 000 YA. This wonderfully convenient mutation rate number (~0.1/site/million years) has since been applied to many human radiations across the globe, spanning temporal scales ranging from ancient continental expansions to recent colonization of South Pacific islands. It also formed the basis for expectations in forensic identity testing laboratories for the frequency with which offspring would differ from their mothers in mtDNA sequence – about once every 600 births. In 1997, we were surprised to discover that, upon actual sequence comparison, mothers and children differed far more frequently.

Our results (Parsons *et al.*, 1997) and others published just before showed an empirically observed mutation rate of the mtDNA control region some 10- to 20-fold higher than the standard molecular clock calibration. As a result, the convenient number was unsettled, and a debate natural to the scientific process ensued. Two predominant themes emerged to counter any disruption of convention: (i) the empirical data are flawed and the convenient number is fine, and (ii) the high observed mtDNA mutation rate is caused by mutational 'hot spots.' Theme one is at an advanced stage of falling by the wayside, due to multiple follow-up studies by various groups that converge on an observed mutation rate that is some 5- to 10-fold higher than the standard clock calibration (e.g., Siguroardottir *et al.*, 2000). Theme two remains predominant, but is elusive in its meaning.

Meanwhile, the field has exercised little additional caution in applying the convenient, conventional number to any given set of mtDNA sequences for dating events in population history.

The issue of mutational hotspots certainly does relate to the discrepancy between observed mutation rate and the standard clock calibration. Different sites in the control region accumulate change at vastly different rates, and about half do not vary at all. Phylogenetic studies show that the fastest variable sites may be as much as 30-fold more likely to mutate than the slowest. It could be that the mutations observed in empirical studies occur at a small number of very fast sites, and these are constantly mutating back and forth, contributing little to sequence divergence over time. Observed mutations do occur at faster-than-average sites, prompting the suggestion that hotspots rescue the convenient number: a great majority of sites accumulate mutations steadily, except the fast sites that virtually immediately saturate with mutations and have no further effect on divergence or the clock. Unfortunately, this rescue doesn't hold up. There are many more slow/moderate sites than fast ones, and sites of directly observed mutations represent the full range of relative rates, at proportions consistent with the rate classes scaled by the number of sites of that class. (The sites of observed mutations are still 'faster than average' because many sites have a mutation rate of zero.)

There is, however, a component of the hotspot issue that likely is very significant – but it doesn't necessarily leave the convenient number unscathed. We seek an explanation for sequences that appear, over a long time, to diverge more slowly than mutations are known to occur between generations. A plausible explanation is that mutations occur quickly enough at many sites that, over time, reversion mutations obscure the total number of mutations that have occurred (and that phylogenetic analyses have under-estimated the extent of this 'homoplasy'). If this is the hotspot theory, hats off – but there is nothing about it that guarantees the convenient number can be applied uniformly to any population sample of mtDNA sequences. The transition curve from 'fast' to 'apparently slower' rates over time has not been charted with any rigor, and users of the convenient number would be wise to wonder where their population sample is on this trajectory.

Thomas J. Parsons, The Armed Forces DNA Identification Laboratory, Rockville, MD, USA

(*Figure 3.14*). The result is that a mutant mtDNA representing a small minority of molecules in the soma of a mother can come to represent a range of proportions from zero to a large majority in her children. This phenomenon, known as **cytoplasmic segregation**, accounts for the wide variation in severity of mtDNA-associated diseases from one generation to the next, and also leads to interpretative problems when mtDNA mutation rates are considered (see *Box 3.6*).

The term **heteroplasmy** is used to refer to a situation where more than one mtDNA type occurs in a cell, and **homoplasmy** when all mtDNAs are identical. Clearly, it is difficult to apply these terms rigorously, since in a large population of mtDNA molecules within a cell at any one time there may be several mutant varieties present at undetectably low concentrations; the degree of heteroplasmy may also be tissue-specific (Lightowlers *et al.*, 1997). In practice, heteroplasmy can only be recognized when a certain proportion of a particular class of mutant molecules has accumulated − this is usually > 1%, but depends on the method of analysis. Note that the sensitivity of PCR does allow the detection of a *specific* class of mutant molecule even at extremely low copy numbers.

3.3 VNTR loci

Another class of genetic variation, much more dynamic that that described in the previous sections, is common in eukaryotic genomes. This variation involves changes in the numbers of repeated DNA sequences arranged in tandem arrays, and the highly heterogeneous classes of loci undergoing these changes are collectively known as **variable number of tandem repeats** loci, or **VNTRs**. While the high variability of these **multiallelic markers** is a useful property in many respects, the underlying high mutation rates mean that, in contrast to SNPs:

▶ alleles with the same size and sequence may not reflect identity by descent, but identity by state (coincidental resemblance, or 'convergent evolution');

▶ ancestral state cannot be determined by reference to great ape DNAs.

VNTRs are classified according to the size of their repeat units, the typical number of units in arrays, and sometimes their level of variability. *Figure 3.15* provides an overview of the scales of the different common classes of VNTRs, and their generally used names. Nomenclature is not systematic, and can be mystifying (see *Box 3.7*).

Many VNTRs are considered as neutral markers. However, there are well known examples in every class of VNTRs that play functional roles, and in which variation in repeat copy number can, in principle or practice, have phenotypic effects:

▶ some mini- and microsatellites lie within the coding regions of genes, or in regulatory regions, and variation can affect gene expression or the function of gene products; telomere repeat arrays at the ends of chromosomes are sometimes classed as a kind of minisatellite, but have essential functions;

▶ some satellites lying at centromeres, and telomere repeat arrays at chromosomal termini, constitute important functional components of chromosomes;

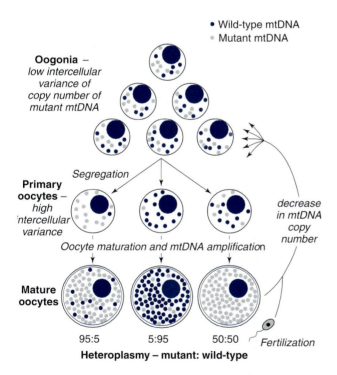

Figure 3.14: Intracellular population genetics: the fate of mutant mtDNAs.

A female developing from a fertilized oocyte that is 50% heteroplasmic for a mutant mtDNA type produces primordial germ cells (oogonia) which show similar levels of heteroplasmy (low variance). During the production of primary oocytes from these cells cytoplasmic segregation of mitochondria leads to a high variance in copy number of the mutant mtDNA type.

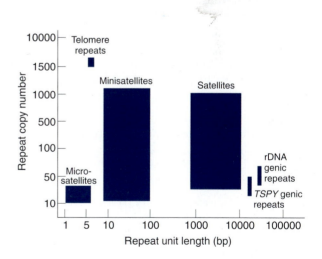

Figure 3.15: Overview of repeat unit length and repeat array length at VNTR loci.

Approximate repeat unit length and repeat copy number are shown on nonuniform scales for microsatellites, minisatellites, satellites and for telomere repeat arrays, which contain about 1500–2500 copies of a hexanucleotide repeat. Microsatellites and minisatellites with fewer repeat units are very abundant but not usually used because of their low mutability and level of polymorphism. For satellites the repeat unit length represents the higher-order repeat; many satellites have a lower-order repeat unit length of a few bp to a few hundred bp. rDNA arrays are on the short arms of the five acrocentric chromosomes (see Section 3.5); the array of the 20-kb unit containing the *TSPY* gene is on Yp.

> **BOX 3.7 Why 'satellite'?**
>
> One term that often crops up in VNTR nomenclature is 'satellite'; since this is nonintuitive, it deserves some historical explanation. Early attempts to fractionate physically sheared DNA used buoyant-density gradient ultracentrifugation, in which separation was on the basis of density in cesium chloride density gradients, often in the presence of binding agents which accentuated differences in density between sequences of different compositions. In these experiments, first reported in 1961 when satellites like Sputnik were in the news, certain minor fractions formed separate bands from the bulk of the DNA, and were therefore called 'satellites' of the major fraction. It later became clear that these fractions represented very high copy number tandemly repeated sequences, including those present at the centromeres of chromosomes in some species (see Section 2.5). The term satellite later became adopted as a term for a wide range of tandemly repeated sequences including micro- and minisatellites, even though (unlike true satellites – see Section 3.3.3) these kinds of sequences are not expected to be separable from the rest in an ultracentrifuge.

▶ some VNTRs contain entire genes, and arrays of different numbers of repeats may affect levels of gene expression; a good example is the tandem arrays of rDNA genes.

VNTR loci cover a vast range of different scales from microsatellites of a few tens of base pairs to satellites covering several megabases, and which are cytogenetically visible. It is important to note that, despite the common element of variation in tandem repeat number, these different loci have little in common in the mechanisms that generate them and maintain their high variability. Their properties are described in the following sections.

3.3.1 Microsatellites

Microsatellites, also widely known as **short tandem repeats** (**STRs**), and sometimes as simple sequence repeats (SSRs), are tandem arrays of repeat units 1–6 bp in length, and those that are useful as genetic markers have a typical copy number of 10–30. *Table 3.3* lists some general properties of microsatellites by repeat unit size, and *Figure 3.16* shows some examples. Some compound microsatellite loci contain repeat units of more than one length, such as adjacent di- and tetranucleotide arrays. Microsatellites with some specific repeat units show clustering, but most are distributed throughout the genome (*Figure 3.17*), which has been a useful property in **linkage analysis** (*Box 3.5*). While most microsatellites are considered to be selectively neutral, some can clearly have phenotypic effects (*Box 3.8*).

The mutation rates of microsatellites have been estimated by direct pedigree analysis (e.g., Brinkmann *et al.*, 1998; Xu *et al.*, 2000), by examining descendants of deep-rooting pedigrees who are separated by many generations (Heyer *et al.*, 1997), and by attempts to detect mutants in small populations of sperm DNA molecules (Holtkemper *et al.*, 2001). Characteristics of the mutation process have also been studied in colon cancer patients who display elevated microsatellite instability (Di Rienzo *et al.*, 1998), and by large-scale comparisons of human and chimpanzee microsatellites (Webster *et al.*, 2002). Typical figures for mutation rate estimates are around 10^{-3}–10^{-4} per locus per

TABLE 3.3: PROPERTIES OF MICROSATELLITES BY REPEAT UNIT SIZE.

Repeated unit/bp	Properties and distribution	Utility
1	Mostly poly(A)/poly (T), associated with *Alu*, LINE and other retroelements	Not used, due to small differences in allele size and problem of allele-calling due to PCR 'stutter' – as well as a band of the expected size, additional bands are seen which are typically one repeat unit larger and smaller, and which result from slippage synthesis errors by the PCR polymerase
2	$(AC)_n/(GT)_n$ most common, representing 0.5% of genome; $(GC)_n$ extremely rare	Widely used in early studies because of ease of discovery; 'stutter' a problem
3	Wide range of different repeat units; some arrays are within or close to genes and can cause diseases through expansion (see *Box 3.8*). $(AAT)_n$ and $(AAC)_n$ most common	Widely used. Alleles easily discriminated, and little 'stutter'
4	Wide range of different repeat units. $(GATA)_n / (GACA)_n$ particularly frequent, and clustered near centromeres, $(AAAC)_n$ and $(AAAT)_n$ again most common	Widely used. Alleles easily discriminated, and little 'stutter'
5	Range of different repeat units	Not widely used because of relative scarcity

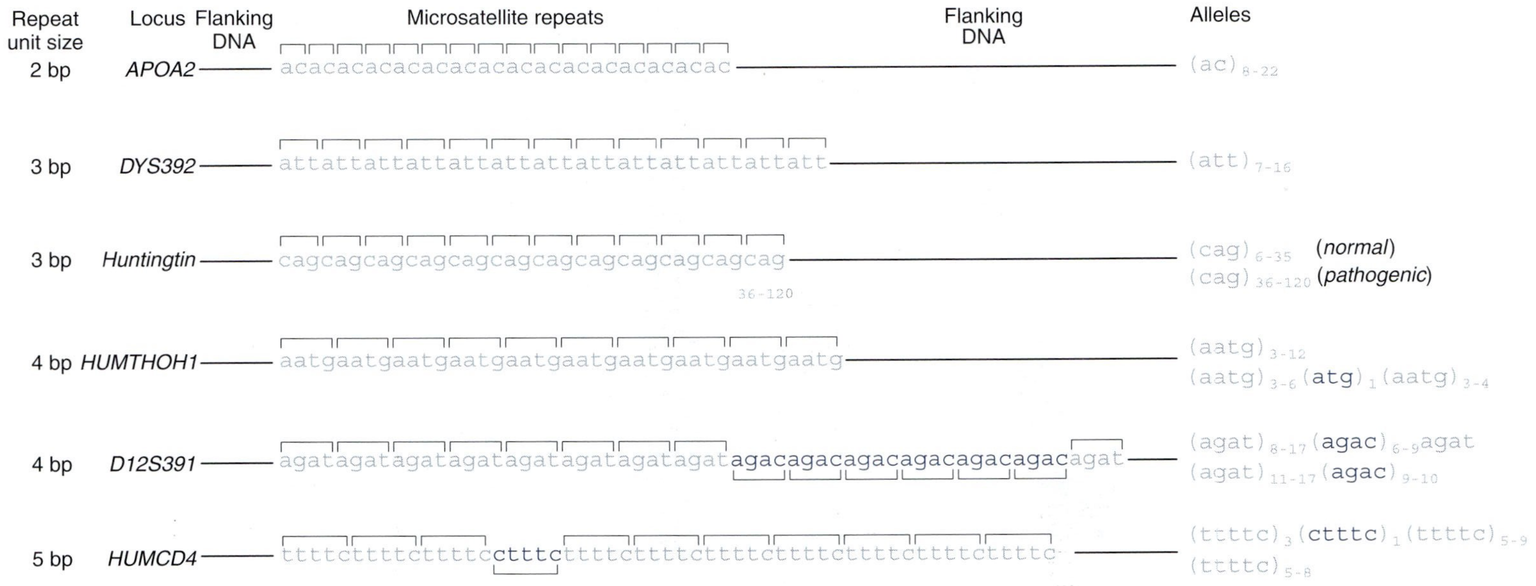

Figure 3.16: Examples of microsatellite structures.

Normal and pathogenic size ranges for the CAG repeat in the *Huntingtin* open reading frame (expanded in Huntington disease) are given.

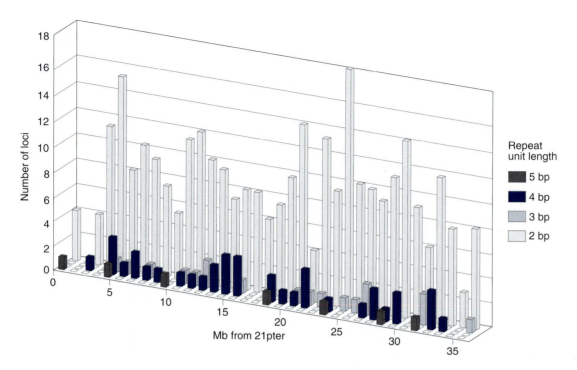

Figure 3.17: Distribution of microsatellites along a human chromosome.

Di-, tri-, tetra-, and pentanucleotide repeats do not show clustering. Microsatellites representing a minimum of 10 perfect repeats of any di-, tri-, tetra- or pentanucleotide motif are indicated by bars along the 35 680 kb of available sequence of chromosome 21. Data were analyzed at the Minisatellite Database website (http://minisatellites.u-psud.fr/), using Tandem Repeats Finder (http://tandem.biomath.mssm.edu/trf/trf.html).

BOX 3.8 Microsatellites with phenotypes

Variation at some microsatellites is not neutral. **Dynamic mutations** at a small subset of trinucleotide microsatellites, in which the loci undergo dramatic expansion in repeat number, are the causes of a number of genetic diseases:

▶ a CAG repeat within the coding region of a gene, encoding a polyglutamine tract, expands from typically 10–30 repeats to typically 40–200 repeats. The resulting polyglutamine expansions cause proteins to aggregate within certain cells and kill them. These CAG expansion disorders, for example Huntington disease (OMIM 143100) and spinocerebellar ataxia type I (OMIM 164400), are mostly dominant;

▶ a trinucleotide repeat (e.g., CGG, CCG, CTG, GAA) in the promoter, untranslated region or an intron of a gene expands from typically 5–50 repeats to typically 50–4000 repeats. This can abolish transcription, as in the case of Fragile X syndrome (OMIM 309550) or Friedreich ataxia (OMIM 229300).

Other classes of microsatellites may also have subtle effects on gene expression. One example is the TCAT tetranucleotide repeat within an intron of the tyrosine hydroxylase gene, where certain alleles can act to enhance transcription (Albanèse *et al.*, 2001).

---

generation. A number of properties of the mutation process emerge from these studies and from research which models mutation to explain diversity data:

▶ Most (> 85%; Brinkmann *et al.*, 1998; Xu *et al.*, 2000) mutations involve an increase or decrease of a single repeat unit.

▶ Overall mutation rate increases as array length increases. However, mutation type, as well as mutation rate, is length-dependent: below a certain number of repeats, mutation becomes undetectably slow;

expansion mutations occur equally throughout the array size range, but the rate of contraction mutations increases as alleles become larger (Xu *et al.*, 2000). This last property explains why microsatellite allele lengths have a stable distribution, and why very large (> 50 repeats) microsatellites are very rare (*Figure 3.18*).

▶ Dinucleotide repeat loci mutate more rapidly than tri- and tetranucleotide repeat loci (Chakraborty *et al.*, 1997; Webster *et al.*, 2002) (note that some direct observations of mutations seem to contradict this, but are based on small sample sizes).

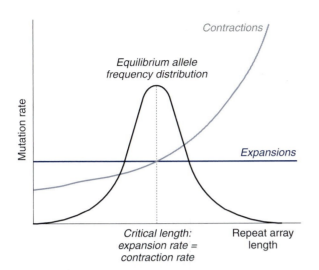

Figure 3.18: Relationship between microsatellite mutation rate and repeat length for expansion and contraction mutations.

Repeat arrays shorter than the critical length tend to expand, while longer arrays tend to contract. The result is a bell-shaped allele frequency distribution at equilibrium, with the modal allele at the critical length. Redrawn from Xu *et al.* (2000).

▶ Uninterrupted ('pure') repeat arrays mutate faster than interrupted arrays containing variant repeats.

The **stepwise mutation model** (SMM – see Section 5.2.3) is often used to model microsatellite evolution. According to this model, the length of a microsatellite varies at a fixed rate independent of repeat length and with the same probability of expansion and contraction. As is clear from the above list of properties, this model is over-simplistic.

Although direct evidence on the mechanism of mutation of microsatellites remains elusive, it is widely accepted that mutation occurs as a result of DNA **replication slippage** (*Figure 3.19*). Experiments show that slippage occurs *in vitro* (Schlötterer and Tautz, 1992), and the similar mutational properties of haploid, Y-chromosomal microsatellites to their diploid autosomal counterparts suggests the absence of a role for inter-chromosomal recombination (Kayser *et al.*, 2000). Despite this, there is evidence of a mutational bias towards longer microsatellites in humans compared to chimpanzees, and this may reflect an underlying influence of heterozygosity – in other words some kind of inter-allelic interaction (see *Box 3.9*).

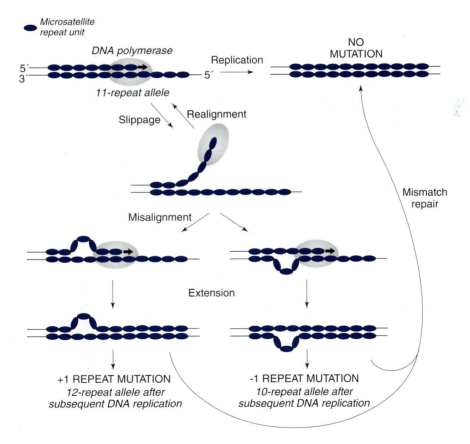

Figure 3.19: Mutation at microsatellites.

Model of the effects of slippage and misalignment in microsatellite mutation. Defects in mismatch repair, such as those in some human cancers, lead to elevated mutation rates in microsatellites. Mutations shown here are a single-step increase or decrease; mutations involving greater repeat numbers occur, but are about 10-fold rarer, and may involve other, nonslippage-based processes.

BOX 3.9 Opinion: Directionality and variation in rate of microsatellite evolution between species

Some time ago, I noticed a number of patterns in genetic data that, although not convincing on their own, together suggested that evolution might be occurring faster in larger populations. If true, this would be strongly at odds with standard population genetic theory, where evolutionary change does not vary with population size. After all, how could a gene sitting on a chromosome 'know' the size of the population in which it existed? In fact, there is a way.

Large populations carry more genetic variability than equivalent smaller populations, such that more individuals will be heterozygous in a larger population. If heterozygous sites were more mutable than homozygous sites, then the required mechanism would exist. Key evidence comes from studies of DNA replication in yeast. When homologous pairs of chromosomes come together prior to the formation of crossovers, large stretches of hybrid DNA are formed, in which each strand of DNA pairs, not with its true partner, but with its equivalent partner on the other chromosome. Then, heterozygous sites show up as mispaired bases in this heteroduplex, and many are recognized and 'repaired' by the mismatch repair machinery present in every cell. If these inter-chromosome repairs are anything but completely error-free, they will provide a source of new mutations that will occur preferentially at or around heterozygous sites. Interestingly, the key enzymes involved in the yeast system have direct homologs in all higher organisms.

To test this exciting possibility I turned to microsatellites. These loci mutate at high frequency by gaining and losing repeats. A second important quality is that the mutation process is usually biased towards expansion, with gain-in-length mutations occurring more often than contractions. Such a property is useful because it means that any change in mutation rate will in time be translated into a difference in mean length. I was now ready to test the hypothesis that population size can affect the rate of evolution. I had to find two related species where one had expanded greatly, then measure the lengths of homologous microsatellites to see whether microsatellites in the expanded species were longer.

The obvious test case is to compare humans (hugely expanded) with chimpanzees (more or less stable). The experiment was conducted by a colleague, David Rubinsztein, who works in medical genetics and conveniently had DNA samples from both humans and other primates. We could not have been more pleased with the results, since the vast majority of all loci were longer in humans, and even under the most stringent statistical tests, mean length overall was longer in humans. A paper was duly published (Rubinsztein et al., 1995).

However, it was not long before a possible flaw was exposed in the form of an ascertainment bias. All the microsatellites we tested were derived from humans and had been selected to be as long as possible. Consequently, the greater length in humans may be due more to the selection process than to any genuine length difference. To address this problem we went through the rigmarole of developing a new panel of microsatellites, this time using chimpanzee DNA and again selecting for maximum length. Once again, we found that the equivalent loci in humans were significantly longer, even though the size difference was reduced (Cooper et al., 1998). This experiment allowed us to conclude that the length difference was made up of two parts, one an artifactual component due to locus selection, and a second part due to a genuine difference in length.

Since our original study, several other reports of consistent microsatellite length differences between related species have emerged, and a few of these have included the vital reciprocal testing mentioned above. Wherever such differences are found, two properties must also exist. First, the mutation process must be asymmetric, otherwise the mean difference in length will remain zero regardless of any change in mutation rate. Second, the two species must differ in the average microsatellite mutation rate over all loci studied. Since microsatellites are selected essentially randomly, and some studies had looked at literally hundreds of markers, the implication is that the genomewide mutation rate has changed.

Currently, the population size hypothesis remains unproven. Genomewide differences in mutation rate could be explained by evolution of enzymes involved with the mutation process itself. Personally, I believe it will prove correct. All elements of the mechanism have been found in yeast, and homologs of these elements are present in higher organisms. Where consistent length differences have been found, as with our human–chimpanzee comparison, it always seems to be the species with the largest population size that has the longer microsatellites. For example, microsatellites in sheep and barn swallows are longer than their homologs in cattle and other *Hirundinae* respectively. If this trend extends to many more species pairs, the pattern may become undeniable.

Bill Amos, Department of Zoology, University of Cambridge, UK

3.3.2 Minisatellites

Minisatellites consist of repeat units from about 8 to 100 bp in length, with copy numbers from as low as 5 to well over 1000. Minisatellites are not simply microsatellites writ large – they are qualitatively different in their variability, mutation rates, mutation processes and chromosomal locations. They are among the most dynamic loci in our genome, some displaying hypervariability, with very large numbers of alleles of different lengths and structures, mutation rates as high as 14% per generation, and complex mutation processes involving both inter- and intra-allelic events. As is the case

BOX 3.10 Minisatellites with phenotypes

As is the case with microsatellites, minisatellites are usually considered to be selectively neutral loci, but there are examples that do play functional roles.

▶ The vast majority of minisatellites are not transcribed, but a few lie in coding regions, encoding repetitive and highly variable proteins. One example is the mucin 1 gene, which encodes a highly polymorphic glycoprotein containing 20-amino acid repeat units (Swallow *et al.*, 1987) varying in copy number from about 20 to over 120 in Europeans. Another example is a minisatellite within the *DRD4* gene, encoding a dopamine receptor. The 48-bp repeat unit encodes a 16-amino acid motif lying in an intracellular loop of the protein, and has a copy number varying from 2 to 11. One allele has been associated with attention-deficit/hyperactivity disorder in children (see *Box 13.1*).

▶ Some minisatellites lie in or close to promoter regions of genes, and different alleles can have an effect on gene expression. An example is the insulin minisatellite, which lies upstream of the insulin gene and has been associated with a complex effect on susceptibility to type 1 and type 2 diabetes mellitus (OMIM 222100 and 125853), among other diseases, possibly by affecting transcription (Kennedy *et al.*, 1995). Expansions of a 12-bp minisatellite repeat unit in the promoter of the cystatin B gene from two to three copies to over 40 copies cause progressive myoclonus epilepsy type 1 (OMIM 254800; Lalioti *et al.*, 1997) by downregulating transcription. Certain alleles of the *HRAS1* minisatellite, lying 1 kb downstream of the proto-oncogene *H*ras-1, are associated with an increased risk of a variety of cancers, through an unknown mechanism (Krontiris *et al.*, 1993).

with microsatellites, most minisatellites are treated as neutral markers. Again, however, there are exceptions (*Box 3.10*).

Minisatellites owe their early discovery, and their exploitation in a wide range of different fields, to the finding (Jeffreys *et al.*, 1985) that a set of 10 or more of these highly variable loci could be detected simultaneously in DNA filter hybridization experiments using a single radio-labeled DNA probe (see *Figure 15.1*). This produced a **DNA fingerprint**, which could be used in individual identification and paternity testing (see Chapter 15). Many other species have minisatellites too, which cross-hybridize with human-derived probes. The useful properties of minisatellites can therefore also be exploited in molecular ecology and in forensic work involving animals. In a recent and high-profile example, Dolly the sheep was shown by traditional DNA fingerprinting to be genetically identical to the mammary cell from which she was derived (Signer *et al.*, 1998). The sequences detected in DNA fingerprinting experiments are actually a subset of minisatellites (about 1000 in number) having repeat units rich in G and C, one strand particularly G-rich, and shared 'core' sequences. Minisatellites of other sequence compositions exist, and behave differently in a number of important respects. Because they cannot be detected *en masse* by DNA fingerprinting, they have been under-investigated compared to their GC-rich counterparts. Examples of minisatellite repeat sequences are shown in *Table 3.4*.

When allele length variation is considered, minisatellites show high levels of diversity, with typical **heterozygosity** values (the probability that two alleles sampled at random from the population are different in length) of well over 90%. Sequence analysis reveals an additional level of diversity – all minisatellites examined contain not homogeneous repeat units, but variant repeats differing by base substitutions and small indels. Note that, unlike microsatellites, these variant repeats do not impair the mutation process. The method of minisatellite variant repeat PCR (**MVR-PCR;** see Section 4.7) allows these variant repeats to be mapped within arrays

conveniently. The resulting fine structures of alleles reveal the spectacularly high degree of variability, and allow access to the details of the mutation process in sperm DNA (Jeffreys *et al.*, 1994). This yields a far greater number of mutant molecules than does pedigree analysis, but is restricted to an analysis of male mutation – all information about female mutation remains pedigree-based, and so limited to a relatively few observations. Mutation at many minisatellites is in fact highly elevated in males compared to females (see Section 3.6); as an extreme example, the minisatellite CEB1 has a mutation rate of 15% per sperm, but only 0.2% per oocyte (Vergnaud *et al.*, 1991).

Germ-line mutation at minisatellites can be highly complex (*Figure 3.20*), involving interactions between alleles in which one allele acquires blocks of repeat units from the other in a nonreciprocal process resembling **gene conversion** (see Section 3.8; reviewed by Jeffreys *et al.*, 1999) resulting in a general bias towards repeat unit gains over losses. The process for most minisatellites is polar, with mutations occurring preferentially at one end of the array. Somatic mutation, in contrast, is a slow and relatively simple, intra-allelic process. The primacy of interallelic processes is emphasized by the absence of variable, GC-rich minisatellites from the nonrecombining region of the Y chromosome, with their concomitant presence both in the recombining pseudoautosomal region, and in other XY-homologous regions of the X chromosome. The great differences between mutation at minisatellites and microsatellites are summarized in *Table 3.5*.

GC-rich minisatellites tend to be clustered towards the ends of chromosomes (Royle *et al.*, 1988), and this, together with some properties of their sequences, suggested early on that they might be associated with recombination hotspots – either as cause or consequence (reviewed by Jarman and Wells, 1989). This has now been confirmed, at least for the minisatellite MS32, by analysis of mutants in sperm DNA. A recombination hotspot exists in the DNA flanking this locus,

TABLE 3.4: GENERAL PROPERTIES OF SIX SELECTED MINISATELLITES.

Mini-satellite	Chromosomal location	Repeat unit length (bp)	GC content	Repeat copy number	Repeat sequences	Sex-averaged mutation rate/generation
MS32	1q42–1q43	29	62%	12–>800	tgactcagaatggagcaggtggccagggg ....c.a.........	0.8%
MS31	7p22–pter	20	60%	14–>500	acccacctcccacagacact gt...............	0.75%
MS205	16pter	45–54	75%	8–87	tgcatgccgacccgtctactcgcccccccacgtacccccgcccc-ta c-----................................c.........c..	0.4%
CEB1	2q37	31–43	70%	5–>300	tcagcccaggggacctcgcaggccacctcctccctcccccc ....a......t......--......t.......t.-	7.6%
B6.7	20qter	34	56%	6–540	gtggacagtgaggggtctctacaggccatgagg ....--.t..a.....a	5.4%
MSY1	Yp11	25	25%	22–114	cataatatacatgatgtatattata ..c....c.....t.c......a.....	3%

The sequence of one repeat unit is given, with positions of variation underneath ('·' same base as above; '-' deleted base). Variant positions exist in combination to give up to 20 different repeat types (CEB1, B6.7). Information from Jeffreys *et al.*, 1991; Armour *et al.*, 1993; Neil and Jeffreys, 1993; Buard and Vergnaud, 1994; May *et al.*, 1996; Jobling *et al.*, 1998; Tamaki *et al.*, 1999.

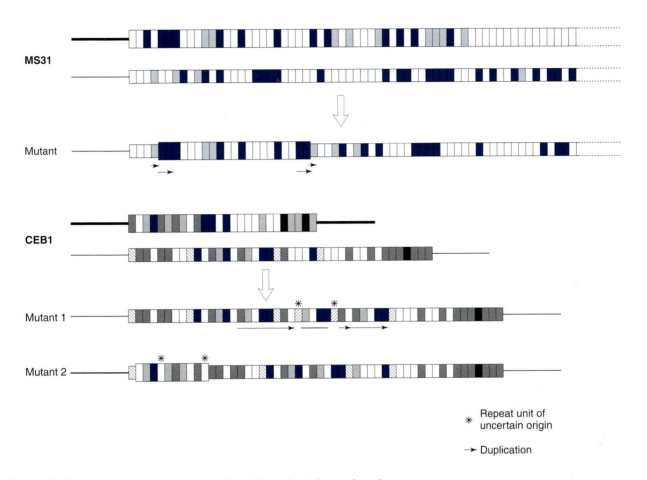

* Repeat unit of uncertain origin

→ Duplication

Figure 3.20: The complex structures of minisatellite alleles and mutation products in sperm.

Structures of two MS31 alleles and two CEB1 alleles in sperm donors are shown – different repeat units are indicated by shading. Mutant alleles were recovered from sperm DNA: the mutant MS31 allele has undergone in-register transfer of repeats from the upper allele, together with target-site duplication of a single repeat, plus a duplication at each end of the transferred segment; the CEB1 mutant 1 allele has undergone complex intra-allelic reduplications, with no evidence for interallelic exchange; the CEB1 mutant 2 has undergone transfer of repeats from the upper allele, together with rearrangements of repeat units in the transferred segment. Redrawn from Jeffreys *et al.* (1999).

TABLE 3.5: MUTATION AT MICRO- AND MINISATELLITES COMPARED.		
	Microsatellites	**Minisatellites**
Germ-line specificity of major mutation process?	No	Yes – somatic mutation is much slower and intra-allelic
Lower threshold effect of array size on mutation rate?	Yes	No – e.g., a five-repeat allele at the CEB1 minisatellite maintains a mutation rate of 0.4%
Mutation rate increases with array homogeneity?	Yes	No – some of the most variable loci have the most heterogeneous repeat units
Interallelic processes?	No, but see *Box 3.9*	Yes
Bias to repeat unit gains over losses?	For small alleles	Yes
Effects on mutation from flanking DNA variation?	No evidence	Yes

and drives polar mutation in the adjacent repeat array (Jeffreys *et al.*, 1998). It seems likely that this class of minisatellites at least exists as a by-product of the meiotic recombination process.

3.3.3 Satellites

Satellites, sometimes dubbed macrosatellites, are large tandem arrays spanning hundreds of kilobases to megabases, and composed of repeat units of a wide range of sizes that can display a higher-order structure (*Figure 3.21*). A good example is alpha satellite, or alphoid DNA, which forms a component of centromeres. The alphoid repeat monomer is 170 bp in length, but in many chromosomes it is arranged in higher-order repeat units of a few kilobases. This higher-order structure can be repeated hundreds or thousands of times to form an array several megabases in size. The detailed structures of satellite arrays are very difficult to investigate because of their size and repetitive nature. The 'complete' human genome sequence will actually contain gaps where major blocks of satellite lie, as the sequences of these regions cannot be assembled from individual sequence reads using standard methods.

The mutation processes at these loci cannot be studied directly, but probably involve unequal crossing over between homologous chromosomes or sister chromatids misaligned by integral numbers of higher order units (*Figure 3.21*). Historically, some satellite polymorphisms have been used in human evolutionary studies (e.g., Oakey and Tyler-Smith, 1990), but nowadays they have been superseded by loci which are easier to type (using PCR), analyze and understand.

3.3.4 Telomere repeat arrays

Telomeres are the structures at the ends of chromosomes, and include as their DNA component tandem arrays of the hexanucleotide repeat TTAGGG, typically 10–15 kb in length. They are sometimes included in the class of minisatellites, but fall outside the definition adopted here, and should be considered on their own. One end of the array is contiguous with subtelomeric DNA, while the other is the

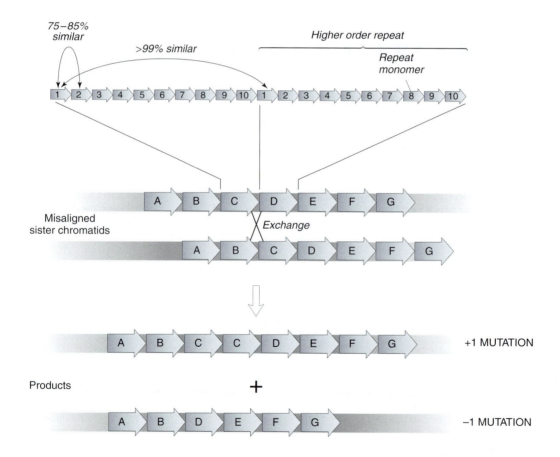

Figure 3.21: Structures of satellites, and mutation by unequal sister chromatid exchange.

This schematic diagram (upper part) shows an array of repeat monomers (for alphoid DNA these are 170 bp in length) and their higher-order structure: individual subunits are about 75–85% similar to each other, but higher-order repeats, which, for alphoid DNA, can be as large as 6 kb depending on the chromosome, are over 99% similar. This indicates that the mode of evolution has been as higher-order repeats, probably in an unequal exchange between sister chromatids of a bivalent in meiosis (see *Figure 2.15*), shown in the lower part of the figure. Mutations may involve changes of more than one higher-order repeat unit.

true chromosomal terminus, and so has no flanking DNA. The most proximal parts of telomere arrays contain variant repeats (e.g., TCAGGG), and this source of variation has been used to study the dynamics of telomeres (Baird *et al.*, 1995). The distal majority of the telomere array appears to be homogeneous, and repeats are added to it by a specialized DNA polymerase, telomerase, which is active in the germ-line and 'regrows' the telomere, which has been losing repeat units from its end at every prior DNA replication.

3.4 Transposable element insertions

Forty-five percent of the human genome is composed of dispersed repeat elements with copy numbers ranging from a few hundred to several hundred thousand (see Section 2.3), including the 868 000 **LINEs** (long interspersed nuclear elements) and 1 558 000 **SINEs** (short interspersed nuclear elements). The best-studied examples of these types are respectively the L1 and *Alu* **retrotransposons** (see Section 2.3). *Figure 3.22* shows the mechanisms of insertion of L1 and *Alu* elements and *Table 3.6* summarizes their transpositional properties. The **reverse transcriptase** enzyme encoded in full-length L1 elements is related to that used by **retroviruses** (like HIV) to integrate a DNA copy of their RNA genome into the host genome. Other dispersed repeat elements are more closely related to retroviruses. Human endogenous retroviruses (HERVs) have retained the ability to replicate themselves throughout a genome but do not have the ability to construct the protein coat that allows them to leave the cell. If these HERVs enter the germ-line and replicate themselves, integrated 'proviral' DNA copies of their genome are inherited by future generations. Over time, HERVs accumulate mutations that prevent them from replicating themselves; almost all HERVs have suffered this fate.

Some of these elements are human-specific – for example, about 2000 human *Alu* insertions are not found in chimpanzee or gorilla DNA (reviewed by Deininger and Batzer, 1999). These human-specific elements belong to particular subfamilies (as defined by shared sequence variants) and a subset of them is still active in transposition. This is demonstrated by their ability to disrupt genes by insertional mutagenesis, thus causing genetic diseases such as hemophilia A (OMIM 306700) and Duchenne muscular dystrophy (OMIM 310200; reviewed by Ostertag and Kazazian, 2001). As well as this direct mechanism for disease causation, *Alu* elements in particular can cause disease by acting as substrates for illegitimate recombination, leading to duplication or deletion of segments between a pair of *Alu*s, or to translocations when the elements are on different chromosomes (see Section 3.5). Thirty-three published examples of this disease-causing mechanism include hypercholesterolemia (OMIM 606945), Duchenne muscular dystrophy and Tay–Sachs disease (OMIM 272800) (Deininger and Batzer, 1999). It is worth noting that illegitimate recombination between L1 elements, HERVs, and particularly between *Alu* elements, actually plays at least as important a role in genetic disease as retrotransposition.

A large number of *Alu* and L1 element insertions have been discovered that are polymorphic in human populations, and useful as molecular markers (Watkins *et al.*, 2001; Myers *et al.*, 2002); a much smaller number of polymorphic HERVs have been found (Turner *et al.*, 2001). All these elements have the advantage that they can insert, but having once inserted, no specific mechanism exists for their removal. It is therefore trivial to decide upon the ancestral state of the polymorphism – it is invariably absence of the element. Also, since there is little or no sequence specificity for insertion sites, when two chromosomes share an inserted element at exactly the same position, it is clearly an example of identity by descent, rather than identity by state, and therefore an example of a '**unique event polymorphism**'.

3.5 Structural polymorphisms and the dynamics of nonallelic homologous recombination

Cytogenetic examination of chromosomes in individuals with genetic disorders has been a useful way to pinpoint disease genes – for example, cytogenetically visible deletions and translocations helped to localize the Duchenne muscular dystrophy gene (one of the first disease genes to be isolated by positional cloning; OMIM 310200) to Xp21. At this level there are also polymorphisms in normal individuals that can be recognized by cytogenetic analysis of banded chromosomes. These include inversions, deletions, duplications, length polymorphisms, and length variations in heterochromatin (as large as several tens of megabases, in the case of the Y chromosome; Verma *et al.*, 1978). There are also some asymptomatic translocations – for example, those generating 'satellited' chromosomes by the translocation of material from the short arms of acrocentric chromosomes such as chromosome 15. Because they are difficult to detect by molecular, PCR-based approaches, these chromosomal polymorphisms are probably under-ascertained, and have not been exploited as a major source of useful variation in human evolutionary genetic studies.

However, recent analysis of DNA sequence data emerging from the genome sequencing projects has revealed the remarkable extent to which our genome comprises recent **segmental duplications**. One estimate is that 5.2% of the genome exists as duplicated sequence which has arisen within the last 40 million years (Bailey *et al.*, 2002). This very high degree of **paralogy**, where paralogs can exhibit sequence identity in excess of 98%, and can stretch for tens or even hundreds of kilobases, has profound implications for the evolution of our genome. This is because paralogous repeats have the potential to sponsor **nonallelic homologous recombination** (NAHR) – they are so similar that the stringent recombination mechanism does not recognize their nonallelic nature. In particular, segments of absolute sequence identity exceed a length threshold, known as the **minimum efficient processing segment** (MEPS), which in mitotically dividing mammalian cells has been estimated at about 200 bp (Lukacsovich and Waldman, 1999)

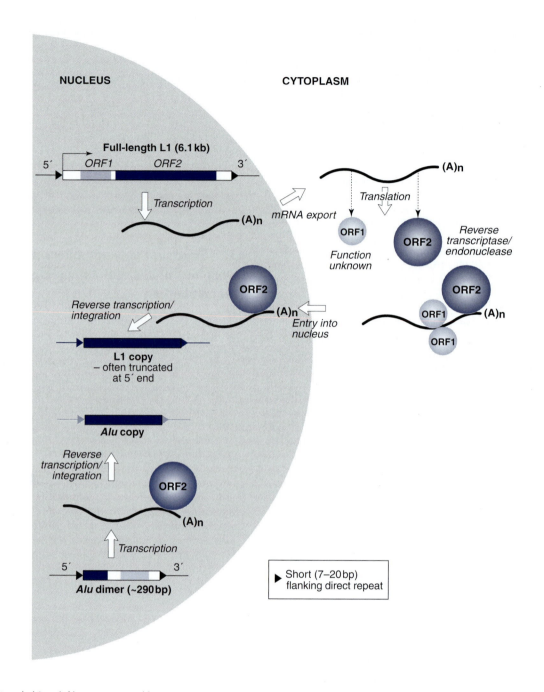

Figure 3.22: Steps in L1 and *Alu* retrotransposition.

A full-length L1 element is transcribed and the transcript exits the nucleus. In the cytoplasm it is translated to yield *ORF* (open reading frame) *1* and *2* proteins; proteins and mRNA may assemble into a ribonucleoprotein particle for re-entry to the nucleus. There, the L1 RNA is reverse-transcribed and integrated into genomic DNA by a process called target-primed reverse transcription, catalyzed by *ORF2*; this is often accompanied by truncation. *Alu* elements are transcribed in the nucleus and their reverse transcription and integration is thought to require the L1 *ORF2* protein. Figure based on Ostertag and Kazazian (2001).

(though little is known about the MEPS length during meiosis). As a result of NAHR the genome is far from the essentially invariant structure suggested by the reference sequence in online databases, onto which the polymorphisms described above (SNPs, VNTRs, retroelement insertions, etc.) can be mapped. Instead it is highly dynamic – structurally polymorphic in different individuals, and prone to disease-causing rearrangements (reviewed by Stankiewicz and Lupski, 2002):

TABLE 3.6: TRANSPOSITIONAL PROPERTIES OF L1 AND *ALU* ELEMENTS.

	L1	*Alu*
Copy number/genome	516 000	1 090 000
Phylogenetic distribution	Mammals	Higher primates
Transcriptionally competent copies	40–50	~2000
Autonomous transposition?	Yes	No
Number of known transpositions causing disease	14[a]	19
Active subfamilies	Ta	Y, Ya5, Ya8, Yb8
Number of polymorphic elements	115 known	> 1000 estimated
Approximation transposition rate (number of transpositions per diploid genome per generation)	0.004–0.1	0.01–0.25

Data from Deininger and Batzer, 1999; Lander *et al.*, 2001; Ostertag and Kazazian, 2001; Myers *et al.*, 2002.

[a]One of the L1 insertions is not pathogenic, but represents a *de novo* insertion in a family segregating hemophilia A.

▶ There are many instances of human diseases which are caused by NAHR between repeats (sometimes known as **low-copy repeats**, or **LCRs**; Stankiewicz and Lupski, 2002), leading to deletion or duplication of dosage-sensitive genes, and subsequent disease phenotypes. One example is rearrangement sponsored by recombination between the *CMT1A* repeats on chromosome 17p – 24-kb direct (i.e., same orientation) repeats flanking a 1.5-Mb region containing the *PMP22* gene. Duplication of this region (i.e., three doses instead of the normal two) leads to Charcot-Marie-Tooth disease (OMIM 118220), while the reciprocal deletion (one dose rather than two) leads to hereditary neuropathy with pressure palsies (OMIM 162500). Diseases caused by rearrangements due to underlying structural features of the genome, rather than to relatively random events such as base substitutions, have been called **genomic disorders** (Lupski, 1998).

▶ Nonpathogenic rearrangements can alter the copy numbers of genes, such as the opsin genes on Xq28, which encode the protein moieties of the visual pigments. Haploid copy numbers vary between 2 and at least 9 (Neitz and Neitz, 1995); while men with these copy numbers have normal color vision, this variation is associated with subtle differences in color perception. Copy number of the ribosomal DNA (rDNA) genes, contained within a 43-kb tandemly repeated unit on the short arms of the five acrocentric chromosomes, varies between 390 and 580 per diploid genome in normal individuals (Veiko *et al.*, 1996).

▶ NAHR between inverted repeats can lead to inversion of the region lying between them. For example, recombination between inverted ~400-kb clusters of olfactory receptor genes on 8p is the cause of a common benign inversion polymorphism. Heterozygous mothers can then undergo aberrant recombination in the inverted region, leading to deletions and **supernumerary** (i.e., in excess of the normal diploid number of 46 chromosomes) inverted duplication chromosomes in offspring (Giglio *et al.*, 2001).

▶ Other examples of supernumerary chromosomes, including an asymptomatic inverted duplicated segment of chromosome 15, and a marker chromosome derived from chromosome 22, associated with cat eye syndrome, also have their origins in aberrant recombination events between paralogous repeats.

3.6 Age and sex effects on mutation rate

Mutations at many of the classes of loci described above show increasing rates with parental age. Furthermore, when the parent of origin of a *de novo* mutant chromosome is determined, by typing polymorphic markers linked to the mutation, a strong bias towards paternal origin is often observed (see *Box 5.8*). Much of the evidence for the sex bias comes from the analysis of mutations associated with genetic disease (reviewed by Crow, 2000): 154 new mutations analyzed in a set of six diseases including achondroplasia (OMIM 100800) all had a paternal origin, and showed a strong paternal age effect (*Figure 3.23*). While these examples (in the three genes *FGFR2*, *FGFR3*, and *RET*) are extreme, studies of a larger number of diseases give a mean male : female mutation ratio of about 10 : 1, and confirm the age effect. The mutations in these cases are base substitutions. When diseases caused more often by deletions (such as Duchenne muscular dystrophy) are considered, the paternal effect is much reduced. What could underlie this paternal-specific effect upon the base substitution rate?

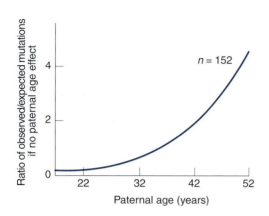

Figure 3.23: Relative frequency of new achondroplasia mutations with increasing paternal age.

Best-fitting exponential curve for real data on 152 new achondroplasia mutations. Adapted from Crow (2000).

One striking difference between the sexes is the number of cell divisions in gametogenesis (*Figure 3.24*) – there are many more in making a sperm than in making an oocyte. What is more, the number of cell divisions in spermatogenesis increases with paternal age, which is not the case for oogenesis, in which all the cell divisions are completed before a woman is born. The number of cell divisions in oogenesis is 22, while the number in spermatogenesis at puberty is 35, increasing by 23 per annum thereafter: the octogenarian father's sperm have undergone more than 40 times as many cell divisions as the 15-year old's (*Figure 3.25*). Since base substitutions are associated with DNA replication through misincorporations, this seems a plausible explanation for much of the excess of mutations in men compared to women, and for the paternal age effect. However, for the *FGFR2*, *FGFR3*, and *RET* genes the extreme male-specific mutation effect seems too large to be accounted for by male–female differences in cell division number, while for other genes the male–female differences in mutation rate are too small: additional factors must be involved.

This difference in base-substitutional mutation rate between males and females is approximated by a figure known as the alpha factor, explained in *Box 5.8*, and has important implications for the evolutionary divergence rates between chromosomes that have different inheritance patterns (see Section 7.4.3).

Sex differences in mutation rates are not confined to base substitutions. Microsatellite mutations occur three to five times more frequently in fathers rather than in mothers (Brinkmann *et al.*, 1998; Ellegren, 2000; Xu *et al.*, 2000). A paternal age effect is also observed (Brinkmann *et al.*, 1998), but the relatively small magnitude of this male excess suggests that there is no simple relationship between cell division number and mutation rate at these loci. Many minisatellites also show a large excess of male mutation (see Section 3.3.2), but in most cases this is likely to be a reflection of sex differences in the behavior of meiotic recombination (see

Section 2.6), rather than of differences in cell division number during gametogenesis. While most genomic disorders (see Section 3.5) show no parent-of-origin bias in their mutations, a few do: for example, about 87% of *CMT1A*-mediated duplications, causing Charcot-Marie-Tooth disease, arise during spermatogenesis, while around 80% of deletions causing neurofibromatosis type 1 (OMIM 162200) are maternal in origin (reviewed by Stankiewicz and Lupski, 2002). Again, differences between the sexes in these cases could reflect differences in meiotic recombination behaviors. An apparent excess of paternal mutation for X-linked genomic disorders has been ascribed to the absence of a pairing partner for most of the length of the X chromosome in male meiosis (Giglio *et al.*, 2000).

3.7 Haplotypes, haplotype diversity and linkage disequilibrium

A **haplotype** refers to the combination of allelic states of polymorphic markers along the same DNA molecule, in other words on the same chromosome or on mtDNA (*Figure 3.26*). These sites can include any class of DNA polymorphism, from a base substitution to a satellite length polymorphism. If we type Y-chromosomal or mtDNA markers we immediately derive a haplotype, since these molecules are themselves haploid. Typing X-chromosomal markers in a male also yields a haplotype, since a male carries a single X chromosome. If we type the same markers in a female, or type autosomal markers in either sex, we do not obtain haplotypes directly, except in the case where both chromosomes have the same haplotype (homozygosity). Instead, a diploid **genotype** – two haplotypes combined – is obtained (*Figure 3.26*). Methods described in the next chapter can then be used to deduce a pair of haplotypes from the genotype (see Section 4.10).

Because mtDNA and the majority of the Y chromosome are nonrecombining, haplotype diversity on these molecules is due only to mutation. In the rest of the genome, in addition to mutation, haplotype diversity is influenced by recombination: a haplotype present in one generation can be broken up to yield a new haplotype in the next (see *Figure 2.17*). The effect of recombination is to increase haplotype diversity. If we consider 10 binary nonrecurrent polymorphisms on a Y chromosome, for example, then the maximum number of haplotypes we expect to observe is 11: the ancestral state is the absence of all the derived alleles, and then each allele is derived in turn to yield an extra haplotype. We may in fact observe fewer than 11 haplotypes, because some could be lost by drift or selection. Now consider the same number of polymorphisms on an autosome: here, if recombination could take place between any pair of markers, the maximum number of haplotypes would be 2^{10}, or 1024.

In reality, the number of observed haplotypes, though likely to exceed 11, tends to be less than this maximum value. The likelihood of recombination between two autosomal markers depends on their relative positions: markers lying far apart are very often separated by

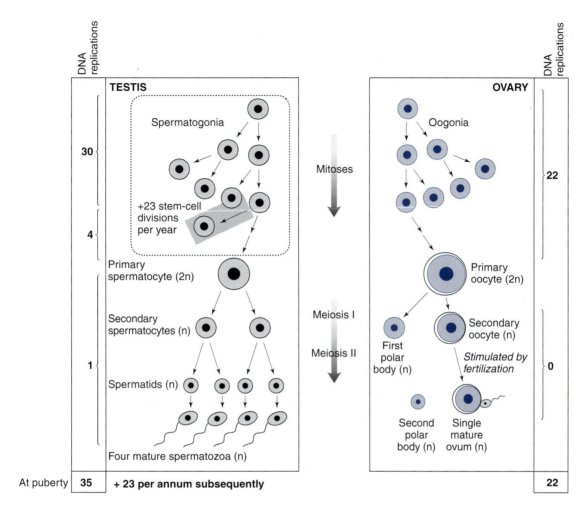

Figure 3.24: Gametogenesis involves more DNA replications in males than in females.

Gametogenesis proceeds in males by mitotic divisions of each diploid spermatogonium (stem cell) to give a diploid primary spermatocyte that proceeds into meiosis, and a new spermatogonium that can divide further in the same way. The total number of cell divisions in males at puberty is 36 (equating to 35 DNA replications because there is only a single replication during the two meiotic divisions). Following puberty there is one stem-cell division every 16 days, or about 23 in a year. In females, diploid oogonia divide to give primary oocytes, which can undergo meiosis. The whole process involves 23 cell divisions, or 22 DNA replications.

recombination, while markers close together are rarely separated (*Figure 3.27a*). In practice, it is not only simple physical distance that governs recombinational distance, since the two are not linearly related (see Section 2.6). Markers only 1 kb apart, but lying either side of a **recombination hotspot** (reviewed by Petes, 2001) will be separated very frequently by recombination; alternatively, markers lying 100 kb apart but in a recombinationally inert region (a 'cold domain') will be very rarely separated (*Figure 3.27b*). *Figure 3.28* illustrates the 'punctate' recombination behavior observed in part of the human MHC locus (Jeffreys *et al.*, 2001), where clusters of hotspots were seen in male meiosis: six hotspots with recombination rates as high as 160 cM/Mb, separate 'cold' domains of **linkage disequilibrium (LD)** showing recombination rates three or four orders of magnitude lower. This study determined haplotypes in sperm

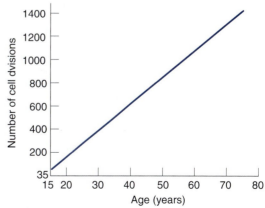

Figure 3.25: Increase of cell division number in sperm with paternal age.

Drawn from data in Crow (2000).

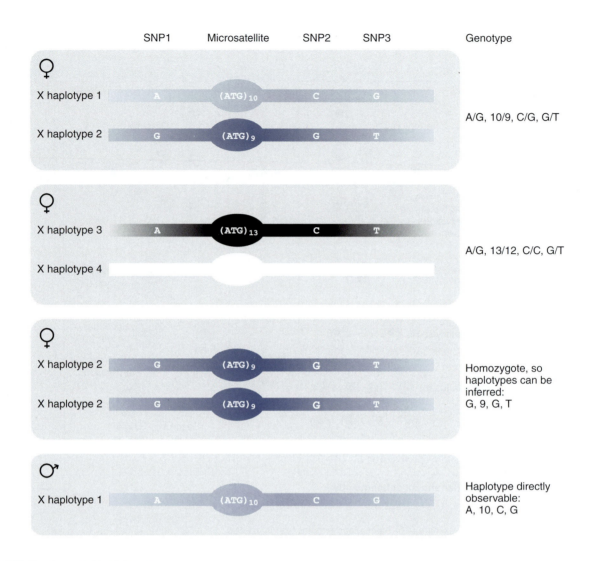

Figure 3.26: Genotypes and haplotypes.

Hypothetical haplotypes involving three SNPs and a microsatellite linked on the X chromosome in females and a male. Direct analysis of these polymorphisms in females will yield genotypes, from which the haplotypes cannot be directly inferred except in the case of a homozygote. Males are hemizygous for the X chromosome, and therefore yield haplotypes directly; autosomal loci will yield genotypes in both sexes.

donors, and then sought rare crossover molecules by PCR using primers directed at alleles lying on *different* haplotypes in sperm DNA from the donor (see Section 4.5.3). Note that the presence of hotspots is difficult to ascertain in pedigree studies because of the small sizes of human families, and the consequent small numbers of recombination events observed.

The tendency of particular alleles at separate loci to be coinherited because of reduced recombination between them can lead to associations between alleles in a population. This property is known as **linkage disequilibrium** (**LD**). The hope that genomewide surveys might identify associations between anonymous markers and genic variation contributing to common disorders has resulted in intense interest in genomewide patterns of LD (reviewed by Ardlie *et al.*, 2002). In particular, LD is said to occur when two alleles are found together on the same chromosome more often than expected if the alleles were segregating at random. There are various different measures for assessing this that differ in their properties and usefulness (*Box 3.11*).

Recent studies examining LD by analyzing SNP haplotype structures of extensive autosomal regions have suggested a block-like structure for parts, at least, of the genome. The haplotype blocks appear as segments of consecutive alleles that seem to be co-inherited: LD is high within blocks but is much lower between blocks (*Figure 3.29*). Conversely, haplotype diversity is low within blocks but high across block boundaries (Cardon and Abecasis,

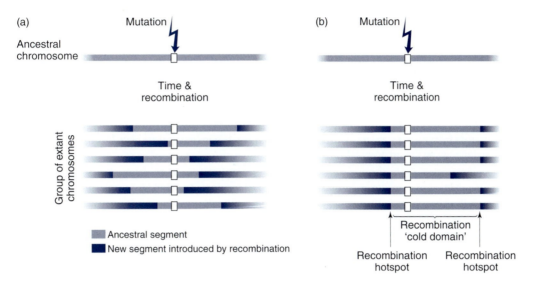

Figure 3.27: Linkage disequilibrium around an ancestral mutation.

(a) Markers which are physically close to the ancestral mutation will tend to remain associated with it despite the passage of time, and recombination events, which are here shown as occurring randomly along the chromosomal segment. (b) When recombination is highly nonuniform, markers that are physically close to the mutation but flanking a hotspot will often be separated from it, and markers far away (in a 'cold domain') will remain associated.

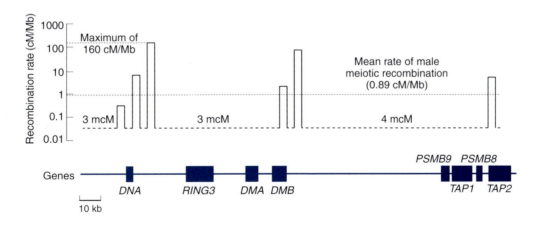

Figure 3.28: Punctate recombination in male meiosis in the class II region of the MHC.

Sperm crossover activity is shown over a 216-kb segment lying in the MHC class II region on chromosome 6. The black line (top) shows the mean male recombination rate for between two and six sperm donors. The background recombination rate of 0.3–0.4 mcM (millicentiMorgans)/Mb is approximate. Adapted from Jeffreys *et al.* (2001).

2003; Stumpf and Goldstein, 2003). A block-like structure has been observed in high-resolution SNP haplotyping of 500 kb of chromosome 5q: discrete blocks of 10–100 kb, each having a few common haplotypes (Daly *et al.*, 2001), randomly-chosen regions of the genome (Gabriel *et al.*, 2002) and in studies of whole chromosomes (Patil *et al.*, 2001). Clearly, if this block-like structure (*Figure 3.29*) is a general feature of the genome, then understanding it will have important implications for the design of studies aiming to identify genes underlying complex disorders and in treating regions of autosomes as small versions of mtDNA or the Y chromosome in phylogeographic studies. It is tempting to equate regions of low recombination with blocks, and hotspots with boundaries between blocks: a genetic 'blockworld'. However, the available data can be explained without invoking recombination hotspots, and this matter is discussed further in Section 14.3.3.

BOX 3.11 Measures of linkage disequilibrium

The oldest and simplest measure of LD, D (Lewontin, 1964), defines it as the difference between the observed frequency of a two-locus haplotype and the expected frequency (based on observed allele frequencies) if the alleles were randomly segregating. For two loci, **A** (alleles A and a) and **B** (alleles B and b), the observed frequency of the haplotype AB (allele A in combination with allele B) is given by P_{AB}. Under random segregation, the expected frequency is the product of the allele frequencies, or $P_A \times P_B$, where P_A is the observed frequency of allele A, and P_B the frequency of allele B. Lewontin's measure of LD is therefore:

$$D = P_{AB} - P_A P_B$$

If D is significantly different from zero, as assessed using a statistical test (Fisher exact test), then LD is said to exist; whether it is positive or negative depends on the arbitrary labeling of alleles.

Though this simple measure is intuitively appealing, its dependence on allele frequencies means that comparisons of different values of D are of limited utility. The measure $|D'|$, which is the absolute value of D divided by its maximum possible value given the allele frequencies at the two loci, is better. It has the useful property that $|D'| = 1$ if, and only if, two alleles have not been separated by recombination during the history of the sample being analyzed. This case is known as complete LD, and at most three out of the four possible two-locus haplotypes are observed (see *Figure*). It is not clear how values of $|D'| < 1$ should be interpreted, though they indicate disruption of LD. $|D'|$ values can be inflated in small samples, and when minor allele frequencies are low, they can give artifactual indications of LD.

Another measure, r^2 (sometimes given as Δ^2), the square of the correlation coefficient between the two loci, is derived by dividing D^2 by the product of the four allele frequencies at the two loci. It has some useful properties that have made it popular in comparing LD in disease association studies. $r^2 = 1$ (known as perfect disequilibrium) if, and only if, the alleles have not been separated by recombination, and have the same allele frequency. r^2 is less inflated by small sample sizes than is $|D'|$.

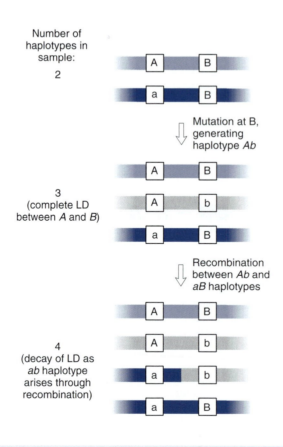

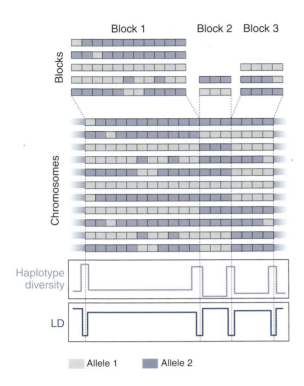

Figure 3.29: Haplotype blocks in the genome.

Haplotype blocks are segments of the genome in which little or no evidence of recombination exists, and only a small proportion of possible haplotypes is found. Here, Allele 1 indicates the first allele found at a locus (i.e., the reference sequence), and Allele 2 the second allele found. In a 'block-like' model of the genome, chromosomes in the population are patchworks of these haplotype blocks, separated by boundaries that may correspond to recombination hotspots; note that hotspots need not necessarily be invoked to explain the existence of blocks (see Section 14.3.3 for a discussion).

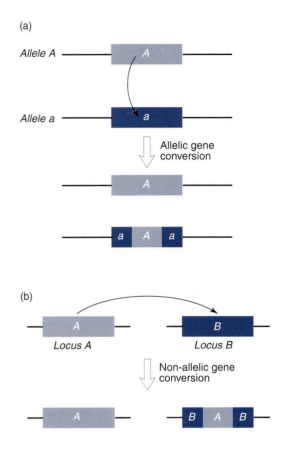

Figure 3.30: Allelic and nonallelic gene conversion.

Nonreciprocal transfer of genetic information between: (a) alleles, and (b) different loci (paralogs). In the latter case a segment of sequence identity of at least about 200 bp (the MEPS) is required for conversion to initiate.

3.8 Nonreciprocal exchange through gene conversion

Mutation and recombination are not the only processes acting to diversify haplotypes. Information transfer can also occur between haplotypes through the process of **gene conversion** (despite its name, this process acts upon all DNA sequences, and not only genes). Unlike classical recombinational cross-over, this is a nonreciprocal transfer of genetic information, in which one allele remains unchanged, but the other is converted to the allelic state of the first (*Figure 3.30a*). Gene conversion has been best studied in yeast, and is relatively poorly characterized in humans, where it cannot be formally distinguished from a double crossover. Conversion tract lengths are in the range of 10–1000 bp. Rather than being seen as a specialized process, gene conversion should be regarded as one of the consequences of homologous recombination via the four-stranded intermediate, the **Holliday junction** (*Figure 3.31*, and see *Figure 2.17*).

Gene conversion has a number of consequences for genome evolution, gene mapping and disease:

▶ When a short stretch of one chromosome is transferred into its homolog by gene conversion this is equivalent to two closely spaced recombination events, and can break down LD in a similar way to recombination. In an LD analysis of randomly chosen sequences having an average distance between SNPs of only 124 bp, a significant fraction of SNP pairs showed incomplete LD (Ardlie *et al.*, 2001). This unexpected finding is best explained by gene conversion, occurring at a rate around 3–10-fold higher than that of recombination.

▶ It acts not only between allelic sequences (homologs), but also between **paralogs** (*Figure 3.30b*) – highly similar sequences which are nonallelic (Section 3.5). In doing so it can increase the length of paralogous sequence identity, and thus provide a more readily utilized substrate for NAHR. The *CMT1A* repeats (see Section 3.5) have been shown to have been homogenized by gene conversion (Hurles, 2001).

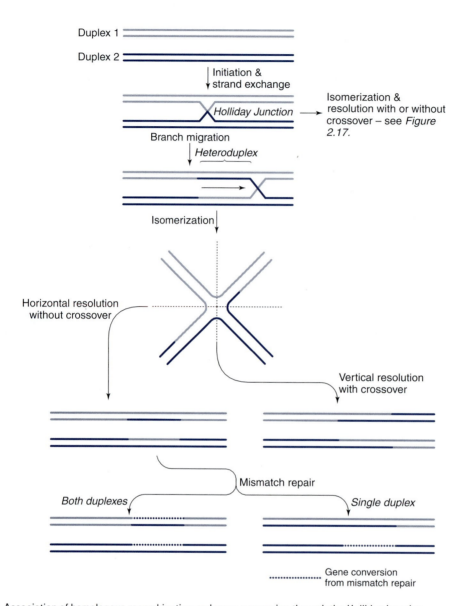

Figure 3.31: Association of homologous recombination and gene conversion through the Holliday junction.

▶ Gene conversion between paralogous genes can act to maintain sequence homology (concerted evolution). An example is the extended sequence similarity between the δ- and β-globin genes, despite their having diverged through gene duplication 85–100 MYA (reviewed by Papadakis and Patrinos, 1999). On the other hand, gene conversion can have deleterious effects. It may be responsible for the spread of some β-thalassemia (OMIM 141900) mutations into different chromosomal haplotypes, and is the cause of 75% of mutations in the *CYP21* (steroid 21-hydroxylase) gene, causing congenital adrenal hyperplasia (OMIM 201910), by transferring sequence from a *CYP21* pseudogene into the nearby active copy (Collier *et al.*, 1993).

▶ As well as generating haplotype diversity, gene conversion can generate excess sequence diversity in paralogous sequences. One example is the **promoter** of the growth hormone gene *GH1*, which shows ~20-fold greater nucleotide diversity than other autosomal loci because of gene conversions with neighboring paralogous copies of the gene (Giordano *et al.*, 1997).

Summary

▶ Genetic differences between people are caused by germ-line mutation. While any DNA sequence change is a mutation, underlying rates and processes differ greatly.

▶ Base substitutions (SNPs) are binary mutations with a low average nuclear rate of $\sim 2 \times 10^{-8}$ per base per generation. There is a bias towards transitions over transversions, and a 10-fold elevated rate at CpG dinucleotides. Most of these mutations show identity by descent, and ancestral state can be determined from great ape sequences.

▶ Base substitutions in genes can adversely affect gene function through amino acid substitution, introduction of stop codons, and regulatory or splicing changes. However, most base substitutions, and other mutations, are selectively neutral.

▶ Mitochondrial DNA base substitutions occur at a higher rate than nuclear: 10-fold outside the control region, and 100-fold within it. Mutation segregation from one generation to the next is complex.

▶ Multiallelic VNTR mutations vary in scale and process, but have relatively high rates – there is little identity by descent, and ancestral state determination is difficult.

▶ Microsatellites have repeats of 1–6 bp and typical repeat numbers of 10–30. Mutation probably occurs through replication slippage. Minisatellites have repeat units of 8–100 bp and repeat numbers of 5 to > 1000, often with complex interallelic mutation processes, and sex-averaged germ-line rates up to 7%. Satellites are large variable tandem arrays that can span hundreds of kilobases, but are difficult to investigate.

▶ Retrotransposons such as *Alu* and L1 elements are very abundant, and some are polymorphic, with very low transposition rates. They show perfect identity by descent, and the ancestral state is always absence.

▶ Structural polymorphisms are widespread, but underinvestigated. Over 5% of the genome comprises recent segmental duplications, and these very similar paralogs can sponsor nonallelic homologous recombination causing disease or nonpathogenic polymorphism through rearrangement.

▶ Many mutations show elevated rates in paternal meiosis, and increase with paternal age. This is only partly explained by increased numbers of germ-line DNA replications in males compared to females.

▶ A haplotype is the combination of different allelic states along the same DNA molecule, and its diversity is increased by recombination. Linkage disequilibrium (LD) is the association of alleles because of reduced recombination between them, which can reflect either physical distance, or variable recombination activity.

▶ Gene conversion, the nonreciprocal transfer of sequence between alleles or between paralogs, can reduce LD, homogenize repeats, and increase nucleotide diversity.

Further reading

Ellegren H (2000) Microsatellite mutations in the germline: implications for evolutionary inference. *Trends Genet.* **16**, 551–558.

Goldstein DB, Schlötterer C (1999) *Microsatellites: Evolution and Applications.* Oxford University Press, Oxford.

Scriver CR, Waters PJ (1999) Monogenic traits are not so simple: lessons from phenylketonuria. *Trends Genet.* **15**, 267–272.

Strachan T, Read AP (2003) *Human Molecular Genetics 3.* Garland Publishing, Inc., New York.

Weatherall DJ (2001) Phenotype–genotype relationships in monogenic disease: lessons from the thalassaemias. *Nature Rev. Genet.* **2**, 245–255.

Electronic references

Human Gene Mutation Database:
http://archive.uwcm.ac.uk/uwcm/mg/hgmd0.html

References

Albanèse V, Biguet NF, Kiefer H, Bayard E, Mallet J, Meloni R (2001) Quantitative effects on gene silencing by allelic variation at a tetranucleotide microsatellite. *Hum. Mol. Genet.* **10**, 1785–1792.

Ardlie KG, Liu-Cordero SN, Eberle MA *et al.* (2001) Lower-than-expected linkage disequilibrium between tightly linked markers in humans suggests a role for gene conversion. *Am. J. Hum. Genet.* **69**, 582–589.

Ardlie KG, Kruglyak L, Seielstad M (2002) Patterns of linkage disequilibrium in the human genome. *Nature Rev. Genet.* **3**, 299–309.

Armour JAL, Harris PC, Jeffreys AJ (1993) Allelic diversity at minisatellite MS205 (D16S309) – evidence for polarized variability. *Hum. Mol. Genet.* **2**, 1137–1145.

Badano JL, Katsanis N (2002) Beyond Mendel: an evolving view of human genetic disease transmission. *Nature Rev. Genet.* **3**, 779–789.

Bailey JA, Gu Z, Clark RA *et al.* (2002) Recent segmental duplications in the human genome. *Science* **297**, 1003–1007.

Baird DM, Jeffreys AJ, Royle NJ (1995) Mechanisms underlying telomere repeat turnover, revealed by hypervariable variant repeat distribution patterns in the human Xp/Yp telomere. *EMBO J.* **14**, 5433–5443.

Bartek J, Lukas J (2001) Mammalian G1- and S-phase checkpoints in response to DNA damage. *Curr. Opin. Cell. Biol.* **13**, 738–747.

Brinkmann B, Klintschar M, Neuhuber F, Hühne J, Rolf B (1998) Mutation rate in human microsatellites: influence of the structure and length of the tandem repeat. *Am. J. Hum. Genet.* **62**, 1408–1415.

Brown WM, George M, Wilson AC (1979) Rapid evolution of animal mitochondrial DNA. *Proc. Natl Acad. Sci. USA* **76**, 1967–1971.

Buard J, Vergnaud G (1994) Complex recombination events at the hypermutable minisatellite CEB1 (D2S90). *EMBO J.* **13**, 3203–3210.

Cardon LR, Abecasis GR (2003) Using haplotype blocks to map human complex trait loci. *Trends Genet.* **19**, 135–140.

Cartegni L, Chew SL, Krainer AR (2002) Listening to silence and understanding nonsense: exonic mutations that affect splicing. *Nature Rev. Genet.* **3**, 285–298.

Cavalli-Sforza LL, Menozzi P, Piazza A (1994) *The History and Geography of Human Genes.* Princeton University Press, Princeton, New Jersey.

Chakraborty R, Kimmel M, Stivers DN, Davison LJ, Deka R (1997) Relative mutation rates at di-, tri-, and tetranucleotide microsatellite loci. *Proc. Natl Acad. Sci. USA* **94**, 1041–1046.

Chao H-K, Hsiao K-J, Su T-S (2001) A silent mutation induces exon skipping in the phenylalanine hydroxylase gene in phenylketonuria. *Hum. Genet.* **108**, 14–19.

Collier S, Tassabehji M, Sinnott P, Strachan T (1993) A de novo pathological point mutation at the 21-hydroxylase locus: implications for gene conversion in the human genome. *Nature Genet.* **3**, 260–265.

Condit CM, Achter PJ, Lauer I, Sefcovic E (2002) The changing meanings of "mutation": a contextualized study of public discourse. *Hum. Mutat.* **19**, 69–75.

Cooper DN, Krawczak M, Antonarakis SE (1995) In: *Metabolic and Molecular Bases of Inherited Disease* (eds CR Scriver, AL Beaudet WS, Sly, D Valle). McGraw-Hill, New York, pp. 259–291.

Cooper G, Rubinsztein DC, Amos W (1998) Ascertainment bias cannot entirely account for human microsatellites being longer than their chimpanzee homologues. *Hum. Mol. Genet.* **7**, 1425–1429.

Cotton RGH, Scriver CR (1998) Proof of 'disease causing' mutation. *Hum. Mutat.* **12**, 1–3.

Crow JF (2000) The origins, patterns and implications of human spontaneous mutation. *Nature Rev. Genet.* **1**, 40–47.

Cummings MR (2000) *Human Heredity: Principles and Issues,* 5th Edn. Brooks/Cole, Pacific Grove.

Daly MJ, Rioux JD, Schaffner SF, Hudson TJ, Lander ES (2001) High-resolution haplotype structure in the human genome. *Nature Genet.* **29**, 229–232.

Deininger PL, Batzer MA (1999) Alu repeats and human disease. *Mol. Genet. Metabol.* **67**, 183–193.

Di Rienzo A, Donnelly P, Toomajian C et al. (1998) Heterogeneity of microsatellite mutations within and between loci, and implications for human demographic histories. *Genetics* **148**, 1269–1284.

Ebersberger I, Metzler D, Schwarz C, Pääbo S (2002) Genomewide comparison of DNA sequences between humans and chimpanzees. *Am. J. Hum. Genet.* **70**, 1490–1497.

Ellegren H (2000) Heterogeneous mutation processes in human microsatellite DNA sequences. *Nature Genet.* **24**, 400–402.

Estivill X, Cheung J, Pujana MA, Nakabayashi K, Scherer SW, Tsui L-C (2002) Chromosomal regions containing high-density and ambiguously mapped putative single nucleotide polymorphisms (SNPs) correlate with segmental duplications in the human genome. *Hum. Mol. Genet.* **11**, 1987–1995.

Gabriel SB, Schaffner SF, Nguyen H et al. (2002) The structure of haplotype blocks in the human genome. *Science* **296**, 2225–2229.

Gardner RJ (1977) A new estimate of the achondroplasia mutation rate. *Clin. Genet.* **11**, 31–38.

Giglio S, Pirola B, Arrigo G et al. (2000) Opposite deletions/duplications of the X chromosome: two novel reciprocal rearrangements. *Eur. J. Hum. Genet.* **8**, 63–70.

Giglio S, Broman KW, Matsumoto N et al. (2001) Olfactory receptor-gene clusters, genomic-inversion polymorphisms, and common chromosome rearrangements. *Am. J. Hum. Genet.* **68**, 874–883.

Giordano M, Marchetti C, Chiorboli E, Bona G, Richiardi PM (1997) Evidence for gene conversion in the generation of extensive polymorphism in the promoter of the growth hormone gene. *Hum. Genet.* **100**, 249–255.

Heyer E, Puymirat J, Dieltjes P, Bakker E, de Knijff P (1997) Estimating Y chromosome specific microsatellite mutation frequencies using deep rooting pedigrees. *Hum. Mol. Genet.* **6**, 799–803.

Hirszfeld L, Hirszfeld H (1919) Essai d'application des methods au probléme des races. *Anthropologie* **29**, 505–537.

Holmquist GP (1998) Endogenous lesions, S-phase-independent spontaneous mutations, and evolutionary strategies for base excision repair. *Mut. Res.* **400**, 59–68.

Holtkemper U, Rolf B, Hohoff C, Forster P, Brinkmann B (2001) Mutation rates at two human Y-chromosomal microsatellite loci using small pool PCR techniques. *Hum. Mol. Genet.* **10**, 629–633.

Hurles ME (2001) Gene conversion homogenizes the CMT1A paralogous repeats. *BMC Genomics* **2**, 11.

Ingman M, Kaessmann H, Pääbo S, Gyllensten U (2000) Mitochondrial genome variation and the origin of modern humans. *Nature* **408**, 708–713.

International SNP Map Working Group (2001) A map of human genome sequence variation containing 1.42 million single nucleotide polymorphisms. *Nature* **409**, 928–933.

Jarman AP, Wells RA (1989) Hypervariable minisatellites: recombinators or innocent bystanders? *Trends Genet.* **5**, 367–371.

Jeffreys AJ, Wilson V, Thein SL (1985) Hypervariable 'minisatellite' regions in human DNA. *Nature* **314**, 67–73.

Jeffreys AJ, MacLeod A, Tamaki K, Neil DL, Monckton DG (1991) Minisatellite repeat coding as a digital approach to DNA typing. *Nature* **354**, 204–209.

Jeffreys AJ, Tamaki K, MacLeod A, Monckton DG, Neil DL, Armour JAL (1994) Complex gene conversion events in germline mutation at human minisatellites. *Nature Genet.* **6**, 136–145.

Jeffreys AJ, Murray J, Neumann R (1998) High-resolution mapping of crossovers in human sperm defines a minisatellite-associated recombination hot spot. *Mol. Cell* **2**, 267–273.

Jeffreys AJ, Barber R, Bois P *et al.* (1999) Human minisatellites, repeat DNA instability and meiotic recombination. *Electrophoresis* **20**, 1665–1675.

Jeffreys AJ, Kauppi L, Neumann R (2001) Intensely punctate meiotic recombination in the class II region of the major histocompatibility complex. *Nature Genet.* **29**, 217–222.

Jobling MA, Bouzekri N, Taylor PG (1998) Hypervariable digital DNA codes for human paternal lineages: MVR-PCR at the Y-specific minisatellite, MSY1 (DYF155S1). *Hum. Mol. Genet.* **7**, 643–653.

Kayser M, Roewer L, Hedman M *et al.* (2000) Characteristics and frequency of germline mutations at microsatellite loci from the human Y chromosome, as revealed by direct observation in father/son pairs. *Am. J. Hum. Genet.* **66**, 1580–1588.

Kennedy GC, German MS, Rutter WJ (1995) The minisatellite in the diabetes susceptibility locus IDDM2 regulates insulin transcription. *Nature Genet.* **9**, 293–298.

Kimura M (1968) Evolutionary rate at the molecular level. *Nature* **217**, 624–626.

Knight RD, Freeland SJ, Landweber LF (1999) Selection, history and chemistry: the three faces of the genetic code. *Trends Biochem. Sci.* **24**, 241–247.

Krawczak M, Ball EV, Cooper DN (1998) Neighboring-nucleotide effects on the rates of germ-line single-base-pair substitution in human genes. *Am. J. Hum. Genet.* **63**, 474–488.

Krontiris TG, Devlin B, Karp DD, Robert NJ, Risch N (1993) An association between the risk of cancer and mutations in the HRAS1 minisatellite locus. *New Engl. J. Med.* **329**, 517–523.

Kunkel TA, Bebebek K (2000) DNA replication fidelity. *Ann. Rev. Biochem.* **69**, 497–529.

Lalioti MD, Scott HS, Buresi C. *et al.* (1997) Dodecamer repeat expansion in cystatin B gene in progressive myoclonus epilepsy. *Nature* **386**, 847–851.

Lander ES, Linton LM, Birren B *et al.* (2001) Initial sequencing and analysis of the human genome. *Nature* **409**, 860–921.

Landsteiner K (1900) Zur Kenntnis der antifermentativen, lytisichen und agglutinierenden Wirkungen des Blutserums und der Lymphe. *Zbl. Bakt. I Abt.* **27**, 357–362.

Lewontin RC (1964) The interaction of selection and linkage. I. General considerations; heterotic models. *Genetics* **49**, 49–67.

Lightowlers RN, Chinnery PF, Turnbull DM, Howell N (1997) Mammalian mitochondrial genetics: heredity, heteroplasmy and disease. *Trends Genet.* **13**, 450–455.

Lukacsovich T, Waldman AS (1999) Suppression of intrachromosomal gene conversion in mammalian cells by small degrees of sequence divergence. *Genetics* **151**, 1559–1568.

Lupski JR (1998) Genomic disorders: structural features of the genome can lead to DNA rearrangements and human disease traits. *Trends Genet.* **14**, 417–422.

May CA, Jeffreys AJ, Armour JAL (1996) Mutation rate heterogeneity and the generation of allele diversity at the human minisatellite MS205 (D16S309). *Hum. Mol. Genet.* **5**, 1823–1833.

Myers JS, Vincent BJ, Udall H *et al.* (2002) A comprehensive analysis of recently integrated human Ta L1 elements. *Am. J. Hum. Genet.* **71**, 312–326.

Nachman MW, Crowell SL (2000) Estimate of the mutation rate per nucleotide in humans. *Genetics* **156**, 297–304.

Neil DL, Jeffreys AJ (1993) Digital DNA typing at a second hypervariable locus by minisatellite variant repeat mapping. *Hum. Mol. Genet.* **2**, 1129–1135.

Neitz M, Neitz J (1995) Numbers and ratios of visual pigment genes for normal red-green color vision. *Science* **267**, 1013–1016.

Oakey R, Tyler-Smith C (1990) Y chromosome DNA haplotyping suggests that most European and Asian men are descended from one of two males. *Genomics* **7**, 325–330.

Osheroff WP, Beard WA, Wilson SH, Kunkel TK (1999) Base substitution specificity of DNA polymerase beta depends on interactions in the DNA minor groove. *J. Biol. Chem.* **274**, 20749–20752.

Ostertag EM, Kazazian HH (2001) Biology of mammalian L1 retrotranspositions. *Ann. Rev. Genet.* **35**, 501–538.

Papadakis MN, Patrinos GP (1999) Contribution of gene conversion in the evolution of the human b-like globin gene family. *Hum. Genet.* **104**, 117–125.

Parsons TJ, Muniec DS, Sullivan K *et al.* (1997) A high observed substitution rate in the human mitochondrial DNA control region. *Nature Genet.* **15**, 363–367.

Patil N, Berno AJ, Hinds DA *et al.* (2001) Blocks of limited haplotype diversity revealed by high-resolution scanning of human chromosome 21. *Science*, **294**, 1719–1723.

Pauling L, Itano HA, Singer SJ, Wells IC (1949) Sickle-cell anemia, a molecular disease. *Science* **110**, 543–548.

Petes TD (2001) Meiotic recombination hot spots and cold spots. *Nature Rev. Genet.* **2**, 360–369.

Przeworski M, Hudson RR, Di Rienzo A (2000) Adjusting the focus on human variation. *Trends Genet.* **16**, 296–302.

Reich DE, Schaffner SF, Daly MJ *et al.* (2002) Human genome sequence variation and the influence of gene history, mutation and recombination. *Nature Genet.* **32**, 135–142.

Richards M, Macaulay V, Hickey E *et al.* (2000) Tracing European founder lineages in the near eastern mtDNA pool. *Am. J. Hum. Genet.* **67**, 1251–1276.

Robertson KD, Wolffe AP (2000) DNA methylation in health and disease. *Nature Rev. Genet.* **1**, 11–19.

Royle NJ, Clarkson RE, Wong Z, Jeffreys AJ (1988) Clustering of hypervariable minisatellites in the proterminal regions of human autosomes. *Genomics* **3**, 352–360.

Rubinsztein DC, Amos W, Leggo J *et al.* (1995) Microsatellite evolution – evidence for directionality and variation in rate between species. *Nature Genet.* **10**, 337–343.

Schlötterer C, Tautz D (1992) Slippage synthesis of simple sequence DNA. *Nucl. Acids Res.* **20**, 211–215.

Shadel GS, Clayton DA (1997) Mitochondrial DNA maintenance in vertebrates. *Ann. Rev. Biochem.* **66**, 409–435.

Signer EN, Dubrova YE, Jeffreys AJ, Wilde C, Finch LM, Wells M, Peaker M (1998) DNA fingerprinting Dolly. *Nature* **394**, 329–330.

Siguroardottir S, Helgason A, Gulcher JR, Stefansson K, Donnelly P (2000) The mutation rate in the human mtDNA control region. *Am. J. Hum. Genet.* **66**, 1599–1609.

Stankiewicz P, Lupski JR (2002) Molecular-evolutionary mechanisms for genomic disorders. *Curr. Opin. Genet. Dev.* **12**, 312–319.

Stumpf MPH, Goldstein DB (2003) Demography, recombination hot spot intensity, and the block structure of linkage disequilibrium. *Curr. Biol.* **13**, 1–8.

Swallow DM, Gendler S, Griffiths B, Corney G, Taylor-Papadimitrou J, Bramwell ME (1987) The human tumour-associated epithelial mucins are coded by an expressed hypervariable gene locus PUM. *Nature* **328**, 82–84.

Tamaki K, May CA, Dubrova YE, Jeffreys AJ (1999) Extremely complex repeat shuffling during germline mutation at human minisatellite B6.7. *Hum. Mol. Genet.* **8**, 879–888.

Turner G, Barbulescu M, Su M, Jensen-Seaman MI, Kidd KK, Lenz J (2001) Insertional polymorphisms of full-length endogenous retroviruses in humans. *Curr. Biol.* **11**, 1531–1535.

Vajo Z, Francomano CA, Wilkin DJ (2000) The molecular and genetic basis of fibroblast growth factor receptor 3 disorders: the achondroplasia family of skeletal dysplasias, Muenke craniosynostosis, and Crouzon syndrome with acanthosis nigricans. *Endocrinol. Rev.* **21**, 23–39.

van Gent DC, Hoeijmakers JHJ, Kanaar R (2001) Chromosomal stability and the DNA double-stranded break connection. *Nature Rev. Genet.* **2**, 196–206.

Veiko NN, Lyapunova NA, Bogush AI, Tsvetkova TG, Gromova EV (1996) Ribosomal gene number in individual human genomes: data from comparative molecular and cytogenetic analysis. *Mol. Biol.* **30**, 641–647.

Venter JC, Adams MD, Myers EW *et al.* (2001) The sequence of the human genome. *Science* **291**, 1304–1351.

Vergnaud G, Mariat D, Apiou F, Aurias A, Lathrop M, Lauthier V (1991) The use of synthetic tandem repeats to isolate new VNTR loci – cloning of a human hypermutable sequence. *Genomics* **11**, 135–144.

Verma RS, Dosik H, Scharf T, Lubs HA (1978) Length heteromorphisms of fluorescent (f) and non-fluorescent (nf) segments of human Y chromosome: classification, frequencies, and incidence in normal Caucasians. *J. Med. Genet.* **15**, 277–281.

Vogel F, Motulsky AG (1997) *Human Genetics: Problems and Approaches.* Springer, Berlin.

Watkins WS, Ricker CE, Bamshad MJ *et al.* (2001) Patterns of ancestral human diversity: an analysis of Alu-insertion and restriction-site polymorphisms. *Am. J. Hum. Genet.* **68**, 738–752.

Webster MK, Donoghue DJ (1997) FGFR activation in skeletal disorders: too much of a good thing. *Trends Genet.* **13**, 178–182.

Webster MT, Smith NGC, Ellegren H (2002) Microsatellite evolution inferred from human–chimpanzee genomic sequence alignments. *Proc. Natl Acad. Sci. USA* **99**, 8748–8753.

Whitfield LS, Sulston JE, Goodfellow PN (1995) Sequence variation of the human Y chromosome. *Nature* **378**, 379–380.

Xu X, Peng M, Fang Z, Xu X (2000) The direction of microsatellite mutations is dependent upon allele length. *Nature Genet.* **24**, 396–399.

Yamamoto F, Clausen H, White T, Marken J, Hakomori S (1990) Molecular genetic basis of the histo-blood group ABO system. *Nature* **345**, 229–233.

Zerjal T, Pandya A, Santos FR *et al.* (1999) In: *Genomic Diversity: Applications in Human Population Genetics* (eds SS Papiha, R Deka, R Chakraborty). Plenum, New York, pp. 91–102.

CHAPTER FOUR

Discovering and assaying genome diversity

CHAPTER CONTENTS

BOXES

4.1 Introduction

Human genetic diversity has been studied since the first genetic markers – blood groups – were discovered (Landsteiner, 1900). Early methods for studying these differences were indirect, based on immunological reactions (see *Box 3.3*), and later, electrophoretic analysis of gene products; methods evolved towards the direct analysis of DNA in the 1970s and 1980s, including Southern blotting analysis (see *Box 4.3*, and *Figure 15.1*). but it was with the invention of the truly revolutionary technique, the **polymerase chain reaction** (**PCR**) (Saiki *et al.*, 1985; Mullis and Faloona, 1987) that the analysis of human genetic variation (and indeed the genetic variation of very many organisms) took a quantum leap forward. Since this book is chiefly concerned with polymorphic markers which can be studied using PCR, we begin with a description of this method; other, older techniques for studying genetic variation will be mentioned elsewhere in this chapter, and in other chapters.

4.2 The polymerase chain reaction (PCR)

PCR has become such a universal technique in genetics and molecular biology that it is difficult for those of us who were around before its advent to remember what life in the lab was like without it. There will be many readers of this book who may not have known the 'pre-PCR' era at all, and will not appreciate how extraordinary a revolution its invention precipitated. Why is PCR so special?

Suppose we are interested in studying the sequence and variation of a human gene, for example, the β-globin gene. This small gene spans about 1.6 kb of DNA, and is therefore of a manageable size for DNA sequencing in a number of different individuals. The problem we are faced with, however, is that in any human DNA sample the β-globin gene represents only 0.00005% of the 3200-Mb genome. The specific sequence of interest is at such low concentration that direct analysis is impossible – its 'signal' is lost in the overwhelming 'noise' of the rest of the genome. The traditional approach to this might have been to use recombinant DNA technology to construct very large 'genomic libraries' of millions of cloned DNA fragments within bacterial plasmids, representing the entire genomes of different individuals, then to identify individual clones containing the gene of interest and analyze these purified gene copies. This would represent an extremely laborious and expensive way of isolating and amplifying (through the medium of bacterial multiplication) individual β-globin gene copies.

What PCR does instead is to provide a way to amplify individual gene copies (DNA sequences) directly, cheaply and quickly (*Figure 4.1*): short (typically 18–24 nucleotide) **oligonucleotide primers** are designed which flank the region of interest (these primers are usually synthesized by one of many commercial suppliers). In the PCR (*Figure 4.2*), which is carried out in a programmable heating block called a thermal cycler, the genomic template DNA is **denatured**

by heating, then the reaction is cooled to a specific temperature to allow the primers to **anneal** to their specific target sequence. The temperature is then raised in the **extension** phase, and a thermostable DNA polymerase, isolated from a thermophilic ('heat loving') bacterium such as *Thermus aquaticus* (*Taq*), carries out DNA synthesis from the primers, using the genomic DNA as template. This cycle of denaturation, annealing and extension is repeated, typically 30 times, and in each cycle previously synthesized product, as well as the original genomic DNA, acts as template: if the reaction were 100% efficient, each cycle would double the amount of target sequence (the **amplicon**). This would mean that, starting with 100 ng (nanograms) of genomic DNA (about 30 000 genome copies), the β-globin gene would be amplified to a copy number of over a billion, with a mass of over 50 μg. In practice, the amplification is around a thousand times less than this, but can yield typically 100 ng of the specific target sequence, which is more than enough DNA for sequencing or other molecular analyses.

The specificity and sensitivity of PCR, combined with the ingenuity of researchers, has led to a vast range of applications, including:

▸ direct determination of the DNA sequence of PCR products;

▸ selective amplification of a specific allele from a diploid genotype (Section 4.10), and typing of polymorphisms within amplicons (see Section 4.6–4.8);

▸ detection of very rare mutant DNA molecules within cell populations in mutation assays, and similarly rare recombinant molecules in recombination assays;

▸ quantification of template molecules in genomic DNA, allowing measurement of copy number of genomic sequences;

▸ detection of specific transcripts in cell extracts, after reverse-transcription of RNA into DNA;

▸ quantification of transcripts and thus monitoring of gene expression;

▸ detection of methylated sites within DNA templates;

▸ simultaneous amplification of all, or almost all, sequences within an individual genome (*Box 4.1*), thus multiplying the amount of DNA available for analysis;

▸ introduction of specific mutations into DNA molecules (*in vitro* mutagenesis);

▸ amplification of minute amounts of DNA from fossil material (ancient DNA; Section 4.11);

▸ amplification of trace amounts of DNA present in biological samples from crime scenes in forensic analysis (see Chapter 15).

Some of these methods will be discussed more fully in this chapter – in particular, the use of PCR in typing polymorphisms in DNA, and the use of PCR in examining DNA sequences in ancient samples. We will first concentrate on analyzing diversity in modern samples, and ancient DNA studies are discussed in Section 4.11.

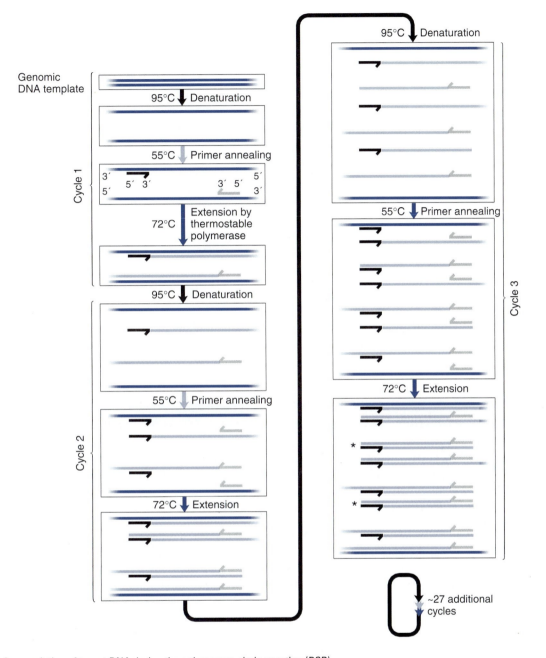

Figure 4.1: Accumulation of target DNA during the polymerase chain reaction (PCR).

In principle the number of copies of the targeted region, delineated by the primers (black arrows), doubles each cycle. A 30-cycle PCR reaction would therefore yield 2^{30} (over 1 billion) copies of the target, but in practice the reaction is not 100% efficient and the yield is less. Note that products delimited at both ends by the primers (marked *) do not start to accumulate until cycle 3. Denaturation and extension temperatures are those used typically; annealing temperature depends on the primer sequences and reagent concentrations.

4.3 First, find your DNA

Any study of human genetic diversity requires DNA samples from a number of different individuals. In this chapter, we set aside the ethical considerations which influence the availability of samples from particular populations (these are discussed in Section 9.2.1), and address only the practical questions. There are several different tissue sources for

human DNA samples, which differ in the amount of DNA that they yield, and in other aspects which can be important. Since the advent of PCR the amount of DNA needed for analysis has dropped dramatically: a single Southern blot typically used 5–10 μg of DNA, while a typical PCR reaction can make do with 1000-fold less. Methods for **'whole genome amplification'** can in principle take a few nanograms of genomic DNA or less, and create a pool of

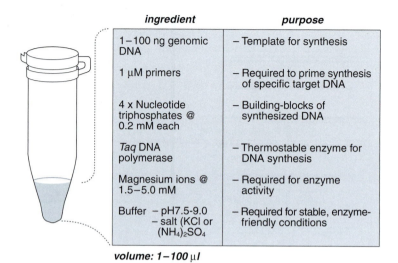

ingredient	purpose
1–100 ng genomic DNA	– Template for synthesis
1 µM primers	– Required to prime synthesis of specific target DNA
4 x Nucleotide triphosphates @ 0.2 mM each	– Building-blocks of synthesized DNA
Taq DNA polymerase	– Thermostable enzyme for DNA synthesis
Magnesium ions @ 1.5–5.0 mM	– Required for enzyme activity
Buffer – pH7.5-9.0 – salt (KCl or (NH₄)₂SO₄	– Required for stable, enzyme-friendly conditions

volume: 1–100 µl

Figure 4.2: Components of a typical PCR reaction.

PCR reactions are often carried out in 96-well plates (in the standard 8 × 12 format), or in 384-well plates (16 × 24), as well as in 0.5-ml or 0.2-ml tubes.

BOX 4.1 Something from next-to-nothing: Methods for whole-genome amplification

Even though one PCR reaction consumes very little DNA, a DNA sample obtained from one individual is finite in amount, and it is often difficult or impossible to return for resampling should the DNA run out. A number of methods have been developed to bypass this problem by amplifying, in as representative a way as possible, sequences from the entire genome in a sample. These **whole-genome amplification (WGA)** methods include:

▶ **degenerate oligonucleotide-primed PCR (DOP-PCR**; Telenius *et al.*, 1992). Short primers with random sequences (degenerate primers) bind to many different sites throughout the genome and provide indiscriminate PCR amplification. Problems associated with this are nonuniform amplification of different loci, and a small average size of products (≤ 1 kb).

▶ **multiple displacement amplification (MDA**; Dean *et al.*, 2002): this non-PCR-based method again employs degenerate primers, but uses a polymerase from a bacterial virus (φ29) to synthesize long (> 10-kb) molecules from the template. Each new strand is displaced from its 5′ end by a following strand, and then new strands themselves become templates for further synthesis. This method can apparently produce 20–30 µg of representative, high-fidelity, high-molecular weight product from as little as a few billionths of a microgram of starting DNA, and has enormous promise for studies where only small amounts of DNA are available.

---

template molecules allowing hundreds or thousands of PCR reactions to be done (see *Box 4.1*).

▶ *Blood* is the most often used source for human DNA. Blood can be taken relatively easily from a vein in the arm, and 5 ml yields 50–200 µg of DNA. While it is usual for DNA to be extracted from blood and analyzed in aqueous solution, there are alternatives; a blood drop (usually produced from the thumb with a lancet) on a specialized paper (FTA paper) allows PCR to be done using DNA immobilized *in situ*. Disadvantages of blood are that venipuncture requires appropriately qualified personnel, is invasive, and involves some risk of blood-borne diseases. There are also practical and administrative difficulties in transporting fresh blood from the field, or even between labs.

▶ *Buccal (cheek) cells* offer a less invasive alternative to blood, with fewer risks of disease. Qualified personnel are unnecessary, and self-sampling is possible – this method is favored by many police forces when sampling individuals for genetic profiling (see Chapter 15). Usually a brush or swab is rubbed in the inside of the cheek and placed in preservative buffer, but alternatively a solution of saline or sucrose can be washed inside the mouth and used as starting material for DNA extraction. A disadvantage of buccal cells is that they yield a small amount of DNA relative to blood.

▶ *Semen* is a good source of DNA (although available from only half the population) and is sampled noninvasively by donation. However, recruitment of semen donors can be sensitive, and semen is not widely used as a starting material in genetic diversity studies. In studies of mutation, particularly at minisatellites and in meiotic recombination, semen samples have been of great importance (see Sections 3.3 and 3.7).

▶ *Hair* can be sampled noninvasively, and has been successfully used in some studies of human genetic diversity. The major issue is whether the hair includes the root or not; the root contains much cellular material which can be extracted and used for PCR amplification of sequences from throughout the genome, though the small amount of DNA obtained makes it best suited to analysis of the high-copy number mtDNA (e.g., Vigilant *et al.*, 1989). The hair shaft contains much less DNA, and has only been used for mtDNA-based forensic investigations (Wilson *et al.*, 1995). Because of these problems, hair is not a material of choice for any sampling of human DNAs.

▶ *Placenta* is a large tissue which is normally discarded, and can be the source of a very large amount of DNA. A disadvantage is that although the placenta is genetically identical to the child who has relied upon it during gestation, it contains maternal blood and so does not yield pure DNA from a single individual. Placenta is not widely used as a source of DNA in diversity studies, but has been useful where a lot of human DNA is needed, such as in reassociation of labeled probe DNAs prior to hybridization, to remove repetitive sequences.

▶ *Cell lines* are established from primary sampled tissues, and offer the advantage that they can be cultured, sometimes indefinitely, yielding effectively unlimited amounts of DNA. Cell lines can be established from fibroblasts taken from skin biopsies, but the primary sampling procedure is unpleasant, and the resulting cell lines are slow to establish and not immortal, eventually senescing and dying. Another common source is lymphocytes, which can be isolated from blood and treated with Epstein–Barr virus, leading to immortalized lymphoblastoid cell lines that grow rapidly and indefinitely. This method (**transformation**) has been used to establish the widely used CEPH (Centre d'Etude du Polymorphisme Humain) pedigree cell lines which were an important resource in linkage analysis, and has been proposed as the method of choice for immortalizing samples taken as part of the Human Genome Diversity Project – 1064 cell lines from different populations are so far available through CEPH (Cann *et al.*, 2002). One major disadvantage of this method is the cost of transformation and cell-line maintenance. Another potential problem is somatic mutations which occur during the propagation of the cells; these have been observed at both mini- and microsatellite loci (Royle *et al.*, 1993; Banchs *et al.*, 1994).

4.4 Finding and typing polymorphisms – general considerations

If we are interested in analyzing genetic variation in a sample of individual genomes, the ideal way to proceed is to analyze all the variation in all the samples – in other words to carry out comprehensive **resequencing**. This is an unbiased way to assess diversity, and will detect DNA variation of all kinds, including SNPs, microsatellites, minisatellites and retroelement insertions. Unfortunately, limitations of technology and attendant cost implications mean that this is not adopted in many cases. Therefore we must often discover variation in a subset of samples, and go on to type the discovered markers in a larger set. This approach will be described here, but it is important to realize that it is suboptimal.

Aside from the invention of PCR, the other quantum leap in human genetic diversity studies has been the availability of the draft genome sequence of our species, itself made possible by the development of automated fluorescent dideoxy DNA sequencing (*Figures 4.3 and 4.4*). Before this, the discovery of the various classes of polymorphic markers relied upon chance detection, the development of specific and arcane methods, and extremely hard work. Genome sequencing from several individuals has automatically made many SNPs available, while **bioinformatic** methods for detecting VNTRs and identifying retroelements belonging to recently transposing families have enormously facilitated the discovery of useful markers. This chapter will concentrate on the 'postgenomic' methods, but will refer to traditional methods as well; note that, for scientists working on less fashionable organisms than humans, which do not have genome projects, these traditional methods are still highly relevant. Molecular ecologists, in particular, have exploited a number of methods of detecting genetic diversity which are not used by those studying human genetics; these methods are summarized in *Box 4.2*.

4.5 Finding and typing single nucleotide polymorphisms (SNPs)

The drive to discover base-substitutional variation in the human genome (SNPs), has come from:

▶ a desire to find the causative mutations underlying human monogenic disorders;

▶ an interest in the evolutionary history of individual loci, including searches for the impact of natural selection;

▶ the idea that a high-density SNP map of the genome will be a useful tool for the mapping of genes involved in complex disorders, via association studies (see Section 14.3).

4.5.1 Finding SNPs the easy way – using databases

There are many different ways for discovering SNPs, but a researcher's first course of action should be the expedient one

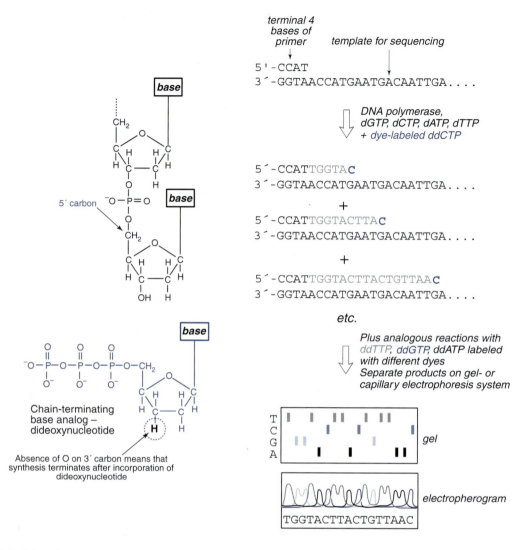

Figure 4.3: Principle of fluorescent dideoxy DNA sequencing.

When a 2′,3′-dideoxy nucleotide analog is incorporated into the growing DNA chain, it terminates synthesis because the 3′-carbon atom does not carry the hydroxyl (–OH) group necessary for the next phosphodiester bond to be made (Sanger *et al.*, 1977). Inclusion of a dye-labeled dideoxy analog (for example, ddCTP) in a reaction will give rise to a series of molecules terminated at C bases in the synthesized strand. Separation of these molecules 'maps' the positions of C bases; similar reactions for the other bases, each labeled with a different dye, allows the sequence to be read from a gel or **electropherogram** after specific dye excitation using a laser. Before fluorescent detection became available, radioactive detection methods were used.

of asking whether someone else has already found SNPs in a sequence of interest, by consulting databases. As we saw in Chapter 3 (Section 3.2.7), the strategy for sequencing the human genome itself has been designed so that SNPs are identified: libraries were constructed from the DNAs of different individuals, and since the amount of sequence obtained covers several genome equivalents, many putative SNPs have been discovered (International SNP Map Working Group, 2001). Other SNPs have been identified through the more directed resequencing efforts of the SNP Consortium. In principle, DNA sequencing is the 'gold standard' for SNP discovery, and should find all the variation

within a sequence. In practice this is almost true, although there are some regions which, because of high GC content or secondary structure, prove difficult to sequence reliably.

Suppose we are interested in finding known SNPs within a particular gene, say, the *APOE* gene. A good starting point is the National Center for Biotechnology Information (NCBI) dbSNP web-page (http://www.ncbi.nlm.nih.gov/SNP/; *Figure 4.5*). Under the heading 'Locus Information', clicking on 'Gene Name or Symbol' takes us to the Locus Information Query page (*Figure 4.5a*), where we can enter the gene name and specify the organism as 'Human'. The resulting Locus Link page (not shown) includes the *APOE*

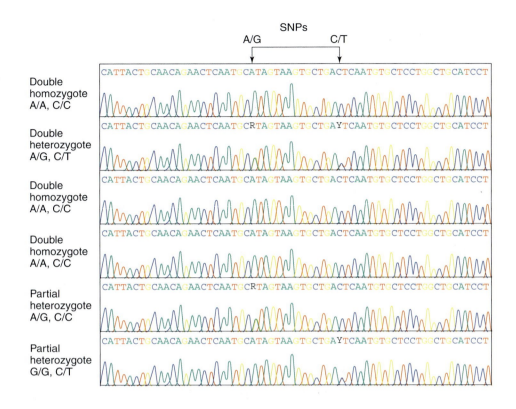

Figure 4.4: Sequencing electropherograms showing heterozygous and homozygous SNPs.

Two base substitutions in the pseudoautosomal region of the sex chromosomes are shown, separated by 13 bp. Each electropherogram shows the diploid sequence obtained from a different individual. Three individuals are double homozygotes, while three contain heterozygous bases, denoted 'R' (A and G) and 'Y' (C and T). Electropherograms kindly supplied by Celia May, Department of Genetics, University of Leicester.

gene, and clicking on the 'V' box gives a list of SNPs (*Figure 4.5B*). If a specific gene is not the object of interest, SNP searches can be done from the same starting point using a genomic clone sequence (via its accession number), or **contig** (set of contiguous clone sequences) instead.

SNPs in the NCBI database were discovered either as a byproduct of BAC sequencing (see Section 3.2.7), or by the SNP Consortium, which used for resequencing a panel of DNAs from 24 ethnically diverse individuals resident in the USA (International SNP Map Working Group, 2001) – a subset of a larger available panel (Collins *et al.*, 1998). These different strategies influence what kinds of SNPs are discovered. Note also that even well-known SNPs resulting from small-scale publications often do not find their way into the databases at all, and so could be easy to miss. An alternative, hand-curated, and more anthropologically based, route to SNPs is via the database ALFRED (for ALlele FREquency Database; http://alfred.med.yale.edu/alfred/index.asp; Osier *et al.*, 2002).

What do we learn about the SNPs that are listed in public databases, and how reliable and useful is the information? There are a number of potential issues:

▶ As has already been discussed in Section 3.2.7, many listed SNPs are in fact PSVs (paralogous sequence variants; Bailey *et al.*, 2002), while other listed SNPs may represent sequencing or assembly errors. In an intensive resequencing survey, 12% of SNPs in the public and private databases could not be confirmed (Reich *et al.*, 2003). Validation of a SNP is therefore important. In the SNP list (*Figure 4.5B*), some SNPs have heterozygosity values given, while some have no data (N.D.). Clicking on a SNP name gives more information: those markers with heterozygosity values have been tested in a population sample(s), and where more than one chromosome carries the SNP, this is good evidence that the SNP is 'real'. Many SNPs, however, remain 'unconfirmed', and represent a difference between only two tested chromosomes. Control DNAs, samples known to have a particular allele for a given published or databased SNP, are very useful in the validation of assays, but are often unavailable.

▶ SNPs vary greatly in allele frequency, and this affects their usefulness in diversity studies. Sometimes frequency information is available in the database, but often not.

BOX 4.2 Methods for molecular ecology: RAPDs and AFLPs

For scientists interested in the genetic diversity of pine trees, mushrooms, spiders, minnows, sea anemones and the like, the absence of genome sequencing projects for these organisms is a major hindrance to developing useful genetic markers. While microsatellites can be developed for many species (reviewed by Zane *et al.*, 2002), as they have been for humans, special methods, not normally applied to human diversity, are also employed to isolate useful markers:

▶ **RAPD (randomly amplified polymorphic DNA)** markers are generated by using short (8–12-bases long) primers to amplify random fragments of DNA. Variation between individuals is seen as presence or absence of bands, or length variation. The advantage of this method is that markers can be quickly identified for any genome, without the need for knowledge of sequence, or for special primer design. The disadvantage is that many (perhaps 60%) of the amplified fragments are nonspecific and nonreproducible.

▶ **AFLPs** (amplified fragment length polymorphisms; reviewed by Mueller and Wolfenbarger, 1999) are generated by digesting genomic DNA with restriction enzymes which produce 5′ single-stranded overhangs, then ligating (using the enzyme DNA ligase) a short adaptor molecule (see *Figure*). Primers targeted to the adaptors give products whose length is determined by the position of the genomic restriction sites, and polymorphism in these sites is thus identified. AFLPs have the advantages of RAPDs, without the problem of specificity (see Section 10.7.2 for an example of the use of AFLPs). Sequencing an AFLP fragment and designing locus-specific primers converts it into a sequence-characterized amplified region (SCAR).

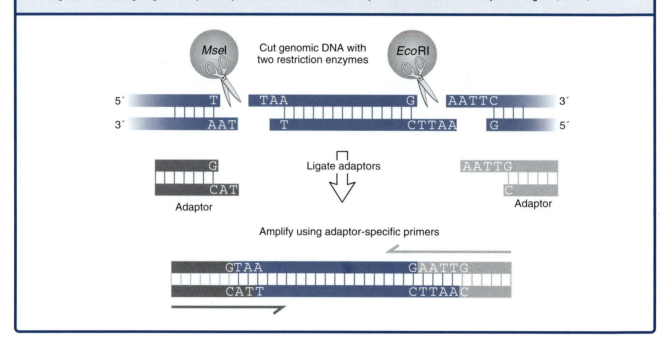

Many SNPs are expected to have different allele frequencies in different populations, and even where heterozygosity values are high (i.e., there is not one rare and one common allele), finding useful information about which populations have been tested is difficult. The ALFRED database is a promising source of such information.

4.5.2 Other methods for SNP discovery

Before PCR and high throughput sequencing became available, SNPs were both discovered and assayed in a limited way using restriction endonucleases [enzymes which cleave DNA at specific short (normally 4–8-bp) sequences], and the detection of cleavage using specific radiolabeled probes (*Box 4.3*). There is now a wide variety of more efficient PCR-based methods for directed SNP discovery, developed to

seek causative mutations in monogenic diseases. Several of these methods rely on the formation of **heteroduplexes**: when alleles in a heterozygote are amplified, then denatured and allowed to reanneal, double-stranded DNA is formed in which one strand is from one allele, and the complementary strand from the other allele. It thus contains a **mismatch** at the SNP site, and has abnormal properties of migration in gels, or of denaturation, which can be exploited in detection systems. The method of **denaturing high performance liquid chromatography** (DHPLC; reviewed by Xiao and Oefner, 2001) detects heteroduplexes by their altered retention time on chromatography columns under near-denaturing conditions (*Figure 4.6*). A popular, simple and cheap method of mutation detection not relying on heteroduplex formation is **single-strand conformation polymorphism** (SSCP) analysis, in which changed mobilities of single-stranded mutant molecules are detected

(a)

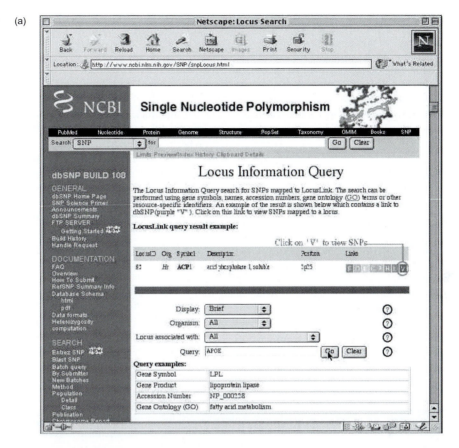

(b)

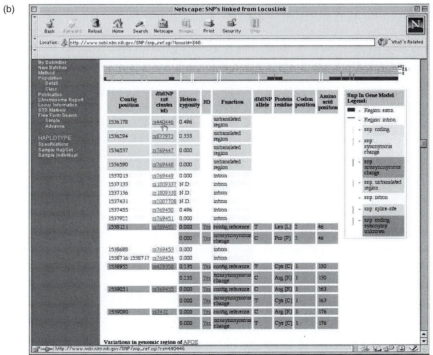

Figure 4.5: Finding SNPs in the NCBI database.

(a) Locus Information Query page; (b) SNPs in the *APOE* gene (page accessed 11/29/02).

on nondenaturing gels (*Figure 4.6*). The various methods differ greatly in their cost, throughput, and in the percentage of SNPs which they detect.

While directed methods such as DHPLC and SSCP analysis are cost effective and can be carried out in a high throughput format, the putative SNPs that they uncover must still be verified by sequencing; the cost of sequencing is likely to decrease in the near future, and will eventually reach a level which renders the other SNP discovery methods obsolete.

4.5.3 SNP typing methods

Once SNPs have been discovered, and if cost considerations preclude resequencing all samples, they must be typed. There are now many different methods for typing any SNP (reviewed by Gut, 2001; Syvänen, 2001), which can be separated into two fundamental steps: interrogation of the SNP, and detection of the allele-specific result of the interrogation. Interrogation rests on one of six basic biochemical principles summarized in *Figure 4.7*. Most of these principles exploit the sequence specificity of different enzymes (ligases, polymerases, special endonucleases and restriction enzymes), while oligonucleotide hybridization utilizes only the specificity of base pairing. Detection involves a wide variety of different technologies, including electrophoretic gels or capillaries, microtiter plate fluorescent readers, oligonucleotide microarrays ('DNA chips'), and mass spectrometry. These technologies detect differences in electrophoretic mobility, fluorescence and mass between the products derived from the different alleles by the interrogation process. *Figure 4.8* shows some ways in which the interrogation methods and the detection methods are connected to constitute typing systems.

Oligonucleotide hybridization methods are available in many different formats, from very 'low-tech' approaches where individual allele-specific oligonucleotides are radiolabeled and hybridized to PCR products on dot-blots (e.g., Jeffreys *et al.*, 2001) to expensive and high-tech DNA-'chip'-based assays (*Figure 4.9a*) involving hybridization of fluorescently labeled PCR products to arrays bearing up to 250 000 oligos per square centimeter (Chee *et al.*, 1996). Another format, with the advantage of absence of post-PCR processing and real-time detection, involves the separation of a fluorophore on one end of an allele-specific oligonucleotide from a quencher on the other, upon hybridization (e.g., molecular beacons; *Figure 4.9b*; Tyagi and Kramer, 1996). A general disadvantage of many methods based on the oligonucleotide hybridization principle (with the exception of the chip-based assay) is the necessity to establish conditions empirically for each SNP, which makes **multiplexing** (typing multiple polymorphisms in a single assay) difficult.

BOX 4.3 Traditional methods for finding and assaying base-substitutional variation

As was discussed in Section 4.2, prior to the PCR era there was a major problem of sensitivity in analyzing single-copy sequences in the large and complex human genome. The problem of specific sequence detection, at least, was overcome by the use of DNA **probes**, specific cloned sequences labeled with the radioisotope phosphorus-32. In this method (see *Figure 15.1*), genomic DNA is cut by restriction endonucleases – enzymes recognizing and cleaving DNA at specific short sequences. These DNA fragments are then fractionated on agarose gels, transferred to a special membrane, and hybridized with the denatured, radiolabeled probe. Washing to remove excess probe, followed by exposure of the filter to X-ray film (**autoradiography**), can reveal specific hybridizing restriction fragments as bands on the film. This process is known as Southern blotting, after its inventor (Southern, 1975).

Because restriction enzymes cleave specific sequences, mutations of any of the bases in these sequences will lead to a failure to cleave. Conversely, a mutation in a sequence closely related to that of a restriction site can create a new site. Such changes are recognized by changes in the lengths of restriction fragments detected by DNA probes: **restriction fragment length polymorphisms (RFLPs)**. RFLPs, marking sites of sequence variation, have been enormously important markers in constructing genetic maps and locating disease genes through linkage analysis. However, as detectors of sequence variation, restriction enzymes have limitations:

▸ many sequences representing sites of variation are unrecognized by known restriction enzymes;

▸ while cleavage of a site, say, for *Bam*HI, indicates that a specific hexanucleotide sequence is present (GGATCC), failure of the same enzyme to cleave this site in another allele does not specify the mutation, but only indicates than any one of the six bases has undergone one of three possible mutations (discounting the possibility of indels or multiple base substitutions). More than one mutation could give the same cleavage pattern;

▸ cleavage by some restriction enzymes is inhibited by DNA methylation, so failure to cleave might be misinterpreted.

In one special case, human genetic variation has been studied directly using restriction enzymes, without the necessity of Southern blotting. This is RFLP variation in mtDNA: the mtDNA molecule is at high copy number and circular (Section 2.7.2), and can therefore be separated from nuclear DNA using differential centrifugation methods developed for isolating circular bacterial plasmids. Isolated mtDNA can then be digested with restriction enzymes, and restriction fragments analyzed directly on agarose gels. This kind of analysis has now been displaced by PCR, but was used to assay mtDNA variation in some early and influential studies (e.g., Cann *et al.*, 1987).

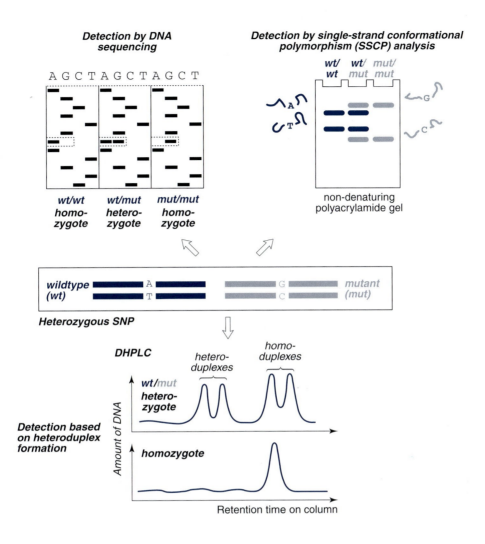

Figure 4.6: Methods for SNP discovery.

A hypothetical autosomal A to G transition can be found by sequencing, SSCP analysis, or a method relying on the detection of heteroduplexes, such as DHPLC. Note that the heteroduplex-based detection of mutations in haploid sequences (or in a diploid homozygote) requires each sample to be mixed with a reference sample. For a more extensive discussion of methods, see Strachan and Read (2003).

The primer extension principle is simple and the reaction robust, so reactions can be carried out under similar conditions, thus minimizing the time needed for optimization. A number of detection methods can be coupled to primer extension. For example, fluorescent dideoxynucleotides can be incorporated and products separated by gel or capillary electrophoresis on a sequencing apparatus (*Figure 4.10*). In another method, a specially modified nucleotide is incorporated which is bound by an antibody conjugated to an enzyme which then forms a colored product, allowing colorimetric detection of primer extension. The method of pyrosequencing (Fakhrai-Rad *et al.*, 2002) monitors release of pyrophosphate when a new base is incorporated.

Two methods of detection showing promise in high throughput applications are mass spectrometry and fluorescence polarization:

▶ Mass spectrometry (**MS**) is used to measure the mass/charge (m/z) ratio of ions, from which their molecular weight can be accurately inferred. The specific technique of **MALDI (matrix–assisted laser desorption-ionization;** reviewed by Jackson *et al.*, 2000) allows the analysis of large, nonvolatile, thermolabile intact molecules by MS, and so is suited to examining changes in molecular weight of primers caused by base incorporation (*Figure 4.11a*). The MALDI process uses a laser pulse to produce ions from the extended primer, which has been co-crystallized

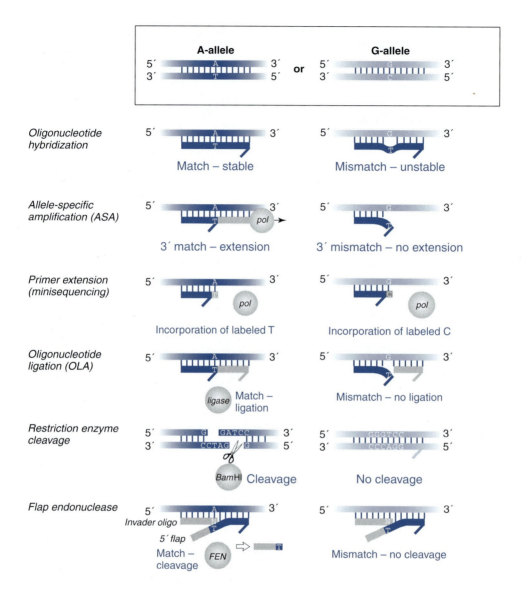

Figure 4.7: Biochemical principles of SNP typing methods.

Different methods of detecting the A-allele of an A/G transition SNP are shown; the G-allele would also be detected in an analogous reaction, except for the restriction enzyme cleavage, where absence of cleavage is the norm, unless a different restriction site exists in the G-allele. ASA is also sometimes known as amplification refractory mutation system (ARMS)-PCR.

with a laser-absorbent chemical matrix. The time taken by these ions to traverse a **time-of-flight** (**TOF**) detector is inversely proportional to m/z, and allows mass to be calculated. There are a number of ways of using this method (MALDI-TOF) for SNP typing; the PinPoint assay (Haff and Smirnov, 1997) incorporates a dideoxynucleotide complementary to the SNP in both alleles, and detects the small resulting mass difference (*Figure 4.11A*). The DNA MassARRAY system detects a larger mass difference by generating products of different lengths for each allele, and can be used for high throughput analysis – 9115 SNPs were analyzed in a recent study (Buetow *et al.*, 2001).

▶ **Fluorescence polarization** (**FP**; *Figure 4.11B*) detects incorporation of a specific fluorescently labeled base by virtue of the ~10-fold increase in size of the molecule to which the fluorophore is attached when it is incorporated into the extended primer, and the consequent reduction in its motion (Chen *et al.*, 1999). When a fluorescent molecule is excited by plane-polarized light, it emits fluorescent light which is also polarized, provided the molecule remains stationary. If the molecule is mobile, fluorescence will be depolarized, and this can be measured using polarization filters. This detection method does not require separation of starting materials from products

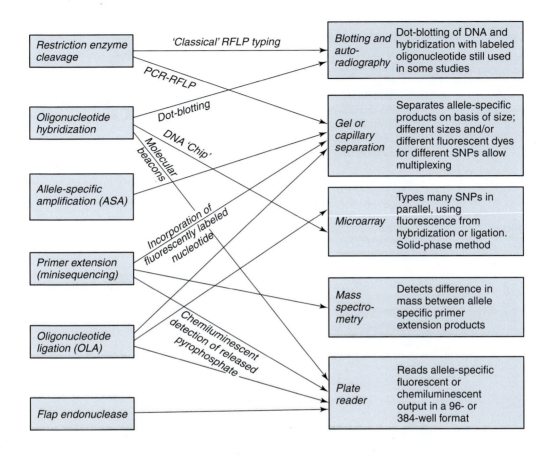

Figure 4.8: Connecting SNP interrogation with allele-specific detection.

The six principles outlined in *Figure 4.7* are shown connected to possible methods for detecting allele-specific products or reactions. The diagram is not meant to be comprehensive, and more information can be found in Syvänen (2001) and Gut (2001).

(it is a 'homogeneous' method), which is advantageous, and it can be used in conjunction with a number of different interrogation methods (reviewed by Kwok, 2002).

Choosing a SNP-typing platform is not easy, because there are so many different systems, with different properties, available; there are several factors to be taken into account:

▶ **throughput**: The numbers of SNPs in recently-published genotyping studies typically varies from tens to a few hundred, and the sample sizes from hundreds to a few thousand individuals; for example, a large survey typed 312 SNPs in 681 individuals – a total of 210 000 genotypes (Rioux *et al.*, 2001). Proposals to study genome-wide LD using SNPs vary widely in the number of SNPs that are suggested for successful analysis, from tens of thousands to a million. However, whatever the figure, an increase in throughput of at least two orders of magnitude is necessary if the aims are to be achieved, given sample sizes of about 1000. This drive towards high throughput analysis is largely responsible for the plethora of different systems for

SNP typing. A key issue in developing high throughput methods is the difficulty of multiplexing – because of primer–primer interactions, nonspecific amplification products, and the requirement of each primer pair for specific optimal conditions, carrying out PCR amplification of more than ten different loci simultaneously has been difficult and only now are large multiplexes coming into use;

▶ **simplicity and robustness of assay**. Some assays are simple, and can type many different SNPs under very similar reaction conditions. They require little optimization – a costly and time-consuming part of producing useful assays. Other systems require individual reactions to be optimized and run under different conditions;

▶ **success rate and accuracy**. Rates of failure or mistyping vary widely (see *Box 4.4*), and acceptable levels depend upon the individual study and its aims. Missing data are a problem in some studies, but not in others; some statistical methods for analyzing LD data, for example, can cope well with missing data;

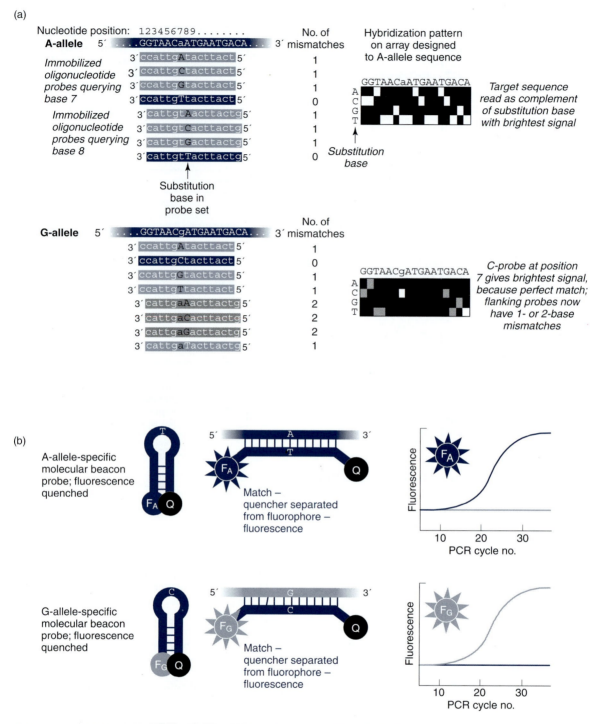

Figure 4.9: Examples of SNP typing methods using the principle of oligonucleotide hybridization.

(a) Oligonucleotide microarray (DNA chip): each position in a target sequence is queried by a set of four oligonucleotide probes, immobilized on a 'chip'. Each probe has a different base at an internal position (the substitution position). Here, the probe array is designed to the A-allele sequence. Target DNA is fluorescently labeled and hybridized to the array. A-allele DNA gives a set of bright signals on the chip corresponding to the complement of the target. G-allele DNA gives only one bright signal, corresponding to the oligonucleotide which perfectly matches the target. Flanking oligonucleotides contain one or two mismatches, and so do not give a strong signal. Figure based on Chee *et al.* (1996). Chips are manufactured by *in situ* light-directed synthesis of oligonucleotides using a photolithographic masking technique.

(b) Molecular beacons (Tyagi and Kramer, 1996): the probes contain a segment complementary to the target sequence, flanked by short self-complementary stretches. When free in solution, the self-complementary stretches hybridize, bringing together a fluorophore on one end of the probe and a quencher on the other, thus quenching fluorescence. In the presence of complementary target sequence, the probe hybridizes and the fluorophore is free to fluoresce. Fluorescence can be monitored in real-time (shown as a schematic graph) or at the PCR end-point.

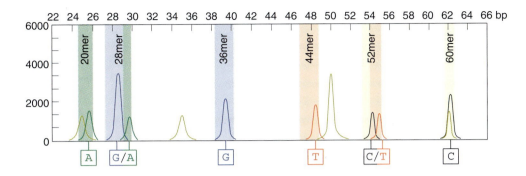

Figure 4.10: Typing SNPs by primer extension ('minisequencing') using the SNaPshot system.

A multiplex PCR amplifying six loci is analyzed using six different extension primers, each of which anneals immediately adjacent to the SNP. Incorporation of a single base according to the allelic state of the SNP is detected by separating the extended primers on the basis of size, and noting the color of fluorescence associated with each. SNPs queried with two primers (the 28-mer and 52-mer) are heterozygous in this example. SNaPshot is a trademark of Applied Biosystems.

▶ **track record**. Since many technologies are new, they do not yet have a long and well-validated track record. In a competitive market, some companies may fail and it may become impossible to maintain and run their systems;

▶ **DNA consumption**. Most methods require only a few nanograms to type a SNP; multiplexing reduces the amount of DNA required per reaction, while methods designed to avoid a PCR step, such as the basic Flap endonuclease assay, require much more DNA;

▶ **cost**. Some platforms, such as those with mass-spectrometric detection, require an expensive piece of apparatus, while others can be done with standard laboratory equipment. Reagent costs vary greatly too: methods requiring large numbers of primers, or specially modified primers carrying fluorophores, quenchers or biotin for immobilization can be very expensive.

4.5.4 Sources of error and bias in SNP typing

There are a number of possible sources of bias and error in SNP typing studies. Some modern surveys of genetic variation sequence the locus of interest in every individual in the sample, thus uniting SNP discovery and SNP typing in an relatively unbiased assessment of diversity (e.g., lipoprotein lipase gene, Nickerson *et al.*, 1998; Xq13.3, Kaessmann *et al.*, 1999). However, as has been discussed above, SNPs are often discovered in a relatively small sample of chromosomes (the screening set), and then typed in a much larger set, sometimes using a different technology. In these cases, the number and choice of samples in the screening set is important, and varies from study to study. For example, an investigation of *APOE* locus variation used sequencing to find 21 SNPs in 144 chromosomes from three populations of interest, then went on to type them by OLA (*Figure 4.7*) in 4358 chromosomes from the same populations

(Nickerson *et al.*, 2000). By contrast, in a study of Y chromosome diversity, DHPLC (*Figure 4.6*) was used to find SNPs in 53 chromosomes chosen for their representation of world populations, then was again used to type the same SNPs in an additional 1009 more geographically diverse chromosomes (Underhill *et al.*, 2000). In the latter strategy, excluding a population from the screening set but including it in the typing set can lead to **ascertainment bias**, since it could exhibit apparently low diversity, particularly if the loci under study show extensive differentiation between populations. A **Eurocentric** bias was introduced into many early studies, where the screening set was made up mostly of easily available 'local' samples, but the typing set included many non-Europeans. This leads to an underestimation of diversity outside European and European-derived populations.

Detection of a single specific base in the 6400 Mb of a diploid genome is a difficult task, and no SNP typing method is 100% accurate. The error rate depends on the method chosen, and on other factors such as DNA quality; as has already been mentioned, its importance depends on the objectives of an individual study. Error rates in published studies of human genetic diversity are often not explicitly discussed (*Box 4.4*), which makes the comparison of different studies difficult (reviewed by Przeworski *et al.*, 2000).

4.6 Finding and typing microsatellites

Now that the human genome sequence is almost complete and easily available, finding microsatellites (and indeed minisatellites) is relatively trivial. This is thanks to computer programs which search DNA sequences systematically for tandem repeats. The best example is Tandem Repeats Finder (Benson, 1999), which moves a 'sliding window' along the sequence to seek candidate matched adjacent repeats of any size in DNA, including repeats containing mismatches and indels. Segments of DNA several hundreds of kilobases in

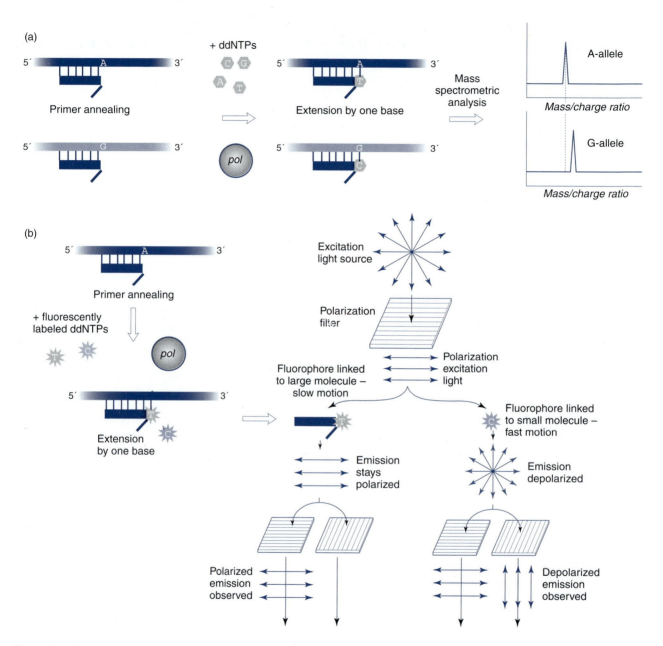

Figure 4.11: Examples of detection technologies used in high-throughput SNP typing via primer extension.

(a) The PinPoint assay (Haff and Smirnov, 1997) detects the incorporation of a specific base complementary to the SNP by measuring the mass/charge (m/z) ratio of the product by mass spectrometry (**MALDI-TOF** – see text).

(b) Fluorescence polarization (FP) detects incorporation of a specific fluorescently labeled base by virtue of the increase in size of the molecule to which the fluorophore is attached when it is incorporated into the extended primer, and the consequent reduction in its motion (Chen *et al.*, 1999). See text for a fuller description.

size can be rapidly searched, and tri-, tetra- or penta-nucleotide loci identified having between 10 and 30 uninterrupted copies of a single repeat type (see, for an example of its results, *Figure 3.17*). Experimental analysis of several genomic DNA samples for length variation, using flanking PCR primers, will then demonstrate whether or not a locus is polymorphic in the sample. An example of this approach is the isolation of novel microsatellites from a

1.33-Mb segment of the Y chromosome (Ayub *et al.*, 2000). Geneticists seeking microsatellites before the human genome sequence data became available had to use different methods to isolate loci of interest, for example hybridization of repeat sequence probes to genomic libraries, followed by subcloning of repeat-containing fragments and DNA sequencing (reviewed by Zane *et al.*, 2002).

BOX 4.4 The extent and impact of experimental error in genetic diversity studies

Any experimental analysis of a large number of samples contains mistakes. The most simple errors are in labeling tubes, and anecdotal evidence from clinical contexts (in which there are occasional catastrophic consequences, such as blood mistransfusion, or administration of the wrong drug) suggests that such errors are in the order of 1%. Efficient laboratory procedures, automation and bar-coding can reduce such errors to a minimum. Other errors occur in the transcription of data, or even in the formatting of data tables during the publication process (an example is mentioned in Section 10.5.5).

The finished portion of the draft genome sequence has an error rate of < 0.01% (Lander *et al.*, 2001), but in most SNP genotyping by sequencing the rate is likely to be much higher than this. Unlike the primary sequencing project, analysis is usually of diploid sequences, so it is important to be able to detect heterozygous nucleotide positions reliably (see *Figure 4.4*). Software such as Polyphred (Nickerson *et al.*, 1997) has been specifically designed for pinpointing such sites in sequence traces, and can reduce error rates in genotyping to < 0.1% (Nickerson *et al.*, 1998). Error rates in other methods of SNP typing are often not discussed, but when they are, they vary widely. For example, retyping of samples already typed at the *APOE* locus using the oligonucleotide ligation assay (see *Figure 4.7*) revealed a primary error rate of < 1.4% (Nickerson *et al.*, 2000), while a DNA-chip based study (*Figure 4.9*) reported an error rate of 7% (Hacia *et al.*, 1999). Commercial companies promoting a particular system may be unwilling to publicize error rates for the systems they produce. For microsatellites, a reported error rate estimated by sample retyping was 0.7% (Wilson *et al.*, 2001).

These random errors are a nuisance, and add an unwelcome aspect of 'noise' to disease association studies, for example, in which the 'signal' of association may already be weak. What is more, the different typing systems and different kinds of errors may confound **meta-analyses** that attempt to draw conclusions by analysis across different studies.

Potentially more serious than random errors are systematic errors which can lead to erroneous conclusions. A high-profile example is a systematic sequencing error which indicated that the same, rare base substitution existed on three very different background haplotypes in HVS I (see *Figure 2.20*) of mtDNA (Hagelberg *et al.*, 1999); this was taken as evidence of recombination in mtDNA, but turned out to be due to, in effect, an artifactual 10-bp deletion within one end of the sequenced region in a number of samples (Hagelberg *et al.*, 2000), and consequent sequence misalignment. The use of phylogenetic methods to detect sequencing errors in mtDNA HVS I sequences has been proposed (Bandelt *et al.*, 2002); these methods suggest that some data-sets have alarmingly high error rates. A survey (Forster, 2003) of 137 papers containing mtDNA data published between 1981 and 2002 showed that ~58% contained confirmed errors that should have been noticed by authors; disturbingly, 14 of these papers were in forensic journals. Claims (Gulcher *et al.*, 2000) that the Icelanders are a highly homogeneous European population have been apparently supported by work with mtDNA (Helgason *et al.*, 2000), but a reanalysis of the primary mtDNA data (Arnason, 2003) reveals plentiful errors in databases, which, when corrected, show the Icelanders to be a highly heterogeneous population, consistent with a history of admixture (see Chapter 12).

Variation at microsatellites is analyzed by PCR, using flanking primers close enough to the array that single-repeat-unit size differences between alleles can be discerned readily (*Figure 4.12a*). Originally, detection was by autoradiography after polyacrylamide gel electrophoresis of radiolabeled PCR products, or by **silver-staining**, another fairly sensitive method of detection. More recently the availability of primers labeled at their 5′ ends with different fluorescent dyes has meant that PCR products can be detected on gel- or capillary-based sequencing platforms using laser technology. This allows the coamplification of several microsatellites (**multiplexing**) labeled with different dyes, and their simultaneous separation and detection (*Figure 4.12b*; see also *Figure 15.4*). These methods have been very important in linkage analysis and association studies, as well as in DNA profiling in forensic work (see Chapter 15). Methods have also been developed which allow the typing of microsatellites on DNA chips (reviewed by Mir and Southern, 2000), though these are not widely used. Some microsatellites have complex structures including variant repeats, or more than one internal block which vary independently; such loci are sometimes typed by sequencing,

since this gives more information than the mere length of a flanking PCR product (Forster *et al.*, 1998).

4.7 Finding and typing minisatellites

As discussed in Section 3.3.2, the best-studied minisatellites are the GC-rich sequences identified by hybridization with DNA probes in DNA fingerprints (Jeffreys *et al.*, 1985). As with microsatellites, identification of new minisatellites is today made easy by the use of Tandem Repeats Finder: this *in silico* method has the advantage that all minisatellites can be identified, regardless of their sequence composition.

Length variation of minisatellites can be measured after Southern blotting: digestion of genomic DNA with a restriction enzyme which does not cut in the minisatellite repeat unit, followed by hybridization with a radiolabeled repeat probe, will reveal an RFLP due not to sequence change in a restriction site, but to variation in repeat number between constant restriction sites. PCR using flanking primers should allow the size of the repeat array to be measured more easily, but some alleles are very large, which

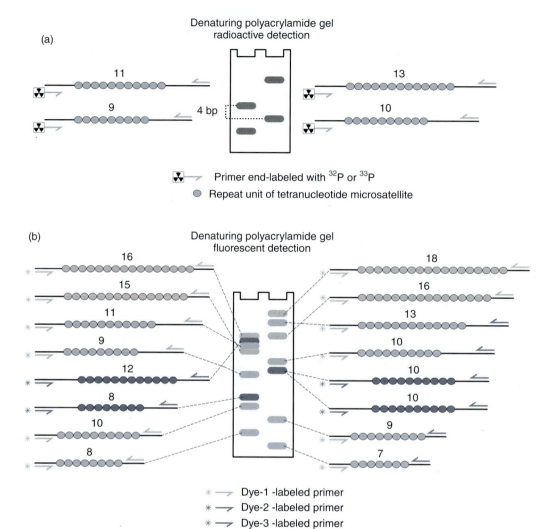

Figure 4.12: Typing microsatellites.

(a) Detection of a diploid tetranucleotide microsatellite by radioactive end-labeling of a primer followed by autoradiography. Silver-staining of the PCR products is an alternative means of detection, but gives a more complex pattern because both strands are detected.
(b) Primers labeled with fluorescent dyes allow multiplexing and simultaneous detection of several loci (here, four) on a sequencing platform. Primers amplifying loci with nonoverlapping size ranges (pale blue) can be labeled with the same dye. A fourth dye is used for a size standard (not shown) which is run in every lane. Capillary electrophoresis systems, as well as gel-based systems, can be used.

initially made this difficult. Specific modifications to the PCR protocol to allow long (up to 35-kb) fragments to be amplified (long-PCR; Barnes, 1994) have overcome this problem for all but the largest alleles.

Discrimination of different allele lengths by Southern blotting or by PCR is limited in resolution. Longer alleles with short repeat units may appear identical by length ('isoalleles'), but actually differ by a small number of repeat units. Even alleles having identical numbers of repeat units might not be identical by descent. Sequence analysis of minisatellites reveals an additional source of variation other than repeat number, which can resolve this: repeat units themselves differ from each other, usually by base substitutions

but sometimes by indels too (see Section 3.3.2). Using the principle of allele-specific PCR, the positions of these 'variant repeats' can be mapped along alleles, in the technique of **minisatellite variant repeat-PCR** (MVR-PCR; Jeffreys *et al.*, 1991). The principle is illustrated in *Figure 4.13*: in essence, PCR products are generated from a fixed flanking primer to repeat-type-specific primers binding to one repeat type only within the array. The lengths of these PCR products thus define the positions of a particular repeat type; separate PCR reactions employing different repeat-type-specific primers define the positions of other repeat types, and an MVR 'code,' or 'map' can be constructed after separation of the PCR products and their detection. In practice technical measures must be taken to reduce the effects of 'PCR collapse', in which repeat-type-specific primers anneal within the products of previous PCR cycles, leading to a reduction of product size and the loss of the longer amplicons from each ladder of products. Most minisatellites are diploid, and require the separation of alleles (Section 4.10.1), either on the basis of size prior to analysis, or by use of an allele-specific flanking primer during MVR-PCR. For some loci, for example MS32 (Jeffreys *et al.*, 1991), both alleles can be analyzed simultaneously ('diploid coding') and still yield much information about structure. Small alleles at any locus can be fully coded using only one flanking primer, some longer alleles can be analyzed in their entirety with both 'forward' and 'reverse' mapping, while the largest alleles resist analysis except at their ends.

One special case in which MVR coding has been useful is in the study of variation within telomere repeats at the ends of chromosomes. Towards their ends, telomere repeat arrays are pure stretches of repeated TTAGGG, but more proximally they include variant repeats whose positions can be mapped from proximal flanking DNA in a variant of MVR-PCR called TVR-PCR (Baird *et al.*, 1995). Since telomeres share many repeated sequences proximal to the telomere repeat array itself, design of assays for specific chromosomes is difficult, and has been so far possible only for chromosomes Xp/Yp (Baird *et al.*, 1995) and 12q (Baird *et al.*, 2000).

4.8 Finding and typing *Alu*/LINE polymorphisms

Early in the history of studies of *Alu* element variation, polymorphic examples of these retroelements were encountered by chance during studies of various genes, and their number was small (in one early study, just four; Batzer *et al.*, 1994). Availability of the genome sequence, together with an improved understanding of the evolutionary relationships of different families of *Alu* elements, as defined

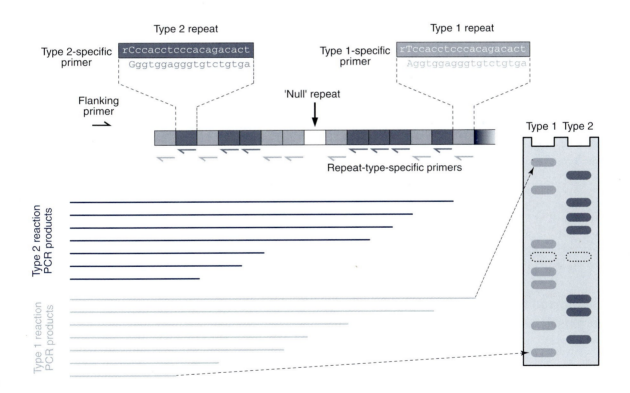

Figure 4.13: Analyzing minisatellite allele structures by MVR-PCR.

MVR-PCR of a single allele at a diploid locus (MS31A; Neil and Jeffreys, 1993) is shown here; alleles can be physically separated on the basis of size, or an allele-specific flanking primer targeted to an adjacent SNP can be used to amplify one allele only. Here, two repeat types are shown, differing by a C/T transition. A second position of variation is adjacent to this, but not assayed here (A/G, shown as 'r'). In two separate reactions, PCR products extending from a fixed primer in the flanking DNA to repeat-type-specific primers are generated. These are separated on a gel, and the MVR code read from the two complementary discontinuous ladders of bands. 'Null' repeats contain additional sites of variation and do not amplify. Gel systems for MVR detection vary from minisatellite to minisatellite. In practice, repeat-specific primers have 5′ extensions ('tags') which are omitted here (Jeffreys *et al.*, 1991).

by sequence variants, has now allowed more systematic attempts to isolate *Alu* polymorphisms, although these are still biased by the small number of individuals who have been sequenced at any locus. Members of the subfamilies related to the *Alu* Y subfamily are 'young', and more prone to polymorphism than other subfamilies, and can be targeted by virtue of their specific sequences during human genome sequence database searches (Roy-Engel *et al.*, 2001). The approach to isolating polymorphic elements is outlined in *Figure 4.14*. Once isolated, typing is straightforward, using PCR primers flanking the *Alu*, which yield a fragment ~300 bp larger from an '*Alu+*' allele than from an '*Alu−*' allele.

Polymorphisms of L1 elements have been less well studied than those of *Alu* elements; they may have a lower rate of transposition (see *Table 3.6*), and the large size of full-length elements, together with their frequent truncation on transposition, makes them a less tractable and uniform target for systematic surveys. As with *Alu* elements, a 'young' subfamily [known as Ta (transcribed, subset a)] has been identified that is prone to tranposition and hence polymorphism – all but one of the 14 known disease-causing L1 transpositions belong to this class (reviewed by Ostertag and Kazazian, 2001). Survey of the genome sequence has revealed 468 of these L1-Ta elements (Myers *et al.*, 2002), and these can then be tested for absence in population samples using an L1-Ta-specific primer together with a specific flanking primer (see *Figure 4.14*). Such an assay requires primers to be designed flanking the L1 insertion point to provide a positive test for the 'L1-' allele.

Using sequence databases to identify potential transposable element polymorphisms will miss those loci where the sequenced genomes possess the 'empty' allele. Consequently, this method will not produce a comprehensive picture of insertional polymorphism in the human genome. Other techniques for detecting such polymorphisms have been developed that do not suffer this constraint, but these often entail complex laboratory procedures to compare pairs of unsequenced genomes (Turner *et al.*, 2001; Buzdin *et al.*, 2002).

4.9 Finding and typing satellite and structural polymorphisms

Satellite polymorphisms represent long-range variation which cannot readily be studied using PCR. Some satellites, such as the alpha satellite at centromeres, are well characterized, and length variation has been studied by cleavage with restriction enzymes cutting outside the repeat unit, followed by **pulsed-field gel electrophoresis** (**PFGE**), a method for separating large (up to megabase-scale) molecules, and Southern blotting using a chromosome-specific alphoid repeat probe. Other satellites are not well characterized, and may remain so given their under-representation in genome sequence databases. PFGE can be used to survey chromosomes for long-range length

variation, which can reflect variation at satellites, or other structural polymorphisms (e.g., Jobling, 1994).

Assays of genomic rearrangements have tended to focus on pathogenic chromosomal structures, rather than structural polymorphism. However, provided they are on a large enough scale, structural polymorphisms of chromosomes can be identified by cytogenetic examination of chromosomal banding patterns (Section 7.3). Submicroscopic rearrangements can also be assayed using non-PCR assays (e.g., Giglio *et al.*, 2001) that often rely on PFGE and/or cytogenetic methods with greater resolution.

Once identified by other means, chromosomal rearrangements can be assayed by PCR-based methods, either by the detection of novel junction sequences, or by changes in copy number. Because inversion polymorphisms do not entail copy number changes, PCR-based methods of these rearrangements are restricted to junction sequence analysis. However, this can be difficult because of the repetitive nature of the sequences sponsoring inversions through nonallelic homologous recombination (see Section 3.5).

Copy number changes that result from deletions and duplications can be assayed by a number of quantitative methods (reviewed by Armour *et al.*, 2002), most of which rely upon PCR procedures that have been designed to amplify products in the same ratios as genomic template molecules. However, while it is comparatively easy to detect differences between one and two copies of a sequence, it is far harder to differentiate between, for example, four and five copies. Many of the known structural polymorphisms exist in tandem repeats of five or more repeats.

4.10 From genotype to haplotype

Knowledge of haplotypes is essential for the construction of 'gene trees' – phylogenies of individual loci – which, as we shall see throughout this book, are important in many areas of human evolutionary genetics. Haplotypes are also the starting point for attempts to locate genes involved in disease, and for understanding the pattern of recombination and LD in the genome. Some loci – the Y chromosome, mtDNA, and the X chromosome of a male – are haploid, and typing markers on these molecules immediately yields haplotypes (*Figure 4.15a*; and see Section 3.7). For most of the genome, however, this is not so (see *Figure 3.26*). It is easy enough to determine which alleles are present, but not how they are associated together on a chromosome – a property sometimes called **'gametic phase'**, or simply **'phase'**. The problem is how to obtain haplotypes from genotype data in diploid DNA. Essentially, there are three different approaches:

▸ molecular separation of one allele from the other allowing individual analysis (see *Figure 4.15*);

▸ statistical methods for estimating haplotypes from genotypes (see *Figure 4.16*);

▸ pedigree studies (see *Figure 4.17*).

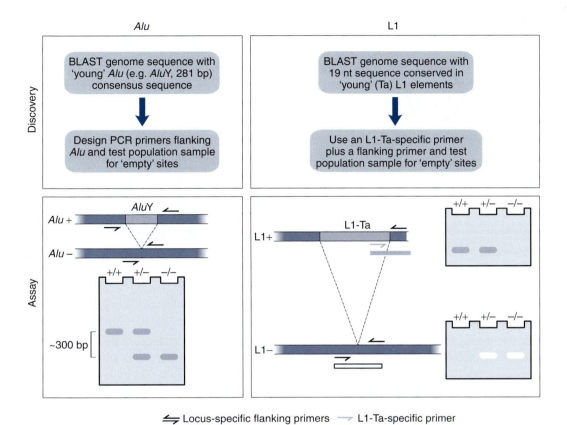

Figure 4.14: Discovery and assay of *Alu* and L1 element polymorphisms.

Methods are described in detail in Roy-Engel *et al.* (2001) and Myers *et al.* (2002). In the L1 assay, flanking PCR to detect 'occupied' sites is not done because expected products are often inconveniently large.

4.10.1 Determining haplotypes by physical separation

Physical separation of one allele from another allows haplotype determination (*Figure 4.15*). There are several ways to do this, which are summarized in *Table 4.1*. In practice, most studies where experimental haplotype determination has been done at individual loci have used allele-specific PCR, since it is technologically simple, generally reliable, and (using 'long PCR' methods) can yield haplotypes covering about 10 kb (Michalatos-Beloin *et al.*, 1996).

Determining the haplotype of one chromosome usually implies the haplotype of the other by subtraction; this can be misleading if there are 'null' or deletion alleles in the population, however, since these cannot be distinguished from homozygosity. Indeed, one of the reasons for inventing the somatic-cell hybrid method (diploid to haploid 'conversion') (*Figure 4.15c*; Yan *et al.*, 2000) was to allow complete analysis of both chromosomes in cases of genetic disorders where the underlying mutation had not been discovered; several mutations were found to be deletion alleles, whose presence is ordinarily masked by the presence of a nondeleted homolog.

4.10.2 Determining haplotypes by statistical methods

Given the practical difficulties of determining haplotypes experimentally, particularly on a large scale in studies of linkage disequilibrium and allele association, statisticians have turned their attention to the possibility of deducing haplotypes from genotypic data. There are three main classes of method:

▶ Clark's algorithm (Clark, 1990) first searches a sample of genotypes for homozygotes and single-site heterozygotes, thus identifying a set of underlying haplotypes unambiguously (*Figure 4.16*). The remaining unresolved genotypes are then examined to ask whether one of the possible underlying haplotypes is one which has already been found. If it has, this haplotype is assumed to be correct, and the other haplotype in the unresolved genotype is deduced by subtraction. In this way the pool of observed and deduced haplotypes increases in size, until, ideally, it contains all haplotypes. If the sample of genotypes is small, it may be that no homozygotes or single-site heterozygotes can be found, in which case the algorithm cannot begin; also, some 'orphan' haplotypes

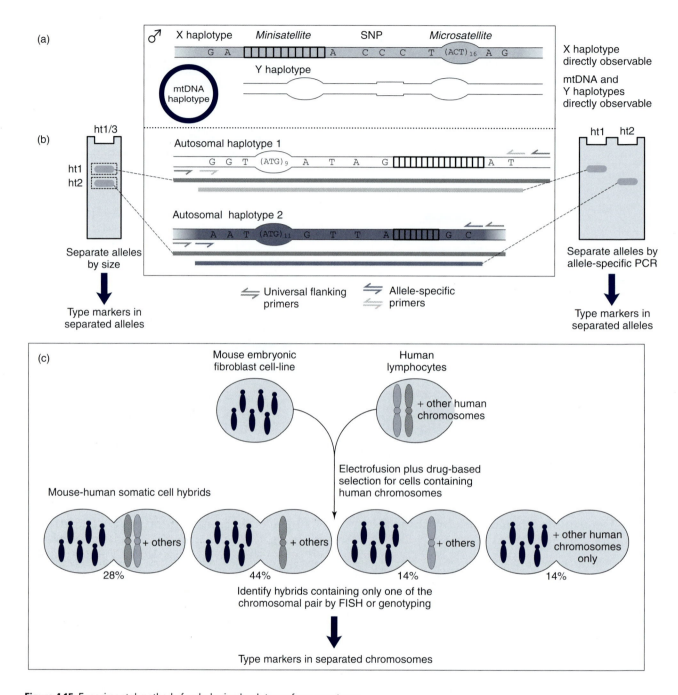

Figure 4.15: Experimental methods for deducing haplotypes from genotypes.

Hypothetical haplotypes composed of SNPs, microsatellites, a minisatellite, and an *Alu* insertion are shown. (a) Haploid molecules (mtDNA, the Y chromosome, and the X chromosome in a man) yield haplotypes directly. (b) Autosomal haplotypes can be deduced by allele-specific amplification of the two alleles followed by typing of markers on each allele individually, or, if alleles differ in size (here, due to a difference in minisatellite array length) they can be physically separated by gel electrophoresis before being typed. Typing of one allele should allow the haplotype of the other to be deduced by subtraction. (c) One chromosome can be isolated from the other by the 'conversion' technique, where somatic cell hybrids containing subsets of human chromosomes are constructed. Identification of hybrids containing only one of the chromosome pair of interest allows a haplotype to be determined (adapted from Yan *et al.*, 2000).

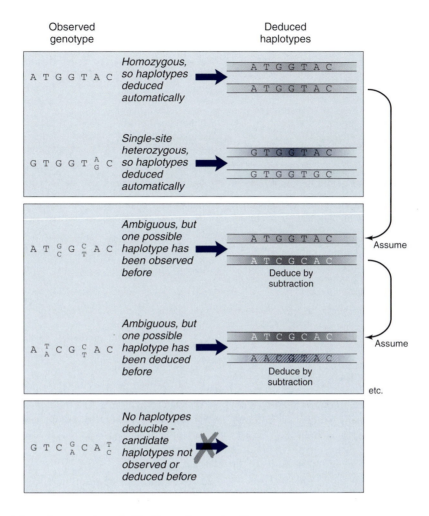

Figure 4.16: Deducing haplotypes from genotypes by Clark's parsimony algorithm.

A hypothetical diploid locus containing seven SNPs is shown. Sequencing reveals homozygous and heterozygous genotypes, and underlying haplotypes are deducible using a parsimony principle; one genotype (bottom) has four possible underlying haplotypes, and a choice cannot be made between them.

may remain undeducible. Nonetheless, this simple algorithm performs remarkably well, for both short segments of DNA and for genomewide patterns of LD. Clark's algorithm is sometimes called a 'parsimony' approach, because it attempts to minimize the number of haplotypes in the sample.

- A maximum-likelihood approach (see *Box 6.2*), implemented through the expectation-maximization (EM) algorithm (Excoffier and Slatkin, 1995) uses an iterative procedure to find the unknown set of population haplotype frequencies that maximizes the likelihood of observing the known genotype frequencies. The method starts with an arbitrary guess at haplotype frequencies, and this guess can influence the final result, so in practice a number of different starting points must be used. While much more computationally intensive than Clark's algorithm, this method has the advantage of estimating haplotype frequencies.

- **Bayesian** phase reconstruction methods (Stephens *et al.*, 2001; Niu *et al.*, 2002) treat haplotypes as unobserved random quantities, and attempt to estimate their distributions given the known genotypes. To do this, a statistical method (the Markov chain-**Monte Carlo method**) makes an initial guess at the haplotype pairs in the sample, then chooses an individual at random and estimates their haplotypes on the assumption that all other haplotypes are correctly reconstructed. Sufficient repetition of this process gives a distribution of the probability of the reconstructed haplotypes given the genotypes. This method has been shown to be superior to the others, and to be able to deal well with missing data on some markers.

TABLE 4.1 METHODS FOR DETERMINING HAPLOTYPE BY ISOLATION OF ALLELES.

Method	Advantages	Disadvantages
Clone genomic DNA or PCR products and type individual cloned copies of alleles	Simple technology	Time-consuming, low throughput
Dilute genomic DNA so that, on average, there is only one molecule per reaction (Ruano *et al.*, 1990)	No manipulations or allele-specific primer design required	Technically demanding single-molecule PCR; sensitive to contamination
If there is a size difference between alleles, separate PCR products on the basis of size (*Figure 4.15b*)	No allele-specific primer design required	Special cases only
Use heterozygous SNP sites as targets for allele-specific PCR primers which allow selective amplification of alleles (*Figure 4.15b*) (Michalatos-Beloin *et al.*, 1996)	Reliable PCR-based method with reasonable throughput	Size of haplotypes limited by PCR range – about 10 kb
Make mouse–human somatic cell hybrids, containing subsets of human chromosomes; identify cells containing only one copy of the chromosome of interest (*Figure 4.15c*) (Yan *et al.*, 2000; Douglas *et al.*, 2001)	Makes whole-chromosome haplotypes available	Time-consuming, low throughput, expensive

4.10.3 Determining haplotypes by pedigree analysis

The traditional method for determining haplotypes is by examination of pedigrees – the principle is illustrated in *Figure 4.17*. While some large pedigrees are assembled for studies of monogenic disease, many studies of complex disorders assemble small pedigrees, such as father–mother–child trios. Determination of haplotypes in children (**'phase'**) in such cases depends on the informativeness of the parental haplotypes, and can often fail (*Figure 4.17b*): for example, haplotypes containing only two polymorphic loci can be ambiguous in up to 11% of cases, depending on allele frequencies (Hodge *et al.*, 1999).

4.11 Studying genetic variation in ancient samples

The length of this book is testimony to the difficulty of making inferences about the past from the genetic diversity of modern human populations. Imagine how much simpler our task would be if we could go back in time and sample the DNA of humans and pre-humans directly: this is the promise of ancient DNA (aDNA) studies. Provided human remains could be found from the appropriate population and era, and DNA isolated from them (*Figure 4.18*), the sensitivity of amplification by PCR could allow us to address directly many of the questions with which this book is concerned, for example:

▶ determination of phylogenetic relationships of ancient hominids from DNA sequence data;

▶ analysis of the ancient sequences of candidate genes involved in important human-specific traits such as language;

▶ direct understanding of mutational dynamics by sampling loci at different time-points;

▶ analysis of the diversities of ancient populations, and thus the reconstruction of their histories;

▶ diagnosis of genetic disorders and measurements of allele frequencies in ancient populations;

▶ determination of past frequencies for alleles involved in phenotypes such as pigmentation, dietary adaptations linked to agriculture, and responses to particular pathogens;

▶ determination of sex of remains, and deduction of kin relationships in group burials.

Further issues could be investigated using DNA not from ancient humans themselves, but from their pathogens, their domesticated animals and plants, and their feces (fossilized as **coprolites**), which would give information about diet.

Unfortunately, there are formidable practical difficulties with ancient DNA analysis in general (reviewed by Hofreiter *et al.*, 2001), and analysis of human samples in particular, which mean that, despite a few notable and spectacular successes, its application has not been very widespread or fruitful:

▶ DNA degrades postmortem, and can rapidly become unamplifiable;

▶ ancient samples can easily become contaminated with ubiquitous modern DNA, which makes proof of the authenticity of 'ancient' sequences difficult or impossible.

4.11.1 The fate of DNA after death

After death, DNA is usually rapidly degraded by the action of **endonucleases**. Particular conditions can allow escape from these enzymes, but over time other processes act to damage DNA. In the cells of a living person, as we saw in Section 3.2.3, DNA is continually being protected from damage by sophisticated monitoring and repair systems. After

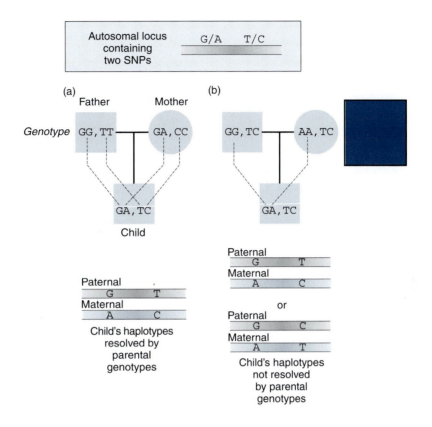

Figure 4.17: Deducing haplotypes from genotypes by pedigree analysis.

A hypothetical locus containing two SNPs (G/A and T/C) is shown, and it is assumed that no recombination can occur between the two SNPs – this is likely if they are physically close. (a) The traditional method of determining haplotype ('gametic phase') by pedigree analysis is successful in this father–mother–child trio because the parental genotypes are informative, allowing deduction of the child's haplotypes. (b) Uninformative parental genotypes do not allow the child's haplotypes to be deduced. Dotted lines show allele inheritance where this is deducible.

death, these systems cease functioning, and the physico-chemical assault can go on unimpeded. The result of this is that the DNA recoverable from bones or other tissues of long-dead humans is severely damaged by cleavage of the sugar-phosphate backbone, resulting in short DNA fragments; loss of bases (abasic sites); chemical modification of bases; and inter- or intramolecular cross-linking of sugar-phosphate backbones. Theoretical considerations based on the chemistry and physics of DNA indicate that it cannot survive any longer than one million years, and probably not longer than 100 000 years.

Cleavage leads to a very short average length of isolated DNA – only a few hundred base-pairs. Lost bases or base modification causes misincorporation by the polymerase during PCR, while cross-linked sites, or oxidized derivatives of cytosine or thymine called **hydantoins** will block the polymerase and truncate synthesis. The longest fragments that can be amplified are only 100–200 bp, and yield of DNA is extremely low. Because of this, mtDNA is the only part of the genome that can be reliably investigated in aDNA studies: its high copy number per cell means that it has a better chance of surviving than does nuclear DNA. The yield

of amplifiable mtDNA molecules in 'good' ancient specimens is about 2000–3000 molecules per gram of tissue, representing a reduction of about six orders of magnitude compared with fresh tissue (Handt *et al.*, 1994; Krings *et al.*, 1997).

One problem for researchers wishing to study ancient human samples is that fossils are rare and precious: museum-curators and paleontologists will be unwilling to submit their specimens to destructive analysis if they feel that the chances of success are low. DNA survival in an ancient sample is influenced by the conditions under which it has existed since it was deposited: temperature, pH, humidity and salt concentration affect the rates of the modifications that DNA undergoes postmortem. The idea that DNA would be protected from degradation by entombment in amber was behind the 'Jurassic Park' scenario of resurrecting dinosaurs from DNA preserved in the guts of blood-sucking insects. Despite early claims, this particular postmortem milieu has since been shown to provide poor protection for DNA (Austin *et al.*, 1997); however, it is clear that some conditions, in particular desiccation and low temperatures, are better than others. Knowledge of the paleoclimatic

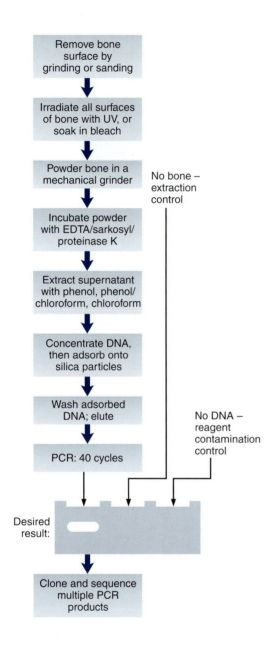

Figure 4.18: Procedure for isolation and analysis of DNA from bone samples.

The flow chart shows typical procedures for preparing a bone sample for extraction, DNA extraction itself, PCR amplification, and control experiments to exclude contamination of reagents. Additional precautions necessary for aDNA work are given in *Box 4.5*.

conditions at a sample site allows an estimation of the sample's 'thermal age' (Smith *et al.*, 2001) – the effective number of years it has spent at 10°C. This analysis suggests that 17 000 years at 10°C may be a practical upper limit for survival, and hence that some Neanderthal sample sites are unlikely to yield amplifiable DNA.

However, since many factors are involved in DNA survival, temperature alone may not be a satisfactory indicator. To address this issue an indirect practical test was introduced (Poinar *et al.*, 1996) which requires very little of a specimen to be analyzed (about 50 times less than would be used for a DNA extraction; Krings *et al.*, 1997). This monitors changes in the three-dimensional structures of amino acids which occur at a condition-dependent rate after death (*Figure 4.19*). In all amino acids except glycine there are four different chemical groups attached to one carbon atom known as the alpha carbon. These groups are arranged in a tetrahedral shape, with the carbon at the center. There are two different ways to arrange the groups, which are chemically identical, but mirror images of each other, called **stereoisomers**. The metabolic processes of amino acid synthesis, and their critical roles in protein structure, constrain natural amino acids in living systems to exist in one stereoisomeric form only, called the L-form. After death, however, transitions to the D-form occur (a process known as **racemization**), and eventually a dynamic equilibrium is reached in which the proportions of L- and D-forms are equal. The rate of racemization depends on the conditions under which a specimen exists, and thus monitoring the **D/L ratio** gives an indication of these conditions, and an indirect estimate of the chance of amplifiable DNA surviving. Empirically, a D/L ratio of about 0.1 for aspartic acid is taken to indicate likely DNA survival. An alternative method that shows promise is flash pyrolysis with gas chromatography and mass spectrometry, which monitors the condition of proteins in a sample via the extent of hydrolysis of peptide bonds between amino acids (Poinar and Stankiewicz, 1999).

4.11.2 The problem of contamination

The most serious problem that bedevils aDNA work is contamination. As we have seen, ancient DNA is present in very small amounts, is seriously damaged and is difficult to amplify; in contrast, modern DNA is plentiful, ubiquitous, in relatively good condition, and very easy to amplify. Efforts to surmount the problem of contamination must be rigorous (*Figure 4.18*, and *Box 4.5*). The problem applies to all aDNA studies, but most particularly to studies of hominid samples. Imagine that you are working on samples from a woolly mammoth: you might inadvertently contribute your own DNA to the sample between the time that it is collected and the time that PCR is carried out, but choice of appropriate PCR primers should mean that your DNA will not be amplified, while the mammoth's DNA will. Even if your DNA *were* amplified and sequenced, using 'universal' mtDNA primers, it would be easily distinguishable from that of the ancient sample because of differences in sequence. In other words, there is a phylogenetic signal in the mammoth DNA which distinguishes it from likely modern contaminating DNAs, and thus provides support for its authenticity. Now consider an attempt to amplify DNA from a 5000-year-old anatomically modern human skeleton. The sequence you obtain might be similar to, or even identical to, many modern human sequences – this was the case with 'Oetzi', the Tyrolean Ice Man (Handt *et al.*, 1994; *Box 4.6*).

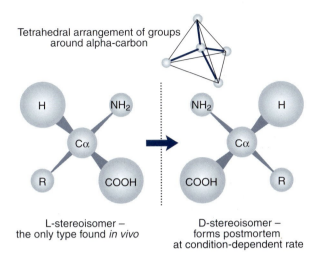

Tetrahedral arrangement of groups around alpha-carbon

L-stereoisomer – the only type found *in vivo*

D-stereoisomer – forms postmortem at condition-dependent rate

Figure 4.19: Amino acid racemization.

After death the L-stereoisomer converts into its mirror-image, the D-stereoisomer, until eventually a dynamic equilibrium is reached at a 1 : 1 ratio. The rate of conversion is dependent upon conditions which also favor DNA damage, and so the D : L ratio provides an indirect measure of the chance of DNA survival in a specimen. 'R' indicates the amino acid side chain, which varies between amino acids.

There is no phylogenetic signal here, and proof of authenticity becomes much more difficult. In the landmark study of the Neanderthal **type specimen** (Krings *et al.*, 1997; see Section 8.7) many appropriate controls were done, but the finding that the Neanderthal sequence lay outside the spectrum of variation of modern human mtDNA sequences was nonetheless of great importance in proving authenticity. The existence of human **numts** (nuclear mitochondrial DNA insertions; see *Box 2.8*) makes life even more difficult, since they are diverged from human mtDNA itself and might easily be confused with ancient, divergent mtDNA sequences (see *Box 4.7*).

Cloning of PCR products provides a defense both against contamination and misincorporation during PCR on damaged templates. Sequencing of multiple overlapping clones allows the reconstruction of the original sequence, and the identification of bases introduced by polymerase errors, as well as the identification of foreign sequences introduced by contamination (*Figure 4.20*).

Application of the rules for authenticity in aDNA work will result in a slow but steady accumulation of data on ancient hominids. Technological advances may improve the recoverability of DNA from ancient samples: for example, the exploitation of knowledge about cellular DNA repair systems may lead to the development of methods for the *in vitro* repair of some of the damage caused to DNA over time. This work has already begun, with the demonstration that enzymatic repair of aDNA templates prior to PCR increases the success of amplification of nuclear DNA sequences about fivefold (Di Bernardo *et al.*, 2002).

Summary

▶ Searches for human genetic variation have been revolutionized by two developments – the invention of the polymerase chain reaction (PCR), and the publication of the draft genome sequence.

▶ Blood, buccal cells and immortalized cell lines are the most useful sources of DNA in population studies.

▶ Resequencing the locus of interest in all individuals is the best way to study its diversity, but is often too expensive.

▶ There are many SNPs in databases, and this should be the first place to look for variation in a specific sequence. SNP discovery has been largely driven by mutation detection in genetic disease studies, and by the desire to build a high density SNP map for seeking genes underlying complex disorders via association studies.

▶ SNP typing systems are very varied in complexity, cost, throughput and accuracy, but all rest on six fundamental biochemical principles for distinguishing one allele from another: oligonucleotide hybridization, allele-specific amplification, primer extension, oligo-nucleotide ligation, restriction enzyme cleavage, and Flap endonuclease cleavage.

▶ Microsatellites are easily discovered using bioinformatic methods, and can be typed in multiplex PCRs using flanking primers and size separation by gel or capillary electrophoresis.

▶ Minisatellites can be identified *in silico* in the same way as microsatellites; length variation can be assessed using flanking PCR, and internal repeat sequence variation measured using minisatellite-variant-repeat PCR.

▶ Polymorphic *Alu* and LINE repeats belong to 'young', recently transposing subfamilies, and knowledge of the sequence of these subfamilies allows potentially polymorphic elements to be targeted in sequence searches. Flanking PCR provides a simple assay for *Alu* insertion polymorphisms.

▶ Since most of human DNA is diploid, methods are needed to deduce haplotypes from genotypes. These include molecular separation of one allele from another (for example by allele-specific PCR), statistical methods, and pedigree analysis.

▶ Ancient DNA studies offer the prospect of studying past populations, their kinship relationships and diets directly. The work is hampered by the degradation of DNA postmortem, and by ubiquitous modern DNA contamination. Stringent criteria for authenticity must be applied to all studies, but especially those on hominids, where contamination is a particularly difficult problem.

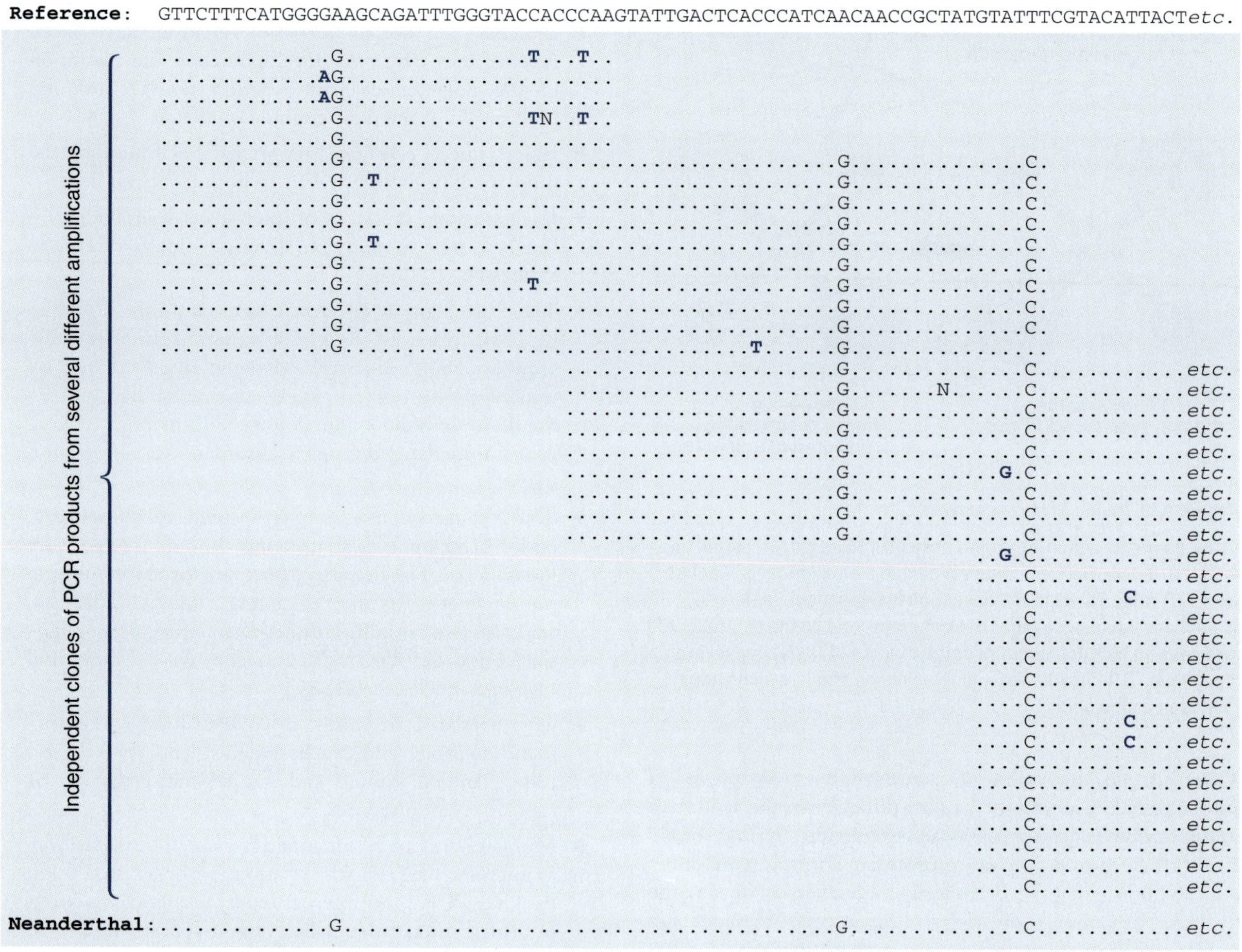

Reference: GTTCTTTCATGGGGAAGCAGATTTGGGTACCACCCAAGTATTGACTCACCCATCAACAACCGCTATGTATTTCGTACATTAC*etc.*

Independent clones of PCR products from several different amplifications

```
..............G...................T..T..
..............AG.....................
..............AG.....................
..............G...................TN..T..
..............G.....................
..............G.......................................G.............C.
..............G..T....................................G.............C.
..............G.......................................G.............C.
..............G.......................................G.............C.
..............G..T....................................G.............C.
..............G.......................................G.............C.
..............G.............................T.........G.............C.
..............G.......................................G.............C.
..............G.......................................G.............C.
..............G.................................T.....G.............C.
                        ................................G.............C............etc.
                        ................................G.....N......C............etc.
                        ................................G.............C............etc.
                        ................................G.............C............etc.
                        ................................G.............C............etc.
                        ................................G...........G.C............etc.
                        ................................G.............C............etc.
                        ................................G.............C............etc.
                        ................................G.............C............etc.
                                                        ..G.C............etc.
                                                        ....C............etc.
                                                        ....C.....C....etc.
                                                        ....C............etc.
                                                        ....C............etc.
                                                        ....C............etc.
                                                        ....C............etc.
                                                        ....C.....C....etc.
                                                        ....C.....C....etc.
                                                        ....C............etc.
                                                        ....C............etc.
                                                        ....C............etc.
                                                        ....C............etc.
                                                        ....C............etc.
                                                        ....C............etc.
```

Neanderthal: ..............G.............................G.............C............*etc.*

Figure 4.20: Inferring an ancient DNA sequence from sequences of overlapping clones.

Bases 16 023–16 106 of hypervariable segment I of mtDNA from the modern human reference sequence (top) and the Neanderthal sequence (bottom) are shown, with a dot indicating identity to the reference. In between these are the sequences of overlapping clones used to infer the Neanderthal sequence. Bases shared between all, or almost all, clones are taken to be genuine differences between the modern and Neanderthal sequence; other bases different from the reference are probably sequence misincorporations during PCR. Note that most of these are present in more than one clone. Adapted from Krings *et al.* (1997), where 123 clones from 13 different PCR amplifications were used to infer a total of 379 bp of HVS I sequence.

BOX 4.5 Opinion: Criteria for authenticity in ancient DNA work

The possibility of tracing ancient lineages back over time, hunting down the sources of ancient diseases and attempting to resolve issues such as whether or not Neanderthals and modern humans interbred is why the hunt for ancient DNA is such an exciting activity.

Unfortunately for the evolutionary biologist, DNA is a relatively weak molecule that degrades rapidly in an environment- and time-dependent manner. DNA which can escape endogenous nuclease degradation is still subject to the bacterial and fungal onslaught. As time passes the molecule is prone to chemical attack, most importantly hydrolysis, oxidation, alkylation and cross-linking reactions, all limiting its half-life *in situ* (see *Figure*). Despite these formidable opponents, the DNA molecule does persist in geological settings: short DNA fragments survive for about 10^4 years in temperate climates, and 10^5 years in colder ones. However, as all the factors which limit and prolong the molecule's survival are not known, these times remain best guesses.

Despite published claims to the contrary, the oldest reproduced and credible sequences to date stem from frozen mammoth carcasses ~50 KY old (see *Box 4.7*). To make ancient DNA a respectable and credible field, some researchers have outlined the necessary controls that should be followed by all. Here, I review these criteria as if they were being carried out by a researcher, on their first ancient DNA quest.

I. **Collect and subsection your samples.** Paleontologists should collect samples meticulously, taking care to avoid any additional contamination. In the lab, a piece of your sample is split into two sections, which will be sent to two independent labs. Upon receipt of the sample, each lab should subsection again for future reproduction within their own lab.

II. **Biochemical preservation.** The bulk molecular component of bone and teeth is protein, namely collagen. Proteins and DNA are subject to similar types of chemical reactions, such as hydrolysis and oxidation. By analyzing the extent of diagenetic alteration and the amount of protein remaining in fossil bone, researchers can screen samples for conditions conducive to DNA preservation. Samples devoid of their endogenous protein are not sources of ancient DNA. This step allows the researcher to focus their efforts on samples which appear chemically more promising.

III. **Extraction and amplification controls.** All ancient DNA work should be carried out in a facility dedicated to the recovery of DNA from low copy number samples (fossils and

Sites vulnerable to:

→ Hydrolytic attack
→ Oxidative damage
→ Alkylation damage
□ Condensation reaction

The DNA molecule and its weakest links.

subfossils). All work should include multiple blank extractions and extensive PCR controls to monitor the level of contamination in all buffers and in the lab. All levels of contamination should be reported in the literature.

IV. **Molecular behavior.** The DNA in fossils is degraded into short fragments (~150 bp). Thus one way to provide indirect evidence for authenticity of an ancient extract is to show, via the strength of amplification products, the presence of more short, than medium (300-bp) or long (500-bp) DNA templates.

V. **Quantify.** Knowing the number of DNA templates initiating your PCR provides valuable information on the state of DNA preservation and whether or not to proceed with your extract. It has been shown empirically that extracts containing ~1000 template molecules can be reproducibly amplified, whereas extracts with < 300 templates yield novel DNA sequences with each new PCR attempt and are thus irreproducible. Quantification should be a requirement for all ancient human samples.

VI. **Amplify and clone.** Amplify small (< 150-bp) overlapping fragments of your target. This ensures to the greatest extent possible that preference is given to your degraded DNA, over generally lower copy number, longer contaminating DNA molecules. It also allows the detection of nuclear mitochondrial pseudogenes (**numts**). Amplifications should be performed twice, cloned and multiple clones from each amplicon sequenced. This allows different sequences present in the extract to be identified, and the reconstruction of a consensus sequence despite PCR- and damage-induced errors in the DNA.

VII. **Intra- and interlab reproducibility.** Reproduce your own findings first from your subsectioned sample (see Step I). It then becomes the daunting task of your collaborating lab to reproduce a small portion of your work.

VIII. **Faunal analysis.** In cases of hominid remains, where elucidating the endogenous sequence can be difficult, it is wise to have faunal samples from the same site to provide indirect evidence for DNA preservation. The DNA of these remains can be targeted with specific DNA primers to avoid human contamination. However, it is wise, at least once, to amplify human DNA from your faunal extracts to learn the ubiquitous nature of human contamination.

IX. **Phylogenetic confirmation.** Analyze your DNA sequences phylogenetically. The presence of novel DNA sequences, specifically human mitochondrial DNA sequences, that do not occur in the database is not a criterion for their authenticity!

Hendrik N. Poinar, Max Planck Institute for Evolutionary Anthropology, Leipzig, Germany

---

BOX 4.6 Oetzi, the Tyrolean Ice Man

The Tyrolean Ice Man, popularly known as Oetzi, was discovered in September 1991. He had emerged in a naturally mummified state from the ice of the Schnalstal glacier in the Oetzal Alps, over 5000 years after his death. Investigations of Oetzi, together with his beautifully preserved clothing, weapons and equipment, offered a rare opportunity to learn about the life of a late Neolithic human. The original view was that Oetzi had been caught in bad weather and frozen to death; yet despite intensive anatomical and pathological studies of the body, it took until 2001 to reveal a flint arrowhead in his back.

Ancient DNA studies (Handt *et al.*, 1994) were undertaken with the hope of investigating Oetzi's affiliations with modern European populations; these studies were extremely difficult, since the DNA was badly degraded (mostly < 150 bp) and there was demonstrable contamination with modern DNA. Short lengths of endogenous mtDNA could be amplified, and it was shown by painstaking analysis and replication of results in two different laboratories that a fragment of HVS I was closely related to the mtDNA reference sequence (see *Box 8.5*). In the 394-bp region analyzed, Oetzi's sequence differed only at two positions: T to C at 16 224 and T to C at 16 331. It is therefore typical of sequences found in many modern European populations, including those from the Mediterranean, Alpine and Northern regions where it represented about 2% of the sample – an unsurprising, but hard-won result.

More recent aDNA analysis has been carried out on the contents of Oetzi's colon and ileum (Rollo *et al.*, 2002). PCR analysis was done using primers directed at species-informative segments of mammalian mtDNA, plant chloroplast DNA, and multi-copy plant/fungal nuclear genes. Phylogenetic analysis of the sequences, together with sequences of known species, suggested that Oetzi's last meals had included the meat of ibex and red deer, and possibly cereals.

BOX 4.7 Opinion: Ancient human DNA – a lesson in doubt

Over the past decade, the field of ancient DNA research has suffered a number of high-profile failures, such as claims of DNA from dinosaur bones, 100 million-year-old amber, and 400 million-year-old salt crystals. This record has resulted from the mismatch between the enormous amplifying power of PCR, the ubiquitous nature of contaminating DNA, and the vanishingly small traces of DNA fragments left behind in old specimens. A further complication has been the lax standards applied by reviewers, journal editors, and the researchers themselves. As a result of the increasingly ancient spiral of speculative DNA claims (see *Figure*), a series of criteria have been invoked as a minimum standard for ancient DNA research (Cooper and Poinar, 2000). While even the basic requirement of independent replication is sometimes still ignored, such an omission is now an obvious sign of unreliable results.

The risk of contamination is best comprehended by considering the amounts of DNA generated in buildings performing molecular biological research. Each of the hundreds of PCR reactions performed daily will generate well over 10^{12} molecules of the targeted sequence in a volume of only 10–100 µl. Aerosol droplets generated when this liquid is transferred during laboratory analysis will routinely contain upwards of a million copies of the template, which are

technically small enough to pass through even ultra-HEPA filtered air supplies. Consequently, normal activities will quickly distribute the PCR product throughout the building, and personnel movement will pass it on to other facilities via clothing, shoes, hair, skin etc. This extensive contamination is rarely detected because modern genetic research is performed at DNA concentrations orders of magnitude above the background level. However, ancient samples may have only a few thousand template copies in 0.5–1 g of bone (the normal sample size), i.e., less than a thousandth of the DNA in a single dried aerosol droplet. As a consequence, ancient DNA work must be physically isolated from molecular biological research areas (or buildings), using strict procedures such as positive air-pressure, and control of personnel movement, to prevent it from being swamped in a sea of contamination.

Human specimens are one of the most difficult areas of ancient DNA research, because the contamination risk includes the above problems, plus the researchers themselves and all those who have come into contact with the specimen before analysis. While it is possible to genetically profile the current researchers, it is generally impossible to characterize the excavating team, archaeologists, museum curators and researchers that have

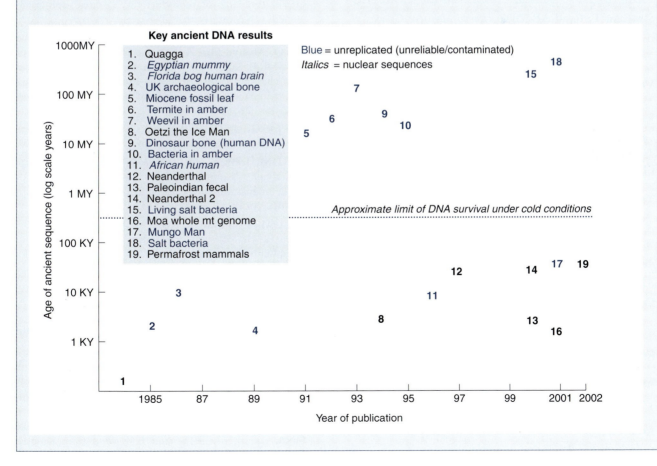

handled a specimen since recovery. Experiments show that modern DNA deposited on specimens during handling is rapidly adsorbed deep inside archaeological bone or teeth specimens, and cannot be removed by surface decontamination procedures. This problem is worse for archaeological specimens which are often washed after recovery, allowing DNA from the (generally re-used) water to permeate throughout the specimen. For example, during work on a well-preserved Viking specimen over 20 different individual sequences were detected in the DNA extracted from a single tooth. Alarmingly, this contamination can easily remain undetected because most ancient DNA research involves mtDNA (due to the high copy number of this organelle in cells), or nuclear loci such as amelogenin (for sexing; see Section 15.2.1), where there may be little, or no, phylogenetic signal to produce an incongruous result. In some cases, for example European Cro-Magnon specimens, it is difficult to imagine how a retrieved sequence could be authenticated.

Another major issue is the modified bases produced by oxidative and hydrolytic damage in ancient DNA strands. The distribution of these damaged sites was long presumed to be relatively random, with little impact on PCR reactions starting from a number of templates. However, recent research (Gilbert *et al.*, 2003) has shown that ancient DNA damage is concentrated in 'hotspots', so that direct sequencing results can easily be misleading. Furthermore, the hotspot positions are also often highly mutable rapidly *in vivo*, and therefore damage causes substitutions to occur primarily at positions likely to be phylogenetic markers in modern human populations. The full extent of the difficulties associated with template damage (and contamination) are generally still not fully appreciated. As a consequence, much of the published research relies on direct sequencing without independent replication, and must be considered suspect. Unfortunately, this includes a large portion of the published ancient human DNA sequences.

Alan Cooper, Department of Zoology, University of Oxford, UK

Further reading

Strachan T, Read AP (2003) *Human Molecular Genetics 3.* Garland Publishing Inc., New York.

Various authors (1999) The Chipping Forecast. *Nature Genet.* **21** (suppl.) 1–60.

Various authors (2002) The Chipping Forecast II. *Nature Genet.* **32** (suppl.) 461–552.

Various authors (2002) A user's guide to the human genome. *Nature Genet.* **32** (suppl.) 1–77.

Electronic references

Tandem Repeats Finder:
http://c3.biomath.mssm.edu/trf.html

PolyPhred:
http://droog.mbt.washington.edu/PolyPhred.html

dbSNP: http://www.ncbi.nlm.nih.gov/SNP/

ALFRED: http://alfred.med.yale.edu/alfred/index.asp

Census of nuclear mitochondrial pseudogenes (Numts): http://www.pseudogene.net/

References

Armour JA, Barton DE, Cockburn DJ, Taylor GR (2002) The detection of large deletions or duplications in genomic DNA. *Hum. Mutat.* **20**, 325–337.

Arnason E (2003) Genetic heterogeneity of Icelanders. *Ann. Hum. Genet.* **67**, 5–16.

Austin JJ, Ross AJ, Smith AB, Fortey RA, Thomas RH (1997) Problems of reproducibility – does geologically ancient DNA survive in amber-preserved insects? *Proc. R. Soc. Lond. Ser. B-Biol. Sci.* **264**, 467–474.

Ayub Q, Mohyuddin A, Qamar R, Mazhar K, Zerjal T, Mehdi SQ, Tyler-Smith C (2000) Identification and characterisation of novel human Y-chromosomal microsatellites from sequence database information. *Nucl. Acids Res.* **28**, 8e.

Bailey JA, Gu Z, Clark RA *et al.* (2002) Recent segmental duplications in the human genome. *Science* **297**, 1003–1007.

Baird DM, Jeffreys AJ, Royle NJ (1995) Mechanisms underlying telomere repeat turnover, revealed by hypervariable variant repeat distribution patterns in the human Xp/Yp telomere. *EMBO J.* **14**, 5433–5443.

Baird DM, Coleman J, Rosser ZH, Royle NJ (2000) High levels of sequence polymorphism and linkage disequilibrium at the telomere of 12q: Implications for telomere biology and human evolution. *Am. J. Hum. Genet.* **66**, 235–250.

Banchs I, Bosch A, Guimera J, Lazaro C, Puig A, Estivill X (1994) New alleles at microsatellite loci in CEPH families mainly arise from somatic mutations in the lymphoblastoid cell-lines. *Hum. Mut.* **3**, 365–372.

Bandelt H-J, Quintana-Murci L, Salas A, Macaulay V (2002) The fingerprint of phantom mutations in mitochondrial DNA data. *Am. J. Hum. Genet.* **71**, 1150–1160.

Barnes WM (1994) PCR amplification of up to 35-kb DNA with high-fidelity and high-yield from lambda-bacteriophage templates. *Proc. Natl Acad. Sci. USA* **91**, 2216–2220.

Batzer MA, Stoneking M, Alegria-Hartman M *et al.* (1994) African origin of human-specific polymorphic Alu insertions. *Proc. Natl Acad. Sci. USA* **91**, 12288–12292.

Benson G (1999) Tandem repeats finder: a program to analyze DNA sequences. *Nucl. Acids Res.* **27**, 573–580.

Buetow KH, Edmonson M, MacDonald R *et al.* (2001) High-throughput development and characterization of a genomewide collection of gene-based single nucleotide polymorphism markers by chip-based matrix-assisted laser desorption/ionization time-of-flight mass spectrometry. *Proc. Natl Acad. Sci. USA* **98**, 581–584.

Buzdin A, Khodosevich K, Mamedov I, Vinogradova T, Lebedev Y, Hunsmann G, Sverdlov E (2002) A technique for genome-wide identification of differences in the interspersed repeats integrations between closely related genomes and its application to detection of human-specific integrations of HERV-K LTRs. *Genomics* **79**, 413–422.

Cann HM, de Toma C, Cazes L *et al.* (2002) A human genome diversity cell line panel. *Science* **296**, 261–262.

Cann RL, Stoneking M, Wilson AC (1987) Mitochondrial DNA and human evolution. *Nature* **325**, 31–36.

Chee M, Yang R, Hubbell E *et al.* (1996) Accessing genetic information with high-density DNA arrays. *Science* **274**, 610–614.

Chen XN, Levine L, Kwok PY (1999) Fluorescence polarization in homogeneous nucleic acid analysis. *Genome Res.* **9**, 492–498.

Clark AG (1990) Inference of haplotypes from PCR-amplified samples of diploid populations. *Mol. Biol. Evol.* **7**, 111–122.

Collins FS, Brooks LD, Chakravarti A (1998) A DNA polymorphism discovery resource for research on human genetic variation. *Genome Res.* 8, 1229–1231.

Cooper A, Poinar HN (2000) Ancient DNA: Do it right or not at ALL. *Science* **289**, 1139–1139.

Dean FB, Hosono S, Fang LH *et al.* (2002) Comprehensive human genome amplification using multiple displacement amplification. *Proc. Natl Acad. Sci. USA* **99**, 5261–5266.

Di Bernardo G, Del Gaudio S, Cammarota M, Galderisi U, Cascino A, Cipollaro M (2002) Enzymatic repair of selected cross-linked homoduplex molecules enhances nuclear gene rescue from Pompeii and Herculaneum remains. *Nucl. Acids Res.* **30**, e16.

Douglas JA, Boehnke M, Gillanders E, Trent JA, Gruber SB (2001) Experimentally-derived haplotypes substantially increase the efficiency of linkage disequilibrium studies. *Nature Genet.* **28**, 361–364.

Excoffier L, Slatkin M (1995) Maximum-likelihood-estimation of molecular haplotype frequencies in a diploid population. *Mol. Biol. Evol.* **12**, 921–927.

Fakhrai-Rad H, Pourmand N, Ronaghi M (2002) Pyrosequencing: an accurate detection platform for single nucleotide polymorphisms. *Hum. Mutat.* **19**, 479–485.

Forster P, Kayser M, Meyer E, Roewer L, Pfeiffer H, Benkmann H, Brinkmann B (1998) Phylogenetic resolution of complex mutational features at Y-STR DYS390 in Aboriginal Australians and Papuans. *Mol. Biol. Evol.* **15**, 1108–1114.

Forster P (2003) To err is human. *Ann. Hum. Genet.* **67**, 2–4.

Giglio S, Broman KW, Matsumoto N *et al.* (2001) Olfactory receptor-gene clusters, genomic-inversion polymorphisms, and common chromosome rearrangements. *Am. J. Hum. Genet.* **68**, 874–883.

Gilbert MTP, Willerslev E, Hansen AJ, Barnes I, Rudbeck L, Lynnerup N, Cooper A (2003) Distribution patterns of postmortem damage in human mitochondrial DNA. *Am. J. Hum. Genet.* **72**, 32–47.

Gulcher JR, Helgason A, Stefansson K (2000) Genetic homogeneity of Icelanders. *Nature Genet.* **26**, 395.

Gut IG (2001) Automation in genotyping of single nucleotide polymorphisms. *Hum. Mutat.* **17**, 475–492.

Hacia JG, Fan J-B, Ryder O *et al.* (1999) Determination of ancestral alleles for human single-nucleotide polymorphisms using high-density oligonucleotide arrays. *Nature Genet.* **22**, 164–167.

Haff LA, Smirnov IP (1997) Single-nucleotide polymorphism identification assays using a thermostable DNA polymerase and delayed extraction MALDI-TOF mass spectrometry. *Genome Res.* **7**, 378–388.

Hagelberg E, Goldman N, Lio P, Whelan S, Schiefenhovel W, Clegg JB, Bowden DK (1999) Evidence for mitochondrial DNA recombination in a human population of island Melanesia. *Proc. R. Soc. Lond. Ser. B-Biol. Sci.* **266**, 485–492.

Hagelberg E, Goldman N, Lio P, Whelan S, Schiefenhovel W, Clegg JB, Bowden DK (2000) Evidence for mitochondrial DNA recombination in a human population of island Melanesia: correction. *Proc. R. Soc. Lond. Ser. B-Biol. Sci.* **267**, 1595–1596.

Handt O, Richards M, Trommsdorf M *et al.* (1994) Molecular genetic analyses of the Tyrolean Ice Man. *Science* **264**, 1775–1778.

Helgason A, Sigurdardóttir S, Nicholson J *et al.* (2000) Estimating Scandinavian and Gaelic ancestry in the male settlers of Iceland. *Am. J. Hum. Genet.* **67**, 697–717.

Hodge SE, Boehnke M, Spence MA (1999) Loss of information due to ambiguous haplotyping of SNPs. *Nature Genet.* **21**, 360–361.

Hofreiter M, Serre D, Poinar HN, Kuch M, Pääbo S (2001) Ancient DNA. *Nature Rev. Genet.* **2**, 353–359.

International SNP Map Working Group (2001) A map of human genome sequence variation containing 1.42 million single nucleotide polymorphisms. *Nature* **409**, 928–933.

Jackson PE, Scholl PF, Groopman JD (2000) Mass spectrometry for genotyping: an emerging tool for molecular medicine. *Mol. Med. Today* **6**, 271–276.

Jeffreys AJ, Wilson V, Thein SL (1985) Hypervariable 'minisatellite' regions in human DNA. *Nature* **314**, 67–73.

Jeffreys AJ, MacLeod A, Tamaki K, Neil DL, Monckton DG (1991) Minisatellite repeat coding as a digital approach to DNA typing. *Nature* **354**, 204–209.

Jeffreys AJ, Kauppi L, Neumann R (2001) Intensely punctate meiotic recombination in the class II region of the major histocompatibility complex. *Nature Genet.* **29**, 217–222.

Jobling MA (1994) A survey of long-range DNA polymorphisms on the human Y chromosome. *Hum. Mol. Genet.* **3**, 107–114.

Kaessmann H, Heissig F, von Haeseler A, Pääbo S (1999) DNA sequence variation in a non-coding region of low recombination on the human X chromosome. *Nature Genet.* **22**, 78–81.

Krings M, Stone A, Schmitz RW, Krainitzki H, Stoneking M, Pääbo S (1997) Neandertal DNA sequences and the origin of modern humans. *Cell* **90**, 19–30.

Kwok PY (2002) SNP genotyping with fluorescence polarization detection. *Hum. Mutat.* **19**, 315–323.

Lander ES, Linton LM, Birren B et al. (2001) Initial sequencing and analysis of the human genome. *Nature* **409**, 860–921.

Landsteiner K (1900) Zur Kenntnis der antifermentativen, lytisichen und agglutinierenden Wirkungen des Blutserums und der Lymphe. *Zbl. Bakt. I Abt.* **27**, 357–362.

Michalatos-Beloin S, Tishkoff SA, Bentley KL, Kidd KK, Ruano G (1996) Molecular haplotyping of genetic markers 10 kb apart by allele-specific long-range PCR. *Nucl. Acids Res.* **24**, 4841–4843.

Mir KU, Southern EM (2000) Sequence variation in genes and genomic DNA: methods for large-scale analysis. *Ann. Rev. Genomics Hum. Genet.* **1**, 329–360.

Mueller UG, Wolfenbarger LL (1999) AFLP genotyping and fingerprinting. *Trends Ecol. Evol.* **14**, 389–394.

Mullis KB, Faloona FA (1987) Specific synthesis of DNA in vitro via a polymerase-catalyzed chain-reaction. *Methods Enzymol.* **155**, 335–350.

Myers JS, Vincent BJ, Udall H et al. (2002) A comprehensive analysis of recently integrated human Ta L1 elements. *Am. J. Hum. Genet.* **71**, 312–326.

Neil DL, Jeffreys AJ (1993) Digital DNA typing at a second hypervariable locus by minisatellite variant repeat mapping. *Hum. Mol. Genet.* **2**, 1129–1135.

Nickerson DA, Tobe VO, Taylor SL (1997) Polyphred: automating the detection and genotyping of single nucleotide substitutions using fluorescence-based resequencing. *Nucl. Acids Res.* **14**, 2745–2751.

Nickerson DA, Taylor SL, Weiss KM et al. (1998) DNA sequence diversity in a 9.7-kb region of the human lipoprotein lipase gene. *Nature Genet.* **19**, 233–240.

Nickerson DA, Taylor SL, Fullerton SM et al. (2000) Sequence diversity and large-scale typing of SNPs in the human apolipoprotein E gene. *Genome Res.* **10**, 1532–1545.

Niu T, Qin ZS, Xu X, Liu JS (2002) Bayesian haplotype inference for multiple linked single-nuceotide polymorphisms. *Am. J. Hum. Genet.* **70**, 157–169.

Osier MV, Cheung K-H, Kidd JR, Pakstis AJ, Miller PL, Kidd KK (2002) An allele frequency database for anthropology. *Am. J. Phys. Anthropol.* **119**, 77–83.

Ostertag EM, Kazazian HH (2001) Biology of mammalian L1 retrotranspositions. *Ann. Rev. Genet.* **35**, 501–538.

Poinar HN, Hoss M, Bada JL, Pääbo S (1996) Amino acid racemization and the preservation of ancient DNA. *Science* **272**, 864–866.

Poinar HN, Stankiewicz BA (1999) Protein preservation and DNA retrieval from ancient tissues. *Proc. Natl Acad. Sci. USA* 96, 8426–8431.

Przeworski M, Hudson RR, Di Rienzo A (2000) Adjusting the focus on human variation. *Trends Genet.* **16**, 296–302.

Reich DE, Gabriel SB, Altshuler D (2003) Quality and completeness of SNP databases. *Nature Genet.* **33**, 457–458.

Rioux JD, Daly MJ, Silverberg MS et al. (2001) Genetic variation in the 5q31 cytokine gene cluster confers susceptibility to Crohn disease. *Nature Genet.* **29**, 223–228.

Rollo F, Ubaldi M, Ermini L, Marota I (2002) Otzi's last meals: DNA analysis of the intestinal content of the Neolithic glacier mummy from the Alps. *Proc. Natl Acad. Sci. USA* **99**, 12594–12599.

Roy-Engel AM, Carroll ML, Vogel E et al. (2001) Alu insertion polymorphisms for the study of human genomic diversity. *Genetics* **159**, 279–290.

Royle NJ, Armour JAL, Crosier M, Jeffreys AJ (1993) Abnormal segregation of alleles in CEPH pedigree DNAs arising from allele loss in lymphoblastoid DNA. *Genomics* **15**, 119–122.

Ruano G, Kidd KK, Stephens JC (1990) Haplotype of multiple polymorphisms resolved by enzymatic amplification of single DNA molecules. *Proc. Natl Acad. Sci. USA* **87**, 6296–6300.

Saiki RK, Scharf S, Faloona F, Mullis KB, Horn GT, Erlich HA, Arnheim N (1985) Enzymatic amplification of beta-globin genomic sequences and restriction site analysis for diagnosis of sickle-cell anemia. *Science* **230**, 1350–1354.

Sanger F, Nicklen S, Coulson AR (1977) DNA sequencing with chain-terminating inhibitors. *Proc. Natl Acad. Sci. USA* **74**, 5463–5467.

Smith CI, Chamberlain AT, Riley MS, Cooper A, Stringer CB, Collins MJ (2001) Not just old but old and cold? *Nature* **410**, 771–772.

Southern EM (1975) Detection of specific sequences among DNA fragments separated by gel electrophoresis. *J. Mol. Biol.* **98**, 503–517.

Stephens M, Smith NJ, Donnelly P (2001) A new statistical method for haplotype reconstruction from population data. *Am. J. Hum. Genet.* **68**, 978–989.

Syvänen A-C (2001) Accessing genetic variation: genotyping single nucleotide polymorphisms. *Nature Rev. Genet.* **2**, 930–942.

Telenius H, Carter NP, Bebb CE, Nordenskjold M, Ponder BAJ, Tunnacliffe A (1992) Degenerate oligonucleotide-primed PCR – general amplification of target DNA by a single degenerate primer. *Genomics* **13**, 718–725.

Turner G, Barbulescu M, Su M, Jensen-Seaman MI, Kidd KK, Lenz J (2001) Insertional polymorphisms of full-length endogenous retroviruses in humans. *Curr. Biol.* **11**, 1531–1535.

Tyagi S, Kramer FR (1996) Molecular beacons: probes that fluoresce on hybridization. *Nature Biotechnol.* **14**, 303–308.

Underhill PA, Shen P, Lin AA *et al.* (2000) Y chromosome sequence variation and the history of human populations. *Nature Genet.* **26**, 358–361.

Vigilant L, Pennington R, Harpending H, Kocher TD, Wilson AC (1989) Mitochondrial DNA sequences in single hairs from a Southern African population. *Proc. Natl Acad. Sci. USA* **86**, 9350–9354.

Wilson JF, Weiss DA, Richards M, Thomas MG, Bradman N, Goldstein DB (2001) Genetic evidence for different male and female roles during cultural transitions in the British Isles. *Proc. Natl Acad. Sci. USA* **98**, 5078–5083.

Wilson MR, Polanskey D, Butler J, DiZinno JA, Replogle J, Budowle B (1995) Extraction, PCR amplification and sequencing of mitochondrial DNA from human hair shafts. *Biotechniques* **18**, 662–669.

Xiao W, Oefner PJ (2001) Denaturing high-performance liquid chromatography: a review. *Hum. Mutat.* **17**, 439–474.

Yan H, Papadopoulos N, Marra G *et al.* (2000) Conversion of diploidy to haploidy. *Nature* **403**, 723–724.

Zane L, Bargelloni L, Patarnello T (2002) Strategies for microsatellite isolation: a review. *Mol. Ecol.* **11**, 1–16.

SECTION THREE

How do we interpret genetic variation?

Insights into the human past can be generated from analyzing modern genetic diversity because genetic variation, described in the preceding section, has not arisen independently in each individual, but has a shared history which has been shaped by several different 'evolutionary forces'. This section introduces the fundamental concepts of population genetics, molecular evolution and phylogenetics that allow these inferences to be drawn, and are consequently central to the rest of this book.

CHAPTER 5   *Processes shaping diversity*

This chapter demonstrates that genetic diversity within a species is shaped by a number of 'evolutionary forces'. These processes include mutation and recombination which generate new genetic variation, genetic drift which removes variation and selection which shapes pre-existing variation. The interplay of these different forces is complex and requires us to develop mathematical models that describe simplified but fundamental features of how populations and molecules change over time.

CHAPTER 6   *Making inferences from diversity*

The knowledge gained from modeling evolutionary forces allows us to develop analytical methods to extract information about the past from genetic diversity data. In this chapter we meet methods for analyzing genetic diversity that allow us to investigate the role that each of these 'evolutionary forces' has played in shaping variation. Case studies to illustrate the application of these inferential methods are found in subsequent chapters and readers are referred forward to the relevant sections. Similarly, in the later chapters readers are directed back to this section for methodological detail. Some may wish to refer to the contents of *Chapters 5* and *6* only as necessary – as if they were a guidebook for the journey through the complexities of human genetic diversity.

CHAPTER FIVE

Processes shaping diversity

5.1 Overview

In the 'Origin of Species' Charles Darwin was primarily interested in the evolution of species over geological time. Others have been more concerned with processes operating on genetic diversity, within a single species, over a time scale of generations. These two scales of evolutionary change are often referred to as **macro-** and **micro-evolution**. While it is often assumed that species-level evolution is just an extrapolation of population-level evolution, the project of reconciling these two fundamental evolutionary levels is by no means complete.

In this chapter, we will show that the microevolutionary processes that shape genetic diversity can be measured by changes in allele and haplotype frequencies within populations.

5.1.1 Why do we need models of populations and of molecules?

Allele frequency changes through time and space are the clues by which evolutionary processes can be investigated. By understanding the mechanisms by which the 'forces of evolution' act on allele frequencies we can produce mathematical models that approximate reality. These models are necessary to understand the subtle interplay between these 'forces', and permit inferences on *past* processes from *modern* diversity.

Using models we can derive equations that allow us to estimate parameters of interest from the data, for example, population growth rate, the age of an allele, or the migration rate between two populations. Models also allow us to test different hypotheses about the past. Alternative models can be compared on the basis of which provides the best fit to the data (for example whether a straight or curved line through a set of plotted points is a better approximation of an apparent trend). There is a variety of methods for testing **'goodness-of-fit'**, some of which are explored in the next chapter (see *Box 6.2* on the concept of **likelihood**).

One of the strengths of many population genetic models is their generality: they can be applied to data from any species that share broad characteristics. For example, some models applied to humans might be equally applicable to all other species that reproduce sexually and do not self-fertilize. However, we must bear in mind that models are 'lies that lead to the truth', and that they require us to make assumptions that may not be true of all species. In certain situations these assumptions deviate so far from reality that they can become untenable and our inferences meaningless. This problem drives mathematical models of evolution to become ever more sophisticated, abandoning simplifying assumptions one by one, and introducing new parameters that provide a better fit to biological reality. Nonetheless, even in the data-rich field of modern human genetics, no amount of data can make up for an inappropriate model. During this chapter we highlight some of the classical population genetic assumptions that deviate significantly from reality.

This book is not the place to find a sophisticated mathematical treatment of modern evolutionary theory; the reader will find sources of more detailed information in the references.

The concept of a **population** is central. By definition, a population must be defined before an allele's frequency within it can be determined. In addition, we are often interested in reconstructing past demographic events; demography is a property of populations, not of individuals. It is for these reasons that this discipline is known as **'population genetics'**. Furthermore, many studies of human genetic diversity aggregate individuals from a number of closely situated, but distinct, locations into a single 'population'. An ecological approach to sampling, such as using regular grid squares, is rarely, if ever, adopted for humans, although, interestingly, it was advocated by Allan Wilson, an early proponent of the ill-fated Human Genome Diversity Project (Roberts, 1991). This sampling of groups, rather than of individuals, leads to their being considered a natural unit of investigation.

One kind of model we will encounter is a mathematical approximation of populations, their interactions and mating structures. When the term 'population' is being used it is important to be clear how the population was defined and whether it refers to individuals grouped together for the sake of analysis, or an idealized group, assumed to be adhering to the assumptions of a mathematical model (e.g., randomly mating). In other words, does the term 'population' refer to a practical or theoretical entity?

The other kinds of mathematical model we will encounter are those describing molecular processes (i.e., **mutation** and **recombination**), which, as we saw in Chapter 3, differ between DNA sequences and genomic regions. These enable us to go beyond allelic definitions and allow us to make the connection between molecular diversity and population processes.

Population genetics matured around 1930, primarily with the work of Ronald Fisher and Sewall Wright. It is worth bearing in mind that these advances took place before DNA was shown to be the molecule of inheritance. Fisher and Wright united the previously incompatible concepts of Darwinian **natural selection** of **phenotypic** characters and Mendel's ideas of particulate inheritance by showing that discrete alleles can underlie continuous traits (see Chapter 13). In the ensuing years, the description of genetic diversity at the molecular level, and the inadequacy of selection as the sole process explaining levels of polymorphism, have required substantial developments to population genetic theory.

5.1.2 Hardy–Weinberg equilibrium

The first major success of population genetics was to explain how allele frequencies in one generation could be used to calculate **genotype** proportions in the next generation of a randomly mating population. In diploid organisms such as humans, two alleles, A and a, at the same locus, with frequency p and q respectively, can be combined to make three genotypes: AA, Aa and aa. If we know the frequency of these two alleles in an idealized population we can predict the proportions of the genotypes in the succeeding generation by combining **gametes** (which contain single alleles) at random – a postulate known as the Hardy–Weinberg principle (Hardy, 1908). Thus the proportion of each genotype in the next generation is:

$$AA = p^2, Aa = 2pq \text{ and } aa = q^2$$

If the genotype proportions in the succeeding generation are calculated in this manner, and are found to be indistinguishable from those in the parental generation, no evolution (defined as change in allele frequencies) is occurring, and the population is said to be at **Hardy–Weinberg equilibrium**. At the time of its discovery, the existence of this equilibrium was important as it showed that mating alone need not reduce variation.

Here is an example of the application of the Hardy–Weinberg principle: a recessive single gene disorder has an incidence of 1 in 2500. What is the frequency in the population of heterozygous carriers of this condition?

▶ The frequency of the recessive allele = q^2, therefore $q = (0.0004)^{0.5} = 0.02$.

▶ The sum of the frequency of the two alleles, the disease and the wild-type allele, must sum to 1. Therefore $p = 1 - q = 0.98$.

▶ The frequency of the heterozygous genotype is $2pq = 2 \times 0.02 \times 0.98 = 0.0392$, or about 1 in 25.

For us to be able to estimate genotype proportions from one generation to the next in this way, the population must be made up of an infinite number of randomly mating, sexually reproducing **diploid** organisms. However, for Hardy–Weinberg equilibrium to be observed the idealized population must have certain additional properties, including:

▶ no selection;
▶ no mutation;
▶ no migration.

In other words, an absence of any factors that might change allele frequencies.

By contrast, if the calculated genotype proportions are not in Hardy–Weinberg equilibrium we might reasonably conclude that evolution (a change in allele frequencies) is indeed occurring, and that one, or a combination, of the above factors is in operation. The rest of this chapter is devoted to examining the processes that influence allele frequencies.

5.2 Generating diversity: Mutation and recombination

5.2.1 Germ-line vs. somatic mutations

Mutation is the sole process generating new alleles: indeed, by definition any change producing a new allele is called a mutation. It provides the raw material on which selection and the other forces of evolution can act. There is a broad variety of mutational changes, and these occur at widely varying rates (see Chapter 3 for more details). Each mutation is a single change occurring in a single cell. Evolutionary consequences only follow from those changes that occur in the **germ-line**, and not those in somatic tissues, as somatic mutations are not heritable. The dynamics of many types of mutations vary between the **soma** and the germ-line. Because of the fidelity of DNA polymerases and the operation of DNA repair mechanisms, germ-line mutations occur at low rates for individual nucleotides, although (given the size of the human genome), they are inevitable in every replication cycle. It has been estimated that every human carries, on average, 128 new mutations (Giannelli *et al.*, 1999).

5.2.2 Mutation changes allele frequencies

In the absence of other processes, an allele will decrease in frequency as it accumulates mutations, a phenomenon known as **mutation pressure**. By knowing the mutation rate for the whole gene (μ), the initial allele (p_0) frequency, assuming no back mutation and ignoring **stochastic** processes, we can calculate this allele's frequency (p_t) t generations later, by:

$$p_t = p_0 e^{-\mu t}$$

At low mutation rates, mutation pressure is a weak force that can only have appreciable impact over long time scales. After 1000 generations, the wild-type allele of a gene 1 kb in size with a per-generation nucleotide mutation rate of 2×10^{-9} will only decrease in frequency from 1.0 to 0.998.

5.2.3 Models of the mutational process

The above example introduced the concept of a gene in which each mutation creates a new allele; in other words we discounted the possibility of **back mutations** and **recurrent mutations**. If we consider a gene 1 kb in length then the number of possible alleles is enormous, 4^{1000}. The probability of back mutations and recurrent mutations is correspondingly small. This model is known as the **infinite alleles model** (Crow and Kimura, 1970), and at first sight seems to be a reasonable approximation of reality for the evolution of DNA sequences, although we will meet more complex models of DNA sequence evolution later.

If we consider the evolution of a polymorphic microsatellite, oscillating in size by whole numbers of repeats, we can see that the opportunity for back mutation and recurrent mutation is much greater than for SNPs. Thus the infinite alleles model does not always appear to be a close approximation of biological reality. We need different models for different types of mutation. The **stepwise mutation model** (SMM) (Ohta and Kimura, 1973) provides a better fit to microsatellite evolution. According to this model, mutations increase and decrease allele length by one unit with equal probability (*Figure 5.1*).

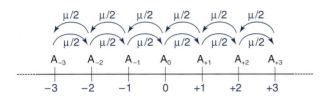

Figure 5.1: The stepwise mutation model.

The model considers only single-step mutations, and regards an increase or decrease as equally probable, and independent of allele length. The average mutation rate is μ, and any allele mutates to a smaller or larger allele with rate μ/2.

Initially, the SMM considered single-step changes only, but there is good empirical evidence for a lower, but nevertheless appreciable, rate for multiple step mutations and the model can be adapted to account for these (Di Rienzo *et al.*, 1994). There are, however, other known aspects of microsatellite evolution not incorporated within the SMM model (see Section 3.3.1):

▶ a positive correlation between allele length and mutability;

▶ a lower length threshold under which mutation rate becomes undetectable;

▶ a possible small bias towards expansions of short alleles, resulting in an increase in size of the microsatellite;

▶ a possible preference for deletions rather than expansions in longer alleles; together with the previous point, this produces an equilibrium allele length distribution;

▶ massive expansions in triplet repeat diseases, and consequent negative selection in these and other examples.

Other types of mutations, such as chromosomal rearrangements and GC–rich minisatellite mutations, fit neither of the above models.

5.2.4 More complex models of DNA sequence evolution

If we are interested in aspects of sequence evolution that require us to suppose that several changes might have occurred at the same site, then we need more complex models of mutation – for example, we may need to consider the probability that an A will mutate to a C and then subsequently back from a C to an A again. These models come into play when considering sequence evolution over long time scales, where back mutations result in the observed **sequence divergence** being an underestimate of the real number of mutational changes. We will come to applications of these models in Chapter 6.

In the simplest model all nucleotide substitutions occur at the same rate, while the most complex model allows a

different rate for each nucleotide change. These models can be represented as a substitution scheme, and as a probability matrix, shown in *Figure 5.2*. The simplest example is known as the **Jukes–Cantor model (JC)** (Jukes and Cantor, 1969), and one of the more complex models is the **general reversible model (REV)**. There are a number of intermediate models that contain some, but not all, of the complexity of the REV model.

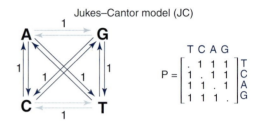

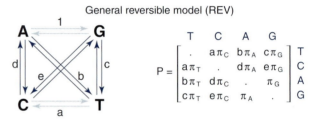

Figure 5.2: Models of sequence evolution.

The probability matrix is shown for two models of sequence evolution. This matrix contains the relative rates of the different possible base substitutions, which are also shown on a substitution scheme that shows transitions in light blue and transversions in dark blue. The REV model also includes the π_i parameter which is the frequency of that base in the sequence.

The frequency of each nucleotide clearly influences the probability of nucleotide changes averaged over an entire sequence. For example, an A to G transition may have the same rate as a C to T transition, but if there are twice as many As as Cs in a sequence then the probability of an A to G occurring within the sequence as a whole is not the same as that of a C to T. The JC model does not take potential bias in **base composition** into account, but the REV model does.

There are aspects of sequence evolution known from empirical studies that are not accounted for in these models (see Chapter 3 for details):

▶ Small (1–20-bp) insertion or deletion events (**indels**) are known to make up about 10% of all sequence changes in noncoding sequences of the human genome (Nachman and Crowell, 2000). Discounting this kind of mutational change can have a large impact; for example, whether or not indels are removed prior to sequence analysis makes a fourfold difference to the apparent sequence divergence between humans and chimpanzees according to one way of measuring it (see Section 7.4.3). The probability of a small indel event occurring is largely determined by the repetitive nature of the surrounding sequence, in a manner that is poorly understood and, therefore, difficult to model. Such changes are rarely found as polymorphisms in coding regions because they often disrupt the open reading frame.

▶ The phenomenon of the increased mutability of **CpG** dinucleotides (Bird, 1980; Krawczak *et al.*, 1998) departs significantly from the REV model (see Section 3.2.5). The mutability of a nucleotide depends on its neighbor, so that not all Cs and Gs have the same probability of mutating. Both transitions and transversions have increased probability at CpGs (Nachman and Crowell, 2000).

Models have been developed that can accommodate rate variation among sites within a sequence. These fit such variations in rate to a statistical distribution. Some, like the gamma distribution, have a single modal value, whereas other models allow multimodal distributions that may be a better fit to the true rate variation among sites, as suggested by the increased mutability of CpGs described above.

5.2.5 Recombination

Meiotic recombination is a consequence of sexual reproduction, and enhances the ability of populations to adapt to their environment through the combining of advantageous alleles at different loci (*Figure 5.3*). By contrast, asexually reproducing species and nonrecombining portions of the human genome are prone to the operation of **Muller's ratchet**, the slow but inexorable accumulation of deleterious mutations. This process of degeneration may explain the low density of functional genes on the nonrecombining portion of the Y chromosome (see *Box 8.8*, and Section 13.7.1)

Recombination generates new combinations of alleles on the same DNA molecule, known as **haplotypes** (see Sections 3.7 and 4.10), and in this way increases haplotype diversity. Consequently, recombination is capable of breaking up advantageous allelic combinations. This results in the theoretical possibility that by increasing the likelihood of disrupting a beneficial haplotype, outbreeding can result in a drop in fitness known as **outbreeding depression**.

While alleles at loci on different chromosomes are randomly segregated during meiosis, alleles at loci closely linked on the same chromosome are not, as recombination between them occurs infrequently. Linked loci share a common evolutionary heritage; selection operating on one locus will affect diversity at the other. An allele that rises to high frequency through **positive selection** at a linked locus is said to be 'hitchhiking'. The reduction in diversity at loci linked to a recently fixed allele is dubbed a **selective sweep**. Reduced diversity of the *tb1* gene in maize represents a clear example of a selective sweep that has arisen through selective breeding practices (see Section 10.7.1). Conversely, negative selection at a locus also reduces diversity at linked loci, albeit

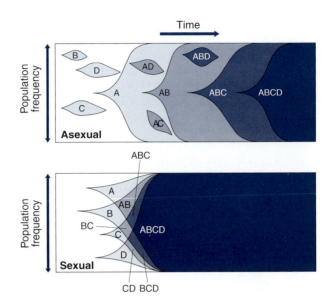

Figure 5.3: The advantage of sexual reproduction (redrawn and adapted from Crow and Kimura, 1970).

Four alleles (A–D) all increase the fitness of the organism, with the fittest having all four alleles. Only one allele at a time can prevail in an asexual organism, so they must be combined serially. By comparison, in a sexually reproducing organism these beneficial alleles can be combined in parallel. Thus it takes much less time to assemble the fittest genotype.

at a slow rate, by a process known as **background selection** (Charlesworth *et al.*, 1993).

5.2.6 Recombination in populations: Linkage disequilibrium

Recombination can be studied at the population level by investigating whether specific alleles at different loci are associated with one another more or less often than would be expected by chance. This nonrandom association is known as **linkage disequilibrium (LD)**. If we consider two neighboring loci **A** and **B** with two alleles – A and a, and B and b – at each locus, and assume that there is no association between them, then we would expect that the only factor determining whether we find A and B on the same haplotype would be their relative frequencies. In other words, the frequency of the haplotype AB should be the product of the frequencies of alleles A and B. If this is indeed the case, then the two loci are said to be in linkage equilibrium. The difference between the *observed* frequency of haplotype AB (P_{AB}) and the *expected* frequency of haplotype AB under linkage equilibrium ($P_A \times P_B$) is known as D, and is used as the basis for more complex measures of LD such as $|D'|$ or r^2 (see *Box 3.11* for more details on these measures of LD). At linkage equilibrium D is zero. If D is significantly different than zero, then the markers are in LD (reviewed by Ardlie *et al.*, 2002).

$$D = P_{AB} - P_A \times P_B$$

In an analogous fashion to the reduction in frequency of an allele by 'mutation pressure', recombination can reduce the frequency of a haplotype. Rather than monitor this process through the decline in frequency of the haplotype itself, we can follow the decay of LD using D as follows. When a new mutation arises on a chromosome, it is linked to all other polymorphic markers on the same chromosome forming a single haplotype. In other words, it will only be found associated with one allele at all of those other loci, and so is in complete LD with them (D is at its maximal possible value). However, over several generations the frequency of the new mutation may grow; if so, recombination events introduce the new mutant allele onto chromosomes with different alleles at the other polymorphic markers (see the figure in *Box 3.11*). As a consequence, LD starts to decay. If we know the recombination rate per generation (r) between the newly mutated locus and a given locus, after a certain number of generations (t) we can track the decay of LD over time, by relating the present value of D (D_t) to the initial value of D (D_0) using the equation:

$$D_t = (1 - r)^t \times D_0$$

From this equation we can see that as time increases, D_t tends to zero (linkage equilibrium). In addition, as we move along the chromosome away from our newly mutated locus, the interlocus recombination rate increases, meaning that D_t will tend to zero even sooner. In an infinitely large population, LD would continue to decay over time as a result of an ever-increasing frequency of recombination between the newly mutated locus and any other locus. However, real populations are not infinitely large and in Section 5.6 we will explore why an inexorable decay of LD is an unrealistic expectation.

5.2.7 Models of recombination

In comparison to models of mutation, models of recombination have traditionally been fairly simple. The simplest model is that the rate of recombination is uniform. In other words, the probability of a **crossover** occurring between a pair of markers is determined only by the physical distance that separates them. The products of this type of recombination event are two new haplotypes containing contiguous stretches of alleles from each ancestral haplotype (see *Figure 5.4*).

Empirical studies of recombination in humans and model organisms have revealed two biological properties of recombination that conflict with the simple model of recombination described above:

▶ not every recombination event results in a crossover (see Section 3.8). A recombination intermediate can be resolved in one of two ways: a crossover, or a **gene conversion** event that converts a small segment of DNA (typically less than a kilobase) in one haplotype in a nonreciprocal way so that it is identical to that same segment in the other haplotype;

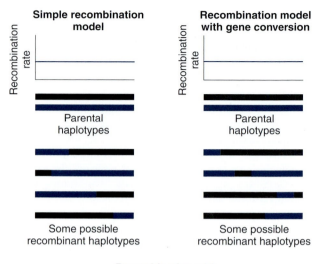

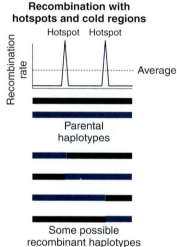

Figure 5.4: Three models of recombination.

The recombination rate along the haplotype is shown above the parental haplotype and four typical recombined haplotypes for three models of recombination discussed in the text.

▶ recombination rates are not uniform along a segment of DNA (see Sections 2.6 and 3.7). Crossovers appear to be concentrated in **hotspots** between which lie recombinationally inert, 'cold' regions. At larger scales, recombination rates vary along the chromosome in ways that are only now being elucidated, but are often low near centromeres and high near telomeres.

It is only recently that the second of these two attributes of recombination at the smallest scale has been discovered. As a consequence, while models of recombination have been constructed that incorporate either one of these two additional complexities, few if any models have combined the two. Incorporating gene conversion into recombination models requires two additional factors (Wiuf and Hein, 2000; Frisse *et al.*, 2001):

▶ the ratio of gene conversions to crossover events;

▶ the length of the gene-converted segment (see *Figure 5.4*).

Recombination rate heterogeneity can be modeled (Stumpf and Goldstein, 2003) by considering:

▶ the size and spacing of recombination hotspots;

▶ the ratio of the recombination rates in hotspots and in cold regions (see *Figure 5.4*).

5.3 Eliminating diversity: Genetic drift

5.3.1 The concept of effective population size

No population is infinitely large, as is assumed by Hardy–Weinberg theorem. Each generation represents a finite sample from the previous one. Thus variation in allele frequency between generations can occur solely through this **stochastic** process of sampling. This source of variation is known as random genetic drift (Wright, 1931).

Intuitively, we might expect that the magnitude of genetic drift relates to the size of the population being sampled, and we can show this to be true. *Figure 5.5* illustrates the change in allele frequency over 100 generations in simulated populations, starting with an initial allele frequency of 0.5. The allele rapidly becomes either fixed or lost from the populations of constant size 20, whereas both alleles persist in the populations of constant size 1000 with more subtle variations in frequency.

The population model described above is known as the **Wright–Fisher model**, and is fundamental to many aspects

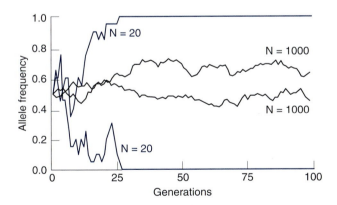

Figure 5.5: Genetic drift in populations of different sizes.

The results of simulations over 100 generations of allele frequencies of a binary marker in populations of size 20 or 1000 are shown, each starting from a frequency of 0.5. The populations are of constant size and have nonoverlapping generations; each generation is sampled randomly from the previous one (so individuals have an equal probability of contributing to the next generation). The allele rapidly becomes either fixed or lost from the populations of constant size 20, whereas more subtle variations are seen in the populations of constant size 1000.

of population genetics. However, this model contains many unrealistic assumptions when compared to real populations. First, generations overlap in real human populations. Second, populations are rarely constant in size and third, large populations do not exhibit random mating. These three factors are of varying importance for any given population. Wright's concept of **effective population size** (N_e) allows us to compare the amount of genetic drift experienced by different populations (Wright, 1931). The effective population size for any population represents the size of an idealized, Wright–Fisher, population that experiences the same amount of genetic drift as the one under study. Effective population size measures the magnitude of genetic drift: the smaller the effective population size, the greater the drift. We can understand the impact of different properties of real populations on genetic drift through the changes they cause in effective population size.

There are, in fact, two ways of defining effective population sizes: one is based on the sampling variance of allele frequencies (i.e., how an allele's frequency might vary from one generation to the next), and the other utilizes the concept of **inbreeding** (i.e., the probability that the two alleles within an individual are identical by descent from a common ancestor). Both of these properties of a finite population depend on the size of that population. For the sake of simplicity in this chapter we treat these two definitions interchangeably, but the reader should be aware that while under most simple population scenarios these two definitions of effective population size give identical values for N_e, in more complex situations this is not the case (Crow and Kimura, 1970).

It is not easy to relate the effective population size (N_e) to the census size of a population (N), as there are many parameters that can affect this relationship, only some of which are relevant to humans. These are discussed later in this section. The effective population size is almost always substantially less than the actual population size. For example, the introduction of overlapping generations alone into the population model reduces N_e to 25%–75% of N (Felsenstein, 1971; Nunney, 1993).

5.3.2 Genetic drift causes the fixation and elimination of new alleles

The concept of effective population size allows us to calculate the probability and rate of **fixation** for a new allele in the absence of selection and mutation. Fixation itself is a rare event – a far more likely outcome for a new allele is that it will be lost. As intuition would suggest, with no favoring of either outcome, the fixation probability of an allele in the absence of selection is equal to its frequency in the population; a new allele would have a frequency of 1/2N. Thus the smaller the population, the greater chance a new mutant has of becoming fixed (see *Figure 5.6*). The average time to fixation (t) in generations has been shown to be:

$$t = 4N_e$$

Therefore a new allele in a smaller population will not only have a higher probability of becoming fixed, but it will also

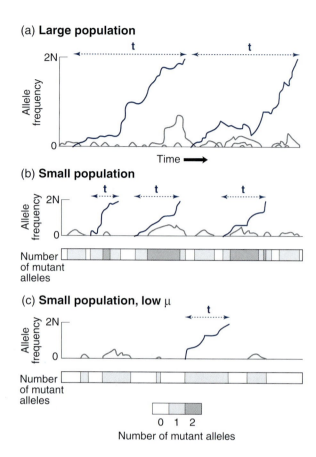

Figure 5.6: Schematic view of the fixation of new alleles in three different populations (redrawn and adapted from Crow and Kimura, 1970).

New alleles that are fixed or eliminated are shown by blue and gray lines respectively. The time to fixation (t) is longer in the larger population (a) than the smaller population (b). A greater proportion of new alleles are fixed in the smaller population than in the larger population. (c) A population of the same population size as (b), but with a lower mutation rate (μ). The time to fixation in (c) is no different from that in (b), but the time between the fixation of new alleles is greater, as is the proportion of time spent with no polymorphism.

be fixed more rapidly than it would in a larger population. Nonetheless, fixation under the influence of drift alone is substantially slower than if selection were acting (see Section 5.4.2).

5.3.3 The impact of size fluctuations on effective population size

Few populations are constant in size for many generations, so what happens to the effective population size during these fluctuations? The long-term effective population size has been shown to be approximately equal to the harmonic mean (Wright, 1938; Crow and Kimura, 1970), rather than the arithmetic mean of the population sizes over time (*Figure 5.7*). The harmonic mean is the reciprocal of the mean of the reciprocals:

$1/N_e = (1/t)\sum_{i=1}^{t}(1/N_i)$ for t generations

In practice, this means that the effective population size is disproportionately affected by the smaller population sizes. So in the recently expanded human population, the effective population size (and hence the amount of neutral variation) is still largely determined by the smaller ancestral population sizes in our past (*Table 6.1* gives estimates of N_e in humans, and *Figure 10.5* shows estimates of the population growth over the past 100 000 years).

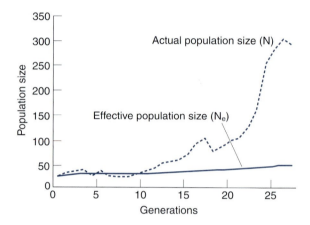

Figure 5.7: N_e in a population of variable size.

The harmonic mean of the census size barely changes despite a recent population expansion.

This dependence of present day variation on past small population sizes brings us to two important population processes that shape the genetic diversity apparent in many human populations, **bottlenecks** and **founder effects**. In many respects the two processes are similar since both result in a reduced ancestral population size, but the difference between them can be seen in *Figure 5.8*. Founder effects relate to the process of **colonization** and the genetic separation of a subset of the diversity present within the source population (e.g., the peopling of the Americas; see Section 11.2). In contrast, bottlenecks refer to the reduction in size of a single, previously larger, population and a loss of prior diversity.

5.3.4 Variation in reproductive success and effective population size

The Wright–Fisher model assumes that all parents have an equal chance of contributing to the next generation. This results in a **Poisson distribution** of number of offspring. However in real human populations there is often substantial variation in the contribution of individuals to the succeeding generation. To put it another way, there is a high variance in the number of offspring, higher than that expected under a Poisson distribution (where the variance equals the mean). This can be due to social causes, and need not be attributed solely to differences in fertility. In general, the higher the **reproductive variance**, the lower the effective population

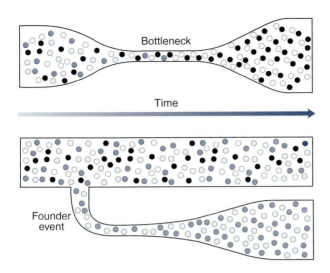

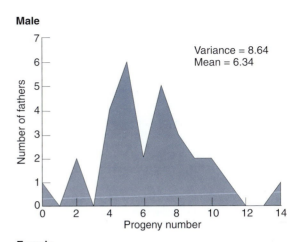

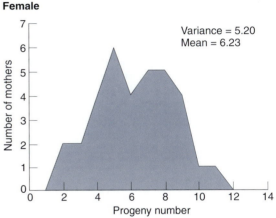

Figure 5.8: Bottlenecks and founder events.

Circles of different colors represent different alleles. Both bottlenecks and founder events result in a loss of allelic diversity.

Figure 5.9: Difference in reproductive variance for male and female Aka pygmies.

The two sexes have almost the same mean value but different variances (data from Hewlett, 1988).

size, because parental contributions become more and more unequal (Crow and Kimura, 1970; Caballero, 1994). It is worth noting that when reproductive variance is less than that expected under a Poisson distribution then N_e can be greater than N. This phenomenon is utilized to minimize loss of genetic diversity in the breeding of endangered species by equalizing the number of offspring from each parent.

Because reproductive variance often differs between the sexes (see *Figure 5.9*), males and females may have different effective population sizes. Most anthropological studies show males to have higher reproductive variance than females, which is expected to result in a lower male effective population size. This has implications for the effective population sizes of portions of the genome with different inheritance patterns (see *Box 5.1*).

Further reductions in effective population size are noted when reproductive variance is correlated between generations, for example, when children of large families tend to have large families of their own. Biologically, this inheritance of **fecundity** could happen when a gene conferring greater fertility is polymorphic within a population. Alternatively, social mechanisms of inherited fertility may operate in structured societies where access to resources is both unequal and inherited. Whatever the cause, inheritance of family size has been noted in many human populations, from different types of demographic data (see *Box 5.2*). In 1932, Huestis and Maxwell used completed questionnaires from University of Oregon students to demonstrate a significant correlation between the number of siblings and the number of children that an individual has (Huestis and Maxwell, 1932). Alternatively, genealogical records can detail past inheritance of family size, as has been demonstrated with the Saguenay-Lac Saint Jean population in Quebec (Austerlitz and Heyer, 1998), and the British nobility (Pearson *et al.*, 1899).

5.3.5 Population subdivision and effective population size

Previously we have considered only randomly mating populations; however, most human populations are not so homogeneous. In one respect, all human mating is nonrandom because it usually involves a conscious choice, but in this context 'random' means only that mating is random with respect to the genetic make-up of each individual. A population may not be randomly mating because it consists of smaller, partially isolated **sub-populations**, also known as **demes**. Alternatively, nonrandom mating may also occur because mate choice is not blind to genetic relatedness.

Population subdivision is often modeled in terms of a **meta-population** that comprises partially isolated sub-populations. This isolation eventually leads to partial **genetic differentiation** as genetic drift operates independently within each subpopulation. Members of the same subpopulation are therefore more closely related, on average, than are members of different subpopulations. Depending on the

BOX 5.1 Effective population sizes of different regions of the genome

Up to this point, effective population size has been considered at the level of individuals; however, not all genomic loci are equally represented in all individuals. If we consider a single mating couple as a microcosm of a species with equal sex ratios, they have four copies of each autosome, three copies of the X chromosome, two copies of mitochondrial DNA (mtDNA), only one of which will be inherited by the succeeding generation, and a single Y chromosome. Thus, given a 1 : 1 sex ratio, the effective population size of the Y chromosome and mtDNA will be only a quarter that of the autosomes, and a third that of the X chromosome. This assumes that the reproductive variances of males and females are equal, as in the Wright–Fisher model.

If, however, we also take into account the differences in effective population sizes between the sexes the relationships for the different loci become more complex. As we saw in Chapter 2, the Y chromosome is inherited paternally and mtDNA is inherited maternally, while the X chromosome is inherited twice as often from females as it is from males. Thus, a higher reproductive variance of males than females reduces the effective population size of the Y chromosome relative to that of mtDNA, the X chromosome and the autosomes, and increases the effective population size of mtDNA and the X chromosome relative to that of the autosomes. In cases of extreme male reproductive variance it is possible that the effective population size of the X chromosome may exceed that of the autosomes, up to a limit of 9/8 that of autosomal N_e (see *Figure*). In such extreme cases the effective population size of the Y chromosome approaches its lower limit of 1/8 that of the autosomes (Caballero, 1995; Charlesworth, 2001). Such considerations may explain why the Y chromosome exhibits such a high degree of population differentiation (see *Box 8.8* and *Figure 9.18*). However, discrepancies in generation times between the sexes also cause their effective population sizes to differ (Helgason *et al.,* 2003). The sex with the shorter generation time will experience more genetic drift (all other factors being equal) as a result of more frequent episodes of sampling a new generation from the previous one. In humans, females appear to have the shorter generation time (see Section 6.6.2), which should lower the effective population size of mtDNA relative to biparentally- and paternally-inherited loci. The relative importance of these opposing factors may differ from population to population. Recent work on detailed Icelandic genealogies seems to indicate that in the last few centuries generation time discrepancies between the sexes have outweighed any differences in reproductive variance, with the consequence that the effective population size of Icelandic mtDNA is less than that of the Y chromosome (Helgason *et al.,* 2003).

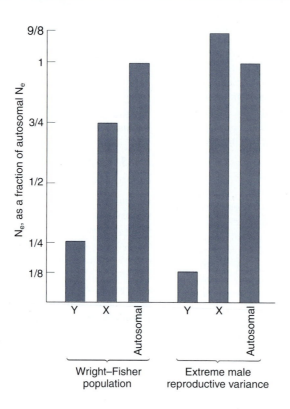

Relative effective population sizes for different regions of the genome.

BOX 5.2 Opinion: From N_e to N in Quebec

In population genetics, the intensity of genetic drift is measured by the effective population size (N_e). This N_e accounts for the rate of stochastic changes in gene frequencies from one generation to the next. A smaller N_e gives a higher rate of stochastic change: the fate of an allele will depend mostly on drift. This N_e equals the census size of an 'ideal population' (Wright–Fisher model) defined by a constant size, nonoverlapping generation, equal sex ratio, and where each individual has the same probability of reproduction. When the population is not of constant size, N_e is equal to the harmonic mean of the population sizes at different times. When generations overlap (as in the case of human populations), a good approximation of N_e is N/3, where N is the census size. When there is variance in reproductive success, N_e is roughly proportional to the inverse of the variance in progeny number.

In recent studies, we have detected a new factor that has a strong impact on N_e (Austerlitz and Heyer, 1998). We showed that the **transmission of reproductive success** from parent to offspring causes a 10- to 20-fold reduction of N_e. By transmission of reproductive success, we mean that members of families with a large number of effective children (children that have a genetic contribution to the population), tend themselves to have a large number of effective children (see *Figure*). This social factor was measured in a population of Quebec and explains its high frequency of some inherited disorders. This population (current census size N = 300 000) was founded 12 generations ago by a limited number of immigrants. The reduction in N_e relative to modern census size can be calculated as follows:

▶ N_e reduces to 17 000 as a consequence of population expansion;

▶ variance in offspring number reduces N_e to 12 000;

▶ transmission of reproductive success reduces N_e to 900.

The main consequences are that transmission of reproductive success:

▶ enhances the effect of drift more than 10 times;

▶ increases linkage disequilibrium.

This transmission of reproductive success is not limited to the Quebec population since this kind of correlation has been measured in other human groups (in Iceland, and the Pyrenees). To what extent can this phenomenon be generalized to other populations? The problem is that demographic data on three generations are necessary to compute this correlation; to this end we have designed an ethno-demographic questionnaire for its measure in populations where vital records are not available and we are using it now for our field investigations.

In parallel, we have recently developed a theoretical framework for this phenomenon in a constant size population. We confirm by simulation that due to the transmission of reproductive success, N_e is divided by one order of magnitude (Sibert *et al.*, 2002) when levels of transmission are equal to those measured in the Quebec population. This formalization also enables us to describe the impact of this fertility transmission on **gene genealogies**. It shows the genetic impact of what has been described as vertical transmission (Cavalli-Sforza and Feldman, 1991). Therefore, the detection of transmission of reproductive success highlights the strong impact of social transmission of demographic behaviors on human genetic diversity: not taking it into account can yield considerable underestimation (typically by one order of magnitude) of the census size of past populations. This is a major issue in almost all studies on the past history of the human species.

Evelyne Heyer

Eco-Anthropology, Muséum National d'Histoire Naturelle, Paris, France

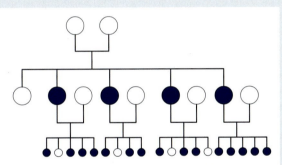

Children who belong to families with a high number of effective children tend to have a high number of effective children

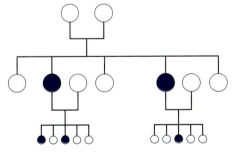

Children who belong to families with a small number of effective children tend to have a small number of effective children

● **Effective child:** one who reproduces in the population

○ **Non-effective child:** death, emigration, celibacy

Inheritance of effective family size.

nature of the *population structure*, the effective population size of the meta-population can be increased or decreased relative to a randomly mating population of the same size. If there are substantial levels of extinction and recolonization of subpopulations then the effective population size of the meta-population can be dramatically reduced relative to the census size.

These subpopulations are not completely isolated and the migration of individuals between them results in gene flow between subpopulations, reducing differentiation. Thus to understand the impact of population subdivision on genetic drift we must model: (i) the number, size and spatial arrangement of the subpopulations; and (ii) gene flow by migration (Crow and Kimura, 1970; Wang and Caballero, 1999). These models are considered in greater depth in Section 5.5. One aspect shared by all these models is a need for some measure of population structure that can be used to estimate parameters of the population subdivision model, for example, the rate of gene flow, or the effective population size of the meta-population. The accuracy of these estimates depends upon how closely the chosen model approximates to reality. Some have argued that such estimates have little, if any, relevance to reality, because all current models are over-simplistic, and contain important assumptions that are violated by all human populations.

Perhaps the best known measure of population structure is F_{ST}. This derives from the **Fixation indices** proposed by Sewall Wright to measure the deviation of observed heterozygote frequencies from those expected under Hardy–Weinberg theorem (Wright, 1951). Subpopulation divergence results in an excess of **homozygotes** and a corresponding deficiency of **heterozygotes** in the meta-population, a phenomenon known as the Wahlund effect. F_{ST} measures the apportionment of genetic variation between subpopulations; in other words, it compares the mean amount of genetic diversity found within subpopulations to the genetic diversity of the meta-population.

F_{ST} can be defined in a number of different ways, and can be estimated from genetic diversity data by a variety of methods, most commonly:

$$F_{ST} = (H_T - H_S)/H_T$$

where H_T is the expected heterozygosity of the meta-population and H_S is the mean expected heterozygosity across subpopulations.

F_{ST} varies between 0 and 1. When gene flow is high and there is little differentiation between subpopulations F_{ST} is close to zero. When subpopulations are highly differentiated then genetic diversity of the meta-population is much greater than in any subpopulation and F_{ST} is close to 1. As we shall see in Section 6.4.2, F_{ST} values between pairs of populations can also be considered to be a measure of genetic distance between them.

5.3.6 Mating choices and effective population size

Nonrandom mating also results from individuals choosing their mates via some assessment of their mutual similarity. If individuals choose partners on the basis of shared phenotypic characteristics such as socio-economic status, IQ or skin color, this is known as **assortative mating**. **Disassortative mating** (or negative assortative mating) results when partners are chosen on the basis of their phenotypic differences rather than similarities (see *Boxes 5.3* and *5.4*). Disassortative mating at a locus generates a greater heterozygote frequency, and assortative mating a lower heterozygote frequency, than that expected under random mating. Assortative mating augments genetic drift by decreasing the effective population size, whereas the opposite is true for disassortative mating.

Inbreeding and **outbreeding** are the corresponding terms for when partners are genetically related, rather than simply sharing phenotypic similarity. The more closely related two partners are, the higher the chance that they will pass on the same deleterious recessive allele to their offspring. Thus there is a fitness cost to inbreeding known as **inbreeding depression.** Incest represents the extreme of inbreeding, and refers to a set of couplings between relatives that are proscribed by human societies. Incest taboos are universal and probably represent a strategy to minimize inbreeding depression. While marriages between first-degree relatives are almost always proscribed, variations exist among different societies as to the rules of marriage with more distant relations.

BOX 5.3 Evidence for (dis)assortative mating in humans

Assortative mating can be based on physical or psychometric traits. Physical traits that are selected include similar attractiveness, age and ethnicity. The last can be clearly demonstrated by the statistical analysis of census data, whereas the first is trickier as it relies upon a subjective notion of beauty. Nevertheless, it has been argued that individuals do choose to mate with those of a similar level of attractiveness to themselves. Psychometric traits thought to have been selected for during assortative mating include IQ, and the presence of a mental disorder. Less easily classified traits that influence assortative mating include religion, deafness and educational qualifications.

One of the best known traits proposed as a candidate for disassortative mating is resistance to infectious diseases (Hamilton and Zuk, 1982). Much of an individual's resistance to infectious disease is encoded in the MHC region of the human genome (see *Box 5.5*). This region contains several closely linked and highly polymorphic genes that are involved in immunological recognition and response. Disassortative mating is one of a number of plausible explanations for the surprisingly high degree of polymorphism in the MHC region (see *Box 5.5*). It has been suggested that individuals can maximize the protection of their offspring from infectious diseases by increasing their heterozygosity at this locus, in other words, by mating with someone with a different set of alleles in this region (see *Box 5.4*).

BOX 5.4 Opinion: Choosing mates – should we follow our noses?

From an evolutionary perspective, perhaps the most important decision that individuals make is their choice of mates. After all, the fitness of our offspring and their descendants determine to a large extent the proportion of our genes that survive through the generations. Thus it is not unexpected that complex mating strategies have evolved to ensure that we select the right mate. In this context, it may not be particularly surprising that immune genes of the **major histocompatibility complex** (MHC; *see Box 5.5*), which play a central role in our defense against pathogens, are involved in mate selection in many species (Penn and Potts, 1999). Although the particular strategy may vary from species to species, MHC-linked olfactory-mediated cues seem to influence preferences for mates in species as diverse as rats, mice, stickleback, salmon, and humans (Potts, 2002). Species-specific preferences may be for MHC differences, MHC heterozygosity, or rare MHC alleles or haplotypes. In all cases, however, the result is the maintenance of MHC polymorphism and disease-resistant progeny.

The human MHC genes, called human leukocyte antigens (or HLA), are among the most polymorphic loci in the human genome, with more than 250 alleles each at the HLA-B and HLA-DRB1 loci. The HLA loci encode at least six highly polymorphic cell surface glycoproteins that participate in the critical process of self vs. nonself recognition, and which ultimately determine whether a cell is infected (and needs to be destroyed) or healthy (and allowed to survive). The extensive polymorphisms that characterize these proteins presumably allow for an efficient immune response against a wide array of pathogens and, ultimately, survival of the species. How this extraordinary level of variation is maintained in populations is still the subject of debate, but it is becoming increasingly clear that these molecules participate in a variety of biological and social processes in addition to their well-established role in the immune response. In fact, the molecular diversity that allows for the recognition of self vs. nonself in an immunologic context may also confer unique odor profiles to individuals that allow for kin recognition. This would provide a biological mechanism for avoiding mating with close kin (or inbreeding) who would have familiar or similar HLA types.

Because of the enormous number of alleles at the HLA loci and the many factors that influence mate selection in most modern populations, it is difficult to study HLA-based mate choice in humans. However, founder populations that are derived from a relatively small number of ancestors and have remained reproductively isolated from other populations will have a limited repertoire of HLA alleles and haplotypes and often simpler and more homogeneous social structures. For example, an Anabaptist group living in South Dakota is derived from only 64 ancestors who lived in the early 1700s to early 1800s. This group has only 22 and 14 alleles at the HLA-B and HLA-DRB1 loci, respectively, and only 67 unique HLA haplotypes. Furthermore, their traditional and homogeneous lifestyle minimizes the number of factors that might influence mate selection in other populations. A study of HLA and mate choice in this population revealed a deficiency of couples matching for an entire HLA haplotype compared with expectations based on random mating with respect to HLA genotypes (Ober et al., 1997). This suggested that individuals in this population were not marrying individuals who had HLA types that were very similar to their own. Such a mating strategy would result in avoiding marrying close kin and ensure the production of heterozygous offspring, despite the limited number of HLA haplotypes present in this population. Although probably not relevant in most modern populations today, this may have played an important role during most of our evolutionary past, when humans lived in relatively small social groups. Further, the mechanism that allows us to discriminate among individuals based on their HLA types may still be present in modern populations (Wedekind and Furi, 1997). Indeed, and contrary to popular belief that humans no longer rely on their sense of smell for social cues, a recent study showed that women's preferences for male odors were strongly influenced by the number of HLA allele matches between them, with a preference for odors from men with a small number of matches (Jacob et al., 2002). Thus, close kin with a large number of matches, as well as 'strangers' with no matches, would be least preferred by these women. Although there is still much more to learn about HLA-mediated odor cues and human behavior, it is likely that many more of our behaviors are influenced by olfaction than previously thought.

Carole Ober, Department of Human Genetics, University of Chicago, USA

5.3.7 Genetic drift and disease heritages of isolated populations

Some human populations exhibit high incidences of multiple genetic diseases that are rare in surrounding populations. They appear to have a distinct heritage of genetic disease. These populations also show high frequencies of usually rare, but neutral, alleles. Often these groups are known to have undergone population processes that have resulted in small effective population sizes: for example founder effects (e.g., Finns [see *Box 14.1*], Afrikaners in South Africa) and

endogamy (within-group marriage, e.g., Roma, see Section 12.6.2).

However, in some cases, where there is good evidence that a disease allele has been imported into a population by a single founder, it appears that insufficient time has elapsed for genetic drift alone to account for the high frequency. An example is the increase in carrier frequency for the disorder of amino acid metabolism, Tyrosinemia I (OMIM 276700), from 1/5000 to 1/22 within 12 generations in the Saguenay-Lac Saint Jean (SLSJ) population of Quebec (Austerlitz and Heyer, 1998). In such cases it is tempting to invoke some form of selective process. However, a more sophisticated appreciation of the demographic factors underpinning genetic drift often provides an adequate explanation. In the SLSJ case, inheritance of family size, which was well documented in the genealogical records for this population, increases genetic drift sufficiently to account for the observed carrier frequencies (see *Box 5.2*).

5.4 Manipulating diversity: Selection

5.4.1 How does selection change allele frequencies?

Natural selection, as defined by Darwin and elaborated by Fisher, is the differential reproduction of genotypes in succeeding generations. Genotypic variation produces individuals with varying capacities to survive and reproduce in different environments. Selection can occur at any stage on the long journey from the formation of a genotype at fertilization to that individual generating viable progeny, including:

▸ survival into reproductive age – viability and mortality;

▸ success in attracting a mate – **sexual selection**;

▸ ability to fertilize – fertility and **gamete selection** (meiotic drive);

▸ number of progeny – fecundity.

The sum of these is the ability of an individual genotype to survive and reproduce, its **fitness**, which is partly dependent on the environment. The important factor is the relative fitness of a genotype compared to other genotypes competing for the same resources. Relative fitness is measured by a **selection coefficient** (s), which compares a genotype to the fittest genotype in the population. A selection coefficient of 0.1 represents a 10% decrease in fitness compared to the fittest genotype.

Mutations that reduce the fitness of the carrier are subject to **negative selection**, also known as purifying selection, whereas mutations that increase fitness undergo **positive selection**. However, to understand the dynamics of selection at diploid loci we must consider the impact of mutants on the fitness of the genotypes, and not on the individual alleles. The two alleles within a diploid genotype can interact to determine the phenotypic fitness of an organism in different

ways. This in turn affects the efficiency of natural selection in fixing or eliminating novel alleles. For example, a novel deleterious allele will be eliminated more rapidly from the population if it reduces the fitness of a heterozygote as well as the homozygote. This situation is known as **codominant selection**. Alternatively, a new allele may increase the fitness of a heterozygote relative to that of both homozygotes. The two homozygous genotypes may exhibit different reductions in fitness (s_1 and s_2). Such selection is known as **overdominant selection** and creates a balanced polymorphism. By contrast, **underdominant selection** operates where new alleles reduce the fitness of the heterozygote alone. These different selective regimes are summarized in *Table 5.1*.

Overdominant selection is not the only mechanism by which balanced polymorphisms can be generated, but is one of a number of processes described collectively as **balancing selection**. One example of an alternative mechanism for balancing selection is that of **frequency-dependent selection**, whereby the frequency of a genotype determines its fitness. If a genotype has higher fitness at low frequencies but lower fitness at higher frequencies, an intermediate equilibrium value will be reached over time. *Box 5.5* describes the *MHC* locus, which has been suggested to be under both frequency-dependent and overdominant selection (also known as **heterozygote advantage**).

Other classic examples of balanced polymorphisms in humans are those that protect against malaria when heterozygous but have a reduced fitness compared to wild-type when homozygous, as a result of red blood cell disorders. A number of these types of balanced polymorphisms have arisen in different areas of malarial **endemicity**, the best known of which is the sickle cell anemia (OMIM 603903) allele Hb^S (Haldane, 1949). This allele dramatically reduces fitness when homozygous. Malarial endemicity is not spread equally across the world, and as a consequence these balanced polymorphisms exhibit a limited range that closely parallels that of malaria (*Figure 5.10*). Arguments have been made that alleles causing other common recessive genetic disorders have reached high frequencies as a result of overdominant selection. For example, the cystic fibrosis (OMIM 219700) recessive allele may provide protection against cholera and other diarrheal diseases (Section 14.2.1). Many of these claims remain difficult to prove, since direct experiments cannot be easily done. Proof in the case of malaria came from attempts to

TABLE 5.1: DIFFERENT TYPES OF SELECTION, AND THEIR EFFECT ON GENOTYPE FITNESS.

Type of selection	Genotype fitness		
	AA	Aa	aa
Simple negative/positive selection	1	1	1−s
Codominant (genic) selection	1	1−s	1−2s
Overdominant selection	1−s_2	1	1−s_1
Underdominant selection	1	1−s	1

BOX 5.5 Locus briefing: The major histocompatibility locus

What is it?

When a tissue is transferred from one individual to another, it may be rejected or accepted by the host immune system; this is known as **histocompatibility**. Although a number of loci throughout the genome are involved in histocompatibility, in humans the major determinants are found in a large gene cluster on chromosome 6 known as the 'major histocompatibility (MHC) locus'. The different MHC-encoded proteins that can be recognized by the immune system are cell-surface proteins that are known collectively as the human leukocyte antigen (HLA). The HLA includes proteins that are expressed on all nucleated cells.

How is the locus arranged?

Due to its medical importance, the gene-dense MHC locus was one of the first large regions to be sequenced during the human genome project (MHC Sequencing Consortium, 1999). The 3.6-Mb locus is divided into three regions, called classes (see *Figure*). Ancient gene duplication events have generated several expressed HLA genes and many pseudogenes within the class I and II regions (Beck and Trowsdale, 2000). These HLA genes are involved in the development of **adaptive immunity** through the presentation of bacterial and viral **antigens** to T lymphocytes. Different alleles at each individual gene vary in their ability to present antigens from different pathogens.

How diverse is it?

Perhaps the most unusual feature of the MHC is the huge amount of variation contained within it. The HLA Sequence Database (http://www.ebi.ac.uk/imgt/hla/) currently contains 1620 allele sequences from 19 different loci within the MHC, see *Table*.

As well as the sheer number of alleles, the differences between them are often many times greater than at other loci (many alleles at the same HLA locus differ by 5–17% of all nucleotides, whereas most alleles at other loci differ by less than 1%), indicating that their common ancestry is ancient. Many of these alleles are so old that they predate the human–chimpanzee split: that is, a human allele may be more closely related to a chimpanzee allele than to an alternative human allele, a characteristic known as trans-species polymorphism. In addition, there is also variation in the copy number of HLA-DRB genes.

Why is it so diverse?

High MHC diversity within modern humans could be explained by selection for diversity or by an elevated mutation rate. However, trans-species polymorphism can only be explained by the operation of selection in preventing the fixation of alleles over time. Traditionally, two alternative selection pressures have been proposed, although they are not mutually exclusive (Cooke and Hill, 2001):

▶ heterozygote advantage – individuals with heterozygous MHC haplotypes are better able to resist infectious disease as a result of having a broader spectrum of antigen binding specificities;

▶ Frequency-dependent selection – low frequency alleles are favored if pathogens have evolved to evade immune detection in individuals carrying the higher frequency alleles.

A role for selection is supported by the concentration of variation in the exons coding for the antigen-binding groove of the protein. Noncoding variation peaks around variable genes, and is therefore **hitch-hiking** along with the balancing selection operating at these loci (Beck and Trowsdale, 2000).

What role does recombination play in generating diversity?

Recombination within heterozygous HLA genes creates new alleles and interallelic and intergenic gene conversion generates additional variation. The MHC exhibits a high degree of **linkage disequilibrium** (LD) that most likely results from the localization

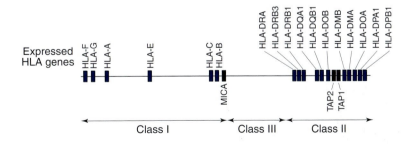

Structure of the MHC region on chromosome 6, showing the division into three classes and the location of the genes found in the HLA sequence database.

of recombination events to certain hotspots within the locus, between which lie long regions of low recombination (Kauppi *et al.*, 2003). As a consequence, linked MHC genes are frequently co-inherited in **haplotype blocks** (see Section 14.3.3), making it easy to identify disease-related haplotypes, but difficult to locate disease-related alleles to a single gene. For example, the tightly linked class II loci, DRB1, DQA1 and DQB1, are often found to be in complete LD.

But what about selection?

Through linkage analysis and **association studies**, numerous MHC haplotypes have been associated with susceptibility to, and protection against, different diseases. These include infectious diseases (Cooke and Hill, 2001) (e.g., the protective effect of HLA-B*5301 against malaria), autoimmune disorders [e.g., susceptibility to multiple sclerosis (OMIM 126200) conferred by HLA-DRB1*1501] and other diseases [e.g., HLA-DQB1*0602 predisposes to narcolepsy (OMIM 605841)]. Because of the medical importance of these associations and the difficulties in disentangling them, the MHC haplotype project (http://www.sanger.ac.uk/HGP/Chr6/MHC/) aims to sequence eight full MHC haplotypes that are commonly associated with type 1 diabetes (OMIM 222100) and multiple sclerosis.

Can HLA variation be used to explore the human past?

The current geographical distribution of HLA alleles is shaped to some degree by events in the human past. High diversity at the protein level facilitated extensive study of these loci prior to the advent of DNA-based methods; HLA variation was the most important marker system in Cavalli-Sforza and coauthors' compendium of protein variation (Cavalli-Sforza *et al.*, 1994; see *Table 10.5*). However, the association of different MHC haplotypes with many different diseases raises the possibility that the spatial distribution of HLA alleles may be shaped, not by population history, but by different selective environments. Selection can be expected to skew the frequencies not only of disease-related alleles, but also of alleles at any linked loci. Consequently, anthropological studies of HLA variation have become less popular in recent years.

How are different alleles named?

The **nomenclature** for the different alleles has been complicated by the use of two different methods to define alleles. Initially, serological methods that detect some but not all variation at the protein level were used to identify alleles. More recently, direct analysis of DNA sequences at this locus has revealed that multiple alternative DNA sequences can encode the same serologically-detected allele. Thus the nomenclature has evolved to include information on the gene at which the allele is found, the serological allele, and the underlying DNA sequence (Marsh *et al.*, 2002). For example, the serologically-defined allele HLA-A1 (the first allele at the A locus within the HLA) can be encoded by the DNA allele HLA-A***01**01 or HLA-A***01**02. The first two numbers (in bold) define the serological allele, and the second two numbers define different nonsynonymous changes that yield different protein alleles with the same **immunoreactivity**. A third level of numbers can be added to indicate any synonymous changes that might be present. So, for example, the two alleles that give the same HLA-A*0101 protein sequence but differ by a mutation that does not cause an amino acid change, are defined as HLA-A*010101 and HLA-A*010102. As an added complication, some of the serologically-defined alleles have been given new names for the purposes of the DNA naming system; so HLA-DR17 has become HLA-DRB1*03, of which there are 25 nonsynonymous alleles (HLA-DRB1*0301 to HLA-DRB1*0325), some of which have synonymous variants (e.g., HLA-DRB1*030501 and HLA-DRB1*030502).

Class I		Class II		Class III	
Gene	Number of alleles	Gene	Number of alleles	Gene	Number of alleles
HLA-A	266	HLA-DRA	3	TAP1	6
HLA-B	511	HLA-DRB(1–9)	403	TAP2	4
HLA-C	128	HLA-DQA1	23	MICA	54
HLA-E	6	HLA-DQB1	53		
HLA-F	1	HLA-DPA1	20		
HLA-G	15	HLA-DPB1	101		
		HLA-DMA	4		
		HLA-DMB	6		
		HLA-DOA	8		
		HLA-DOB	8		

infect volunteers of different genotypes with the malaria parasite, supplemented by hospital observations of malarial mortality. The relative incidence of fatal infections is significantly lower amongst individuals carrying the sickle cell allele.

5.4.2 The dynamics of selection

Even small selective forces are capable of causing appreciable changes in allele frequencies over many generations. The intergenerational change in allele frequencies can be calculated by incorporating the selective coefficients described in *Table 5.1* into the Hardy–Weinberg theorem. *Figure 5.11* compares the selection dynamics of a low frequency, advantageous, allele under positive and codominant selection. It can be seen that selection achieves the most rapid changes in allele frequencies when alleles are at intermediate frequencies. However, in Section 5.6 we shall see that other forces acting on allele frequencies may outweigh small selective forces.

5.4.3 Sexual selection

In cases of limiting numbers of mating partners, selection can operate at the level of **mate choice**. For humans, where the levels of investment of males and females in their offspring are unbalanced, the availability of females is the factor limiting male reproduction and therefore females can exercise mate choice. Desirable traits may be those that indicate health, access to resources and ability or willingness to invest in offspring. An alternative mechanism of sexual selection in response to limited female mating resources is that of competition between males:

'We may conclude that the greater size, strength, courage, pugnacity, and energy in man, in comparison with woman, were acquired during

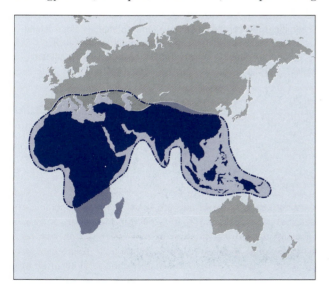

Figure 5.10: The overlapping geographical distributions of red blood cell disorders and malaria.

Blue indicates regions of current malarial incidence, and the shading shows the distribution of red blood cell disorders (adapted from Cooke and Hill, 2001).

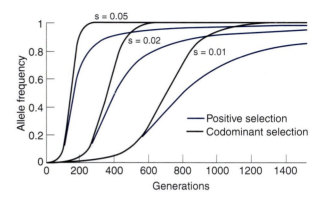

Figure 5.11: Positive and codominant selection for an advantageous allele.

The selection dynamics of a low frequency, advantageous, allele are compared under positive and codominant selection. Selection achieves the most rapid changes in allele frequencies when alleles are at intermediate frequencies.

primeval times, and have subsequently been augmented, chiefly through the contests of rival males for the possession of the females.'
Charles Darwin, Descent of Man (1871)

If these attractive or competitive traits are to some degree genetically determined then these loci are said to be under **sexual selection**. Mate choice can be differentiated from assortative mating on the basis that specific preferences are shared among all members of the same sex.

Darwin invoked sexual selection (see Section 13.3.5) to explain the presence of secondary sexual characteristics among humans. Others have proposed that the human mind itself is largely a result of this selective process (Miller, 2000). Human traits that may be under sexual selection are the result of complex multigenic inheritance that has yet to be elucidated. Consequently, molecular investigations of the relevance of this selective mechanism to human evolution have not yet begun. Studies of sexual selection in humans have limited themselves to observing indirect correlations between phenotypes and reproductive success, a recent example being the apparent greater number of offspring born to taller men (Pawlowski *et al.*, 2000). Sexual selection, whether genetically or culturally determined, could be expected to lower the effective population size of the selected sex by increasing the reproductive variance.

5.5 Migration

Unlike genetic drift, mutation and selection, migration cannot change species-wide allele frequencies, but it is capable of changing allele frequencies within a sub-population. It thus belongs to a second tier of population processes that shape human genetic diversity. As noted previously, gene flow counteracts genetic differentiation and is modeled within the framework of a larger, subdivided, meta-population.

First, we must be clear on some definitions, because they are often used interchangeably in the literature. **Colonization** is the process of movement into previously unoccupied land, thus entailing a founder effect. By contrast, **migration** is the movement from one occupied area to another. **Gene flow** is the *outcome* when a migrant contributes to the next generation in their new location. Thus, to observe gene flow directly we not only need to monitor the movement of migrants but also their reproductive success. Estimates of gene flow have, therefore, relied upon indirect methods that relate measures of population subdivision to gene flow via a model for the population structure. Often migration is used as a synonym for gene flow.

5.5.1 Models of migration

Perhaps the simplest model of gene flow is the **island model** devised by Sewall Wright. A meta-population is split into 'islands' of equal size N, which exchange genes at the same rate, *m*, per generation. Under the assumptions of this model the rate of migrant exchange can be related directly to a measure of population subdivision (F_{ST}), by the equation:

$$F_{ST} = 1/(1 + 4Nm)$$

The assumptions of the island model include:

- no geographical substructure apart from the division into islands: all islands are equivalent;
- each population persists indefinitely;
- no mutation;
- no selection;
- each population has reached equilibrium between mutation and drift;
- the migrants are a random sample from the source 'island'.

The **stepping-stone model** (Kimura and Weiss, 1964) seeks to remove one obvious flaw of the island model – the lack of geographic substructure. The stepping-stone model introduces the idea of geographical distance by only allowing the exchange of genes between adjacent discrete sub-populations. *Figure 5.12* shows a comparison of the island and stepping-stone models. The stepping-stone model also assumes equal rates of migration between subpopulations. Both kinds of model have been used to show that even very low rates of migration between subpopulations are capable of retarding their genetic differentiation.

Migration can be modeled as occurring within a continuous population, rather than discrete subpopulations, by considering that mating choices are limited by distance, and that these distances are typically less than the overall range of the population. This is the basis for **isolation by distance** (IBD) models (Wright, 1943; Malécot, 1948). Within such models, genetic similarity develops in neighborhoods as a function of dispersal distances. These can be thought of either as the difference between birthplaces of parent and offspring, or marital distances. Different mathematical functions have been used to relate the decline

Island model

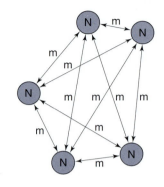

Stepping-stone model

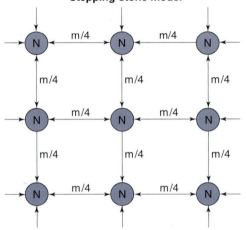

Figure 5.12: The island and stepping-stone models of gene flow.

Each diagram represents one of a family of models: the n-island model, and the two-dimensional stepping-stone model, also known for obvious reasons as the lattice model. N, Population size; m, rate of exchange of genes per generation.

in frequency of these dispersals over geographical distance. Once the system has reached equilibrium between gene flow and the differentiation caused by genetic drift, genetic similarity declines over distance in a predictable fashion. The stepping-stone model described above is a discontinuous example of IBD.

These migration models are mathematically tractable and can be generalized to many species. However, for many human populations (unlike those of other species) we often have detailed data on parameters such as migration rates, migration distances and marital distances. The migration matrix model uses this detailed information and thus can incorporate different migration rates and asymmetric migration between subpopulations. In this way, a more complex and realistic relationship between distance and migration is obtained. Nevertheless, it seems unlikely that present day migration rates have been constant for long enough to allow the system to reach the state of equilibrium required by such models.

5.5.2 Population structure and migration in real populations

Clearly, the clumped pattern of most human habitation across the planet falls between the models of discrete subpopulations and uniform continuity assumed by the stepping-stone and isolation by distance models respectively. Furthermore, migration processes are far more complex than the current models allow. Migration processes often include long-distance movements as well as smaller scale mating choices. The choice to migrate is taken by individuals on the basis of multiple 'push' and 'pull' factors, so that migration rates are rarely, if ever, symmetric between two populations. Migrants are seldom a random sample of their source population; they are often age-structured, sex-biased (see *Box 5.6*) and related to one another. The latter property of migrants is known as **kin-structured migration** and is well documented both ethnographically (Fix, 1999) and archaeologically (Anthony, 1990). In light of these complications, we should be cautious in attempting to estimate parameters of population structure and be skeptical of their relationship to reality.

5.6 Interplay among the different forces of evolution

5.6.1 Equilibria between opposing forces

Thus far, we have examined individually some of the important factors influencing the level of variation in a population. Mutation, recombination and migration increase diversity, random genetic drift decreases it, and selection can do either. In this section, we investigate how these opposing forces interact with one another.

Migration drift equilibrium
In the previous section we saw that in a subdivided population the opposing forces of migration and drift can reach an equilibrium state whereby differentiation among subpopulations, as measured by F_{ST}, remains constant over time. It is only by assuming that this equilibrium has been attained that we can estimate migration rates from F_{ST} values in real populations. There are other similarly important equilibria in population genetics.

Mutation drift equilibrium
In the simplest model of a population with no selection or migration and the usual assumptions of constant size and random mating, diversity will reach an equilibrium value where the number of novel variants (generated by mutation) entering the population is balanced by the number lost by drift. This is known as **mutation drift equilibrium.** John Relethford uses a simple analogy to illustrate this point (Relethford, 2001): imagine a paper cup with a hole in the bottom under a dripping tap (*Figure 5.13*). Water will accumulate in the cup until the amount entering from the tap (= mutations) is balanced by the amount lost through the

hole (= drift), leading to a stable water level (= diversity). If the mutation rate increases (more water in) or decreases (less water in), diversity at equilibrium will increase or decrease correspondingly. Similarly, if drift increases (larger hole) or decreases (smaller hole), diversity will decrease or increase. This equilibrium value of diversity is known as the **population mutation parameter (θ or 'theta')**, and is discussed in more detail in Chapter 6.

Recombination drift equilibrium
Earlier in this chapter we saw that in an infinitely large population linkage disequilibrium decays over time as a result of recombination generating new haplotypes. However, random genetic drift is continually removing haplotypes from the population. As a consequence LD can reach an equilibrium value in finite populations. This equilibrium value of LD is determined by the **population recombination parameter (ρ or 'rho')**. This parameter combines information on effective population size and recombination rate (c) using the equation:

$$\rho = 4N_e c$$

The precise relationship between measures of LD (such as D, $|D'|$ and r^2) and ρ is complex, but under certain conditions:

$$r^2 = 1/1 + \rho$$

so that when ρ is large

$$r^2 \approx 1/\rho$$

Thus we can see that LD decreases as ρ increases, for example as a result of a larger effective population size (lower genetic drift) or higher recombination rate.

In real populations, patterns of LD are only beginning to be described. Yet, it has already become apparent that LD is not simply an equilibrium between recombination and drift, but can be greatly affected by selection, mutation, gene conversion and demography. These influences on LD are discussed in greater depth in *Box 5.7*. Software for analyzing LD is shown in *Table 5.2*.

Because demography influences LD, analysis of LD within a population can allow inferences to be made on the prehistoric demography of that population (for an example of how patterns of LD can reveal ancient admixture events, see Section 12.5). These relationships between LD and demography are of use for medical genetics studies, which are discussed in more detail in Chapter 14. LD can be used to track disease alleles through the nonrandom association of a neutral marker with the disease. The greater the LD within a population, the lower the density of markers that need to be screened to identify a chromosomal region linked to the disease allele. However, the fine mapping of a disease gene within a previously defined interval requires much smaller amounts of LD. The population to be screened can be chosen on the basis of the desired extent of LD, through knowledge of its demographic prehistory.

If we are examining LD over a large genomic region containing many polymorphic markers, it is unclear how best to combine the information from measures of LD based on

BOX 5.6 Sex-specific differences in migration

If we consider possible differences between the sexes in their migration behavior, an intuitive hypothesis on observing the modern world might be that men tend to migrate over longer distances than women: intercontinental migrants tend to be male-biased and recent history documents explorers, traders and soldiers as being almost exclusively male. However, when considering the impact of migration on genetic diversity, we must not only examine long distance migration patterns but also small-scale local migrations.

Marital residence patterns are critical to investigating local migration patterns. **Patrilocality** describes the phenomenon by which a female from one village, when marrying a man from a different village, take up residence in the man's village. In contrast, **matrilocality** describes the situation when the husband moves to the wife's village. It has been estimated that roughly 70% of modern societies are patrilocal (Murdock, 1967; Burton *et al.*, 1996). In other words, in the majority of societies, mtDNAs are moving between villages each generation, whereas Y chromosomes are staying put. Similarly, the X chromosome is more mobile than the autosomes, as it passes down the female line twice as often as it does down the male line.

How might we determine which of the above current phenomena – the apparent male bias of long-distance migration or the female bias of intergeneration marital movement – has had a greater role in shaping modern genetic diversity?

To resolve this issue, we can compare the geographical patterning of genetic diversity at loci that have different inheritance patterns. Migration reduces population differentiation, and so by studying the relationship between geographical distances and genetic distances among the same set of populations, we can identify which loci appear to have been experiencing greater gene flow. Panel a of the *Figure* shows a schematic representation of the genetic patterning of mtDNA and Y-chromosomal diversity found in two of the best studied global regions: Europe (Seielstad *et al.*, 1998) and the islands of South-East Asia, Near Oceania and Australia (Kayser *et al.*, 2001). In both of these regions the Y chromosome exhibits greater genetic differentiation than mtDNA, indicating a higher female migration rate in the past. This suggests that the more local sex biases outlined above have outweighed the long-distance processes.

(a)

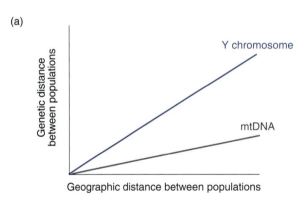

(b)

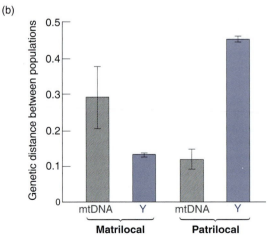

This kind of interlocus comparison assumes that migration rate is the only factor determining population differentiation, however, genetic drift also influences levels of population subdivision, and, as we saw in *Box 5.1*, the effective population sizes of the Y chromosome and mtDNA need not be equal. Consequently it has been argued that these differences between loci reflect differences in genetic drift as a result of sex differences in reproductive variance, and not migration rates. To resolve this issue, it has been demonstrated that populations with a matrilocal pattern of marital residence do indeed exhibit the converse pattern to that described above (Oota *et al.*, 2001), namely of greater mtDNA than Y-chromosomal genetic differentiation (see panel b of the *Figure*). Thus it is suggested that a sex-specific difference in migration rate and not drift is the primary factor influencing the discrepant patterns of genetic differentiation between loci with opposing sex-specific patterns of inheritance.

comparisons between individual pairs of markers (i.e., D, |D′| or r^2). Therefore recent attention has focused on estimating ρ itself for these kinds of data, as this gives a single measure of LD for the entire region. Estimating ρ requires the use of population models and is computationally intense, but it allows the other forces that shape LD (e.g., demography and mutation) to be taken into account (reviewed by Pritchard and Przeworski, 2001). An additional

advantage of studying ρ is that it allows c – the recombination rate across the region – to be estimated, and compared between different regions of the genome.

Mutation selection equilibrium

Some deleterious mutational events, such as certain chromosomal rearrangements (see Section 3.5), have sufficiently high mutation rates that within a large population they occur several times within a single generation, and can be considered recurrent mutations. Mutation and selection are opposing forces determining the frequency of such mutant alleles in the population. The rate at which new mutations are generated can be balanced by the eventual elimination of each mutant by selection so that the average number of a given mutation reaches an equilibrium value within the population. Lowering the selective cost of a mutation, or increasing its mutation rate, will increase this equilibrium value.

In addition, rather than considering a single recurrent mutation in isolation, if we consider all deleterious alleles together, a balance between mutation and selection may operate over the genome as a whole, such that at equilibrium each genome contains a certain number of deleterious alleles.

5.6.2 Does selection or drift determine the future of an allele?

So far in this section we have not considered the interplay of selection and drift, and their relative weight in influencing allele frequencies. For example, the selection dynamics described in Section 5.4.2 assume an infinitely large population. What happens in finite populations where random genetic drift is also operating?

Because drift operates more effectively in smaller populations, stronger selection is required to influence fixation or elimination. Whether drift or selection predominates depends on a number of factors, which include:

▸ the effective population size;
▸ the selection coefficient;
▸ the type of selection;
▸ the frequency of the allele under selection.

Equations relating these parameters exist for different types of selection. They can be used to determine whether an allele is

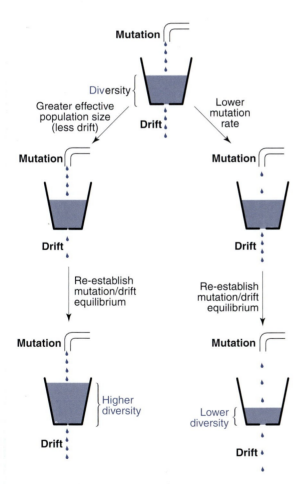

Figure 5.13: A metaphorical depiction of the relationship between mutation rate, drift and diversity.

A change in either the mutation rate or effective population size changes the diversity at mutation-drift equilibrium – see text.

BOX 5.7 Factors affecting patterns of LD

Mutation

For slowly mutating markers such as SNPs, recurrence and reversion of mutations is too rare to have any appreciable effect; for markers such as microsatellites, mutation rate is higher and the lack of identity by descent could mean that patterns of apparent association would need careful interpretation.

Recombination rate heterogeneity

Recombination across the genome is not uniform (see Section 3.7). The presence of recombination hotspots in human DNA has long been suspected, and has now been demonstrated empirically from direct studies in sperm DNA close to a minisatellite (Jeffreys *et al.*, 1998) and within the major histocompatibility complex (Jeffreys *et al.*, 2001). The extent to which this is a general property of the genome remains unclear. Nonetheless, recombination rate heterogeneity can be expected to result in large differences in LD among different genomic regions.

Demography

Stochastic changes in allele and haplotype frequencies occur every generation through the random sampling of the gametes of one generation to form the next (see Section 5.3). In small populations, these changes are magnified, and should lead to an increase of LD as more haplotypes are lost. Isolated populations with small effective population sizes have been shown to have high LD, compared to older, larger populations. Conversely, rapid population growth should decrease LD, and recently expanded populations have been demonstrated to exhibit decreased LD relative to constant sized populations.

Admixture

Immediately after admixture of two populations (see Chapter 12) having different allele frequencies, LD is related not to the distance between the markers, but to the interpopulation allele frequency differences. 'Spurious' LD can exist in the early stages after admixture, even between unlinked alleles. Admixture between populations should also elevate LD among linked alleles in the resulting mixed population, as a result of different haplotype frequencies in the ancestral populations.

Natural selection

Positive selection for an allele (see Section 5.4) can lead to 'hitch-hiking' of a segment of DNA with it, and a 'selective sweep' of markers in the segment through the population, with associated elevated LD. Effects of negative selection are more subtle.

Gene conversion

The nonreciprocal transfer of genetic information from one allele to another, gene conversion (see Section 3.8), can also affect LD. For example, LD between markers separated by very short physical distances (~ 100 bp) is less than expected, and this has been ascribed to gene conversion (Frisse *et al.*, 2001).

---

likely to be under the influence of selection. Relating these parameters together allows us to draw four important conclusions:

1. Although selection often substantially increases the probability that an advantageous allele becomes fixed compared to a neutral allele, in humans, most new advantageous alleles are still far more likely to be eliminated than fixed.

2. If new alleles are almost exclusively deleterious, the optimal allele can persist unchanged over very long time scales. This conforms to the hypothesis that functional constraint on important proteins such as **histones** underlies their extreme lack of variability among diverse species.

3. The time taken to fix an advantageous allele is much shorter than that to fix a neutral allele.

4. A general rule for diploid loci is that for selection to be operating then the following relationship should hold:

$$s > 1/2N_e$$

For **haploid** loci with one-quarter the effective population size of diploid loci, the relevant rule is:

$$s > 2/N_e$$

The use of this last rule can be seen in the following example. A polymorphic inversion on the human Y chromosome, present in roughly 70% of British males, protects against XY translocations during meiosis. The offspring resulting from

---

TABLE 5.2: SOFTWARE: METHODS FOR ANALYZING LD.

Purpose	Software	URL
Estimating pairwise measures of LD (D' and r^2)	DNASP	http://www.ub.es/dnasp/
Estimating the population recombination parameter, ρ	LDhat	http://www.stats.ox.ac.uk/~mcvean/

these rearrangements (XX males and XY females) are infertile; infertility is evolutionary death for the individual. However, these translocations occur only at low rates, such that the selective advantage (s) of this inversion has been calculated as 1/90 000 (Jobling *et al.*, 1998). Given that the Y chromosome is a haploid locus and the effective population size (N_e) of humans is 10 000, the selective advantage of this inversion is not sufficiently large to overcome the effects of drift (1/90 000 < 1/5000).

5.6.3 The neutral theory

The removal of individuals with deleterious alleles from a population imposes a cost in the form of selective deaths. If these costs are cumulative then it suggests that there is a limit to the amount of selection a population can support without going extinct. The presence of deleterious alleles imposes a **genetic load**. Overdominant selection, as well as negative selection, contributes to the genetic load. Before information on molecular diversity became available it was hypothesized that genetic load would mean that only a limited amount of polymorphism would be compatible with a sustainable population. Muller famously predicted that only one in a thousand genes would be heterozygous. According to this view, polymorphisms were stable entities, maintained by balancing selection. By contrast, substitutional processes were independent, being the result of positive selection.

Starting with protein electrophoresis, developed in the 1950s, and culminating in the SNP consortium data of today, the amount of polymorphism uncovered within human genomes, and those of other species, is many times greater than expected by this 'selectionist' viewpoint. One possible explanation is that the costs of harboring deleterious alleles have been overestimated, because the genetic load is minimized by the presence of recombination and **epistatic** interactions between deleterious alleles. Recombination enables deleterious mutations to be pooled in the same individual genome without changing the number of mutations in the population, and subsequently the deleterious mutations can be eliminated with a minimal number of selective deaths. This is enhanced by interactions among the deleterious mutations, resulting in a reduction in fitness that is more than simply multiplying together the negative effects of the individual alleles. Gene products often operate in metabolic, signaling and developmental pathways. A given pathway may be buffered against a single deleterious mutation, but if this allele is combined with other mutations in the same pathway the buffering capacity can be overwhelmed, resulting in a dramatic decrease in fitness. There is, however, an alternative explanation, the **neutral theory of molecular evolution** (Kimura, 1968).

Neutral theory states that rather than being the product of different types of selection, most polymorphisms and substitutions represent the same process, namely the random change in frequency by genetic drift of alleles that are neutral (i.e., have no effect on fitness). According to this view, polymorphisms are transient entities, awaiting eventual fixation or elimination by genetic drift. A role for negative selection is allowed within neutral theory to eliminate the deleterious minority of mutations that arise in regions of functional constraint. However, cases of positive and balancing selection are regarded as rare events. The neutral hypothesis led to many predictions for both inter- and intraspecific molecular data, which could be compared with selectionist expectations. These comparisons have produced qualified successes for both viewpoints. The substantial developments resulting from neutral theory have been the greater appreciation of the role of genetic drift and a body of analytical methods designed to detect the imprint of selection (see Section 6.3). In addition, neutral theory suggests that the rate of sequence evolution may be driven solely by the rate of mutation, which provides support for the molecular clock hypothesis that is central to our ability to gain information on the timing of evolutionary events.

5.6.4 The molecular clock hypothesis

So far we have considered mutation as a random, low probability, event. The radioactive decay of the isotope carbon-14 is an analogous event, and has been used to relate carbon isotope ratios to time in archaeological dating. Can measures of genetic diversity be used similarly to date the divergence of lineages (Zukerkandl and Pauling, 1965)?

The putative regularity of mutation might not translate into a regularity of evolution if selection plays a dominant role in determining the survival of new mutations in some lineages. Alternatively, if drift predominates it can be shown that the rate of substitution is equal to the rate of mutation. This is an important finding from neutral theory. The rate of nucleotide substitution (k) is equal to the rate at which new mutations are generated, multiplied by their probability of fixation (u). In a population of size N, diploid loci have a population size of 2N, and the rate of nucleotide substitution (k) is therefore:

$$k = 2N\mu u$$

Remember that for a neutral mutation the probability of fixation is its frequency, which for a new mutation is the reciprocal of the population size (1/2N).

$$k = 2N*1/2N*\mu \qquad \text{therefore} \qquad k = \mu$$

Nonetheless, perhaps mutation rates measured per year are not equal across all lineages. This alone would cause lineages to diverge at different rates.

The molecular clock hypothesis holds that for any given DNA sequence, the rate of evolution is approximately constant over all evolutionary lineages. This regularity of molecular evolution would stand in direct contrast to the nonuniform change of morphological evolution. We will address applications of the molecular clock hypothesis and genetic dating in the next chapter; however, here we need to consider whether this is a reasonable approximation of the evolutionary process.

In principle, this hypothesis could be tested using known calibration points of dated lineage divergences. This requires accurate, independent dating of the lineages being investigated by other disciplines, most notably paleontology. However, fossil dates are often unreliable so a different test is commonly used. The **relative rates test** does not require absolute divergence times, but only knowledge of the order in which a number of lineages diverged from one another (Sarich and Wilson, 1973). This test (*Figure 5.14*) compares the rate of evolution in two lineages by relating them to a third, which is known to be an outgroup (a lineage known to be more distantly related to the other two lineages than they are to one another). Although the lineages in question are often individual species, the test can also be used on nonrecombining regions of the genome within a species (where the phylogeny is known). The number of mutational events in each branch of the tree relating the three lineages is calculated. The significance of any differences between the mutational distances shown in *Figure 5.14* can be assessed by a number of different methods. The relative rates test has been used to demonstrate a mutation rate slowdown in hominoid species as compared to other higher primates (see Section 7.4.4).

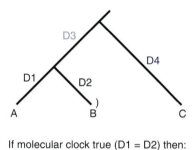

If molecular clock true (D1 = D2) then:

$$D_{AC} = D_{BC} (D1 + D3 + D4 = D2 + D3 + D4$$

Thus, test to see if $D_{AC} - D_{BC} = 0$

Figure 5.14: Testing the molecular clock with the relative rates test.

The rate of evolution in two lineages, leading to A and B, is compared with that in an outgroup lineage, C. D1 to D4 are mutational distances on different branches of the phylogeny relating the three species.

Proving that the molecular clock hypothesis is false for a given set of homologous sequences does not mean that these sequences cannot be used as a means of relating genetic divergence to time. Rather, the clock requires some calibration if it is to be accurate. In an analogous fashion, radiocarbon dating requires calibration for the amount of carbon-14 in the atmosphere at the time that the dated sample was absorbing carbon (*Box 9.4*).

5.6.5 Problems with the molecular clock hypothesis

Comparisons of rates of evolution between species have often shown significantly different rates of nucleotide substitution evolution. When the same rates are found across

multiple loci it suggests that different annual mutation rates rather than selection are the cause of rate inequalities. There are several possible explanations, which relate the mutation rate to biochemical processes within the cell (**lineage effects**). The different processes that have been proposed to cause lineage effects need not be mutually exclusive, but could operate in concert (*Figure 5.15*).

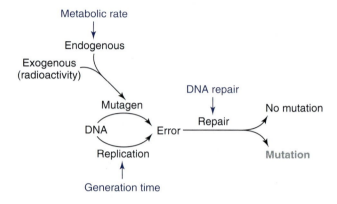

Figure 5.15: Different sources of lineage effects.

A number of processes are involved in generating a new mutation. Potential sources of lineage effects are shown in blue.

The lineage effect that has received most attention is the **generation time hypothesis**. This assumes that most mutations occur during DNA replication in the germ-line. As a consequence, the mutation rate is determined by the number of replications during a certain period of time. If species have the same number of cell divisions per generation then the mutation rate becomes dependent on the generation time. This hypothesis of replication errors causing mutations also underpins the hypothesis of **male-driven evolution** (see *Box 5.8*).

The generation time hypothesis has gained support from studies of mammalian species that show a better correlation of mutational distance with generation time than with calendar time. Many studies have demonstrated that while generation times within mammalian groups have the order rodents < monkeys < humans, evolutionary rates have the order rodents > monkeys > humans, both for indels and substitutions. Similarly the mutation rate among higher primates seems to be negatively correlated with **generation time**. The mutation rate increase in rodents does not appear to be linear with respect to generation length as there is only a two- to fourfold increase relative to humans despite a 40-fold difference in generation times; however, the difference in replications per generation can account for some of this discrepancy. Human males have four times more replications per generation than do male mice, and this would be expected to lessen rate differences. Further problems for the generation time hypothesis appear in the difference in the substitution rate ratio of synonymous and

BOX 5.8 Male-driven evolution

J.B.S. Haldane first proposed that the greater number of genome replications in the male germ-line should result in the male mutation rate being higher than that of the female (Haldane, 1947). His hypothesis is based on the same assumption – that mutation is caused by errors in replication – as the generation time hypothesis discussed in the text (Hurst and Ellegren, 1998).

In humans, the number of replications required during oogenesis is constant, at about 22, whereas the number of replications in the male germ-line increases with age (see *Figure 3.24*). Thirty replication events are required to generate stem spermatogonia at puberty (~13 years old), which then go through ~23 replications per year, before the final five replications required to make mature spermatozoa. Thus a male reproducing at 25 will be using DNA that has gone through $30 + 23*(25 - 13) + 5 = 311$ replication cycles. This is about 14 times as many as for the oocyte DNA.

The ratio of male to female mutation rates (α – also known as the **alpha-factor**) can be calculated by comparing the number of mutations that have accrued in homologous autosomal, Y-chromosomal and X-chromosomal sequences over the same time period. The ratios of these amounts can be related to α by the following equations:

$X/A = 2/3(2 + \alpha)/(1 + \alpha)$

$Y/A = 2\alpha/(1 + \alpha)$

$Y/X = 3\alpha/(2 + \alpha)$

Many estimates for α have been generated for higher primates, and they vary substantially between studies as shown in the *Table*. Nonetheless, they all indicate that the mutation rate is significantly higher in the male than in the female germ-line. More recent estimates have tighter confidence limits as a result of being based on longer sequences. The most recent study in the *Table* also incorporates a consideration of the diversity apparent in the human-chimpanzee common ancestor, which allows more accurate estimation of α than the equations given above.

α (95% CI)	Comparisons	Reference
∞	Y/A and X/A	Miyata *et al.*, 1987
6 (2–84)	Y/X	Shimmin *et al.*, 1993
2.5	Y/A, X/A and Y/X	Erlandsson *et al.*, 2000
1.7 (1.2–2.9)	Y/X	Bohossian *et al.*, 2000
2.8 (2.3–3.4)	Y/X	Ebersberger *et al.*, 2002
2.6 (2.2–3.2)	Y/A	Ebersberger *et al.*, 2002

nonsynonymous mutations between primates and rodents, suggesting a potential role for selection. In addition, there remain a number of cases where calendar time, rather than generation time, is better correlated with rates of evolution, even after correcting for the differing number of replications per generation among the different species.

The **metabolic rate hypothesis** is predicated on the idea that most mutations result from the presence of endogenous mutagens (Martin and Palumbi, 1993). Free radical byproducts of aerobic respiration are the prime suspects, and, therefore, organisms with higher metabolic rates should produce more mutagenic free radicals. Differences in metabolic rates have been used to explain rate differences that were previously difficult to reconcile with generation time differences, notably the shorter generation time but slower mutation rate of whales compared to primates. In addition, considerations of metabolic mutagens may explain why rate differences are generally more pronounced among mitochondrial sequences than nuclear ones, as most oxidative free radicals are produced within mitochondria themselves.

Alternative sources of lineage effects could lie in the enzymatic mechanisms that act to repair the effects of mutagenic processes, rather than the processes themselves. While most attention has focused on varying efficiencies of **DNA repair**, another source could be differences in the many pathways that mop up mutagens of different kinds before they are able to create DNA lesions. At present, the relative efficiencies of these pathways in different lineages are too poorly characterized to allow these hypotheses to be tested.

In addition to rate variation amongst lineages there are often also differences between the rates of nonsynonymous and synonymous substitution (see Section 3.2.6). Ohta and Gillespie have proposed different, nonbiochemical, hypotheses to explain this weakness with the molecular clock hypothesis. Gillespie believes that episodic selective pressures at nonsynonymous sites, created by environmental changes, distort the molecular clock (Gillespie, 1984) whereas Ohta has proposed that nonsynonymous changes are slightly deleterious rather than neutral, such that the interplay of drift and selection is sensitive to fluctuations in population sizes (Ohta, 1992). The idea is that negative correlation between population size and generation time allows the rate of nonsynonymous changes to operate largely independently of any generation-time effect. Small populations tend to have longer generation times and drift predominates, frequently fixing nonsynonymous mutations that occur at low rates. By contrast, negative selection predominates at nonsynonymous sites in large populations, so fixation occurs infrequently although mutations are generated more rapidly as a result of shorter generation times. These factors cancel out such that large and small populations have similar rates of evolution with respect to calendar time. This debate has not been resolved, although the two processes of **episodic selection** and **nearly neutral** mutations could operate in conjunction.

Summary

- To understand the forces that influence diversity we need to construct mathematical models of both populations and molecular evolution.

- Mutation and recombination increase diversity by generating new alleles and new haplotypes respectively.

- Genetic drift results from the random sampling of one generation from the preceeding one, causes stochastic change in allele frequencies, and is measured by the effective population size (N_e).

- The effective population size varies for loci with different patterns of inheritance (i.e., Y chromosome, X chromosome, mtDNA and autosomes).

- The effective population size can be very different from the census population size and is affected by population size fluctuations, variance in reproductive success and population structure.

- Selection can operate in a number of different ways to increase, decrease or maintain an allele's frequency:

  - purifying selection removes deleterious alleles and acts on most genes;

  - balancing selection maintains more than one allele by mechanisms such as heterozygote advantage or frequency-dependent selection;

  - directional selection changes the frequency of an allele over time, such as when a new advantageous mutation occurs and increases in frequency.

- The MHC is the most diverse locus in the human genome, probably as a result of balancing selection acting to maintain polymorphism.

- Migration increases diversity by introducing new alleles into a population, and can be modeled using a meta-population split into partially isolated sub-populations.

- The different forces acting on allele frequencies will balance one another out so that with sufficient time diversity within the population reaches an equilibrium value. Human populations, however, are not at equilibrium.

- The discovery of far higher than expected levels of natural polymorphism led to the development of the 'neutral theory', which states that most mutations in the human genome have no effect on fitness.

- The molecular clock hypothesis is generally useful but must be modified by the finding of different mutation rates in different phylogenetic lineages, which are explained by a number of factors, known as lineage effects.

Further reading:

Greater detail on population genetic theory and models of molecular evolution can be found in:

Li W-H (1997) *Molecular Evolution.* Sinauer Associates, Inc., Sunderland, MA.

Page DM, Holmes EC (1998) *Molecular Evolution: A Phylogenetic Approach.* Blackwell, Oxford.

For further mathematical detail, interested readers should consult:

Cavalli Sforza LL, Bodmer WF (1971) *The Genetics of Human Populations.* W.H. Freeman, San Francisco.

Hartl DL, Clark AG (1989) *Principles of Population Genetics.* Sinauer Associates, Inc., Sunderland, MA.

Nei M (1987) *Molecular Evolutionary Genetics.* Columbia University Press, New York.

References

Anthony D (1990) Migration in archaeology: The baby and the bathwater. *American Anthropologist* **92**, 895–914.

Ardlie KG, Kruglyak L, Seielstad M (2002) Patterns of linkage disequilibrium in the human genome. *Nature Rev. Genet.* **3**, 299–309.

Austerlitz F, Heyer E (1998) Social transmission of reproductive behavior increases frequency of inherited disorders in a young expanding population. *Proc. Natl Acad. Sci. USA* **95**, 15140–15144.

Beck S, Trowsdale J (2000) The human major histocompatibility complex: lessons from the DNA sequence. *Annu. Rev. Genomics Hum. Genet.* **1**, 117–137.

Bird AP (1980) DNA methylation and the frequency of CpG in animal DNA. *Nucleic Acids Res.* **8**, 1499–1504.

Bohossian HB, Skaletsky H, Page DC (2000) Unexpectedly similar rates of nucleotide substitution found in male and female hominids. *Nature* **406**, 622–625.

Burton ML, Moore CC, Whiting JWM, Romney AK (1996) Regions based on social structure. *Curr. Anthropol.* **37**, 87–123.

Caballero A (1994) Developments in the prediction of effective population size. *Heredity* **73**, 657–679.

Caballero A (1995) On the effective size of a population with separate sexes, with particular reference to sex-linked genes. *Genetics* **139**, 1007–1011.

Cavalli-Sforza LL, Feldman MW (1991) *Cultural Transmission and Evolution: a Quantitative Approach.* Princeton University Press, Princeton.

Cavalli-Sforza LL, Menozzi P, Piazza A (1994) *The History and Geography of Human Genes.* Princeton University Press, Princeton, N.J.

Charlesworth B, Morgan MT, Charlesworth D (1993) The effect of deleterious mutations on neutral molecular variation. *Genetics* **134**, 1289–1303.

Charlesworth B (2001) The effect of life history and mode of inheritance on neutral genetic variability. *Genet. Res.* **77**, 153–166.

Cooke GS, Hill AV (2001) Genetics of susceptibility to human infectious disease. *Nature Rev. Genet.* **2**, 967–977.

Crow JF, Kimura M (1970) *An Introduction to Population Genetics Theory.* Harper and Row, New York.

Di Rienzo A, Peterson AC, Garza JC, Valdes AM, Slatkin M, Freimer NB (1994) Mutational processes of simple-sequence repeat loci in human populations. *Proc. Natl Acad. Sci. USA* **91**, 3166–3170.

Ebersberger I, Metzler D, Schwarz C, Paabo S (2002) Genomewide comparison of DNA sequences between humans and chimpanzees. *Am. J. Hum. Genet.* **70**, 1490–1497.

Erlandsson R, Wilson JF, Paabo S (2000) Sex chromosomal transposable element accumulation and male-driven evolution. *Mol. Biol. Evol.* **17**, 804–812.

Felsenstein J (1971) Inbreeding and variance effective numbers in populations with overlapping generations. *Genetics* **68**, 581–597.

Fix AG (1999) *Migration and Colonization in Human Microevolution.* Cambridge University Press, Cambridge.

Frisse L, Hudson RR, Bartoszewicz A, Wall JD, Donfack J, Di Rienzo A (2001) Gene conversion and different population histories may explain the contrast between polymorphism and linkage disequilibrium levels. *Am. J. Hum. Genet.* **69**, 831–843.

Giannelli F, Anagnostopoulos T, Green PM (1999) Mutation rates in humans. II. Sporadic mutation-specific rates and rate of detrimental human mutations inferred from hemophilia B. *Am. J. Hum. Genet.* **65**, 1580–1587.

Gillespie JH (1984) The molecular clock may be episodic. *Proc. Natl Acad. Sci. USA* **81**, 8009–8013.

Haldane JBS (1947) The mutation rate of the gene for haemophilia, and its segregation ratios in males and females. *Annals Eugenics* **13**, 262–271.

Haldane JBS (1949) Disease and evolution. *Ricerca Scientifica* **19**:(Suppl. 1), 3–10.

Hamilton WD, Zuk M (1982) Heritable true fitness and bright birds: a role for parasites? *Science* **218**, 384–387.

Hardy GH (1908) Mendelian proportions in a mixed population. *Science* **28**, 49–50.

Helgason A, Hrafnkelsson B, Gulcher JR, Ward R, Stefansson K (2003) A populationwide coalescent analysis of Icelandic matrilineal and patrilineal genealogies: evidence for a faster evolutionary rate of mtDNA lineages than Y chromosomes. *Am. J. Hum. Genet.* **72**, 1370–1388.

Hewlett BS (1988) In: Sexual selection and paternal investment among Aka pygmies. *Human Reproductive Behaviour: A Darwinian Perspective* (eds L Betzig, M Borgerhoff Mulder, P Turke). CUP, Cambridge, pp. 263–276.

Huestis RR, Maxwell A (1932) Does family size run in families? *J. Hered.* **23**, 77–79.

Hurst LD, Ellegren H (1998) Sex biases in the mutation rate. *Trends Genet.* **14**, 446–452.

Jacob S, McClintock MK, Zelano B, Ober C (2002) Paternally inherited HLA alleles are associated with women's choice of male odor. *Nature Genet.* **30**, 175–179.

Jeffreys AJ, Murray J, Neumann R (1998) High-resolution mapping of crossovers in human sperm defines a minisatellite-associated recombination hotspot. *Mol. Cell* **2**, 267–273.

Jeffreys AJ, Kauppi L, Neumann R (2001) Intensely punctate meiotic recombination in the class II region of the major histocompatibility complex. *Nature Genet.* **29**, 217–222.

Jobling MA, Williams G, Schiebel K *et al.* (1998) A selective difference between human Y-chromosomal DNA haplotypes. *Curr. Biol.* **8**, 1391–1394.

Jukes TH, Cantor CR (1969) Evolution of protein molecules. In: *Mammalian Protein Metabolism* (ed. NH Munro). Academic Press, New York, pp. 21–123.

Kauppi L, Sajantila A, Jeffreys AJ (2003) Recombination hotspots rather than population history dominate linkage disequilibrium in the MHC class II region. *Hum. Mol. Genet.* **12**, 33–40.

Kayser M, Brauer S, Weiss G, Schiefenhovel W, Underhill P, Stoneking M (2001) Independent histories of human Y chromosomes from Melanesia and Australia. *Am. J. Hum. Genet.* **68**, 173–190.

Kimura M, Weiss GH (1964) The stepping stone model of population structure and the decrease of genetic correlation with distance. *Genetics* **49**, 561–576.

Kimura M (1968) Evolutionary rate at the molecular level. *Nature* **217**, 624–626.

Krawczak M, Ball EV, Cooper DN (1998) Neighbouring nucleotide effects on the rates of germ-line single-base-pair substitution in human genes. *Am. J. Hum. Genet.* **63**, 474–488.

Malécot G (1948) *Les Mathématiques de l'hérédité.* Masson, Paris.

Marsh SG, Albert ED, Bodmer WF *et al.* (2002) Nomenclature for factors of the HLA system, 2002. *Eur. J. Immunogenet.* **29**, 463–515.

Martin AP, Palumbi SR (1993) Body size, metabolic rate, generation time, and the molecular clock. *Proc. Natl Acad. Sci. USA* **90**, 4087–4091.

MHC Sequencing Consortium (1999) Complete sequence and gene map of a human major histocompatibility complex. The MHC sequencing consortium. *Nature* **401**, 921–923.

Miller GF (2000) *The Mating Mind: How Sexual Choice Shaped the Evolution of Human Nature.* Doubleday, New York.

Miyata T, Hayashida H, Kuma K, Mitsuyasa K, Yasunaga T (1987) Male-driven molecular evolution: A model and nucleotide sequence analysis. *Cold Spring Harb. Symp. Quant. Biol.* **52**, 863–867.

Murdock GP (1967) *Ethnographic Atlas.* University of Pittsburgh Press, Pittsburgh.

Nachman MW, Crowell SL (2000) Estimate of the mutation rate per nucleotide in humans. *Genetics* **156**, 297–304.

Nunney L (1993) The influence of mating system and overlapping generations on effective population size. *Evolution* **47**, 1329–1341.

Ober C, Weitkamp LR, Cox N, Dytch H, Kostyu D, Elias S (1997) HLA and mate choice in humans. *Am. J. Hum. Genet.* **61**, 497–504.

Ohta T, Kimura M (1973) A model of mutation appropriate to estimate the number of electrophoretically detectable molecules in a finite population. *Genet. Res.* **22**, 201–204.

Ohta T (1992) The nearly neutral theory of molecular evolution. *Annu. Rev. Ecol. Syst.* **23**, 263–286.

Oota H, Settheetham-Ishida W, Tiwawech D, Ishida T, Stoneking M (2001) Human mtDNA and Y-chromosome variation is correlated with matrilocal versus patrilocal residence. *Nature Genet.* **29**, 20–21.

Pawlowski B, Dunbar RI, Lipowicz A (2000) Tall men have more reproductive success. *Nature* **403**, 156.

Pearson K, Lee A, Bramley-Moore L (1899) On the inheritance of fertility in mankind. *Phil. Trans. R. Soc. Lond.* **192**, 257–330.

Penn DJ, Potts WK (1999) The evolution of mating preferences and Major Histocompatibility Complex genes. *The American Naturalist* **153**, 145–164.

Potts WK (2002) Wisdom through immunogenetics. *Nature Genet.* **30**, 130–131.

Pritchard JK, Przeworski M (2001) Linkage disequilibrium in humans: models and data. *Am. J. Hum. Genet.* **69**, 1–14.

Relethford JH (2001) *Genetics and the Search for Modern Human Origins.* Wiley, New York.

Roberts L (1991) Scientific split over sampling strategy. *Science* **252**, 1615.

Sarich VM, Wilson AC (1973) Generation time and genomic evolution in primates. *Science* **179**, 1144–1147.

Seielstad MT, Minch E, Cavalli-Sforza LL (1998) Genetic evidence for a higher female migration rate in humans. *Nature Genet.* **20**, 278–280.

Shimmin LC, Chang BH-J, Li W-H (1993) Male-driven evolution of DNA sequences. *Nature* **362**, 745–747.

Sibert A, Austerlitz F, Heyer E (2002) Wright–Fisher revisited: The case of fertility correlation. *Theor. Popul. Biol.* **62**, 181–197.

Stumpf MP, Goldstein DB (2003) Demography, recombination hotspot intensity, and the block structure of linkage disequilibrium. *Curr. Biol.* **13**, 1–8.

Wang J, Caballero A (1999) Developments in predicting the effective size of subdivided populations. *Heredity* **82**, 212–226.

Wedekind C, Furi S (1997) Body odour preferences in men and women: do they aim for specific MHC combinations or simply heterozygosity? *Proc. Roy. Soc. Lond. Ser. B* **264**, 1471–1479.

Wiuf C, Hein J (2000) The coalescent with gene conversion. *Genetics* **155**, 451–462.

Wright S (1931) Evolution in Mendelian populations. *Genetics* **16**, 97–159.

Wright S (1938) Size of population and breeding structure in relation to evolution. *Science* **87**, 430–431.

Wright S (1943) Isolation by distance. *Genetics* **28**, 114–138.

Wright S (1951) The genetical structure of populations. *Annals Eugenics* **15**, 323–354.

Zukerkandl E, Pauling L (1965) Evolutionary divergence and convergence in proteins. In: *Evolving Genes and Proteins* (eds V Bryson, HJ Vogel). Academic Press, New York, pp. 97–166.

CHAPTER SIX

Making inferences from diversity

CHAPTER CONTENTS

BOXES

6.1 Introduction

6.1.1 What are our objectives?

Current genetic diversity contains information on past population parameters, and on the history of human adaptation to changing environments. In particular, analysis of genetic variation among contemporary individuals has been used to clarify important aspects of the human past. But, what do we want to know? And how much of this can be inferred from modern genetic diversity? There are two related goals:

(1) A *description* of the distribution of present diversity, which allows comparisons between species or between populations within a species. These include comparisons of genetic diversity, and its apportionment between subpopulations.

(2) *Inferences* about how modern diversity evolved. Studies of human genetic diversity are usually limited to a single time-slice, such as analyzing diversity among modern populations, which constrains our ability to investigate the past. Prehistorical and historical processes must therefore be inferred from modern diversity. Such **inferential methods** require explicit or implicit models of the evolutionary processes, some of which were described in the previous chapter. Inferences on past processes are motivated by:

▶ an anthropological interest in the prehistory of populations, their origins, movements and demographies;

▶ an interest in the evolutionary history of specific segments of DNA, be they individual genes, chromosomes or entire genomes.

As we saw in the previous chapter, these interests, although conceptually distinct, often lead to related questions, because population processes affect molecular diversity.

In this chapter, we will come across both descriptive and inferential methods. There is no formal distinction between the two; descriptions of present diversity almost inevitably lead to discussions of how it might have arisen. It is worth noting that there is often no simple and unique answer to complex questions, such as 'How did a particular pattern of genetic diversity arise?' Therefore, a combination of several analytical approaches, and the validation of the results over as many loci as possible, are advisable.

6.1.2 The nature of the data

The nature of the genetic data from which we can infer past processes has changed radically over the past 30 years. Initially, polymorphic genetic markers comprised different protein 'types' (alleles), characterized immunologically or by protein **electrophoresis**. Because the molecular basis of these polymorphisms was unknown, evolutionary inference could only be based on the analysis of allele frequencies. These 'pre-DNA' markers are typified by blood groups, and

are commonly referred to today as **'classical markers'** (see *Box 3.3* on the ABO blood group system and *Box 5.5* on HLA alleles).

The advent of direct methods for investigating the sequence of protein and DNA molecules led to the development of **'molecular markers'**. By defining the underlying molecular differences between alleles at a locus, these methods allow the introduction of the concept of evolutionary distances between alleles – for example, the number of repeat units by which two **microsatellite** alleles differ, or the number of variant bases between two sequences. In this way, evolutionary inference can be based both on the analysis of allele frequencies, *and* on the molecular comparison of different alleles. In addition, these methods allowed diversity in noncoding regions of the genome to be investigated.

Some analytical methods described here are applicable to both types of data, classical and molecular, whereas others are suitable only for one type. Several methods devised for classical data have been adapted to take account of the extra information within molecular data, and this may render a method suitable for only some kinds of molecular data. For example, a method devised for analyzing DNA sequences may not be applicable to microsatellite diversity.

It is worth noting that a single locus contains less information on our evolutionary past than do many loci, no matter how informative that individual locus is. A single locus gives but a single account of the evolutionary process. As any historian knows, the collation of several corroborative sources is vital to an accurate reconstruction of the past; any single account may be biased whether by chance (drift) or by design (selection). Until now, most studies of DNA diversity have focused on individual, often nonrecombining, loci. As genetic diversity among humans is uncovered in more and more regions of the genome, combining this information together in a coherent fashion will become a major challenge.

6.2 Measuring and summarizing genetic variation

6.2.1 Measures of identity

Statistics that summarize the amount of variation do not encapsulate all information present in the data; in addition, different evolutionary processes may lead to the same value of one or more statistics. Nonetheless, these **summary statistics** allow comparisons between populations and between loci.

Perhaps the simplest way to describe the amount of diversity is to count the number of alleles present. This can be done either within many populations for a single locus, allowing a comparison of diversity between populations, or at many loci within a single population, allowing a comparison among loci. Clearly this measure does not account for molecular distances between alleles, and so is applicable to both classical and molecular markers. It is, however, highly dependent on sample sizes, which is usually a serious disadvantage.

A commonly used measure of diversity that is also blind to molecular distance between alleles is Nei's **gene diversity** statistic (Nei, 1987). Despite its name, this measure is suitable for both coding and noncoding markers. This statistic measures the probability that two alleles drawn at random from the population will be different from each other. Consequently, for **diploid** loci this statistic is sometimes referred to as a measure of **heterozygosity**, which becomes **'virtual heterozygosity'** at haploid loci. The unbiased estimator used for gene diversity is given below:

$$H = n(1 - \Sigma x_i^2)/(n-1)$$

Where n is the number of gene copies and x_i is the frequency of the *i*th allele.

In situations where there is a high degree of polymorphism (such as when complex minisatellites or many linked microsatellites are being considered), almost all alleles (or haplotypes) are different from one another, and thus gene diversity ceases to be useful as it is close to 1 in all populations. In such situations, measures of diversity are required that take account not of the *identity* of the allele, but the *distances* between alleles. Such measures must take account of the molecular nature of allelic variation, and therefore are often specific to different types of data.

6.2.2 Measures of nucleotide diversity

We have seen in Chapter 5 that under neutral evolution the level of diversity within a population can reach an equilibrium value whereby the generation of new alleles by mutation is canceled out by the elimination of alleles by drift (**mutation drift equilibrium**). It is therefore possible to define the *expected* level of diversity (θ – 'theta') in a population in terms of the mutation rate (μ per site per generation) and drift. In practice, since drift is inversely proportional to **effective population size** (Section 5.3), N_e is used and the equation relating these parameters for diploid loci is:

$$\theta = 4N_e\mu$$

Thus by knowing the mutation rate and θ, and assuming an equilibrium state, we can estimate the effective population size of a diploid population from DNA sequence diversity (see *Table 6.1*). θ is a fundamental parameter of molecular evolution, sometimes referred to as the 'neutral parameter' or the **'population mutation parameter'** and, in different forms, crops up in many different analytical methods. The above equation is specific for diploid loci; however, a more general form of the equation that considers loci with other inheritance patterns can be considered. Here, n represents the number of heritable copies of the locus per individual, which is 2 for diploid loci, 0.5 for the Y chromosome and mtDNA and 1.5 for the X chromosome.

$$\theta = 2nN_e\mu$$

We can see that assuming the same mutation rate, the reduced number of heritable copies of the sex chromosomes

should result in their having lower diversity at equilibrium than autosomal loci.

How can we summarize the diversity within a set of nucleotide sequences? We could count the number of nucleotide sites that vary within the entire set of aligned sequences (see *Box 6.1* and *Table 6.2*), known as **'segregating sites'**. However, such a measure is clearly dependent on the length of sequence analyzed: the longer the sequence, the greater the number of segregating sites. Measuring the proportion of all sites that are segregating would surmount this problem, but this itself depends on the number of sequences sampled – as more sequences are studied, more segregating sites are found. A sample size-independent measure known as nucleotide diversity, π, is analogous to Nei's gene diversity. Nucleotide diversity describes the probability that two copies of the same nucleotide drawn at random from a set of sequences will be different from one another. The commonly used estimator for π is shown below:

$$\pi = n(\Sigma x_i x_j \pi_{ij})/(n-1)$$

Where n is the number of sequences, x_i and x_j the frequencies of the *i*th and *j*th sequences respectively and π_{ij} the proportion of different nucleotides between them.

As we have seen, the amount of variation expected at each nucleotide site under neutral evolution is given by θ. Thus, if selection is absent (and, of course, the other assumptions are met), π and θ should be equal.

There are several different methods for estimating θ from sequence data. These methods make different assumptions and are calculated using different parameters derived from the observed diversity, including:

▶ the number of alleles;
▶ the number of segregating sites (S);
▶ the number of **singletons** (η);
▶ the observed homozygosity;
▶ the mean number of **pairwise** differences (π).

These can be represented as θ_S etc. In an idealized neutrally evolving population these different estimators of θ should give the same value.

Comparisons of genetic diversity between continental human populations have been instructive in deciding between alternative models for the origins of modern humans (see Section 8.5 and *Table 8.1*). Interestingly, humans appear to have lower levels of nucleotide diversity than other apes (see Section 7.5 and *Table 7.3*) despite their much larger present population size.

6.2.3 Other measures of molecular diversity

A commonly used way of representing diversity for different types of molecular data is the **mismatch distribution**, also known as the distribution of pairwise differences. The mismatch distribution is appropriate for data where discrete differences between alleles can be counted; these differences can be base substitutions, RFLP sites, or microsatellite repeat

TABLE 6.1: NUCLEOTIDE DIVERSITY, NEUTRAL EXPECTATION OF θ AND EFFECTIVE POPULATION SIZE ESTIMATES FOR HUMANS.

Locus (Length)	π ($\times 10^{-4}$)	θ ($\times 10^{-4}$)	μ ($\times 10^{-9}$)	N_e	Reference
APOE (5.5 kb)	5.3	6.87 (S)	23.5	7300	(Fullerton *et al.*, 2000)
Chr. 1 (10 kb)	5.8	9.51 (S)	14.8	16 000	(Yu *et al.*, 2001)
Chr. 22 (10 kb)	8.8	13.2 (S)	23	14 400	(Zhao *et al.*, 2000)
X chr. (10.2 kb) Xq13.3	3.6	6.8 (S)	18.4	12 300	(Kaessmann *et al.*, 1999;)
X chr. (4.2 kb) *PDHA1*	–	4.41 (ML)	19.2	7700	(Harris and Hey, 1999)
Y chr. (64 kb)	0.74	2.01 (S)	24.8	8100	(Thomson *et al.*, 2000)
mtDNA (15.4 kb) excluding control region	28	28 (π)	340	8200	(Ingman *et al.*, 2000)
Alu insertions	–	–	–	17 500	(Sherry *et al.*, 1997)

N_e is calculated using locus-specific per-generation nucleotide mutation rates (μ). Among the different studies, θ per nucleotide was calculated using estimators based on a variety of sequence characteristics: S (segregating sites), π (pairwise differences) and ML (a maximum likelihood estimator). These sequence-derived estimates are compared with an estimate from *Alu* insertion polymorphisms.

units. The distribution of the number of such differences between each allele and every other allele summarizes the discernible genetic diversity. *Figure 6.1* shows schematically how the mismatch distribution is calculated from DNA sequences (Rogers and Harpending, 1992).

The mismatch distribution is a good example of a descriptive summary statistic that overlaps with inferential methods. As well as describing the diversity apparent within a sample, the shape of the distribution has been shown to be indicative of population history, in particular being influenced by episodes of population expansion. While the mean of the distribution provides a simple description of the overall diversity, the shape of the distribution is also informative. A smooth, bell-shaped mismatch distribution indicates a period of rapid population growth from a single haplotype, whereas a ragged, multimodal distribution indicates a population whose size has been constant over a long period (Rogers and Harpending, 1992), see *Figure 6.2*. To distinguish between these two types of distribution a **raggedness** statistic (r) is used. This is simply the sum of the squared difference between neighboring peaks, and is estimated by the equation below:

$$r = \sum_{I=1}^{d+1} (x_i - x_{i-1})^2$$

Where *d* is the greatest number of differences between alleles, and x_i is the relative frequency of *I* pairwise differences.

Smooth distributions typically have lower raggedness values (less than 0.03 for sequence data), than multimodal distributions (as illustrated in *Figure 6.2*), and indicate a past population expansion. The age of this expansion can be estimated in a number of ways, most of which are related to the distance of the mean of the distribution from the y axis. As time elapses since the population expansion, the mean moves further away from the y axis (see *Figure 6.3*). Note that for sequence data, the mean of the mismatch distribution is similar to π, as defined above, except that distances between alleles are not divided by sequence length. Thus it is a measure of sequence diversity rather than nucleotide diversity.

Mismatch distributions of mitochondrial DNA diversity have been used to detect population expansions in humans that have been dated to 34–62 KYA (see Section 9.5.5 for a discussion of evidence for early human population expansions).

TABLE 6.2: SOFTWARE: SEQUENCE ALIGNMENT.

Alignment	Software	URL
Pairwise	Needle	http://www.ebi.ac.uk/emboss/align/
Multiple	ClustalW	http://www.ebi.ac.uk/clustalw/
Database homology search	BLAST	http://www.ncbi.nlm.nih.gov/blast/
Pairwise homology search	BLAST 2	http://www.ncbi.nlm.nih.gov/blast/bl2seq/bl2.html

---

BOX 6.1 Sequence alignment and BLAST searches

Having obtained sequence data from a number of individuals, we want to identify the evolutionary changes between two homologous sequences (i.e., sequences that share a common ancestor). To do this, we align the sequences to ensure that the bases that are compared derive from the same nucleotide position in the common ancestor. Sequences must be aligned before we carry out any comparative analysis, for example: calculating sequence diversity, constructing phylogenetic trees, or testing for selection.

Several different methods produce **sequence alignments**. Some methods maximize the number of matching aligned bases (similarity methods), while others minimize the number of mismatched aligned bases (distance methods). Gaps within the alignment where a base in one sequence is not matched by a base in the other (perhaps as a result of an insertion or a deletion) are also kept to a minimum. Scores are calculated to compare alternative sequence alignments, which are based on arbitrary weighting of gaps, mismatches and matches. For example, consider the two alternative alignments below; if we score 1 for a mismatch and 5 for a gap, then the top alignment has the lowest score (in bold) and is therefore preferable. However, if we score 1 for a mismatch and 0.5 for a gap then the bottom alignment has the lower score (again in bold). In this example we have not taken into account the size of the gap; most popular alignment methods give greater weighting to larger gaps than to smaller gaps.

	mismatch score 1 gap score 5	mismatch score 1 gap score 0.5
Alignment 1	TACTCTGATC | | || | TAGTC--GCC 1 +5 +1+1 = **8**	TACTCTGATC ||| || | TAGTC--GCC 1 +0.5+1+1 = 3.5
Alignment 2	TACTCTGATC || || | | TAGTC-GC-C 1 +5 +1+5 = 12	TACTCTGATC || || | | TAGTC-GC-C 1 +0.5 +1+0.5 = **3**

The methods described above are for 'pairwise' sequence alignments. Aligning multiple sequences is more complex, and there exists a variety of different methods for comparing the quality of different alignments. Alignments of human sequences are generally relatively simple given the small number of changes between homologs. However, as evolutionary distance between sequences increases, it becomes more difficult to choose between alternative alignments.

A related problem to sequence alignment is that of homology searching. For example, one might want to find a homologous sequence within a large database of genomic sequence, or find the position of a small sequence within a single much longer sequence. Homology search methods attempt to identify regions of *local* similarity, rather than optimize an alignment over the entire length of all sequences (known as *global* methods). Many of the popular methods are based on an algorithm known as BLAST – Basic Local Alignment Search Tool. *Table 6.2* lists software for performing different types of sequence alignment.

A BLAST search usually returns a number of homology matches between the query sequence and the database. Each match comes associated with a 'score' and an 'E value' that determine how much significance should be ascribed to each individual match. The score is a measure of the match identified, which is assessed in a manner similar to the scoring example for pairwise alignments given above, except that a similarity method is used such that the highest score represents the best alignment. The E (Expect) value describes the number of matches with that score that could be expected given the size of the database. As the E value gets closer to zero, the significance of the associated match increases.

A number of specific measures have been developed that summarize the molecular diversity among micro- and minisatellites. Molecular distances are calculated between individual alleles, or between haplotypes at **nonrecombining loci**. Many of these measures have been developed specifically for dating purposes, often based upon a single-step mutation mechanism (SMM; see Section 5.2.3). In other words, diversity expressed using these measures accrues linearly with respect to time, and thus provides a molecular clock. These diversity measures – which include the **variance** in allele frequencies and the average squared distance between alleles – are discussed later in this chapter.

6.2.4 Measures of apportionment of diversity

If genotype frequencies among our sampled individuals are not in **Hardy–Weinberg equilibrium**, but exhibit a deficiency of heterozygotes, we should question whether

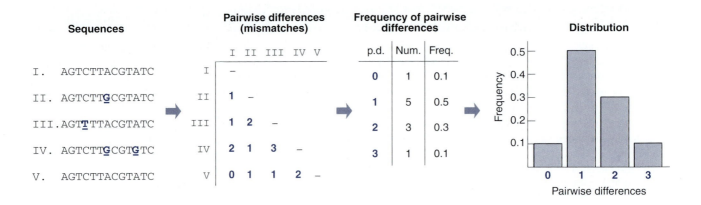

Figure 6.1: Generation of a mismatch distribution from a matrix of pairwise distances.

A distance matrix is constructed using pairwise differences between a set of five sequences. Mutations in the sequences are shown in blue and underlined to facilitate comparisons between sequences. The mismatch distribution is a histogram obtained by counting the number of pairwise comparisons that share the same number of differences between the two sequences.

these individuals come from a single randomly mating population (see Section 5.1.2). Subdividing the population into partially isolated subpopulations alone may cause an identical deficiency in heterozygotes. This often arbitrary subdivision process generates a **hierarchical population structure**. A single meta-population composed of a number of subpopulations is the simplest, two-tiered, example of such a structure. Once data have been shown not to conform to randomly mating (panmictic) expectations, then it can be said that there is population structure, and we can measure the apportionment of diversity amongst these different tiers of the hierarchy.

In the previous chapter we encountered F_{ST}, a popular statistic that describes the degree to which variation at classical markers within a meta-population is apportioned among subpopulations (see Section 5.3.5). F_{ST} can be thought of as measuring the proportion of the total variance in allele frequencies that occurs between subpopulations. If genetic drift results in the subpopulations being highly differentiated this proportion (and F_{ST}) will be large, while if large amounts of gene flow between subpopulations maintain their similarity, this proportion will be much smaller ($F_{ST} \approx 0$). F_{ST} values vary between 0 and 1: for example an F_{ST} of 0.3 means that 30% of the total allele frequency variance is found between subpopulations, with the corollary that 70% of allele frequency exists within the subpopulations themselves.

In humans 9–13% of allele frequency variance is typically found between continental groups (see *Table 9.1*), which is low compared to other large-bodied mammals with large geographical ranges (Section 9.3). However, values of population subdivision at some loci are significantly elevated or depressed when compared to this average. In some cases this is due to the impact of selection, whereas in others it is a consequence of a lower effective population size (see Sections 9.3, 9.4 and 13.3.4).

There are several other similar measures that apportion diversity (see *Table 6.3*). As indicated in the previous section, many of these measures can also be used to infer how such diversity arose. Again, some of these measures have been adapted, or specifically designed, to accommodate molecular data.

A critical aspect of all of these measures is some kind of significance test to demonstrate that any population subdivision is greater than could be expected by chance. We must exclude the possibility that subpopulations are not differentiated, and that any apparent differences in allele frequencies result from sampling effects alone. Excluding a null hypothesis of random mating is a natural first step to all analyses of population subdivision; why attempt to explain something that may not exist?

One method for determining the significance of population subdivision is to use a **permutation** test. These are common in population genetics where the variation of a given measure cannot be easily predicted by standard statistical distributions (e.g., normal, Poisson, binomial). These tests are also known as **Monte-Carlo methods**, a name resulting from use of random numbers to emulate a casino-like situation.

Permutation tests randomize the empirical data many times and calculate the measure of interest from each randomization (*Figure 6.4*). The 'real' measure from the observed data is then compared against the 'simulated' measures to see if it is significantly different (Roff and Bentzen, 1989).

Alternative properties of the apportionment of alleles among subpopulations, other than allele frequency variance, can be used to generate measures of population subdivision. For example, G_{ST} considers the proportion of total gene diversity that occurs between subpopulations. By incorporating a molecular measure, **nucleotide diversity**,

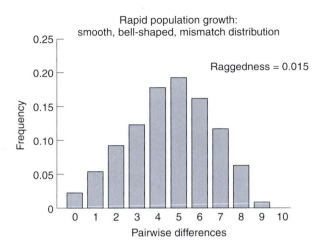

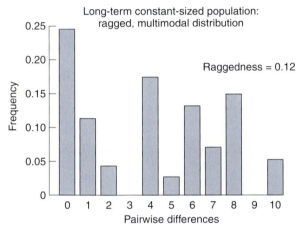

Figure 6.2: Ragged and smooth mismatch distributions.

Two mismatch distributions are shown together with their associated raggedness values calculated according to the equation given in the text. The smoother, unimodal distribution has a much smaller raggedness value than the multimodal distribution.

rather than gene diversity, N_{ST} represents a G_{ST} analog for use with sequence data (Lynch and Crease, 1990). R_{ST} is another molecular analog of G_{ST}, derived specifically for microsatellite data (Slatkin, 1995). As discussed above, methods incorporating molecular information become useful when mutation rates are high and most alleles are rare.

Another alternative to allele frequency-based methods is to use the Analysis of Molecular Variance (**AMOVA**), a method that considers, for example, the variance in the number of microsatellite repeat units at a given locus (Excoffier *et al.*, 1992). This method takes into account the molecular relationship of alleles, rather than just their frequency, when apportioning variance between tiers of the hierarchical population structure, and can calculate Φ_{ST} (Φ is pronounced 'phi'), a molecular analog of F_{ST}. The AMOVA method can be applied to any data where genetic distances between alleles can be calculated (see *Table 6.4* for software

to calculate some of the measures of population subdivision discussed above).

If we assume that one of the simple models for population substructure outlined in the previous chapter (e.g., an **island** or **stepping-stone model**) is a reasonable approximation of the meta-population being studied, we can calculate certain parameters of that structure, for example the migration rate per generation between subpopulations. To do this we have to assume that the meta-population structure has reached **migration drift equilibrium** (see Section 5.6.1), in other words present rates of migration and genetic drift have remained unchanged for long enough that the level of population subdivision has reached a stable equilibrium value. These assumptions might seem untenable in real populations; nevertheless, comparison of such parameters between different populations is capable of revealing plausible patterns.

6.3 Testing for selection

6.3.1 Introduction

A low level of genetic diversity within a population may reflect limited immigration, extensive drift (for example, a low population size), or selective pressures against a certain set of alleles. Likewise, a high level of genetic diversity may result from extensive immigration, a large population size (reducing drift), or selection favoring the increase of genetic diversity. In each case these factors can be combined together. Therefore, before trying to explain diversity at a locus in terms of prehistorical population processes, we need to account for any effect of selection. As we have seen in the previous chapter, selection comes in many guises, each of which is expected to have a different effect on genetic diversity (see Section 5.4). Negative or **purifying selection** removes new, deleterious, variants from the population whereas positive, or **diversifying, selection** increases the probability that a new variant will become fixed. Additionally, selection need not be acting on the locus itself, but could be operating on a linked locus.

How can we test for the imprint of past selection? There are many methods (Wayne and Simonsen, 1998; Kreitman, 2000; Yang and Bielawski, 2000), which typically compare some feature of the observed diversity to that expected under neutral evolution. Such methods are also known as **'neutrality tests'**. No single neutrality test has predominated, which is partly because the alternative methods require different sorts of data and vary in their ability to detect the influence of different selective regimes. Most of these methods are oriented towards summary statistics of DNA sequence diversity. Although the use of summary statistics does not incorporate all of the information contained within the data, alternative methods remain too computationally intense at present to be used widely.

A significant difference between any test statistic and neutral expectations need not result solely from selection. The equilibrium model of neutral evolution also assumes that

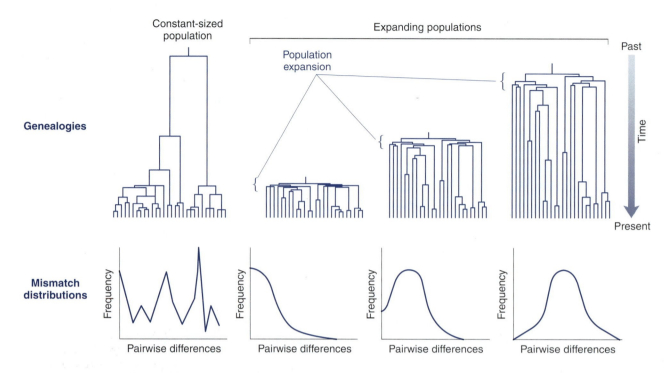

Figure 6.3: Genealogies and mismatch distributions for a constant-sized population and three populations that have undergone population expansions at different times.

The longer branches in the genealogy of the constant population are lineages ancestral to many individuals, whereas the longer branches in the expanding populations are often specific to individuals. Because branch length is indicative of time and therefore the accumulation of mutations, more mutations are shared among individuals in the constant population than in the expanding population, and greater numbers of mutations differentiate individuals in populations with more ancient expansions. Thus it can be seen how the mismatch distributions summarize this information.

the population has certain demographic characteristics. The test statistic may differ from neutral expectation if these demographic assumptions alone are violated. In fact, given that humans have grown in number from thousands to billions over the past hundred millennia (see *Figure 10.5*), we can be certain that our species is *not* in a state of constantly sized neutral equilibrium.

We consider a number of different neutrality tests below, classified on the basis of the different effects of selection each attempts to detect. The power of these tests to detect

selection depends on, amongst other factors, the characteristics of the selective regime:

▶ the type of selection operating;

▶ the strength of selection;

▶ the period during which selection occurred or is occurring.

The results of a neutrality test will often depend on whether the region being analyzed is itself under selection or is linked to a region under selection.

TABLE 6.3: MEASURES OF POPULATION SUBDIVISION.

Measure	Considers	Data type	Reference
F_{ST}	Allele frequency variance	Classical	(Wright, 1951)
G_{ST}	Gene diversity	Classical	(Nei, 1973)
N_{ST}	Nucleotide diversity	Molecular	(Lynch and Crease, 1990)
R_{ST}	Microsatellite diversity	Molecular	(Slatkin, 1995)
φ_{ST}	Molecular variance	Molecular	(Excoffier *et al.*, 1992)

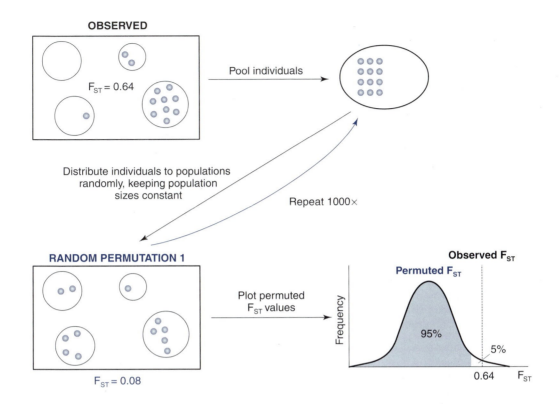

Figure 6.4: Example of a permutation test for population differentiation.

A measure of population subdivision (F_{ST}) is calculated from the observed frequency of two alleles in four subpopulations. To generate simulated datasets for comparison, each allele is randomly assigned to a subpopulation, such that the meta-population frequency of each allele remains constant, as does the sample size of each subpopulation. F_{ST} is then calculated for each of the 1000 simulated datasets. For the empirical value of F_{ST}, or indeed any other measure of population subdivision, to be significantly different from zero, it must be greater than a certain proportion (X) of the simulated values, where 1–X is the significance level being tested. For example, if the observed F_{ST} is greater than it is in more than 950 out of 1000 simulated datasets, then it is significant at the 5% level.

6.3.2 Codon-based selection tests

Nucleotides within coding sequence can be classified into those that result in a change in amino acid when mutated (**nonsynonymous**) and those that do not (**synonymous** – see Section 3.2.6). Synonymous sites can be assumed to be selectively neutral and so the proportion of such sites that are variable within a set of sequences (d_S or K_s) is independent of selection, and therefore a product of neutral evolution. The proportion of the nonsynonymous sites that are variable within the same set of sequences (d_N or K_a) will be greater

TABLE 6.4: SOFTWARE: MEASURING DIVERSITY AND ITS APPORTIONMENT.		
Statistic	**Software**	**URL**
Nei's diversity	ARLEQUIN	http://lgb.unige.ch/arlequin/
π and ϑ	"	"
Mismatch distribution	"	"
F_{ST}	"	"
AMOVA	"	"
G_{ST}	DNASP	http://www.ub.es/dnasp/
N_{ST}	"	"
R_{ST}	RSTCALC	http://helios.bto.ed.ac.uk/evolgen/rst/rst.html

than d_S under diversifying selection, and less than d_S under purifying selection. The ratio d_N/d_S (K_a/K_s) is often known as ω. By testing if ω is significantly different from 1, we are in effect testing the d_N statistic against the neutral expectation d_S – making comparisons from data with a single gene. If ω is significantly greater than 1, diversifying selection would appear to be acting. If ω is significantly less than 1 (a more common occurrence), then purifying selection predominates.

If sequences are closely related, there is little information in the data to give us statistical power to detect differences between d_N and d_S. However, if there are large differences between the sequences then the possibility arises that there are parallel mutations or **reversions** between sequences. The observed number of differences between sequences becomes an underestimate of the true number of changes and this value must be corrected using a model of sequence evolution (Yang and Bielawski, 2000). *Figure 6.5* shows how, over time, saturation of the number of sites in a nucleotide sequence leads the observed genetic distance to underestimate the actual genetic distance. Methods for correcting these underestimates of distances between molecules are discussed in greater depth in Section 6.4.3.

Using the synonymous substitution rate as a proxy for neutral evolution avoids possible problems associated with differential mutation rates at different loci. In Chapter 3 we saw that rates of neutral evolution vary from region to region throughout the genome, perhaps because of differences in sequence composition and genealogical history. However, a further complication arises from the fact that many organisms

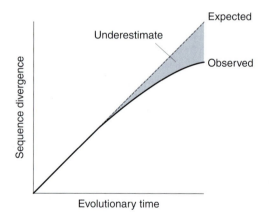

Figure 6.5: Saturation in sequence evolution leads to an underestimate of distances between molecules.

Over long evolutionary time periods some sites within a sequence will have mutated more then once, which can lead to problems when comparing two distantly-related sequences. For example, if in one sequence an A mutates to a T and then to a C, while in the distantly-related homologous sequence no mutations occurred at that site, it will appear as if only a single mutational event has occurred: either an A to a C in the first sequence, or a C to an A in the second. The number of actual mutational events is underestimated by the observed differences between the sequences.

show biases in their codon usage resulting from both **mutational bias** and selection for the efficient translation of proteins. As a consequence, evolution at synonymous sites may not be truly neutral. Fortunately, these can be accounted for by using a more complex evolutionary model that considers such biases. *Table 6.5* lists some of the genes that have been shown to be under positive selection in primates from a consideration of ω.

As described above, ω represents an average over many nucleotides. However, selective pressures are likely to be different among nucleotides within the same gene (some sites might be under negative selection while mutations at others may be neutral, or even undergo positive selection), and thus it could be argued that an average of these is relatively meaningless. However, when many sequences can be compared, it does become possible to detect selective pressures at individual sites using codon-based tests, and ω can also be used to compare different portions of the same gene to detect whether certain functional modules have been under different selective regimes.

The comparative analysis of whole genome sequences provides an opportunity to use ω to detect selection patterns over a huge number of genes. The recent publication of a draft sequence of the mouse genome contained an analysis of 12 845 genes **orthologous** between mouse and human (Mouse Genome Sequencing Consortium, 2002). The median value of ω among these genes was 0.115, indicating that most are under purifying selection. Values of ω were particularly low in regions of proteins containing recognizable protein domains. This observation suggests that most protein domains are under greater functional constraint than domain-free regions. However, a few of these protein domains had significantly elevated ω values, indicating that they are under reduced purifying selection, or that a subset of sites are under diversifying selection. Strikingly, many of these protein domains represent the extra-cellular domains of proteins involved in the immune response, which accords with the hypothesis that co-evolution between pathogens and their hosts drives the rapid evolution of immune-related proteins.

Positive selection may be episodic, especially when considered over long time periods, and selection pressures may have been limited to certain lineages. We know, for example, that selection pressures on globin genes have not been equal among all human populations: resistance to malaria in regions of **endemicity** has resulted in localized, distorted, patterns of diversity. By reconstructing ancestral sequences within a **phylogeny** relating the different observed sequences, codon-based selection tests can indicate whether selection has been acting equally in all regions of the tree, or has been episodic, and so exhibits lineage-specific evolutionary pressures.

Rather than comparing the amount of change at synonymous and nonsynonymous sites between two species, an alternative codon-based test compares diversity within a species to divergence between species. The McDonald–Kreitman test (McDonald and Kreitman, 1991)

TABLE 6.5: GENES UNDER POSITIVE SELECTION IN PRIMATES FROM ω.

Gene	System of function	Reference
Eosinophil cationic protein	Immunity/Defense	(Zhang *et al.*, 1998)
Protamine P1	Reproduction	(Rooney and Zhang, 1999; Wyckoff *et al.*, 2000)
SRY	Reproduction	(Pamilo and O'Neill, 1997)
Lysozyme	Digestion	(Messier and Stewart, 1997)
Morpheus	Unknown	(Johnson *et al.*, 2001)

compares the amount of nonsynonymous and synonymous polymorphism within a species, to the amount of nonsynonymous and synonymous fixed differences between species. Under neutral evolution, intraspecific polymorphism levels and interspecific substitutions are both determined by the rate of mutation, and thus the d_N/d_S ratios should be equal between and within species. Positive selection in the lineage leading to a species could be expected to increase interspecific nonsynonymous substitutions relative to intraspecific nonsynonymous polymorphisms (*Figure 6.6*).

The McDonald–Kreitman test has been used to suggest that purifying selection has been acting on human mitochondrial genes (Nachman *et al.*, 1996). At this locus, intraspecific nonsynonymous polymorphisms are more frequent than their interspecific counterparts. Under purifying selection, weakly deleterious alleles become polymorphic but are fixed less often then they should be under neutral expectations.

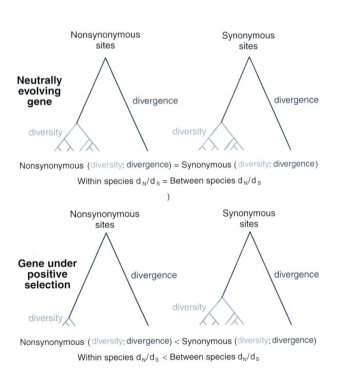

Figure 6.6: The basis of the McDonald–Kreitman neutrality test.

Intraspecific diversity and interspecific divergence is shown for two genes, one that is evolving in a neutral fashion and the other that is under positive selection. Under neutral evolution the ratio of diversity to divergence should be the same at synonymous and nonsynonymous sites. If the gene has undergone adaptive evolution, since divergence from the common ancestor many more advantageous alleles will have become fixed, than might otherwise have been expected. This positive selection can be detected as a reduction in the ratio of diversity to divergence at nonsynonymous, but not synonymous, sites.

6.3.3 Selection tests based on the frequencies of variant sites

Segregating sites within a set of sequences are present at different frequencies; some are found in only a single sequence (singletons), while others are in multiple sequences. Both selection and demography shape the spectrum of these frequencies (known as the **site frequency spectrum**). For example, in a population that has undergone a recent expansion many variant sites will have arisen relatively recently and so will be present only at low frequencies at all loci. Such a population can be considered to have an excess of rare alleles. Similarly, if a specific lineage at a locus is undergoing positive selection, it will be increasing in frequency relative to other lineages, and so a lineage-specific excess of rare alleles may be observed at that locus, but not at others. By contrast, population subdivision and balancing (overdominant) selection maintain multiple lineages in the population for longer than would be expected under neutral evolution, and so produce an excess of intermediate frequency alleles (*Figure 6.7*). Population subdivision increases the number of intermediate frequency alleles at all loci, whereas balancing selection only influences the site frequency spectrum at the locus at which selection is acting.

Methods addressing the frequencies of variant sites are often based on the expectation that under neutral evolution different estimates of θ should be equal. Only some estimates of θ incorporate information on allele frequencies; the number of segregating sites (S) is independent of frequencies, but nucleotide diversity (π) is not. Consequently, discrepancies between estimates of θ that incorporate frequency information differently (or not at all) detect departures from neutral expectations of the allele frequency spectrum. One commonly used statistic, known as Tajima's D, compares two estimates of θ, based, respectively, on S and π. Under neutrality, Tajima's D is expected to be zero. Significantly positive values of this statistic indicate population subdivision or balancing selection, whereas negative values indicate positive selection or population growth.

A related statistic takes into account the ancestral state of the variants (to see how the ancestral state is identified, see Section 3.2.4 and *Figure 3.9*). Under neutrality, very few high-frequency derived alleles are present in a population. However, directional selection leads to a **selective sweep** that increases the proportion of linked high-frequency derived alleles as the sweep is nearing completion, or in adjacent regions after fixation. This pattern can be measured by Fay and Wu's *H* statistic (Fay and Wu, 2000). Both this *H* statistic and Tajima's D statistic have been used to detect positive selection at the *FOXP2* gene that has been proposed to be involved in language evolution (see Section 7.6.2 for more details).

These tests are particularly useful because they can be applied to data from a single species, irrespective of whether the sequences being considered are coding or not. However, their inability to distinguish between demographic effects and selection is a disadvantage. These methods lack power to

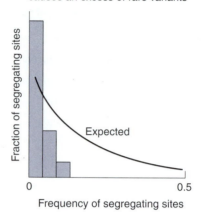

Positive selection or population growth causes an excess of rare variants

y-axis: Fraction of segregating sites

Expected

x-axis: Frequency of segregating sites (0 to 0.5)

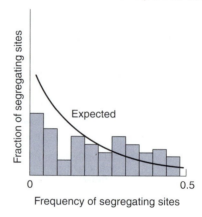

Balancing selection or population subdivision causes an excess of more frequent variants

y-axis: Fraction of segregating sites

Expected

x-axis: Frequency of segregating sites (0 to 0.5)

Figure 6.7: Frequency spectrum of segregating sites under different types of selection.

Two types of deviation from the site frequency spectrum expected in a constant size neutrally evolving population (smooth line) are shown. One spectrum shows rare alleles (those at low population frequency) being more prevalent than expected, and the other shows intermediate alleles being over-represented. Both scenarios can be caused by either selection or demographic factors.

detect very recent selection pressures when the selected lineage may be in the minority.

6.3.4 Selection tests based on comparing multiple loci

The Hudson–Kreitman-Aguadé (HKA) test (Hudson *et al.*, 1987) compares within-species polymorphism, and between-species divergence at two (or more) loci. Under neutrality, the level of within-species polymorphism should be correlated with between-species divergence. This degree of correlation should be the same at both loci if they are evolving in a neutral fashion. Thus the null hypothesis of neutrality can be tested at a chosen locus by comparing it with sequence polymorphism and divergence at a neutral control locus in the same two species.

Under neutrality, three parameters are expected to relate observed diversity within and between the two species: θ; the time since divergence; and the ratio of effective population sizes. In essence, these are estimated from the observed diversity at *all* loci, and are then used to generate expected polymorphism and divergence for *each* locus. A statistical method known as a **goodness-of-fit test** is then used to compare these locus-specific measures of observed and expected diversity. If neutral evolution is operating at both loci, the expected and observed measures of diversity should agree reasonably well. *Table 6.6* lists some of the instances where the HKA test has been used to demonstrate positive selection in primates.

While originally formulated for comparing two unlinked loci, this test has been further developed to compare linked regions within the same gene. A more recent adaptation slides a 'window' along a sequence, to explore selective differences within a single gene. This type of analysis can reveal situations where one protein domain, but not others, has been under selection.

6.3.5 Selection tests comparing allelic frequency and intra-allelic haplotypic diversity

A discrepancy between the frequency of an allele and its intra-allelic diversity expected under neutrality provides evidence of past selection (Slatkin and Bertorelle, 2001). Intra-allelic diversity can be measured from haplotype data in a number of ways, including:

▸ the number of nonrecombinants at a linked binary marker;

▸ the length of a conserved haplotype;

▸ the number of mutations at a linked multiallelic marker.

High frequency alleles are expected to be old, because time is required for an allele to increase in frequency by drift. They therefore have accumulated high levels of intra-allelic diversity. Low frequency alleles can be either old or young, and may therefore have high or low diversity. Positive selection rapidly increases the frequency of an allele, so it becomes common without achieving a high level of intra-allelic diversity.

The relationship between frequency and intra-allelic diversity allows the growth rate for each allele to be estimated. Comparing these growth rates represents a test for selection. If the growth rates among different alleles are statistically indistinguishable then it is likely that this represents the growth rate of the population at large. However, if these rates are not the same, then one allele may have been preferentially amplified in frequency.

The power of these haplotype-based methods for detecting past selection has been recently exemplified by analysis of haplotypes encompassing the glucose-6-phosphate dehydrogenase (*G6PD*) and CD40 ligand (*TNFSF5*) genes (Sabeti *et al.*, 2002). One allele at each of these genes is thought to provide protection against malaria and as a consequence these alleles may well be under positive selection (Section 14.4). The authors identified the local haplotypes on which these alleles lie and then examined extended haplotypes by incorporating more distant SNPs. They showed that the local haplotypes containing the protective alleles had unusual properties when compared to other haplotypes at the same locus:

▸ at each locus the extended haplotypes of the haplotype containing the protective allele were conserved over significantly longer distances than those of other local haplotypes with similar frequencies (*Figure 6.8*);

▸ the only extended haplotypes that were significantly different from haplotypes simulated under a model of neutral evolution were those containing the protective alleles.

These unusual properties are indicative of positive selection, yet all of the other selection tests described in this section were unable to detect evidence of selection from these same data. This suggests that haplotype-based tests may well be more sensitive for detecting recent positive selection than other methods.

6.3.6 The future of selection tests

As data on the diversity of individual genes within the human genome grow exponentially, attempts to detect the possible imprint of selection will increase. It should be noted that the methods discussed here have little power to detect selection when intragenic recombination is operating (Wall, 1999) – many incidences of selection would not be detected by these methods. **Population structure** may also play an important role in reducing our ability to detect positive selection (Przeworski, 2002).

TABLE 6.6: GENES UNDER POSITIVE SELECTION IN PRIMATES FROM THE HKA TEST.

Gene	System of function	Reference
FY-Duffy locus	Immunity	(Hamblin *et al.*, 2002)
Factor IX	Blood clotting	(Harris and Hey, 2001)
MC1R	Pigmentation	(Makova *et al.*, 2001)
Monoamine oxidase A	Behavior	(Gilad *et al.*, 2002)

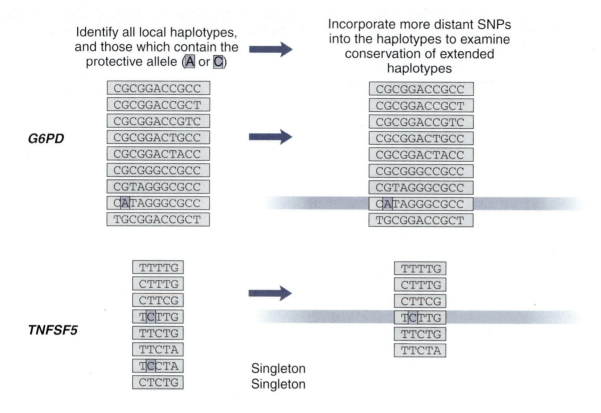

Figure 6.8: Evidence for positive selection in haplotypes at *G6PD* and *TNFSF5*.

The degree of extended haplotype conservation is shown by bars extending on either flank from local haplotypes (within the box) at two genes identified using closely spaced SNPs. Longer bars indicate greater haplotype conservation. The protective alleles at either locus are shown in blue. The conservation of haplotypes extending out from the core region is much greater for local haplotypes containing the protective alleles. These bars are drawn schematically, and are not to scale. The conservation of extended haplotypes cannot be determined for local haplotypes that are only found once in the population.

When interpreting departures from neutrality, as indicated by significant values for many of the neutrality test statistics described above, there are often two equally plausible explanations: one invoking demographic factors, and the other selective ones. Given that much is known about human demography from other disciplines, or, at the very least, is knowable from future multilocus analysis, how can this information be used to discriminate between alternative explanations for departures from neutrality? Rather than adopting a simple neutral null hypothesis it is possible to compare many non-neutral hypotheses, using a **likelihood** framework (see *Box 6.2*). Although they are computationally intensive, these methods provide perhaps our best hope for the future disentanglement of demographic and selective histories.

No doubt, future selection tests will continue to compare patterns of variation across loci. Selection tends to affect different genes differently, whereas the likely impact of demographic processes is the same all over the genome. This time-honored criterion to disentangle selection from its alternatives is bound to play an ever increasing role in future comparative studies of genetic diversity at multiple loci.

Software for performing some of the selection tests discussed in this chapter is listed in *Table 6.7*.

6.4 Genetic distance measures

6.4.1 Why do we need measures of genetic distance?

Measures of **genetic distance** are statistics that allow us to compare the relatedness of populations or molecules. The greater the evolutionary distance between them, the greater the value of the statistic. If a measure is greater between population A and B than between C and D, we can say that C and D are more closely related than are A and B.

Such measures allow us to explore population structure and molecular diversity in greater detail, by pairwise comparisons, rather than by averaging over all populations or molecules. As we shall see, by making certain assumptions, it becomes possible to convert distance measures to an evolutionary time-scale. This might allow us to say, for example, not only that C and D share a more recent common ancestor than do A and B, but that the common

BOX 6.2 Likelihood, maximum likelihood and the likelihood ratio test

There is a different philosophical interest in probability as opposed to **likelihood**. Probability is concerned with making predictions of outcomes (data) from a solid set of hypotheses (model plus parameters). For example, with a balanced coin (hypothesis), what is the probability of getting five 'heads' from 10 spins (outcome)? Likelihood, on the other hand, is concerned with generating better hypotheses having observed the outcome. For example, how likely is it that the coin is balanced (hypothesis) given that five 'heads' were observed in 10 spins (outcome)?

The *likelihood* of the hypothesis (H) given the outcome (O), is proportional to the *probability* of the outcome given the hypothesis.

$$L(H|O) \propto P(O|H)$$

The hypothesis can be varied by changing the parameters within a model, or by changing the model itself, and comparing the likelihoods. Commonly, individual likelihoods are very small and so are often expressed as log-likelihoods.

Evolutionary models tend to have many parameters. Often, the true values of the model parameters are unknown, but we would like to estimate them. Typical parameters within evolutionary models include: mutation rate, population size, population growth rate and ages of alleles. **Maximum likelihood** can be used to estimate the parameters of these models. The criterion used to choose one value for a parameter over another is that a given value maximizes the likelihood of the chosen model.

In the above scenario, the investigator is at the mercy of the evolutionary model. An alternative model may very well give an alternative maximum likelihood estimate for the same parameter. How can we decide which model is the more appropriate? A **likelihood ratio test** (LRT) compares the ability of alternative models to explain the data, by considering the significance of the test statistic below:

$$2\log \frac{\text{maximum likelihood under alternative hypothesis}}{\text{maximum likelihood under null hypothesis}}$$

This test statistic often approximates to the χ^2 distribution with one degree of freedom, allowing easy assessment of the relative merits of different hypotheses. Generally, specifying more parameters gives a better fit of the model to the data. The principle of the LRT is to make sure that the increase in the model's complexity does significantly improve our ability to account for the observed data.

As evolutionary models become more complex, more parameters are required, and the information present in the data can be spread more thinly amongst them. Consequently, more data are often required to maintain similar levels of certainty when more complex models introduce new parameters.

ancestor of A and B is twice as old as that of C and D. Although initially devised for classifications of populations, genetic distance measures also allow the construction of phylogenies of populations (or molecules).

Historically, genetic distances between populations were based solely on allele frequencies, but as with the measures of population subdivisions discussed above, recent developments allow the inclusion of molecular information, in the form of distances between alleles. Indeed, some measures of genetic distances are merely pairwise analogs of measures of population subdivision, and share the same names.

6.4.2 Distances between populations

There are a number of commonly used measures of genetic distances between populations, both 'classical' and 'molecular' in nature. Despite the plethora of different measures, most are highly correlated. This abundance of measures has arisen in response to different data types and different expectations about the underlying evolutionary processes. For example, diversity data from markers with a high mutation rate may be analyzed with a genetic distance measure that emphasizes the contribution of mutational processes to population divergence. Alternatively, genetic drift may be thought to be

TABLE 6.7: SOFTWARE: TESTING FOR NEUTRALITY.

Method	Software	URL
McDonald–Kreitman test	DNASP	http://www.ub.es/dnasp/
Tajima's D	''	''
HKA test	''	''
ω	PAML	http://abacus.gene.ucl.ac.uk/software/paml.html
H	htest	http://crimp.lbl.gov/htest.html

the predominant process causing population divergence at slowly mutating markers, and the genetic distance measures chosen accordingly.

If we consider two populations **X** and **Y** with the frequency of the ith allele being x_i and y_i respectively, the simplest measure of genetic distance between two populations sums the difference between the allele frequencies, $\Sigma(x_i - y_i)$. This needs to be squared to avoid differences in sign canceling each other out, $\Sigma(x_i - y_i)^2$. However, this quantity fails to give sufficient weight to alleles with frequencies close to 0% or 100%.

Two commonly used classical measures of genetic distance are F_{ST} and Nei's standard genetic distance, D (Nei, 1987). Both of these vary between 0, for identical populations, and 1 for populations that share no alleles. For use as a genetic distance, F_{ST} is specifically formulated for two populations and can be defined as:

$$F_{ST} = V_p / p(1-p)$$

where p and V_p are the mean and variance of gene frequencies between the two populations respectively. This is just a weighted form of the simple measure considered above, that increases the weight given to alleles that are almost fixed ($p \sim 100\%$) or barely polymorphic ($p \sim 0\%$).

As before, there is a variety of different methods for estimating F_{ST} from empirical data. Thus F_{ST} can be regarded as a family of distances, rather than a single quantity.

Nei's standard genetic distance, D, relates the probability of drawing two identical alleles from the two different populations (which is $\Sigma x_i y_i$) to the probability of drawing identical alleles from the same population $\Sigma(x_i^2$ and $\Sigma y_i^2)$ by the following equation:

$$D = -\ln(\Sigma x_i y_i / (\Sigma x_i^2 (y_i^2)^{1/2})$$

By making assumptions about the processes that are driving the divergence of populations, we can relate distance measures to absolute time. This relationship can then be used to generate a 'corrected' (or 'transformed') version of the statistic that can be shown (under certain assumptions) to be linear with respect to evolutionary time.

For example, Wright showed that under the action of drift alone, F_{ST} varies with time (t) according to the equation

$$F_{ST} = 1 - e^{(-t/2N)}$$

Rearranging this gives:

$$t = -2N \ln(1 - F_{ST})$$

Thus the measure $-\ln(1 - F_{ST})$ is linear with respect to time.

Similarly, Nei's standard genetic distance, D, under certain assumptions, should also be linearly related to time by:

$$D = 2\alpha t$$

Where α is the rate of nucleotide substitution. The implications of these equations for genetic dating are discussed in greater detail in Section 6.6.

However, some population episodes can disrupt the linear relationship between a given genetic distance measure and time. Commonly cited examples include population bottlenecks and even minimal amounts of migration between diverging populations.

Linearity of the genetic distance measure is a useful property especially when constructing phylogenies. The other major property that affects the usefulness of a measure is its variance. How accurately does the estimate from an empirical sample reflect the true population value? If hundreds of individuals are needed from each population for an estimated measure to be accurate, this will seriously affect its usefulness. Our confidence in saying that two populations are genetically different depends on the variance of the statistic we use – the lower the variance of the statistic, the higher the confidence.

A number of measures of population genetic distance have been specifically developed for use with microsatellite loci. As discussed in the previous chapter, the mutational dynamics of these loci differ significantly from the standard **infinite alleles model** of molecular evolution. Genetic distances incorporating the **stepwise mutational model** (**SMM**) contain information on the molecular distances between alleles. *Table 6.8* lists a number of these measures.

While the linearity of these measures is an improvement on methods that do not account for molecular distances between alleles, their performance for phylogeny construction does not seem as powerful (Perez-Lezaun *et al.*, 1997). This apparent weakness is probably due to the fact that mutations accumulate slowly through evolutionary time. Most questions of anthropological interest involve processes occurring over relatively short time periods, during which few mutations accumulated, but genetic drift and migration may have been substantial.

Whatever measure of genetic distance between populations is used, we must test its significance: i.e., determine if the distance is significantly different from zero. This is especially important for human populations, which are often closely related. *Table 6.9* lists software for calculating some of the population distance measures discussed above.

6.4.3 Distances between molecules

In principle, once the molecular basis of allelic variation has been defined, it is trivial to estimate distances between alleles.

TABLE 6.8: POPULATION GENETIC DISTANCE MEASURES FOR MICROSATELLITE LOCI.

Measure	Reference
$\delta\mu^2$	(Goldstein *et al.*, 1995)
R_{ST}	(Slatkin, 1995)
D_{SW}	(Shriver *et al.*, 1995)

TABLE 6.9: SOFTWARE: CALCULATING GENETIC DISTANCES BETWEEN POPULATIONS.

Distance	Software	URL
F_{ST} (sequences & STRs)	ARLEQUIN	http://lgb.unige.ch/arlequin/
D (Nei's distance)	GDA	http://lewis.eeb.uconn.edu/lewishome/software.html
F_{ST} (STRs)	Microsat	http://hpgl.stanford.edu/projects/microsat/
$\delta\mu2$ (STRs)	"	"
R_{ST} (STRs)	"	"
D_{SW} (STRs)	"	"

Two homologous sequences will differ at a discrete number of sites, and two microsatellite alleles will differ by a discrete number of repeat units. By assuming that such differences accumulate in a stepwise fashion these statistics in themselves represent simple genetic distances.

As we saw previously, linearity with respect to evolutionary time is an important property of genetic distances. The simple measures described above can be improved upon in this regard. As evolution proceeds, parallel mutations and reversions disrupt the linearity of molecular genetic distances. As a result, the number of differences observed between two molecules is an underestimate of the actual number of mutational steps differentiating them. In Section 6.3.2 we saw how this process operates on nucleotide sequences. A similar process operates at microsatellites, except that linearity decays much more rapidly as result of the higher mutation rate and stepwise mutations of these loci, where a second mutation is as likely to produce a return to the original allele as to result in a new allele.

Genetic distances between nucleotide sequences can be 'corrected' using a specific model of sequence evolution. This estimates the number of actual mutational steps between the sequences by inflating the observed number of sequence differences. There are a number of different models of sequence evolution, varying in complexity, which were discussed in greater depth in Section 5.2.4. The simplest model considers all bases to be equally mutable and is known as the **Jukes–Cantor model**. A more complex model incorporates the common finding that transitions are more frequent than transversions (see Section 3.2), and thus assigns them separate mutation rates. The most complex model assumes that each base substitution has its own mutation rate. However, the amount of sequence divergence observed between homologous human sequences is so low, typically less than 0.1%, that sites are not close to saturation, and so the observed sequence differences are a reasonable approximation of the actual number of mutational changes. Consequently, as long as mutation rates at all sites are similar, little if any correction is required. It becomes more important to correct the observed sequence divergences when comparing sequences from different primate species, especially if they are distantly related. Jukes–Cantor distances between different ape species have been found to vary among genomic regions as a result of differences in the underlying mutation rate (see *Table 7.1*).

Similarly, we can use models of microsatellite mutation to correct for the observed divergence between two alleles being an underestimation of their true divergence. The model most frequently used is the stepwise mutational model (SMM) discussed in the previous section. This model has been used to derive a measure of microsatellite genetic distance, which is the average of the squared distance between alleles (known as **ASD**). This measure has been shown to be linear over longer time periods than other measures. Nevertheless, once more complex, and realistic, models of microsatellite evolution are considered, such as those that incorporate constraints on allele size, the linearity of all measures is substantially reduced. *Table 6.10* lists software for calculating various genetic distances between molecules.

TABLE 6.10: SOFTWARE: CALCULATING GENETIC DISTANCES BETWEEN MOLECULES.

Distance	Software	URL
Jukes–Cantor	PHYLIP	http://evolution.genetics.washington.edu/phylip.html
More complex models of sequence evolution	PAML	http://abacus.gene.ucl.ac.uk/software/paml.html
More complex models of sequence evolution	TREE-PUZZLE	http://www.tree-puzzle.de
ASD	Microsat	http://hpgl.stanford.edu/projects/microsat/

6.4.4 Visual representations of genetic distance

Having obtained a set of pairwise genetic distances between a set of populations or molecules, how can we display this information in a comprehensible manner? It is difficult to detect patterns from a table of pairwise differences; graphical displays of diversity are preferable.

If we have n populations, we require n−1 dimensions to fully display their pairwise genetic distances as graphical, or Euclidean distances (*Figure 6.9*). However, we often have more than four populations or molecules that we want to compare, yet we are unable of conceiving of, or representing, the multidimensional spaces required to display these data. **Multivariate analyses** allow us to reduce this multidimensional space to the two or three dimensions we can comprehend, while reducing the inevitable loss of information. **Multidimensional Scaling** (MDS) is one of these methods. It is also possible to use multivariate analysis to generate two- or three-dimensional graphical representations of interpopulation or intermolecular distances using the raw data of allele frequencies, rather than genetic distances. **Principal Components Analysis** (PCA) is a commonly used example of this approach. Individual axes, known as principal components (PCs), are extracted sequentially, with each PC encapsulating as much of the remaining variation as possible. Using PCA it is possible to estimate the proportion of the total variance in the total dataset that has been summarized within these reduced dimensions. *Figure 6.10* shows how PCA reveals global relationships from a set of pairwise distances between populations. An example of a PCA plot derived from real data can be seen in *Figure 10.21*, which relates different African populations using mitochondrial DNA diversity. PC data can also be used to construct **'synthetic' maps** that summarize information from several alleles with similar geographic distributions (see Sections 6.7.1 and 10.5.3).

Alternative methods for the graphical display of interpopulation and intermolecule relationships attempt to reconstruct the ancestral relationships between all the entities being investigated. These are known as **phylogenetic methods**, and their outputs are **phylogenies**. This most important class of methods is the subject of our next section.

6.5 Phylogenetic methods

6.5.1 Introduction to tree terminology

The tree is an intuitively attractive method for displaying the relationships between many kinds of variant entities. Sometimes the tree itself describes the actual ancestral relationships of these entities; it encapsulates the *mechanism* by which diversity arose. Such is the case for trees of separate species or nonrecombining haplotypes of unique markers – for example, human Y-chromosomal SNPs (see *Box 8.8*). Often however, the tree is a convenient tool for the graphical display of diversity. In such cases, it may imply a *model* for how diversity arose, but does not itself represent the *mechanism*. For example, a tree of populations implies a model whereby populations split (fission) from common ancestors, and subsequently do not mix. This certainly does *not* represent the reality of population evolution.

Due to their inherent attractiveness as graphical tools, trees are used across a wide variety of disciplines, for a broad range of purposes. This has led to a set of partially redundant terminologies that can easily confuse the uninitiated (*Figure 6.11*).

A tree consists of **branches** (also known as 'edges') between **nodes**. The ultimate aim is to relate groups of

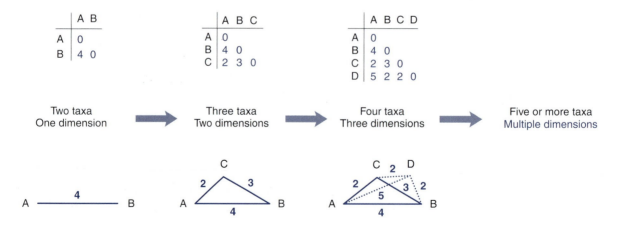

Figure 6.9: How many dimensions are needed to display population relationships?

Distance matrices are shown relating genetic distances between increasing numbers of populations. As the number of populations increases, the number of dimensions needed to display these distances visually such that the length of the lines represent the genetic distances between them quantitatively also increases. A single line of a given length can relate two populations. Three populations can be related by a triangle, whose sides have lengths proportional to the genetic distance. Four populations can be represented by a pyramid. With five or more populations it is no longer possible to represent population relationships in three-dimensional space.

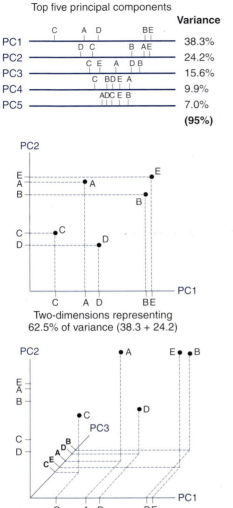

Top five principal components

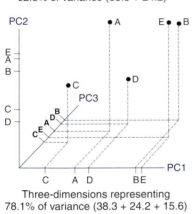

Two-dimensions representing
62.5% of variance (38.3 + 24.2)

Three-dimensions representing
78.1% of variance (38.3 + 24.2 + 15.6)

Figure 6.10: Graphical representations of principal components analysis of five populations in both two and three dimensions.

Principal components (PC) are extracted from multivariate data from five populations (A–E) such that each successive PC contains a smaller proportion of the overall variance. Each PC is displayed as a single axis. The first five principal components account for 95% of the variance within the dataset. These PCs can then be used as axes for graphs that maximize the amount of variation displayed in a fixed number of dimensions.

populations or molecules, known as **taxa** or operational taxonomic units (OTUs). These taxa occupy special nodes. Nodes unoccupied by taxa represent hypothetical ancestors (or hypothetical taxonomic units – HTUs). In a **cladogram**, all terminal nodes (known as leaves) represent taxa, whereas all internal nodes are hypothetical ancestors. However, other types of evolutionary trees also allow taxa to occupy internal nodes.

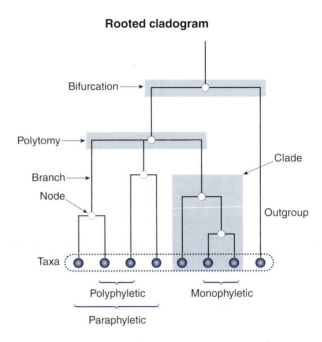

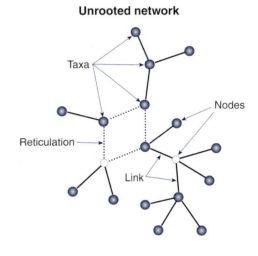

Figure 6.11: Terminologies for trees and networks.

A number of different features of evolutionary trees and networks described in the text are highlighted.

Trees can be **rooted** or **unrooted**. Rooted trees contain one taxon that can be defined as having the most ancestral divergence compared to all other taxa. This taxon is also known as an **outgroup**. The root of the tree lies between the outgroup and all other taxa. This property orientates rooted trees with respect to evolutionary time, meaning that evolutionary changes on the tree have a defined direction of change, from ancestral to derived. An unrooted tree can be rooted either by assuming that the root falls midway along the longest branch on the tree (mid-point rooting), or by incorporating a taxon known to be an outgroup to all other taxa, and seeing where it joins the unrooted tree.

The proximity of an internal node to the root of the tree determines the relative antiquity of the divergence event it represents. Nodes that are closer to the root are more ancient than those further from the root. Unrooted trees are not able to relate the ancestry of different nodes in this way; it is not immediately obvious which nodes are descendants and which are ancestors. The number of possible trees for any given number of taxa increases rapidly and is different for rooted and unrooted trees. For nine taxa, there are over 135 000 possible unrooted trees but over 2 000 000 rooted ones.

The branching pattern of a tree is known as its **topology**, and the descendants of a single node form a **clade**. Some trees only allow two branches to descend from each node, a process known as **bifurcation**. Alternatively, more than two branches can descend from the same internal node, forming a **polytomy** (or **multifurcation**). A cladogram represents solely the relationships between taxa – branch lengths between nodes are irrelevant. In contrast, **additive** trees use branch lengths to reflect evolutionary distance quantitatively. Thus additive trees can vary not only topologically, but also quantitatively in the length of their branches.

The relationship between a group of taxa and clades allows taxon groupings to be classified on the basis of the tree. A grouping of taxa that fall into a single clade is **monophyletic**. However, if this grouping excludes other members of the same clade it is **paraphyletic**. Taxa groups that span multiple clades are **polyphyletic**. Only a monophyletic grouping of taxa is considered to be a coherent evolutionary lineage.

An important property of trees is that as evolutionary time progresses towards the present, branches diverge but never coalesce. However, some biological processes (e.g., recombination), can cause lineages to merge, while others (e.g., parallel mutation), cause them to *appear* to merge. In either case the result can be represented as a four-sided closed structure known as a **reticulation**, or cycle. Trees that incorporate such structures, in an attempt to include these biological processes, are known as **networks**. Some network methods represent taxa solely as terminal nodes, while others also allow taxa to occupy internal nodes. Lines connecting nodes are known as **links**.

6.5.2 Different approaches to phylogeny reconstruction

There are many different methods for constructing trees and networks from genetic data. Not all are suitable for the different types of data available, and no single method predominates. Tree construction ('phylogenetic') methods are generally classified using two important criteria: first, the type of data used as input, and second, the means by which a tree is constructed (see *Table 6.11*).

Input data fall into two major classes: *distances* and *characters*. Genetic distances between populations or molecules must first be calculated from raw data, as described in the above section, and are represented in the form of a **distance matrix**. Characters are discrete units of evolution, whether single base changes in a nucleotide sequence, or

changes in numbers of repeat units of a micro- or minisatellite. Character-based methods allow us to infer the character content of ancestors.

There are two main classes of phylogeny construction methods. The first, known as *'clustering'* methods, uses an iterative algorithm to combine taxa together in a hierarchical fashion (one-by-one). The second class, known as *'searching'* methods, considers the whole range of possible trees and chooses that which best fits the data according to some **'optimality criteria'**. In practice, the range of possible trees is often so large that it becomes computationally unfeasible to compare all trees. It has been shown that if a million trees could be compared every second, it would still take 10 million years to compare all trees formed by only 20 taxa. Consequently these methods seek to sample a representative subset of all trees. In practice this is done by: (i) jumping between separated regions of the entire range; (ii) finding the best tree in each locale; and (iii) comparing the best trees from each locale.

How can we choose one phylogenetic method over another? A good phylogenetic method should have five characteristics:

▸ efficient – a tree is constructed rapidly;

▸ consistent – the same tree is obtained as more data are added;

▸ robust – the tree is insensitive to violations of the method's assumptions;

▸ powerful – few data are required to get the correct tree;

▸ **falsifiable** – the validity of the method's assumptions can be tested.

As we shall see below, no single method has all five characteristics – each emphasizes some desirable properties over the others. Fast methods are not always robust, and powerful methods are often slow.

6.5.3 Distance matrix phylogenetic methods

The Unweighted Pair-Group Method with Arithmetic mean (UPGMA) is perhaps the simplest phylogenetic method. The tree is built by an iterative clustering process that combines the two taxa that have the least genetic distance between them. When taxa are combined they form a new taxon. The genetic distances between this composite taxon and other taxa are the average of the distances from the individual constituent taxa. The UPGMA produces a special form of additive tree, known as an **ultrametric** tree. This tree has the interesting property that the distance between each taxon and the root is the same. If we draw a scaled version of this tree, we find that all the terminal nodes are aligned. Therefore, the UPGMA method has the advantage being a convenient representation of taxa living in the same moment in time. However, the UPGMA method assumes equal rates of evolution for all taxa, and if evolutionary rates are unequal among lineages, topological errors may result (Sneath and Sokal, 1973).

TABLE 6.11: CLASSIFICATION OF PHYLOGENETIC METHODS.

| Methodology | Type of data | |
	Characters	Distances
Clustering	–	Neighbor joining, UPGMA
Searching	Maximum Parsimony, Maximum Likelihood	Minimum evolution
Networks	Minimum spanning, Median Networks	Split decomposition

Cavalli-Sforza and Edwards suggested in 1967 that the best tree based on a distance matrix would be that which gave the shortest sum of branch lengths (S). This is known as the principle of 'minimum evolution' (Cavalli-Sforza and Edwards, 1967). Whilst this is a good example of an optimality criterion that can be used as the basis for a 'searching' phylogenetic method, it also provides a rationale for designing alternative clustering methods to UPGMA. Clustering methods are much faster than search-based methods.

Neighbor-Joining (NJ; *Figure 6.12*) is a clustering method that attempts to find the tree with the minimal value for S (Saitou and Nei, 1987). The iterative procedure used to reconstruct the phylogeny is very fast to compute and often produces trees that are very close to the minimum evolution tree.

A tree of Pacific populations constructed using the Neighbor-Joining algorithm can be seen in *Figure 11.19*. The genetic distances used to construct this tree were derived from classical markers.

6.5.4 Character-based phylogenetic methods

The principle of **Maximum Parsimony** (MP) defines the best tree as the one that requires the smallest number of

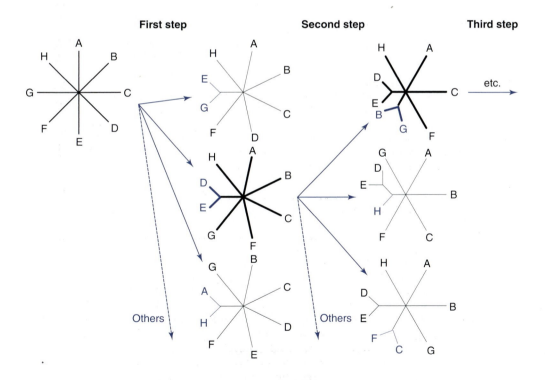

Figure 6.12: Schematic representation of phylogeny reconstruction by the Neighbor-Joining algorithm.

Initially, all taxa (A–H) are related by a single polytomy, forming a tree known as a **star phylogeny**. Then, all possible pairs of taxa are pulled out of this tree in turn, and the pair that gives the shortest overall tree (lowest value for S) is selected (shown with thicker branches). As with UPGMA, this pair is combined into a single taxon, and distances from the composite taxon to all other taxa recalculated as the average of distances from the constituent taxa. Pairs of taxa are sequentially selected in this manner until all interior branches have been found.

evolutionary changes to account for the data. The branch lengths of trees produced from character-based methods are the numbers of individual evolutionary changes along that branch. Thus MP is for character data what minimum evolution is for distance matrix methods. In the case that two (or more) trees are equally parsimonious, there is no criterion for choosing between them, and no unique tree can be inferred.

Having defined the optimality criterion for MP, how is this ideal tree sought? For the number of taxa commonly used in studies of human genetic diversity, it is often not possible to examine all possible trees because it would be computationally too laborious. The method need only consider the subset of nucleotide sites known as **informative sites**. These are polymorphic sites at which at least two alleles are present in two or more individuals. This reduces some of the computational load, but a search strategy is still required, both for jumping between different locales within the range of possible trees, and for identifying the most parsimonious tree in each locale. There are some famous examples where an initially published tree has been shown in subsequent analyses not to be the most parsimonious (Vigilant *et al.*, 1991; Penny *et al.*, 1995). The initial search strategies missed more globally optimal trees.

MP methods can incorporate information about the relative rate of different mutations, for example if transversions are known to occur less frequently than transitions, or if certain sites are known to be hypermutable. These mutational events can be weighted accordingly, such that the rarer changes carry more influence.

MP methods can be sensitive to unequal rates of evolution, via a process known as **'long branch attraction'** (*Figure 6.13*). Two lineages that have been mutating at a faster rate have longer branches. As a result, they may by chance share parallel mutations. If these mutations are more numerous than those that distinguish their common ancestors, MP will produce the wrong tree.

A tree can be regarded as a hypothesis that attempts to explain an outcome, the data. Thus alternative trees can be considered to be competing hypotheses which can be compared through a likelihood framework, as described in *Box 6.2*. The optimality criterion used is that the chosen tree should be the most likely one. Under a given evolutionary model, the best tree is that which has the **maximum likelihood** (ML) of producing the data. Moreover, alternative evolutionary models can be compared by applying the likelihood ratio test. This lends falsifiability, which was one of the desirable properties for a phylogenetic method considered earlier. Both tree topology and branch lengths should be inferred at the same time using a ML approach, as both affect the likelihood of the tree. As a result, ML methods are computationally intensive.

ML methods require models of sequence evolution (discussed in Chapter 5). The more complex of these models can incorporate the differential rates of different types of sequence changes, as well as variable mutabilities among

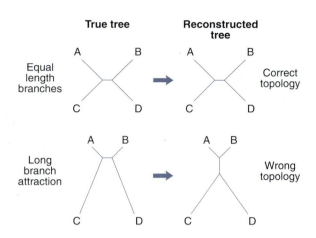

Figure 6.13: Long branch attraction.

Two trees relating the evolutionary relationships of four taxa are shown, one in which the rate of evolution is equal for all lineages and the other for which taxa C and D have experienced much faster mutation rates, which gives them longer branch lengths. When MP is used to reconstruct the phylogeny relating these four taxa from their sequence divergences, it identifies the correct phylogeny in the first scenario but not in the second.

different sites. However, these models incorporate parameters (e.g., how much more likely transitions are than transversions), that are unknown for the dataset in question. This introduces a 'chicken and egg' problem. If we knew the tree relating a set of sequences we could estimate the parameters of the model more accurately, but we need the parameters of the model to get the tree in the first place. This means that trees and model parameters must both be varied, which again increases the computational load enormously.

Although maximum parsimony was originally proposed as an approximation to maximum likelihood methods, it has been shown that the shortest tree (the MP tree) is not always the most likely tree (the ML tree), although in practice they are often very similar. Some conditions under which MP and ML trees are prone to differ are known: when there has been a large amount of evolutionary change within the tree, or substantial rate variation among lineages. *Table 6.12* lists some of the available software for **phylogeny reconstruction**.

Examples of phylogenies based on character state data can be found in many chapters in this book, but they have been especially important in investigating the origins and earliest migrations of modern humans (Sections 8.6 and 9.5.4), and identifying our closest primate relative (see Section 7.4.3). The most detailed phylogeny of a single locus is that relating all Y-chromosomal SNP haplotypes shown in *Box 8.8*. This phylogeny was constructed using MP, and can be shown to be the single most parsimonious tree relating these haplotypes.

TABLE 6.12: SOFTWARE: PHYLOGENETIC METHODS.

Method	Software	URL
Most tree methods	Phylip	http://evolution.genetics.washington.edu/phylip.html
Most tree methods	PAUP*	http://paup.csit.fsu.edu/index.html
Minimum spanning network	ARLEQUIN	http://lgb.unige.ch/arlequin/
Reduced median networks	Network	http://www.fluxus-engineering.com/sharenet.htm
Median-joining networks	"	"
Split decomposition	Splitstree	http://www-ab.informatik.uni-tuebingen.de/software/splits/welcome_en.html

6.5.5 Assessing the significance of phylogenies

Having obtained a phylogeny for our data, how confident can we be that we have obtained an *accurate* and *precise* phylogeny? After all, given a dataset of unknown quality, every phylogenetic method will reconstruct at least one phylogeny. 'Accuracy' refers to the proximity of the tree obtained to the true tree, while 'precision' refers to the number of alternative trees that can be excluded.

The accuracy of phylogenetic methods can be tested by generating data from a known phylogeny, and then asking which phylogenetic method reconstructs the tree closest to the known tree. Such datasets can be generated either by simulating sequence divergence **in silico** along the branches of a predetermined tree, or by manipulating experimental organisms or molecules in the laboratory in a controlled fashion, and subsequently analyzing polymorphic markers within the different lineages.

The creation of *in silico* datasets allows many different models and parameters of sequence evolution to be explored. 'Zones' of evolutionary processes in which specific methods are vulnerable to reconstructing erroneous trees can be identified. For example, it has been shown that when mutation rates vary substantially along different lineages, UPGMA becomes inaccurate. Thus methods can be chosen according to whether real data appear to fit into any of these 'zones'.

What are the potential sources of error that might lead to the wrong phylogeny?

▶ Tree structure – a convoluted evolutionary history relates the taxa.
▶ Too few data.
▶ An incorrect model of sequence evolution.

The first of these sources of error is unavoidable; certain evolutionary histories simply do not lend themselves to phylogenetic reconstruction. For example, rapid species divergences and large population sizes will lead to substantial numbers of incongruent gene trees due to the maintenance of ancestral polymorphisms between consecutive speciation events and their random **assortment** into the different lineages. In such cases, the reconstruction of species relationships by phylogenetic methods is doomed to confusion.

The level of confidence we should have in a reconstructed phylogeny can be assessed statistically using the '**bootstrap**'. This method is based on the idea that if a dataset strongly supports a certain statistical result (in this case a tree), then randomly chosen subsets of the data should support the same result (tree) (Efron, 1982). In practice, the phylogenetic bootstrap is performed by resampling sites from the sequence alignment *with replacement*. The same number of sites as were present in the original alignment is selected, but each time a site is selected from the alignment it remains able to be selected again. Therefore, some of the sites in the original alignment are present more than once in these synthetically replicated datasets, and others are absent. Each of a large number of synthetic datasets (typically 100–1000), is used to reconstruct a phylogeny, and each time an internal node in the original tree is precisely replicated in a bootstrapped dataset this is noted. Bootstrap values are usually displayed on the original phylogeny in the form of percentages next to the nodes to which they refer. Thus, a value of 92 means that the same node was reconstructed from 92% of all synthetic datasets. There is no agreement over what constitutes a good bootstrap value, although bootstrapping is generally regarded as being relatively conservative and values over 70% are often considered reasonably reliable.

6.5.6 Networks and split decomposition

Some biological processes are not adequately represented by a phylogeny in which taxa are forever splitting but never joining. These reticulate processes include recombination, which reunites and merges two previously divergent haplotypes. Similarly, gene flow between populations can result in their sharing young alleles despite a more ancient fission from a common ancestor. These kinds of processes generate loops within phylogenies known as **reticulations** or **cycles**. Such phylogenies are called **networks**.

A single network contains within it several trees. Thus if two trees are similarly well supported by the data, we do not have to choose one over the other, but can summarize them in a network. This network represents more of the information present in the data than does either tree. As with trees, networks can be constructed from distance or character data, and there are a number of alternative methods of construction.

The method of *split decomposition* detects antagonistic signals favoring different phylogenies within genetic distance data. In contrast, *minimum spanning networks* and *median networks* are based on character-based data.

A minimum spanning network can easily be constructed by hand (illustrated for microsatellite haplotypes in *Figure 6.14*), by following a simple procedure:

▶ a character-based distance matrix between all taxa is computed;

▶ links are drawn between all taxa separated by single mutational steps;

▶ mutational steps of increasing size are considered until all taxa are linked into a single network and all most parsimonious links of equal length from any given taxa have been reconstructed.

The minimum spanning network shown in *Figure 6.14* is by no means the most parsimonious method for linking all taxa. The overall length of a network can often be shortened by adding hypothetical ancestral, unobserved nodes to the network (*Figure 6.15a*). Median networks provide a systematic method for reconstructing these ancestral nodes whilst guaranteeing that the resultant network contains all the most parsimonious trees (Bandelt *et al.*, 1995). Median networks are also constructed by a simple procedure, although it is often performed computationally due to its repetitious nature:

▶ variant sites within sequences or microsatellite haplotypes are converted to binary haplotypes;

▶ sites showing perfectly correlated variation across all haplotypes are combined into a single character, which is weighted for the number of sites incorporated within it;

▶ for each triplet of haplotypes in turn, a median haplotype is calculated, which is the consensus of the three haplotypes, and, if novel, is added to the set of haplotypes;

▶ the observed haplotypes plus the median haplotypes are then linked by a single-step network, and branches representing characters represented by multiple sites are weighted accordingly.

Median networks are commonly used when **homoplasies** resulting from parallel mutations or reversions are frequent. This type of data includes mitochondrial DNA control region sequences (where the base substitution mutation rate is high) and microsatellite haplotypes. Median networks are prone to producing hyperdimensional cubes of reticulations when the number of taxa becomes large, which quickly make the network unintelligible. A coherent set of rules has therefore been developed to 'reduce' the network by removing some reticulations through elimination of the least likely links (Bandelt *et al.*, 1995).

An alternative method for constructing networks with limited levels of reticulation, known as median joining, has also been developed (Bandelt *et al.*, 1999). The algorithm

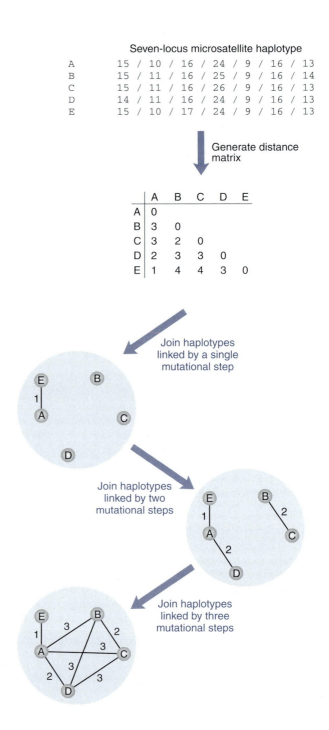

Figure 6.14: Constructing a minimum spanning network from microsatellite haplotypes.

The data consist of five haplotypes (A–E) comprising numbers of repeat units at seven linked microsatellites. Assuming that microsatellite mutations occur in single steps, a distance matrix can be calculated. For example, haplotypes A and C differ by one repeat unit at the second locus, and two repeat units at the fourth locus, making a total of three repeat unit differences between these two haplotypes. The network is then constructed from this distance matrix using the procedure described in the text.

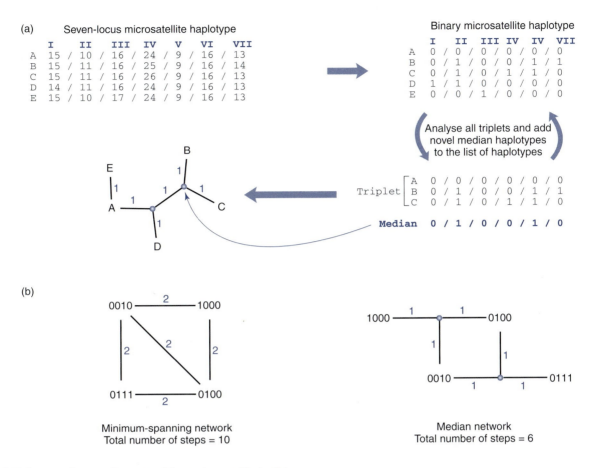

Figure 6.15: Constructing a median network from microsatellite haplotypes.

(a) The haplotype data used are the same as for the minimum spanning network example in *Figure 6.14*. First, haplotypes of allele repeat numbers are converted to binary format. Invariant loci (V and VI) are ignored. The smallest allele at a locus is designated 0. An allele one repeat larger is designated 1. For locus IV, where three alleles are present, the smallest is 00, the allele one repeat unit longer is 01 and the allele two repeats longer than the shortest is 11. This takes account of a single-step mutational mechanism. Ancestral nodes not present in the sampled data are represented by small filled circles. In this example, a single most parsimonious tree is produced by the median network algorithm. (b) A minimum spanning network and median network for the same four binary haplotypes, both networks contain reticulations (cycles). Again, the median network is shorter.

used to construct these median-joining networks is based on the limited introduction of likely ancestral sequences/ haplotypes into a minimum spanning network of the observed sequences. Again, these likely ancestral sequences are identified through the calculation of median haplotypes. The median joining algorithm has the advantage of being applicable to multi-state markers and is useful for large datasets, but is more unreliable for phylogenies with long branches. Consequently, the median joining algorithm is most often used for closely related, intraspecific haplotypes.

Thus reduced median and median joining networks represent alternative methods for obtaining intelligible networks with limited amounts of reticulation. The former method generates all possible ancestral sequences and then eliminates the least likely, whereas the latter method introduces limited numbers of the most likely ancestral sequences into a phylogeny of the observed sequences.

An example of a network based on HVS I sequences can be seen in *Figure 10.13*, which shows the major European maternal lineages.

6.6 Dating

A rooted phylogeny provides a *relative* chronology for genetic changes; changes close to the tips of a tree must have occurred after those closer to the root within the same clade. However, when integrating genetic data with those from other disciplines of prehistory, it is highly desirable to produce an *absolute* chronology, in other words, to provide actual time estimates of when such changes occurred. In this way, the timing of a change, be it in a population or a molecule, can be placed in a wider context, perhaps related to an archaeological culture or a paleoclimatological event. Such contextualization of genetic data is vital for our reconstruction of human prehistory.

How might we reconstruct this chronological record? We need two tools:

▸ a molecular clock – a process generating variation that changes predictably with time. This process could be genetic drift, mutation or recombination;

▸ a means of determining how fast the clock ticks – calibrating the predictable rate of change.

We will consider dating methods that attempt to date population splits and molecular changes separately. For both of these classes of methods, it is usually assumed that selection has not been acting on the loci being studied, as this would complicate the relationship between variation and time (see Section 5.6.3 on the 'neutral theory of evolution').

6.6.1 Population splits

Dating population splits revealed by population phylogenies requires the assumption that populations are akin to species: that no gene flow has occurred between them after they split. Classical population genetics allows us to relate the difference in gene frequencies between two populations to the time since they shared a common ancestor. In principle, the measures of genetic distance discussed above (e.g., F_{ST}) should give us some measure of the time, as they increase as the time passes since two populations split. However, many of these distances do not show linear relationships with time, or if they are linear, only exhibit linearity over short time spans. Accordingly, various transformations of these standard statistics have been proposed that improve their linearity.

Sometimes it is assumed that any differences in gene frequencies are due solely to genetic drift, with no mutation or selection. In such cases, the rate of population divergence depends only on the effective population sizes. When drift is operating as the molecular clock, to calculate the time at which two populations diverged we need to know their effective population sizes, and assume that they have been constant in size. This approach is necessary when using F_{ST} to estimate divergence times.

Under other assumptions, mutational processes, rather than drift, drive the molecular clock. A commonly used method adopting this approach considers the genetic distance measure, D, which as we saw above is related linearly to time via the parameter α, the rate of fixation of nucleotide substitution. In Chapter 5, we saw that the rate of fixation of neutral mutations is independent of the population size, and is equal to the neutral mutation rate. In the 1970s and early 1980s, much work went into assessing α through changes in the electrophoretic mobility of a number of different proteins between populations (or species) (Nei and Roychoudhury, 1982).

The introduction of molecular data in recent years has necessitated the use of more sophisticated mutational models for dating population splits. In addition, greater attention has been paid to mutation rate calibration. For example, the $\delta\mu^2$ measure has been devised for use with microsatellite data. This measure incorporates the stepwise mutation model of microsatellite evolution, and has the useful property of being independent of population size (Goldstein *et al.*, 1995).

Although much initial attention focused on the dating of population splits, such putative events rarely if ever represent the reality of population evolution; an absence of post-fission gene flow seems unlikely for most human populations. The advent of molecular data has led to a greater focus on the dating of molecular events, and it is these methods that we will consider now.

6.6.2 Most recent common ancestors (MRCA) of nonrecombining lineages

Nonrecombining portions of the genome contain haplotypes of unique markers that can be related by a single most parsimonious phylogeny. These portions of the genome include the Y chromosome and mitochondrial DNA, as well as short stretches of the autosomes within which no recombination events have occurred since the MRCA of all extant sequences. It is possible to date all nodes within such a phylogeny.

A number of different methods exist which estimate the time to the most recent common ancestor (TMRCA) of a set of chromosomes sharing a common mutational change at a unique marker. Note that this is different from dating the mutational change itself. There may be a substantial time lag between a mutation and the MRCA of the sampled chromosomes that carry it (*Figure 6.16*).

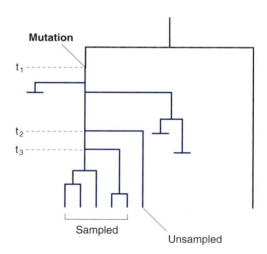

Figure 6.16: The age of an MRCA is not necessarily that of the mutation it carries.

Not all chromosomes carrying the mutation (blue branches) have survived to the present day, and not all chromosomes carrying the mutation in the extant population have been sampled. Consequently, the age (t_3) of the most recent common ancestor of sampled chromosomes carrying a specific allele may be younger than the age of the most recent common ancestor of all extant chromosomes carrying that allele (t_2), and is certainly younger than the age of the mutation generating that allele (t_1).

These methods can be usefully classified into those that invoke a population model and those free from any such model (Stumpf and Goldstein, 2001). This latter class uses summary statistics of **intra-allelic diversity** to date an allele. A new haplotype defined by a unique marker comes into being on a single chromosome with zero diversity at other markers linked to it. As this haplotype grows in frequency and becomes a population of closely related haplotypes, diversity accumulates through mutation at linked markers. The amount of intra-allelic diversity among the members of this population can be related to the age since they last shared a single common ancestor. These linked markers can be base substitutions, or multiallelic systems such as microsatellites.

An intra-allelic diversity summary statistic collapses many data points into a single value. These summary statistics increase linearly with time, in a similar manner to the genetic distances discussed in the previous section. The crucial parameter required to relate these summary statistics to time is the mutation rate. For nonrecombining haplotypes, mutation drives diversification, and so it alone represents the molecular clock.

Mutation rate calibration can either be performed *directly* on individual meioses, or *indirectly* through the observation of a certain amount of divergence across a known time span. The latter approach often uses divergence between species whose divergence is well dated in the fossil record. The slow rate of DNA sequence evolution is not generally amenable to direct calibration of mutation rates (although mutations in the hypermutable mtDNA control region can be detected directly – see Section 3.2.8), and so, typically, estimation requires divergence comparisons between closely related species. By contrast, the faster mutation rates of microsatellite loci can be measured using direct methods (Brinkmann *et al.*, 1998; Kayser *et al.*, 2000). The estimation of mutation rates is discussed more fully in Chapter 3. Different studies often appear to use different mutation rates for the same locus; however, sometimes these rates are in fact identical, but are expressed in different ways (see *Box 6.3* for more details).

Direct calibration can be performed by observing individual mutations in pedigrees or in pools of known numbers of sperm, and consequently generates a per-generation mutation rate. This introduces an additional

---

BOX 6.3 Why are there so many ways of expressing mutation rates?

One of the most confusing aspects of the scientific literature is that mutation rates quoted for some loci appear to be different between different publications. Given the importance of mutation rates for calibrating the molecular clock, these potential sources of conflict between interpretations must be understood. Some mutation rates are different because they genuinely reflect different rates of evolution, while other rates only appear to be different; they are in fact the same rate expressed in different ways.

A mutation rate is essentially the number of mutations that can be expected to occur within a segment of DNA over a certain period of time. Typically, the segment of DNA considered is often either the entire locus under study, or a nucleotide within that locus. Similarly, there are two time periods that are frequently considered, a year, and a generation (a million years is also sometimes encountered in the field of mtDNA research; see Section 3.2.8). If we consider the simple case of a locus 1500 bp in size, and a generation time of 25 years there are four ways of expressing the same mutation rate:

▶ 0.6 mutations per locus per year;

▶ 15 mutations per locus per generation;

▶ 0.0004 mutation per nucleotide per year;

▶ 0.01 mutations per nucleotide per generation.

Whether mutation rates are expressed per year or per generation often depends on how that mutation rate is calibrated. If the mutation rate is estimated from pedigrees (as for microsatellites), it is often given per generation, whereas if it is calculated using calibration points from the fossil record (more commonly used for sequence evolution), an annual rate is more frequently used. We can be more confident about the number of years to a dated common ancestor than we can be about the number of generations, which depends in part on the reproductive behavior of all intermediate ancestors.

The earliest mutation rate estimates were based on the incidence of dominant disorders that must have occurred **de novo**, rather than been inherited (see *Box 3.4*). These estimates first appeared prior to the advent of DNA sequencing technologies, and are only observable through the pathological outcome. As a consequence, the length of the gene could not be known, and these mutation rates could only be expressed per locus. There are other population genetics parameters which can similarly be expressed in per nucleotide or per locus form, most notably the population mutation parameter (also known as 'θ' – see Section 6.2.2).

The most confusing situation regarding apparently conflicting mutation rates occurs when diversity at a locus is assayed in different ways, most of which do not capture the full diversity. For example, mtDNA diversity can be assayed using HVS I sequences, RFLP analysis (covering ~20% of the mitochondrial genome) or whole genome sequences. Additional confusion results from our ability to calibrate the mitochondrial molecular clock using either pedigrees or fossil calibration points (see *Box 3.6* and Section 3.2.8).

uncertainty in making a conversion to calendar time, namely: what has been the average generation time over the period of time being considered? In this context, generation time is the average age of a parent at the birth of their offspring.

While genealogical records can provide us with accurate estimates of generation time in recent centuries (Tremblay and Vezina, 2000), dramatic demographic transitions during this period mean that these estimates are of little relevance to the more distant past. Although the age at which humans first become capable of reproduction has been decreasing over recent decades, this is more than outweighed by the much shorter life expectancies in prehistory. As a result, we assume that prehistoric generation times were shorter than they are in the present (Weiss, 1973).

One striking finding from studies of relatively modern genealogical records is that males tend to have significantly longer generation times than females. One large study of French–Canadians revealed an average female generation time of 29 years, but an average male generation time of 35 years (Tremblay and Vezina, 2000). This is due in part to menopause in women, with no analogous reproductive limit for men. Nevertheless, there are clearly other factors involved. For example, female fertility declines rapidly between the ages of 20 and 40 years, with no such dramatic change in male fertility. In addition, social pressures have resulted in a situation in many societies where men tend to be older, on average, than their partners. Whatever the causal mechanisms in modern societies, the possibility that male and female generation times in prehistory may also have been significantly different cannot be excluded. This would mean that different generation times should be used for loci with different patterns of inheritance. *Table 6.13* shows generation times proposed for different loci on the basis of the French–Canadian data.

Due to the increase in generation times over the past 5 million years, the generation time chosen depends to some degree on the time-depth being investigated. Generation time estimates used for modern humans tend to vary between 20 and 30 years, whereas, due to the expected shorter generation time of our common ancestor, 15 years has often been taken as the average generation time of hominids since our split from chimpanzees.

Table 6.14 lists three summary statistics that have been used to relate intra-allelic diversity to time, together with the type of data for which they are applicable.

Dating using the statistic ρ ('rho') requires the construction of a phylogeny relating the intra-allelic diversity. This phylogeny must contain the root haplotype, and so median networks are often used, as they allow the reconstruction of ancestral haplotypes, even if they are not observed within the sample. The ρ statistic represents the average number of mutational changes between the root haplotype and every individual in the sample. These mutational changes are counted from the network itself, rather than by estimation from the observed number of mutation differences between two haplotypes; this takes account of possible reversions and parallelisms at sites with greater mutation rates (*Figure 6.17*). This statistic is simply related to time by the equation:

$$\rho = \mu t$$

The Average Squared Distance (ASD) between a root microsatellite haplotype and all other haplotypes has also been shown to be linearly related to the time since the MRCA:

$$ASD = \mu t$$

Unlike the ρ statistic, ASD is calculated from the data without the need to construct a phylogeny. How are parallelisms and reversions accounted for? The above equation relating ASD to time was derived mathematically such that correction for reversions and parallelisms is automatic. This mathematical derivation assumes certain dynamics of mutation, namely, the stepwise mutation model, and consequently can only be applied to markers that accord with this model. Therefore, while ρ dating can be performed on haplotypic data from any type of marker, ASD is restricted to microsatellite haplotypes.

TABLE 6.13: GENERATION TIMES FOR LOCI WITH DIFFERENT PATTERNS OF INHERITANCE.

Locus	Generation time (in years)
Y chromosome	35
Mitochondrial DNA	29
X chromosome	31
Autosomes	32

Derived from French–Canadian data (from Tremblay and Vezina, 2000).

TABLE 6.14: SUMMARY STATISTICS USE FOR GENETIC DATING.

Statistic	Data	Specify root	Reference
ρ (Rho)	DNA sequence and microsatellite	Yes	(Forster *et al.*, 1996)
Variance	microsatellite	No	(Di Rienzo *et al.*, 1994)
ASD	microsatellite	Yes	(Goldstein *et al.*, 1995)

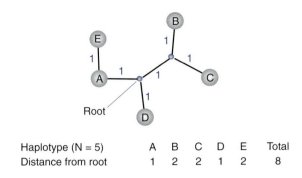

Haplotype (N = 5)	A	B	C	D	E	Total
Distance from root	1	2	2	1	2	8

Mean distance from root (ρ) = 8/5 = 1.4
Mutation rate (μ) = 1 mutation every 20000 years
Time to MRCA (ρ/μ)= 28000 years

Figure 6.17: Basis of dating using intra-allelic diversity (rho dating).

The phylogeny used is identical to that constructed using the median network algorithm in *Figure 6.15*. The root is assigned to one of the unobserved internal nodes, and the distance of each individual taxon from the root is calculated from the phylogeny. Eight mutational changes are observed between the root and the five observed haplotypes, giving a ρ value of 1.4. The time to the MRCA of these haplotypes is ρ/μ, from rearranging the equation given in the text.

Rho and ASD dating require that a root haplotype is specified, as they depend on comparing mutational distances between all sampled chromosomes and this root. Incorrect specification of this root will give rise to erroneous date estimates.

The variance-based method is only applicable where a smooth mismatch distribution can be demonstrated. This method detects population expansion, and a smooth mismatch distribution of intra-allelic diversity suggests a continuous expansion of a lineage since the MRCA, such that the date of the MRCA and the date of the expansion are one and the same. *Table 6.15* shows estimates of the age of the MRCA of a Y-chromosomal lineage using the above three different summary statistic methods on the same data.

One drawback of these summary statistic-based methods is that whilst they produce unbiased point estimates, it is

difficult to get a true estimate of their 95% confidence limits. The sources of error in these estimates come largely from:

▶ uncertainty in the parameters, specifically mutation rates and generation times;

▶ uncertainty about the demographic history of the population.

Whilst it is relatively simple to include confidence limits based on uncertainty in the parameters, it is only by assuming certain idealized demographies that the latter source of error can be appreciated. One such demography is that represented by the so-called '**star phylogeny**' (*Figure 6.18*). In this situation, each observed haplotype is expected to have had an independent history since the MRCA of all haplotypes. This expectation is only realistic of populations that have been growing very rapidly in size, starting from a small group of founders. These idealized demographies tend to underestimate the range within confidence limits when compared to more realistic demographies.

By contrast, model-based methods are computationally much more intense, requiring substantial simulations to estimate statistical confidence. These simulations can be designed such that both of the above potential sources of error can contribute to the overall confidence in the final dating estimates. These methods are 'model based' because a model is required to direct the simulations. The predominant class of such methods does not simulate the evolution of haplotypes forward in time (from the past to the present), but rather backwards, focusing on the 'coalescence' of observed haplotypes in ancestral generations (from the present to the past). These methods are called 'coalescent simulations' and they form an integral part of '**coalescent theory**', a recent development in population genetics that has many potential applications. This class of methods is discussed in greater detail in *Box 6.4*. *Table 6.16* lists some of the available software for estimating the time to the MRCA for nonrecombining lineages.

One problem with model-based methods is that the accuracy of the estimates they produce is prone to bias resulting from incorrect specification of the model. For example, a constant population size model may not be appropriate for a population that has been rapidly expanding. While alternative models are often available, it is not yet possible to include all the complexities of real populations within such models.

TABLE 6.15: DIFFERENT ESTIMATES OF THE DATE OF THE MRCA OF CHROMOSOMES CARRYING THE SRY$_{-2627}$ ALLELE ON THE NONRECOMBINING PORTION OF THE HUMAN Y CHROMOSOME.

Method	Age (years)	95% confidence limits (years)
Rho	2693	1154–9425
ASD	3452	1480–12083
Variance	3076	1318–10766

Confidence limits represent uncertainty in the mutation rate (data taken from Hurles *et al.*, 1999).

Figure 6.18: Three representations of a star phylogeny.

Different rooted and unrooted examples of a star phylogeny all share the defining feature that each lineage descends from a single multifurcation, and so each has evolved independently since the MRCA.

The sensitivity of coalescent and other inferential methods can be tested by using **forward simulations** to generate artificial datasets. In this way it is possible to implement more complex evolutionary models than can be presently accommodated by coalescent simulations. If the inferences from a subsequent coalescent analysis of the artificial dataset produces parameter estimates outside those used to generate the data, then something is amiss: the coalescent model is not robust enough to cope with some property of the model used to generate the data. If this property is likely to be shared by real populations, then any estimates produced by the coalescent method cannot be taken at face value.

Forward simulations have also been used to show that a given evolutionary model, containing assumptions about both populations and molecules, can generate data that closely resemble the observed data. However, this is an inconclusive method for making inferences. It suffers from the fundamental weakness that showing the observed data can be produced by a particular model does not guarantee that the model represents the reality (Dyke, 1981); alternative, untested, models may replicate the observed patterns of diversity equally well. However, it does allow the rejection of models that provide a poorer fit to the data.

Having obtained a date for the MRCA of a set of extant haplotypes, what relevance does this have for population processes that we might be interested in? This is a thorny and highly contentious issue, on the cusp of population genetics and phylogenetics. The date of a MRCA is a purely genetic estimate and need not have any obvious correlate in the other prehistorical records. This date does not represent the timing of a migration that spread the chromosomes carrying those haplotypes, although it does represent an upper bound

to the age of the migration. The TMRCA of a set of alleles is typically much older than the population in which they are found. Only when there is a strong bottleneck or founder effect is the TMRCA similar to the age of the population split.

Guido Barbujani uses the following analogy to demonstrate this point (Barbujani *et al.*, 1998; and see Section 10.5):

> '*...suppose that some Europeans colonize Mars next year: If they successfully establish a population, the common mitochondrial ancestor of their descendants will be Paleolithic. But it would not be wise for a population geneticist of the future to infer from that a Paleolithic colonization of Mars.*'

Having made this point, some attempts have been made to date population movements through a consideration of intra-allelic diversity, using a method known as 'founder analysis' (*Box 6.5*; and see Section 10.5). Overall, dates from genetic diversity have provided powerful evidence for a recent African origin for modern humans (see Section 8.6), and have also been useful in regional studies of human diversity (see Section 11.3.4 for an example examining the origins of Pacific peoples).

6.6.3 Most recent common ancestors of recombining lineages

The process of recombination complicates the dating of alleles at autosomal and X-chromosomal loci. Intra-allelic diversity at markers linked to a specific allele is generated not only through the production of new alleles from the ancestral allele by mutation, but also by replacement of the ancestral allele at the linked locus by recombination. Thus any molecular clock must take into account both recombination and mutation.

As a result, some approaches adopted for dating these alleles have focused not on intra-allelic diversity, but on the time taken to expand to the observed frequency. This is a relatively crude approach: an allele found at an appreciable frequency might have continuously increased in frequency since its origin over a relatively short period of time, or it might be declining from a previous higher frequency, and thus have been in existence for much longer. Consequently, estimates of allele age based solely on frequency come with very wide confidence limits. In addition, positive selection is likely to increase the frequency of an allele far more rapidly than is expected under the neutral models that are typically used for such estimations.

TABLE 6.16: SOFTWARE: METHODS FOR DATING THE TMRCA.

Method	Software	URL
Rho	Network	http://www.fluxus-engineering.com/sharenet.htm
Coalescent (Likelihood)	Genetree	http://www.stats.ox.ac.uk/~griff/software.html
Coalescent (Bayesian)	Batwing	http://www.maths.abdn.ac.uk/~ijw/downloads/download.htm

BOX 6.4 Coalescent theory

In the past two decades, a new approach to describing how patterns of genetic variation are generated has revolutionized population genetics. This approach is known as **'coalescent theory'** and it has enabled significant new insights into the evolution of human genetic diversity to be obtained.

Coalescent theory focuses on genealogical descriptions of the ancestry of a number of copies of the same haplotype (Kingman, 1982; Hudson, 1990). This set of ancestral relationships is known as the **gene genealogy** (see *Figure a*). As we move backwards in time through the generations, we will start to encounter haplotypes that are ancestral to two existing haplotypes. This process of the merging of lineages as we go backwards in time is known as 'coalescence'. As we go further back these ancestral haplotypes continue to coalesce until a single common ancestor of all modern haplotypes is encountered. Note how in *Figure a*, although only a subset of lineages are sampled, they have the same most recent common ancestor as the entire population.

It turns out that mathematical models describing this process of coalescence going backwards in time are simpler and more efficient (although perhaps less intuitively obvious) than the evolution of genetic variation in populations forward in time. For one thing, coalescent methods need only consider those haplotypes that have been sampled, not all of the haplotypes present in the population as a whole. Huge computational savings are achieved through not having to consider those haplotypes in previous generations that did not contribute to the current sample.

The patterns of genetic diversity apparent in the observed haplotypes depend entirely on the shape of the gene genealogy that recounts their ancestry. For example, the deeper the tree, the more mutational differences will be observed between the two most differentiated haplotypes. Thus the effect that different evolutionary processes have on genetic diversity (e.g., migration, recombination, etc.) can be considered via their effect on the gene genealogy. *Figure b* shows how different population demographies change the shape of the gene genealogy.

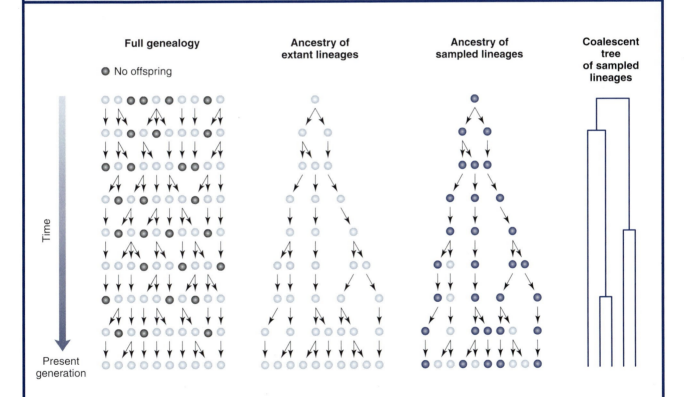

(a) Genealogical relations among nine generations of a population with constant size 10.

Considering only those individuals who have contributed to the present generation reveals a most recent common ancestor of all lineages. If only a subset of lineages are sampled (sampled lineages are shaded in dark blue) they share the same common ancestor as that shared by all extant lineages.

continued

There are many applications of coalescent theory (reviewed by Rosenberg and Nordborg, 2002); those most pertinent to the study of human genetic diversity include:

Mathematical modeling

Mathematical models of the coalescent process can be used to derive equations that estimate population parameters from genetic diversity. This application represents a natural extension of classical population genetics, which often sought to achieve the same goal.

A simulation tool for hypothesis testing

Coalescent simulations can be used to test hypotheses by asking: 'Are the genetic data compatible with a certain model of evolution?' Gene genealogies are repeatedly simulated under a defined population model (the null hypothesis) to generate independent simulated data. Then the observed data are compared to these replicates to see if they are significantly different. Selection tests are one example of hypothesis testing using coalescent simulations. In these tests, a neutrally evolving population model is used to generate simulated data for comparison with observed genetic diversity. If the observed data are significantly different from the simulated data then the null hypothesis (neutral evolution) can be rejected. Other examples of coalescent-based hypothesis testing use different models to guide the coalescent process, and therefore allow different null hypotheses to be tested. Coalescent simulations can also be used to test the performance of different statistics that purport to detect the same phenomenon.

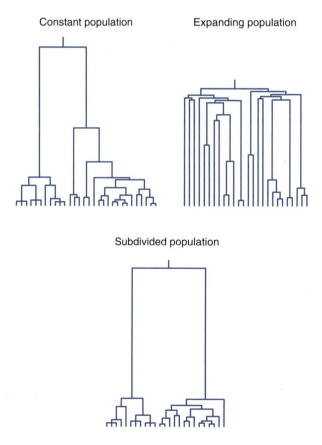

(b) Typical genealogies under different population demographies.

In a population of constant size, about half of the gene genealogy consists of waiting for the last two ancestral lineages to coalesce into the MRCA. In a population that has undergone a recent expansion, many of the lineages coalesce at a single time point that marks the beginning of the population expansion. In a subdivided population, two ancestral lineages persist for the majority of the gene genealogy.

continued

Statistical inference and parameter estimation

Coalescent methods can be used to efficiently extract information contained within diversity data to draw inferences from the data and estimate parameters of the evolutionary model (for example, mutation rate, or effective population size). This information is contained within the underlying gene genealogy relating a set of haplotypes. However, it is impossible to precisely reconstruct this actual genealogy, as this would require us to know in which generation each mutation occurred, and in which generation any pair of alleles shared their most recent common ancestor. Consequently, the coalescence of lineages is simulated stochastically according to a specified evolutionary model and thousands of plausible gene genealogies are generated. This set of genealogies can then be used to estimate the parameters of the evolutionary model, often through a likelihood framework (see *Box 6.1*).

The major strengths of this class of inferential methods are that they use all of the information present in the data and that a wide range of parameters can be estimated. The complexity of the evolutionary model used to guide the simulation of coalescence events determines the parameters that can be estimated, but these typically include the age of different mutations within the phylogeny, migration rates and population growth rates. Because past population processes alter the shape of the gene genealogy (see *Figure b*), confidence limits around parameter estimates depend greatly on demographic history.

Coalescent-based methods are growing in sophistication, incorporating ever more complex processes such as recombination, selection and gene conversion into the genealogical process, and can be expected to be a mainstay of future studies of human genetic diversity.

It is, however, possible to devise dating methods that utilize intra-allelic diversity in the presence of recombination. The 'decay' of the **ancestral haplotype** resulting from both recombination and mutation can be modeled, and the age of the allele calculated, assuming that the mutation rate of the marker locus and the rate of recombination between it and the allele being dated are known (Stephens *et al.*, 1998; Goldstein *et al.*, 1999). Ideally, the recombination rate is estimated by studying linkage of the two loci through pedigrees. However, in the absence of such data, the recombination rate of the entire chromosomal region, as determined from **genetic maps**, has been used. This latter approach is not ideal given the substantial local heterogeneity in recombination rates that appears to be a feature of the human genome (see Section 3.7).

In order to monitor the decay of the ancestral haplotype, it is necessary to be able to define the ancestral state of a linked marker upon which the allele in question arose. If numerous recombination events have occurred between the linked marker and the allele then it will be difficult to identify the ancestral state. If, however, very few recombination events have taken place, there is very little information in the data with which to make precise dating estimates. Consequently, the linked marker should be chosen to balance these two factors – close enough for easy identification of the ancestral state, but far enough away to give statistical power. If the most frequent allele at the linked marker locus in chromosomes carrying the allele to be dated is different from the most frequent allele in the population, it is likely that this allele represents the ancestral state on which the allele arose (*Figure 6.19*).

6.7 Phylogeography and other geographical analyses

Until now our consideration of population subdivision has focused solely on a qualitative measure – the population affiliation of a given individual. This neglects some of our quantitative knowledge about each data point, namely its geographical location. Individuals 1, 2 and 3 may belong to populations A, B and C respectively; however, if populations A and B are neighbors, whereas C is further away, information has been overlooked. By incorporating this information we can start to integrate genetic data with information about the landscape from which they were sampled. For example, high levels of differentiation between two nearby populations may result from the presence of a mountain range separating them. By relating genetic data to specific geographical locations we can also attempt to integrate patterns of modern diversity with past geographical processes, for example, sea-level changes that result in the formation of land-bridges. Geographical analyses of genetic data allow us to partially disentangle the relative contributions of history and geography to modern genetic diversity. In the absence of external, nongenetic, information, it can be difficult to discern whether geographical constraints or historical episodes account for observed patterns of genetic diversity.

In many cases, the first step is to test whether the genetic data show any form of geographic structure. There must be demonstrable geographical patterning before any explanation need be sought. This is analogous to the tests for population subdivision discussed earlier in this chapter. Once patterning has been detected, it is necessary to determine the nature of the pattern; are certain alleles found in patches? Or do

BOX 6.5 Dating movements by founder analysis

Founder analysis (Richards *et al.*, 2000) relies on the principle that by comparing the molecular diversity of a source population and of a population derived from it by migration, the timing of this migration can be determined. Founder chromosomes are identified within the derived population, and the amount of diversity that has arisen on those founder chromosomes, since migration, is quantified. This is achieved by subtracting the diversity seen in the source population from that in the derived population, as shown in the *Figure*. Typically the ρ statistic described above is used to relate diversity to time.

However, the actual source population is an ancient one, so the present day population occupying the same region must be taken as a surrogate. This approach assumes that the modern day descendants of the actual source population:

▶ are present in the same geographical location as the ancestral source population;

▶ contain all of the diversity present within the ancestral population.

This method also assumes that any recent back migrants from the 'derived' population to the 'source' population can be identified and excluded from the analysis. Including back-migrants in the analysis would underestimate the age of the migration. An additional source of error comes from postmigration recurrent mutation in both the source and the derived population. This can generate shared haplotypes between the populations that might be assumed to result from migration rather than the accumulation of diversity postmigration. Again, this would cause the postmigration diversity to be underestimated, with the age of the migration being similarly biased. The method is also sensitive to multiple dispersals of a haplotype. Application of this method has been highly contentious because of these factors (see Section 10.5.5), although recent studies have attempted to take them into account.

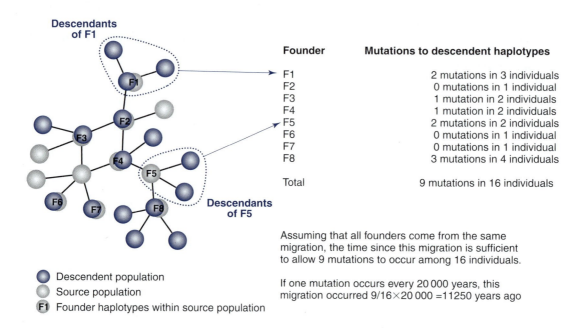

Founder	Mutations to descendent haplotypes
F1	2 mutations in 3 individuals
F2	0 mutations in 1 individual
F3	1 mutation in 2 individuals
F4	1 mutation in 2 individuals
F5	2 mutations in 2 individuals
F6	0 mutations in 1 individual
F7	0 mutations in 1 individual
F8	3 mutations in 4 individuals
Total	9 mutations in 16 individuals

Assuming that all founders come from the same migration, the time since this migration is sufficient to allow 9 mutations to occur among 16 individuals.

If one mutation occurs every 20 000 years, this migration occurred 9/16×20 000 = 11250 years ago

🔵 Descendent population
⚪ Source population
Ⓕ¹ Founder haplotypes within source population

Schematic representation of founder analysis.

A network showing the relationship between haplotypes sampled in the descendent (blue circles) and source (grey circles) populations allows putative founders to be identified. If it is assumed that these founders all arrived at the same time, the time since the migration can be calculated. The mutation rate used in the calculation is that commonly used when studying HVS I diversity in mtDNA (see Section 3.2.8).

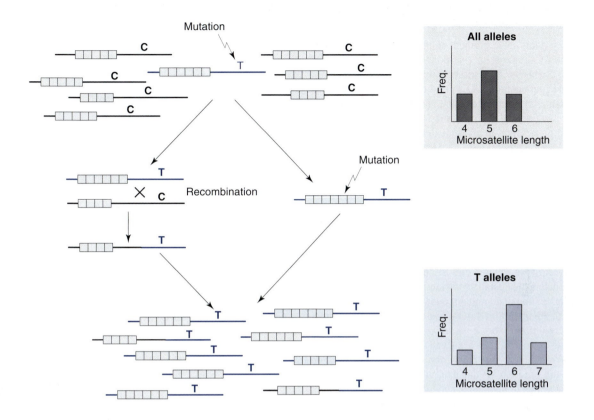

Figure 6.19: Mutation and recombination cause decay of the ancestral haplotype of a specific allele.

A C to T transition occurs on a haplotypic background of a six-repeat allele at a linked microsatellite. This ancestral haplotype breaks down as recombination transfers this T allele onto chromosomes that have microsatellites with different numbers of repeats, and as mutation also alters the copy number of microsatellite repeat units. Over time the population of T-allele chromosomes acquires diversity at the linked microsatellite. Nonetheless, the ancestral six-repeat allele remains the most frequent, in contrast to the five-repeat allele that is most frequent in the general population.

smooth gradients of allele frequencies span the sampled area? Such gradients of allele frequencies are known as **clines**, and they can be generated by a number of evolutionary processes (see *Box 6.6*).

What kinds of data are required for these kinds of analyses? It is not enough to have a few sites that are sparsely sampled over a wide geographical region. Any landscape being investigated in this manner should be extensively sampled, with as regular as possible a distribution of sample sites. Some analyses require a population at each sample site, whereas others can cope with a separate geographical location for each sampled individual.

Phylogeography refers to the study of the geographical distribution of the clades within a phylogeny. The phylogeny provides a temporal dimension to be combined with the spatial dimension of geography. However, the incorporation of genetic and geographic data within a common analytical framework need not require a phylogeny. The term 'phylogeography' was itself only coined in 1987 (Avise *et al.*,

1987), and so a statistical methodology is still being developed. By contrast, the combining of allele frequency data and geography has a longer history. Many analyses rely on visual inspection of the allele frequency data overlaid onto maps. One disadvantage of such methods is the lack of temporal information that often leaves geographic patterns open to alternative interpretations (Barbujani, 2000). Nevertheless, the use of densely sampled genetic data throughout a geographical region allows the reconstruction of a 'genetic landscape', the features of which have been shaped by past processes, be they historical or geographical.

6.7.1 Interpolated maps

How can patterns of allele frequency be shown on maps? The simplest approach is to plot graphical representations of allele frequencies within a population onto the geographical location from whence it came. Pie charts allow the frequencies of several (but not many) alleles to be shown in the same map.

BOX 6.6 Opinion: Clines can be generated by several processes

We label a geographic gradient in gene frequencies a **cline**. Sometimes the cause for this pattern will be obvious. For example, the further from the equator, the lighter the color of human skin pigmentation. This pattern is almost surely due to natural selection – darker skin in the tropics protects against the harmful effects of solar radiation while lighter skin allows sufficient penetration of UV light to stimulate production of vitamin D (see Section 13.3).

Another long-cited example of a cline is the gradual lessening of the frequency of the red blood cell antigen B (see *Box 3.3*) from east to west across Europe. The historical incursions of Asiatic peoples who generally have high frequencies of the B allele was thought to have diffused the gene to Europeans. In this case, the causal factor for the cline is gene flow.

In theory, however, clines due to gene flow such as the B gene should not be limited to one locus but should exist for *all* loci. This is because mating passes the entire parental genome to the offspring. Natural selection, on the other hand, is expected to act on a *specific* locus as a particular phenotype is selected by a particular environmental factor (e.g., UV radiation). This difference in expectation is often cited as the criterion for deciding whether a cline is due to natural selection or gene flow: one locus = natural selection; many loci = gene flow. Correct inference in human population genetics, however, is not necessarily so simple.

Consider the clinal pattern for (apparently) many loci that extends across Europe from the southeast to the northwest (see Section 10.5). This cline parallels a major event in European history, the diffusion of agriculture from its origin in the Near East throughout the continent over the course of several thousand years. This process has been well documented by archaeologists through the analysis of the cultural remains associated with farming. A long held view was that this diffusion of crops and techniques was *cultural*; that is, indigenous hunting gathering peoples of Europe borrowed farming practices from their neighbors so that farming spread as a slow wave from its center of origin. More recently, it has been proposed that the spread of agriculture was *demic*, that is through the expansion of the human farming population, beginning in the Near East and intermarrying with resident hunter-gatherers, gradually spreading farming techniques *and* their genes. The data to support this hypothesis included many loci but the majority of the alleles showing the southeast–northwest cline were from the human leukocyte antigen (**HLA**) system, genes forming part of the immune system (see *Box 5.5*). The potential role of these genes in disease resistance raises the possibility that they were not merely genetic markers passively transported by the expanding farmer populations. Instead, it may be that the cline in HLA allele frequencies, though representing many alleles, is due to natural selection rather than gene flow (**demic diffusion**).

One way that the cultural diffusion of agriculture could have produced the cline would be through the changed way of life

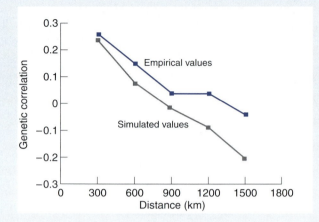

Comparison of simulated spatial genetic correlations with those based on actual European gene frequencies.

Similar spatial autocorrelations are shown in the empirical data and simulated data generated using the model described in the text.

due to farming. Domestic animals such as cattle, sheep and goats, are the likely source of many human diseases (**zoonoses**). As Europeans began to tend these animals, they would have been exposed to a variety of new diseases exerting new selective pressures on HLA genes. People closest to the Near East would have adopted animal husbandry and experienced the new disease environment first. As successive European populations borrowed animal domesticates from their neighbors, they too began to experience the new diseases. The farther from this point of origin, the more recent the changed selective environment. Thus in the southeast disease selection would have had many more generations to affect HLA frequencies with progressively fewer generations of selection in populations to the northwest. This time gradient in natural selection can produce a geographic gradient in gene frequencies (cline).

The *Figure* compares the results of a computer simulation of this process with the actual gene frequencies in a transect across Europe. Both the simulated and real data show the clinal pattern of steadily declining genetic similarity with increasing distance between populations.

Thus a clinal distribution of gene frequencies by itself does not specify the evolutionary mechanism that produced it. Even the rule that a cline in many loci is caused by gene flow may not be infallible as the HLA case illustrates. To resolve the issue, additional genetic data unlikely to be affected by disease selection as well as better archaeological information on the process of agricultural spread are needed.

Alan Fix

Department of Anthropology, University of California Riverside, USA

[Further information on these issues may be found in Fix, 1996 and 1999].

Alternatively, a contour map of allele frequencies can be produced. The contour lines (known as isogenic lines) join points of equal allele frequency, like lines of altitude on a topological map, or lines of equal air pressure on a weather map. In principle, the drawing of these contour lines could be done by hand; for example, a 35% contour line would lie between two sample sites with frequencies of 30% and 40%. In practice, the sample densities are rarely sufficient to allow any accuracy by this method. Alternatively, the method of **interpolation** shown in *Figure 6.20* can be used to generate a synthetic dataset of regularly spaced pseudo-sites. This regular grid of estimated gene frequencies can then be used to generate contour maps more easily. For an example of a contour map showing the global distribution of the *CCR5* Δ*32* allele, see *Figure 14.15*.

Interpolation is performed by laying a fine grid on top of a map and estimating the gene frequency at each point where the grid lines intersect ('pseudo-sites'). Several different methods are available for this purpose. These methods rely on identifying the nearest sampled locations to the pseudo-site, and estimating the gene frequency at the pseudo-site from the allele frequencies at these locations. The distance between the sampled locations and the pseudo-site is often taken into account by weighting the allele frequency at the observed site according to its proximity. Thus, the gene frequency at the closest observed site has more impact on the gene frequency at the pseudo-site than that at the furthest observed site (see *Figure 6.20*). Other interpolation methods can be far more statistically involved. One such example, known as 'kriging', estimates a mathematical function that describes the genetic surface, rather than estimating single pseudo-values on the grid. When the geographical region being investigated is large, geographical distance calculations between sites need to take account of the Earth's curvature. This is done using a trigonometric equation that calculates the **'great circle distance'**.

One disadvantage of these gene frequency maps is that it is only possible to show information on a single allele on a single map. How can information from multiple alleles, and indeed multiple loci, be combined on the same map? Cavalli-Sforza and his colleagues developed an ingenious method to circumvent this problem. Their approach is to use principal components analysis to integrate information from multiple alleles. Rather than producing maps of gene frequencies, they produce interpolated maps of principal components, which account for quantifiable amounts of the total variation apparent in the data (see *Figure 10.11* for an example). As we saw earlier, principal components condense information from multiple alleles into a reduced number of dimensions with minimal loss of information. In this manner, frequencies of tens or hundreds of alleles can be used to generate maps of the first and second principal components that often account for the majority of the variation. This is because many alleles share similar distributions. The geographical distribution shared by most alleles is often represented faithfully as the first principal component. The next most common distribution is represented in the second principal component, and so on. One complication not taken into account by this method is

that without a temporal perspective, alleles with similar geographical distributions that result from migrations at different times, but along a similar geographical axis, will be conflated into a single signal. If this signal becomes represented in one of the higher principal components it might be tempting to conclude a single migration, overestimating the impact of this prehistorical episode on modern diversity. These issues are discussed in the context of the origins of agriculture in Europe, in Section 10.5.

A more serious problem with gene frequency maps is that the process of interpolation is capable of generating clines, and yet the identification of clines may be one objective of the analysis. After interpolation it can be impossible to tell the difference between genuine clines present in the data and artificial clines that were generated during interpolation (Rendine *et al.*, 1999; Sokal *et al.*, 1999).

6.7.2 Genetic boundary analysis

It is often of interest to identify the zones of greatest allele frequency change within a genetic landscape. A zone of abrupt genetic change results from populations dwelling on either side of a **'genetic boundary'** being more differentiated than might otherwise be expected, and points to limited gene flow between them. That may mean that the populations are currently isolated because of some **'genetic barrier'** to mating and/or dispersal, physical or cultural, or that the populations evolved independently and came into contact only recently. The former explanation emphasizes the effects of geography, the latter the effects of history. Once identified, genetic barriers can be correlated with physical or cultural barriers to gene flow, to identify which of these has had the greatest impact on the distribution of modern diversity.

A method to detect genetic boundaries by combining data from all alleles was devised by Womble (1951). By calculating the gradient (first derivative) of allele frequency change at a number of locations for each allele separately, regions of greatest change (i.e., steepest gradients) have high values (represented by peaks in three dimensions) and regions of least change (flattest gradients) have lower values. Consequently, these **'surfaces'** of allele frequency change can be summed from all alleles. Peaks present in the same location for multiple alleles reinforce once another. Only peaks over a certain height are deemed significant, and these represent the genetic boundaries within the genetic landscape. This method of analysis has become known as 'Wombling' (Barbujani *et al.*, 1989; Rosser *et al.*, 2000). *Figure 6.21* shows a simple example of the Wombling procedure summing information on genetic boundaries from two alleles.

6.7.3 Spatial autocorrelation

Processes whereby populations exchange genes in a geographically restricted fashion (such as isolation by distance; see Section 5.5.1) leads to the nonindependence of genetic variation throughout space. In other words, the frequency of an allele within one population is to some

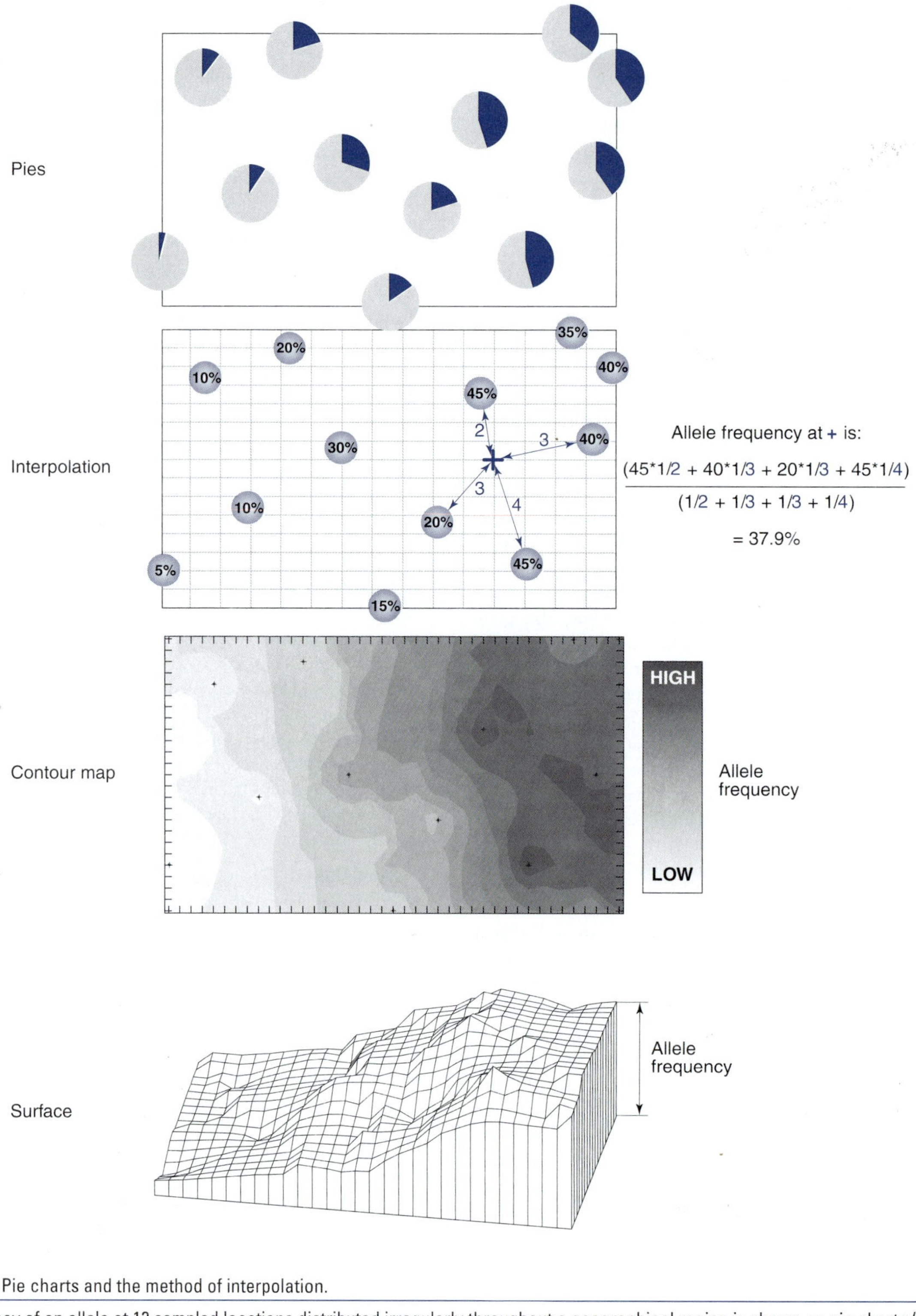

Pies

Interpolation

Allele frequency at **+** is:

$$\frac{(45*1/2 + 40*1/3 + 20*1/3 + 45*1/4)}{(1/2 + 1/3 + 1/3 + 1/4)}$$

$$= 37.9\%$$

Contour map

HIGH

Allele frequency

LOW

Surface

Allele frequency

Figure 6.20: Pie charts and the method of interpolation.

The frequency of an allele at 12 sampled locations distributed irregularly throughout a geographical region is shown as pie charts (blue slice). These locations can be used to estimate allele frequencies at regularly spaced pseudo-sites defined by intersecting gridlines by the process of interpolation. Allele frequencies at the four closest sampled locations to a pseudo-site are identified, and the gene frequency is determined as the average of the gene frequencies at these observed sites, weighted according to the inverse of distance between them. Other distance weighting functions could also be used, for example weighting by the inverse distance squared. Once allele frequencies have been estimated at all pseudo-sites, allele frequencies across the region can be represented by a two-dimensional contour map, where the shading indicates the frequency, or a three-dimensional surface, where the height indicates the frequency.

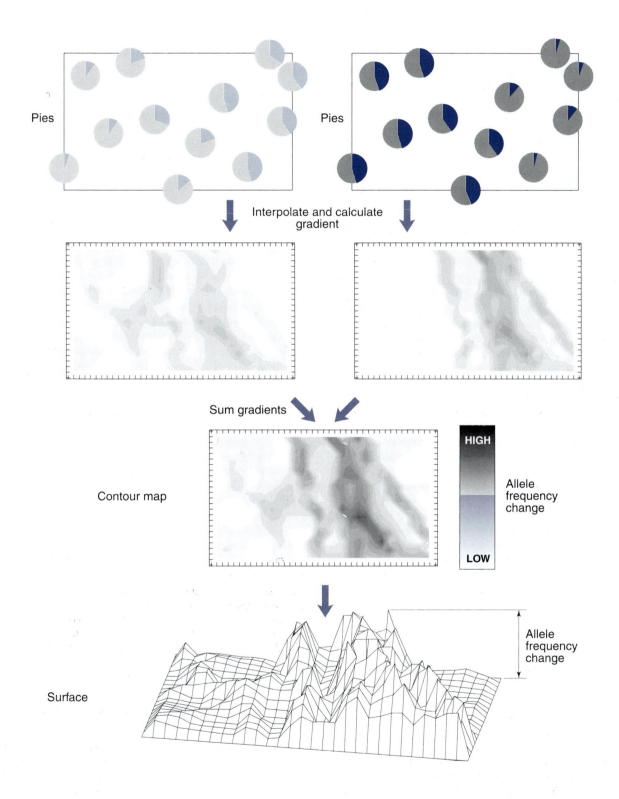

Figure 6.21: Detecting genetic boundaries by 'Wombling'.

The frequencies of two alleles throughout the same geographical region are shown as pie charts (blue sectors). Interpolation is performed for each allele separately, and the rate of change of allele frequencies determined at each pseudo-site. Contour maps of the rate of allele frequency change are shown for each allele individually. These values of allele frequency change at each pseudo-site can be summed, and the composite displayed as either a two-dimensional contour map or three-dimensional surface.

degree dependent on the frequency of the same allele in neighboring populations. This relationship of genetics and geography can be quantified using **'spatial autocorrelation'**. It has been shown that the nature and extent of this **autocorrelation** can be used to investigate the processes causing it. Different patterns of spatial autocorrelation are expected under:

▸ random distribution of alleles in space (a rather unlikely condition for humans);

▸ isolation by distance;

▸ clinal variation;

▸ the presence of multiple nonoverlapping clines in the region of interest.

While the latter three processes generate significant geographical structure, it can be difficult to determine which has been in operation from observation of this structure alone.

So, how is spatial autocorrelation quantified? Pairs of sampled localities are pooled into different arbitrary distance classes. For each distance class the level of spatial autocorrelation is calculated using a measure of genetic similarity that varies between 1 (strong positive correlation) and −1 (strong negative correlation). Plotting the level of spatial autocorrelation against the distance class generates a **correlogram**, which describes quantitatively the geographical pattern of genetic variation, and tests for its departure from spatial randomness. From its shape, inferences on the likely processes generating the geographical structure can often be drawn. There are different measures of genetic similarity that can be used for spatial autocorrelation analysis. Some are applicable to classical allele frequency data, whereas others have been specially formulated for molecular data, taking into account genetic distances between alleles (Bertorelle and Barbujani, 1995). *Figure 6.22* shows the patterns of spatial autocorrelation, in the form of correlograms, expected under random mating, isolation by distance and clinal variation. Spatial autocorrelation analysis has been especially prominent in analyses of European genetic diversity, where different interpretations of clinal patterns of diversity have been hotly debated (see Sections 10.5.4–10.5.6).

6.7.4 Mantel testing

For a set of human populations, we often have a number of distance matrices that contain pairwise comparisons of different aspects of these populations. These distances can be genetic distances, linguistic distances, geographical distances – in fact any form of anthropological measurement. Often we want to detect correlations between these matrices, which might tell us something about their evolutionary past. However, we have a problem, because each pairwise distance within a matrix is not independent, and consequently, the assumptions of classical statistical methods for detecting correlations between variables (e.g., bivariate regression methods) are violated. A Mantel test (Mantel, 1967) detects correspondence between matrices, and

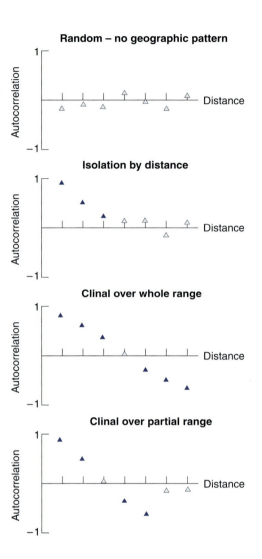

Figure 6.22: Spatial autocorrelation shows different patterns for clines, isolation by distance and random mating.

Pairwise comparisons of autocorrelation between populations are binned into different classes on the basis of the distance between them. Triangles represent the average levels of autocorrelation observed within each distance class. Filled triangles indicate levels of autocorrelation that are significantly different from that obtained when haplotypes are distributed randomly across the geographical space.

provides an alternative method for disentangling history and geography.

Correspondence between any two distance matrices may result from a causal relationship between them, for example a migration that disperses both genes and languages – a historical explanation. Alternatively, correspondence between them may result from both being correlated with a third matrix. In this case superficial correspondence between genetic and linguistic distances may result from both being correlated with geographical distances rather than any direct relationship between them. In such situations, a geographical process, for example isolation by distance, generates geographic structuring of both languages and genes. Thus both historical and geographical processes may underlie correspondences between any two matrices.

Extensions to the Mantel test allow correspondences between three distance matrices to be tested. In this way, a common historical process causing similarity between genetic and linguistic datasets need only be invoked once it has been demonstrated that geographical distances are not themselves the determining factor. In such a case, genetic and linguistic distances would be better correlated with each other, than either would be to geographical distances (Smouse *et al.*, 1986).

Mantel tests have also formed part of the debate on the origins of European diversity and the potential impact of the spread of agriculture, which is discussed in greater detail in Section 10.5.3.

6.7.5 Nested cladistic analysis

The geographical distribution of different parts of a phylogeny may result from different processes: historical events may have occurred in one clade while geographical factors predominate in others. The tree cannot be analyzed as a whole, because different haplotypes have different evolutionary histories. Templeton has devised a truly phylogeographic method that attempts to disentangle historical and geographical explanations of patterns of genetic diversity. This method, called **nested cladistic analysis**, investigates patterns of genetic diversity within different segments of the phylogeny separately (Templeton, 1998).

The first step is to subdivide the known phylogeny into its component parts, which can then be analyzed in turn. This process of subdivision apportions the phylogeny into nested clades, starting at the tips. An example of such a nesting scheme is shown in *Figure 6.23*. Each clade comprises two parts, an ancestral, interior haplotype and tip haplotypes one mutational step from the ancestral haplotype. Once all tip haplotypes have been incorporated into nested clades, unincorporated interior clades are similarly assigned into nested clades of the same level. Once all haplotypes have been incorporated into nested clades, the next level of clades is defined, which nests the clades defined in the prior level. This process is repeated until the entire tree is incorporated within a single nested clade.

The second step is to identify those nested clades that show significant geographic differentiation. Once identified, these nested clades are the only ones for which explanations are sought. Templeton envisages that three main competing explanations are possible:

▸ restricted gene flow i.e., isolation by distance;

▸ past fragmentation of populations;

▸ range expansion, carrying haplotypes into previously unoccupied territories. This can either result from contiguous range expansion or long distance colonization.

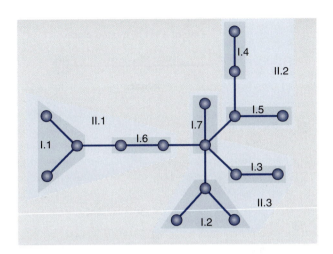

Figure 6.23: Nesting scheme for Nested Cladistic Analysis.

A haplotype phylogeny is shown in blue with haplotypes represented by circles and lines indicating mutations. This network is apportioned into nested clades, as discussed in the text. Lighter shades indicate higher level nested clades. There are seven nested clades at the lowest level (I.1–I.7) and three at the second level (II.1–II.3).

Statistics are calculated for each nested clade to enable clades exhibiting significant geographic differentiation to be identified and the above different scenarios to be distinguished. These statistics incorporate geographical distances (*Figure 6.24*) and are:

▸ D_c – *the clade distance* – measures the geographical range of a particular clade by averaging the distance between each individual carrying a specific haplotype and the geographic center of every individual carrying that haplotype;

▸ D_n – *the nested clade distance* – compares the geographical range of a particular clade to that of its sister clades, i.e., those clades descending from the same ancestral haplotype. This is performed by measuring the 'center of gravity' between individuals carrying a specific haplotype and the geographic center of every individual carrying haplotypes within the next highest level clade (this includes individuals carrying the haplotype in question).

The difference between D_c (or D_n) of a tip clade (T) and that of its interior clade (I) is also calculated (I−T). The significance of all these statistics is determined by permutation tests. By comparing the values and significance of these statistics, using a set of predetermined rules, it can be determined whether or not enough data are present for inferences to be drawn for any given clade, and if so, which biological process most likely underpins the present distribution of genetic diversity within that clade. For example, in the situation shown in *Figure 6.24*, if D_c of the tip clade is the only significant statistic, and is larger than expected by chance, contiguous range expansion is the most likely explanation.

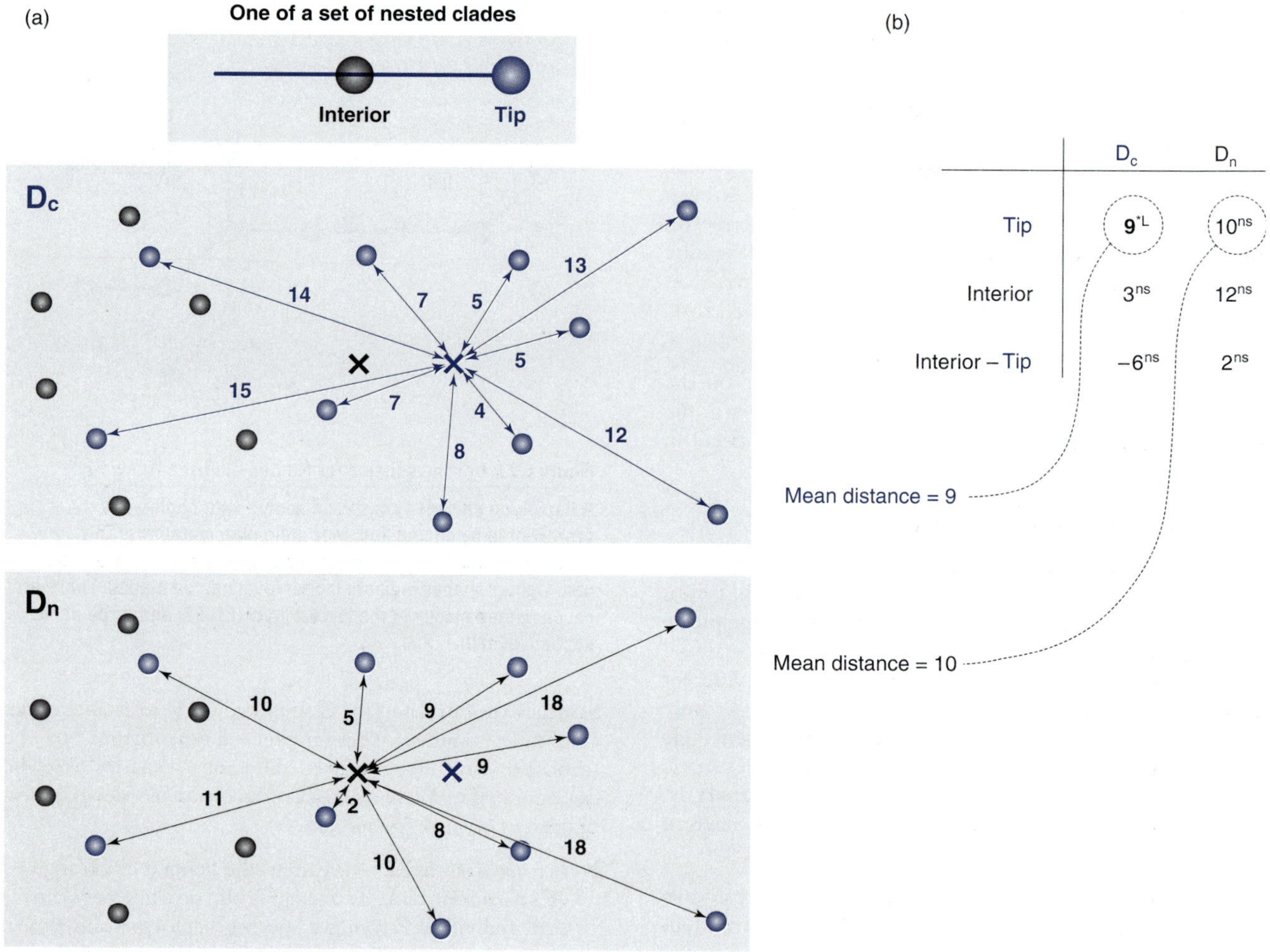

(a)

(b)

Figure 6.24: Calculation of D_c and D_n.

(a) A single nested clade is shown and the tip and interior haplotypes defined. Shaded rectangles indicate the same geographical region. Blue and black circles within this region indicate sites at which individuals belonging to the tip clade and the nested clade respectively, were found. Crosses indicate centers of gravity for the tip clade and the nested clade. The geographic distances used to calculate D_c and D_n for the tip clade are shown in blue and black respectively. (b) D_c and D_n values for the tip and nested clade shown previously, $*^L$ indicates a significantly large value (shown in bold) as determined by a permutation test, and ^{ns} indicates a nonsignificant value.

TABLE 6.17: SOFTWARE: PHYLOGEOGRAPHIC METHODS.

Method	Software	URL
Interpolation	Surface III	http://www.kgs.ukans.edu/Tis/surf3/surf3Home.html
Spatial Autocorrelation	AIDA	http://web.unife.it/progetti/genetica/Giorgio/giorgio_soft.html
Mantel Testing	ARLEQUIN	http://lgb.unige.ch/arlequin/
Nested Cladistic Analysis	GeoDis	http://InBio.byu.edu/Faculty/kac/crandall_lab/geodis.htm

The set of rules, known as an 'inference key', has been criticized as being an *ad hoc* method. Templeton has invested much effort into taking examples of known geographical or historical processes and indicating that his method can be used with genetic data to make the correct prehistorical inferences. Nevertheless, fully statistical methods of phylogeographic inference remain to be developed.

Nested cladistic analysis has been used to support the multi-regional theory of modern human origins (see Section 8.6) and suggest more recent back migrations to Africa (see *Box 9.6*).

Table 6.17 lists software for performing some of the phylogeographic analyses described in this section.

Summary

- ▶ Although describing present genetic diversity and inferring how it arose are two separate analytical objectives, the methods by which they are accomplished are often interrelated.

- ▶ The nature of genetic diversity data has changed substantially since the first half of the twentieth century. The superseding of 'classical' protein variants by a variety of DNA-based 'molecular' markers with different mutational dynamics has necessitated changes to analytical methods to accommodate the additional information of mutational distances between alleles.

- ▶ In the near future, the integrated study of data from many loci will become more commonplace, as it offers greater demographic information than single-locus studies.

- ▶ Measures of genetic diversity have been developed which are independent of both the size of the population, and the length of DNA being studied. These measures allow unbiased comparisons to be made between diversity in different populations.

- ▶ Apportionment of diversity between subpopulations can be measured by various statistics, many of which are related to F_{ST} and some of which are restricted to markers with specific mutational dynamics.

- ▶ Selection can shape genetic diversity in a number of ways, and consequently there is a variety of selection tests that vary in their ability to detect different types of selection. Selection can be detected in genic and nongenic sequences and its imprint can also be seen in markers linked to the locus under selection.

- ▶ Phylogenies are a natural way to think about the evolution of genetic diversity, both of populations and of molecules. There are many methods for reconstructing phylogenies from genetic data, but all will output a phylogeny regardless of whether the genetic diversity evolves in a tree-like manner. Consequently a clear distinction should be drawn between trees as mechanisms and trees as metaphors. Alternative methods (such as principal components analysis) that do not impose a bifurcating pattern on evolution are usually more appropriate for revealing population relationships.

- ▶ Chronological information can be obtained from genetic data, and is very useful for integrating genetic evidence with that from other disciplines. However, relating diversity to time requires many assumptions to be made and confidence limits around age estimates are often very broad.

- ▶ Coalescent theory and its associated analytical methods are a significant recent development in population genetics and have a variety of applications including hypothesis testing and parameter estimation.

- ▶ Integrating geographic information with genetic data is a powerful approach for making inferences from genetic diversity, and will only increase in the future as sampling density becomes denser. Analyzing specific features within densely sampled genetic landscapes, such as genetic boundaries, clines and the distributions of the different clades within a phylogeny, improve understanding of the spatial impact of evolutionary processes.

Further reading

Measures of diversity and subdivision: **Nei M** (1987) *Molecular Evolutionary Genetics*. Columbia University Press, New York.

Selection/Neutrality testing: **Kreitman M** (2000) Methods to detect selection in populations with applications to the human. *Annu. Rev. Genomics Hum. Genet.* **1**, 539–559.

Dating of alleles: **Slatkin M, Rannala B** (2000) Estimating allele age. *Annu. Rev. Genomics Hum. Genet.* **1**, 225–249.

Melding genetic and geographical information: **Barbujani G** (2000) Geographical patterns: how to identify them, and why. *Hum. Biol.* **72**, 133–153.

Catalog of useful software for phylogenetic analysis, distance calculation and dating:

http://evolution.genetics.washington.edu/phylip/software.html

References

Avise JC, Arnold J, Ball RM *et al.* (1987) Intraspecific phylogeography: the mitochondrial DNA bridge between population genetics and systematics. *Annu. Rev. Ecol. Syst.* **18**, 489–522.

Bandelt H-J, Forster P, Sykes BC, Richards MB (1995) Mitochondrial portraits of human populations using median networks. *Genetics* **141**, 743–753.

Bandelt H-J, Forster P, Röhl A (1999) Median-joining networks for inferring intraspecific phylogenies. *Mol. Biol. Evol.* **16**, 37–48.

Barbujani G, Bertorelle G, Chikhi L (1998) Evidence for Paleolithic and Neolithic gene flow in Europe. *Am. J. Hum. Genet.* **62**, 488–491.

Barbujani G, Oden NL, Sokal RR (1989) Detecting regions of abrupt change in maps of biological variables. *Syst. Zool.* **38**, 376–389.

Barbujani G (2000) Geographical patterns: how to identify them, and why. *Hum. Biol.* **72**, 133–153.

Bertorelle G, Barbujani G (1995) Analysis of DNA diversity by spatial autocorrelation. *Genetics* **140**, 811–819.

Brinkmann B, Klintschar M, Neuhuber F, Huhne J, Rolf B (1998) Mutation rate in human microsatellites: influence of the structure and length of the tandem repeat. *Am. J. Hum. Genet.* **62**, 1408–1415.

Cavalli-Sforza L, Edwards AWF (1967) Phylogenetic analysis: models and estimation procedures. *Evolution* **21**, 550–570.

Di Rienzo A, Peterson AC, Garza JC, Valdes AM, Slatkin M, Freimer NB (1994) Mutational processes of simple-sequence repeat loci in human populations. *Proc. Natl Acad. Sci. USA* **91**, 3166–3170.

Dyke B (1981) Computer simulation in anthropology. *Annu. Rev. Anthropol.* **10**, 193–207.

Efron B (1982) The jacknife, the bootstrap and other resampling plans. Regional conference series in applied mathematics, Philidelphia.

Excoffier L, Smouse PE, Quattro JM (1992) Analysis of molecular variance inferred from metric distances among DNA haplotypes: application to human mitochondrial DNA restriction data. *Genetics* **131**, 479–491.

Fay JC, Wu CI (2000) Hitchhiking under positive Darwinian selection. *Genetics* **155**, 1405–1413.

Fix AG (1996) Gene frequency clines in Europe: demic diffusion or natural selection? *J. Roy. Anthropol. Inst. (N.S.)* **2**, 625–643.

Fix AG (1999) *Migration and Colonization in Human Microevolution.* Cambridge University Press, Cambridge.

Forster P, Harding R, Torroni A, Bandelt H-J (1996) Origin and evolution of Native American mtDNA variation: a reappraisal. *Am. J. Hum. Genet.* **59**, 935–945.

Fullerton SM, Clark AG, Weiss KM *et al.* (2000) Apolipoprotein E variation at the sequence haplotype level: implications for the origin and maintenance of a major human polymorphism. *Am. J. Hum. Genet.* **67**, 881–900.

Gilad Y, Rosenberg S, Przeworski M, Lancet D, Skorecki K (2002) Evidence for positive selection and population structure at the human MAO-A gene. *Proc. Natl Acad. Sci. USA* **99**, 862–867.

Goldstein DB, Linares AR, Cavalli-Sforza LL, Feldman MW (1995) An evaluation of genetic distances for use with microsatellite loci. *Genetics* **139**, 463–471.

Goldstein DB, Reich DE, Bradman N, Usher S, Seligsohn U, Peretz H (1999) Age estimates of two common mutations causing factor XI deficiency: recent genetic drift is not necessary for elevated disease incidence among Ashkenazi Jews. *Am. J. Hum. Genet.* **64**, 1071–1075.

Hamblin MT, Thompson EE, Di Rienzo A (2002) Complex signatures of natural selection at the Duffy blood group locus. *Am. J. Hum. Genet.* **70**, 369–383.

Harris EE, Hey J (1999) X chromosome evidence for ancient human histories. *Proc. Natl Acad. Sci. USA* **96**, 3320–3324.

Harris EE, Hey J (2001) Human populations show reduced DNA sequence variation at the factor IX locus. *Curr. Biol.* **11**, 774–778.

Hudson RR, Kreitman M, Aguade M (1987) A test of neutral molecular evolution based on nucleotide data. *Genetics* **116**, 153–159.

Hudson RR (1990) Gene genealogies and the coalescent process. *Oxf. Surv. Evol. Biol.* **7**, 1–44.

Hurles ME, Veitia R, Arroyo E *et al.* (1999) Recent male-mediated gene flow over a linguistic barrier in Iberia suggested by analysis of a Y-chromosomal DNA polymorphism. *Am. J. Hum. Genet.* **65**, 1437–1448.

Ingman M, Kaessman H, Pääbo H, Gyllensten U (2000) Mitochondrial genome variation and the origin of modern humans. *Nature* **408**, 708–719.

Johnson ME, Viggiano L, Bailey JA, Abdul-Rauf M, Goodwin G, Rocchi M, Eichler EE (2001) Positive selection of a gene family during the emergence of humans and African apes. *Nature* **413**, 514–519.

Kaessmann H, Heissig F, vonHaeseler A, Pääbo S (1999) DNA sequence variation in a non-coding region of low recombination on the human X chromosome. *Nature Genet.* **22**, 78–81.

Kayser M, Roewer L, Hedman M *et al.* (2000) Characteristics and frequency of germline mutations at microsatellite loci from the human Y chromosome, as revealed by direct observation in father/son pairs. *Am. J. Hum. Genet.* **66**, 1580–1588.

Kingman JFC (1982) On the genealogy of large populations. *J. Appl. Prob.* **19A**, 27–43.

Kreitman M (2000) Methods to detect selection in populations with applications to the human. *Annu. Rev. Genomics Hum. Genet.* **1**, 539–559.

Lynch M, Crease TJ (1990) The analysis of population survey data on DNA sequence variation. *Mol. Biol. Evol.* **7**, 377–394.

Makova KD, Ramsay M, Jenkins T, Li WH (2001) Human DNA sequence variation in a 6.6-kb region containing the melanocortin 1 receptor promoter. *Genetics* **158**, 1253–1268.

Mantel N (1967) The detection of disease clustering and a generalised regression approach. *Cancer Res.* **27**, 209–220.

McDonald JH, Kreitman M (1991) Adaptive protein evolution at the Adh locus in *Drosophila*. *Nature* **351**, 652–654.

Messier W, Stewart CB (1997) Episodic adaptive evolution of primate lysozymes. *Nature* **385**, 151–154.

Mouse Genome Sequencing Consortium (2002) Initial sequencing and comparative analysis of the mouse genome. *Nature* **420**, 520–562.

Nachman MW, Brown WM, Stoneking M, Aquadro CF (1996) Nonneutral mitochondrial DNA variation in humans and chimpanzees. *Genetics* **142**, 953–963.

Nei M (1973) Analysis of gene diversity in subdivided populations. *Proc. Natl Acad. Sci. USA* **70**, 3321–3323.

Nei M, Roychoudhury AK (1982) Genetic relationship and the evolution of human races. *Evol. Biol.* **14**, 1–59.

Nei M (1987) *Molecular Evolutionary Genetics*. Columbia University Press, New York.

Pamilo P, O'Neill RJ (1997) Evolution of the *Sry* genes. *Mol. Biol. Evol.* **14**, 49–55.

Penny D, Steel M, Waddell PJ, Hendy MD (1995) Improved analyses of human mtDNA sequences support a recent African origin for *Homo sapiens*. *Mol. Biol. Evol.* **12**, 863–882.

Perez-Lezaun A, Calafell F, Mateu E, Comas D, Ruiz-Pacheco R, Bertranpetit J (1997) Microsatellite variation and the differentiation of modern humans. *Hum. Genet.* **99**, 1–7.

Przeworski M (2002) The signature of positive selection at randomly chosen loci. *Genetics* **160**, 1179–1189.

Rendine S, Piazza A, Menozzi P, Cavalli-Sforza LL (1999) A problem with synthetic maps: Reply to Sokal *et al*. *Hum. Biol.* **71**, 15–25.

Richards M, Macaulay V, Hickey E *et al*. (2000) Tracing European founder lineages in the near eastern mtDNA pool. *Am. J. Hum. Genet.* **67**, 1251–1276.

Roff DA, Bentzen P (1989) The Statistical-analysis of mitochondrial-DNA polymorphisms – Chi-2 and the problem of small samples. *Mol. Biol. Evol.* **6**, 539–545.

Rogers AR, Harpending H (1992) Population growth makes waves in the distribution of pairwise genetic differences. *Mol. Biol. Evol.* **9**, 552–569.

Rooney AP, Zhang J (1999) Rapid evolution of a primate sperm protein: relaxation of functional constraint or positive Darwinian selection? *Mol. Biol. Evol.* **16**, 706–710.

Rosenberg NA, Nordborg M (2002) Genealogical trees, coalescent theory and the analysis of genetic polymorphisms. *Nature Rev. Genet.* **3**, 380–390.

Rosser ZH, Zerjal T, Hurles ME *et al*. (2000) Y-chromosomal diversity in Europe is clinal and influenced primarily by geography, rather than by language. *Am. J. Hum. Genet.* **67**, 1526–1543.

Sabeti PC, Reich DE, Higgins JM *et al*. (2002) Detecting recent positive selection in the human genome from haplotype structure. *Nature* **419**, 832–837.

Saitou N, Nei M (1987) The neighbor-joining method: a new method for reconstructing phylogenetic trees. *Mol. Biol. Evol.* **4**, 406–425.

Sherry ST, Harpending HC, Batzer MA, Stoneking M (1997) *Alu* evolution in human populations: using the coalescent to estimate effective population size. *Genetics* **147**, 1977–1982.

Shriver MD, Jin L, Boerwinkle E, Deka R, Ferrell RE, Chakraborty R (1995) A novel measure of genetic distance for highly polymorphic tandem repeat loci. *Mol. Biol. Evol.* **12**, 914–920.

Slatkin M (1995) A measure of population subdivision based on microsatellite allele frequencies. *Genetics* **139**, 457–462.

Slatkin M, Bertorelle G (2001) The use of intraallelic variability for testing neutrality and estimating population growth rate. *Genetics* **158**, 865–874.

Smouse PE, Long JC, Sokal RR (1986) Multiple-regression and correlation extensions of the Mantel test of matrix correspondence. *Syst. Zool.* **35**, 627–632.

Sneath PHA, Sokal RR (1973) *Numerical Taxonomy*. W.H. Freeman, San Francisco.

Sokal RR, Oden NL, Thomson BA (1999) A problem with synthetic maps. *Hum. Biol.* **71**, 1–13.

Stephens JC, Reich DE, Goldstein DB *et al*. (1998) Dating the origin of the CCR5–Δ32 AIDS-resistance allele by the coalescence of haplotypes. *Am. J. Hum. Genet.* **62**, 1507–1515.

Stumpf MP, Goldstein DB (2001) Genealogical and evolutionary inference with the human Y chromosome. *Science* **291**, 1738–1742.

Templeton AR (1998) Nested clade analyses of phylogeographic data: testing hypotheses about gene flow and population history. *Mol. Ecol.* **7**, 381–397.

Thomson R, Pritchard JK, Shen PD, Oefner PJ, Feldman MW (2000) Recent common ancestry of human Y chromosomes: Evidence from DNA sequence data. *Proc. Natl Acad. Sci. USA* **97**, 7360–7365.

Tremblay M, Vezina H (2000) New estimates of intergenerational time intervals for the calculation of age and origins of mutations. *Am. J. Hum. Genet.* **66**, 651–658.

Vigilant L, Stoneking M, Harpending H, Hawkes K, Wilson AC (1991) African populations and the evolution of human mitochondrial DNA. *Science* **253**, 1503–1507.

Wall JD (1999) Recombination and the power of statistical tests of neutrality. *Genet. Res.* **74**, 65–79.

Wayne ML, Simonsen KL (1998) Statistical tests of neutrality in the age of weak selection. *Trends Ecol. Evol.* **13**, 236–240.

Weiss K (1973) Demographic models for anthropology. *Am. Antiq.* **38**, 1–186.

Womble WH (1951) Differential systematics. *Science* **114**, 315–322.

Wright S (1951) The genetical structure of populations. *Ann. Eugen.* **15**, 323–354.

Wyckoff GJ, Wang W, Wu CI (2000) Rapid evolution of male reproductive genes in the descent of man. *Nature* **403**, 304–309.

Yang Z, Bielawski JP (2000) Statistical methods for detecting molecular adaptation. *Trends Ecol. Evol.* **15**, 496–503.

Yu N, Zhao Z, Fu YX *et al.* (2001) Global patterns of human DNA sequence variation in a 10-kb region on chromosome 1. *Mol. Biol. Evol.* **18**, 214–222.

Zhang J, Rosenberg HF, Nei M (1998) Positive Darwinian selection after gene duplication in primate ribonuclease genes. *Proc. Natl Acad. Sci. USA* **95**, 3708–3713.

Zhao Z, Jin L, Fu YX *et al.* (2000) Worldwide DNA sequence variation in a 10-kilobase noncoding region on human chromosome 22. *Proc. Natl Acad. Sci. USA* **97**, 11354–11358.

SECTION FOUR

Where and when did humans originate?

We now turn from considerations of what questions we might ask, what sources of genetic information are available and what analytical tools we can use to address these questions, to a discussion of the answers that can be provided. We will follow a loosely chronological course in the next two sections, and begin in this section by examining the path leading to the origin of modern humans.

CHAPTER 7   *Humans as apes*

First we consider humans in the context of other apes. How similar are we to other apes? Which living apes are our closest relatives? What changes have occurred to make us human? Evidence comes from the structure of our bodies and our chromosomes, but in its fullest form from molecular analysis, especially of DNA. We will see that our closest living relatives are chimpanzees and bonobos, but that we differ from them substantially in our large population size coupled with unexpectedly low genetic diversity. However, we are only just beginning to understand the genetic changes that have allowed these differences to develop.

CHAPTER 8   *Origins of modern humans*

This chapter turns to the human-specific or 'hominid' line, and we follow its development from the time of the chimpanzee–human split, probably more than seven million years ago, to the appearance of anatomically modern humans about 130 000 years ago. Crucial evidence in this highly contentious field comes from fossils and archaeological remains, and also from the extent and distribution of genetic diversity within humans, with contributions from the analysis of ancient DNA. For most of this time, our ancestors were just one of several hominid species living in Africa, expanding into the rest of the world when conditions allowed. These early expansions, however, seem to have contributed little to modern human genomes, although this conclusion is disputed by some.

CHAPTER SEVEN

Humans as apes

CHAPTER CONTENTS

BOXES

7.1 Overview

We are apes, but a unique form of ape. The similarities and differences between humans and other living apes are the subject of this chapter. There are three major themes.

Who are our closest living relatives?
This question has been considered within an evolutionary framework since Darwin wrote that 'much light will be thrown on the origin of man and his history' in *The Origin of Species*. Initial work used morphological methods (Section 7.2); subsequently, studies of chromosomes (Section 7.3) and molecules (Section 7.4) have come to dominate the field. These studies address two main questions:

▶ what was the branching order of the species?

▶ what was the time-scale?

A crucial point to appreciate is the distinction between the **species tree** and what is usually called a **gene tree** (even if it represents a noncoding DNA sequence rather than a gene). There is an independent gene tree for each DNA segment, so both branching order and time-scale may differ when gene trees are compared with one another and with the species tree, of which there is just one (*Box 7.1*).

Are humans typical apes?
Despite the rudimentary information available on ape **population genetics**, it is clear that humans are atypical in many respects: diversity, **effective population size**, and population structure. These unusual features of our population genetics result from our evolutionary history and underlie the remainder of this book.

What genetic changes have made us human?
We do not know the answer, but work that is beginning now will form one of the most exciting areas of research in the future.

The studies we discuss in the next three sections compare humans with apes, mostly using the simple terms 'chimpanzees', 'gorillas' and 'orangutans'. However, these three groups can be subdivided into at least five different species and many subspecies – this is discussed in Section 7.5.

7.2 Evidence from morphology

Before the development of molecular techniques during the past 40 years, morphological characters were the major source of evidence available to place humans in the tree of life. Furthermore, given the probable limits of ancient DNA techniques to within the past 100 KY (see Section 4.11.1), morphology is the only tool we have for reconciling most fossils with living species. As with other **cladistic** enterprises, the emphasis is on identifying characters that are phylogenetically informative, in an attempt to avoid the impact of **convergent evolution**, and reveal the underlying tree of ancestral relationships.

Once it has been decided that a morphological character is **homologous** rather than **analogous**, and that it varies among the taxa being classified, it remains to be determined whether it is a 'primitive' (**plesiomorphic**) character that was present in the common ancestor or is 'derived' (**apomorphic**) and arose later. If derived characters are shared among multiple taxa (**synapomorphic**) then they can be used to infer phylogenetic branchings.

The work of Carl Linnaeus in the eighteenth century laid the foundation for all further attempts to classify organisms by producing a hierarchical system in which species could be lumped together into ever more inclusive groupings. A current classification of humans within this hierarchy is shown below:

Kingdom: Animalia

Phylum: Chordata

Class: Mammalia

Infraclass: Eutheria

Order: Primates

Superfamily Hominoidea

Family: Hominidae

Genus: *Homo*

Species: *sapiens*

But how does this linear progression relate to those of other species, both extant and extinct?

7.2.1 Primates as mammals

Whilst some levels in the taxonomy of mammals are clearly resolved, the **phylogenetic** relationships of the groupings within these levels often remain hotly disputed. There are over 20 living Orders of Mammals (of which Primates is one) which themselves fall into three accepted groupings on the basis of reproductive mechanisms: the egg-laying monotremes (Prototheria) found only in Australia and New Guinea (e.g., platypus, echidna), the marsupials (Metatheria) found in Australia and the Americas, and the placental mammals (Eutheria) found over most of the globe. In contrast to these relative certainties, the phylogenetic tree relating the 18 extant Eutherian orders is difficult to resolve using morphological data. Recent analyses of nuclear and mitochondrial sequences have shown that these Eutherian orders fall into four clades: Afrotheria, Xenarthra, Laurasiatheria and Euarchontoglires (also known as Supraprimates, Lin *et al.*, 2002; Murphy *et al.*, 2001a, 2001b). These superordinal groups correspond to the biogeographic distribution of early mammals, with Afrotheria and Xenarthra being restricted to Africa and South America respectively, and Laurasiatheria and Euarchontoglires having origins in the northern hemisphere. The 356 living species of primates (Groves, 2001) belong to the latter clade, along with rodents, lagomorphs (e.g., rabbit and pika), flying lemurs and tree shrews (*Figure 7.1*). The grouping of rodents, primates and tree shrews is supported by these being the only Eutherian orders known to have *Alu*-like **SINEs** (see Section 2.3) derived from 7SL RNA (Nishihara *et al.*, 2002).

BOX 7.1 Gene trees and species trees

Populations contain many genes and many copies of each gene. As a result of recombination and **independent assortment** of chromosomes, each segment of DNA (defined as a region between recombination positions) has an independent history, (see *Figure*). DNA phylogenies differ because of many factors, including the stochastic nature of evolution and selection; for further discussion, see Section 7.4.3.

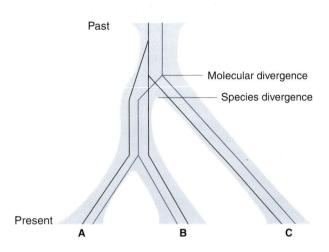

Gene trees and species trees.

A species tree shows the evolutionary relationships of three species: species A and B are most closely related. One gene tree (blue line) has the same branching order as the species tree, although the divergence times are earlier. A second gene tree (black line), shows a different branching order and divergence times: species B and C are the most closely related.

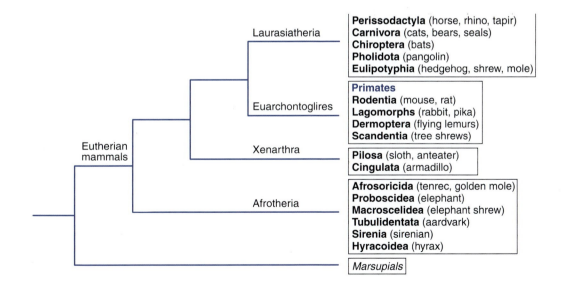

Figure 7.1: Phylogeny of the four major clades of Eutherian mammals.

The higher order taxonomic groupings within each of the four superordinal mammalian clades are shown, together with representatives of that group. Marsupials are placed as an outgroup.

The term 'Primate' (from the Latin for 'first') was coined by Linnaeus himself and remains well accepted, yet compared to many of the other Eutherian orders the Primates are not very clearly defined by a suite of derived morphological characters. For example, pentadactyly (five digits) is a primitive character that has been frequently modified in other mammalian orders. However, shared derived characters include the binocular vision and shortened muzzle or snout that indicate a greater reliance on sight than on smell. In addition the Primates have grasping hands and feet and appreciably larger brains relative to their body size compared to other mammals.

7.2.2 Apes as primates

Traditional morphological classifications of primates identified two groupings: the lower primates (Prosimians) which included the tarsiers, lemurs and lorises and the higher primates (Anthropoidea) which included the Apes and Monkeys. More recently it has been argued on both morphological and molecular grounds that the tarsiers share a common ancestor with the Anthropoidea rather than with the other Prosimians (*Figure 7.2*).

The extant Anthropoidea can be clearly resolved morphologically into three groups. The new world monkeys (Ceboidea), old world monkeys ('OWM', Cercopithecoidea) and the Hominoidea which includes people and apes. Whilst the old and new world monkeys share numerous primitive traits, a number of clearly derived traits indicate that the OWM and hominoids share a more recent common ancestor. These include nostril orientation and common aspects of dentition.

7.2.3 Humans as apes

It is difficult to consider morphological classifications of the Hominoidea in isolation from molecular evidence, as this superfamily has undergone substantial revision over the past 40 years. However, three families have been traditionally recognized: the Hylobatidae or lesser apes (gibbons and siamang), the Pongidae or 'great apes' (orangutans, chimpanzees, bonobos and gorillas) and Hominidae (humans). On gross anatomical grounds the Pongidae and Hominidae are clearly differentiated from the Hylobatidae, and all are differentiated from the OWM by, amongst other characters, the absence of a tail (see *Box 7.2*). Note that if humans and only a subset of Pongidae share a most recent common ancestor then the classification Pongidae is **paraphyletic** (as is the term 'apes' if it does not include humans).

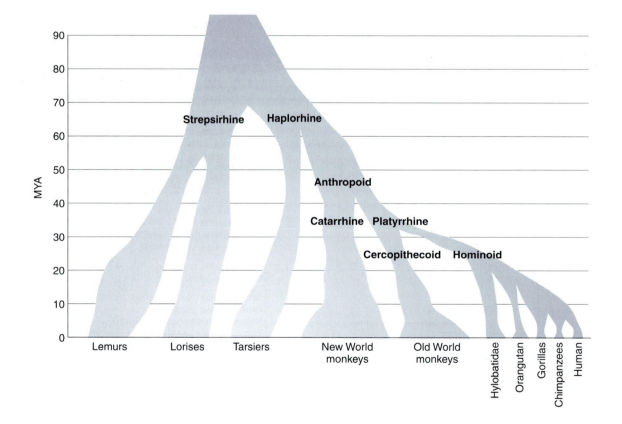

Figure 7.2: A phylogeny of extant primate groups.

No consensus has yet been reached on divergence timings. Whereas the oldest known fossil primates date to 54 MYA, recent statistical modeling of species preservation (Tavaré *et al.*, 2002) suggests the last common ancestor of primates lived about 82 MYA, a date that agrees better with the initial divergence time inferred from molecular data.

BOX 7.2 A history of the classification of humans

Whilst classifications of nature and our place within it were doubtless made earlier in human history, the first person known to have recognized the similarities between ourselves and other primates was the Roman gladiatorial physician Galen of Pergamum. He noted similar morphologies between the Barbary Apes (OWM) he dissected and the unfortunate gladiatorial recipients of his unanesthetized surgery. After his death (c. 200 AD) Galen's anatomical work remained largely unchallenged for over 1000 years.

The Swedish botanist Linnaeus (1707–1778) devised the binomial system of nomenclature (*Genus species*), recognized the grouping 'primates', and coined the name *Homo sapiens* in his *Systema Naturae* (1758). Interestingly, this is one of only two of Linnaeus' 44 primate binomial names which is retained today. Linnaeus boldly included chimpanzees in the same genus (as '*Homo troglodytes*'), but did not believe in evolution.

Thomas Henry Huxley, by contrast a fervent believer in Darwin's evolution by natural selection, published 'Evidence as to Man's place in Nature' in 1863, and in it dispelled the then popular idea that humans should be classified within their own order, stating:

'It is quite certain that the Ape which most nearly approaches man, in the totality of its organization, is either the Chimpanzee or the Gorilla.'

He also noted that:

'Thus, whatever system of organs be studied, the comparison of their modifications in the ape series leads to one and the same result – that the structural differences which separate Man from the Gorilla and the Chimpanzee are not so great as those which separate the Gorilla from the lower apes.'

Nonetheless, Huxley still classified *Homo sapiens* within its own family, Anthropini, and later even within its own suborder, Anthropidae. Charles Darwin, however, disagreed in *The Descent of Man* (1871):

'. . . from a genealogical point of view it appears that this rank is too high, and that man ought to form merely a family, or possibly even only a sub-family.'

Many current taxonomies distinguish humans as a distinct family. However, a recent consideration of species separation times, rather than the more abstract concept of morphological discontinuity, considers that humans should share the *Homo* genus with chimpanzees and bonobos, and that all living apes fall into the family Hominidae (Goodman, 1999). It remains to be seen whether this idea gains wide acceptance.

The ancestral relationships of the living species within the Pongidae and Hominidae are hard to resolve on morphological evidence. **Bonobos** (sometimes called 'pygmy chimpanzees') and common chimpanzees clearly share a common ancestor to the exclusion of the others, and, on morphological grounds, there are a number of seemingly derived features that group these two species with the gorillas. These features include anatomical similarities of the wrist and hand that are linked to a common mode of locomotion, **'knuckle walking'**, and thin dental enamel. However, the derived nature of thin enamel has been questioned, and recent detailed anatomical studies of fossil hominids reveal features associated with knuckle walking (Richmond and Strait, 2000).

In light of these difficulties, perhaps 'fossil apes' could shed light on the evolution of the important morphological characters? The global fossil record contains a few representatives of many hominoid forms over the past 25 million years. However, confounded by the problems of small sample sizes, **sexual dimorphism** and uncertainty over the developmental stage attained, the fossil hominoids have undergone substantial taxonomic tinkering. This phylogenetic shuffling may also have been intensified by a tendency among researchers to assert their own discoveries as direct hominid ancestors. Analyses of the dentition of the fossil hominoid *Ramapithecus* in the 1960s and 1970s placed it on a lineage towards humans, and the early dates suggested a split of humans from other hominoids ~12 million years ago (MYA). However, more recent discoveries link *Ramapithecus* to *Sivapithecus* (an ancestral orangutan) through the identification of a number of derived facial characteristics that they share solely with orangutans. Thus it appears that these species, and not humans, split first from the other great apes (Andrews and Cronin, 1982).

Although many morphological characters are shared between humans and the African apes (chimpanzees and gorillas) most are also present in gibbons, and are therefore primitive (Andrews and Cronin, 1982). However, the derived broad nasal aperture shared between the African apes and humans indicates a shared ancestor. The hominoid fossil record in Africa between 10 and 5 MYA is notoriously sparse (Section 8.2.1), hindering resolution of the gorilla/chimpanzee/human **trichotomy** so that any conclusion remains weakly supported by morphological data alone. Not all sources of morphological data are equally reliable, however (see *Box 7.3*).

7.2.4 Morphological changes en route to *Homo sapiens*

What, in terms of our morphology, separates us from our closest primate relatives? In answering this question we can

BOX 7.3 Not all morphological characters are equal

Cladistics requires a distinction to be made between shared characters which have arisen once and those that have arisen independently, perhaps due to common environmental circumstances. If the true phylogeny is known from independent evidence, a suite of characters can be used to construct an alternative phylogeny which can be tested for accuracy against the known one. In this manner, obtaining a phylogenetic consensus is not just an end in itself, but illuminates the evolutionary processes that have been operating on any given suite of characters.

This approach has been adopted to ask whether or not all kinds of morphological evidence are equally reliable. Phylogenies of living hominoids constructed from either hard tissue (craniodental) or soft tissue characters are different. The soft tissue characters reconstruct the molecular phylogeny (regarded as the 'true' one) much better than the craniodental characters. The hominoid phylogeny produced from craniodental data not only support an incorrect tree, see *Figure*, but certain erroneous clades within the tree attain **bootstrap** support of 95% (Collard and Wood, 2000; Gibbs *et al.*, 2000). This is unfortunate, because soft tissue characteristics are rarely preserved in the fossil record, and so their use is limited to existing species.

Phylogeny from craniodental data

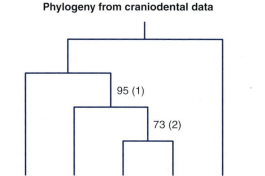

Phylogeny from soft tissue data

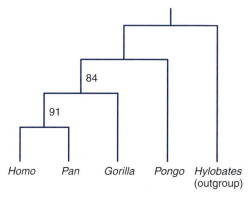

Comparisons of hominoid phylogenies from two sources of morphological characters.

Numbers next to branches indicate percentage bootstrap support. Numbers in parentheses indicate bootstrap support for that clade from the alternative dataset. The chimpanzee–gorilla–orangutan (Pan–Gorilla–Pongo) clade, supported by 95% of bootstraps of the craniodental data, is supported by only 1% of bootstraps from the soft tissue data.

attempt to draw upon the increasingly detailed fossil record of the hominids, to discern not only the changes but the tempo and pattern of change. However, there is an element of circularity in this, as what determines a fossil's classification as a hominid (as opposed to nonhominid) is determined by the prevailing theory of what features define hominid status. Another potential problem is that the hominid phylogeny appears to be very bushy, with all but one lineage going extinct. The ancestral line of modern humans through this tangled heritage remains open to interpretation.

Evidence for a **bipedal** mode of locomotion appears amongst the earliest hominids, whereas a rapid expansion in relative brain size appears much later, starting around 2 MYA. Accompanying this latter change is the first archaeological evidence for stone tool use (although the use of tools derived from other materials would undoubtedly have predated the existence of worked stone). A sharp reduction in sexual dimorphism, which may reflect a change in mating practices, similarly appears later in the hominid fossil record, around 1.8 MYA.

Anatomically modern humans differ from other *Homo* taxa in a number of cranial features and by having a less robust skeleton. It has been claimed that many of the cranial features, including a globular braincase with a vertical forehead, small brow ridges, a face that does not protrude, a canine fossa (a depression on the superior maxillary bone caused by the socket of the canine tooth) and a large chin are not independent, and relate to a reduction in the length of the **sphenoid**, the central bone of the cranial base (Lieberman, 1998). This has been suggested to be an adaptation for speech. However, language does not fossilize, and the origins of a capacity for speech are highly controversial. The descent of the larynx is seen by many as being pivotal to the development of a capability for language; however, given that this organ does not fossilize either, whether archaic *Homo* had essentially modern vocal tracts, let alone the neurological ability to use them, remains open to debate. The recent finding that deer, too, show laryngeal descent demonstrates that this feature is not uniquely human, and suggests that it may exist as an adaptation allowing a species to exaggerate its perceived body size by allowing lower frequency resonances in the vocal tract (Fitch and Reby, 2001).

7.3 Evidence from chromosomes

Like the morphologies of the animals themselves, the morphologies of their chromosomes provide information about the relationships of primate species. Study of the number and physical appearance of mitotic metaphase or late prophase chromosomes under the microscope (the **karyotype**) is the province of cytogeneticists, who use a number of different staining methods to reveal underlying features of chromosomal DNA structure (see Section 2.4).

Karyotypes are useful in phylogenetic comparisons for a number of reasons:

▸ they change relatively slowly, since most chromosomes are constrained by the requirement to pair with homologs, limiting their extent of gross rearrangement;

▸ they provide a 'global' comparison of genomes, rather than of specific segments which may be under the influence of particular selective effects;

▸ methods of study are technically straightforward, requiring universal reagents, and so are readily applicable across a wide range of species.

Because of their relative stability, comparisons of karyotypes among the higher primates can give quite robust information about branching order. However, no 'cytogenetic clock' is available to calibrate the gross changes which chromosomes undergo, and so karyotypic analysis gives no information about dates within the phylogeny.

7.3.1 Human and great ape karyotypes look similar, but not identical

The most obvious difference between the chromosomes of humans and the great apes is in their number: while humans have only 46 chromosomes, chimpanzees, gorillas and orangutans have 48. Comparison of **G-banded** karyotypes, however (Yunis and Prakash, 1982), demonstrates a very high degree of similarity between the four species, and shows that the difference in chromosome number results from an apparently simple end-to-end fusion of two small chromosomes seen in the great apes to form the large **metacentric** chromosome 2 in humans (*Figure 7.3*). Alignment of G-banded chromosomes allows other less dramatic rearrangements to be seen. Most of these are inversions of chromosomal segments, variations in **heterochromatin** adjacent to centromeres, or the presence of G-bands at the ends of chimpanzee and gorilla chromosomes which are absent from those of humans and orangutans. Translocations seem to be limited to a reciprocal event between the equivalents of human chromosomes 5 and 17 in the gorilla with respect to the other species.

Careful consideration of these differences allowed possible ancestral karyotypes to be reconstructed, and a phylogeny to be deduced (*Figure 7.4*). This has chimpanzees as our closest relative, chimpanzee and human as a sister-group to gorilla, and chimpanzee–human–gorilla as a sister-group to the orangutan. Although this branching order agrees with the more recent consensus from molecular data (see Section 7.4), this was not a truly objective attempt at tree building, since the status of orangutan as outgroup was assumed from the beginning.

Unsurprisingly, sharing of chromosomal banding patterns between primate chromosomes is reflected in a high degree of conserved **synteny** – location of a set of genes on the same chromosome. One striking illustration of this at the phenotypic level is the fact that the commonest human autosomal **aneuploidy**, **trisomy 21**, has also been observed in chimpanzees (McClure *et al.*, 1969) gorillas (Turleau *et al.*, 1972) and orangutans (Andrle *et al.*, 1979), and in each species confers a similar phenotype to human Down syndrome (OMIM 190685).

7.3.2 Molecular cytogenetic analyses support the picture from karyotype comparisons

Molecular techniques allow a more fine-grained analysis of the organization of DNA sequences on chromosomes, and so can show whether or not the picture revealed by simple inspection of G-banded karyotypes is accurate.

In **chromosome painting** (also known as chromosome *in situ* suppression hybridization), DNA from a single human chromosome, for example from a chromosome-specific

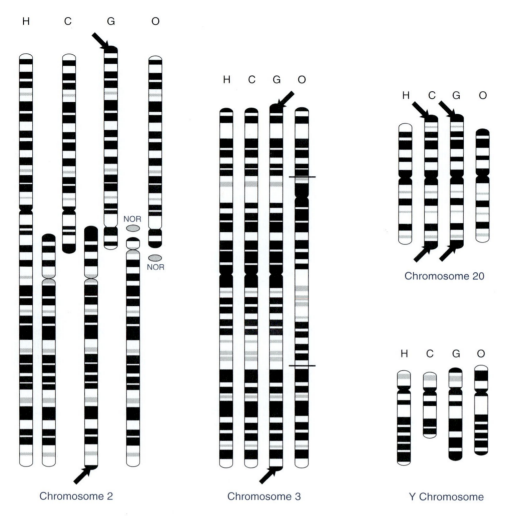

Figure 7.3: Examples of human and great ape chromosomes.

Ideograms of G-banded chromosomes 2, 3, 20 and Y from humans (H), with their orthologs in chimpanzees (C), gorillas (G) and orangutans (O). Conservation of banding pattern is evident. Human chromosome 2 results from a terminal fusion of two chromosomes since the last common ancestor with chimpanzees. One centromere was inactivated. Orangutan orthologs have nucleolar organizer regions (NOR) at the tips of their short arms, where rRNA genes lie. Gorilla- and chimpanzee-specific terminal satellites are indicated by arrows; horizontal lines mark the positions of complex inversion breakpoints on the orangutan ortholog to human chromosome 3. The Y chromosome is poorly resolved and highly variable between species. Redrawn from Yunis and Prakash (1982).

library or alternatively PCR products amplified from a specific flow-sorted chromosome, is labeled and hybridized back to a **metaphase spread** under conditions that allow only unique sequences to generate a signal. This method reveals a picture consistent with that gained from inspection of G-banded chromosomes (Jauch *et al.*, 1992). The origin of human chromosome 2 appears to be confirmed, as does the 5;17 translocation in gorilla.

These chromosome painting methods cannot detect inversions or other intrachromosomal rearrangements; however, this can be done using a particularly high resolution method known as 'chromosomal bar-coding', that uses chromosome paints derived by amplifying human-

specific PCR products from somatic hybrid rodent cell lines containing fragments of a subset of human chromosomes. This process allows 160 molecular cytogenetic landmarks to be identified (Müller and Wienberg, 2001), and detailed ancestral karyotype reconstruction. Even more detailed analysis of particular regions can also be done by **fluorescence *in situ* hybridization** (**FISH**) using specific clones, rather than DNA from whole chromosomes. For example, yeast artificial chromosome (**YAC**) clones forming a **contig** over a region thought to be involved in inversions are hybridized to metaphase chromosomes: a clone which spans a breakpoint gives a single signal on the human chromosome, but a 'split' signal on the ape chromosome. Such analyses of human chromosomes 4, 9 and 12 have

X chromosomes are particularly well conserved in gene content, and also in banding pattern – very similar patterns are seen not only in great apes, but also in all apes and monkeys (Dutrillaux, 1979). In contrast, traditional comparative analysis of Y chromosomes has been rather uninformative, since the chromosome is small, with a poorly defined G-banded appearance, and is also polymorphic in length and organization among humans. The Y chromosome is unique in being **constitutively** haploid: this absence of a homolog, and, therefore, of a requirement to pair along the whole of its length, has apparently freed it from the constraints borne by other chromosomes, and it is therefore cytogenetically more dynamic. FISH analysis using YAC clones (Archidiacono *et al.*, 1998) shows that inversions have been frequent, and confirms the presence of a block of sequence (known to be about 4 Mb in size) on the short arm which has **transposed** from the long arm of the X chromosome since humans and chimpanzees last shared an ancestor (Page *et al.*, 1984). While the short-arm XY-homologous **pseudoautosomal** region, in which recombination in male meiosis takes place, is highly conserved among humans and great apes, molecular analysis reveals another evolutionary novelty in humans – a second pseudoautosomal region (PAR2) on the tip of the long arms of the sex chromosomes (Freije *et al.*, 1992), which again is X-specific in great apes.

The origin of the most striking landmark in the human karyotype, chromosome 2, has been investigated in the greatest detail (*Figure 7.5*). *In situ* hybridization showed that telomeric DNA could be found at band 2q13, consistent with a terminal fusion event. **Cosmids** were therefore isolated which came from this region and contained (TTAGGG)$_n$ telomere repeats (Ijdo *et al.*, 1991). Sequencing identified arrays of the telomere repeat in a head-to-head arrangement – as would be expected for a simple ancestral telomere–telomere fusion – and flanking sequences which are typical of human subterminal repeats, and which cross-hybridize with other human telomeres. FISH analysis (Kasai *et al.*, 2000) has shown that the order of 38 cosmid clones which span the human chromosome 2 fusion point is conserved between human and chimpanzee, where they are divided between the tips of the short arms of chimpanzee chromosomes 12 and 13.

This **anthropocentric** perspective does not allow the investigation of features of great ape karyotypes that do not exist in humans. One such feature is the presence of chimpanzee- and gorilla-specific terminal G-bands. Using a PCR-based strategy including one primer directed at the (TTAGGG)$_n$ repeats of the telomere, material from these bands has been isolated and shown to be composed of a 32-bp AT-rich satellite sequence which is absent from the genomes of humans and orangutans (Royle *et al.*, 1994). The very different organization of subtelomeric sequences between these closely related species reflects the evolutionarily dynamic nature of these regions of the genome.

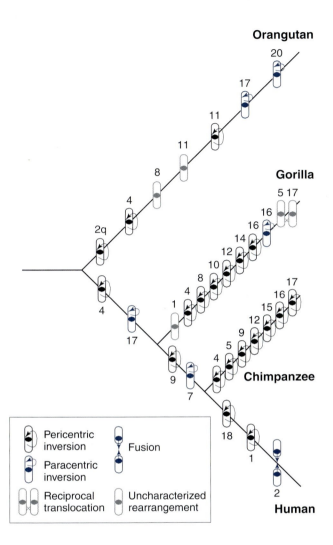

Figure 7.4: Phylogeny of humans and great apes from karyotypic evidence.

Numbers above or below the symbols identify the chromosomes involved, using the human nomenclature. Rearrangements involving heterochromatin are not shown. Drawn from data in Yunis and Prakash (1982).

pinpointed the breakpoints of **pericentric** inversions in the chimpanzee **orthologs** precisely (Nickerson and Nelson, 1998), and again confirm the picture from earlier work. One example of such comparative FISH studies identified a 100-kb segment of DNA sequence originating from chromosome 1 that has inserted into the Y chromosome (Wimmer *et al.*, 2002). Interestingly, this insertion is shared by humans, chimpanzees and bonobos, but absent from three gorillas, as well as orangutans and members of some other ape species. This chromosomal **synapomorphy** (shared derived trait) provides good evidence supporting the *Homo-Pan* clade.

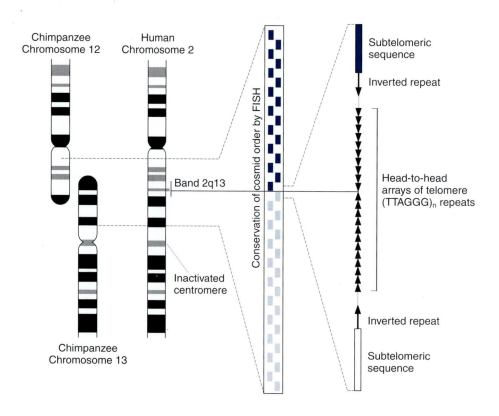

Conservation of cosmid order by FISH

Chimpanzee
Chromosome 12

Human
Chromosome 2

Chimpanzee
Chromosome 13

Band 2q13

Inactivated
centromere

Subtelomeric
sequence

Inverted repeat

Head-to-head
arrays of telomere
$(TTAGGG)_n$ repeats

Inverted repeat

Subtelomeric
sequence

Figure 7.5: Structure of the ancestral fusion point on human chromosome 2.

Drawn from data in Ijdo *et al.* (1991) and Kasai *et al.* (2000).

7.3.3 What is the role of chromosomal changes in speciation?

Karyotypic differences between species might contribute to a reproductive barrier by leading to a hybrid of reduced viability, or a hybrid that is viable, but sterile. A good example of the latter is the production of sterile mules or hinnies from horse (*Equus caballus*) and donkey (*Equus asinus*) matings; these species have different numbers of chromosomes (64 and 62 respectively), and also show numerous apparent **pericentric** inversions and other rearrangements. Mechanisms which have been suggested for the reduced fertility of hybrids include failure of chromosomal pairing in meiosis, general gametic imbalance of functionally equivalent genes from the parents due to inversions and other rearrangements, and dosage problems of genes specific to one parent species or the other. Whether or not hybrids could arise between great apes and humans is a matter of somewhat prurient speculation, but there exists little hard evidence: a notorious alleged chimpanzee–human hybrid, 'Oliver', who showed an atypical chimpanzee phenotype, has been demonstrated to possess 48 normal chimpanzee chromosomes, rather than 47 mixed ones (Ely *et al.*, 1998). By contrast, bonobo–chimpanzee matings are known to have produced viable hybrids in captivity (Vervaecke and Van Elsacker, 1992).

The often reduced fertility of hybrids has led to a view that chromosomal rearrangements may play a major causative role in speciation (King, 1993). However, the apparent paradox that hybrids must be infertile, and yet the rearrangement must come to fixation in the population, means that most chromosomal speciation models involve small, isolated and inbred populations, or a selective advantage conferred by the rearrangement. The opposite view is that chromosome rearrangements are irrelevant to speciation (Rieseberg, 2001), and arise after a premating barrier (such as geographical separation) has become established. In support of this, some species have very different karyotypes, but often form fertile hybrids – an example among the primates is the collared and the white-collared brown lemur (Dutrillaux and Rumpler, 1977), which have 52 and 48 chromosomes respectively. In the absence of evidence for a selective advantage arising from the chromosomal changes, they are most readily interpreted as **neutral** changes that drifted to fixation in small inbred populations. However, see *Box 7.10* for an alternative view on a particular rearrangement – the transposition of X-specific material onto Yp.

7.4 Evidence from molecules

Molecular analyses provide the most extensive set of information about the order and timing of the branches in the great ape/human tree. Studies of individual proteins or genes result in gene trees (*Box 7.1*); the species tree is likely to be the consensus of these. Whole genome **hybridization** produces such a consensus in a single experiment, but has its own limitations. In the last few decades there has been a marked improvement in the methods available, progressing from immunological distance measurements in the 1960s to DNA hybridization studies and now extensive DNA sequence information. Interpretation of all of these datasets requires assumptions about whether and how selection is acting on the loci; most changes are assumed to be neutral, or nearly so, but it is important to remember that selection can give rise to quite different trees, for example at the **HLA** region (*Box 5.5*).

7.4.1 A molecular date for the ape/human divergence

Molecular anthropology can perhaps be said to have begun in 1967 with the publication of a key paper entitled 'Immunological time-scale for Hominid evolution' (Sarich and Wilson, 1967). Sarich and Wilson presented the first use of molecular methods to estimate a date for the great ape/human split and, to the surprise of many, claimed that this date was as recent as 5 MYA; at that time *Ramapithecus* fossils dating to ~ 12–15 MYA were classified as hominids (Section 7.2.3). The basis of this work was:

▸ quantitative measurement of the structural differences between orthologous proteins (serum albumins) in old world monkeys, great apes and humans using an immunological method, microcomplement fixation (*Box 7.4*);

▸ calibration of these measurements using a date of 30 MYA for the split between old world monkeys and hominoids.

The results were expressed as an immunological distance (ID), a measure which, by definition, is 1.0 for **antigen** and antiserum from the same species, and increases for more distantly related antigens as more antiserum is required to produce the same immunological response (see *Box 7.4*). Using an antiserum to human albumin, an ID of 1.09 was obtained for the gorilla, 1.14 for both the chimpanzee and

bonobo, and an average of 2.46 for six species of old world monkey. Taking into account the IDs obtained when antisera to chimpanzee or gibbon albumin were used, and assuming that the log of ID is proportional to time, a time of 5 MYA was calculated for the split between gorillas, chimpanzees and humans (*Figure 7.6*).

While the results with antiserum against human albumin, quoted above, and antiserum against gibbon albumin, placed humans closer to gorillas than chimpanzees, the measurements with antiserum to chimpanzee albumin identified humans as the closest relative of chimpanzees. Thus the method could not resolve the gorilla/chimpanzee/human split in a consistent way and this was presented as an unresolved **trichotomy**. Note that this is a failure to resolve a gene tree and is distinct from the gene tree/species tree difference. The resolution of this trichotomy formed the focus of much subsequent work. Other limitations were:

▸ only a single locus was assayed;

▸ immunological distance is an indirect measure of evolutionary distance.

7.4.2 Resolution of the gorilla/chimpanzee/human trichotomy using DNA–DNA hybridization

An attractive approach to determining the branching order of the species tree in a single experiment is DNA–DNA hybridization, which effectively compares the entire single-copy components of two genomes with one another and thus produces an average tree. If most regions of the genome are evolving neutrally, this is expected to correspond well to the species tree.

In DNA–DNA hybridization experiments, two DNA samples (either from the same species or different species) are reassociated and the thermal stability of the hybrid is measured (see *Box 7.5*). **Heteroduplexes** formed between molecules of different species are less stable than homoduplexes formed between molecules of the same species, and this reduction in stability provides a measure of the divergence between the species. The method is well suited to comparing species that have diverged for greater than 10 MY, but for closely related species, the small differences can be masked by random experimental error. For

BOX 7.4 The microcomplement fixation assay

▸ Mix dilutions of antigen (albumin, or serum containing albumin as a major component), a standard quantity of antibody raised against purified albumin, and guinea pig complement; incubate for 16–18 hours at 4°C.

▸ Add sheep erythrocytes (red blood cells), sensitized by treatment with rabbit anti-sheep erythrocyte antibody; incubate at 37°C for 40–60 min to allow lysis of erythrocytes.

▸ Measure OD_{415} of supernatant. This provides a quantitative measure of the amount of lysis and thus the amount of antigen.

▸ Express results as ID (immunological distance, or index of dissimilarity), the relative concentration of antigen required to produce a similar reaction to the homologous albumin.

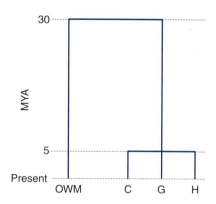

Figure 7.6: Phylogeny based on immunological distance.

The phylogenetic relationship between human (H), chimpanzee (C) and gorilla (G) is unresolved in this tree, and presented as a trichotomy. The outgroup is old world monkeys (OWM). Data from Sarich and Wilson (1967).

gorillas, chimpanzees and humans the differences are small and the subject has been unusually contentious, even by the standards of human evolutionary studies, leading to accusations of data fudging.

The experiments of Sibley and Ahlquist (Sibley and Ahlquist, 1984) presented evidence, in the form of their ΔT50H values, for a closer relationship between chimpanzees and humans than between either species and gorillas (*Figure 7.7*). This work was criticized because it ignored some of the complexities of the hybridization process (a fraction of DNA remains unhybridized; does this significantly affect the conclusions?) and because the T50H statistic is not the only, and possibly not the best, way to analyze the data (Lewin, 1988a; Lewin, 1988b). Thus although the work appears to definitively resolve the trichotomy, concerns about its interpretation meant that Sibley and Ahlquist's conclusion was not universally accepted. Nevertheless, subsequent work has supported the view that chimpanzees and humans are the most closely related of these three ape species.

7.4.3 Resolution of the trichotomy using DNA sequence data

Gene trees do not necessarily have the same topology as species trees (*Box 7.1*). The simplest explanation of this difference is that they do not share the same evolutionary history. If polymorphism survives from one speciation event to the next then there is a possibility that alleles may assort themselves in such a way that the most similar sequences are not necessarily in the most closely related species (*Figure 7.8*). Unsurprisingly, for any pair of speciation events, the degree to which we might expect allele misassortment depends on the probability of alleles becoming fixed over a given length of time. For alleles not under selection this depends on the effective population size. Thus, on average, there is a higher probability that gene trees from haploid mitochondrial and Y-chromosomal data more accurately reflect the species tree, due to their lower effective population sizes (see *Box 5.1*), than trees from autosomal data.

Some loci within the human genome have polymorphisms that are maintained by **balancing selection**, over much longer time spans that would be expected of neutral alleles. This explains why alleles at some human HLA loci are more closely related to their chimpanzee counterparts than they are to other human alleles (*Box 5.5*). Such loci are to be avoided when reconstructing species trees.

In addition there is the added complication that even when a gene tree shares the same underlying topology as the species tree, reconstructing the true phylogeny from sequence data can be hampered by other evolutionary processes. These processes include selection, unequal rates of evolution amongst lineages, and biased **base composition** (see Sections 5.2 and 6.5 for more details).

Consequently the best approach to resolving the gorilla/chimpanzee/human trichotomy lies with examining gene trees from as many neutral, single-copy, **orthologous** loci as possible in the three species (Chen and Li, 2001; Ruvolo, 1997). These data can subsequently be analyzed in a number of ways. The sequences can be concatenated (combined into a single, long sequence) and **bootstrap** support (see Section 6.5.5) obtained for clades within the

---

BOX 7.5 **DNA–DNA hybridization**

▶ Shear DNA to an average length of ~ 500 bp.

▶ Prepare 'tracer' fraction by denaturing, renaturing for a limited time to reassociate repeated sequences but not unique sequences, followed by purification of the remaining single-stranded, single-copy DNA. Label with [125]I.

▶ Mix tracer DNA with 1000-fold excess of the sample to be used for comparison, the 'driver'. This is unlabeled, whole-genome DNA. Denature and allow to reassociate extensively (5 days) so that even single-copy sequences form duplexes.

▶ Bind reassociated, double-stranded DNA to a hydroxyapatite column at 55°C. Raise the temperature in steps of 2.5°C and measure the amount of driver DNA that elutes at each step by counting the radioactivity washed off.

▶ Plot percentage single-stranded DNA against temperature: the 'melting curve'.

▶ Determine the temperature at which 50% of the DNA has melted, the T50H value, and compare between homoduplex and different heteroduplexes, the ΔT50H value.

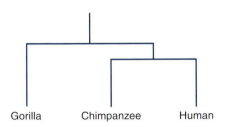

Figure 7.7: Resolution of the gorilla–chimpanzee–human trichotomy.

DNA–DNA hybridization data (Sibley and Ahlquist, 1984) support a closer relationship between chimpanzees and humans than between either species and gorillas.

single resulting phylogeny. Alternatively, phylogenies can be reconstructed for individual loci and multi-locus tests applied to determine whether any topology is significantly predominant. *Figure 7.9* shows recent data demonstrating unequivocal support for a human–chimpanzee clade to the exclusion of the gorilla.

Once a definitive phylogeny has been obtained, and the branch-points dated, the proportion of gene trees which do not support this phylogeny can be used to estimate the effective population size of the ancestral population that existed between the species' splits. In the above scenario, this ancestral population corresponds to the human–chimpanzee common ancestor. Although, as we shall see in the next section, dating branch-points is not without controversy, it appears that the effective population size of this common ancestor was at least five times greater than that of humans. There are many possible explanations for this dramatic reduction, and it may not be possible to arrive at a conclusive answer.

Different loci within the human genome exhibit different degrees of **sequence divergence**. Regions under greater selective constraint will exhibit less sequence divergence; consequently, **nonsynonymous** changes within coding regions (K_a) are less frequent than **synonymous** changes (K_s). The reasons underlying other differences in sequence divergences shown in *Table 7.1* are less immediately apparent.

Inheritance pattern has a dramatic effect on sequence divergence. It has long been suspected that the larger number of mitotic cell divisions in the male germ-line than in the female increase mutation rates in males relative to females, and that consequently evolution is 'male driven' (see *Box 5.8* and *Figure 3.24*). The ratio of these sex-specific mutation rates is known as the '**alpha factor**', estimates for which are given in *Box 5.8*. Y chromosomes are inherited solely through the male germ-line, whereas X chromosomes pass through twice as many female meioses as male. Thus, male-driven evolution explains why sequence divergence amongst noncoding regions is greatest among Y-linked loci, and least among X-linked loci.

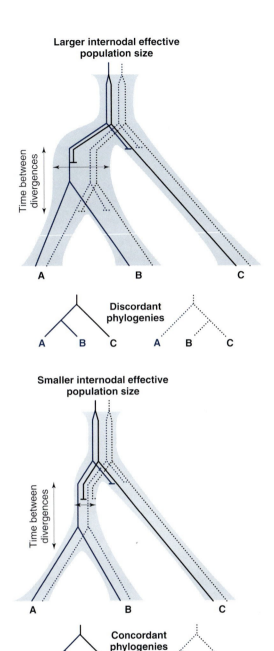

Figure 7.8: The effect of ancestral effective population size on the frequency of discordant gene trees.

The degree to which gene trees of neutral loci fail to reflect the same phylogeny as the species tree depends on the time between successive speciation events and the effective population size of the ancestral population. In these two examples, the time between speciation events is the same in the two scenarios. Two gene trees are shown within each species tree (in blue) by the solid and dashed lines. In the first example, the greater width of the species ancestral to A and B represents a larger effective population size that allows a polymorphism in one of the gene trees (dotted line) to persist between the two speciation events. The subsequent lineage assortment of this polymorphism results in discordant phylogenies.

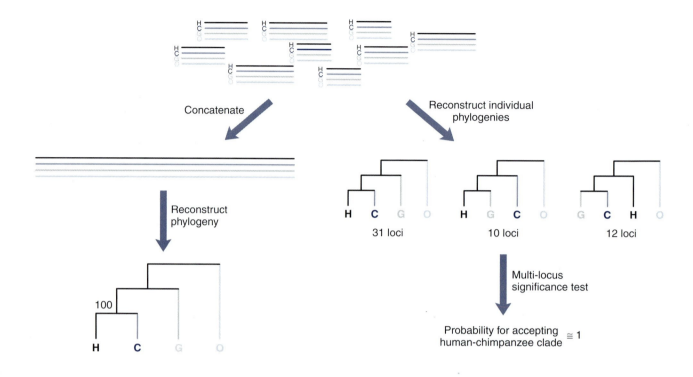

Figure 7.9: Combining data from multiple orthologous loci to resolve the gorilla–chimpanzee–human trichotomy.

The number (100) next to the human–chimpanzee clade indicates percentage bootstrap support. H, Human; C, chimpanzee; G, gorilla; O, orangutan. Data are from Chen and Li (2001).

The comparatively greater sequence divergence of *Alu* elements and pseudogenes relative to noncoding regions which have the same inheritance pattern relates to a higher average frequency of **CpG dinucleotides** in these sequences. CpG dinucleotides are roughly 10-fold more mutable than the genome average (see Section 3.2.5).

Other studies have revealed further significant differences among loci in human/chimpanzee sequence divergences that can not be explained by the above factors (Ebersberger *et al.*, 2002) These sequence divergence disparities appear to be chromosome-wide. For example, chromosomes 4, 19 and 21 appear to have accumulated substitutions at a greater rate

TABLE 7.1: PERCENTAGE SEQUENCE DIVERGENCES (JUKES–CANTOR DISTANCES) BETWEEN HOMINOIDS IN DESCENDING ORDER.

Locus	H-C	H-G	C-G	H-O	C-O	G-O
Alu elements	2	–	–	–	–	–
Non-coding (Chr. Y)	1.68 ± 0.19	2.33 ± 0.2	2.78 ± 0.25	5.63 ± 0.35	6.02 ± 0.37	6.17 ± 037
Pseudogenes (autosomal)	1.64 ± 10	1.87 ± 11	2.14 ± 0.11	–	–	–
Pseudogenes (Chr. X)	1.47 ± 0.17	–	–	–	–	–
Noncoding (autosomal)	1.24 ± 0.07	1.62 ± 0.08	1.63 ± 0.08	3.08 ± 0.11	3.12 ± 0.11	3.09 ± 0.11
Genes (K_s)	1.11	1.48	1.64	2.98	3.05	2.95
Introns	0.93 ± 0.08	1.23 ± 0.09	1.21 ± 0.09	–	–	–
Noncoding (Chr. X)	0.92 ± 0.10	1.42 ± 0.12	1.41 ± 0.12	3.00 ± 0.18	2.99 ± 0.17	2.96 ± 0.17
Genes (K_a)	0.80	0.93	0.90	1.96	1.93	1.77

Human, H; Chimpanzee, C; Gorilla, G; Orangutan, O. Data from Chen and Li (2001) and references therein.

than chromosomes 14 and 17. Unlike interlocus differences in *intraspecific* sequence diversity, which may be explained by the different genealogical histories of each locus (see Section 3.2.7), discrepant *interspecific* sequence divergences among loci require alternative explanations to account for the different rates at which substitutions have accumulated. These explanations could include unknown selection pressures, or variable chromosomal mutation rates as a result of different recombination rates, but whatever the reason, there is evidence that the underlying inequalities have been maintained over long time periods (Lercher *et al.*, 2001; Ebersberger *et al.*, 2002).

It is worth noting that the above measures of sequence divergence only take into account base substitutional differences between orthologous loci. DNA sequences may also differ by insertions or deletions (**indels**) of small numbers of bases (see Section 3.2.6). Typically these indels are stripped from alignments before the calculation of sequence divergence measures such as the **Jukes–Cantor** distances (see Section 6.4.3) used in *Table 7.1*. A different measure of sequence divergence takes indels and base substitutions into account by considering what fraction of bases in one species are exactly matched in another species. This measure has been applied to 779 kb of aligned human and chimp sequence in which ∼ 106 000 base substitutions were supplemented by 1000 indels (Britten, 2002). The rate of base substitutional sequence divergence in these alignments (1.4%) was similar to that described above, yet, despite their lower frequency, indels contribute an additional 3.4% of sequence divergence, generating an overall divergence that approaches 5%! This method for calculating sequence divergence is only accurate for comparisons of closely related species, because the mutational mechanism of indel events is poorly known, which makes it impossible to make the necessary correction for parallel and reversion mutations in comparisons of more distantly related species.

7.4.4 Dating hominoid divergences

Dating divergences among hominoids is important for a number of reasons, given in *Box 7.6*. The global **molecular clock** is the simplest method for estimating species divergence dates from molecular diversity. It assumes that, for a given locus, mutation is occurring at a constant rate across all taxa. Thus species bifurcations that are well resolved in the fossil record can be used to date less well-characterized divergences. However, a number of factors may alter rates of sequence change over evolutionary time, including differences in generation times, in DNA repair mechanisms and in metabolic rates (see Section 5.6.5). Empirical data indicate that rates do vary between mammalian lineages, and unsurprisingly the molecular clock becomes less reliable the more deeply rooting the phylogeny. For example, it has been demonstrated that mitochondrial evolution has proceeded faster in primates than in many other mammalian orders, and that within primates there has been an evolutionary slowdown in the hominoids, perhaps due to their longer generation time (see *Figure 7.10*). In addition, a recent analysis of 15.3 kb of intronic autosomal sequences has demonstrated that the slowdown among higher primates has been strong enough to result in both hominoids and old world monkeys having slower mutation rates than the mammalian average (Yi *et al.*, 2002).

Because of these rate differences, a number of statistical methods have been devised to detect mutation rate heterogeneity. Most are based on the **relative rates test**, which compares molecular distances between pairs of species that are equally related: two ingroup species compared with a known outgroup (see *Figure 5.14*). Under a constant rate these distances should be equal. The statistical significance of any difference between the two distances can be assessed in different ways.

BOX 7.6 Why do we need to know the date of the human/chimpanzee split?

It might seem only of academic importance to know how recently we split from our nearest living primate relative. However, because the rate of sequence evolution at individual loci is too slow to be measured directly by current methods, we must rely on calculating these mutation rates indirectly. Often this is done by calibrating the amount of sequence divergence between humans and chimpanzees with their estimated age of divergence (other divergence points can be used but require greater correction for multiple hits and are more likely to encounter rate heterogeneity). The rate of sequence evolution is central to evolutionary genetics, and underpins a great number of important evolutionary calculations. For example it is used to estimate:

▶ the age of the most recent common ancestor for a given locus;

▶ the ages of individual lineages, which can provide information about the timings of prehistoric migrations;

▶ the ages of demographic events, such as population expansions and bottlenecks;

▶ the deleterious mutation rate in the human genome.

In addition, as was discussed in Section 7.4.3 (see *Figure 7.8*), the effective population size of the ancestral population and the time between species' divergences determine the proportion of gene trees which do not have the same topology as the species tree. Conversely, if we know this proportion and the time between species divergences we can estimate the effective population size of the ancestral population.

(a) ML tree of mtDNA amino acid sequences

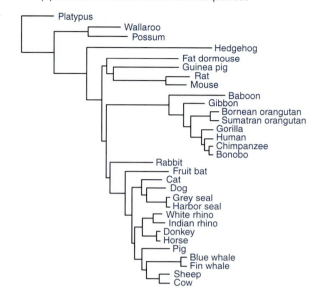

(b) MP tree of 15 aligned sequences of the ψη-globin gene locus

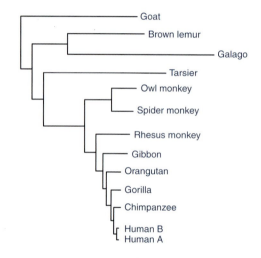

Figure 7.10: Phylogenies exhibiting rate heterogeneity among lineages.

If mutation rates are constant across lineages, the sum of branch lengths leading to each taxon in an outrooted tree should be equal, and thus the tips should be level. Longer branch lengths indicate a faster rate and shorter branch lengths a slower rate. (a) The mtDNA tree (redrawn from Yoder and Yang, 2000) shows a higher rate of mtDNA mutation in primates compared to other mammals. (b) The ψη-globin tree (redrawn from Bailey *et al.*, 1991) shows a decrease in mutation rate at the ψη-globin gene locus in hominoids compared to other primates.

If rate heterogeneity is detected, various analytical techniques allow it to be taken into account when calculating divergence times. These techniques require accurate estimates of branch lengths, and in the case of faster mutating loci or deeper divergences, this involves a correction of observed sequence distance measures for multiple mutations at the same site. As with other applications of **phylogenetic reconstruction** methods, changing the model of sequence evolution often gives different inferences.

There are three potential sources of error in molecular divergence dates:

▶ the chosen calibration point;

▶ the methods used to reconstruct the phylogeny and to take into account rate heterogeneity;

▶ the choice of data.

These three factors are not unrelated – for example, the calibration point needs to be one for which data are available.

This explains the predominance of mitochondrial studies when more ancient, nonprimate calibration points are used, as few other loci have been studied in so many different species. It also explains why, in many non-mtDNA studies, contentious, more recent, primate divergence calibration points are used. In other words, for a given locus, the fewer data that are available, the more contentious the calibration point.

One way to validate a calibration point is to have a number of paleontologically well dated divergences incorporated within the data: this allows the reciprocal calibration of one well dated divergence by another. If these reciprocal calibrations do not agree, then either the data, the method or the calibration must be changed. By using different variations of all three factors, reciprocal calibrations can reveal which combinations are least accurate. A number of recent studies have questioned the use of primate divergence calibration points (Arnason et al., 1998; Yoder and Yang, 2000). However, note that genetic estimates of clade age will always be older than estimates made using the fossil record. This is because genetic data identify the earliest stages of divergence, which predate the important morphological changes.

Studies of many individual loci have estimated the date of the human/chimpanzee split as 4–7 MYA, with the gorilla splitting off earlier, between 6 and 9 MYA. Specific studies often present more precise estimates; however, varying the models, calibration points and the form of the data within accepted bounds results in the range given above. These ranges therefore represent the best picture of the lack of precision currently available (Yoder and Yang, 2000). This range has become the consensus view, even though most of these estimates rely on the patchy primate fossil record for calibration. Often a date of 12–16 MYA for the split of the orangutan from the other extant hominoids has been used, or a date of 25–30 MYA for the Cercopithecoidea/Hominoidea split. If these dates are revised upwards, as suggested by recent results (Arnason et al., 1998; Yoder and Yang, 2000) (see Box 7.7), then all estimates based on these calibrations will be similarly increased. This includes the estimates based on immunological and DNA–DNA hybridization data (Sarich and Wilson, 1967; Sibley and Ahlquist, 1987).

7.5 Genetic diversity among the great apes

Having arrived at a consensus about the phylogenetic relationship of humans with our closest nonhuman relatives, we can ask whether the pattern of genetic diversity in the great apes resembles that of our own species. Polymorphism maintained between species may indicate balancing selection, and discrepancies of levels of diversity within and between species may signal the action of natural selection, or different population structures and histories, and different **effective population sizes**.

Certainly there are enormous differences in actual population sizes today. These are difficult to determine for the great apes since they live in forest environments which are not easy to survey, but estimates are given in *Figure 7.11*. Currently, it takes the human population less that 2 days to increase by a number equal to the total world population of great apes. While our species is currently distributed over almost all the land surface of the earth, distributions of great apes are highly restricted (*Figure 7.11*), and are becoming more so through our own activities. The World Conservation Union lists all great ape species as either endangered or critically endangered, and a survey of western equatorial Africa (Walsh et al., 2003) documents a recent catastrophic decline in numbers due to hunting and Ebola hemorrhagic fever. This sad fact gives the understanding of great ape diversity a vital and urgent purpose beyond the academic activities of primate taxonomists (*Box 7.8*).

7.5.1 How many genera, species and subspecies are there?

Describing the genetic diversity within a species or subspecies first requires these categories to be defined. This is not straightforward (Mallet, 2001), and indeed there is some circularity when genetic data themselves are sometimes used as criteria for defining species or subspecies, as is the case for chimpanzees (e.g., Morin et al., 1994; Gonder et al., 1997).

Original ideas of a species as possessing a perfect 'essence', and of variations within species as imperfections, were associated with the belief that each species was separately created by a deity. Darwin's rejection of this notion was part of his understanding that variation was fundamental to living organisms, and allowed species to evolve. Since Darwin's time, a wide range of species definitions has been proposed. Traditional versions have been based on taxonomically useful characters such as morphology, while more recent 'species concepts' have focused on phenomena such as shared ecology (and geographical distributions), or genetic similarity. A prominent model has been the **biological species concept** (Mayr, 1970), which defines a species in terms of an interbreeding natural population that is reproductively isolated from others by a number of possible mechanisms (*Box 7.9*). Models such as this regard species as 'real' entities, and have been challenged by more recent ideas which suggest that species have no objective reality, but are man-made constructs designed for the purposes of categorization (Hey, 2001; Mallet, 2001). One view points out that it is the local population, united by gene flow within it, and not the species, which is the evolutionary unit. This ambiguity can have a profound effect on the number of species that we recognize – for example, what were once considered as four species of baboon are now regarded as only one.

Recent changes in great ape classification (*Table 7.2*) have introduced a new subspecies of chimpanzee (P. t. vellerosus), and have divided gorillas into two species (G. gorilla and G. beringei). Arguments based on the large genetic differences between Sumatran and Bornean orangutans (approaching or exceeding differences between bonobos and chimpanzees; Zhi et al., 1996) may lead to their being classified as distinct species in the future.

BOX 7.7 Opinion: Molecular estimates of primate divergences

It is traditionally accepted that *Pan* and *Homo* diverged ~ 5 MYA. Consequently it is also generally held that the earliest divergences among modern humans took place ~ 170 000 years ago (essentially 5 MY divided by 30, since the greatest molecular differences among recent humans are about 1/30 of that between *Pan* and *Homo*, even in cases when more complicated models of molecular evolution are used). In the absence of conclusive paleontological support for the timing of the *Pan/Homo* split, this divergence has commonly been dated using three primate calibration points: (i) the divergence between Catarrhini (old world monkeys and apes) and Platyrrhini (new world monkeys) set at ~ 35 MYA; (ii) that between Cercopithecoidea (old world monkeys) and Hominoidea (lesser and great apes and *Homo*) set at 25–30 MYA; and (iii) that between *Pongo* (orangutan) and *Gorilla/Pan/Homo* set at ~ 15 MYA. Use of each of these calibration points gives a dating of ~ 5 MYA for the *Pan/Homo* split. However, the datings for the three calibration points are not well supported by fossil data and, indeed, are even paleontologically refutable (Arnason *et al.*, 2000b). Furthermore, they yield paleontologically untenable estimates for various mammalian divergences – placing eutherian origin at 65–70 MYA, cetacean (whale) origin at ~ 30 MYA and the divergence between *Equidae* (horses) and *Rhinocerotidae* at ~ 25 MYA. The first dating is incompatible with the Albian (98–113 MYA) age of recognizable marsupial and eutherian fossils, while the latter two are refuted *inter alia* by 53.5-MY-old cetacean fossils and by 46–48-MY-old equid and rhinocerotid fossils.

The dubious validity of the three primate calibration points has necessitated the establishment of references with a more solid paleontological support. The fossil record of artiodactyls (even-toed ungulates, e.g., cows), perissodactyls (odd-toed ungulates, e.g., horses, rhinos) and cetaceans is more extensive than that of most other mammalian orders. Therefore we have used the fossil record of these groups in conjunction with complete mitochondrial (mt) DNA sequences to establish three nonprimate calibration points. These calibration points are: A/C-60, the interordinal divergence between ruminant artiodactyls and cetaceans 60 MYA; E/R-50, the intraordinal perissodactyl divergence between equids and rhinoceroses 50 MYA; and O/M-33, the intraordinal divergence between odontocetes (toothed whales) and mysticetes (baleen whales) 33 MYA. The three references yield congruent and credible datings when tested both against each other and the mammalian fossil record (Arnason *et al.*, 2000a). *Inter alia* they suggest that primates originated ~ 95 MYA and that Eutheria originated 130–140 MYA (a dating generally accepted by mammalian paleontologists).

Molecular datasets allow extrapolation between different lineages – for example, from lineages with a good fossil record to lineages for which the record is poor or nonexistent. However, one of the problems in all dating estimates is that the rate of molecular evolution can vary

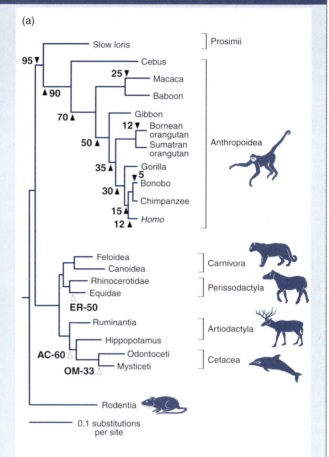

(a)

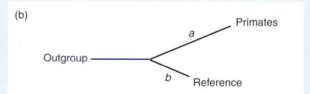

(b)

Age and position of the three nonprimate calibration points and the estimated divergence time of various primate lineages.

The tree is based on ML analysis of amino acid sequences from complete mitochondrial genomes.

among taxa. Therefore a universal molecular clock cannot be assumed. In the tree shown in the *Figure* (part a), the branches leading to the terminal primate taxa are longer than those leading to the artiodactyls and cetaceans. This is because the evolutionary rate of the primates is faster than that of the other groups. Rate differences must be taken into account and corrected for in any dating extrapolation between different lineages. This is done by estimating the evolutionary rate of, for example, the primates *relative* to that of artiodactyls and cetaceans using an indisputable outgroup to all of these taxa. In this particular case, the rodents or marsupials would constitute such an outgroup. The rate differences between different lineages can be calculated from the branch lengths of maximum likelihood, neighbor joining or maximum parsimony trees or can be calculated from the distances according to the following simple scheme:

$$\Delta R = a/b = \frac{(D_{OG\text{-}prim} + D_{prim\text{-}ref} - D_{OG\text{-}ref})/2}{(D_{OG\text{-}ref} + D_{prim\text{-}ref} - D_{OG\text{-}prim})/2}$$

ΔR denotes the relative rate difference estimated from distances (D) between the outgroup (OG), primates (prim) and the calibration point taxa or reference (ref). $D_{OG\text{-}prim}$ is the distance between the outgroup and primates, $D_{prim\text{-}ref}$ is that between primates and the reference and $D_{OG\text{-}ref}$ is that between the outgroup and the reference. In this way, the molecular differences which have accumulated on branch a (primates – see *Figure*, part b) and branch b (the reference) can be calculated and the ratio between a and b established without involving the primate fossil record. After this procedure ΔR can be used to rectify distances (or branch lengths) between any of the three references mentioned above and the primates.

Using the amino acid dataset of complete mtDNA sequences, the above scheme suggests that primates evolve 1.64 times faster than the species behind the A/C-60 reference. After establishing this ΔR it can be used to estimate the divergence time for different splits among the primates. For example, the molecular distance between ruminant artiodactyls and cetaceans is 0.141, corresponding to 60 MY. The molecular distance between Cercopithecoidea (baboon) and Hominoidea (gibbon, great apes and *Homo*) is 0.207, but their time of divergence cannot be paleontologically established. Based on A/C-60, the molecular estimate of the time of divergence between Cercopithecoidea (OWM) and Hominoidea becomes 53.6 MYA:

Distance artiodactyls–cetaceans
0.141, corresponding to 60 MYA

Calibrated distance Cercopithecoidea–Hominoidea
0.207/1.64 = 0.126

Time of divergence Cercopithecoidea/Hominoidea
(0.126/0.141) × 60 MYA = 53.6 MYA

Our estimate of the time of divergence between Cercopithecoidea and Hominoidea is about twice as old as the date arbitrarily attributed to it as one of the three local primate calibration points. The situation regarding the other two local primate calibration points is entirely analogous.

As mentioned above, use of the Cercopithecoidea/Hominoidea calibration point set at 25–30 MYA and of the other two local primate references gives an estimate of ~ 5 MYA for the *Pan/Homo* divergence. That dating is refuted by recent fossil finds in Kenya (*Orrorin tugenensis, Samburupithecus kiptalami*) and Ethiopia (*Ardipithecus ramidus kadabba*). Our estimates (Arnason *et al.*, 2000b) on the other hand, place the *Pan/Homo* divergence at 10.5–13 MYA (depending on the evolutionary model used).

Ulfur Arnason and Axel Janke

Division of Evolutionary Molecular Systematics, Department of Cell and Organism Biology, University of Lund, Sweden

7.5.2 Levels of intraspecific diversity in apes and humans

Phylogenies for two loci, mtDNA HVS I (Gagneux *et al.*, 1999), and a 10-kb segment of Xq13.3 (Kaessmann *et al.*, 2001) (*Figure 7.12*) illustrate dramatically reduced diversity in humans compared to great apes. One explanation for lower diversity in Xq11.3 and mtDNA might be selection, but this would have to act upon both loci similarly, which seems unlikely, and statistical tests for selection (see Section 6.3) at Xq13.3 are negative. Alternatively, the difference may represent a difference in population history, such as a recent expansion from a founder population in humans, and tests based on the Xq13.3 sequences support this. Analysis of the same data suggests an effective population size of 35 000 for chimpanzees compared with 11 000 for humans, and a chimpanzee TMRCA of 2100 KYA (95% confidence interval 1160–3350 KYA) compared with a human TMRCA of 645 KYA (319–1150 KYA). Y-chromosomal sequence diversity has also been shown to be greater in chimpanzees and bonobos than in humans (Stone *et al.*, 2002). Sequence analysis of an ~ 3-kb segment in 101 chimpanzee, and seven bonobo Y chromosomes revealed 23 variant sites, but the humans were monomorphic. The estimated effective male population size was 24 000 for chimpanzee, and 21 000 for bonobo; from other studies, the value for humans is 7500

BOX 7.8 Practical reasons for studying great ape diversity

▸ Defining species and subspecies is important, as these are the units of protection and captive breeding.

▸ Understanding the diversity of wild gene pools allows the preservation of as much diversity as possible within captive populations. This is important for maintaining fitness in response to selective pressures.

▸ Defining population genetic subdivision is important for relocation and reintroduction programs, which need to consider the genetic distinctiveness of isolated populations.

▸ Defining population substructure aids in identifying the geographical origins of confiscated illegal pets.

(a) Chimpanzees and bonobos

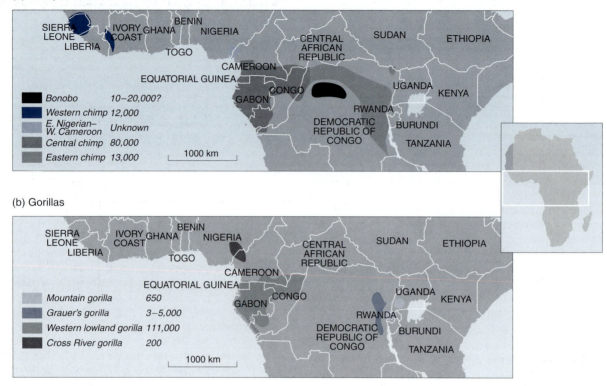

(b) Gorillas

(c) Orangutans

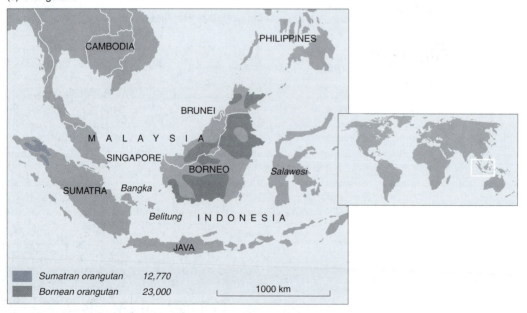

Figure 7.11: Current distributions and approximate population sizes of the great apes.

Numbers within boxes in each part of the figure are approximate population sizes, taken from the WWF International website. Many of these population sizes are the subject of debate and continuing revision. Distributions are taken from the same website, and from Zhi *et al.* (1996).

BOX 7.9 Reproductive barriers between species

The reproductive isolation of species can be divided into two parts:

▶ prezygotic – mechanisms preventing fusion of egg and sperm, so no zygote can form – due to species differences in traits such as:

  ▶ sexual behavior;

  ▶ geographical range or habitat preference;

  ▶ seasonal breeding;

  ▶ gamete compatibility (post-mating, but still prezygotic).

▶ postzygotic – mechanisms preventing a zygote from developing into a fertile offspring – due to:

  ▶ hybrid inviability;

  ▶ hybrid sterility.

(Thomson *et al.*, 2000). Respective estimates for TMRCA are 720 KYA (370–1070 KYA), 500 KYA (230–730 KYA), and 190 KYA (90–290 KYA).

Several independent studies have confirmed that reduced diversity is a general finding for human mtDNA; but despite the concordance of the Xq13.3 data with mtDNA data, the general picture for the nuclear genome is more complex. Early studies of **electrophoretic** variation among red blood cell enzyme and serum proteins indicated that chimpanzees were much less variable than humans, and this picture has been supported for a number of nuclear loci by DNA sequence data (*Table 7.3*). The fact that blood group genes

(Sumiyama *et al.*, 2000) and HLA-A genes (Adams *et al.*, 2000) are less variable in chimpanzees than in humans may reflect selection, through infectious diseases, for increased allelic variation in the human lineage; rigorous tests for selection have not been carried out in these cases, however. Recent data showing reduced chimpanzee diversity at the HLA-A, -B, and -C loci suggest that a selective sweep has occurred in chimpanzees, prior to subspeciation; because chimpanzees are resistant to **AIDS**, it is claimed that the sweep may have been caused by simian immunodeficiency virus SIVcpz, a close relative of **HIV** (de Groot *et al.*, 2002). Lower variability in great apes at micro- (Wise *et al.*, 1997) and mini-satellites (Ely *et al.*, 1992) may be the result of

---

TABLE 7. 2: SPECIES AND SUBSPECIES DISTINCTIONS AMONG THE GREAT APES.

	Species	Taxonomic issues
Humans	*Homo sapiens sapiens*	• Should *Pan* and *Homo* be united as the same genus?
Common chimpanzees	*Pan troglodytes verus* (Western); *P.t. troglodytes* (Central); *P.t. schweinfurthi* (Eastern); *P.t. vellerosus* (Eastern Nigerian–West Cameroon)	• Subspecies distinctions genetic • Behavioral differences do not follow subspecies lines • Subspecies are interfertile in captivity, but no data on hybrid inviability
Bonobo (aka Pygmy chimpanzee)	*Pan paniscus*	• Behavioral (including social), morphological, genetic and geographical distinctions from *P. troglodytes* • Interfertile with common chimpanzees in captivity
Eastern gorillas	*Gorilla beringei graueri* (Grauer's) *G.b. beringei* (Mountain)	• Species distinction morphological, geographic and genetic • Subspecies distinction morphological and geographic
Western gorillas	*Gorilla gorilla gorilla* (Western lowland) *G.g. diehli* (Cross River)	• Subspecies distinction craniodental and geographic
Orangutans	*Pongo pygmaeus abelii* (Sumatran) *P.p. pygmaeus* (Bornean)	• Distinctions morphological, behavioral, cytogenetic (pericentric inversion of chromosome 2), genetic • Genetic differences large enough to warrant different species; other differences relatively small • Subspecies interfertile in captivity; hybrids fertile

Based on several sources, including Uchida (1996), Zhi et al. (1996), and Kemf and Wilson (1997).

ascertainment bias in the choice of highly variable human loci for analysis (see *Box 3.9*), or could reflect a higher mutation rate or mutational bias in humans at these tandem repeat sequences. The finding of lower sequence variability in gorilla Y chromosomes compared to human ones (Gibbons, 2001) is likely to be a reflection of the social structure of gorillas, where a dominant male can father many offspring. Further systematic studies of long sequences, together with appropriate tests for selection, are required to provide a proper understanding of relative nuclear genome diversity.

The distribution of different sequence types at various loci among different species and subspecies within the genus *Pan* have been examined in a number of studies. In a tree of mtDNA HVS I haplotypes (*Figure 7.13a*), bonobo haplotypes lie on one branch, and chimpanzee subspecies

each also occupy single branches, with no haplotypes shared between subspecies. This clear separation includes *P.t. vellerosus*, and indeed the distinctiveness of this clade plus its geographical range is effectively the definition of this subspecies. Separation is similarly clear in a tree of Y-chromosomal haplotypes based on sequence variation (*Figure 7.13b*), though the number of different haplotypes is much smaller (Stone *et al.*, 2002). In contrast, in a tree of haplotypes based on variation at Xq13.3 (Kaessmann *et al.*, 1999), while bonobos are separated from chimpanzees, subspecies are not well separated (*Figure 7.13c*). This apparent discrepancy can be explained by the threefold greater effective population size of the X chromosome compared to mtDNA, which results in a greater time-depth. If subspecies divergence occurs between these two coalescent times then the mtDNA gene tree will exhibit subspecies-specific clades, whereas the X chromosome gene tree will not (*Figure 7.14*).

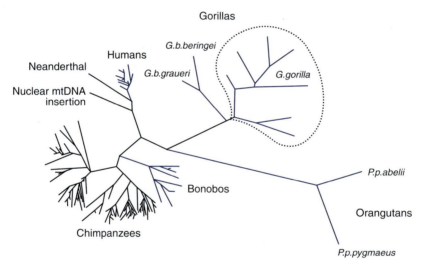

(a) mtDNA HVS I; unrooted and pruned

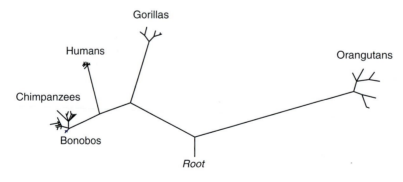

(b) Xq13.3 sequences; rooted using gibbon outgroup

Figure 7.12: Phylogenies showing relative diversities among apes.

(a) mtDNA phylogeny from HVS I sequences, redrawn from Gagneux *et al.* (1999). The tree ('pruned' to remove homoplasies) includes sequences from a Neanderthal, and from a human nuclear mtDNA insertion. Gorilla species are differentiated; for more on chimpanzee subspecies, see *Figure 7.12*. (b) Xq13.3 sequence phylogeny, redrawn from Kaessmann *et al.* (2001).

TABLE 7. 3: RELATIVE DIVERSITIES OF VARIOUS LOCI IN HUMANS AND GREAT APES.

Locus	Chimpanzee vs. human	Bonobo vs. human	Gorilla vs. human	Orangutan vs. human
Xq13.3	3-fold greater	n.d.	2-fold greater	3.5-fold greater
mtDNA	3–4-fold greater	Greater	Greater	Greater
Y chromosome	Greater	n.d.	Less	Greater
MHC class I genes	Greater, but less in HLA-A comparison	n.d.	n.d.	n.d.
ABO blood group genes	2–3-fold less	4–7-fold less	n.d.	n.d.
Microsatellites	Less	n.d.	n.d.	n.d.
Minisatellites	Less	n.d.	n.d.	n.d.

Compiled from data in Ely *et al.* (1992); Rubinsztein *et al.* (1995); Wise *et al.* (1997); Gagneux *et al.* (1999); Adams *et al.* (2000); Sumiyama *et al.* (2000); Gibbons (2001); Kaessmann *et al.* (2001). n.d. - no data. Many studies do not give quantitative measures, and comparison between them is complicated by the use of different measures and different sample sizes.

7.6 What genetic changes have made us human?

This section focuses on the genetic changes that have made us definitively human, which has involved morphological and behavioral modifications, among others. We do not yet know which the crucial adaptations were, but we can approach the question in two complementary ways:

▸ identify the genetic changes that have occurred and evaluate their potential significance ('bottom-up');

▸ identify the important phenotypic changes and search for the genes responsible ('top-down').

Comparing the entire DNA sequences of a human and another ape should reveal much about the differences between us, and might also have some medical benefits (*Box 7.10*); however, as we write, the chimpanzee sequencing project is far from complete and relatively few sequence data from any great ape are currently available.

7.6.1 Genetic differences between apes and humans

If chimpanzee and human have diverged by 1–2% in their single copy DNA, and the differences are independent point mutations, there will be 30–60 million of them. About half of these mutations will have occurred on the human lineage, and most are likely to be neutral. It is therefore a formidable task to identify the few among the 15–30 million human-specific mutations that have been functionally important. How many important changes were there? Simple reasoning suggests that there must be many important changes. If there was just one or a few, we would expect to observe back-mutations; but among all of the mutant human phenotypes

known, there are none that result in an ape-like phenotype. Nevertheless, it is unclear whether there have been 10, 100, 1000, or more functionally important changes.

What changes are found?
All the classes of mutation discussed in Chapter 3 can potentially contribute to great ape/human differences. The largest scale, cytogenetically visible, changes have been discussed in Section 7.3. Tandemly repeated DNA sequences evolve rapidly; indeed, the most rapidly changing regions of the genome are the blocks of **heterochromatin** near the centromeres and telomeres (Section 7.3.1). These have low gene densities and are therefore poor candidates for functionally important changes. Examples of **retroposon** (LINE and SINE) and **endogenous retrovirus** insertions fixed in the human lineage have been discovered. These include the human-specific endogenous retrovirus family HERV-K cluster 9 and the retrovirus-derived long terminal repeat (LTR) family LTR13. LTRs carry powerful **enhancers**, so these insertions could potentially increase the expression of nearby genes, but no such effect has yet been documented. Minisatellites and microsatellites evolve rapidly, and can influence the expression of neighboring genes, but again no consequences for human-specific gene expression have been demonstrated. Attention therefore focuses on duplications, deletions and point mutations affecting the copy number, structure or expression pattern of genes. *A priori* expectations have been that 'regulatory mutations' will be the most relevant; these may change the transcription pattern of individual genes, or if they occur in genes such as transcription factors, they may affect large classes of genes. An alternative view is that we are in some respects 'degenerate apes' and some of our specific characteristics such as slow development, or loss of hair and muscle strength may be due to loss-of-function mutations (Olson and Varki, 2003).

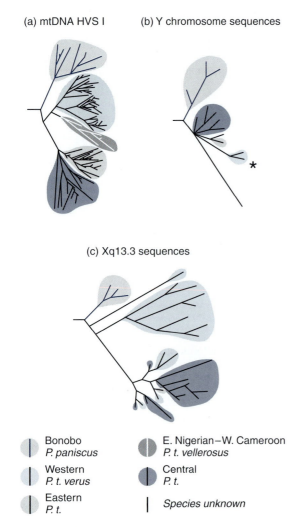

(a) mtDNA HVS I   (b) Y chromosome sequences

(c) Xq13.3 sequences

Bonobo *P. paniscus*	E. Nigerian–W. Cameroon *P. t. vellerosus*
Western *P. t. verus*	Central *P. t.*
Eastern *P. t.*	*Species unknown*

Figure 7.13: mtDNA and Y-chromosome but not X-chromosome sequences, resolve the chimpanzee subspecies.

(a) Phylogeny of bonobos and four chimpanzee subspecies, from mtDNA HVS I sequences (drawn from Gagneux *et al.*, 1999). Subspecies fall neatly into mtDNA clades. (b) Phylogeny of bonobos and three chimpanzee subspecies, from Y chromosome sequences (drawn from Stone *et al.*, 2002). There is no sharing of lineages between subspecies, but some individuals are not assigned to subspecies, including a pair of individuals on a long branch. Note also that one lineage (indicated by an asterisk) accounts for 76/101 of the chimpanzees. (c) Phylogeny of bonobos and three chimpanzee subspecies (*P.t. vellerosus* was not included in this study), from Xq13.3 sequences (drawn from Kaessmann *et al.*, 1999). Members of a subspecies fall into more than one clade. Note that distances between some chimpanzee sequences are greater than the distances between them and bonobo sequences.

Known differences in gene expression

With a catalogue of more than 30 000 human genes now available from the human genome sequencing project, it is, in principle, possible to investigate systematically which ones show differences in expression between apes and humans. A start along these lines has now been made (Enard *et al.*, 2002a). In one set of experiments, **arrays** containing ~ 18 000 human cDNAs were probed with transcripts from three species: rhesus macaque, chimpanzee and human; in each case a pool of several individuals was used to reduce the effects of intraspecific polymorphism. Three tissues were investigated: blood, liver and brain neocortex, and a summary measure of the distance between the relative expression profiles was calculated. In blood and liver, relative expression levels were most similar in chimpanzees and humans, as expected from the phylogeny; but in brain, humans were the most distinct (*Figure 7.15*). Experiments using material from single animals instead of pools revealed substantial differences between individuals within each species. Nevertheless, it seems that the massive size increase of the human brain is accompanied by changes in the expression of many genes, and studies of these individual genes can now begin.

In addition to such systematic studies, investigations initiated for other reasons have fortuitously revealed human-specific differences. A few genes that are active in apes and other mammals are inactive in humans; genes are not created *de novo*, but some have been duplicated in humans and these are of interest because duplication may be followed by divergence and the acquisition of new functions (*Table 7.4* and *Box 7.11*). One of the most extensively studied genes that has been inactivated on the lineage to humans is the CMP-Neu5Ac hydroxylase gene (Chou *et al.*, 1998; Irie *et al.*, 1998) which catalyzes the conversion of the cell surface sugar N-acetylneuraminic acid into N-glycolylneuraminic acid (Neu5Gc). Consequently, human cells largely lack Neu5Gc, and the trace amounts that are present may be derived from the diet. The inactivity of the enzyme is due to a genomic deletion which removes a 92-bp exon and introduces a **frameshift** resulting in a truncated protein. This could be a neutral or nearly neutral change that has been fixed as a result of genetic drift, but there is a possibility that it results from selection. Many pathogens bind to these cell surface sugars, so loss of Neu5Gc could have been selected as a way of providing resistance to a pathogen affecting past populations. A second, more speculative, possibility is suggested by the observation that Neu5Gc is rare in the brains of all mammals: could the inactivation of the gene have contributed to the unusual development of the human brain? Sialic acids such as Neu5Gc survive better in fossils than DNA does, and it has been possible to show that the lack of Neu5Gc is shared by modern humans and Neanderthals, so the inactivation of the gene is likely to predate the common ancestor of these groups ~ 500–600 KYA (Chou *et al.*, 2002). An attempt to date the inactivation from the excess of nonsynonymous substitutions in the human CMP-Neu5Ac hydroxylase gene suggested that it occurred ~ 2.8 MYA (Chou *et al.*, 2002). It is difficult to assess the uncertainty associated with this estimate but, if reliable, it would have occurred just before the increase in brain size associated with the genus *Homo* (Section 8.2.3).

This gene inactivation may be linked to another difference between ape and human genomes. Apes have a

Mitochondrial DNA

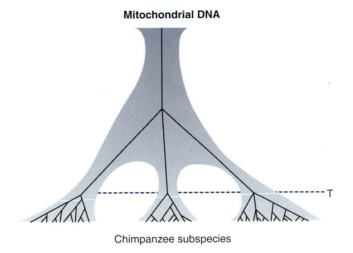

Chimpanzee subspecies

X chromosome

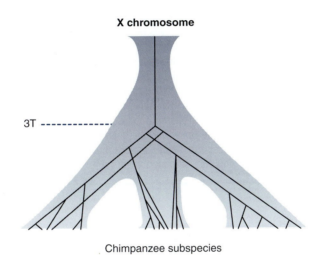

Chimpanzee subspecies

Figure 7.14: Gene trees of different time-depth within a species tree of chimpanzee subspecies.

The three-fold greater effective population size of the X chromosome compared to mtDNA results in a greater time-depth. If the actual subspecies divergence occurs between these two coalescent times then the mtDNA gene tree will exhibit subspecies-specific clades, whereas the X chromosome gene tree will not.

protein, provisionally named Siglec-L1, which binds to Neu5Gc, and this binding requires the presence of a particular arginine residue (Angata *et al.*, 2001). Humans have Siglec-L1, but the human form lacks the arginine residue necessary for Neu5Gc binding. The role of Siglec-L1 is unknown, but there appears to be a set of functionally related genes that are co-evolving in a specific fashion on the human lineage.

The Yq pseudoautosomal region (PAR2) is present in one copy at the tip of Xq in apes, but has two copies in humans as the result of the formation of a second pseudoautosomal region at the tip of Yq. It contains four genes; these remain identical to their X counterparts because of continuing genetic exchange and the information available about them does not obviously suggest a role in the development of

human-specific characteristics (*Table 7.4*). In contrast, a duplication of ~ 4 Mb of Xq sequence on the non-recombining part of Yp represents the largest human-specific duplication known and any genes it contains are of great interest (*Table 7.4* and *Box 7.11*).

7.6.2 Phenotypic differences between apes and humans

How can the genetic basis of phenotypic changes be investigated?

As with the investigation of a genetic disease, two approaches can be used to investigate the genetic basis of the differences between ape and human phenotypes. One is to understand the developmental, cellular and/or biochemical pathways leading to a phenotype and look for differences in the genes

---

BOX 7.10 Benefits of an ape genome project

The understanding of ape/human differences will be greatly facilitated by a directed ape genome-sequencing project, which will replace fortuitous discoveries and small-scale analyses with a systematic catalog of differences. Which ape genome would be most useful? If the aim is to understand what genetic characteristics make us human, the most suitable one would be the most closely related: either chimpanzee or bonobo. Of these, chimpanzees are probably the better choice because they are commoner and more often studied than bonobos.

Because of the great similarity of the chimpanzee genome to the human, the sequence can for the most part be assembled using the human sequence as a template. Thus a modest level of random shotgun sequencing will provide much of the data needed. The current aim is to achieve 'rough draft' 4 × coverage by the end of 2003. It is unclear whether more complete and accurate sequencing will be carried out; if it is not, there is a danger that some of the most rapidly-evolving and thus most interesting regions, containing duplicated, deleted or multiply-substituted sequences, will be misassembled or misinterpreted (Olson and Varki, 2003).

There are, potentially, significant medical benefits because the sequence may provide insights into the genetic basis of phenotypes that differ between chimpanzees and humans (Varki, 2000), and some of these phenotypes are medically important. These include fundamental characteristics like lifespan, which is rarely longer than 60 years in chimpanzees, even in captivity, and the lack of menopause, which has not been observed in long-lived female chimpanzees. Major human diseases like the common carcinomas and Alzheimer disease are rarely reported in chimpanzees. It is difficult to know to what extent this is due to poor ascertainment, shorter lifespan, or differences in diet or other environmental factors, but the reported high frequency of leukemias and lymphomas, and the absence of neurofibrillary tangles associated with Alzheimer disease suggest that the differences cannot be entirely due to problems in ascertainment and diagnosis. An ape genome project would throw light on their genetic basis. Susceptibility to infectious diseases can differ substantially. Chimpanzees are largely immune to infection by *Plasmodium falciparum*, the agent which causes the most severe form of malaria in humans: this was clearly illustrated when captive chimpanzees and their keepers were all exposed to *falciparum* malaria in Gabon, but only the keepers became infected. Late complications in hepatitis B and C, which are common in humans, are rare in apes. The best-known example of all is the consequence of HIV infection. In chimpanzees the symptoms are mild, while in humans infection progresses to AIDS, which is now devastating large parts of the world. It is thought that HIV-1 has been present in chimpanzees for a long time, but has been transferred to humans only recently. Thus an evolutionary explanation for the difference is available, but an understanding of the molecular and cellular basis for the difference could be of value for the treatment of humans infected with HIV.

---

involved; the second is to find mutants affecting the phenotype and identify the genetic changes by **positional cloning**, without necessarily understanding the molecular basis of the phenotype at all.

There are striking morphological differences between the skeletons of apes and humans and these have received considerable attention, partly because bones, particularly crania and teeth, survive as fossils (Section 7.2). The morphology of modern humans differs significantly even from our ancestors 200 000 years ago. It is not immediately obvious how we can investigate the genetic basis of these characteristics, but a careful study comparing chimpanzee and human development and fossils (Lieberman, 1998) has suggested that many of them can be explained by a single underlying change: a reduction in the length of the **sphenoid** bone at the base of the cranium (Section 7.2.4). The genes that control development of the sphenoid are unknown, but provide a focus for investigating the cranial changes and a model for studying the genetic basis of other morphological changes.

An example of a phenotypic difference between humans and great apes at the biochemical level is the lowered plasma concentration of transthyretin (prealbumin) in humans (Gagneux *et al.*, 2001). This blood plasma protein regulates

thyroid hormone metabolism and correspondingly, consistent differences between human and chimp thyroid hormone status have also been observed. In addition, transthyretin has also been implicated in brain development.

The search for language genes

A phenotype of particular interest is language, but it is difficult to deduce what molecular changes underlie human language ability. An ideal way to investigate the genetics of language ability would thus be to identify families in which a language disorder gene is segregating and use positional cloning to identify the gene. Specific language impairment, where children have significant difficulties in acquiring language despite adequate intelligence and opportunity, is found in 2–5% of children. It has a strong genetic component, showing **concordance** in about 70% of **monozygotic** (identical) twins and about 45% of **dizygotic** (nonidentical) twins, leading to a heritability estimate of 0.45 (Folstein and Mankoski, 2000). However, most families segregating such disorders do not show a simple **Mendelian** pattern of inheritance.

One exceptional family, KE, is a large, three-generation pedigree described in 1990 (Hurst *et al.*, 1990) and consisting of 15 affected members, 16 unaffected members and six unaffected spouses. The pedigree indicates autosomal

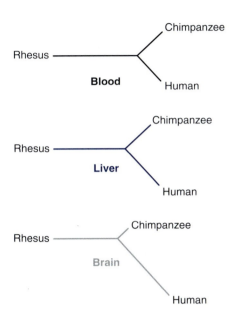

Figure 7.15: Comparison of gene expression patterns in humans, chimpanzees and rhesus macaques.

Expression levels of ~18 000 human genes were measured using RNA from different tissues (blood, liver or brain) from the three species, and a distance measure between the profiles was calculated. In blood and liver, humans and chimpanzee expression patterns are the most similar, as would be expected from their evolutionary relationship. However, in brain, chimpanzee and rhesus are the most similar, interpreted as indicating particularly pronounced changes in gene expression in the human brain (data from Enard *et al.*, 2002a).

dominant inheritance with full **penetrance**. There has been debate about the nature and specificity of the disorder. Some investigators have emphasized the specificity of the grammatical defect, citing findings such as the affected members' ability to distinguish correctly between pairs of phrases like 'He washes him'/'He washes himself' and 'The mother's baby'/'The baby's mother', but not to develop general grammatical rules such as that for creating plurals by adding an *s* to a word, so that they could not deduce that the plural of the imaginary animal the 'wug' would be 'wugs' (Gopnik, 1990). Others, however, have emphasized the complex nature of the phenotype: affected members have low IQs and their speech is often incomprehensible to naive listeners due, at least in part, to poor mobility of the lower face, including the mouth and upper lip (Vargha-Khadem *et al.*, 1998). The phenotype is thus better described as a combined speech and language disorder than a specific language disorder alone.

Linkage analysis in the KE family was used to localize the gene, designated *SPCH1*, to 7q31 within a critical region of 5.6 cM (Lai *et al.*, 2000). This region was > 6.1 Mb in length and contains 17 known genes. It is a considerable challenge

to determine whether one of these, or an unrecognized gene within this region, is responsible for the phenotype. Lai *et al.* therefore investigated a short-cut: a balanced translocation found in CS, a 5-year-old boy with a somewhat similar language impairment who, according to his mother, has never been able to sneeze or laugh spontaneously. The translocation involves chromosomes 5 and 7 and the breakpoint on chromosome 7 lies within the *SPCH1* critical region. It disrupts the gene *FOXP2*, which also contains a G to A transition leading to an arginine to histidine substitution in the KE family (*Figure 7.16*) (Lai *et al.*, 2001). This changes the conserved 'forkhead' domain of the protein and thus could inactivate it. The protein is predicted to be a transcription factor, and **haploinsufficiency** could lead to abnormal development of neural structures important for speech and language.

It might be expected that such a gene would have undergone selective changes during human evolution, and this possibility was investigated by comparing human, other primate and more distant mammalian sequences (Enard *et al.*, 2002b; Zhang *et al.*, 2002). The amino acid sequence is highly conserved and, apart from variation in the length of the polyglutamine tracts (*Figure 7.16*), only three amino acid differences were found between the human and mouse sequences. Two of these differences, however, had arisen on the human lineage after its divergence from other apes, and it was suggested that the N325S substitution might create a new phosphorylation site, but this residue is also found in carnivores such as cats and dogs, and so alone is unlikely to account for speech and language ability (Zhang *et al.*, 2002). Such a pattern of change is consistent with positive selection on the human lineage, but also with a relaxation of constraints. In order to distinguish between these possibilities, and investigate the timing of any selection, the variability within humans was examined by sequencing an ~ 14-kb region adjacent to the two amino acid changes in 20 geographically diverse individuals (Enard *et al.*, 2002b). The value of Tajima's *D* (Section 6.3.3) was negative and unusually low (−2.20): in a survey of 313 genes, only one lower value had been found (*Figure 9.8*). This reveals an excess of rare alleles. An excess of high-frequency derived alleles, measured by Fay and Wu's *H* (Section 6.3.3), was also found. Although a range of explanations for these statistics is possible, a selective sweep at the amino acid changes in the *FOXP2* locus provides a simple explanation for both of them. The excess of high-frequency derived alleles is expected to be transient: they will soon drift to fixation, and modeling in an expanding population suggested that such a sweep should have occurred within the last 200 KY. Similarly, a study of ~ 10 kb spanning the two amino acid changes in 10 individuals found a reduced diversity and negative value (−1.36) for Tajima's *D*, as expected after a sweep (Zhang *et al.*, 2002). This work provides a starting point for further investigations, such as an analysis of the phenotype associated with reversion of the two amino acid changes, and a model for identifying other genes that have been selected during recent human evolution.

TABLE 7.4: SELECTED EXAMPLES OF APE/HUMAN DIFFERENCES IN GENE STRUCTURE AND EXPRESSION.

Gene	Function	Change	Consequence
CMP-Neu5Ac hydroxylase	Cell surface sugar (Neu5Gc) synthesis	Deletion of exon	Loss of activity in humans
Phi hHaA	Type I hair keratin gene	Base substitution	Loss of gene product in humans
TCRG-V10	T-cell receptor	Base substitution in splice site	Loss of gene product in humans
Tropoelastin (ELN)	Component of elastic fibers	Deletion of exon 34	Activity retained; possible subtle change
OR 912–93	Olfactory receptor	Nonsense mutation	Loss of gene product, possible reduction in sense of smell
ProtocadherinY (PCDHY)	Cell–cell interaction in nervous system	Duplication X to Y	Duplicated + diverged gene in humans
HSPRY3	Growth factor signaling, but inactive on Y	Formation of Yq PAR2	Duplicated in humans
SYBL1	Transport in brain, but inactive on Y	Formation of Yq PAR2	Duplicated in humans
IL9R	T-cell lymphokine	Formation of Yq PAR2	Duplicated in humans
CXYorf1	Unknown	Formation of Yq PAR2	Duplicated in humans
Morpheus	Unknown	Duplication on chromosome 16p	Additional copies in humans

Summary

▶ Morphological evidence from fossils and extant species clearly shows that humans are a mammalian species, belong to the taxonomic order of primates and are most closely related to the chimpanzee, bonobo, gorilla and orangutan.

▶ The 5–8 million years of hominid evolution since humans and chimpanzees last shared a common ancestor was characterized firstly by a move towards bipedal locomotion, and subsequently by a dramatic increase in brain size.

▶ The karyotypes of humans and other apes are similar but differ in a number of respects, the most notable of which is the reduction in chromosomal number from 48 to 46 as a result of a chromosomal fusion on the hominid lineage.

▶ Molecular comparisons among ape species by DNA–DNA hybridization and phylogenetic analysis of DNA sequences reveal that humans and chimpanzees are more closely related to one another than either is to gorillas.

▶ The time since humans and chimpanzees last shared a common ancestor can be estimated from molecular comparisons between the two species, and these calculations can be reconciled with the fossil-derived estimates of the timing of the speciation event, although precision is hampered by difficulties in calibrating the molecular clock.

▶ There is substantial controversy over the number of ape species and subspecies, which is due in no small part to difficulties in defining what a species actually is.

▶ Genetic diversity in other ape species appears to be significantly greater than in humans despite the apes' much smaller, and decreasing, population sizes.

▶ Identifying the important genetic changes involved during hominid evolution can be investigated in one of two ways: defining important human-specific phenotypes (e.g., language) and researching their genetic bases, or identifying all genetic changes and seeing which are of functional importance.

▶ The search for genes involved in language has identified one gene, FOXP2, that when mutated causes a language impairment disorder in humans, and appears to have undergone a recent selective sweep (in the past 200 KY) due to positive selection for a new nonsynonymous variant.

BOX 7.11 Opinion: A theory of the speciation of modern *Homo sapiens*

This theory (Crow, 2002) – that a single gene played a critical role in the transition from a precursor species – is founded upon:

(i) the premise that hemispheric asymmetry is the defining feature of the human brain and the only plausible correlate of language;

(ii) an argument for a specific candidate region (the Xq21.3/Yp11.2 region of homology) based upon the reciprocal deficits associated with the sex chromosome aneuploidies, and the course of chromosomal change in hominid evolution. A gene (*protocadherinXY*) that has been identified within this region is expressed in the brain with the potential to account for a sex difference;

(iii) a particular evolutionary mechanism (sexual selection acting on an X-Y-linked gene) to account for species-specific modification of what initially was a saltational change (in this case a chromosomal rearrangement).

Cerebral asymmetry as the species-defining feature

Most humans have a preference to use the right hand for many activities, particularly those involving fine control; it has been argued that population-based directional asymmetry of handedness is specific to *Homo sapiens* (Annett, 1985). A review of the primate literature concludes that 'nonhuman primate hand function has not been shown to be lateralized at the species level – it is not the norm for any species, task or setting, and so offers no easy model for the evolution of human handedness' (Marchant and McGrew, 1996).

The contrast is illustrated by the comparison of hand usage for the everyday range of activities in chimpanzees and humans (see *Figure*).

Directional asymmetry is present in the human but absent in the great ape population: when can this discontinuity have arisen? Clearly sometime between the separation of the hominid and chimpanzee lineages, that is between approximately 5 MYA and 100 KYA (the minimal estimate for the origin of modern *Homo sapiens*). The change must have had a genetic basis.

The case for an X-Y homologous gene

Where is the gene? There are sex differences in verbal ability – females have an advantage over males. There is also a sex difference in degrees of handedness – females are more strongly right-handed than males and are less likely to be left-handed. It is plausible that these sex differences are related and that, in turn, both are related to the sex difference in brain growth – brain development is faster in females than in males.

The key to the genetics of asymmetry lies in the neuropsychological deficits associated with the sex chromosome aneuploidies. Individuals who lack an X chromosome – XO or Turner syndrome – have relative deficits of nondominant hemisphere capacity ('performance IQ'), whilst individuals with an extra X chromosome – XXY or

Klinefelter syndrome and XXX syndrome – have relative deficits of dominant hemisphere capacity ('verbal IQ'). Since XXY individuals are male and XXX individuals are female these effects cannot be attributed to gonadal hormones. Therefore a gene on the X chromosome influences the relative development of the hemispheres. The fact that deficits comparable to those in Turner syndrome are not present in normal males who, like Turner syndrome individuals, have only one X chromosome, indicates that the gene must be present also on the Y chromosome. This argument generates the hypothesis that the asymmetry factor belongs to the class of X-Y homologous genes.

The significance of the Xq21.3 translocation and the Yp paracentric inversion

Several regions of homology between X and Y chromosomes have arisen as a result of translocations to the Y chromosome from the X and the time course of these translocations can be charted in cross-species comparisons. The regions of greatest interest with respect to evolutionary developments in humans are those that have been subject to change between the chimpanzee and *Homo sapiens*. Two regions – the Xq21.3/Yp region of homology and the 0.4-Mb pseudoautosomal region (PAR2) at the telomeres of the long arms of the X and Y – stand out. Both representations on the Y chromosome were established after the separation of the hominid and chimpanzee lineages. Of these two regions a gene within Xq21.3/Yp more readily explains a sex difference on the basis that sequence divergence can take place in the absence of recombination within the sex-specific regions of the X and Y whereas this is not the case within PAR2.

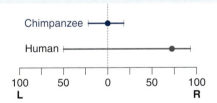

| | Number of | |
	individuals	activities
Chimpanzee	38	46
Human	1960	75

Hand preference in chimpanzees and humans compared.

Data for chimpanzees refer to a community of wild chimpanzees (*P.t. schweinfurthii*) observed in the Gombe National Park (Marchant and McGrew, 1996). Data for humans were collected by questionnaire from populations of undergraduate psychology students in Scotland and Australia (Provins *et al.*, 1982). In each case the data relate to a wide range of everyday activities. Medians (filled circles) and boundary values (horizontal bars) for 95% of the population have been extracted from graphs in the original publications.

The time course of the changes to these X and Y homologous regions is of great interest. The translocation from Xq21.3 to Yp has been estimated on the basis of X-Y sequence divergence at approximately 3 MYA. This was followed by a paracentric inversion, that has not been dated, that split and reversed the block in Yp. Within this distal block on the Y a gene has been located (*protocadherinXY*) that is also present in the Xq21.3 homologous region. Protocadherins are cell surface adhesion molecules that play a role in axonal guidance. The sequence of *protocadherinY* differs from that of *protocadherinX*; both forms are expressed in the brain. *ProtocadherinXY* is thus a gene that has changed in recent hominid evolution, being present on the X and Y in humans but only on the X in other extant primates, that has the potential to bring about changes in connectivity in the brain.

The fact that it is present with some sequence differences on both X and Y chromosomes indicates that it can account for differences between the sexes (e.g., in brain growth and asymmetry). These differences may be subject to sexual selection or differential mate choice, that according to some theories plays a role in speciation. The chromosomal changes that have influenced the form and mode of expression of *protocadherinXY* thus are candidates as speciation events in hominid evolution and the most recent of these, the paracentric inversion, may be directly involved in the origin of modern humans.

Timothy J. Crow

POWIC SANE Research Centre, Warneford Hospital, Oxford, UK

Further reading

Carroll SB (2003) Genetics and the making of *Homo sapiens*. *Nature*, **422**, 849–857

Klein RG (1999) *The Human Career: Human Biological and Cultural Origins*. University of Chicago Press, Chicago.

Electronic references

WWF International: http://www.livingplanet.org/

Ape Alliance: http://www.4apes.com/

Primate Cytogenetics Network: http://www.selu.com/bio/cyto/

Ape Genome Sequencing: http://sayer.lab.nig.ac.jp/~silver/index.html

References

Adams EJ, Cooper S, Thomson G, Parham P (2000) Common chimpanzees have greater diversity than humans at two of the three highly polymorphic MHC class I genes. *Immunogenetics* **51**, 410–424.

Andrews P, Cronin JE (1982) The relationships of Sivapithecus and Ramapithecus and the evolution of the orang-utan. *Nature* **297**, 541–546.

Andrle M, Fiedler W, Rett A, Ambros P, Schweizer D (1979) A case of trisomy 22 in *Pongo pygmaeus*. *Cytogenet. Cell Genet.* **24**, 1–6.

Angata T, Varki NM, Varki A (2001) A second uniquely human mutation affecting sialic acid biology. *J. Biol. Chem.* **276**, 40282–40287.

Annett M (1985) *Left, Right, Hand and Brain: The Right Shift Theory*. Lawrence Erlbaum, London.

Archidiacono N, Storlazzi CT, Spalluto C, Ricco AS, Marzella R, Rocchi M (1998) Evolution of chromosome Y in primates. *Chromosoma* **107**, 241–246.

(a) *FOXP2* gene structure

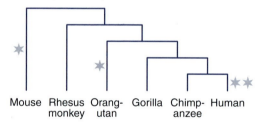

(b) *FOXP2* protein sequence evolution

Figure 7.16: Structure and evolution of the *FOXP2* gene implicated in speech and language development.

(a) Structure of the *FOXP2* transcript showing 17 exons (boxes); note that several variant transcripts resulting from alternative splice patterns are known, and the exon numbering can be different. The open reading frame may extend from exon 2 to 17 (light gray) or from exon 4 to 17 (dark gray). The locations of structural features within the protein, inactivating mutations (translocation, R553H), and evolutionary changes within the human lineage (T303N, N325S) are shown (based on Enard *et al.*, 2002b; Lai *et al.*, 2001).

(b) Amino acid changes during *FOXP2* evolution within mammals. Changes (stars) are located on the phylogeny, which represents the branching order but not the time-scale. Note that the protein is highly conserved: there are very few changes. Nevertheless, two of them have occurred during recent human evolution (data from Enard *et al.*, 2002b).

Arnason U, Gullberg A, Gretarsdottir S, Ursing B, Janke A (2000a) The mitochondrial genome of the sperm whale and a new molecular reference for estimating eutherian divergence dates. *J. Mol. Evol.* **50**, 569–578.

Arnason U, Gullberg A, Janke A (1998) Molecular timing of primate divergences as estimated by two nonprimate calibration points. *J. Mol. Evol.* **47**, 718–727.

Arnason U, Gullberg A, Schweizer Burguete A, Janke A (2000b) Molecular estimates of primate divergences and new hypotheses for primate dispersal and the origin of modern humans. *Hereditas* **133**, 217–228.

Bailey WJ, Fitch DHA, Tagle DA, Czelusniak J, Slightom JL, Goodman M (1991) Molecular evolution of the psi-eta-globin gene locus – gibbon phylogeny and the hominoid slowdown. *Mol. Biol. Evol.* **8**, 155–184.

Britten RJ (2002) Divergence between samples of chimpanzee and human DNA sequences is 5%, counting indels. *Proc. Natl Acad. Sci. USA* **99**, 13633–13635.

Chen FC, Li WH (2001) Genomic divergences between humans and other hominoids and the effective population size of the common ancestor of humans and chimpanzees. *Am. J. Hum. Genet.* **68**, 444–456.

Chou HH, Hayakawa T, Diaz S *et al.* (2002) Inactivation of CMP-N-acetylneuraminic acid hydroxylase occurred prior to brain expansion during human evolution. *Proc. Natl Acad. Sci. USA* **99**, 11736–11741.

Collard M, Wood B (2000) How reliable are human phylogenetic hypotheses? *Proc. Natl Acad. Sci. USA* **97**, 5003–5006.

Crow TJ (2002) Sexual selection, timing and an X-Y homologous gene: did *Homo sapiens* speciate on the Y chromosome? In: *The Speciation of Modern Homo sapiens*, (T Crow, ed.) *Proc. Brit. Acad.* **106**, 197–216.

de Groot NG, Otting N, Doxiadis GG *et al.* (2002) Evidence for an ancient selective sweep in the MHC class I gene repertoire of chimpanzees. *Proc. Natl Acad. Sci. USA* **99**, 11748–11753.

Darwin C (1871) *The Descent of Man, and Selection in Relation to Sex.* J. Murray, London.

Dutrillaux B (1979) Chromosomal evolution in primates: tentative phylogeny from *Microcebus murinus* (prosimian) to man. *Hum. Genet.* **48**, 251–314.

Dutrillaux B, Rumpler Y (1977) Chromosomal evolution in Malagasy lemurs. II. Meiosis in intra- and interspecific hybrids in the genus Lemur. *Cytogenet. Cell Genet.* **18**, 197–211.

Ebersberger I, Metzler D, Schwarz C, Pääbo S (2002) Genomewide comparison of DNA sequences between humans and chimpanzees. *Am. J. Hum. Genet.* **70**, 1490–1497.

Ely J, Deka R, Chakraborty R, Ferrell RE (1992) Comparison of five tandem repeat loci between humans and chimpanzees. *Genomics* **14**, 692–698.

Ely JJ, Leland M, Martino M, Swett W, Moore CM (1998) Technical note: chromosomal and mtDNA analysis of Oliver. *Am. J. Phys. Anthropol.* **105**, 395–403.

Enard W, Khaitovich P, Klose J *et al.* (2002a) Intra- and interspecific variation in primate gene expression patterns. *Science* **296**, 340–343.

Enard W, Przeworski M, Fisher SE *et al.* (2002b) Molecular evolution of *FOXP2*, a gene involved in speech and language. *Nature* **418**, 869–872.

Fitch WT, Reby D (2001) The descended larynx is not uniquely human. *Proc. Roy. Soc. Lond. B* **268**, 1669–1675.

Folstein SE, Mankoski RE (2000) Chromosome 7q: Where autism meets language disorder? *Am. J. Hum. Genet.* **67**, 278–281.

Freije D, Helms C, Watson MS, Donis-Keller H (1992) Identification of a second pseudoautosomal region near the Xq and Yq telomeres. *Science* **258**, 1784–1787.

Gagneux P, Wills C, Gerloff U *et al.* (1999) Mitochondrial sequences show diverse evolutionary histories of African hominoids. *Proc. Natl Acad. Sci. USA* **96**, 5077–5082.

Gagneux P, Amess B, Diaz S *et al.* (2001) Proteomic comparison of human and great ape blood plasma reveals conserved glycosylation and differences in thyroid hormone metabolism. *Am. J. Phys. Anthropol.* **115**, 99–109.

Gibbons A (2001) Studying humans – and their cousins and parasites. *Science* **292**, 627–629.

Gibbs S, Collard M, Wood B (2000) Soft-tissue characters in higher primate phylogenetics. *Proc. Natl Acad. Sci. USA* **97**, 11130–11132.

Gonder MK, Oates JF, Disotell TR, Forstner MRJ, Morales JC, Melnick DJ (1997) A new west African chimpanzee subspecies? *Nature* **388**, 337.

Goodman M (1999) The genomic record of humankind's evolutionary roots. *Am. J. Hum. Genet.* **64**, 31–39.

Gopnik M (1990) Feature-blind grammar and dysphagia. *Nature* **344**, 715.

Groves C (2001) Why taxonomic stability is a bad idea, or why are there so few species of primates (or are there?). *Evol. Anthropol.* **10**, 192–198.

Hey J (2001) The mind of the species problem. *Trends Ecol. Evol.* **16**, 326–329.

Hurst JA, Baraitser M, Auger E, Graham F, Norell S (1990) An extended family with a dominantly inherited speech disorder. *Devel. Med. Child Neurol.* **32**, 352–355.

Huxley TH (1863) *Zoological Evidences as to Man's Place in Nature.* Williams and Norgate, London.

Ijdo JW, Baldini A, Ward DC, Reeders ST, Wells RA (1991) Origin of human chromosome 2: an ancestral telomere–telomere fusion. *Proc. Natl Acad. Sci. USA* **88**, 9051–9055.

Irie A, Koyama S, Kozutsumi Y, Kawasaki T, Suzuki A (1998) The molecular basis for the absence of N-glycolylneuraminic acid in humans. *J. Biol. Chem.* **273**, 15866–15871.

Jauch A, Wienberg J, Stanyon R, Arnold N, Tofanelli S, Ishida T, Cremer T (1992) Reconstruction of genomic rearrangements in great apes and gibbons by chromosome painting. *Proc. Natl Acad. Sci. USA* **89**, 8611–8615.

Kaessmann H, Wiebe V, Pääbo S (1999) Extensive nuclear DNA sequence diversity among chimpanzees. *Science* **286**, 1159–1162.

Kaessmann H, Wiebe V, Weiss G, Pääbo S (2001) Great ape DNA sequences reveal a reduced diversity and an expansion in humans. *Nature Genet.* **27**, 155–156.

Kasai F, Takahashi E, Koyama K, *et al.* (2000) Comparative FISH mapping of the ancestral fusion point of human chromosome 2. *Chrom. Res.* **8**, 727–735.

Kemf E, Wilson A (1997) *Wanted Alive! Great Apes in the Wild.* WWF Species Status Report, Gland, Geneva.

King M (1993) *Species Evolution.* Cambridge University Press, Cambridge.

Lai CSL, Fisher SE, Hurst JA, *et al.* (2000) The *SPCH1* region on human 7q31: Genomic characterization of the critical interval and localization of translocations associated with speech and language disorder. *Am. J. Hum. Genet.* **67**, 357–368.

Lai CSL, Fisher SE, Hurst JA, Vargha-Khadem F, Monaco AP (2001) A forkhead-domain gene is mutated in a severe speech and language disorder. *Nature* **413**, 519–523.

Lercher MJ, Williams EJB, Hurst LD (2001) Local similarity in evolutionary rates extends over whole chromosomes in human–rodent and mouse–rat comparisons: implications for understanding the mechanistic basis of the male mutation bias. *Mol. Biol. Evol.* **18**, 2032–2039.

Lewin R (1988a) Conflict over DNA clock results. *Science* **241**, 1598–1600.

Lewin R (1988b) DNA clock conflict continues. *Science* **241**, 1756–1759.

Lieberman DE (1998) Sphenoid shortening and the evolution of modern human cranial shape. *Nature* **393**, 158–162.

Lin YH, McLenachan PA, Gore AR, Phillips MJ, Ota R, Hendy MD, Penny D (2002) Four new mitochondrial genomes and the increased stability of evolutionary trees of mammals from improved taxon sampling. *Mol. Biol. Evol.* **19**, 2060–2070.

Linnaeus C (1758) *Systema Naturae*, 10th edition.

Mallet J (2001) *Species, Concepts of; in: Encyclopedia of Biodiversity* (Levin SA, ed.), pp 427–440. Academic Press, New York.

Marchant LF, McGrew WC (1996) Laterality of limb function in wild chimpanzees of Gombe National Park: comprehensive study of spontaneous activities. *J. Hum. Evol.* **30**, 427–443.

Mayr E (1970) *Populations, Species and Evolution.* Harvard University Press, Cambridge, MA.

McClure HM, Belden KH, Pieper WA, Jacobson CB (1969) Autosomal trisomy in a chimpanzee: resemblance to Down's syndrome. *Science* **165**, 1010–1012.

Morin PA, Moore JJ, Chakraborty R, Jin L, Goodall J, Woodruff DS (1994) Kin selection, social structure, gene flow, and the evolution of chimpanzees. *Science* **265**, 1193–1201.

Müller S, Wienberg J (2001) "Bar-coding" primate chromosomes: molecular cytogenetic screening for the ancestral hominoid karyotype. *Hum. Genet.* **109**, 85–94.

Murphy WJ, Eizirik E, Johnson WE, Zhang YP, Ryder OA, O'Brien SJ (2001a) Molecular phylogenetics and the origins of placental mammals. *Nature* **409**, 614–618.

Murphy WJ, Eizirik E, O'Brien SJ *et al.* (2001b) Resolution of the early placental mammal radiation using Bayesian phylogenetics. *Science* **294**, 2348–2351.

Nickerson E, Nelson DL (1998) Molecular definition of pericentric inversion breakpoints occurring during the evolution of humans and chimpanzees. *Genomics* **50**, 368–372.

Nishihara H, Terai Y, Okada N (2002) Characterization of novel Alu- and tRNA-related SINES from the Tree Shrew and evolutionary implications of their origins. *Mol. Biol. Evol.* **19**, 1964–1972.

Olson MV, Varki A (2003) Sequencing the chimpanzee genome: insights into human evolution and disease. *Nature Rev. Genet.* **4**, 20–28.

Page DC, Harper ME, Love J, Botstein D (1984) Occurrence of a transposition from the X-chromosome long arm to the Y-chromosome short arm during human evolution. *Nature* **311**, 119–122.

Provins KA, Milner AD, Kerr P (1982) Asymmetry of manual preference and performance. *Percept. Motor Skills* **54**, 179–194.

Richmond BG, Strait DS (2000) Evidence that humans evolved from a knuckle-walking ancestor. *Nature* **404**, 382–385.

Rieseberg LH (2001) Chromosomal rearrangements and speciation. *Trends Ecol. Evol.* **16**, 351–358.

Royle NJ, Baird DM, Jeffreys AJ (1994) A subterminal satellite located adjacent to telomeres in chimpanzees is absent from the human genome. *Nature Genet.* **6**, 52–56.

Rubinsztein DC, Amos W, Leggo J, *et al.* (1995) Microsatellite evolution – evidence for directionality and variation in rate between species. *Nature Genet.* **10**, 337–343.

Ruvolo M (1997) Molecular phylogeny of the hominoids: Inferences from multiple independent DNA sequence data sets. *Mol. Biol. Evol.* **14**, 248–265.

Sarich VM, Wilson AC (1967) Immunological time-scale for hominid evolution. *Science* **158**, 1200–1203.

Sibley CG, Ahlquist JE (1984) The phylogeny of the hominid primates, as indicated by DNA–DNA hybridisation. *J. Mol. Evol.* **20**, 2–15.

Sibley CG, Ahlquist JE (1987) DNA hybridisation evidence of hominoid phylogeny: results from an expanded data set. *J. Mol. Evol.* **26**, 99–121.

Stone AC, Griffiths RC, Zegura SL, Hammer MF (2002) High levels of Y-chromosome nucleotide diversity in the genus Pan. *Proc. Natl Acad. Sci. USA* **99**, 43–48.

Sumiyama K, Kitano T, Noda R, Ferrell RE, Saitou N (2000) Gene diversity of chimpanzee ABO blood group genes elucidated from exon 7 sequences. *Gene* **259**, 75–79.

Tavaré S, Marshall CR, Will O, Soligo C, Martin RD (2002) Using the fossil record to estimate the age of the last common ancestor of extant primates. *Nature* **416**, 726–729.

Thomson R, Pritchard JK, Shen P, Oefner PJ, Feldman MW (2000) Recent common ancestry of human Y chromosomes: Evidence from DNA sequence data. *Proc. Natl Acad. Sci. USA* **97**, 7360–7365.

Turleau C, de Grouchy J, Klein M (1972) Phylogénie chromosomique de l'homme et des primates hominiens (*Pan troglodytes*, *Gorilla gorilla* et *Pongo pygmaeus*): essai de reconstitution du caryotype de l'ancetre commun. *Ann. Génét.* **15**, 225–240.

Uchida A (1996) What we don't know about great ape variation. *Trends Ecol. Evol.* **11**, 163–168.

Vargha-Khadem F, Watkins KE, Price CJ, *et al.* (1998) Neural basis of an inherited speech and language disorder. *Proc. Natl Acad. Sci. USA* **95**, 12695–12700.

Varki A (2000) A chimpanzee genome project is a biomedical imperative. *Genome Res.* **10**, 1065–1070.

Vervaecke H, Van Elsacker L (1992) Hybrids between common chimpanzees (*Pan troglodytes*) and pygmy chimpanzees (*Pan paniscus*) in captivity. *Mammalia* **56**, 667–669.

Walsh PD, Abernethy KA, Bermejo M, Beyersk R, De Wachter P, Akou ME, *et al.* (2003) Catastrophic ape decline in western equatorial Africa. *Nature* **422**, 611–614.

Wimmer R, Kirsch S, Rappold GA, Schempp W (2002) Direct evidence for the Homo-Pan clade. *Chrom. Res.* **10**, 55–61.

Wise CA, Sraml M, Rubinsztein DC, Easteal S (1997) Comparative nuclear and mitochondrial genome diversity in humans and chimpanzees. *Mol. Biol. Evol.* **14**, 707–716.

Yi S, Ellsworth DL, Li W-H (2002) Slow molecular clocks in Old World Monkeys, Apes and Humans. *Mol. Biol. Evol.* **19**, 2191–2198.

Yoder AD, Yang ZH (2000) Estimation of primate speciation dates using local molecular clocks. *Mol. Biol. Evol.* **17**, 1081–1090.

Yunis JJ, Prakash O (1982) The origin of man: a chromosomal pictorial legacy. *Science* **215**, 1525–1530.

Zhang J, Webb DM, Podlaha O (2002) Accelerated protein evolution and origins of human-specific features: *FOXP2* as an example. *Genetics* **162**, 1825–1835.

Zhi L, Karesh WB, Janczewski DN, *et al.* (1996) Genomic differentiation among natural populations of orang-utan (*Pongo pygmaeus*). *Curr. Biol.* **6**, 1326–1336.

CHAPTER EIGHT

Origins of modern humans

CHAPTER CONTENTS

BOXES

8.1 Introduction

This chapter presents our current understanding of where and when modern humans arose. In the last chapter, we saw that the split between humans and our closest living relatives, chimpanzees and **bonobos**, occurred at least 5 MYA and perhaps several million years before this. Here, we will consider events subsequent to that split, leading to the origin of our species, modern *Homo sapiens*, a much more recent event. There can be ambiguity about the meaning of the term 'human': we know that we are human, and that chimpanzees are not, but where in the continuum of evolutionary ancestors that link us should we draw the line between human and nonhuman? Any decision is a matter of opinion; here, we will draw it within the genus *Homo*, and will sometimes refer to 'archaic' and 'modern' humans where clarification is needed; but we will also use the phrase the 'human lineage', meaning the lineage that *led to* humans, and includes all species on the human line since the ape/human split.

We differ from our last common ancestor with other apes in several respects.

▸ Morphology: the structure of our bodies, including our brains.

▸ Behavior: from the way we walk to our social organization and language.

▸ Genetics: many neutral and some selected changes.

Individual changes could have occurred independently, or in packages linked by selection or drift. Two kinds of evidence are important for understanding the likely times and places of the changes.

▸ Fossils and archaeology, which provide information about the environment, morphology and, to some extent, the behavior of our ancestors and the related species present at different times.

▸ Genetics, which reveals the history of the lineages that have survived in modern humans.

Vincent Sarich famously contrasted the two by remarking *'I know my molecules had ancestors, the paleontologist can only hope that his fossils had descendants'*.

8.2 Evidence from fossils

Extinct and extant species more closely related to humans than to chimpanzees are known as **hominids**, although the alternative term **hominin** is sometimes used. Candidate fossils can therefore be examined for characteristics that differ between humans and apes (*Box 8.1*) and their likely position in the phylogeny determined. This aim is hindered by the rarity of fossil hominids: our ancestors appear to have existed at low population densities and were seldom fossilized; and also by the incompleteness of the fossils that are found: teeth are the most frequent finds, then the relatively tough bones of the head, the **cranium** (skull) and **mandible** (lower jaw), but other bones (often collectively called 'post-cranial') are rarely preserved, and soft body tissues never. In addition, there are conceptual problems: there would undoubtedly have been variation within each species due to differences between individuals within the same population including **sexual dimorphism**, and also geographical differences, as well as changes over time. A decision has to be made about which fossils should be grouped under the same name (i.e., genus and species), and this inevitably has an arbitrary element. The person who discovers and names a new hominid species gains considerable credit and even celebrity, especially if it is claimed to be a direct human ancestor. There is thus a danger of excessive 'splitting' and over-emphasis of human similarities at the expense of the ape-like features. The plethora of names can be very confusing and the reader should be aware that the field is subject to continuous revision; opinions about the number of genera and species, and their relationships, differ considerably between experts, and change over time.

8.2.1 The earliest hominids

We saw in Chapter 7 that the human and chimpanzee lineages are generally assumed to have diverged about 5–7 MYA. The earliest hominid fossils should therefore originate from this period, but not before (see *Box 8.2* for a summary of physical dating methods). For many years very few candidate fossils of this age were known, but recently several have been described (*Figures 8.1* and *8.2*). The oldest of these consists of a cranium ('Toumai' meaning 'hope of life'), jaw fragment and teeth from Chad, Africa, designated

---

BOX 8.1 Human characteristics that can be preserved in stone

▸ Brain cavity: absolutely and relatively larger in humans compared with other apes; attachment to spinal cord is more centrally located in the base of the skull.

▸ Teeth: enamel is thicker; canines and incisors are smaller in humans than other apes.

▸ Chest: more cylindrical in humans; that of apes widens towards the base to accommodate larger gut.

▸ Legs longer, feet and pelvis adapted for upright walking in humans.

▸ Bipedalism can also be recognized from tracks of preserved footprints.

▸ Hands adapted for grasping with longer thumb in humans.

▸ Slow development and prolonged childhood in humans, identified from the 'age at death' of fossils, measured e.g., from growth lines on teeth.

▸ Tool use more extensive in humans; complex tools and fire specific to humans.

Sahelanthropus tchadensis (Brunet *et al.*, 2002). Direct dating of the fossil deposits was not possible, but the associated vertebrate fauna suggested a date of 6–7 MYA. It is worth noting that if *S. tchadensis* is truly on the hominid lineage and as old as 7 MY, these findings can barely be reconciled with a 5–7-MYA chimpanzee/human split, and emphasize the importance of critical evaluation of the molecular dating data (see *Boxes 7.6* and *7.7*). Despite having a chimpanzee-sized brain, *S. tchadensis* was assigned to the hominid lineage on the basis of a number of features including the relatively flat face, position of attachment of the spinal cord and intermediate tooth enamel thickness. Not everyone agrees with this assessment, and one researcher has suggested *'This is the skull of a female gorilla'* (Senut, 2002). Unfortunately, an understanding of its mode of locomotion awaits discovery of post-cranial remains. The environment appears to have had a mosaic structure including forest, savanna, desert and lakes.

Two genera dating to 5–6 MYA have been described: *Orronin* and *Ardipithecus*. *Orronin tugenensis*, also called 'Millennium man' is represented by 13 fossils dating to about 5.8–6.1 MYA from the Tugen Hills in Kenya, East Africa (Senut *et al.*, 2001). These fossils include a fragmentary thigh-bone which indicated upright walking, and small thick-enameled molars judged to link *Orronin* to the human lineage; these authors place *Ardipithecus* on a separate lineage leading to chimpanzees (Aiello and Collard, 2001). *Ardipithecus ramidus kadabba* is represented by 11 fossils, including a nearly complete foot, from the Middle Awash in Ethiopia, Africa, dated to between 5.2 and 5.8 MYA (Haile-Selassie, 2001). Despite the presence of thin enamel,

Haile-Selassie considers *Ardipithecus* to lie on the human lineage and the affinity of *Orronin* to be uncertain, but perhaps to have no descendants or represent a chimpanzee ancestor. Others have suggested that the two sets of fossils may be derived from the same creature (Gibbons, 2002). A more thorough comparison of all of these early fossils is needed.

Few fossils dating between 5 MYA and 4 MYA are available, but a small collection from the Middle Awash in Ethiopia dating to about 4.4 MYA, initially called *Australopithecus ramidus* (White *et al.*, 1994) was subsequently designated *Ardipithecus* in a corrigendum (an error, usually implying a printer's error) to the original article (White *et al.*, 1995), and is currently known as *Ardipithecus ramidus ramidus*. Analysis of fragments comprising approximately half of a single skeleton await publication (Gibbons, 2002) and should provide a better understanding of this species.

Where the appropriate features can be identified, all of these early hominid fossils appear to represent chimpanzee-sized, upright-walking species that lived in or near a wooded environment. The evidence for **bipedalism** at an early date is an important finding and raises the question of how the human–chimpanzee common ancestor moved around. The most radical possibility is that bipedalism is the primitive trait and **knuckle-walking** in chimpanzees is derived, although this possibility requires the nonparsimonious assumption that knuckle-walking arose independently in gorillas. The difficulty of distinguishing between human and chimpanzee lineages at times close to their split should not be a surprise.

BOX 8.2 Physical dating methods for fossil sites

Method	Basis	Type[a]	Useful time-span (KYA)	Materials used
Isotopic:	Radioactive decay	Absolute		
Uranium series			1–350	Carbonates
^{40}K-^{40}Ar			> 10	Volcanic rocks
^{14}C			< 40	Organic
Trapped electron dating: Thermoluminescence (TL)	Electrons caught in defects in crystal lattices	Relative	1–200	Quartz, ceramics,
Optically Stimulated Luminescence (OSL)			1–500	quartz, zircon
Electron Spin Resonance (ESR)			1–500	Carbonates, silicates
Amino acid racemization	Chemical instability of natural L amino acids	Relative	40–200	Organic materials
Paleomagnetism	Direction of magnetic field (N or S)	Binary	All	Iron-rich rocks
Paleontology	Comparison of fossils with dated sites	Relative	All	Fossils

[a]Absolute, provides a date in years; relative, allows comparisons with other sites or materials; binary, two possible findings, which allow refinement of an approximate date determined by other methods.

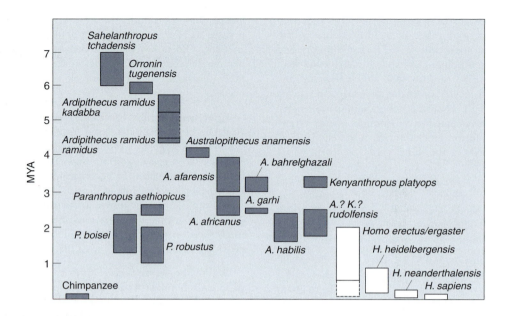

Figure 8.1: Fossil hominids.

The time span of each species indicates either the uncertainty in dating or the times of the earliest and latest fossils, whichever is larger. Dotted lines indicate either a lack of intermediate *Ardipithecus ramidus* fossils, or particular uncertainty about the later dates for *Homo erectus*. Dark blue: found only in Africa. White: found in Africa and elsewhere, or only outside Africa. Many aspects of the classification of these fossils are still debated and are likely to be revised.

8.2.2 Australopithecines and their contemporaries

Most fossils dating after about 4.2 MYA and before the appearance of *Homo* are ascribed to the genus *Australopithecus*, although, as noted, even *Ardipithecus ramidus* has been included in *Australopithecus*. The earliest species now commonly placed in the genus *Australopithecus* is *anamensis* from Lake Turkana and other sites in Kenya at around 3.9–4.2 MYA (Leakey *et al.*, 1995). This species is distinguished from the better-known *Australopithecus afarensis* (Johanson and White, 1979) mainly by the larger size of the males (estimated as ~ 55 kg vs. 45 kg); the females of the two

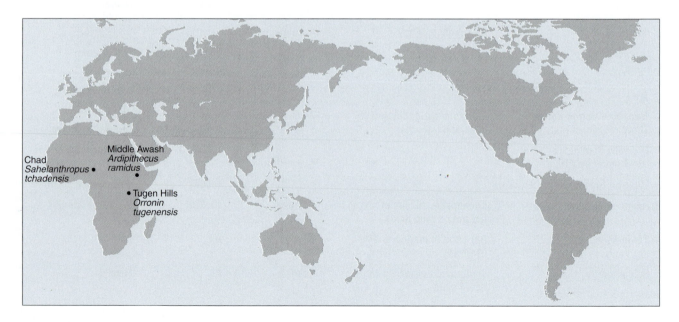

Figure 8.2: Sites of the earliest hominid fossils (dating to 7–4.5 MYA) in Chad and East Africa.

species are similar. *A. afarensis* (ca. 3.0–3.9 MYA) is present at a number of sites in East Africa from Ethiopia to Tanzania, and includes the famous partial skeleton 'Lucy' (3.2 MY; *Figure 8.3*) and the Laetoli footprints (3.5 MY) (Leakey and Hay, 1979), dramatically illustrating its bipedal locomotion. The name *A. bahrelghazali* (3.0–3.5 MY) has been given to fossils from Chad (Brunet *et al.*, 1995), 2500 km to the west of the Eastern Rift Valley, which are morphologically similar to *A. afarensis* and may represent a regional variant of that species. *Australopithecus africanus*, the first member of the genus to be named (the 'Taung child') (Dart, 1925), is also known from a number of sites, all from the south of the continent, and most of these fossils date to between 2.4 and 2.9 MYA. In East Africa, later *Australopithecus* is represented by fossils from the Middle Awash in Ethiopia designated *A. garhi* (~ 2.5 MYA) (Asfaw *et al.*, 1999), characterized by relatively large teeth. Thus **gracile** (lightly built) Australopithecines were present in many areas of Africa from around 4 MYA to less than 2.5 MYA (*Figure 8.4*). The relationships between the five species mentioned here are unclear: the simplest scheme would consider *afarensis* and *bahrelghazali* geographical variants, and similarly *africanus* and *garhi* geographical variants at a later date; *afarensis/ bahrelghazali* would be descendants of *anamensis*, and *africanus/garhi* of *afarensis/bahrelghazali*. The fossil material available from *A. afarensis* is extensive enough to allow many of its characteristics to be deduced. The species is estimated to have been 1–1.5 m tall (and, as mentioned, bipedal); weight was between 25 and 50 kg, with considerable dimorphism between the sexes. Brain size was 400–500 cc: similar, in proportion to body mass, to that of the chimpanzee. The habitat is thought to have been more open than that inhabited by the earlier hominids, perhaps with grassland as well as trees. *A. africanus* was probably a similar species, although with less sexual dimorphism.

Fossils from Lake Turkana in Kenya also dating to this period (3.2–3.5 MYA) show a distinct combination of primitive and derived features and have been placed in a separate genus and species, *Kenyanthropus platyops* (Leakey *et al.*, 2001). Similarities between these and the larger-brained later fossils represented by KNM-ER-1470 (Leakey, 1973) (~ 1.8 MY) from Koobi Fora in Kenya, previously designated *Homo* or *Australopithecus rudolfensis*, have led to *rudolfensis* being assigned to *Kenyanthropus*; but while these specific names are widely used, their generic affiliations remain controversial.

Robust (heavily built) hominids with small brains and large jaws and chewing teeth were originally included in *Australopithecus* but now are commonly placed in a separate genus, *Paranthropus*. *P. aethiopicus* is represented by only a small number of specimens, but these include the 'Black skull', a fairly complete ~ 2.5-MY-old skull from Lake Turkana. *Paranthropus boisei* fossils, including the skull 'Zinj' from Olduvai Gorge, Tanzania, Africa (Leakey, 1959), are found mostly in Ethiopia, Tanzania, and Kenya and span the range 1.2–2.3 MYA. *Paranthropus robustus* remains are known from several sites in South Africa (Swartkrans, Dreimulen, and Kromdraai), and are not reliably dated, but probably lie

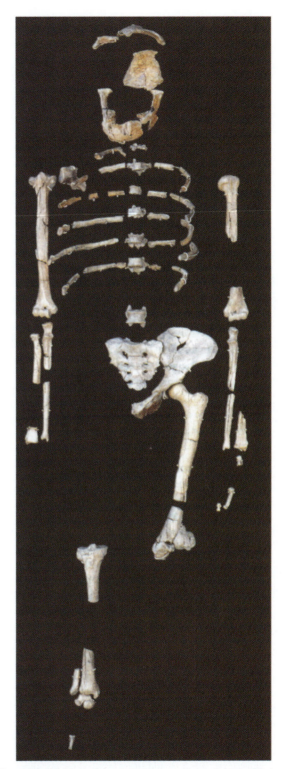

Figure 8.3: Skeleton of *Australopithecus afarensis* (A. L. 288–1 or 'Lucy').

A. afarensis lived in East Africa between 3 and 4 MYA and Lucy, dating to ~ 3.2 MYA and named after the Beatles' song 'Lucy in the sky with diamonds', is its best-known representative. Lucy was a mature adult female when she died, but was only just over 1 m in height. Reproduced with permission from the American Museum of Natural History.

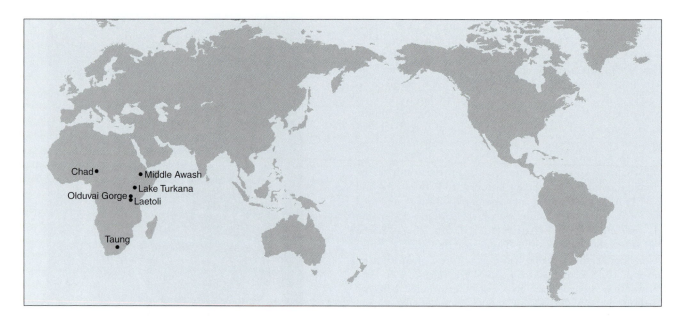

Figure 8.4: Sites of gracile Australopithecine fossils (dating to 4–2.4 MYA), spread over much of Africa but absent from the rest of the world.

in the 1.0–2.0 MYA range. The robust morphology of these species is thought to represent an adaptation to a diet that required heavy chewing, such as low-quality fibrous vegetable food, e.g., roots and nuts.

The question as to which of these hominid species is our direct ancestor has attracted considerable attention. It is widely agreed that the robust Australopithecines form a separate lineage with no surviving descendants, and few would place *A. africanus* on the human line. *A. anamensis* and *A. afarensis* are good candidates for human ancestors before 3 MYA, but the position of the large-brained *rudolfensis* is unclear and there seems to be no consensus about which fossils represent our ancestors between 3 MYA and the emergence of *Homo*.

8.2.3 The genus *Homo*

The reader will not be surprised to learn that there are disagreements about which species should be included within our own genus, *Homo*, and thus about its origin. For many years, the earliest member of the genus was considered to be *H. habilis*, 'handy man', on the basis of a partial skull and jaw, OH 7, from Olduvai Gorge, Tanzania (Leakey *et al.*, 1964); some *habilis* specimens may date back to about 2.5 MYA. However, this species has subsequently been described as '*a mishmash of traits and specimens, whose composition depends upon what researcher one asks*'; in addition, *habilis* does not show the body size and shape, or small teeth, characteristic of humans, while later species do. These features appear shortly after 2 MYA in fossils described as *Homo ergaster* or *erectus*, and it therefore seems reasonable to draw the distinction between *Australopithecus* and *Homo* here (Wood, 1996; Wood and Collard, 1999); thus *habilis* would be assigned to *Australopithecus* and *erectus/ergaster* would be the first *Homo*.

We will therefore refer to '*A. habilis*' in the following sections. The clumsy term '*erectus/ergaster*' is used because there is also debate about the distinctiveness of these two species: it is difficult to find morphological characteristics that separate them reliably. While one view considers *ergaster* to cover African individuals and reserves *erectus* for those found outside Africa, an alternative analysis would include all these specimens as a single widespread and variable species, *erectus*. The latter view is somewhat strengthened by the finding of a ~ 1.0-MY-old specimen resembling Asian *erectus* in Africa (Asfaw *et al.*, 2002). Here, *erectus* will be used for this whole group of fossils. The first ones date to 1.8 or 1.9 MYA and, like all earlier hominids, are found in Africa, demonstrating an African origin for our genus. *H. erectus* fossils include the outstanding Nariokotome Boy (~ 1.6 MY; *Figure 8.5*) (Walker and Leakey, 1993), the most complete early hominid ever found, which provides important insights into this species. He is thought to have been in early adolescence when he died, male, tall and thin at about 1.5 m tall and 47 kg in weight. As an adult he would probably have reached 1.8 m and 68 kg, common figures for modern humans. His limb proportions and tooth size were also similar to modern humans, but his brain size (880 cc, corresponding to 909 cc at maturity) was significantly smaller than the modern human average (around 1450–1500 cc), although just within the modern human range (830–2300 cc).

H. erectus is the earliest hominid to be found outside Africa (*Figure 8.6*). Fossils found in the late nineteenth and early twentieth centuries, 'Java Man' (Indonesia, the site of the **type specimen** 'Trinil 2' discovered in 1891) and 'Peking Man' (China) demonstrate the presence of *erectus* in East Asia at dates that may be as early as 1.8 MYA (Swisher *et al.*, 1994), and more recent discoveries that include Dmanisi

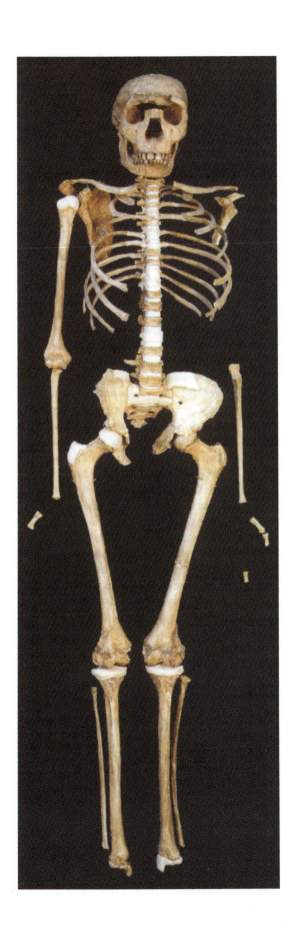

(Georgia) (Gabunia and Vekua, 1995) are similarly dated to 1.6–1.8 MYA. These dates are little younger than the earliest African *H. erectus*; it has been suggested that the large body size providing tolerance to heat stress and dehydration, coupled with improved stone tool kits, may have allowed the species to live in a wide range of environments and thus expand out of Africa (Wood, 1996). In Java, *H. erectus* may have survived until 27 KYA (Swisher *et al.*, 1996); if so, they would have been contemporaries of modern humans.

Later *Homo* from Africa and Europe (*Figure 8.7*) are less robust and have larger brains (~ 1200 cc instead of ~ 900 cc) and are often designated *H. heidelbergensis*, the type specimen of which is the 500-KY-old 'Heidelberg jaw' from Germany (*Figure 8.8*). Related specimens include the massive 'Bodo' cranium (Ethiopia, ~ 600 KY), the tibia (lower leg bone) from Boxgrove (England, ~ 500 KY) and the 'Petralona 1' cranium (Greece, age uncertain but perhaps 200–400 KY), while human footprints in volcanic ash on the Roccamonfina volcano in southern Italy date to 385–325 KYA (Mietto *et al.*, 2003) and could be associated with small individuals, perhaps children, belonging to this species. Many would also place the 780-KY-old specimens from Atapuerca, Spain, designated *Homo antecessor* by their discoverers (Bermudez de Castro *et al.*, 1997), within *heidelbergensis*. According to this view, *heidelbergensis* would have been a widespread and somewhat variable species, perhaps originating from *erectus* in Africa soon after 1 MYA and giving rise to more recent *Homo* species, including *H. sapiens*.

Neanderthals (*Figure 8.9*) form a reasonably distinct group of fossils from Europe and Western Asia between ~ 250 KYA and ~ 28 KYA, relatively robust with large brains (~ 1400 cc, larger than many modern humans) and well-developed brow ridges. Well-known examples include the type specimen of uncertain date from the Neander valley in Germany, the 'old man of La Chapelle-aux-Saints' from France (a 40–50-year-old individual showing evidence of arthritis; ~ 50 KY), 'Kebara 2' (Israel, ~ 60 KY) and Shanidar 4 from Iraq (~ 60 KY), sometimes interpreted as representing a deliberate burial. Neanderthals are thought to be descendants of *H. heidelbergensis* and are usually assigned to a distinct species *Homo neanderthalensis*, but their relationship to modern humans has aroused intense debate and is considered further below.

Needless to say, these classifications are not universally accepted. While the terms '*Homo erectus*' and modern '*Homo sapiens*' are seldom disputed, all intermediate species are sometimes referred to collectively as 'archaic *sapiens*'.

Figure 8.5: Skeleton of *Homo erectus* (WT 15000 or the 'Nariokotome boy').

H. erectus (African specimens are sometimes called *H. ergaster*) is known from Africa at ~ 1.9 MYA and from Asia soon afterwards (~ 1.8 MYA), possibly surviving there until 27 KYA. This specimen, from Lake Turkana, Kenya, dates to about 1.6 MYA. The Nariokotome boy was an adolescent male when he died, with the body size and shape of modern humans but a smaller brain. Reproduced with permission from the American Museum of Natural History.

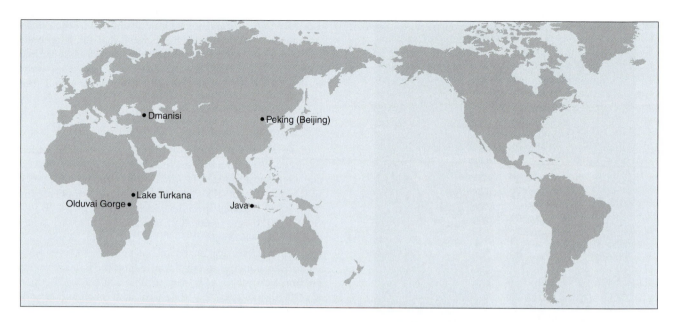

Figure 8.6: Sites of early *Homo* fossils (dating to 2.0–1.6 MYA), including sites outside Africa.

8.2.4 Anatomically modern humans

The origin of modern humans has probably been the most contentious issue in the field over the last 20 years. We will begin by considering modern human *morphology* and ask where and when this first appeared. Anatomically modern humans (often abbreviated AMH) differ from earlier humans ('archaic humans' or 'archaic *sapiens*'), but these differences

are not easy to define: indeed, it is often pointed out that there is no type specimen for *Homo sapiens*. Paleontologists have focused on cranial features, which can be summarized by two characteristics, derived from a comparison of 100 recent humans, 10 fossils classified as anatomically modern, and nine classified as Neanderthals or *heidelbergensis* (Lieberman *et al.*, 2002):

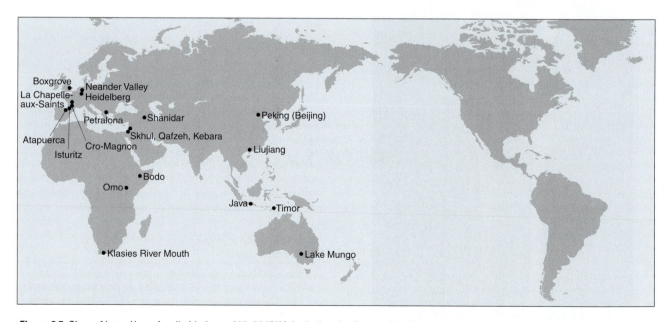

Figure 8.7: Sites of later *Homo* fossils (dating to 800–30 KYA), including the first modern humans.

Sites are found in Africa, Asia, Australia and Europe, but not in the Americas.

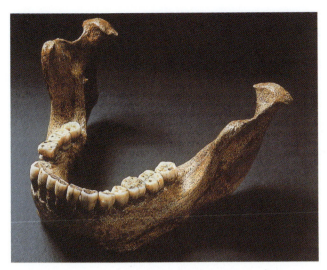

Figure 8.8: *Homo heidelbergensis* mandible (Mauer 1 or the 'Heidelberg jaw').

Opinions differ about which African and European fossils dating from ~ 800–200 KYA should be ascribed to *H. heidelbergensis*, but this mandible is the type specimen and so must belong to *heidelbergensis*. It was found near Heidelberg, Germany, and dates to approximately 300–500 KYA. See also the cover illustration. Reproduced with permission of the Science Photo Library.

▶ extent of the globular shape of the skull;

▶ degree of retraction of the face.

This system allows a clear distinction between AMH and archaic humans, with zero overlap, but has the disadvantage that relatively complete specimens are needed; for fragmentary specimens it is necessary to use less reliable criteria. For an alternative view, see *Box 8.3*.

Fossil crania of one child and two adults from Herto (Ethiopia) date to 154–160 KYA and show many of the features of modern human morphology (White *et al.*, 2003). The most complete is large (1450 cc) and has the globular braincase of modern humans, but retains some more archaic features such as protruding brows. Interestingly, all show evidence of *post mortem* modification, including cut marks, interpreted as resulting from mortuary practices. The authors describe them as '*on the verge of anatomical modernity but not yet fully modern*' and create a new subspecies *Homo sapiens idaltu* to accommodate them. Fully modern human fossils are known from Omo-Kibish (Ethiopia *Figure 8.10*) with a suggested date of 130 KY determined from associated shells, but perhaps younger than this; fragmentary specimens from Klasies River Mouth in South Africa at 90–120 KY; and two sites from Israel dated to 90–100 KY, the cave at Qafzeh with parts of more than 20 skeletons and the rockshelter at Skhul

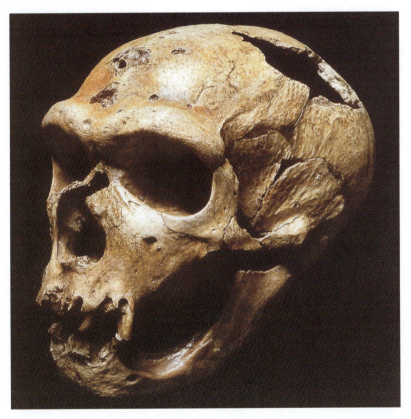

Figure 8.9: *Homo neanderthalensis* skull (the old man of La Chapelle-aux-Saints).

H. neanderthalensis lived in Europe and Western Asia from ~ 250 to 28 KYA. This specimen from France dates to ~ 50 KYA and was derived from a 40–50-year-old man. Note the large brow ridges and small chin. Several pathological features, including arthritis and resorption of the tooth sockets, are also present and contributed to the misinterpretation of Neanderthals as shuffling, brutish cavemen. Reproduced with permission of the Science Photo Library.

BOX 8.3 Opinion: Multiregional origins of modern humans

Did modern humanity first appear recently, in one place, like the speciation at the time of hominid origins when the hominid line parted from our last common ancestor with chimpanzees? Or is it the culmination of many small changes that happened at different times and in different places? Key to answering this question is another: can 'modern human' be defined? This is more than a semantic issue. If modern humanity had a single recent origin, and modern people replaced archaic ones, there should be such a definition because of the recent origin and the shared characteristics that promoted the replacement of the archaic hominids. But in fact there is no unique anatomical or behavioral definition of 'modern human.' Various attempted definitions either include some archaic humans such as Neanderthals as modern, or exclude some members of living populations. Something is wrong.

Multiregionalists believe that what is wrong is the expectation that modern humans can be defined in some unique anatomical or behavioral way. They find that there is fossil evidence supporting the interpretation that modernity was approached over a long time period as successful new features and behaviors appeared in different places and were able to spread across the human range because people migrated and exchanged genes (see *Figure*). If true, this would provide a clear refutation of the alternative (Eve Replacement) theory that there was a singular event of modernization, a **Rubicon** that was crossed by a single population, perhaps a new species. The fossil evidence for differing morphological continuities in various regions and the absence of significant worldwide discontinuities suggests that there was no worldwide replacement of archaic human populations by a 'modern human' species, and no Rubicon was crossed by the widespread populations of humanity.

Instead, the continued contacts between populations, through migrations and mate exchanges, created a network of genetic interchanges that linked even the most far-flung peoples and allowed them to share new alleles (or allele combinations) and behaviors that were adaptive. Modernity describes the way we are now, and the process that brought us here, not an event that happened at a specific time in the past to distinguish modern humans from other kinds of humans.

Multiregionalists consider the human species to be much older than modernity; it includes 'archaic' humans as well as living ones, and is defined by branch points (speciations) and extinctions, not by anatomical differences. The last recognizable branch point was some 2 MYA, at a time when there were dramatic changes in body and brain size (and other anatomy) – a time that is compatible with the coalescence estimates for nuclear genes from recombining regions (implying a population size bottleneck). The pattern of diversity that developed for the human species since then is unique among primates. Modern humanity is very widespread; yet, these far-flung populations are each both

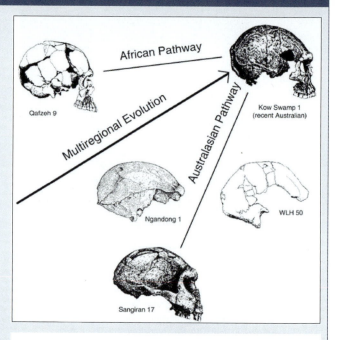

A graphic depiction of the reticular pattern for multiregional evolution.

In this case, the recent Native Australian from Kow Swamp could be a unique descendant of earlier African 'modern humans' such as from the Qafzeh site in Israel (Eve Replacement Theory), or it could be the unique descendant of the earlier Australasians from Indonesia such as Sangiran 17, through intermediates such as the later Indonesians from Ngandong, or Willandra Lake Australians (Candelabra Theory). However, the multiregional explanation is that *both* regional sources are significant in understanding the ancestry in this, and all other parts of the world. It is not clear where any boundary between archaic and modern humans might lie.

internally variable and genetically similar; it is said that differences between the most disparate human populations amount to less than those between adjacent populations of a single frog species! One of the first behavioral changes for these new humans was when they became colonizers and soon came to occupy a vast range of ecological niches around the world. Most geographically disperse mammals occupy an ecological niche that is broadly distributed, but the human pattern is of a widespread single species with many *different* ecological niches. The difference is culture, the unique human capacity to transform experience into symbols for information storage and transmission and complex communications. Culture allows unprecedented behavioral flexibility, critical in allowing human populations to occupy so many diverse niches. The capacity for complex culture is what made us modern.

Multiregionists explain this pattern of diversity as the result of gene flow and natural selection. Genic exchanges between populations, migrations and gene flow across a worldwide network of relationships, let most adaptively advantageous genes, or gene combinations, spread to all groups. Selection, and not recent origin, accounts for the unique similarities across the human species. Differences in selection, of course, also account for some populational variation, but three other factors can also play significant roles in the Multiregional explanation for differences: older genetic variation reflecting differences in ancestry, genetic drift for nonadaptive variation, and isolation by distance. Thus, while the Eve Replacement Theory implies that differences between populations are adaptive, Multiregional evolution implies that many of these differences may result from drift, or in cases of **exaptation**, historical accident, and it is the unique population similarities that are shared – modernity itself – that is adaptive.

Milford Wolpoff

Department of Anthropology, University of Michigan, Ann Arbor, MI, USA

with at least 10 individuals, including some likely burials. The Israeli fauna at this time is interpreted as an extension of the African fauna, and thus all of these early remains can be considered African. Outside Africa, claims for early fossils include the Liujiang skull (China, placed at 67 KY by some, but 10–30 KY by others), Lake Mungo 3 (Australia, 62 ± 6 KYA according to some, but with a more recent date of 40 ± 2 KYA according to others; see Section 9.5.2) and East Timor (30–35 KYA); if reliable, these dates demonstrate that modern humans were present in East Asia and Australia by 40 KYA or perhaps by 60 KYA. In Europe, the earliest unequivocally modern and securely dated remains do not occur before 39 KYA, when sites such as Isturitz in the French Pyrenees are known (Zilhao and d'Errico, 1999); the Cro-Magnon type specimen is later, and appears to represent a deliberate burial.

Evidence for the first modern human presence in different regions of the world will be discussed further in Chapters 9 and 11. Here, we note that, despite uncertainties in classification and dating, even the earliest claimed dates outside Africa are much more recent than widely accepted dates inside Africa: it is clear that modern human morphology appeared considerably earlier in Africa than elsewhere.

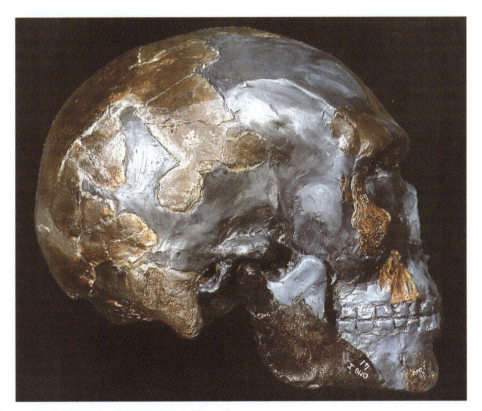

Figure 8.10: The earliest anatomically modern human cranium ('Omo I').

Modern features include the high forehead and developed chin; note that the gray portions are reconstructed. This specimen from Omo-Kibish in southern Ethiopia provides crucial evidence that modern human anatomy had developed in Africa by ~ 130 KYA. Reproduced with permssion by Michael Day.

8.3 **Evidence from archaeology**

Archaeological evidence may be considered as the preserved signs (other than fossils) of hominid activity, although perhaps this definition should be extended to include ape activity. While hominid fossils are very rare, archaeological remains, such as stone tools, are much more common. Assemblies can be classified and associated with particular hominids through rare sites that contain both, and then allow the presence of these hominids to be inferred elsewhere.

Chimpanzees use a range of tools, including sticks to extract termites and stones to break open nuts (Whiten *et al.*, 1999), and orangutans also use tools in a variety of ways, including for seed extraction and autoerotic purposes (van Schaik *et al.*, 2003), so a parsimonious assumption is that our common ancestor used tools as well. Most of these would not be preserved in the archaeological record, and, although the study of chimpanzee stone tool sites is now beginning (Mercader *et al.*, 2002), chimpanzee archaeological sites have not been recognized because the tools are not sufficiently distinct from natural objects. Perhaps, for similar reasons, any tools used by the earliest hominids have not been identified, so archaeology begins only with the **Oldowan** culture about 2.5 MYA (*Figures 8.11* and *8.12a*).

Oldowan tools, named after Olduvai Gorge, Tanzania, East Africa, could be recognized as artifacts by characteristics such as:

▶ **conchoidal fracture patterns** resulting from striking one stone with another, which differ from the fractures seen in naturally cracked stones;

▶ transport of stone over several kilometers, so that tools may be made from lavas or quartzite which do not occur naturally at the site;

▶ concentrations of stone tools, sometimes in association with butchered animal bones.

Tools consist of hammerstones, flakes and cores, and, as implied above, were probably used to scavenge animal carcasses, including breaking open the bones. Even hyenas cannot crack the thick-walled limb bones of large animals, so tool-use would have provided the hominids with a rich and novel food source – bone marrow. It is impossible to be certain of the identity of the toolmakers at Olduvai Gorge, but they are usually assumed to be *A. habilis* or *K. rudolfensis*.

At about 1.6 MYA, a striking change occurred: symmetrical teardrop-shaped **bifaces**, worked around all or most of the margin and often referred to as **handaxes**, started to be made (*Figure 8.12b*). These are called **Acheulean** after the French site St. Acheul, and are often found in association with flake tools and choppers. The uses of these bifaces are poorly understood, but they have been described as the 'Swiss Army knife' of the **Paleolithic** and continued to be used, with little obvious change in structure, until around 150 KYA. They were, perhaps, the most successful of all human tools. They are found throughout Africa, in Europe, and in Asia south of the **Movius Line** which runs from the Caucasus mountains to the Bay of Bengal, but are largely

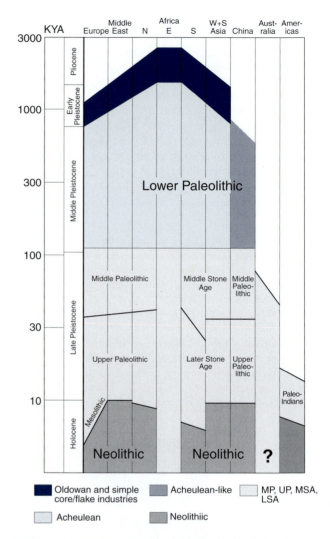

Figure 8.11: Chronology of archaeological stages in different regions of the world.

The first column shows the time-scale (note that the scale is nonlinear), and the second column gives the geological period. Subsequent columns show the archaeological stage in selected world regions. Archaeological remains appear earlier in Africa than elsewhere. Opinions vary about whether or not there was a Neolithic period in Australia.

absent from Eastern Asia. There, well-crafted stone tools are known from sites as early as 803 KYA, such as the Bose basin in Southern China, where hominids apparently exploited the rock exposed by a meteorite impact (Yamei *et al.*, 2000). However, these assemblies are described as 'Acheulean-like' rather than 'Acheulean', and archaeology thus reveals an important cultural difference between regions east and west of the Movius Line. Acheulean and Acheulean-like technologies are likely to have been made by *H. erectus/ergaster*. Some have speculated that their construction required advanced mental capacity, including the ability to visualize their shape in advance.

While stone tools dominate the early archaeological record, we would expect that many other materials would

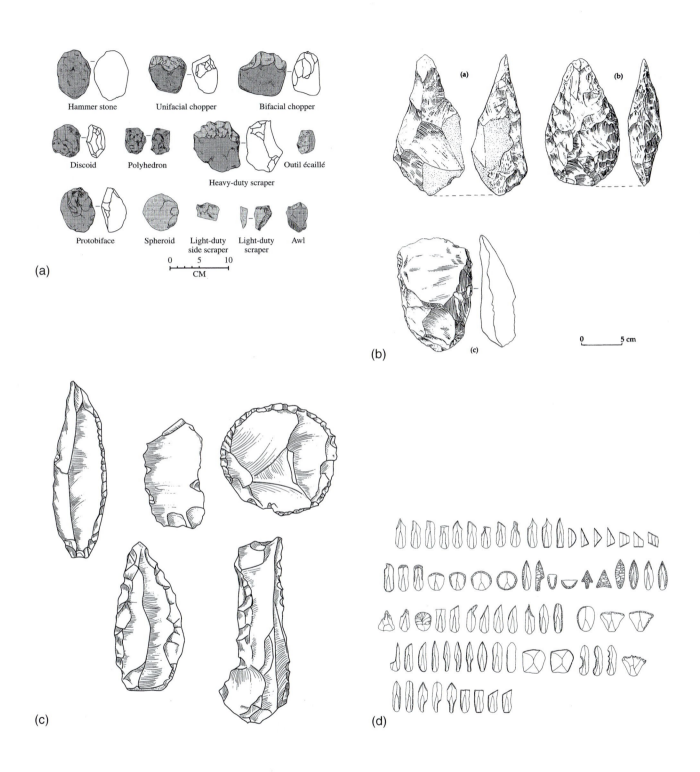

Figure 8.12: Stone tool technologies.

(a) The oldest recognized stone tools are Oldowan, manufactured from pebbles and dating back to ~ 2.5 MYA. (b) After ~ 1.6 MYA, Acheulean tools are found, including bifaces. They continued to be used until ~ 150 KYA. (c) Mousterian tools were manufactured by the Levallois technique after ~ 300 KYA, associated with both Neanderthals and early anatomically modern humans. (d) Upper Paleolithic tools dating after 50 KYA showing the wide range of forms. Reproduced with permission from Blackwell Science Ltd.

have been used, but would seldom have been preserved. A set of throwing spears from Schöningen in Germany was found in association with butchered horses and is dated to ~ 400 KYA (Thieme, 1997), providing evidence for use of multiple materials and sophisticated hunting activity at this time.

Soon after 300 KYA, the **Levallois technique** was developed, in which the shaping of the tool was accomplished by removing flakes from a core, followed by removal of one final shaped flake which would form the tool itself (*Figure 8.12c*). The technique is thought to have originated in Africa, but was being used in Europe by 200 KYA. Among the **Middle Paleolithic** industries in Europe is the **Mousterian**, characterized by flakes described as side-scrapers and points. The human remains associated with Mousterian artifacts are usually Neanderthal, but at Qafzeh and Skhul they are early modern humans, and late Neanderthals may have used different tools (Section 9.5.3). Mousterian-like toolkits are found in Asia as far to the east as Lake Baikal, and in Southern Asia tools have been labeled as 'Mousteroid'; in Africa a range of Middle Stone Age industries has been described.

In the **Upper Paleolithic**, the predominant tools are described as **blades** instead of flakes; blades are long narrow flakes made from specialized cores and then reworked in a number of ways (*Figure 8.12d*). In particular, they may be retouched at the end rather than the side. Objects made from other materials, such as wood and bone, become much more abundant in the Upper Paleolithic and unequivocal art is found. The Upper Paleolithic is often associated with modern humans although, as we have seen above, there is unlikely to be a simple one-to-one correspondence between toolkits and species. Discussion of subsequent developments will be continued in later chapters.

8.4 Hypotheses to explain the origin of modern humans

While many of the fossil discoveries described in the first half of this chapter have taken place in the last few years, and dates have often been refined or revised, the basic pattern of an early exodus of *H. erectus* from Africa to occupy much of the Old World, followed by a much later appearance of modern humans, has been clear for decades, and has conditioned the debate that dominated the field during the second half of the twentieth century. This debate can be most easily appreciated by first considering the extreme views (*Figure 8.13*); then we will consider more complex scenarios.

▸ The 'multiregional' model proposes that the transition from *erectus* to *sapiens* took place in a number of places in the Old World, with the diverse modern human characteristics arising at different times in different places (see *Box 8.3*).

▸ In contrast, the 'out of Africa' model proposes that the transition took place recently (< 200 KYA) in Africa, and that these humans replaced the hominids already present on other continents (*Box 8.4*).

One way of characterizing the difference is that, according to the multiregional model, our ancestors 1 MYA lived on several continents; in contrast, according to the out of Africa model, our ancestors 1 MYA lived only in Africa. Even with these extreme views, it is worth noting that it is more difficult to falsify the multiregional model than the out of Africa one: a single ancestor living in Asia 1 MYA would falsify the strict out of Africa model, but 10 out of 10 ancestors living in Africa would not necessarily falsify the multiregional model. These models were formulated before the classification of many of the species between the times of *H. erectus* and modern *H. sapiens* was adopted, and it is not entirely clear how all the additional species would fit into them.

Intermediate models are obviously possible, involving a recent origin of many human characteristics in Africa, but also interbreeding with archaic populations outside Africa. Relethford (2001) has suggested that it is useful to distinguish between the *mode* of transition (multiregional evolution within a species, or a speciation event followed by replacement) and the *location and timing* of the transition (no single time and place, or recent in Africa). As a result, he classifies models into three groups:

▸ regional coalescence (multiregional evolution within a species, no single time and place);

▸ primary African origin (multiregional evolution within a species, recent African origin);

▸ African replacement (speciation followed by replacement, recent African origin).

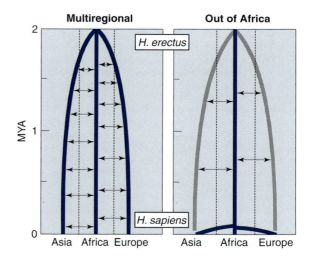

Figure 8.13: Two extreme models for the origins of modern humans.

Both models begin with *H. erectus* shortly after 2 MYA and lead to contemporary humans; many intermediate models could also be proposed. Horizontal arrows indicate gene flow between populations on different continents. In the multiregional model, extensive gene flow is required; the out of Africa model requires less. Blue lines: ancestors of modern humans. Gray lines: lineages that are not ancestors of modern humans.

BOX 8.4 Opinion: Modern human origins – why it's time to move on

While the sentence 'There are two theories of modern human origins' has launched a thousand student essays, this reflects more the historical situation than the current one. When the 'mitochondrial Eve' hypothesis was first developed in the 1980s, giving momentum to the 'single origin, recent origin, out of Africa with replacement' model (originally developed by W.W. Howells), it was based on what we can now see to be relatively little evidence (a small sample, one genetic locus), and on weak statistical analysis. Despite this, it had the effect of galvanizing research across a range of approaches from archaeology to molecular genetics, and also hardening up the alternative into a coherent theory – the **multiregional** model.

However, what is surprising is how robust the essential finding has proved. As more genetic loci have been used, as the techniques for analyzing molecular genetic data have improved, as better and better mathematical models have been developed, the basic finding has been substantiated – that humans lack genetic diversity (reflecting a recent origin in a small, localized population), that global diversity is a subset of African diversity (hence the probability of that population being in Africa), and that Neanderthals (the only archaic species for which we have comparable genetic evidence) lie outside the range of human diversity, and show their own evolutionary history. No genetic evidence specifically supporting multiregional evolution has accrued, and critiques of the single origin model have rather focused on emphasizing that the confidence limits on many data cannot exclude the remote possibility of some form of admixture, and therefore a residual support for a watered down version of multiregionality. Y chromosome data, complete mtDNA sequences, SNPs, and multi-locus analyses all point in the same direction. Furthermore, the fossil and archaeological data do not support multiregionalism, although their interpretation is less resolved than the genetic.

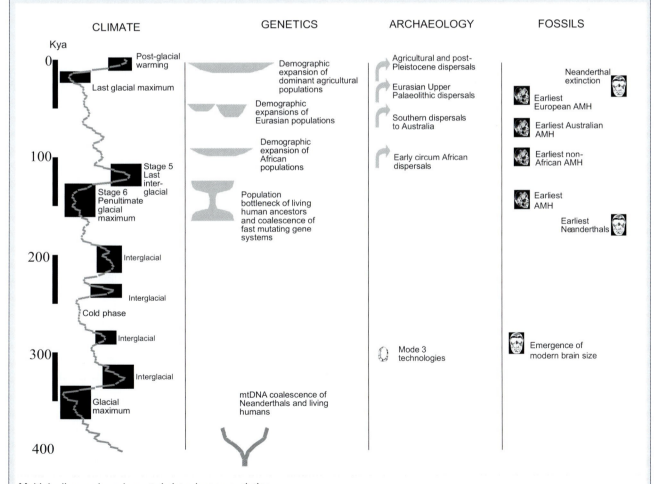

Multiple dispersals and events in later human evolution.

While the out of Africa model is clearly supported by the evidence, the nature of the underlying events and processes is more complex and extended than originally thought.

For most people active in the field the 'modern human origins', debate has moved on – not 'out of Africa' versus 'multiregionality', but, what particular patterns and processes are involved in 'out of Africa'. That is where the action now is, in both genetics and paleoanthropology. The original Cann *et al.* (1987) model was simple, assuming a synchrony of events – a population bottleneck or crash, the extinction of archaic forms, the evolution (by speciation event) of anatomically modern humans, the rapid dispersal and diversification of populations bearing fully modern behavior and capacities. We now know that is not the case, and that what we can think of as the evolution of modern humans is spread over at least 100 000 years (but not 1 000 000). This greater complexity has led to a number of alternative versions of 'out of Africa', and it is debate among these that are now the focus of attention. Multiregionality is no longer one of the serious models.

So what are the major issues of modern human evolution now? We have argued that the model that best describes the evolution of modern humans, and accounts for the data across the range from molecular genetics to fossil morphology to archaeology, is a *multiple dispersals* or *multiple event* one. Much work is now focused on these separate events. For example, if there was a bottleneck producing a small population in Africa around 150 KYA, where in Africa was it (it's a very large continent!), and how many people were involved (current estimates suggest around 10 000 individuals)? What is the relationship between this bottleneck and the drop in global temperatures that occurred at that time (Stage 6 – see *Figure*)? Does this bottleneck coincide with the evolution of modern human features, as may be implied by the earliest known *Homo sapiens* fossils

(Omo Kibbish; see *Figure 8.10*)? What behavioral and cognitive shifts occurred then?

The dispersals associated with modern humans both within and beyond Africa represent another major issue. Archaeological and fossil evidence suggests that these did not occur until around 110 KYA, and then were partial. We can recognize a phase of within African dispersal during the warmer stage 5, but then there seem to have been further contractions and isolation (leading to diversification) as the climate cooled again. The first major 'out of Africa' dispersals, around 70 KYA, were not global, but were confined to the southern parts of Asia and eastward (the southern dispersals). The colonization of Eurasia, so strongly associated with the spread of the Upper Paleolithic and the extinction of the Neanderthals, the last surviving archaic hominins, occurred much later, around 40 KYA. Ecology and behavior, as well as mapping the distribution of fossils and archaeological material, are essential to understanding these and subsequent events in human evolution.

The 'out of Africa' model of human evolution has basically proved to be empirically sound, and the field is now (at last!) moving on. Multiple events are being recognized, and changing the nature of what started as a simple, one horse (or at least, one gene!) model. The next stage will be to focus on what both genes and fossils can tell us about the processes underlying these events. In the end, the genes are only a part of the evolutionary problem, which at heart is understanding why one particular species of hominin – *Homo sapiens* – is so distinctive and successful.

Robert Foley and Marta Mirazón Lahr

Leverhulme Centre for Human Evolutionary Studies, University of Cambridge, UK.

The reader may also come across the terms 'strong garden of Eden hypothesis' and 'weak garden of Eden hypothesis' (Harpending *et al.*, 1993). These refer to two demographic variants of the out of Africa model. According to the 'strong' hypothesis, modern humans appeared as an *H. erectus* subpopulation, perhaps a new species, and spread continuously over much of the Old World. According to the 'weak' hypothesis, modern humans again appeared in one subpopulation, but spread was slow, taking tens of thousands of years and producing separate daughter populations that later re-expanded.

With this background, we can now consider the contemporary genetic data: patterns of diversity should contain information about demographic history.

8.5 Evidence from genetic diversity

We saw in Chapter 6 that the genetic diversity found in a population depends on the balance between a number of factors such as the rate at which variants are introduced by mutation, and the rate at which they are lost by drift, which

in turn is influenced by factors including the effective population size and the amount of subdivision. If a population is not at mutation–drift equilibrium, when genetic diversity remains constant over time, the diversity will also depend on its age: an older population will contain more diversity. It has therefore been of interest to compare the diversity levels in different populations (usually considered as continents, to avoid local population-specific factors); but do the results reflect population age, size, or some other factor?

8.5.1 Levels of human diversity and effective population size

The scenarios in *Figure 8.13* make different predictions about the total level of diversity we might expect to see in humans, because of the different number of ancestors more than 150 KYA contributing to modern populations. According to the multiregional model, our ancestors were widely dispersed across several continents but formed a single population because of gene flow, and this would require a large population which in turn would carry high diversity. In contrast, according to the out of Africa model, our ancestors before the recent expansion were restricted in location and number, and would thus exhibit low diversity. We have seen in Section 7.5 that humans actually have low diversity compared to other apes, and estimates of effective population size are around 10 000. What population size would the two models require? Harpending and Rogers (2000) have argued that distribution in the Old World over an area of 25 million square kilometers at population densities comparable to those of modern foragers would lead to a population size of 125 000–1 000 000. The effective population size would be perhaps one-third of this: over 40 000 and possibly over 300 000, which is considerably greater than 10 000. However, as discussed in Section 5.3, relating effective population size to census population sizes is problematic and some have argued that an effective population size of 10 000 could be compatible with a population large enough to cover the Old World (Harpending *et al.*, 1993). The out of Africa model involves a smaller population and would be consistent with an effective size of 10 000. This line of reasoning thus suggests that the human population was most likely not large enough for the multiregional model to be tenable.

8.5.2 Geographical distribution of diversity

The scenarios in *Figure 8.13* also make different predictions about the *geographical distribution* of genetic variation. According to the multiregional model, there is no *a priori* reason for one geographical region to show more diversity than another; the out of Africa model predicts greater diversity in Africa.

Table 8.1 shows that different studies have detected the highest diversity in different parts of the globe, and thus appear to provide support for the multiregional model. However, this interpretation is too simple. The high RFLP diversity in Europe can be explained by a bias in the **ascertainment** of these markers, most of which were initially chosen because they detected variation in Europeans: it is not surprising that such markers, which have low

mutation rates, detect less variation in other populations. Of the two loci showing their highest diversity in Asia, β-globin may be subject to selection, while the low diversity of the Y chromosome in Africa may be due to the recent expansion of a single haplotype which made up 45% of the sample examined (see Section 10.6.3). Indeed, this study found that a different measure of Y diversity, the number of pairwise differences, identified Africa as the continent with the highest diversity. The studies showing the least ascertainment bias and involving the largest numbers of neutral loci (30 microsatellites, 50 autosomal segments), as well as some studies of individual loci (Xq13.3, mtDNA), identify Africa as the location of the highest diversity. Does this finding therefore support the out of Africa model? It is certainly consistent with it, since higher diversity in Africa could indicate an older population. However, as we have seen, it could also reflect a larger long-term effective population size. More evidence is needed.

A detailed study of a single locus was carried out by Tishkoff *et al.* (1996), who analyzed haplotype diversity at the *CD4* locus on chromosome 12. These haplotypes consisted of two polymorphisms 9.8 kb apart: (i) a binary marker, the presence or absence of a 256-bp deletion; and (ii) a multiallelic marker, a microsatellite containing between four and 15 repeat units. Haplotype diversity was highest in sub-Saharan Africa, and lowest in the Americas and the Pacific (*Figure 8.14*). The deletion could be associated with any one of nine microsatellite alleles within Africa, but 98% of deletion chromosomes found outside Africa were associated with a single allele containing six repeats; among the nondeletion chromosomes, which were associated with 12 different alleles in sub-Saharan Africa, only the 5-repeat and 10-repeat alleles were present at a frequency > 3% outside Africa. Thus three lineages made up 95% of the non-African chromosomes. Moreover, the haplotypes found outside Africa were a subset of those in sub-Saharan Africa. These results were interpreted as indicating that the non-African lineages originated in Africa and experienced a bottleneck as they spread throughout the

TABLE 8.1: DIVERSITY OF SELECTED LOCI IN AFRICA, ASIA AND EUROPE.

Locus (measure of diversity)	Diversity			Reference
	Africa	Asia	Europe	
79 RFLPs (h)	0.297 ± 0.007	0.327 ± 0.012	**0.379** ± 0.015	Bowcock *et al.*, 1994
30 microsatellites (h)	**0.807** ± 0.014	0.685 ± 0.021	0.730 ± 0.016	Bowcock *et al.*, 1994
β-globin, 5′ flanking region[a] (π)	0.323%	**0.331%**	0.232%	Harding *et al.*, 1997
Xq13.3[a] (π)	**0.035%**	0.025%	0.034%	Kaessmann *et al.*, 1999
50 autosomal segments (π)	**0.115** ± 0.016%	0.061 ± 0.010%	0.064 ± 0.010%	Yu *et al.*, 2002
mtDNA control region (p)	**2.08**	1.75	1.08	Vigilant *et al.*, 1991
43 Y-chromosomal binary markers (h)	0.841 ± 0.001	**0.904** ± 0.000	0.852 ± 0.000	Hammer *et al.*, 2001

[a]Value calculated by Yu *et al.* (2002) from the data described in the 'Reference' column. The highest diversity in each study is shown in bold. h, Locus or haplotype diversity; p, pairwise difference/100 bp; π, nucleotide diversity (see Section 6.2.2).

rest of the world. The authors presented an age of 102 KYA for the maximum time of exit of the deletion lineage. This was obtained by assuming that the maximum age of the deletion lineage within Africa corresponded to the chimpanzee/human split, taken as 5 MYA, and using the decay of association between the markers in Africa and outside of Africa. If the age of this lineage within Africa was less than 5 MY (and times of 1 MY are more typical for autosomal loci), the time of exit would be correspondingly later. A coalescence-based reconsideration of this time suggested 154 KYA for the maximum age of the deletion chromosomes outside Africa (Tishkoff *et al.*, 1998).

While the details of the first age calculation can be criticized, the pattern for the *CD4* locus is clear: high haplotype diversity within Africa compared with low diversity and strong **linkage disequilibrium** in the rest of the world. This indicates an African origin for this locus and illustrates the power of even the minimal phylogenetic information derived from two linked polymorphisms, a point that will be explored in the next section. As has been emphasized several times, the history of a single locus does not necessarily represent the history of the entire genome, and many loci must be examined to obtain an overall picture.

8.6 Evidence from genetic phylogeny

The phylogeny of a locus can provide information about the time and place of its origin. To do this we must make some assumptions. These assumptions are:

▶ the phylogeny can be reconstructed accurately;

▶ geographical movement has been limited, so that the modern distribution provides information about the ancient distribution.

Molecular phylogenetic information was first applied to the question of human origins when mtDNA data became available. The features of this locus that make it particularly suitable for such studies are explained in *Box 8.5*, and aspects of the nomenclature of clades in *Box 8.6*. The first studies attracted enormous attention, both from scientists and the public, and some criticism (*Box 8.7*).

8.6.1 Mitochondrial DNA phylogeny

A more recent mtDNA phylogeny based on the complete mtDNA sequence of 53 individuals of diverse geographical origins, rooted by comparison to a chimpanzee sequence, was determined by Ingman *et al.* (2000). All sequences were different and 657 variable positions were found, 516 of which were outside the hypervariable region. Despite the elevated mutation rate of mtDNA compared with nuclear sequences, a robust phylogeny could be obtained from the complete sequence excluding the rapidly mutating control region (*Figure 8.15*). This has some striking features:

▶ complete separation of African and non-African lineages;

▶ the first three branches lead exclusively to African lineages, while the fourth branch contains both African and non-African lineages;

▶ deep branches within African lineages; **starlike** (see *Figure 6.18*) structure within non-African lineages;

▶ **TMRCA** for the entire phylogeny: 172 ± 50 KY;

▶ TMRCA for the branch containing African plus non-African lineages, marked with an asterisk in *Figure 8.15*: 52 ± 28 KY;

▶ expansion time for non-African lineages estimated at 1925 generations or 38 500 years at 20 years/generation.

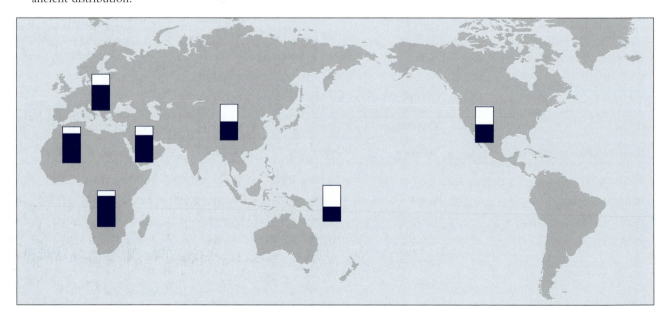

Figure 8.14: *CD4* haplotype diversity in different regions of the world.

Diversity is represented by the height of the blue shading within the rectangle. Note that the highest diversity is found in Africa. Drawn from the data of Tishkoff *et al.* (1996).

BOX 8.5 Locus briefing: the mitochondrial genome

What are its origins?

The cytoplasmic location and energetic function of mitochondria (mt) derive from their origin as **endosymbiotic** bacteria (see *Box 2.8*). Each mitochondrion contains several identical mitochondrial genomes (mtDNA) that appear to be exclusively maternally inherited in humans (see *Figure 2.18*).

What genes are encoded within the mitochondrial genome?

The endosymbiont progenitor of modern mitochondria contained many more genes than are presently found in mtDNA. Most of these genes have been lost, with some being transferred to the nuclear genome. mtDNA retains 37 genes whose products function in energy production (oxidative phosphorylation) and mitochondrial protein synthesis (see Section 2.7.2 and *Figure 2.20*). These genes are tightly packed within the 16.5-kb genome, whose only significant stretch of noncoding sequence contains a dedicated replication origin and is known as the 'D-loop', or 'control region'. The two strands of mtDNA are biased in their base content, with the G-rich and C-rich strands known as the 'heavy' and 'light' strands respectively.

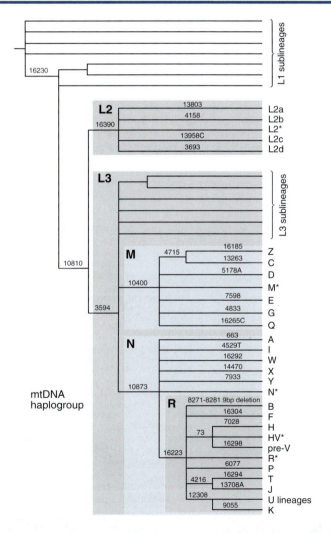

A high-level phylogeny of human mitochondrial DNA.

Clades are labeled with the position within the genome of one of the clade-defining mutations. All mutations are transitions with the exception of a few transversions that are indicated by the nucleotide position being followed by the derived base. No consensus has yet been reached on the nomenclature of L1 and L3 sublineages and so the lineages remain unlabeled here. The topology of L1 and L3 sublineages (taken from Salas *et al.*, 2002) illustrates that L1 is not monophyletic. Geographical distributions of the major clades are shown in *Figure 9.16*. Data are from Macaulay *et al.*, 1999; Schurr *et al.*, 1999; Chen *et al.*, 2000; Ingman *et al.*, 2000; Herrnstadt *et al.*, 2002; Kivisild *et al.*, 2002; Salas *et al.*, 2002; Mishmar *et al.*, 2003.

continued

What diseases are caused by mutations within mtDNA?

The maternal inheritance of mtDNA allows inherited disease-causing mutations to be mapped to this locus on the basis of the segregation of the disease through pedigrees. Most diseases affect organs with a high energy requirement, such as nerves and muscles. Pathogenic mutations include base substitutions, deletions, duplications, insertions and inversions (see *Table* and http://www.mitomap.org/). Rearrangements are often flanked by short repeats of 2–13 bp, suggesting that mitochondria contain machinery for homology recognition and the splicing of DNA ends, despite the apparent lack of homologous recombination. Mitochondrial DNA mutations may also be acquired in somatic tissues, and may be involved in the aging process.

How has the study of mtDNA diversity developed?

The first full mitochondrial genome sequence was a chimera of European placental and African HeLa cell line DNA (Anderson *et al.*, 1981; Arnason *et al.*, 1996; Andrews *et al.*, 1999), and subsequently become known as the Cambridge Reference Sequence (CRS); the nucleotide numbering of mtDNA sequences is based on this. Diversity within mtDNA was initially studied by RFLP analysis of purified mtDNA; high-resolution RFLP typing that uses ~ 18 different restriction enzymes can screen roughly 20% of the mitochondrial genome for mutations. The advent of PCR facilitated sequencing of certain informative portions of mtDNA, with more recent advances in throughput allowing the sequencing of whole mitochondrial genomes (Ingman *et al.*, 2000).

How can we find mtDNA diversity among modern humans?

Polymorphisms are easier to find in mtDNA than in the nuclear genome as a result of an elevated mutation rate. Different parts of mtDNA mutate at different rates, perhaps because of different selective constraints and different DNA structures (see Section 3.2). The control region contains the two most polymorphic regions, known as hypervariable segments (HVS) I and II. Sequencing HVS I has proved the most popular method for assaying mtDNA, with well over 10 000 HVS I sequences available in databases. While the absence of mtDNA recombination (see *Box 2.9*) makes it possible to reconstruct a single phylogeny for mtDNA lineages, the presence of parallel mutations and reversions at fast-mutating sites has hindered this process. There are now more than 500 published whole mtDNA sequences, and combining this information with RFLP data and control region sequences allows the reconstruction of a phylogeny of major mtDNA lineages shown in the *Figure*.

Why are all the deep rooting clades called L?

The present nomenclature (see *Box 8.6*) has been established piecemeal since the advent of high-resolution RFLP typing. These studies focused initially on Native American and Eurasian populations, and by the time large studies of Africa were initiated, few single letter clades remained unassigned. The early African study which defined L1 and L2 (Chen *et al.*, 1995) misplaced the root of the phylogeny, such that it appeared that L1 referred to a monophyletic clade, which is now known not to be the case, since L1 encompasses the root.

Why is mtDNA so useful for exploring the human past?

▶ The high mutation rate makes it easy to assay diversity in many samples efficiently.

▶ The low effective population size leads to increased genetic drift, which generates geographic structure (e.g., continent-specific lineages, see *Figure 9.16*).

▶ Exclusively maternal inheritance allows access to female-specific processes.

▶ The high copy number of mtDNA per cell facilitates the analysis of degraded contemporary samples and ancient DNA (see Sections 4.11 and 15.1.5).

EXAMPLES OF MITOCHONDRIAL DISEASES.

Mutation	Disease
Base substitution in protein gene: 11 778 transition in NADH dehydrogenase 4 (*ND4*) gene	Leber's hereditary optic neuropathy (LHON; OMIM 535000)
Base substitution in tRNA gene: 3243 transition in tRNA Leucine I gene	Mitochondrial encephalomyopathy, Lactic acidosis, and stroke-like episodes (MELAS; OMIM 540000)
Deletion: 4977-bp deletion (8469 to13447) between 13-bp direct repeats	Kearns–Sayre syndrome (OMIM 530000)
Duplication: 266-bp duplication between 7-bp direct repeats in the control region	Mitochondrial myopathy
Inversion: 7-bp inversion (3902–3908) in NADH dehydrogenase 1 (*ND1*) gene	Mitochondrial myopathy
Insertion: C inserted at 7472 in tRNA Serine 1 gene	Progressive encephalopathy (PEM; OMIM 590080)

What about possible selection pressures?

Testing for association between mitochondrial polymorphisms and disease is one way of identifying *current* selective pressures on mtDNA. There is good evidence that haplogroup T is significantly associated with poor sperm motility, and hence male infertility. However, because the negative effects are observed only in males and there is no paternal transmission of mtDNA, these lineages are not subject to selection (Ruiz-Pesini *et al.*, 2000). Some disease-causing mtDNA mutations arise preferentially on specific lineages [e.g., LHON on haplogroup J (Torroni *et al.*, 1997; Brown *et al.*, 2002)], as a result of synergistic interaction between the primary LHON mutations and polymorphisms defining the haplogroup. An association between a mutation in the control region (16189) and diabetes suggests that variability in replication efficiency may play a role in metabolic disease (Poulton *et al.*, 2002). It has also been suggested that this same mutation provided a selective advantage in the past through the efficient use of minimal bioenergetic resources (see *Box 11.7* and Section 13.4.4). Statistical methods have been used to detect the imprint of *past* selection in whole genome sequences from healthy individuals (see Section 6.3). Adaptation to colder climates may have been an important factor in determining patterns of mtDNA variation in non-African populations (Mishmar *et al.*, 2003).

BOX 8.6 Haplogroup nomenclature

Complex nomenclatures are used to describe Y-chromosomal and mtDNA haplogroups (see *Boxes 8.5* and *8.8*). This complexity is required to maintain stability in the face of changes to the phylogeny as a result of additional markers and/or haplogroups. These nomenclatures are based on a **cladistic** appreciation of the underlying phylogeny relating the ancestral relationships of haplogroups. Major clades are identified by single capital letters (e.g., haplogroup C), sublineages within these clades are given numerical suffixes (e.g., haplogroup C1), and this can be continued using alternating lower case letters and numbers until all lineages have been named (e.g., C1a3b2).

The set of chromosomes that share the derived state of a unique event polymorphism are by definition **monophyletic**. By contrast, a set of chromosomes defined by the sharing of the derived state of a deep-rooting marker and ancestral states at the various nested markers that define sublineages, are potentially **paraphyletic**. In other words, a new mutation may be found that is carried by some but not all these chromosomes. These sets of chromosomes are sometimes named paragroups instead of haplogroups, and consequently are highlighted by using a * suffix (e.g., chromosomes belonging to haplogroup C, but not C1 or C2 are called C*). Often not all of the markers defining known sublineages are typed when paragroups are identifed and additional suffixes are added to reflect this. For example, if only markers defining the C1 and not the C2 sublineage are typed, the corresponding paragroup is named C*(xC1).

Both the Y-chromosomal and mtDNA phylogenies contain **multifurcations** that may be resolved into a set of bifurcations in the future. A new marker found to be derived in two lineages descending from a multifurcation defines a new clade which is named by the union of the names of the individual clades (e.g., a clade that unites haplogroups C and D is known as CD).

BOX 8.7 Early controversies about 'mitochondrial Eve'

'All these mitochondrial DNAs stem from one woman who is postulated to have lived about 200 000 years ago, probably in Africa' These words of Cann *et al.* (1987) stimulated the 'mitochondrial Eve' controversy. Cann and colleagues extracted mtDNA from 147 individuals and digested it with 12 restriction enzymes, thus identifying polymorphisms throughout the molecule. They constructed a phylogenetic tree linking the lineages and dated the tree using the mtDNA mutation rate. The tree suggested a root in Africa and a time depth of about 200 KY (140–290 KY). Why did their claim prove so controversial? Of its three parts, the descent of all mtDNAs from one woman surprised some people, but should not have done so because any nonrecombining region must trace back to a single ancestor. More serious criticisms focused on the methodology used, and the second two parts of the claim:

▶ most of the 'African' DNA samples were from African–Americans: only two of the 20 were born in sub-Saharan Africa. Are they representative of African mtDNAs?

▶ the method used to generate the tree was not guaranteed to find the most parsimonious tree. Different, more parsimonious, trees might exist;

▶ the method used to locate the root (midpoint rooting) placed it at the midpoint of the longest branch. This could be unreliable if, for example, the rate of evolution was higher in Africa than elsewhere. Outgroup rooting is preferred.

Despite these criticisms, subsequent studies have supported the major conclusion of Cann *et al.*: that is, a recent African origin for mtDNAs.

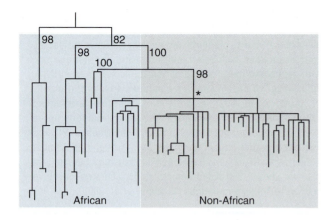

Figure 8.15: Neighbor-joining tree of mtDNA sequences.

The tree was constructed using sequence information from the entire mtDNA genome except the control region. Blue shading: African lineages. Gray shading: non-African lineages. Numbers indicate the percentage of bootstrap replicates. The asterisk is discussed in the text. Based on Ingman *et al.* (2000).

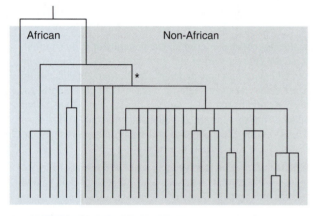

Figure 8.16: Y-chromosomal phylogeny.

Blue shading: African lineages; gray shading: non-African lineages. The asterisk is discussed in the text. Based on Thomson *et al.* (2000).

8.6.2 Y-chromosomal phylogeny

The Y chromosome is also a highly informative locus for such phylogenetic studies (*Box 8.8*). A Y-chromosomal phylogeny was derived from **DHPLC**-based mutation detection in 64 kb of DNA from 43 individuals by Thomson *et al.* (2000). They detected 56 variants which distinguished 32 lineages that fell into the parsimony tree shown in *Figure 8.16*, again rooted by comparison with ape sequences. Although less detailed than the mtDNA phylogeny, its structure shows close parallels:

▶ complete separation of African and non-African lineages;

▶ the first two branches lead exclusively to African lineages, while the third branch contains both African and non-African lineages;

▶ TMRCA for the entire phylogeny: 59 (40–140) KY, assuming 25 years/generation;

▶ TMRCA for the branch containing African plus non-African lineages, marked with an asterisk in *Figure 8.16*: 40 (31–79) KY.

The point estimate (best single estimate, but not taking account of the uncertainty) for the Y phylogeny TMRCA is very recent, and will be discussed further in Chapter 9.

8.6.3 Other phylogenies

Phylogenies from several autosomal or X-chromosomal loci are available. These are potentially complicated by recombination, but by analyzing very closely linked polymorphisms, usually within 10 kb or less, haplotypes showing little recombination can be identified and the effects of recombination minimized or excluded.

One early example of this approach was an analysis of 2.67 kb from the β-globin region on chromosome 11 in 349 chromosomes from around the world. This allowed Harding *et al.* (1997) to construct a haplotype tree accounting for 326 of the sequences (*Figure 8.17*), with the features:

▶ root in Africa, and some lineages exclusive to Africa, but location of many lineages in both African and non-African populations;

▶ TMRCA for the entire phylogeny: 750 (400–1300) KY, assuming 20 years/generation;

▶ lineages such as C2 and C3 (*Figure 8.17*) dating to > 200 KY exclusive to *Asia*.

Some results from this analysis, such as the African root, are similar to those seen with mtDNA and the Y chromosome; others, such as the older TMRCA, are expected because of the larger effective population size of an autosomal locus and are in reasonable agreement with the other data and neutral expectations. A notable finding of this study, however, was the occurrence of lineages > 200-KY-old that were exclusive to Asia. While this could be interpreted as evidence for ancient lineages that had arisen within Asia before 200 KYA and survived in modern populations because of interbreeding between ancient and modern humans, other interpretations are also possible. These lineages could be present in Africa, but just not detected in this study because of the limited sample size used, or they could have been lost recently from African populations. Alternatively, the ages could be overestimated; neutrality was assumed because the data were consistent with this simple possibility, but β-globin has been a target of balancing selection, and this might not have been detected by the tests used, which are renowned for their lack of sensitivity (see Section 6.3).

Several other studies have been carried out and selected examples are summarized in *Table 8.2*. These studies identified a root that was either in Africa, or could not be located precisely but lay in a large region that included Africa. An examination by Takahata and coworkers (Takahata *et al.*, 2001) of the published data found that it was possible to infer the ancestral origin of a total of 10 loci,

BOX 8.8 Locus briefing: The Y chromosome

How has it evolved?

The human sex chromosomes look very different from each other, but were once a pair of homologous autosomes. The process of divergence was initiated when one of them acquired a male sex determining function early in mammalian evolution, followed by a repression of recombination. The nonrecombining region of the Y has expanded over time such that it now constitutes ~ 95% of the chromosome, with the pseudoautosomal regions confined to the termini (see Section 2.7.1).

What does the chromosome contain?

Women survive quite well without Y chromosomes (see *Figure 2.18*), and the Y therefore cannot contain genes important for survival that are not shared with the X chromosome. In fact, the Y is extremely gene poor, with the 23 Mb of non-pseudoautosomal euchromatin coding for only 27 different proteins (1.2/Mb; Skaletsky *et al.*, 2003) compared to the 717 genes on the 160-Mb X chromosome (4.5 genes/Mb). Much of their common ancestral gene content has been lost from the Y chromosome, and it has been suggested that this decline presages an inevitable disappearance of the Y itself (see Section 13.7.1). However, this residue of ancestral genes, which includes the sex-determining gene *SRY*, has been supplemented by newly acquired genes with male-specific functions (e.g., spermatogenesis), many of which have been amplified into multiple copies. Gene loss was due in part to the inability of the Y chromosome to eliminate mutant alleles via recombination with a nonmutant homolog, and as a consequence many genes degenerated (**Muller's ratchet**).

In contrast to the impoverishment of genes, the Y is enriched for many different types of repeats, including **SINEs**, **endogenous retroviruses** and **segmental duplications**. The consequent susceptibility to rearrangements by **nonallelic homologous recombination** results in unusually high levels of structural polymorphism among humans (see Section 3.5) and structural divergence among great apes. The instability of the Y and the nature of its gene content means that the commonest classes of pathogenic mutation on this chromosome are deletions that cause male infertility by removing spermatogenic genes (see ideogram in *Figure 2.19*).

How similar are Y chromosomes within and between species?

The Y chromosome has unusual evolutionary characteristics compared to other nuclear loci: a higher mutation rate, higher *between species* sequence divergence, but lower *within species* sequence diversity. How can these seemingly discordant properties be reconciled?

Higher mutation rates result from the exclusive passage of Y chromosomes through the male germ-line (see *Box 5.8*), which is more mutagenic that the female germ-line. This leads to greater Y-chromosomal sequence divergences between hominoid species than for other loci (see Section 7.4.3). However, there is lower diversity within a species because the Y chromosome is prone to genetic drift by virtue of its smaller effective population size (see *Box 5.1*). This results in a very recent common ancestor, and therefore less time to accrue diversity (see Section 6.2.2). The enhanced genetic drift more than outweighs the increased mutation rate. The increased genetic drift also leads to large differences between populations (see *Box 5.6*), making the Y the most geographically informative locus in the genome.

What molecular polymorphisms are found on the Y chromosome?

Since the discovery of the first polymorphic molecular marker in 1985, the number of Y-chromosomal markers has accumulated at an ever-increasing rate. The discovery of SNPs has been greatly facilitated by the **DHPLC** technique described in Section 4.5. Microsatellite identification has been accelerated by the analysis of genomic sequences (see Section 4.6). Of the polymorphisms found on recombining chromosomes, only GC-rich minisatellites (see Section 3.3.2) appear to be absent from the nonrecombining Y chromosome.

How should the polymorphic information from different markers be combined?

The molecular markers identified above have differing mutation rates; some are sufficiently slow mutating that they can be considered unique in human evolution. These unique event markers are typically binary SNPs or indels and can easily be combined into haplotypes (see Section 3.7), known as haplogroups. The absence of recombination (see *Box 2.7*) means that these monophyletic haplogroups can be related by a single phylogeny (shown on the following page) using the principle of maximum parsimony (see Section 6.5.4), which is the most detailed for any region of the genome. The global distribution of haplogroups is shown in *Figure 9.18*. Diversity of the rapidly mutating multiallelic markers can be examined within haplogroups to gain further haplotypic resolution, and some chronological insight.

What are the applications of studying Y-chromosomal diversity?

▶ Evolutionary studies (see Jobling and Tyler-Smith, 2003, and Chapters 7 to 12);

▶ Genealogical investigations (see Chapter 15);

▶ Forensic work (see Chapter 15);

▶ Medical research, e.g., association studies for disease phenotypes.

continued

Is there any evidence of selection on the Y chromosome?

Drawing inferences on human population history from Y-chromosomal diversity assumes that selection has not played a role in shaping present patterns of diversity. Is there any evidence to the contrary? Some estimates of the worldwide Y TMRCA are recent [e.g., 59 (40–140) KYA; Thomson *et al.*, 2000] which could be due to selection, and several associations between Y haplogroups and phenotypes have been claimed, such as with low sperm count in Danish males (Krausz *et al.*, 2001). If true, these could cause differential selection between haplogroups, but all require further investigation. Evidence for selection is inconclusive.

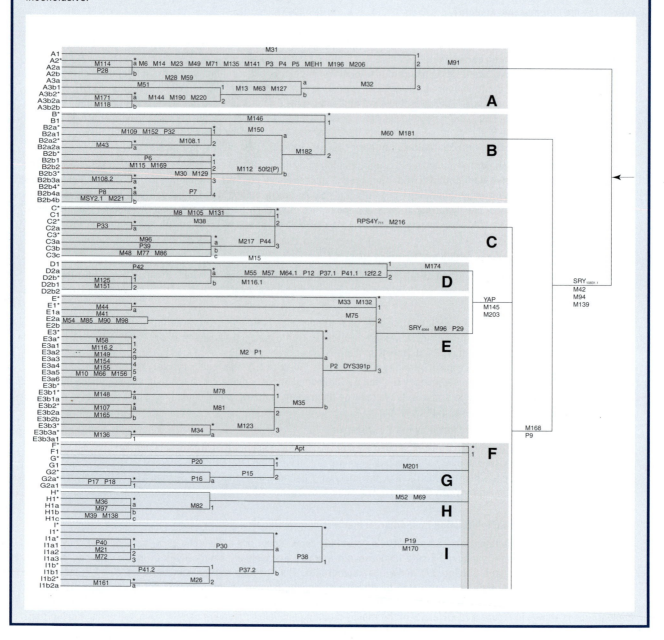

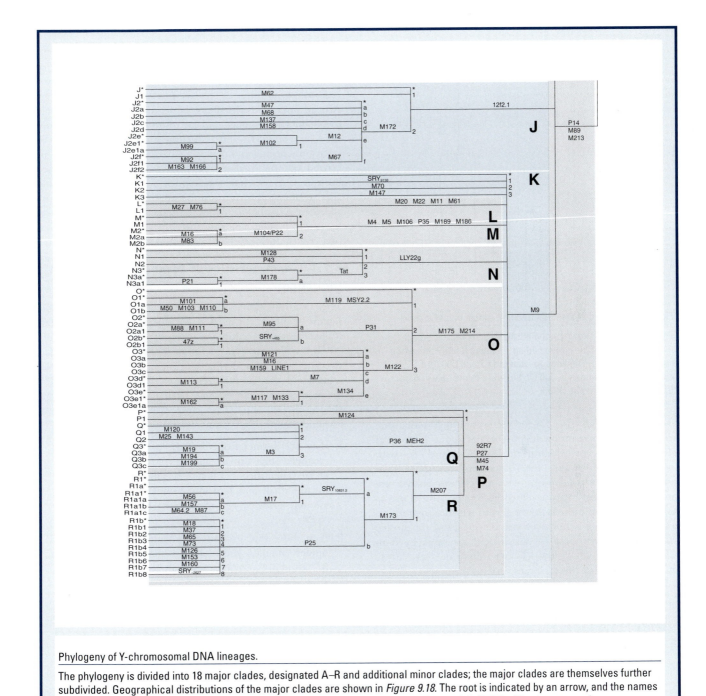

Phylogeny of Y-chromosomal DNA lineages.

The phylogeny is divided into 18 major clades, designated A–R and additional minor clades; the major clades are themselves further subdivided. Geographical distributions of the major clades are shown in *Figure 9.18*. The root is indicated by an arrow, and the names of the binary markers defining the clades are indicated. From the Y Chromosome Consortium (2002) with minor corrections.

including those in *Table 8.2*. Nine of the 10 were in Africa, and one was in Asia. The latter, *GK*, was based on a sample size of 10 and a single variable position, so this conclusion might change if more data were available.

The phylogenetic analysis thus overwhelmingly points to an African origin for most loci. However, we must remember that these are analyses of *loci*, not populations. If the population size were the same on each continent, the finding of an African origin for at least nine out of 10 loci would support the out of Africa model. In reality, population sizes have not been the same and it is likely that African populations were larger than those on other continents for most of the pre-Neolithic period. There is some evidence from phylogeny for ancient non-African contributions to our gene pool (*Figure 8.17*), and **nested cladistic analysis**

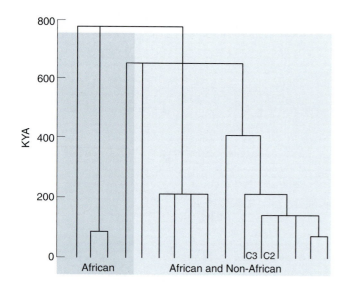

Figure 8.17: Scaled coalescent tree of β-globin haplotypes.

The time-scale is shown on the left-hand side. Blue shading: African lineages. Blue-gray shading: lineages found both inside and outside Africa. The lineages marked C2 and C3 were estimated to have ages > 200 KY, but were found only in Asia. Based on Harding *et al.* (1997).

(Templeton, 2002) (*Box 8.9* and see Section 6.7.5) identifies an out of Africa expansion 420–840 KYA, about the time of *H. heidelbergensis*, in addition to one 80–150 KYA, with interbreeding between the populations rather than replacement. The conclusions of nested cladistic analysis are not, however, universally accepted, in part because of the *ad hoc* criteria used and the lack of rigorous statistical hypothesis testing: the debate is not over yet. Nevertheless, the simple conclusion is that most of our ancient ancestors lived in Africa (*Box 8.10*), although we do not yet know quantitatively what 'most' means.

8.7 Evidence from ancient DNA

The analysis of ancient DNA should be an ideal way to distinguish between different hypotheses of the origins of modern humans: it should tell us directly whether there was regional continuity or replacement of early lineages by African ones in recent times. Unfortunately, it is impossible to obtain DNA sequence data from most fossils: DNA does not survive well and contaminating DNA, from the environment and humans who have handled the fossils or carried out the analysis, provides a high background (see Section 4.11). Ancient DNA work is technically demanding and stringent criteria must be met before results can be accepted as authentic. These criteria are discussed in *Boxes 4.5* and *4.7*. Thus ancient DNA analysis has so far made less of a contribution than was once hoped.

8.7.1 Characteristics of Neanderthal mtDNA sequences

The determination of the partial mtDNA hypervariable region sequence from the Neanderthal **type specimen** (called 'Feldhofer' after the cave of origin) (Krings *et al.*, 1997) was a major triumph for ancient DNA workers. So far, two additional Neanderthal specimens have yielded mtDNA sequence data; a sample size of three is small, but at least the specimens are quite widely separated in time and space (*Figure 8.18*) and some important conclusions about Neanderthal mtDNA sequences are possible:

▶ they are distinct from modern human mtDNAs: for example, the initial analysis of 360 bp by Krings *et al.* identified 27.2 ± 2.2 substitutions between Neanderthals and humans, compared with 8.0 ± 3.1 among a diverse set of modern humans: see *Figure 8.19*;

▶ they show low diversity, comparable to that within modern humans and much less than that within apes: 3.7% and 3.4% for Neanderthals and modern humans,

TABLE 8.2: GEOGRAPHICAL LOCATION OF THE MOST RECENT COMMON ANCESTOR OF SELECTED LOCI.

Chromosome	Locus	Length (bp)	Sample size	Root	Phenotypic association	Reference
11	β-globin	2670	326	Africa	Hemoglobinopathies	Harding *et al.*, 1997
18	*LPL*	9734	71	Africa and elsewhere	Cardiovascular disease	Clark *et al.*, 1998
16	*MC1R*	954	356	Africa and elsewhere	Skin pigmentation	Harding *et al.*, 2000
X	*PDHA1*	4200	35	Africa	Neurological disease	Harris and Hey, 1999
Y	Y	64 120[a]	43	Africa		Thomson *et al.*, 2000
-	mtDNA	15 435	53	Africa		Ingman *et al.*, 2000

[a]Compared by DHPLC rather than resequencing; DHPLC is expected to detect most, but not all, variants.

BOX 8.9 Opinion: Against recent replacement

Cann *et al.* (1987) were the first to use an evolutionary tree of mitochondrial DNA variation to infer that the ancestors of modern humans evolved in Africa, then expanded out of Africa, and drove all other humans in Eurasia to complete genetic extinction (the out-of-Africa replacement hypothesis). This inference was made by noting that the oldest parts of the mtDNA tree were located in Africa. No assessment was made of whether or not this phylogeographic pattern was based upon sufficient data to be statistically significant, nor were any explicit criteria used to discriminate among plausible explanations of the pattern. A procedure known as **nested cladistic analysis (NCA**; Section 6.7.5) circumvents these problems by first testing the null hypothesis that there is no association between the tree of genetic variants and geography, and secondly by providing explicit, *a priori* criteria for the biological interpretation of any statistically significant phylogeographic patterns. The inference criteria have been validated by applying NCA to cases where we know the answer (e.g., validating the criteria for range expansion by analyzing species whose range now includes formerly glaciated areas).

Another constraint on inference from an evolutionary tree is the time it takes all the current genetic variants to coalesce to a common ancestral DNA molecule. There is no possibility of associations between genetic variants and geography once this coalescence has occurred as there is no longer any genetic variation. Coalescence time therefore places an absolute limit on how far back in time inferences can be made. Indeed, the theory of coalescence implies that the first temporal half of the evolutionary tree of a DNA region is expected to consist of only two DNA lineages – usually far too little variation to contain significant phylogeographic information. As a rule of thumb, the evolutionary tree of genetic variation for a particular DNA region is informative only for the second half of the tree's time depth. The time depth of the mtDNA tree has been estimated to be around 200–250 KY, implying that mtDNA is informative about human evolution only to about 100–125 KYA. Given that the out of Africa expansion in the replacement model is estimated at around 100 KYA, this event is at the temporal extreme of informative inference from mtDNA. This in turn raises a serious logical flaw in using mtDNA to test the replacement hypothesis. Replacement requires that genetic patterns in Eurasia older than 100 KY should have been completely erased. MtDNA can be consistent with the replacement model, but it lacks sufficient temporal depth to detect older events and processes in Eurasia. The detection of such older events is critical to testing the replacement hypothesis. Therefore, although mtDNA has been used as the primary genetic evidence for the out of Africa replacement hypothesis, it is in fact inappropriate for testing this hypothesis because it contains little or no information that could possibly falsify the replacement hypothesis.

To avoid the temporal limitation associated with mtDNA, I applied NCA to 10 different DNA regions, including the mtDNA and eight regions in the human genome that have much greater coalescent times (varying from 670 KYA to 8.5 MYA). These analyses (Templeton, 2002) show that the oldest event detected by mtDNA is indeed an out of Africa expansion event around 100 KYA. However, we have to look at the eight DNA regions with greater temporal depth to see if this expansion event was also a replacement event. The NCA of the eight older DNA regions all revealed older events involving Eurasian populations, including an older out of Africa expansion event around 600 KYA and evidence for genetic interchange between African and Eurasian populations that goes back at least 610 KY with 95% statistical confidence. Indeed, all eight of the eight older DNA regions contributed to these statistically significant inferences. Consequently, every potentially informative DNA region without exception tells us that when humans expanded out of Africa they interbred to at least some extent with Eurasian populations. A complete replacement around 100 KYA is strongly and conclusively rejected by all the informative genetic data, without any contradictory genetic inferences at all. This includes the mtDNA data, which of course contains no information whatsoever about the validity of the replacement hypothesis and hence is irrelevant to any test of the replacement hypothesis.

Alan Templeton

Department of Biology, Washington University, St Louis, USA.

respectively, according to one measure (Krings *et al.*, 2000), compared to 14.8% and 18.6% among chimpanzees and gorillas, respectively;

▶ they are no more similar to Europeans than to other modern humans: for example, Neanderthal–European differences = 28.2 ± 1.9 while Neanderthal–African differences = 27.1 ± 2.2 (Krings *et al.*, 1997).

8.7.2 Conclusions from Neanderthal mtDNA sequence data

The Neanderthal mtDNA sequence data have been widely interpreted as supporting the out of Africa model of modern human origins: they show that Neanderthal mtDNAs are distinct from those of living humans, and no more similar to Europeans than to other modern humans. How robust is this conclusion? Nordborg (1998) has argued on the basis of quantitative modeling that the findings exclude models where Neanderthals and modern humans were members of the same randomly mating population. However, they do not exclude all possibility of interbreeding since few of the mtDNAs present before exponential growth of modern humans began have survived; any Neanderthal sequences in a mixed population could just have been lost by chance (see *Box 8.11*). Analysis of additional loci would help to resolve this question, but it will be even more difficult to obtain

BOX 8.10 Opinion: Against the multiregional hypothesis

Despite recent significant fossil and genetic discoveries that directly address modern human origins, the debate has not been resolved. It seems that one reason is due to lack of quantitative arguments. Almost all hypotheses put forward for modern human origins have not been specific enough to develop solid population genetic models and could adjust to new findings after each confrontation (Klein and Takahata, 2002). Inevitably, the debate has been verbal and often incomprehensible.

According to the multiregional hypothesis, the regional populations dispersed worldwide and evolved **anagenetically** (forming a new species without splitting) throughout the past 2 MY, but the hypothesis is nebulous about its specific claims with respect to gene flow, natural selection, and genetic drift. One way of testing the hypothesis in light of the available genetic data is to take into account its essential features only, and examine the ranges of parameters within which the hypothesis is viable. It is essential that there was a worldwide network of genetic exchanges among evolving regional populations throughout the **Pleistocene**. It is also essential that the regional populations diverged from one another in certain characters, but simultaneously they also changed uniformly by acquiring a common set of novel characters. Since these two features are likely to be associated with two types of selected mutations, we must consider not only neutral, but also regionally and globally advantageous mutations.

DNA sequence data allow us to reconstruct the genealogical relationships in a sample of genes at a locus. They contain important information on the demographic history of humans. Specifically, we can infer the time and place (regional population) of the most recent common ancestor (MRCA). For globally advantageous mutations, the genealogy must be shorter than the time of appearance of modern human characters (less than 200 KYA) and the place of the MRCA can be any regional population. For regionally advantageous mutations, the genealogy likely goes beyond the Pleistocene and the place of the MRCA should be Africa. The genealogy of such genes would be reciprocally monophyletic with respect to regional populations. The genealogical behavior of neutral mutations depends heavily on the extent of gene flow and genetic drift within regional populations. If genetic drift is weak (large regional populations) and/or the extent of gene flow is low, the outcome is the same as in regionally advantageous mutations. Under some appropriate conditions, the MRCA is expected to occur during the Pleistocene, but the place of MRCA can again be any regional population.

We can ask if all regional populations made significant genetic contributions to human evolution during the Pleistocene. If we observe a strong bias for one particular regional population, a model of the multiregional hypothesis imposes conditions under which it is viable (Takahata *et al.*, 2001). At present, there are 12 loci at which their respective genealogies can be reconstructed from worldwide samples of DNA sequences. The estimated time of the MRCA at all loci is within the Pleistocene. In addition, there is no indisputable indication that these sampled loci harbor globally or regionally advantageous mutations. If the sample contains such loci, they could have strongly supported or rejected the multiregional hypothesis. However, there is one notable feature in their genealogies. The estimated place of the MRCA is Africa, or the genealogy of Africans is **paraphyletic** to that of non-Africans, at all but one locus, which is consistent with a number of observations that the African population has had a great genetic impact upon the current human gene pool. The imposed condition on the multiregional hypothesis is that the African population must have been much larger in size or more geographically structured within it than the non-African population. The multiregional hypothesis may absorb and adjust to this condition. The result would be a resemblance to the Assimilation hypothesis of African origins of modern humans. Despite an obvious difference in time-scale between these hypotheses, roles of non-African populations must be lessened to a great extent and I wonder if such a modified version can still appropriately be called 'multiregional'.

Naoyuki Takahata

Department of Biosystems Science, Graduate University for Advanced Studies (Sokendai), Japan.

authentic data because nuclear loci are present in much lower copy number than mtDNA.

It is worth bearing in mind that if Neanderthal and modern human sequences *had* been similar, the authenticity of the Neanderthal DNA would have been much more difficult to establish. Of the nine criteria listed for establishing the authenticity of ancient DNA (*Box 4.5*), phylogenetic distinctiveness was decisive in demonstrating that genuine Neanderthal DNA had been identified. Prospects for complementary information from other loci thus look bleak.

Summary

▶ Information about modern human origins is provided by fossils, archaeological remains, studies of modern human genetic variation and, to a limited extent, by analysis of ancient DNA.

▶ Interpretations of almost all sources of evidence are hotly debated and it is difficult to identify a consensus view about many topics.

▶ Fossils that date to the approximate time of the chimpanzee/human split, about 5–7 MYA, have been

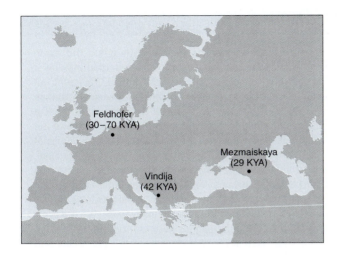

Figure 8.18: Sites of Neanderthal fossils that have provided mtDNA sequence data.

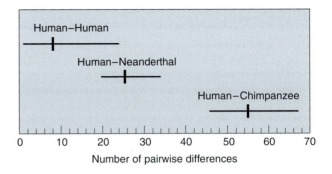

Figure 8.19: Comparison of modern human, Neanderthal and chimpanzee mtDNA sequences.

Pairwise comparisons were carried out using data from 333 bp of HVS I. The mean and range for each comparison is shown. Data are from Krings *et al.* (1997).

described recently from several locations in Africa. They show some features that place them on the human line, including upright walking, but remain to be fully evaluated.

▶ Several hominid species are known from Africa between about 4.2 MYA and the appearance of *Homo* at around 2 MYA. Most of these belong to the genus *Australopithecus*, but there are additional hominids with uncertain affiliations. The later Australopithecines, such as *A. habilis*, are associated with Oldowan stone tools.

▶ The first *Homo* species, *H. erectus*, appeared around 1.9 MYA in Africa, and exhibited a height and weight similar to modern humans, but a smaller brain. *H. erectus*, associated with Acheulean technology in some places, spread rapidly over much of the Old World, including Georgia, Indonesia and China.

Several later *Homo* species are known, including *heidelbergensis*, *neanderthalensis* and *sapiens*, although their relationships are still debated. They are associated with several different archaeological assemblages.

▶ Modern human morphology is first found in Africa at about 130 KYA, but only substantially later (probably after 50 KYA) in other parts of the world.

▶ These observations led to the development of several hypotheses about the origins of modern humans, two of which were the multiregional and out of Africa models.

▶ Genetic diversity for most loci is higher in Africa than on other continents, which is consistent with a longer period of evolution in Africa, or a larger population size.

▶ Most phylogenies of individual loci show a root in Africa and a subset of lineages in other parts of the world; this is seen particularly clearly in the well-resolved phylogenies of mtDNA and the Y chromosome. Such results imply an African origin for most, if not all, of our ancestors.

▶ DNA cannot be extracted from most fossils, but ancient DNA analysis has been used to demonstrate that the mtDNA of *H. neanderthalensis* was distinct from that of living humans.

▶ The consensus view in the field is therefore that the out of Africa model explains the data most effectively.

BOX 8.11 Opinion: Were Neanderthals and anatomically modern humans different species?

The fossil record indicates that Neanderthals existed in Europe and the Near East until approximately 28 KYA. The first traces of anatomically modern humans in these regions appear much earlier, perhaps 100 KYA in Israel (Section 8.2). These two morphologically distinct groups thus co-existed for a very long period of time, and the question of whether they also interbred has gained much attention, ranging from a best-selling series of Paleolithic romance novels, to recent debates among academics about the discovery of a (fossilized) Neanderthal 'love child'.

Many people considered the question answered when Svante Pääbo's group managed to sequence a short piece of mtDNA using ancient DNA extracted from the Neanderthal type specimen (Krings *et al.*, 1997). Phylogenetic analysis of this sequence and more than 900 homologous sequences from modern humans revealed that the Neanderthal sequence was distinct from all the modern ones (in other words, the modern sequences were monophyletic with respect to the Neanderthal sequence). As illustrated in the *Figure*, the time to the MRCA of a modern human mtDNA and the Neanderthal mtDNA was estimated to be approximately four times greater than the time to the MRCA of the most distantly related modern human mtDNAs. Their result was widely interpreted as proving that Neanderthals and anatomically modern humans did not interbreed.

This conclusion is not warranted. Rather, it is a canonical example of the danger of confusing species and gene trees (*Box 7.1*). Clearly, the mtDNA tree in the *Figure* is precisely what we would expect to see if Neanderthals and anatomically modern humans did not interbreed; however, consistency with a hypothesis does not constitute proof of that hypothesis. What is required is a demonstration that alternative hypotheses are rendered less likely by the data. Specifically, what is the likelihood of seeing a gene tree like the one in the *Figure* if Neanderthals and anatomically modern humans did in fact interbreed? Straightforward calculations using standard population genetics models reveal that this probability is very high (Nordborg, 1998). In other words, regardless of whether Neanderthals and anatomically modern humans interbred, it is highly likely that the mtDNA tree would look like the tree in the *Figure*. Thus, whatever beliefs one may have had about the existence of Neanderthal–human affairs (and most people who study human evolution seem to have strong beliefs), there is no reason to change them because of the mtDNA data.

The result that interbreeding may not leave a trace in the mtDNA data may at first seem baffling, but is actually easy to understand intuitively. Consider a large urn full of gold coins. You are concerned that there has been some funny business, and that some of the coins are in fact bronze. To test this, you randomly sample coins and look at them carefully. After looking at more than 900 without finding any bronze coins, you would be reasonably confident that few, if any, of the coins in the urn are bronze. Analogously, the mtDNA data show that no modern human has a sequence like the Neanderthal one. However, the question of interbreeding concerns ancient humans, not modern ones. Did some anatomically modern humans carry Neanderthal mtDNA at the time when Neanderthals were still around? This turns out to be impossible to answer by looking at modern human sequences. The reason is so-called genetic drift, or random changes in allele frequencies over time. Going back to the coin analogy, imagine that a new urn is filled with copies of coins from the old one. The old coins are randomly sampled, one at a time, and returned to the old urn after a copy has been put in the new one. If there were any bronze coins in the old urn, there may be fewer or more in the new one, just by chance. If the process is repeated many times, we are likely to end up with a pure gold (or pure bronze) urn, even if the original urn was mixed. Analogously, it is highly likely that even if Neanderthal mtDNAs existed among anatomically modern humans 50 KYA (say), they would all have been lost by now.

At this point, the reader may think 'OK, so there may have been some admixture, but why does it matter if all traces were lost?' The answer to this is that all traces would not have been lost. Just because all Neanderthal mtDNAs were lost does not mean that Neanderthal alleles at other loci were also lost. In fact, simple calculations show that this is exceedingly unlikely to have happened (Nordborg, 1998). In order to rule out admixture, it is necessary to look at many different (unlinked) loci. Going back to the coins, the genome contains tens of thousands of urns, each with a different type of coin. We need to look at more than one urn before we can say that none of them contains bronze coins. Further studies of mtDNA will tell us nothing: it is necessary to take a genomic approach. At least with current techniques, this is not likely to be feasible with ancient DNA. Statistical analysis of sequence polymorphism from modern humans is a more promising strategy (Wall, 2000), but it is important to recognize that modern DNA contains very limited information about what happened in the past. It is unfortunately true that evolutionary biology contains many questions that are in principle interesting but are in practice unanswerable.

Magnus Nordborg

Molecular & Computational Biology, University of Southern California, Los Angeles, USA

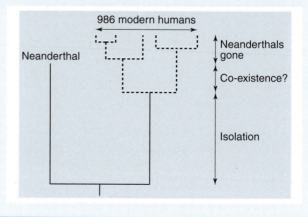

Further reading

Aitken MJ, Stringer CB, Mellars PA, eds (1993) *The Origin of Modern Humans and the Impact of Chronometric Dating.* Princeton University Press, Princeton.

Johanson D, Edgar B (1996) *From Lucy to Language.* Weidenfeld and Nicolson, London.

Klein J, Takahata N (2002) *Where do we Come From? The Molecular Evidence for Human Descent.* Springer, Berlin.

Relethford JH (2001) *Genetics and the Search for Modern Human Origins.* Wiley-Liss, New York.

Electronic references

Human evolution and fossils:
http://www.archaeologyinfo.com

Photographs and discussion of many key fossils:
http://www.modernhumanorigins.com

References

Aiello LC, Collard M (2001) Our newest oldest ancestor? *Nature* **410**, 526–527.

Anderson S, Bankier AT, Barrell BG et al. (1981) Sequence and organization of the human mitochondrial genome. *Nature* **290**, 457–465.

Andrews RM, Kubacka I, Chinnery PF, Lightowlers RN, Turnbull DM, Howell N (1999) Reanalysis and revision of the Cambridge reference sequence for human mitochondrial DNA. *Nature Genet.* **23**, 147.

Arnason U, Xu X, Gullberg A (1996) Comparison between the complete mitochondrial DNA sequences of *Homo* and the common chimpanzee based on nonchimeric sequences. *J. Mol. Evol.* **42**, 145–152.

Asfaw B, Gilbert WH, Beyene Y, Hart WK, Renne PR, WoldeGabriel G, Vrba ES, White TD (2002) Remains of *Homo erectus* from Bouri, Middle Awash, Ethiopia. *Nature* **416**, 317–320.

Asfaw B, White T, Lovejoy O, Latimer B, Simpson S, Suwa G (1999) *Australopithecus garhi*: a new species of early hominid from Ethiopia. *Science* **284**, 629–635.

Bermudez de Castro JM, Arsuaga JL, Carbonell E, Rosas A, Martinez I, Mosquera M (1997) A hominid from the lower Pleistocene of Atapuerca, Spain: possible ancestor to Neandertals and modern humans. *Science* **276**, 1392–1395.

Bowcock AM, Ruiz-Linares A, Tomfohrde J, Minch E, Kidd JR, Cavalli-Sforza LL (1994) High resolution of human evolutionary trees with polymorphic microsatellites. *Nature* **368**, 455–457.

Brown MD, Starikovskaya E, Derbeneva O, Hosseini S, Allen JC, Mikhailovskaya IE, Sukernik RI, Wallace DC (2002) The role of mtDNA background in disease expression: a new primary LHON mutation associated with Western Eurasian haplogroup J. *Hum. Genet.* **110**, 130–138.

Brunet M, Beauvilain A, Coppens Y, Heintz E, Moutaye AH, Pilbeam D (1995) The first australopithecine 2,500 kilometres west of the Rift Valley (Chad). *Nature* **378**, 273–275.

Brunet M, Guy F, Pilbeam D et al. (2002) A new hominid from the Upper Miocene of Chad, Central Africa. *Nature* **418**, 145–151.

Cann RL, Stoneking M, Wilson AC (1987) Mitochondrial DNA and human evolution. *Nature* **325**, 31–36.

Chen YS, Torroni A, Excoffier L, Santachiara-Benerecetti AS, Wallace DC (1995) Analysis of mtDNA variation in African populations reveals the most ancient of all human continent-specific haplogroups. *Am. J. Hum. Genet.* **57**, 133–149.

Chen Y-S, Olckers A, Schurr TG, Kogelnik AM, Huoponen K, Wallace DC (2000) mtDNA variation in the South African Kung and Khwe – and their genetic relationships to other African populations. *Am. J. Hum. Genet.* **66**, 1362–1383.

Clark AG, Weiss KM, Nickerson DA et al. (1998) Haplotype structure and population genetic inferences from nucleotide-sequence variation in human lipoprotein lipase. *Am. J. Hum. Genet.* **63**, 595–612.

Dart R (1925) *Australopithecus africanus*: the man-ape of South Africa. *Nature* **115**, 195–199.

Gabunia L, Vekua A (1995) A Plio-Pleistocene hominid from Dmanisi, East Georgia, Caucasus. *Nature* **373**, 509–512.

Gibbons A (2002) In search of the first hominids. *Science* **295**, 1214–1219.

Haile-Selassie Y (2001) Late Miocene hominids from the Middle Awash, Ethiopia. *Nature* **412**, 178–181.

Hammer MF, Karafet TM, Redd AJ, Jarjanazi H, Santachiara-Benerecetti S, Soodyall H, Zegura SL (2001) Hierarchical patterns of global human Y-chromosome diversity. *Mol. Biol. Evol.* **18**, 1189–1203.

Harding RM, Fullerton SM, Griffiths RC, Bond J, Cox MJ, Schneider JA, Moulin DS, Clegg JB (1997) Archaic African and Asian lineages in the genetic ancestry of modern humans. *Am. J. Hum. Genet.* **60**, 772–789.

Harding RM, Healy E, Ray AJ et al. (2000) Evidence for variable selective pressures at *MC1R*. *Am. J. Hum. Genet.* **66**, 1351–1361.

Harpending H, Rogers A (2000) Genetic perspectives on human origins and differentiation. *Annu. Rev. Genomics Hum. Genet.* **1**, 361–385.

Harpending HC, Sherry ST, Rogers AR, Stoneking M (1993) The genetic structure of ancient human populations. *Curr. Anthropol.* **34**, 483–496.

Harris EE, Hey J (1999) X chromosome evidence for ancient human histories. *Proc. Natl Acad. Sci. USA* **96**, 3320–3324.

Herrnstadt C, Elson JL, Fahy E *et al.* (2002) Reduced-median-network analysis of complete mitochondrial DNA coding-region sequences for the major African, Asian, and European haplogroups. *Am. J. Hum. Genet.* **70**, 1152–1171.

Ingman M, Kaessmann H, Pääbo S, Gyllensten U (2000) Mitochondrial genome variation and the origin of modern humans. *Nature* **408**, 708–713.

Jobling MA, Tyler-Smith C. (2003) The human Y chromosome: an evolutionary marker comes of age. *Nature Rev. Genet.* **4**, 598–612.

Johanson DC, White TD (1979) A systematic assessment of early African hominids. *Science* **203**, 321–330.

Kaessmann H, Heissig F, von Haeseler A, Pääbo S (1999) DNA sequence variation in a noncoding region of low recombination on the human X chromosome. *Nature Genet.* **22**, 78–81.

Kivisild T, Tolk HV, Parik J, Wang Y, Papiha SS, Bandelt HJ, Villems R (2002) The emerging limbs and twigs of the East Asian mtDNA tree. *Mol. Biol. Evol.* **19**, 1737–1751.

Krausz C, Quintana Murci L, Rajpert De Meyts E, Jørgensen N, Jobling MA, Rosser ZH, Skakkebaek NE, McElreavey K (2001) Identification of a Y chromosome haplogroup associated with reduced sperm counts. *Hum. Mol. Genet.* **10**, 1873–1877.

Krings M, Capelli C, Tschentscher F *et al.* (2000) A view of Neandertal genetic diversity. *Nature Genet.* **26**, 144–146.

Krings M, Stone A, Schmitz RW, Krainitzki H, Stoneking M, Pääbo S (1997) Neandertal DNA sequences and the origin of modern humans. *Cell* **90**, 19–30.

Leakey LSB (1959) A new fossil skull from Olduvai. *Nature* **184**, 491–493.

Leakey LSB, Tobias PV, Napier JR (1964) A new species of genus *Homo* from Olduvai Gorge. *Nature* **202**, 7–9.

Leakey MD, Hay RL (1979) Pliocene footprints in the Laetolil Beds at Laetoli, northern Tanzania. *Nature* **278**, 317–323.

Leakey MG, Feibel CS, McDougall I, Walker A (1995) New four-million-year-old hominid species from Kanapoi and Allia Bay, Kenya. *Nature* **376**, 565–571.

Leakey MG, Spoor F, Brown FH, Gathogo PN, Kiarie C, Leakey LN, McDougall I (2001) New hominin genus from eastern Africa shows diverse middle Pliocene lineages. *Nature* **410**, 433–440.

Leakey RE (1973) Evidence for an advanced plio-pleistocene hominid from East Rudolf, Kenya. *Nature* **242**, 447–450.

Lieberman DE, McBratney BM, Krovitz G (2002) The evolution and development of cranial form in *Homo sapiens*. *Proc. Natl Acad. Sci. USA* **99**, 1134–1139.

Macaulay V, Richards M, Hickey E *et al.* (1999) The emerging tree of West Eurasian mtDNAs: a synthesis of control-region sequences and RFLPs. *Am. J. Hum. Genet.* **64**, 232–249.

Mercader J, Panger M, Boesch C (2002) Excavation of a chimpanzee stone tool site in the African rainforest. *Science* **296**, 1452–1455.

Mietto P, Avanzini M, Rolandi G (2003) Human footprints in Pleistocene volcanic ash. *Nature* **422**, 133.

Mishmar D, Ruiz-Pesini E, Golik P *et al.* (2003) Natural selection shaped regional mtDNA variation in humans. *Proc. Natl Acad. Sci. USA* **100**, 171–176.

Nordborg M (1998) On the probability of Neanderthal ancestry. *Am. J. Hum. Genet.* **63**, 1237–1240.

Poulton J, Luan J, Macaulay V, Hennings S, Mitchell J, Wareham NJ (2002) Type 2 diabetes is associated with a common mitochondrial variant: evidence from a population-based case–control study. *Hum. Mol. Genet.* **11**, 1581–1583.

Ruiz-Pesini E, Lapena A-C, Díez-Sánchez C *et al.* (2000) Human mtDNA haplogroups associated with high or reduced spermatozoa motility. *Am. J. Hum. Genet.* **67**, 682–696.

Salas A, Richards M, De la Fe T, Lareu M-V, Sobrino B, Sanchez Diz P, Macaulay V, Carracedo A (2002) The making of the African mtDNA landscape. *Am. J. Hum. Genet.* **71**, 1082–1111.

Schurr TG, Sukernik RI, Starikovskaya YB, Wallace DC (1999) Mitochondrial DNA variation in Koryaks and Itel'men: population replacement in the Okhotsk Sea-Bering Sea region during the Neolithic. *Am. J. Phys. Anthropol.* **108**, 1–39.

Senut B (2002) *The Observer*, 14th July, p. 19.

Senut B, Pickford M, Gommery D, Mein P, Cheboi K, Coppens Y (2001) First hominid from the Miocene (Lukeino Formation, Kenya). *C. R. Acad. Sci.* **332**, 137–144.

Skaletsky H, Kuroda-Kawaguchi T, Minx PJ, Cordum HS, Hillier L, Brown LG, *et al.* (2003) The male-specific region of the human Y chromosome: a mosaic of discrete sequence classes. *Nature* **423**, 825–837.

Swisher CCr, Curtis GH, Jacob T, Getty AG, Suprijo A, Widiasmoro (1994) Age of the earliest known hominids in Java, Indonesia. *Science* **263**, 1118–1121.

Swisher CCr, Rink WJ, Anton SC, Schwarcz HP, Curtis GH, Suprijo A, Widiasmoro (1996) Latest *Homo erectus* of Java: potential contemporaneity with *Homo sapiens* in southeast Asia. *Science* **274**, 1870–1874.

Takahata N, Lee SH, Satta Y (2001) Testing multiregionality of modern human origins. *Mol. Biol. Evol.* **18**, 172–183.

Templeton AR (2002) Out of Africa again and again. *Nature* **416**, 45–51.

Thieme H (1997) Lower Palaeolithic hunting spears from Germany. *Nature* **385**, 807–810.

Thomson R, Pritchard JK, Shen P, Oefner PJ, Feldman MW (2000) Recent common ancestry of human Y chromosomes: evidence from DNA sequence data. *Proc. Natl Acad. Sci. USA* **97**, 7360–7365.

Tishkoff SA, Dietzsch E, Speed W *et al.* (1996) Global patterns of linkage disequilibrium at the *CD4* locus and modern human origins. *Science* **271**, 1380–1387.

Tishkoff SA, Kidd KK, Clark AG (1998) Inferences of modern human origins from variation in *CD4* haplotypes. In: *Proceedings of the Trinational Workshop on Molecular Evolution* (eds MK Uyenoyama, A von Haeseler). Duke University Publications Group, Durham, NC. pp. 181–198.

Torroni A, Petrozzi M, D'Urbano L *et al.* (1997) Haplotype and phylogenetic analyses suggest that one European-specific mtDNA background plays a role in the expression of Leber hereditary optic neuropathy by increasing the penetrance of the primary mutations 11778 and 14484. *Am. J. Hum. Genet.* **60**, 1107–1021.

van Schaik CP, Ancrenaz M, Borgen G *et al.* (2003) Orangutan cultures and the evolution of material culture. *Science* **299**, 102–105.

Vigilant L, Stoneking M, Harpending H, Hawkes K, Wilson AC (1991) African populations and the evolution of human mitochondrial DNA. *Science* **253**, 1503–1507.

Walker A, Leakey REF (Eds) (1993) *The Nariokotome* Homo erectus *Skeleton*. Cambridge, Harvard University Press.

Wall JD (2000) Detecting ancient admixture in humans using sequence polymorphism data. *Genetics* **154**, 1271–1279.

White TD, Suwa G, Asfaw B (1994) *Australopithecus ramidus*, a new species of early hominid from Aramis, Ethiopia. *Nature* **371**, 306–312.

White TD, Suwa G, Asfaw B (1995) *Australopithecus ramidus*, a new species of early hominid from Aramis, Ethiopia. *Nature* **375**, 88.

White TD, Asfaw B, DeGusta D, *et al.* (2003) Pleistocene *Homo sapiens* from Middle Awash, Ethiopia. *Nature* **423**, 742–747.

Whiten A, Goodall J, McGrew WC *et al.* (1999) Cultures in chimpanzees. *Nature* **399**, 682–685.

Wood B (1996) Human evolution. *BioEssays*, **18**, 945–954.

Wood B, Collard M (1999) The human genus. *Science* **284**, 65–71.

Y Chromosome Consortium (2002) A nomenclature system for the tree of human Y-chromosomal binary haplogroups. *Genome Res.*, **12**, 339–348.

Yamei H, Potts R, Baoyin Y *et al.* (2000) Mid-Pleistocene Acheulean-like stone technology of the Bose basin, South China. *Science* **287**, 1622–1626.

Yu N, Chen FC, Ota S *et al.* (2002) Larger genetic differences within Africans than between Africans and Eurasians. *Genetics* **161**, 269–274.

Zilhao J, d'Errico F (1999) The chronology and taphonomy of the earliest Aurignacian and its implications for the understanding of Neanderthal extinction. *J. World Prehistory* **13**, 1–68.

SECTION FIVE

How did humans colonize the world?

The transformation of humans from a rare African species into a numerous one with a worldwide distribution is an unprecedented biological phenomenon, and is central to understanding why humans are genetically so similar to one another, and explaining the small, but appreciable, geographical differences that do exist among human populations. We continue along a path that is approximately, but not precisely, chronological, discussing the early movements of modern humans out of Africa before considering the major effects that have followed the subsequent introduction of farming and the meeting of populations.

CHAPTER 9   *The distribution of diversity – out of Africa and into Asia, Australia and Europe*

The study of human diversity raises important ethical and methodological questions, and we begin this chapter by considering these. Next, we examine how human genetic variation is distributed around the world, and how the patterns found for neutral genetic loci can be affected by natural selection. The evidence for the routes and timing of movements into Asia, Australia and Europe, derived from fossil, archaeological and genetic sources, is then discussed.

CHAPTER 10   *Agricultural expansions*

In this chapter we focus on the appearance of agriculture in Europe and Africa. Did the farmers increase in number at the expense of nearby non-farming populations, or did the neighbors learn farming and then expand themselves? As well as considering the effects of the agricultural revolution on humans themselves, we examine its genetic impact on the plant and animal species that early farmers chose to domesticate.

CHAPTER 11   *Into new found lands*

We now consider the peopling of the Americas and the islands of the Pacific. The peopling of the Americas is highly contentious, but there is agreement that most of the current indigenous gene pool dates back to a migration over 15 000 years ago. The enormous stretch of the Pacific forming Remote Oceania was uninhabited until 3500 years ago, and here there is more agreement that the major migration was from the west, but with genetic contributions from several sources.

CHAPTER 12   *What happens when populations meet?*

In the final chapter of this section, we consider a process that must have been important throughout the spread of *Homo sapiens*: the mixing of populations, or admixture. This can affect a population in many ways, including its culture and language, as well as its genetics. Genetic admixture, however, is not simply an equal or random mixing of genomes from the two populations. It is usually sex-biased, affecting the mtDNA, autosomes and Y chromosome differentially, and establishing linkage disequilibrium: a legacy that persists for many generations.

CHAPTER NINE

The distribution of diversity – out of Africa and into Asia, Australia and Europe

9.1 Introduction

In the last chapter we saw that the human genus (*Homo*) originated in Africa, and that most evidence also supports the idea of a recent (< 150 KYA) origin for modern humans in Africa. However, uncertainty remains about the size, if any, of the contribution of archaic hominids outside Africa to the modern gene pool. In this chapter we will explore two topics:

▸ the distribution of contemporary genetic diversity around the world: is it structured, and, if so, how?

▸ the exodus of anatomically modern humans from Africa leading to the peopling of the Old World and Australia: when and by which routes did humans reach Asia, Australia and Europe?

Human entry to other areas of the world is discussed later (Chapter 11). Sources of evidence are again:

▸ genetics of modern populations;

▸ fossils and archaeology;

▸ environmental conditions.

9.2 Studying human diversity

The study of human diversity has a long history and has in the past been linked to both blatant and subtle forms of **racism**. We will begin by considering some of the historical aspects of the field, and ethical issues raised by such work.

9.2.1 The history and ethics of studying diversity

Should we study human genetic diversity, or is this an area of work where the potential for misuse (see *Figure 9.1* and *Box 9.1*) outweighs the potential benefits to such an extent that it should not be pursued? Such studies already have a long history, as we will see. One pragmatic answer to this question is that information on human genetic diversity is needed and is therefore generated for medical and forensic applications, and so is already available whatever evolutionary geneticists decide, so we must be ready to consider its implications and consequences. Another is that the information in fact refutes any scientific basis for racism and so can be used to combat misuse; furthermore, it is of enormous intrinsic interest to many people, not just scientists.

Linnaeus' classification of human diversity
Our biological classification system originates with Linnaeus (1707–1778), who subdivided humans into two species (*diurnus* and *nocturnus*; see *Figure 9.2*) and a total of seven categories:

Apollo Belvidere

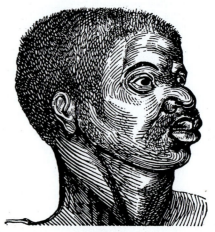

Negro

Young chimpanzee

Figure 9.1: A racist view of humanity

From *Indigenous Races of the Earth* by Nott and Gliddon, 1868. Note not only the hierarchy, but also the falsification of Negro and chimpanzee skulls. Reproduced with permission from Ayer Co Publishing Inc.

BOX 9.1 Race and racism

'On May 7, 1876, Truganini, the last full-blood Black person in Tasmania, died at seventy-three years of age. Her mother had been stabbed to death by a European. Her sister was kidnapped by Europeans. Her intended husband was drowned by two Europeans in her presence, while his murderers raped her.

It might be accurately said that Truganini's numerous personal sufferings typify the tragedy of the Black people of Tasmania as a whole. She was the very last. 'Don't let them cut me up,' she begged the doctor as she lay dying. After her burial, Truganini's body was exhumed, and her skeleton, strung upon wires and placed upright in a box, became for many years the most popular exhibit in the Tasmanian Museum and remained on display until 1947. Finally, in 1976 – the centenary year of Truganini's death – despite the museum's objections, her skeleton was cremated and her ashes scattered at sea.'

From *Black War: the Destruction of the Tasmanian Aborigines* by Runoko Rashidi, http://www.cwo.com/~lucumi/tasmania.html

The genocide of the Tasmanian Aborigines by the European settlers in the nineteenth century provides one of the worst of many examples of racism, and is notable for the 'anthropological' justification of the public exhibition of Truganini's, and other, remains. Yet racism does not consist only of such crude episodes: racist thinking penetrates deeply into Western, and perhaps all, culture, evolutionary thought and genetics. Consider these two quotations:

'I advance it, therefore, as a suspicion only, that the blacks, whether originally a distinct race, or made distinct by time and circumstance, are inferior to the whites in the endowment both of body and mind.'

'We hold these truths to be self-evident: that all men are created equal.'

Both are from Thomas Jefferson, and the next is Charles Darwin, writing about the gap between humans and apes after an anticipated future extinction of gorillas and 'Hottentots':

'The break will then be rendered wider, for it will intervene between man in a more civilized state, as we may hope, than the Caucasian, and some ape as low as a baboon, instead of as at present between the negro or Australian and the gorilla.'

Hierarchies of humans and apes, such as that illustrated in *Figure 9.1* were common in anthropological and biological literature. They were usually based on a small number of visible characteristics such as:

▶ skin color;

▶ hair color and morphology;

▶ facial features.

These characteristics are influenced by both environmental and genetic factors, but even if allowance was made for the environment, the genes affecting these phenotypes have probably been subject to particular selection pressures and perhaps sexual selection (Sections 13.2 and 13.3). Thus they are unrepresentative of the majority of the genome. Inevitably, the compilers put their own group at the top and those they wanted to exploit at the bottom (see *Figure 9.1*). Another notable feature of these schemes was that the number of 'races' identified varied greatly between authors.

'Race' in addition to its everyday usages, is a biological term with a clear meaning: it refers to a group of individuals who can be cleanly distinguished from other groups of the same species. Of course, this requires that we specify what is meant by 'cleanly', and an F_{ST} value ≥ 0.3 is commonly used: that is, 30% or more of the variation needs to be found between groups for these groups to be classified as 'races'. Some species are divided into races; the question of whether or not humans are such a species is an empirical one. We will see that the answer (Section 9.3.1) is 'no'.

▶ *diurnus* (also referred to as *Homo sapiens* by Linnaeus);

   ▶ americanus: red, with black hair and a scanty beard, obstinate, free, painted with fine red lines, regulated by customs;

   ▶ europeus: white, long flowing hair, blue eyes, sanguine, muscular, inventive, covered with tight clothing, governed by laws;

   ▶ asiaticus: yellow, melancholy, black hair and brown eyes, severe, haughty, stingy, wears loose clothing, governed by opinions;

   ▶ afer (i.e., African): black, cunning, phlegmatic, black curly hair, women without shame and lactate profusely, anointed with grease, ruled by impulse;

   ▶ monstrosus: a miscellaneous collection including dwarfs and large, lazy Patagonians.

▶ *nocturnus*

   ▶ troglodytes: nocturnal, hunts only at night, lives underground.

In some classifications he also included ferus: wild, hairy, runs about on all fours. Apart from the misleading notion that humans can be categorized into such groups and the language that now sounds offensive, this classification is notable for its mixing of real and imaginary categories, and physical, intellectual and cultural characteristics.

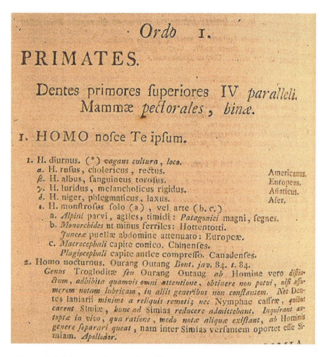

Figure 9.2: Linnaeus' 1756 classification of humans.

Galton's 'Comparative worth of different races'

In *Hereditary Genius: an Inquiry Into its Laws and Consequences* published in 1869, Francis Galton (1822–1911) established a grading system, A, B, C etc., for people within each race, and then compared the grades between races. At the top of the racial hierarchy were the ancient Greeks; Galton was slightly critical of his own race, the English, *'the calibre of whose intellect is easily gauged by a glance at the contents of a railway book-stall'*, and placed them two grades below the Greeks, although *'the average standard of the Lowland Scotch and the English North-country men is decidedly a fraction of a grade superior to that of the ordinary English'*. Inevitably, *'the average intellectual standard of the negro race is some two grades below our own'* and *'the Australian type is at least one grade below the African negro'*.

Modern attitudes to studying diversity

If genetic diversity studies of humans are an acceptable part of science, what are their prerequisites? There is a fundamental requirement, recognized by international law, that research on humans can only be undertaken with **informed consent** from the subject. This means that, with some exceptions for forensic investigations and in limited medical situations, individuals must not only agree to the research, but must do this on the basis of an understanding of the nature and purpose of the research, and of its risks and benefits. The risks associated with the physical procedures of donating cheek cells, hair or blood to provide DNA are minimal (see Section 4.3); debate has focused on the risks associated with the use of the information obtained, discussed in Greely, 2001a, and *Box 9.2*. The results may have implications for:

▶ health and, in some countries, health insurance: what if the donor is found to have a disease predisposition?

▶ stigmatization: what if the donor carries a trait judged to be undesirable?

▶ commercial applications: what if a cell line or DNA sequence leads to a patentable or saleable product?

Additional novel questions are raised by genetic research because we share DNA variants with our relatives, so study of one individual provides information about other members of their family and population. Therefore **group informed consent** is required in many situations and it would be unethical to sample consenting individuals from a group that had not given consent. The appropriate authority to provide such group consent, if there is one, can only be determined on a case-by-case basis for each population.

Benefits from genetic diversity studies are:

▶ increased understanding of genetic history and relationships;

▶ medical advances such as the identification of genes predisposing to disease (see Chapter 14);

▶ accurate paternity testing, victim and assailant identification, and other forensic applications (see Chapter 15);

▶ some immediate benefits to the population such as medical advice or treatment.

Complications exist because the people who receive most of the long-term benefits may not be the donors.

Outstanding issues that have not been fully resolved include:

▶ is informed consent from members of cultures that do not ascribe to Western scientific values truly 'informed'?

▶ how much information about the donor should accompany a cell line or DNA sample?

▶ can samples collected many years, or perhaps decades, ago with no written consent be used?

▶ can samples collected for one study be used in another?

▶ can an individual give general consent for all future studies, which may involve techniques that do not yet exist and have implications that are not currently understood?

It is difficult to give general answers to many of the ethical questions that diversity studies raise; answers may emerge more satisfactorily through the consideration of individual cases. The Human Genome Diversity Project (Section 9.2.3) has drawn up guidelines and these may be found on the website http://www.stanford.edu/group/morrinst/hgdp/protocol.html

9.2.2 Who should be studied?

The starting point for any study of human diversity is a set of humans, and this raises the question: who should be studied? Sampling always creates problems: is the sample representative?

BOX 9.2 Opinion: The Human Genome Diversity Project, a personal view

It was in Quetzaltenango, Guatemala, in late November 1993, that I was first accused of being an agent for the CIA. At least, that's what the translator told me, but I had the distinct impression that she was leaving out other, even less savory, descriptions. I had come by invitation to the General Congress of the World Council of Indigenous Peoples as a representative of the proposed Human Genome Diversity Project (HGDP) to explain the Project; I left having learned at least as much as I had taught.

The HGDP was first proposed in 1991 by a group of human geneticists, led by Luca Cavalli-Sforza and Allan Wilson. The Project's concept was simple – to complement the Human Genome Project, which hoped to sequence the equivalent of one human genome, by collecting, analyzing, and making available for research a broad set of samples of the 6 billion existing human genomes, drawn from throughout the human species. The primary goals of the founders were to advance research into human history and evolution, but they foresaw other possible uses – in medicine, population genetics, anthropology, and other fields. What they did *not* foresee was the ethical – and political – storm ahead for the Project.

Even before the HGDP was launched in September 1993, an NGO called the Rural Advancement Foundation International (RAFI) had begun to excoriate it as a 'bio-pirate', interested in stealing valuable genes from indigenous peoples. RAFI and others accused the HGDP not only of underhanded commercial goals, but of planning to undermine indigenous cultures, to overthrow indigenous land rights, to help the US produce ethnically-targeted biological weapons, and to clone armies of indigenous warrior slaves. RAFI summed up the HGDP with the sobriquet 'the Vampire Project'.

The researchers involved in the HGDP were dumbfounded by this reaction. Academics with great sympathy for indigenous people, they tried to stress that the project was resolutely noncommercial (and even more nonmilitary), that it was not focused on indigenous peoples but was interested in all the world's populations, and that no samples would be taken without full informed consent and local government approval. The North American Committee of the HGDP went farther and, in its Model Ethical Protocol for Collecting DNA Samples, bound itself not to take any samples without both individual and, whenever feasible, *group* informed consent (http://stanford.edu/group/morrinst/hgdp.htm). The HGDP continues to exist but, probably at least in part because of the political controversy attached to it, has never received substantial funding – and has never come close to achieving its goals (Greely, 2001b). Its unhappy experience, however, has illuminated some of the tough ethical and political issues in genetic research on human populations.

Rules for protecting human research subjects have focused on preventing harm to individuals, but groups have interests too (Greely, 2001a). Groups are interested in avoiding disease-related stigma; they may be interested in preserving their own culture's history and traditions; they may be more interested in immediate benefits to their people than participation in academic projects. And whenever genetics is used to look at nationalities or ethnicities, its methods, and its history, raise concerns about how the data might be used, or abused, to support racist or nationalistic views. When the groups involved have been oppressed, they may well fear commercial exploitation or worse harms, up to genocide. Such fears may seem bizarre to the researchers, but they find support in the groups' history. The Cornish or the French may be happy to participate; Australian aboriginal peoples or Native American nations may not be. And, given their history, their reluctance to participate is not, and must not be treated as, unreasonable.

These concerns cause ethical, political, and in some cases, legal problems for researchers. The researchers' answer must not be to try to override or fast talk around the group's concerns. Those concerns must be met openly and forthrightly through building trust. But building trust takes time, effort, and money and its results are uncertain. Sometimes, in spite of a researcher's best efforts, people will say no. One key to ethical human population genetics research is learning to accept that answer.

Some people in Quetzaltenango (though not the ones who thought I was a CIA agent), asked me 'Even if you are not going to hurt us, why should we help you?' It was a good question to the HGDP in 1993; it remains a good question today.

Hank Greely, School of Law, Stanford University, USA

If not, conclusions drawn from the sample may not be applicable to the entire population. Analyzing everyone would avoid the difficulties of sampling, and some have argued that it is fairer (Williamson and Duncan, 2002), but at present it is impractical for DNA studies and even for DNA-free genetics based, for example, on phenotypic traits. This is likely to remain true for the foreseeable future, so the issues raised by sampling must be addressed. It can be useful to consider each issue in two ways: first, how sampling would be organized for a nonhuman species such as a worm; then, how the involvement of humans affects it.

Although human genetic diversity had been investigated for a long time, the early studies aroused little controversy or public interest. Attitudes changed with the launch of the Human Genome Diversity Project (HGDP) in 1991, and this will be discussed next.

9.2.3 The Human Genome Diversity Project

The HGDP was announced in a paper published by Cavalli-Sforza, Wilson and others (Cavalli-Sforza *et al.*, 1991). The authors called for the collection of *'material to record human ethnic and geographic diversity'*, particularly from populations 'that have been isolated for some time [and] are likely to be linguistically and culturally distinct'. One scientific issue that arose immediately was the sampling strategy, and the authors differed among themselves: Wilson favored defining a geographical grid and sampling every 100, 50 or even 10 miles (161, 80 or 16 km), while Cavalli-Sforza favored using cultural and linguistic criteria to identify isolated populations. It may be useful to consider the worm analogy here: in such a case we would use a grid strategy. Yet this has not been used for sampling humans. Why not? Contributing factors may be the expected low degree of sampling as a proportion of the total population and the extent of additional information available about each sample: many wanted to be sure to include populations that had already aroused interest because of their culture or language, and these might, by chance, be omitted if sampling were based on a grid.

A meeting took place in 1992 between the HGDP organizers and anthropologists to choose about 500 key populations out of an estimated 7000 populations in the world. According to one account (Roberts, 1992), the anthropologists began by identifying broad issues that might be addressed; in Africa, for example, these included questions about the origins of modern humans or the Bantu-speaker expansion. Then

> '. . . they first selected isolated populations believed to be relatively unmixed descendants of ancestral populations, like the !Kung of Botswana and Namibia and the Hadza of Tanzania. Next they selected a nearest neighbor or two, to determine if these isolates are as distinct genetically as they are culturally or linguistically. Then, at Cavalli-Sforza's urging, they tried to get a representative sample of the entire continent by using the 1500 or so major language groups as a guide. Finally, the group plotted all these populations on a map to reveal geographic holes and selected 25 additional populations to fill them in.'

The populations who would be included in such a sampling strategy appear not to have been consulted during the early stages of the project and many first learned about it from the press. Response to 'the vampire project' was, understandably, often surprise or horror that it was considered important to sample them because they were in imminent danger of extinction, with little attention paid to preventing their demise! Some of the concerns and reactions are described in *Box 9.2*. Concerns focused on:

▶ commercialization, with the profits going to Western companies or laboratories;

▶ diversion of funds from more immediate needs of the population;

▶ the potential for racism and biological warfare.

One response from Nilo Cayuqueo (http://www.nativenet. uthscsa.edu/archive/nl/9311/0069.html) was

> 'With all the money and effort that is going to be expended to try to further exploit us, we believe the time, energy and money could be better put to much better use by:

▶ *helping our communities in our struggle for self-determination,*

▶ *getting governments to acknowledge our way of life, honor broken treaties and allow us to reclaim our rightful territories,*

▶ *stopping the human rights violations against our people and our land,*

▶ *bringing proper sanitation and medical care to our impoverished communities,*

▶ *stopping the multinational corporations from exploiting and destroying our natural resources on which we all depend.'*

The ethical standards of the HGDP are probably higher than those of many small-scale piecemeal sampling schemes, so perhaps the best hope of achieving ethically responsible sampling lies with the HGDP. Nevertheless, it has not achieved large-scale success; its greatest achievement so far has been the establishment of a panel of 1064 cell lines (Cann *et al.*, 2002). Why did the HGDP fail to attract more widespread support and funding? The contrast with the human genome sequencing project may be informative. Despite initial doubts about its implications for human genetics and the scientific community, the public sequencing project's policy of making data freely and immediately available to all, instead of just to the labs generating the data, may have been a key ingredient in its success. The medical relevance of the sequencing project was obviously an important factor, and its concentration in scientifically advanced countries simplified its organization, but perhaps a more 'open' diversity project would be more successful.

9.2.4 What is a population?

The phrase 'human population' is used widely, including in this book, but we now need to examine more carefully what is meant by the term 'a population'. When sets of individuals are phenotypically distinct and do not interbreed, it is easy to distinguish populations: for example the human population and the chimpanzee population. However, although humans from anywhere in the world are potentially capable of interbreeding, they clearly do not form one worldwide randomly-mating population, but exhibit structure. What criteria can be used to identify this structure? Among the ways in which we can decide whether or not people belong to the same population are:

▶ geographical proximity: individuals from the same population must be able to meet;

▶ a common language: they must be able to communicate with each other;

▶ shared ethnicity, culture, religion: they are more likely to intermarry if they share history and values.

None of these criteria is an absolute: we do not necessarily say that someone belongs to a different population if they move to a different country, if grandparents and grandchildren speak different languages, or if someone converts to a new religion. Nevertheless, after several generations, such changes could lead to the establishment of a new population. If these criteria are considered alone, each individual can be considered to belong to many populations (geographical, linguistic etc.) defined in different ways (see

Box 1.2), and these may change with time. The extent to which these memberships are correlated between individuals is unclear; if they were highly correlated, we could meaningfully identify distinct populations that would summarize the relationships between individuals, but if they were poorly correlated, it would be difficult to identify populations clearly. The criteria used in the studies reported in this book are not always consistent. One common practice is to use self-determination: a person is a member of the group they identify with. The French historian Ernst Renan described a nation as *'a group of people united by a mistaken view of the past and a hatred of their neighbors'*; although perhaps unduly cynical, this encapsulates much of what we mean by 'a population'.

9.2.5 How many people should be analyzed?

We have touched on the HGDP's approach to sampling: the target was 25 people from each of 500 selected populations, a total of 12 500 individuals. While this was never achieved, the questions remain:

▸ how many individuals need to be examined in order to address a particular question?

▸ how should these be distributed among different populations?

There are no simple answers to these questions, because the answer depends on the specific question and also on the number and type of markers used. Generally, the smaller the effect that is sought, the larger the sample that will be needed to investigate it.

A related issue is what kind of weighting scheme should be used in choosing numbers of individuals and populations. It is generally agreed that, except in forensic investigations, weighting according to current population size is a poor option because it biases strongly towards recent expansions and these are not usually the main focus of interest. A geographical scheme is probably the most common, and aims to sample roughly equally from each geographical area. A third possibility is to use linguistic criteria, on the grounds that the major linguistic divisions predate recent demographic expansions. For example, a study of Xq diversity (Kaessmann *et al.*, 1999) examined 69 individuals distributed among 16 out of 17 major language phyla.

In practice, sample availability is often the major criterion. Throughout this book, we will see that sample sizes of 20–50 per population are common in DNA analyses, and 100 would be large. Studies often examine a few hundred individuals in total. Advances in technology (Chapter 4) may allow much larger studies to be undertaken in the near future.

9.3 Apportionment of human diversity

When information on the variation of a set of 'classical' markers (*Box 3.3*) became available from different populations, it was possible to investigate how diversity was apportioned between human populations or groups of populations, and a pioneering study was presented by Lewontin in 1972 (Lewontin, 1972). Despite limitations in technology and dated terminology, this work identified the basic view that is still current, more than 30 years later.

9.3.1 Establishment of the apportionment of diversity

Lewontin (1972) used 17 loci (blood groups, serum proteins and red blood cell enzymes) for which variation had been detected by immunological or electrophoretic methods (see *Box 3.3*), and had allele frequency data available for several populations. The populations were classified into seven 'races' termed **Caucasians**, Black Africans, Mongoloids, South Asian Aborigines, Amerinds, Oceanians and Australian Aborigines, based on morphological, linguistic, historical and cultural criteria. Diversity was measured by H, the Shannon information measure:

$$H = -\sum_{i=1}^{n} p_i \ln 2 \, p_i$$

where p_i is the frequency of the ith allele; H is a somewhat similar measure to Nei's gene diversity (Section 6.2.1) and in this study ranged from 0 (no variation) to 1.9 (high variation). It was calculated at three levels for each locus:

▸ H_{pop}, the value for each individual population averaged over populations within a race;

▸ H_{race}, the value for each race, calculated from the average gene frequency over all populations within that race;

▸ $H_{species}$, calculated from the frequency averaged over all populations within the species.

These values were then used to apportion the diversity:

▸ within populations = $H_{pop}/H_{species}$;

▸ between populations within races = $(H_{race} - H_{pop})/H_{species}$;

▸ between races = $(H_{species} - H_{race})/H_{species}$.

The proportion of variation within populations ranged from 63.6% for the Duffy blood group to 99.7% for Xm, although only four populations were available for the latter marker; the mean proportion within populations was 85.4%. On average, 8.3% (range, 2.1–21.4%) corresponded to differences between populations within races, and only 6.3% (range, 0.2–25.9%) was found between races.

The overwhelming conclusion was that most variation lies within populations, and that 'races' had no genetic reality, a conclusion reinforced by subsequent analyses using independent population samples and DNA markers. Lewontin concluded:

'Human racial classification is of no social value and is positively destructive of social and human relations. Since such racial classification is now seen to be of virtually no genetic or taxonomic significance either, no justification can be offered for its continuance.'

9.3.2 The apportionment of diversity at different loci

Subsequent studies differ in that they used DNA markers of several kinds, analyzed data by AMOVA (Section 6.2.4) and referred to 'continental groups' rather than 'races', but the conclusions are strikingly similar: ~ 83–88% of autosomal variation is found within populations and ~ 9–13% between continental groups (*Table 9.1*).

Results from mtDNA and the Y chromosome are somewhat different, with less of the variation within populations and more between groups, as might be expected from their smaller effective population sizes (see *Boxes 8.5 and 8.8*). The latter is particularly marked for the Y chromosome, which may be partly explained by **patri-locality** (see *Box 5.6*), although there were large differences between studies. One detected no variation between groups, probably because the markers used were a small set of rapidly mutating microsatellites, while another actually found that most of the variation (53%) was between groups (*Table 9.1*).

A comparison of the distribution of diversity in humans with that found in other species of large-bodied mammals provides a useful perspective (Templeton, 1999). Humans, despite their worldwide distribution, show a low F_{ST} value comparable to that seen in either waterbuck or impala, each from a limited geographical region, Kenya; species with wider ranges, such as coyotes from North America or gray wolves from Eurasia have higher values (*Figure 9.3*).

9.3.3 Can the origin of an individual be determined from their genotype?

Although most human variation is found within populations, the small proportion that lies between populations or continents is still significantly greater than zero. A question is therefore whether individuals can be assigned to their population or continent of origin on the basis of their genotype. If all variation were within populations, this would be impossible; if all variation were between populations/continents, it would be trivial. How successful is it in reality?

TABLE 9.1: SELECTED STUDIES OF THE APPORTIONMENT OF HUMAN DIVERSITY.

Locus	Variation (%)			Reference[a]
	Within population	Between population within groups	Between groups	
Autosomal				
17 classical markers	85.4	8.3	6.3	Lewontin, 1972
30 microsatellites	84.5	5.5	10.0	Barbujani *et al.*, 1997
79 RFLPs	84.5	3.9	11.7	Barbujani *et al.*, 1997
60 microsatellites	87.9	1.7	10.4	Jorde *et al.*, 2000
30 SNPs	85.5	1.3	13.2	Jorde *et al.*, 2000
21 *Alu* insertions	82.9	8.2	8.9	Romualdi *et al.*, 2002
Beta-globin	79.4	2.8	17.8	Romualdi *et al.*, 2002
mtDNA				
RFLPs	75.4	3.5	21.1	Excoffier *et al.*, 1992
RFLPs	81.4	6.1	12.5	Seielstad *et al.*, 1998
HVS I	72.0	6.0	22.0	Jorde *et al.*, 2000
Y-chromosomal				
22 binary markers	35.5	11.8	52.7	Seielstad *et al.*, 1998
30 markers, several types	59	25	16	Santos *et al.*, 1999
6 microsatellites	83.3	18.5	−1.8	Jorde *et al.*, 2000
14 binary markers	42.5	17.4	40.1	Romualdi *et al.*, 2002

[a]Reference for apportionment analysis, which may use data first published elsewhere.

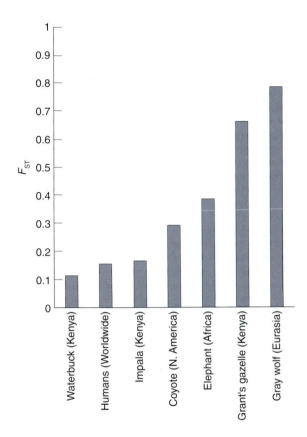

Figure 9.3: F_{ST} values for humans and other large-bodied mammals.

Note the low human F_{ST}, typical of species with a restricted geographical distribution, despite the wide distribution of humans. Data from Templeton (1999).

A study using 30 autosomal microsatellites in ~10 individuals from each of 14 populations (Bowcock *et al.*, 1994) calculated a distance measure between pairs of individuals (1−*Ps*) where *Ps* was the proportion of alleles they shared, averaged over all loci. Distances were then used to construct a tree of individuals, which was thus based on their genotypes without taking information about their origins into account, and could then be used to examine whether or not continent-specific or population-specific clusters of individuals existed. Eighty-eight percent of individuals lay in continent-specific clusters, with 5% not clearly defined, and 7% in a continent-specific cluster that did not correspond to their geographical origin. Population-specificity was less, but 9/14 populations formed subclusters that included > 50% of the individuals from that population. These results imply that genotypes do contain significant information about geographical origin.

Subsequent studies have investigated the power of other methods to cluster genotypes and detect structure within the human population. Wilson and coworkers (Wilson *et al.*, 2001) tested 39 microsatellites in a total of 354 individuals from eight populations, and analyzed the data with the

program STRUCTURE (*Box 9.3*). Four groups were detected, and the proportion of each population assigned to each group is shown in *Figure 9.4*. Group B was largely specific to Papua New Guinea and Group D to China. Most South African Bantu speakers belonged to Group C, and most Ashkenazi, Norwegians and Armenians to Group A. In two populations, Ethiopians and Afro-Caribbeans, individuals were split more evenly between groups: in both cases Groups A and C. The four groups A–D thus correspond quite well to the geographical regions Western Eurasia, New Guinea, sub-Saharan Africa and China, respectively, with allowance for admixture. The authors emphasize, however, that the groups do not correspond well to either skin color or continental categories. A label 'Black' that grouped Ethiopians with South African Bantu speakers would misclassify three-quarters of the Ethiopians, while a label 'Asian' that grouped China and New Guinea together would miss a major distinction. The genetic grouping provided the best prediction of the distribution of variants that influence drug metabolism, a topic discussed further in Section 14.5.1, and not a foregone conclusion since drug-metabolizing genes may have been under differential selection in different regions of the world, so that their variation would not necessarily reflect neutral variation. More recent studies (Rosenberg *et al.*, 2002; Bamshad *et al.*, 2003) have also used STRUCTURE with larger numbers of markers and individuals. Data on 377 autosomal microsatellites in 1056 individuals from 52 populations (Rosenberg *et al.*, 2002) confirmed the high level of within-population variation, accounting for 93–95% of the total. Nonetheless, when STRUCTURE was applied to these data, clusters at K = 5 corresponded largely to the major geographical regions, Africa, Eurasia, East Asia, Oceania and the Americas.

A study using either eight or 21 *Alu* binary insertions as genetic markers (Romualdi *et al.*, 2002) was unable to identify continent-specific groups, perhaps indicating that large numbers of highly informative markers are needed. *Figure 9.5* shows how clustering at K = 3 improves as the number of *Alu* **insertion polymorphisms** (see Section 3.4) in Africans, Asians and Europeans increases from 20 to 100 in a set of 565 individuals (Bamshad *et al.*, 2003). With the full set of markers, accuracy of assignment reached 99–100% for these broad continental categories. The ability to identify, to a significant extent, the geographical origin of an individual has implications for forensic investigations; these are discussed in Section 15.2.2.

How can the finding that it is possible, with a sufficiently large set of markers, to deduce much about the population of origin of an individual be reconciled with the earlier conclusion that most variation exists within populations? Particular alleles were usually not continent- or population-specific, although specific searches for these have had some success (Section 15.2.2); extraction of the geographical information generally required analysis of large numbers of loci. If it is possible to identify genetic groups within humans, are these groups then 'races'? No, because, as we have seen, the groups identified by current genetic techniques do not

---

BOX 9.3 The program STRUCTURE

http://pritch.bsd.uchicago.edu/

STRUCTURE uses genotype data from multiple independent loci to investigate population structure (Pritchard *et al.*, 2000). It is assumed that a certain number, *K*, of genetic groups is present, each characterized by a set of allele frequencies at each locus. Since the number of genetic groups, i.e. the value of *K*, is not usually known in advance, the first step in using the program is to calculate the likelihood of the data, *X*, for a range of values of *K*. The posterior probability of *K*, given *X* (written as *K*|*X*), can then be calculated. Commonly, the probability of one value of *K*|*X* is close to 1 and the others are close to 0, so a clear indication of the number of groups is obtained. The program can then be used in two different ways, according to whether admixture is incorporated into the model or not. If admixture is not allowed, the posterior probability for each individual of belonging to each of the *K* groups is calculated, and the person can be considered a member of the group with the highest probability. This is how the program was used in the articles described in the text. If desired, a cut-off (e.g., 50%) can be used and individuals with a probability of less than 50% of belonging to any one group can be considered unclassified. Alternatively, if admixture is allowed, the fraction of each individual's genotype derived from each group is estimated (see Section 12.3.3).

---

correspond to traditional races, and the differences between them are too small to justify being called races, which would require ≥30% difference between groups.

Why are human populations so similar to one another? In principle, the similarity could exist because of a recent common ancestry, or because of high levels of genetic exchange between populations. The recent origin of modern humans in Africa (Chapter 8) is probably the major factor. Long-distance (intercontinental) travel by individuals was rare in prehistoric times, but gene flow over long distances can result from the cumulative effect of many short migrations by individuals, and a better quantitative understanding of these levels is needed.

9.4 The effect of selection on the apportionment of diversity

In the previous sections of this chapter, we considered studies conducted using markers that were chosen for historical reasons (data were available when the work was carried out) or because they were considered neutral. We saw in Section 9.3.1 that, although the mean between-population component of diversity was 14.6%, there was variation among autosomal loci: it ranged from as little as 0.3% for Xm to as much as 36.4% for the Duffy blood group. How much variation in these values should we expect for neutral markers, and how much do we observe? What effects does selection have? Are all F_{ST} values consistent with neutral expectation?

9.4.1 The expected distribution of F_{ST} values

Under neutral conditions, variation in allele frequencies between populations is determined by drift, which affects all loci in the same population equally, so all are expected to show the same F_{ST} value, although there will be a spread around this value because of stochastic factors. Attempts have been made to calculate the expected distribution of F_{ST} values (Bowcock *et al.*, 1991), although these were hindered by the incomplete understanding of human demographic history, and simplifying assumptions had to be made. Nevertheless, Bowcock *et al.* compared their calculated distribution with the

observed F_{ST} values from 100 DNA polymorphisms, and concluded that there was an excess of both very low F_{ST} values and high F_{ST} values. The observed distribution of classical marker F_{ST} values is shown in *Figure 9.6*.

9.4.2 Low F_{ST}: balancing selection at *CCR5*

Balancing selection, which, for example, can act through **heterozygote advantage** or **frequency-dependent selection** (Section 5.4), favors the maintenance of two or more alleles in the population and thus high diversity, but if the same alleles are favored in all populations, F_{ST} will be low. Among the classical markers, some **HLA** alleles (e.g., *HLAB**37*) show low F_{ST} (see *Box 5.5* for an introduction to the *HLA* locus); outside the *HLA* region, a good example of low F_{ST} is provided by the 5′ region of *CCR5* (*Figure 9.7*). Note that this is a different section of the gene from that containing the *Δ32* mutation, which lies in the coding region and will be discussed in Section 14.3.2.

Bamshad *et al.* (2002) sequenced about 1 kb of DNA from the *CCR5* 5′ end in 400 chromosomes with a worldwide distribution. They found that diversity was high ($\pi = 0.0029 \pm 0.0017$ or 0.0021 ± 0.0013 in two different subsets of their sample): about three times the average value (see Sections 3.2.7 and 6.2), but that the F_{ST} value was 0.016, about 10-fold lower than the average and not significantly different from zero. Evidence for a departure from neutrality was seen in the high positive values of Tajima's *D* (*Figure 9.8*) and Fu's *Fs* (see Section 6.3 for an explanation of these statistics), indicating an excess of intermediate-frequency alleles. A **network** of haplotypes revealed two main haplotype clusters separated by seven mutational steps, which the authors estimated corresponded to a divergence time of about 2.1 MYA.

The functional consequences of the polymorphisms identified in the *CCR5* 5′ region are unknown, and it remains possible that they are in **linkage disequilibrium** with polymorphisms in another gene, but *CCR5* has been a target of recent selection (Section 14.3.2). The most interesting interpretation of these results would therefore be that it has been a target of balancing selection in the past: for example, frequency-dependent pathogen resistance.

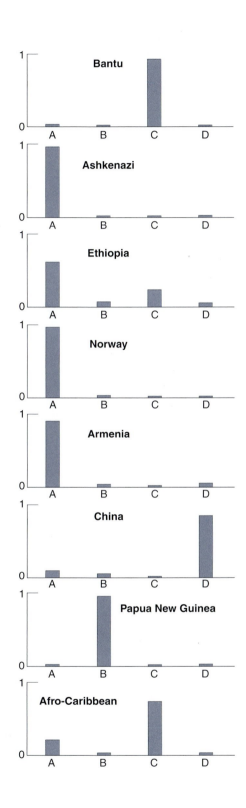

Figure 9.4: Assignment of individuals to groups on the basis of genotypic data.

Individuals from eight populations were assigned to one of the groups A, B, C or D using the program STRUCTURE. From data in Wilson *et al.* (2001).

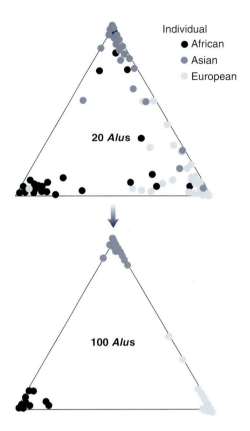

Figure 9.5: Partitioning of individuals from Africa, Asia and Europe using *Alu* insertion polymorphisms.

The proportion of ancestry from each of the three sources, estimated using the program STRUCTURE, increases towards the apex of the triangle. Individuals are partitioned more clearly into the three classes as the number of *Alu* insertion polymorphisms increases from 20 to 100. Redrawn from Bamshad *et al.* (2003).

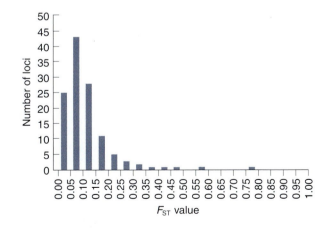

Figure 9.6: Distribution of F_{ST} values for 122 classical loci.

Drawn from data in Table 2.12.1 of Cavalli-Sforza *et al.* (1994).

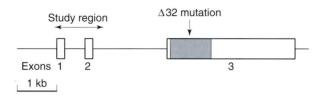

Figure 9.7: The *CCR5* gene, indicating the 5' region analyzed by re-sequencing.

Evidence for balancing selection was found in one region ('study region'). Open boxes represent exons and the gray box represents the coding region. The Δ32 mutation discussed in Chapter 14 is distinct and lies within the coding region of the gene.

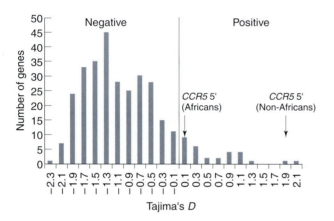

Figure 9.8: High positive value for Tajima's *D* in the *CCR5* 5' region.

The histogram shows the distribution of values for Tajima's *D* from 313 genes measured by Stephens *et al.* (2001), and the arrows indicate the values for the *CCR5* 5' region in Africans and non-Africans. Most values of *D* are negative and are interpreted as reflecting population expansion, but the value for the *CCR5* 5' region in non-Africans is positive and one of the highest known.

9.4.3 High F_{ST}: directional selection at the Duffy blood group locus

In contrast to balancing selection, directional selection acting in a subset of populations will lead to high frequencies of different alleles in different populations. Diversity within any one population may be low, but worldwide F_{ST} will be high. A clear example is provided by the Duffy blood group.

The Duffy (FY) antigen was first detected serologically, and three main alleles, A, B and O, were identified. Note that these are distinct from the ABO blood group system which represents an entirely different locus, and is described in *Box 3.3*. The *FY*O* allele shows the highest F_{ST} value known (0.78; *Figure 9.6*), being at or near fixation in most sub-Saharan African populations and rare outside Africa. In addition, the *FY*A* allele shows near-fixation in eastern Asia and the Pacific, so that the ancestral *FY*B* allele has been lost from large parts of the world. The molecular basis of this variation is now understood. The *Duffy antigen receptor for chemokines* (*DARC*) gene codes for a cell surface protein

expressed in many tissues, including red blood cells, kidney, spleen and brain. The *FY*A*/*FY*B* difference is due to a G/A substitution at position 625 which leads to a Gly/Asp substitution at amino acid 44 (*Figure 9.9*). Individuals with the *FY*O*/O* genotype lack the *DARC* gene product on red blood cells because a T to C substitution 46 bp upstream of the transcription start site in a binding site for the GATA transcription factor prevents binding of this factor, and, as a consequence, abolishes transcription in erythroid cells (the cell lineage leading to red blood cells) but not in other cell types (Tournamille *et al.*, 1995). This mutation has arisen on an *FY*B* background (see *Figure 9.9*). Individuals lacking the *DARC* gene product on their red blood cells are completely resistant to infection by the malaria parasite *Plasmodium vivax*, which suggests that the high frequency of *FY*O* in Africa is a response to selection for resistance to this form of malaria.

The pattern of variation around the *DARC* gene, however, is considerably more complex than would be expected from simple selection for *FY*O* in Africa and not in the rest of the world. An analysis of 1.9 kb of DNA surrounding the *FY*O* mutation in Africans (48 *FY*O* chromosomes) and Italians (14 *FY*A* and 20 *FY*B* chromosomes) has been carried out (Hamblin and Di Rienzo, 2000). It revealed a significant departure from neutrality in Africa using the HKA test (Section 6.3), which was interpreted as due to the lower diversity in Africa resulting from selection for the haplotype carrying the *FY*O* allele and its surrounding sequences. F_{ST} was elevated in the immediately adjacent sequences (Hamblin *et al.*, 2002), but the effect was confined to a region of < 20 kb encompassing the mutation (*Figure 9.10*). An excess of rare alleles would be expected if there had been a recent **selective sweep**, but in fact alleles of intermediate frequency were common (Hamblin and Di Rienzo, 2000). Thus the allelic frequency spectrum did not differ significantly from neutral expectation (measured by Tajima's *D*). This was because the *FY*O* allele occurred on two common haplotypic backgrounds, perhaps due to recurrent mutation, gene conversion or recombination early in the spread of the allele. Thus *FY*O*-bearing chromosomes show several of the features expected in a region of DNA subject to directional selection, and a selection coefficient of 0.002 or larger was estimated, but, unexpectedly, exhibit two different backgrounds.

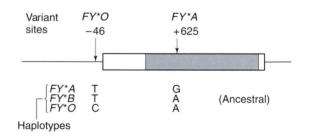

Figure 9.9: Structure of the *DARC* gene.

The gene and variants responsible for the Duffy (*FY*) A, B and O alleles are shown. Box: transcribed region; gray box: coding region.

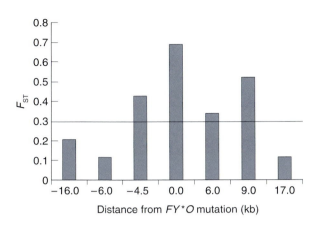

Figure 9.10: F_{ST} values surrounding the *FY*O* mutation.

Italians (carrying the *FY*A* and *FY*B* alleles) are compared with Africans (carrying the *FY*O* allele). The horizontal line represents the maximum value seen in the same population samples for loci assumed to be neutral (Hamblin *et al.*, 2002). Note how the high F_{ST} falls off rapidly with distance away from the *FY*O* mutation along the chromosome.

The near-fixation of the *FY*A* allele in some populations in Asia also leads to a high F_{ST} value for this allele (0.33), raising the possibility that it too may have spread because of selection. In this case the mechanism of selection is not known, and there is a possibility that selection is acting on a nearby locus in LD; nevertheless, it is usually assumed that the *FY*A* mutation is itself the target of selection. A similar decay of F_{ST} with distance along the chromosome was seen and two regions of low diversity were present, one covering the *DARC* gene and the other 3′ to the gene, but they were separated by a region of normal diversity. Values for Tajima's *D* were not significant near the gene, but further in the 3′ direction significantly positive values were found, indicating an excess of intermediate-frequency alleles. As we saw above for the *FY*O* allele in Africa, this is the opposite of the result expected from a simple selective sweep. Thus variation near the *DARC* gene is not consistent with simple directional selection outside Africa, but could perhaps be explained by more complex forms of selection.

It is notable that both of the examples chosen of genes showing unusual F_{ST} values, *CCR5* and *DARC*, appear to have experienced multiple distinct episodes of selection in different parts of the world and at different times. We assume that the variation around most genes has not been shaped by balancing or directional selection, but perhaps the few genes that are selected in these ways have often been targeted many times because of their central importance in interacting with many different pathogens or other environmental factors.

9.5 Colonization of the Old World and Australia

We saw in Chapter 8 that the earliest fossils with modern human morphology found outside Africa date to < 70 KYA,

with the exception of early examples from Israel which are interpreted as a temporary extension of the African fauna. Similarly, archaeological remains indicating modern human behavior appear in Africa by 80 KYA but outside Africa only after 50 KYA (Burenhult, 1993). In this section we will consider in more detail the different lines of evidence for the spread of modern humans into the Old World. Climate had a major influence on these events, and will be considered first.

9.5.1 The environment

Over the last two billion years, the average temperature of the surface of the earth has varied by more than 10°C, oscillating between warm (sometimes entirely ice-free) and cold (**glacial** or **ice age**) conditions. These changes are thought to result from variations in the positions of the continents, the amount of carbon dioxide in the atmosphere, and the earth's orbit. For the last 3 MY, the earth has been in a cool period, but there has been considerable variation within this, including warm **interglacials** at around 250 and 110–130 KYA with temperatures similar to the present; indeed, a ~100 KY warm–cold periodicity can be traced back to ~900 KYA. For the period of particular interest here, the last 100 KY, detailed records of variables such as temperature, precipitation and volcanic activity can be obtained from ice cores, and slightly less detailed records from lake and ocean sediments. The conditions we think of as 'normal', experienced during the last 12 KY, are exceptional for both their high temperature and stability, seen in the moderate-resolution record from the Antarctic ice (*Figure 9.11a*). Before ~12 KYA the temperature was 2–9°C colder in the Antarctic, with a mean of 6.8°C colder than present for the period 18–70 KYA. A higher-resolution record from the Arctic (*Figure 9.11b*) for the last 40 KY shows a similar overall profile where the temperature before ~12 KYA was 10–20°C colder than at present, but with very rapid fluctuations of up to 10°C. It has been suggested that such changes occurred in sudden jumps over periods of just a few decades, perhaps mediated by alterations to the circulation of ocean currents such as the present day Gulf Stream in the North Atlantic. Twenty-three **interstadials**, short-lived warm events, have been identified within the last 100 KY, and at least six **Heinrich events**, short-lived periods of extreme cold. The most recent warming event between 13.0 and 11.5 KYA (the **Younger Dryas** to **Holocene** transition) is the easiest to study and has been interpreted as '*a series of warming steps, each taking less than 5 years. About half of the warming was concentrated into a single period of less than 15 years*' (Adams *et al.*, 1999). Thus extreme climatic changes would have occurred within a human lifetime on many occasions during the spread of modern humans.

Low temperature had several consequences: precipitation was low, so that large areas of the world were cold deserts. Glaciers formed where there was sufficient snowfall at high latitudes, and combined to form huge ice caps that rendered large regions uninhabitable. These ice caps locked up a significant proportion of the world's water, so that sea levels

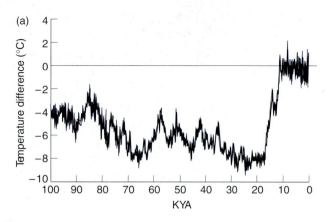

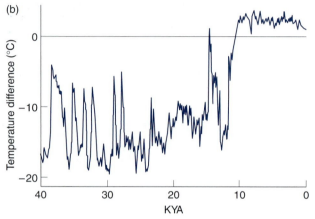

Figure 9.11: Temperature variation over the last 100 KY.

(a) Data from the Vostok ice core in Antarctica (Petit *et al.* 1999 and http://cdiac.esd.ornl.gov/trends/temp/vostok/jouz_tem. htm). Note that for most of this period the temperature has generally been colder than during the last 10 KY, particularly around 17–30 KYA. (b) Higher-resolution data from the Greenland GISP2 core for the last 40 KY showing a similar overall pattern but very rapid fluctuations of up to 10°C (http://www.amap.no/maps-gra/mg-cc.htm).

were lower. While it is difficult to generalize about the climate over the entire Old World during this period, the following scenario can be suggested:

▸ 100–70 KYA. The last interglacial, when temperatures were similar to the present, had ended at around 110 KYA. The temperature subsequently fell, so it was generally slightly cooler than the present, and the period included some intense cold periods in Asia around 91, 83, 75 and 70 KYA. Nevertheless, the African fauna (including humans) which had expanded during the interglacial, remained in areas such as the Levant (Eastern Mediterranean) during this period.

▸ 70–55 KYA. Glacial Maximum. One hypothesis suggests that this event was triggered by the eruption of Mount Toba in Sumatra, Indonesia. This was the

largest explosive volcanic eruption in the last 450 MY (perhaps 3000 times larger than the 1980 Mount St. Helens, USA, eruption) and produced substantial deposits of volcanic ash as far away as India, so it could have had a major environmental impact (Ambrose, 1998a). Fragmentation of the environment occurred within Africa separating sub-Saharan, Northeastern and Northwestern regions; conditions elsewhere, and sea levels, may have been similar to those during the more recent glacial maximum (see below and *Figure 9.12*) with extensive ice caps and a hostile high-latitude environment. Australia was cooler and drier than at present.

▸ 55–25 KYA Somewhat warmer than the glacial maximum, but still colder than the present and highly variable; sea level still as much as 70 m below the current level. There is evidence for wood or forest in Central Siberia in the later part, Taiga (cold coniferous forest) vegetation in the Altai mountains, cold dry steppe in northern China; some areas of Australia were wetter than at present. In Europe, there was an interstadial as warm as the present at 43–41 KYA, followed by a cold Heinrich event 41–39 KYA and then relatively mild temperatures.

▸ 23–14 KYA Last Glacial Maximum (LGM). The sea level fell to as much as 120 m below the present level. In Africa, it was dry and the Sahara extended southwards. Northern Asia was a dry, treeless, polar desert or semi-desert since there was not enough snowfall for ice caps to build up; average temperatures in Siberia were 12–14°C colder than at present. Northern China was semi-desert, but there were more hospitable grasslands and cool temperate forest in the south. Australia was very dry with a large area of extreme desert in the center. Northern Europe was covered by ice sheets or polar desert; to the south there was semi-desert or grassland.

▸ 12 KYA to present. The Holocene (meaning 'wholly recent'). Temperatures rose rapidly to the current level, but as discussed, in a very irregular fashion. For most of the period, they have been unusually stable.

In conclusion, the environment during the period when humans spread from Africa to the rest of the Old World was significantly different from our present environment: mostly colder and drier with lower sea level, accompanied by marked short-term fluctuations in climate.

9.5.2 Fossil evidence

It seems likely that the hominids living in Africa expanded out of Africa whenever conditions permitted. We have seen the early expansion of *H. erectus* soon after 2 MYA, and the suggestion that *H. heidelbergensis* appeared in Africa around 1 MYA and spread into Europe and perhaps Western Asia as well (Section 8.2.3). Securely dated fossils of anatomically modern humans would provide the most direct evidence for the times at which modern humans reached different parts of the Old World. However, human fossils are rare and dating

Figure 9.12: The earth during the Last Glacial Maximum at ~ 18 KYA.

Note the prominent ice caps and lower sea level leading to larger land areas, particularly in Southern Asia/Northern Australia and Beringia, the connection between Siberia and North America. Such a detailed reconstruction is not available for the earlier glacial maximum, but the appearance may have been similar. From http://www.johnstonsarchive.net/spaceart/earthicemap.jpg.

is often disputed, so the age of the earliest widely accepted human fossil from any region may be considerably later than the time of initial entry. Archaeological remains are more plentiful but provide only indirect evidence for modern humans. It can sometimes be difficult to decide whether some finds result from human activity or natural processes; if artificial, we may still not be sure whether they were created by archaic or modern humans. However, the earliest archaeological dates are likely to be closer to the time of entry than the earliest fossil dates.

Humans share their environment and many aspects of their behavior with other animals, so the study of past animal and plant distributions should be a rich source of comparative information. The African fauna expanded into West Asia during the warm climate of the last interglacial. The Qafzeh and Skhul modern human fossil sites in Israel (see Section 8.2.4) are seen as part of this general expansion; similarly, these people were replaced by the cold-adapted Asian fauna, including Neanderthals, when the climate deteriorated. Studies of flora and fauna have also been useful for elucidating the locations of glacial **refugia** and the postglacial recolonization of Europe (Taberlet *et al.*, 1998).

We will now examine the evidence from modern human fossils in different areas of the Old World and Australia. In East Asia, some fossils with modern morphology are claimed to have ages as old as 67 KY (*Table 9.2*), but these dates are not generally accepted (Etler, 1996); some have argued for a true 'hominid fossil gap' from 100 to 40 KYA in the otherwise rich record from China (Jin and Su, 2000). The earliest widely accepted date may be that for the Salawusu material from Inner Mongolia, a region yielding miniscule flaked tools and human remains consisting of two complete frontal bones, two partial mandibles, limb bones and other fragments (Etler, 1996). These are placed at 33 KYA or earlier. Elsewhere in East Asia, fossils from around 17 KYA are known from Japan (*Table 9.2*).

In Australia, the earliest date claimed for a human fossil is ~ 62 KYA for LM3, a red ochre-covered burial of a short man or tall woman discovered in 1974. This specimen has attracted controversy because of the ancient DNA claims associated with it (see Section 9.5.4 below), but the dating has also been criticized and some have argued that it must be younger than the 43-KY-old sediments into which the burial was inserted, with some estimates of 30 KYA or less for the skeleton. A recent re-evaluation using optical dating methods (*Box 8.2*) has suggested a date of 40 ± 2 KYA (Bowler *et al.*, 2003). The nearby LM1, a fragmentary female cremation, was initially dated to 17–26 KYA using a ^{14}C-based method, but was also dated to 40 ± 2 KYA by Bowler *et al.* in the same study. Thus there is plausible evidence for human fossils in Australia by around 40 KYA. There is considerable morphological diversity among the Australian fossils, with both **robust** and **gracile** individuals represented. These could represent different colonization events or variation in a single population due to drift or selection, but the uncertain chronology of many of the specimens makes it difficult to identify a temporal pattern or relationships.

Europe is the most intensively studied region, and as a result, the earliest accepted date of 41 KYA (*Table 9.2*; Section 8.2.4) occurs relatively soon after the entry of modern humans, as judged from their archaeological remains, and, remarkably, is the oldest outside Africa in the Table. The latter observation emphasizes the paucity of the hominid fossil record in the rest of the Old World; that from elsewhere in Asia and the Pacific is at present too fragmentary to provide useful insights.

In conclusion, the most widely accepted dates for modern human fossil material outside Africa are around 40 KYA or later, but the record is so incomplete that these are unlikely to reflect the real time of appearance, and little significance can be attached to the apparent differences between continents.

9.5.3 Archaeological evidence

Archaeological evidence is important for two related reasons. It provides:

▸ evidence for the presence of modern humans at sites or times where no human fossils have been found. This requires that we can distinguish between archaeological deposits left by archaic and modern humans;

▸ evidence for forms of behavior that are considered 'modern', such as art.

Archaeological terms such as '**Upper Paleolithic**' refer to assemblies of remains such as stone or bone tools that represent *cultures*, not time periods. These cultures may have been present in different places at different times, so the date of an archaeological type needs to be related to a particular location. The Upper Paleolithic, for example, may have reached one place at 45 KYA but another only at 35 KYA; in other regions it may never have been present.

TABLE 9.2: SELECTED MODERN HUMAN FOSSILS FROM OLD WORLD AND AUSTRALIAN LOCATIONS OUTSIDE AFRICA.

Region	Date (KYA) Early estimate[a]	Late estimate[a]	Location
China	67	10	Liujiang, Guangxi
	60	28	Laishui, Hebei
	50	**33**	**Salawusu, Inner Mongolia**
	40	7	Ziyang, Sichuan
	34	10	Zhoukoudian (Upper Cave), Beijing
Japan	18	16	Minatogawa, Okinawa
Australia	62	30	Lake Mungo 3 (LM3)[b]
	40	17	**Lake Mungo 1** (cremation)[b]
	13		Kow Swamp
Europe	**41**		**Isturitz, France**
	36		Vogelherd Cave, Germany

Fossils that might represent the ancestors of living populations are listed; the fossil with the oldest widely-accepted date from each region is indicated in bold.

[a]Some dates are disputed, so an indication of the range of possible dates is given.

[b]Both dated to 40 ± 2 KYA by Bowler *et al.* (2003).

Africa

African archaeological deposits dating from around 250 KYA to 40 KYA belong to the **Middle Stone Age** (MSA); similar assemblies outside Africa are described by the alternative, but broadly equivalent, term **Middle Paleolithic** (MP). Archaic humans of this period and the first anatomically modern humans are thus both associated with MSA remains, but differences can be found between the two (Mellars, 2002). At Klasies River Mouth at the extreme south of the African continent, where anatomically modern human fossils have been found, the site contains stone tool assemblages designated 'Howieson's Poort' industries that are comparable in some respects to those from European Upper Paleolithic sites (< 40 KYA, see below) but date to around 70 KYA. The nearby Blombos Cave site may be even older, perhaps closer to 80 KYA, and contains small leaf-shaped projectile points that have been compared by some to those of the much later European Solutrian (20–17 KYA). Bone tools are present at both sites. Most striking of all, lumps of red ochre incised with abstract linear patterns dating to 75–80 KYA have been found at Blombos Cave (Henshilwood *et al.*, 2002) and such images are interpreted as evidence for modern human behavior. Nevertheless, these cultures are classified as MSA. Evidence for their geographical extent comes from Abdur on the Red Sea coast of Eritrea dating to ~ 125 KYA during the last interglacial (Walter *et al.*, 2000), where the artifacts resemble those from Klasies River Mouth from the same period. Thus humans using MSA technologies

seem to have been able to spread along the African coast at this time, and some have speculated that humans adapted to such an environment could readily have extended further to the east. While it is possible that descendants of the Abdur inhabitants did migrate further east, there is no evidence that they survived the glacial period and contributed to modern populations. The Qafzeh and Skhul sites in Israel dating to around 90–100 KYA (Section 8.2.4) also contain modern human fossils in association with MP archaeological deposits. Thus variant MSA remains that include some, but not all, features associated with later periods were present in Africa by the time of the glacial maximum beginning ~ 70 KYA.

It might seem logical to begin a consideration of modern human entry into the rest of the world by examining evidence from sites close to Africa. Unfortunately, little information is available for many of the most relevant places. This may be due, in part, to sea level changes so that the relevant sites lie below the current sea level, but a major factor has been the concentration of archaeological activity in restricted regions (e.g., Europe), which leads to a strong ascertainment bias. Thus the absence of evidence from large areas of the world cannot be interpreted as the absence of modern human sites from these regions (see *Box 10.1* for similar considerations on evidence of later agricultural activity).

Australia

It is nevertheless somewhat surprising that the earliest archaeological evidence for the expansion of the ancestors of modern populations from Africa into the rest of the world comes from Australia. For most of the last 100 KY, the level of the sea has been lower than at present and Australia has been part of a larger continent that included New Guinea and Tasmania, known as **Sahul**. Despite this, Sahul was always separated from Asia (**Sunda**) by a significant barrier: > 90 km (56 miles) of water even when sea level was lowest. The biological significance of this barrier is illustrated by the maintenance of very distinct faunas, with kangaroos, koalas and platypus on the Australian side of the **Wallace Line** and apes, tigers and pigs on the Asian side. Entry into Australia was thus impossible for most animals, apparently including even the *H. erectus* present in Southeast Asia from ~ 1.8 MYA, so the crossing of a considerable distance of water can itself be interpreted as a sign of modern human technology. The evidence suggests that it may have been achieved by 50 KYA. Thermoluminescence (TL) dating of the Malakunanja II site in northern Australia, a 4-m-deep sand deposit containing Middle Paleolithic artifacts such as flakes, red and yellow ochre and a grindstone in the upper 2.6 m, and an absence of such artifacts lower down, provided a date of ~ 50 KYA for the lowest occupation level (Roberts *et al.*, 1990). The study mentioned above which provided dates of ~ 40 KYA for the LM1 and LM3 fossils (Bowler *et al.*, 2003) suggested that the earliest occupation of the Lake Mungo site in southern Australia dated to 50–46 KYA. Older dates, such as those associated with increased charcoal deposits, are difficult to associate definitively with human activity, so there is no widely accepted evidence for human occupation earlier than 50 KYA. There are few radiocarbon dates before ~ 40 KYA, and this observation has led some to question the validity of the early TL dates. By 35 KYA, however, sites are found in Tasmania, so most of the continent except the central desert seems to have been occupied by this time. It is striking that, during the Pleistocene, Australia was populated by 24 genera of large marsupials, including 3-m-high kangaroos, reptiles and birds, and 1-ton carnivorous lizards, but 23 of these **megafauna** genera became extinct at around 46 KYA (95% confidence interval, 40–51 KYA) (Roberts *et al.*, 2001). Although the cause of this massive extinction remains controversial, many would attribute it to humans, either directly through hunting or indirectly through alterations to the environment. If so, this would provide evidence for a significant human presence by this time.

The first Australians must have come from nearby southern Asia, and archaeological sites from this region dating to ≥ 50 KYA would be of great interest. However, the earliest reported sites such as Niah (NW Borneo, ~ 40 KYA) and Lang Rongrien (Thailand, ~ 37 KYA) are considerably later than this. The absence of earlier sites makes it impossible to identify the immediate ancestors of the Australians and trace their movements. Nevertheless, it is often assumed that a route from East or Northeast Africa along the southern coast of the Asian continent was followed; this would have allowed the travelers to occupy a similar ecological niche throughout. Travel at 16 km per year would have covered this distance in just 1000 years.

Asia, Europe and the Upper Paleolithic

The transition from the Middle Stone Age to the **Later Stone Age** may have begun in East Africa, where sites such as the Enkapune Ya Muto rockshelter in the Rift Valley of Kenya are estimated to place it at around 50 KYA (Ambrose, 1998b), although precise dating is difficult because such dates are close to the limit possible with radiocarbon (*Box 9.4*). Flake-based industries were replaced by **blades**; ground bone tools and perforated ornaments became common, and long-distance trading was established. The corresponding transition outside Africa, the Middle to Upper Paleolithic (UP) transition, is sometimes referred to as the 'Upper Paleolithic revolution' because of the rapid and extensive nature of the changes in some regions. The first UP traditions are called the **Aurignacian** and **Châtelperonian**. In Western Asia, the Aurignacian appears at about 47 radiocarbon KYA at sites such as Boker Tachtit and Nahal Avdat (both in Israel) and Temnata (Bulgaria).

In Europe, the MP to UP transition occurred a few thousand years later than in West Asia. Here, it can be interpreted as a transition from Neanderthal to modern humans. Considerable archaeological and fossil data are available, and it is generally agreed that the MP (**Mousterian**) tradition at this time was associated with Neanderthals. Similarly, it is widely believed that the UP Châtelperonian was Neanderthal, while the UP Aurignacian was modern human. Debate continues about whether these were independent transitions, or whether the Neanderthals copied the modern humans, and also involves considerable controversy about dates. A widely held view places the early Aurignacian at locations such as Northern Spain at ~ 40 radiocarbon KYA, perhaps corresponding to 43 calendar KYA (see *Box 9.4*) and the early Châtelperonian about 2 KY later (Mellars, 1998). Proponents of this view regard the coincidental occurrence of two independent Middle to Upper Paleolithic transitions as unlikely, and conclude that the Neanderthals, who had the later transition, copied modern humans. An alternative view (Zilhão and d'Errico, 1999) dates the Aurignacian to 36.5 radiocarbon KYA and the Châtelperonian (as above) to 38 radiocarbon KYA, implying that the Neanderthals had already accomplished their own Middle to Upper Paleolithic transition by the time the modern humans arrived and so could not have copied them.

In Siberia, there is evidence of a Middle Paleolithic dated to 70–40 KYA and extending almost 2000 km east of the limit of known Neanderthal fossils in Uzbekistan. The creators of these sites are unknown, but could have been Neanderthals or other archaic humans. The transition to the UP is dated to around 42 KYA at Kara-Born in the Altai region where UP layers lie above Mousterian, and to older than 39 KYA at Makarovo-4 near Lake Baikal, with a total of 20 sites older than 30 KYA (Goebel, 1999), so modern humans appear to have been widespread soon after 40 KYA. All these sites are south of 55°N latitude and the more northern areas only appear to have been colonized by mammoth hunters after 26 KYA.

BOX 9.4 Radiocarbon (^{14}C) dating and its calibration

Radiocarbon dating is based on the principle that ^{14}C produced in the upper atmosphere by cosmic ray bombardment of nitrogen-14 (^{14}N) is incorporated into biological materials during their lifetime, along with normal ^{12}C. This ceases at death, and the ^{14}C then decays, via a series of 'half lives', with each half-life taking 5730 years. Measurement of the amount of ^{14}C remaining in a biological specimen therefore reveals the amount of time since death. With current techniques for detecting ^{14}C, samples as old as ~ 50 KY or slightly older can be dated using this method, so it is central to understanding the timing of the spread of modern humans. If the amount of ^{14}C in the atmosphere had been constant, it would be simple to convert a measured proportion of ^{14}C into a date: half the present level would give a date of 5.7 KYA, one quarter a date of 11.5 KYA and so on. However, the amount of ^{14}C has varied considerably and there can be substantial differences between '^{14}C years' and 'calendar years'. Thus ^{14}C dates need to be calibrated against an independent standard. For dates $\leq$ 10 KYA, counting of annual growth rings in trees or timbers ('dendrochronology') can be used, but this calibration cannot be extended far back into the glacial period because no good record is available. Alternative calibration sources include annually layered sediments and corals, but the most extensive comparison, extending back to 45 KYA, is derived from a stalagmite that contained both thorium-230 (allowing measurement of the calendar age) and ^{14}C in the calcium carbonate (allowing measurement of the ^{14}C age) (Beck *et al.*, 2001). The resulting comparison curve – it is not yet a universally-agreed calibration curve (see *Figure*) – showed that ^{14}C levels were higher between 30 and 40 KYA, so that ^{14}C ages from this period translate into considerably older calendar ages. For example, 31 KYA (^{14}C) would correspond to ~ 38 KYA (calendar). Thus it is essential to know whether ^{14}C dates encountered in the literature are 'raw' or 'corrected'. Note that dates from methods such as thermoluminescence (TL) and electron spin resonance (ESR) (*Box 8.2*), which are measured in calendar years, are sometimes 'decalibrated' by converting to radiocarbon years.

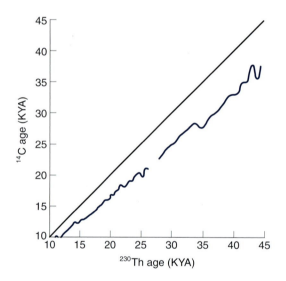

Comparison curve between radiocarbon and calendar dates for the period ~ 45–10 KYA.

Comparisons were made using sections from a stalagmite that could be dated by both methods. Drawn from data kindly provided by David Richards (Beck *et al.*, 2001).

Conclusions

In this section, it will have become apparent that:

▸ the fossil and archaeological records are very sparse in many parts of the world;

▸ our current conclusions are critically dependent on the chronology, which is often disputed. In particular, the evidence for the settlement of Australia by 50 KYA depends on a small number of TL dates.

Thus, future discoveries or recalibrations may drastically change our view of human presence during this crucial period. At present, our conclusions are that hominids probably expanded out of Africa whenever conditions permitted, including during the last interglacial, but that these expansions made little genetic contribution to modern human populations. The expansions that interest us here are those of anatomically modern humans who contributed most to current populations. There seem to have been at least two: the Middle Stone Age people who arrived in Australia by 50

KYA, and the Upper Paleolithic people who were present in West Asia by 47 KYA and brought their technologies to Europe and Siberia a few thousand years later (*Figure 9.13*). Even if humans arrived in Australia ~ 40 KYA rather than ~ 50 KYA, they were clearly using a different technology from the UP expansion, and most likely represent a distinct migration. While this is undoubtedly an over-simplification of the real situation, we will see that a two-migration model, based on archaeological and fossil findings, can readily be reconciled with much of the genetic data.

9.5.4 Genetic evidence from phylogenies

Genetic studies provide information about alleles and lineages, but these give only indirect information about population movements. Early attempts to extract such information were made when data from classical markers became available. As we have seen, genetic distances between populations can be calculated from allele frequencies and displayed as trees or using methods such as **principal component analysis** (Section 6.4). In such analyses, an example of which is given in *Figure 9.14*, the largest distances between populations outside sub-Saharan Africa usually lie between Southeast Asians, including Australians, Chinese and Pacific Islanders, and the rest of the world, including North Asians, Americans, South Asians such as Indians, Europeans and North Africans. We must remember that such trees represent **genetic distances**, which are influenced by many factors including migration and drift, so do not show a phylogeny of the populations. For these reasons, the time depth is also uncertain. Nevertheless, early history is likely to

have had a major influence on the genetic distances, and it is striking that the first split in the non-African part of the tree distinguishes the Australians from a group containing West Asians, Europeans and North Asians: areas that have different fossil and archaeological histories.

Non-Africans generally carry a small subset of African diversity (Chapter 8), and non-African mtDNA and Y phylogenies stem from a common ancestor some distance from the root in the molecular phylogenies. These results are most simply explained by the common origin of all non-African populations from a small subset of African ones. Data from the *COL1A2* locus also illustrate this point. In many human populations, a derived allele carrying a 38-bp deletion is present. In sub-Saharan Africans, the frequency of this was found to be close to zero, while in almost all non-African populations it was present at a substantial frequency, ~ 24% on average (Mitchell *et al.*, 1999). Since this deletion probably has a single origin, it provides evidence for a common origin, at least at this locus, for all non-Africans. Such a conclusion, of course, provides no information about the number of migrations out of Africa: any number could have occurred, as long as they shared a common source. It does, however, argue against any large-scale migration back to Africa, since such an event would introduce the deletion into African populations (*Box 9.6*).

mtDNA analysis
More detailed information, including an indication of time depth, is obtained from nonrecombining haplotypes:

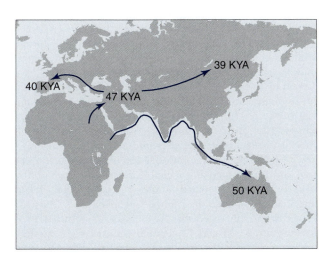

Figure 9.13: The timing of the initial modern human colonizations of Australia and the Old World, based on archaeological evidence.

There is evidence for Middle Paleolithic people in Australia by about 50 KYA, and it is suggested that they followed a southern coastal route out of Africa. Upper Paleolithic people were present in the Near East by about 47 KYA, and in Western Europe and Siberia by 40 KYA or soon after; these people represent a separate migration, following a more northern route. The arrows illustrate the distinct migrations, but not the precise routes of travel, which are unknown.

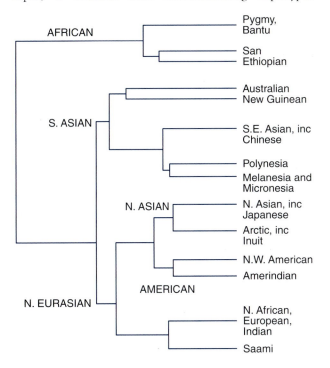

Figure 9.14: Tree of genetic distances between populations based on 120 classical markers.

Redrawn and simplified from data in Cavalli-Sforza *et al.* 1994.

mtDNA and the Y chromosome. A well-resolved mtDNA phylogeny is available, based on sequencing of the complete molecule, or sequencing of the control region combined with typing of selected SNPs (Ingman *et al.*, 2000; Richards and Macaulay, 2001). mtDNAs outside Africa fall into two clades: M and N (*Box 8.5*; *Figure 9.15*). In the commonly used haplogroup nomenclature (Richards and Macaulay, 2001), branches within M are designated C, D, E and G, while the subdivisions within N include A, B, F, H, I, J, K, R, T, U and V. These haplogroups often have different geographical distributions (*Figure 9.16*). M and its subdivisions are found primarily in South and East Asia and the Americas. Some subsets of N are found in these regions, but N makes up most of the mtDNAs of West Asian and European populations. M and N are both rare in sub-Saharan Africa, where the mtDNA types are designated L, but distinct M variants make up about 20% of the mtDNAs in Ethiopia, leading Quintana-Murci and colleagues (Quintana-Murci *et al.*, 1999) to identify Eastern Africa as the source of a migration out of Africa involving the ancestors of the current Indians and other Asian populations. It is possible to estimate dates for the nodes in the mtDNA phylogenetic tree. Quintana-Murci *et al.* use the ρ method (Section 6.6.2) to estimate coalescence times of 36 ± 11 or 48 ± 15 KYA for the Ethiopian haplogroup M and 53 ± 7 or 56 ± 7 KYA for Indian haplogroup M chromosomes. They judge that these ancestors must themselves have diverged for several thousand years, so that the common ancestor of all M mtDNAs would have lived about 60 KYA. Using a similar approach with the complete mtDNA sequence data, Ingman *et al.* estimated a time of 52 ± 27.5 KYA for the MRCA of both the M and N lineages (Ingman *et al.*, 2000). They did not make a separate estimate for M alone, but this TMRCA would necessarily be younger than 52 KYA. There is thus a difference between the two time estimates for the origin of M, but when the confidence intervals are taken into account, they overlap. It is tempting to attempt to link this molecular evidence with the migration that produced archaeological evidence for modern humans in Australia by 50 KYA, and suggest that a second population leaving Africa with a lower frequency of M but a higher frequency of N corresponded to the Upper Paleolithic culture of West Asia, Europe and Siberia. However, many subsequent events have affected the frequencies of the mtDNA haplogroups in these populations, so it is unclear how much information about the initial exodus can be obtained.

Y-chromosomal DNA analysis
The Y chromosome provides the greatest phylogenetic resolution of any locus, and 153 haplogroups defined by binary markers have now been recognized and assembled into a detailed phylogeny (Y Chromosome Consortium, 2002; *Box 8.8*). The first two branches lead almost exclusively to African lineages; all other lineages, which include some African and all non-African ones, carry the derived states of two markers, M168 and P9 (*Figure 9.17*). We would therefore assume that the Y-chromosomal ancestors who left Africa carried chromosomes with the derived states of the M168 and P9 markers. Subsequent

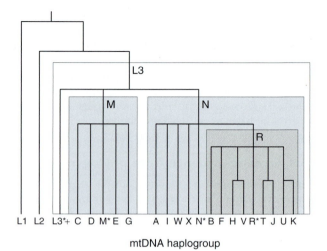

mtDNA haplogroup

Figure 9.15: Simplified mtDNA phylogeny.

Most African mtDNAs belong to haplogroups L1, L2 and L3*+, while most non-African mtDNAs belong to haplogroups M and N. There are, however, overall differences in the distributions of M and N. 'L3*+' includes haplogroups L3b, L3d and L3e as well as L3*; branch lengths have no significance. See *Boxes 8.5* and *8.6* for more detail.

branches of the tree, however, have geographically distinct distributions outside Africa, and these may provide information about the early stages of this exodus (*Figure 9.18*). The phylogenetic tree is incompletely resolved in this region and a trifurcation is seen, leading to haplogroups C (markers RPS4Y$_{711}$, M216), DE (YAP, M145, M203) and F (M89, M213, P14). Haplogroups C and D are largely confined to East Asia; E to Africa, West Asia and Europe; while subdivisions of F are widespread in most regions outside Africa. It has been suggested that haplogroup C was carried by the people responsible for the first colonization of New Guinea and Australia, where it now makes up around half of the Y chromosomes (Underhill *et al.*, 2001), but it appears that most of the haplogroup C chromosomes in Australia have a more recent origin, perhaps only 3–5 KYA and corresponding to the introduction of the dingo and new tool and plant-processing technologies (Redd *et al.*, 2002). Thus it is unclear whether the Y chromosome retains information about the initial peopling of Australia. Nevertheless, the general East Asian concentration of haplogroups C, D and O (a subdivision of F), contrasting with the more Western and Northern distribution of haplogroups E, I, J, N and P (more subdivisions of F) does indicate a SE/NW distinction like that seen in the other genetic data.

The dates of key nodes in the Y phylogeny are thus of interest. The M168 mutation, shared by essentially all non-African Y chromosomes, has been dated to 69 (56–81) (Hammer and Zegura, 2002), 40 (31–79) or 42 (36–109) KYA (Thomson *et al.*, 2000). The last two point estimates are not consistent with the ancestry of modern Y chromosome from two distinct migrations out of Africa

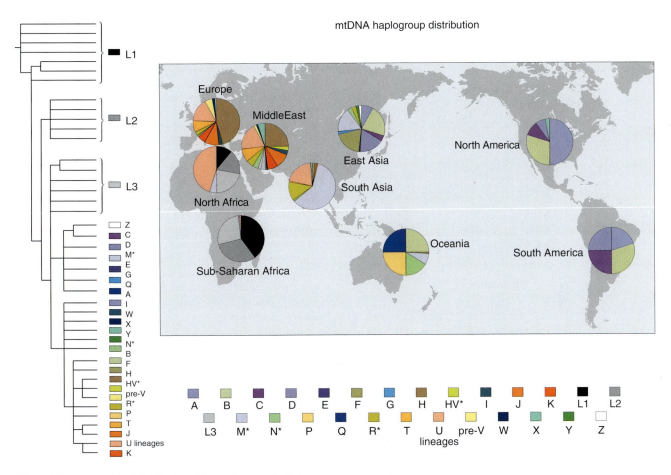

mtDNA haplogroup distribution

Figure 9.16: Geographical distribution of the major mtDNA clades.

Each major clade (haplogroup) is assigned a color reflecting its position in the phylogeny (left), and its frequency in population samples from broad geographical regions is shown in the pie charts. Note the most basal haplogroups (L) in Africa and the contrast between the frequencies of M (and subclades) compared to N (and subclades) in southern Asia compared with more northern parts of Asia and Europe. Color figure kindly sponsored by Oxford Ancestors Ltd.

> 50 KYA and > 47 KYA: a lineage cannot spread before it exists. However, the confidence intervals of all estimates encompass such dates, so there is no major inconsistency.

While it is striking that Australian and European populations are highly differentiated, this finding would be expected under isolation by distance after a single migration. However, the fact that the lineages that reflect this differentiation are the deepest rooting non-African lineages in both Y and mtDNA phylogenies suggests an early (about 40–60 KYA) fission among populations ancestral to all non-Africans, either within Africa or shortly after exit from Africa. Nevertheless, the sharing of common lineages among all non-African populations suggests that separate migrations originating from genetically distinct source populations in Africa are less likely.

Some populations have attracted particular attention because of the insights they may provide into early population movements. These include the Andaman

Islanders, seen by some as isolated descendants of an early migration out of Africa, who are discussed in *Box 9.5*. Movements would not have been one-way, and evidence for migrations back to Africa has also been presented (*Box 9.6*).

Ancient DNA analysis

In principle, analysis of **ancient DNA** is an ideal way to answer questions about which lineages were carried by fossil specimens, or when particular lineages appeared in different parts of the world. In practice, however, its use is limited by the difficulty of deciding whether any sequences recovered are endogenous or contaminants (*Boxes 4.5* and *4.7*): a problem which is even more acute with modern human material than with Neanderthals (Section 8.7.1) since phylogenetic distinctiveness is a less useful criterion. These difficulties are exemplified by the debate over the mtDNA sequence from LM3, 'Mungo Man' (*Figure 9.19*). The date of this fossil is disputed (see above and *Table 9.2*), but if the true date is 40 KYA, it is one of the most ancient Australian

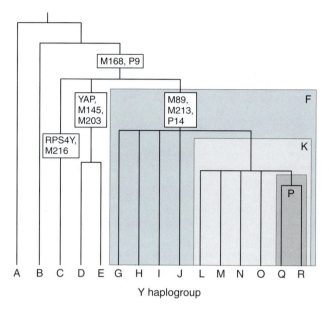

Figure 9.17: Simplified Y-chromosomal DNA phylogeny.

Haplogroups A and B are found mainly in Africa. All other haplogroups carry the derived states of the M168 and P9 markers. These haplogroups are found in almost all Y chromosomes outside Africa, and in some inside Africa. For more detail, see *Box 8.8.*

fossils available. Considerable interest therefore accompanied the report of mtDNA control region sequence from LM3 and nine other ancient Australians dated between 2 and 15 KYA (Adcock *et al.*, 2001). The authors emphasized the distinct outlying position of LM3 in their tree (*Figure 9.20*), lying closer to the nuclear mtDNA insert (**numt**) sequence (see *Boxes 2.8* and *4.5*) than to modern mtDNAs and conclude *'Our results indicate that anatomically modern humans were present in Australia before the complete fixation of the mtDNA lineage now found in all living people'*. While they took precautions to avoid artifacts, their techniques have been criticized (Cooper *et al.*, 2001): of the criteria defined in *Box 4.5*, they did not carry out biochemical studies of bone preservation (II in *Box 4.5*), cloning (VI) or replication in an independent lab (VII). In addition, their phylogenetic analysis has also been questioned. Incorporation of rate heterogeneity or additional sequences into their tree construction model produced a tree (*Figure 9.20b*) in which LM3 lies among modern sequences (Cooper *et al.*, 2001). It is thus not clear that this sequence, whether authentic or not, is very distinct from those of modern humans or other ancient Australians.

9.5.5 Genetic evidence from demographic inferences

Geographical expansion out of Africa has resulted in a large increase in human population size. Demographic changes

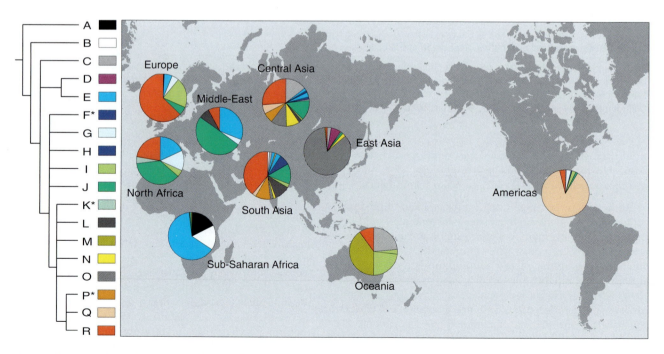

Figure 9.18: Geographical distribution of the major Y-chromosomal DNA clades.

Each major clade (haplogroup) is assigned a color reflecting its position in the phylogeny (left), and its frequency in population samples from broad geographical regions is shown in the pie charts. The basal haplogroups A and B are largely confined to Africa, and there is a broad distinction between East Asia and the rest of Asia and Europe. Data kindly supplied by Peter Underhill, and based on Underhill *et al.* (2001). Colour figure kindly sponsored by Oxford Ancestors Ltd.

BOX 9.5 Genetic analysis of the Andaman Islanders

The Andaman Islands, lying in the Bay of Bengal between India, Burma and Indonesia, are inhabited by hunter–gatherer populations with a characteristic phenotype of dark skin and short stature, sometimes referred to as 'Negrito' that contrasts with their geographical neighbors. They have remained isolated until recently, in part because of their vigorous repulsion of intruders, but are very few in number: perhaps only 400–500 individuals in total. It has been suggested that they may be isolated descendants of the early migrants out of Africa. Two recent DNA analyses have now allowed this possibility to be tested using genetic evidence (Endicott *et al.*, 2003; Thangaraj *et al.*, 2003). MtDNA sequences were obtained from small numbers of individuals from three sources: (i) contemporary populations (Onge, Jarawa and Greater Andamanese); (ii) hair samples collected in 1906–1908 (Greater Andamanese); and (iii) teeth from skeletal material collected in the nineteenth century (Jarawa, Aka-Bea from Greater Andaman). Y data were obtained from males in the contemporary populations. The mtDNAs were predominantly haplogroup M (M2 and M4 subgroups), and some represented novel variants. The Y haplotypes of the Onge and Jarawa (27 individuals in all) belonged exclusively to a novel sublineage within haplogroup D, with very little variation in Y microsatellites; those from Greater Andaman were heterogeneous and may represent recent admixture. In populations with such small census size, genetic diversity is expected to be very low and it is not surprising to find only one or two lineages. The presence of novel Andaman-specific variants provides good internal evidence that the haplotypes represent isolated indigenous populations. The particular deep-rooting lineages found, within mtDNA haplogroup M and Y haplogroup D, are among those previously thought to have been carried by the early southern migration out of Africa, providing support for the hypothesis that the Andaman Islanders represent an independent line of descent derived from this migration.

BOX 9.6 Back to Africa?

People have undoubtedly traveled between Africa and the rest of the world throughout history and prehistory. Here, we are interested in the question of whether any of the ancient movements produced a significant genetic impact on sub-Saharan Africa. The finding of alleles such as the *COL1A2* deletion that are common in almost all non-African populations but essentially absent from sub-Saharan Africa (Mitchell *et al.*, 1999) is one line of evidence against a major genetic impact. Nevertheless, this is a single locus and other parts of the genome may have a different history.

Studies of the Y chromosome, perhaps because of its high power to detect male movements, have led two different groups to formulate hypotheses about male movement back to Africa. One of these proposed an Asian origin for Y chromosomes carrying the *Alu* insertion YAP (haplogroups D and E) (Altheide and Hammer, 1997). YAP+ chromosomes are found in Africa (mostly E), Europe (mostly E) and Asia (mostly D) and actually make up the majority of African Ys. Altheide and Hammer noticed that all the binary markers subdividing this lineage (and available in 1997) were derived in Africa and ancestral in Asia, and favored the explanation that the insertion had taken place in Asia, implying that most African Y chromosomes had subsequently been replaced by Asian Ys. Further work has identified derived mutations specific to the Asian sub-branch D, so this no longer appears 'ancestral', and small numbers of chromosomes from both branches have been found widely throughout the world. It is therefore difficult to decide between the two places of origin on the basis of this evidence, so there is no strong evidence for large-scale replacement of African Y chromosomes. It is worth noting, however, that these observations formed the basis of the inference of a range expansion back to Africa derived from **nested cladistic analysis** (Section 6.7.5).

The second study involved a different section of the Y phylogeny: a lineage within haplogroup R (see *Box 8.8*) (Cruciani *et al.*, 2002). This haplogroup is rare in Africa, being found mainly in Asia and Europe, and is thought to have originated in Asia. However, a distinct R sublineage formed ~ 40% of the Northern Cameroon sample, located in the western central part of the African continent. This lineage was distinguishable from the Asian and European R sublineages, which carried additional derived markers, suggesting that it did not result from back-migration during historical times; the authors calculated a TMRCA for the African lineage from the variation found at three microsatellites and estimated about 4 (2.4–8) KYA, but argued for an earlier migration because other lineages thought to be present in Western Asia 4 KYA were not found in Cameroon. While the source and timing of this migration remain uncertain, it does seem to represent a small-scale and relatively recent migration back to Africa.

It is likely that more such events will be detected in future studies, but will not alter the overall conclusion that back-to-Africa migrations before the colonial era have only had a limited effect on the African gene pool.

Figure 9.19: LM3, a well-preserved early Australian modern human fossil.

This skeleton has been dated to ~ 62 KYA by some but to only around half this age by others; recent estimates put it at ~ 40 KYA. In addition to providing evidence for the timing of the initial entry of modern humans, attempts have been made to analyze ancient DNA extracted from it.

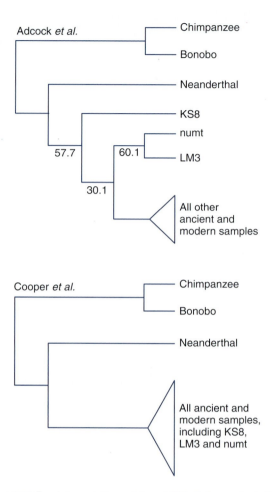

Figure 9.20: Two interpretations of the mtDNA phylogeny including LM3.

In the Adcock *et al.* (2001) interpretation, the ancient sequences KS8 and LM3 are distinct from modern sequences; numbers on the tree represent relative likelihood support for selected branches determined by the authors. In contrast, in the Cooper *et al.* (2001) interpretation, KS8 and LM3 are interspersed among modern sequences. 'numt' is an insertion of mtDNA sequence into the nuclear genome (see Section 4.11.2). Note that only the branching order is represented in this figure: branch lengths have no significance.

affect the pattern of genetic variation: in an expanding population, a locus is expected to show a smooth unimodal distribution of pairwise differences, an excess of rare variants (negative Tajima's D), and a **star phylogeny** (Chapter 6). As we have seen, these patterns can also be produced in other ways, such as by positive selection, but if similar patterns are seen at many unlinked loci, they are more likely to result from demographic changes than selection. Genetic analysis can therefore provide information about past demography. Much interest has focused on the entire human population:

when did it start to increase in numbers? The beginning of the demographic expansion does not necessarily coincide with the range expansion, but the information should be taken into account in our models of human history. In addition, we can ask whether or not there is evidence for different demographic changes in different populations.

Early DNA studies used a single locus, mtDNA, and examined mismatch distributions, Tajima's D or Fu and Li's D^* (Rogers and Harpending, 1992; Wall and Przeworski, 2000). D and D^* were negative, indicating expansion. A constant-sized population produces a ragged mismatch distribution, while in an expanding population the position of the peak reflects the time since the start of the expansion

(Section 6.2.3; *Figure 6.3*). Rogers and Harpending found that data from two African populations produced ragged distributions, while four non-African populations (from the Middle East, Sardinia, Japan and America) produced smooth distributions indicating a start to expansion between 34 KYA (Sardinia) and 62 KYA (Middle East), or earlier if a different estimate for the mutation rate was used. It was not possible to provide confidence intervals for these dates, but there is considerable uncertainty and they are consistent with expansion before, during or after movement out of Africa. Since mtDNA is a single locus, they are also consistent with a selective spread of one mtDNA type in a constant-sized population. An analysis of Y-chromosomal microsatellite data also found evidence for population growth; the start was placed at around 18 (6–44) KYA in the world population (Pritchard *et al.*, 1999) and at similar times in regional populations. These dates do not exclude growth starting after the Neolithic transition (see Chapter 10) and, as with the mtDNA data, are derived from a single locus.

Studies of multiple autosomal microsatellite loci should give more straightforward information about demographic changes, but have yielded conflicting conclusions. One study found evidence for population expansion 640–49 KYA in Africa, but not outside (Reich and Goldstein, 1998); another suggested growth in Asians and Europeans but not in Africans (Kimmel *et al.*, 1998). A re-examination of the available X-chromosomal and autosomal sequence data for consistency with a model of constant size followed by exponential growth was presented in 2000 (Wall and Przeworski, 2000), based on D and D^*. Values at many loci were not consistent with growth starting > 50 KYA, including two out of eight African samples and five out of eight non-African samples. It seems obvious from nongenetic data that there has been enormous demographic growth; failure to detect this in the patterns of genetic variation may reflect selection at the few loci examined, deviations from the mutational model used, or more complex demographic changes. It is hoped that larger datasets and more sophisticated models will resolve these issues, and there are grounds for optimism. One lies in the predominantly negative values of D in *Figure 9.8*, which are most simply explained by demographic expansion, and another comes from the largest study so far, examining 377 autosomal microsatellites in 1056 individuals from the HGDP panel (Section 9.2.3) (Zhivotovsky *et al.*, 2003). A difference was found between hunter-gatherers and farmers in sub-Saharan Africa, with evidence for expansion in the farmers, as well as Eurasians and East Asians, but not in the hunter-gatherers or populations from Oceania or the Americas.

Linkage disequilibrium (LD) is also affected by demography. A study of the extent of LD at 19 randomly chosen loci found it to extend to around 60 kb, on average, in Europeans, but < 5 kb in Africans (Reich *et al.*, 2001). This was interpreted as a result of a population bottleneck followed by an expansion in Europeans, with simulations suggesting that the observed data could be fitted by a bottleneck event 27–53 KYA.

Summary

▸ Studies of human diversity raise important ethical questions which need to be addressed before each individual project is carried out.

▸ The sampling strategy will influence the conclusions and can be based on current population size, geographical or linguistic criteria.

▸ The definition of a 'population' is not simple, but reflects a combination of geographical proximity, a common language and shared ethnicity, culture and religion.

▸ Most (about 85%) autosomal variation is found within populations and only about 10% between different continents; more geographical differentiation is seen for mtDNA and Y-chromosomal sequences.

▸ With a large number of markers, considerable information about the population of origin of an individual can be obtained.

▸ Selection influences the apportionment of diversity for a few loci: balancing selection can lead to low F_{ST} values (e.g., *CCR5*) while directional selection can lead to high F_{ST} values (e.g., *FY*O*).

▸ Climate has had a major influence on the spread of humans: after the last interglacial period (130–110 KYA), the climate cooled and was very unstable, with glacial maxima at 70–55 and 32–14 KYA. Only after 12 KYA did the climate become warm and stable in the way that we consider 'normal'.

▸ Fossil and archaeological evidence suggest that modern humans expanded out of Africa whenever the climate allowed, including during the last interglacial, but the expansions before 60 KYA appear to have contributed little to modern human populations.

▸ By 50 KYA, people using Middle Paleolithic technologies had reached Australia; around the same time Later Stone Age technology was developed in Africa, and soon after the equivalent Upper Paleolithic technology appeared in the Middle East, Siberia and Europe, becoming widespread by 40 KYA. There thus appear to have been at least two distinct (southern and northern) expansions out of Africa.

▸ Genetic evidence from classical marker frequency analyses and phylogeographic studies of mtDNA and the Y chromosome also suggest a major distinction between modern southern and northern populations.

▸ Although the population has expanded considerably over the last 100 KY, genetic evidence for demographic change varies between studies; several do, however, detect a demographic expansion during the Paleolithic.

Further reading

Burenhult G (ed) (1993) *The First Humans: Human Origins and History to 10,000 BC.* Harper San Francisco.

Gould SJ (1981) *The Mismeasure of Man.* Penguin Books, Middlesex, UK.

Electronic references

Human Genome Diversity Project:
http://www.stanford.edu/group/morrinst/hgdp.html

global environments in the last 130 KY:
http://members.cox.net/quaternary/

Asian and Australian paleoanthropology:
http://www-personal.une.edu.au/~pbrown3/palaeo.html

human fossils in China:
http://www.chineseprehistory.org//

References

Adams J, Maslin M, Thomas E (1999) Sudden climate transitions during the Quaternary. *Prog. Physical Geog.* **23**, 1–36.

Adcock GJ, Dennis ES, Easteal S, Huttley GA, Jermiin LS, Peacock WJ, Thorne A (2001) Mitochondrial DNA sequences in ancient Australians: Implications for modern human origins. *Proc. Natl Acad. Sci. USA* **98**, 537–542.

Altheide TK, Hammer MF (1997) Evidence for a possible Asian origin of YAP+ Y chromosomes. *Am. J. Hum. Genet.* **61**, 462–466.

Ambrose SH (1998a) Late Pleistocene human population bottlenecks, volcanic winter, and differentiation of modern humans. *J. Hum. Evol.* **34**, 623–651.

Ambrose SH (1998b) Chronology of the Later Stone Age and food production in East Africa. *J. Archaeol. Sci.* **25**, 377–392.

Bamshad MJ, Mummidi S, Gonzalez E et al. (2002) A strong signature of balancing selection in the 5′ cis-regulatory region of *CCR5*. *Proc. Natl Acad. Sci. USA* **99**, 10539–10544.

Bamshad MJ, Wooding S, Watkins WS, Ostler CT, Batzer MA, Jorde LB (2003) Human population structure and the inference of group membership. *Am. J. Hum. Genet.* **72**, 578–589.

Barbujani G, Magagni A, Minch E, Cavalli-Sforza LL (1997) An apportionment of human DNA diversity. *Proc. Natl Acad. Sci. USA* **94**, 4516–4519.

Beck JW, Richards DA, Edwards RL et al. (2001) Extremely large variations of atmospheric 14C concentration during the last glacial period. *Science* **292**, 2453–2458.

Bowcock AM, Kidd JR, Mountain JL, Hebert JM, Carotenuto L, Kidd KK, Cavalli-Sforza LL (1991)

Drift, admixture, and selection in human evolution: a study with DNA polymorphisms. *Proc. Natl Acad. Sci. USA* **88**, 839–843.

Bowcock AM, Ruiz-Linares A, Tomfohrde J, Minch E, Kidd JR, Cavalli-Sforza LL (1994) High resolution of human evolutionary trees with polymorphic microsatellites. *Nature* **368**, 455–457.

Bowler JM, Johnston H, Olley JM, Prescott JR, Roberts RG, Shawcross W, Spooner NA (2003) New ages for human occupation and climatic change at Lake Mungo, Australia. *Nature* **421**, 837–840.

Cann HM, de Toma C, Cazes L et al. (2002) A human genome diversity cell line panel. *Science* **296**, 261–262.

Cavalli-Sforza LL, Menozzi P, Piazza A (1994) The history and geography of human genes. Princeton, NJ, Princeton University Press.

Cavalli-Sforza LL, Wilson AC, Cantor CR, Cook-Deegan RM, King MC (1991) Call for a worldwide survey of human genetic diversity: a vanishing opportunity for the Human Genome Project. *Genomics* **11**, 490–491.

Cooper A, Rambaut A, Macaulay V, Willerslev E, Hansen AJ, Stringer C (2001) Human origins and ancient human DNA. *Science* **292**, 1655–1656.

Cruciani F, Santolamazza P, Shen P et al. (2002) A back migration from Asia to sub-Saharan Africa is supported by high-resolution analysis of human Y-chromosome haplotypes. *Am. J. Hum. Genet.* **70**, 1197–1214.

Endicott P, Gilbert MT, Stringer C, Lalueza-Fox C, Willerslev E, Hansen AJ, Cooper A (2003) The genetic origins of the Andaman Islanders. *Am. J. Hum. Genet.* **72**, 178–184.

Etler DA (1996) The fossil evidence for human evolution in Asia. *Ann. Rev. Anthropol.* **25**, 275–301.

Excoffier L, Smouse PE, Quattro JM (1992) Analysis of molecular variance inferred from metric distances among DNA haplotypes: application to human mitochondrial DNA restriction data. *Genetics* **131**, 479–491.

Goebel T (1999) Pleistocene human colonization of Siberia and peopling of the Americas: an ecological approach. *Evol. Anthropol.* **8**, 208–227.

Greely HT (2001a) Informed consent and other ethical issues in human population genetics. *Annu. Rev. Genet.* **35**, 785–800.

Greely HT (2001b) Human genome diversity: what about the other human genome project? *Nature Rev. Genet.* **2**, 222–227.

Hamblin MT, Di Rienzo A (2000) Detection of the signature of natural selection in humans: evidence from the Duffy blood group locus. *Am. J. Hum. Genet.* **66**, 1669–1679.

Hamblin MT, Thompson EE, Di Rienzo A (2002) Complex signatures of natural selection at the Duffy blood group locus. *Am. J. Hum. Genet.* **70**, 369–383.

Hammer MF, Zegura SL (2002) The human Y chromosome haplogroup tree: nomenclature and phylogeny of its major divisions. *Annu. Rev. Anthropol.* **31**, 303–321.

Henshilwood CS, d'Errico F, Yates R et al. (2002) Emergence of modern human behavior: Middle Stone Age engravings from South Africa. *Science* **295**, 1278–1280.

Ingman M, Kaessmann H, Pääbo S, Gyllensten U (2000) Mitochondrial genome variation and the origin of modern humans. *Nature* **408**, 708–713.

Jin L, Su B (2000) Natives or immigrants: modern human origin in east Asia. *Nature Rev. Genet.* **1**, 126–133.

Jorde LB, Watkins WS, Bamshad MJ, Dixon ME, Ricker CE, Seielstad MT, Batzer MA (2000) The distribution of human genetic diversity: a comparison of mitochondrial, autosomal, and Y-chromosome data. *Am. J. Hum. Genet.* **66**, 979–988.

Kaessmann H, Heissig F, von Haeseler A, Pääbo S (1999) DNA sequence variation in a non-coding region of low recombination on the human X chromosome. *Nature Genet.* **22**, 78–81.

Kimmel M, Chakraborty R, King JP, Bamshad M, Watkins WS, Jorde LB (1998) Signatures of population expansion in microsatellite repeat data. *Genetics* **148**, 1921–1930.

Lewontin RC (1972) The apportionment of human diversity. *Evol. Biol.* **6**, 381–398.

Mellars P (1998) The fate of the Neanderthals. *Nature* **395**, 539–540.

Mellars P (2002) In: *The Speciation of Modern Homo sapiens* (ed. Crow TJ). Oxford University Press, Oxford, UK. pp. 31–47.

Mitchell RJ, Howlett S, White NG et al. (1999) Deletion polymorphism in the human *COL1A2* gene: genetic evidence of a non-African population whose descendants spread to all continents. *Hum. Biol.* **71**, 901–914.

Petit JR, Jouzel J, Raynaud D et al. (1999) Climate and atmospheric history of the past 420,000 years from the Vostok ice core, Antarctica. *Nature* **399**, 429–436.

Pritchard JK, Seielstad MT, Perez-Lezaun A, Feldman MW (1999) Population growth of human Y chromosomes: a study of Y chromosome microsatellites. *Mol. Biol. Evol.* **16**, 1791–1798.

Pritchard JK, Stephens M, Donnelly P (2000) Inference of population structure using multilocus genotype data. *Genetics* **155**, 945–959.

Quintana-Murci L, Semino O, Bandelt HJ, Passarino G, McElreavey K, Santachiara-Benerecetti AS (1999) Genetic evidence of an early exit of *Homo sapiens sapiens* from Africa through eastern Africa. *Nature Genet.* **23**, 437–441.

Redd AJ, Roberts-Thomson J, Karafet T et al. (2002) Gene flow from the Indian subcontinent to Australia: evidence from the Y chromosome. *Curr. Biol.* **12**, 673–677.

Reich DE, Goldstein DB (1998) Genetic evidence for a Paleolithic human population expansion in Africa. *Proc. Natl Acad. Sci. USA* **95**, 8119–8123.

Reich DE, Cargill M, Bolk S et al. (2001) Linkage disequilibrium in the human genome. *Nature* **411**, 199–204.

Richards M, Macaulay V (2001) The mitochondrial gene tree comes of age. *Am. J. Hum. Genet.* **68**, 1315–1320.

Roberts L (1992) Genome Diversity Project: Anthropologists climb (gingerly) on board. *Science* **258**, 1300–1301.

Roberts RG, Jones R, Smith MA (1990) Thermoluminescence dating of a 50,000-year-old human occupation site in northern Australia. *Nature* **345**, 153–156.

Roberts RG, Flannery TF, Ayliffe LK et al. (2001) New ages for the last Australian megafauna: continent-wide extinction about 46,000 years ago. *Science* **292**, 1888–1892.

Rogers AR, Harpending H (1992) Population growth makes waves in the distribution of pairwise genetic differences. *Mol. Biol. Evol.* **9**, 552–569.

Romualdi C, Balding D, Nasidze IS et al. (2002) Patterns of human diversity, within and among continents, inferred from biallelic DNA polymorphisms. *Genome Res.* **12**, 602–612.

Rosenberg NA, Pritchard JK, Weber JL, Cann HM, Kidd KK, Zhivotovsky LA, Feldman MW (2002) Genetic structure of human populations. *Science* **298**, 2381–2385.

Santos FR, Pandya A, Tyler-Smith C et al. (1999) The central Siberian origin for native American Y chromosomes. *Am. J. Hum. Genet.* **64**, 619–628.

Seielstad MT, Minch E, Cavalli-Sforza LL (1998) Genetic evidence for a higher female migration rate in humans. *Nature Genet.* **20**, 278–280.

Taberlet P, Fumagalli L, Wust-Saucy AG, Cosson JF (1998) Comparative phylogeography and postglacial colonization routes in Europe. *Mol. Ecol* **7**, 453–464.

Templeton AR (1999) Human races: a genetic and evolutionary perspective. *Am. J. Anthropol.* **100**, 632–650.

Thangaraj K, Singh L, Reddy AG et al. (2003) Genetic affinities of the Andaman Islanders, a vanishing human population. *Curr. Biol.* **13**, 86–93.

Thomson R, Pritchard JK, Shen P, Oefner PJ, Feldman MW (2000) Recent common ancestry of human Y chromosomes: evidence from DNA sequence data. *Proc. Natl Acad. Sci. USA* **97**, 7360–7365.

Tournamille C, Colin Y, Cartron JP, Le Van Kim C (1995) Disruption of a GATA motif in the Duffy gene promoter abolishes erythroid gene expression in Duffy-negative individuals. *Nature Genet.* **10**, 224–228.

Underhill PA, Passarino G, Lin AA *et al.* (2001) The phylogeography of Y chromosome binary haplotypes and the origins of modern human populations. *Ann. Hum. Genet.* **65**, 43–62.

Wall JD, Przeworski M (2000) When did the human population size start increasing? *Genetics* **155**, 1865–1874.

Walter RC, Buffler RT, Bruggemann JH *et al.* (2000) Early human occupation of the Red Sea coast of Eritrea during the last interglacial. *Nature* **405**, 65–69.

Williamson R, Duncan R (2002) DNA testing for all. *Nature,* **418**, 585–586.

Wilson JF, Weale ME, Smith AC *et al.* (2001) Population genetic structure of variable drug response. *Nature Genet.* **29**, 265–269.

Y-Chromosome-Consortium (2002) A nomenclature system for the tree of human Y-chromosomal binary haplogroups. *Genome Res.* **12**, 339–348.

Zilhão J, d'Errico F (1999) The chronology and taphonomy of the earliest Aurignacian and its implications for the understanding of the Neandertal extinction. *J. World Prehist.* **13**, 1–68.

Zhivotovsky LA, Rosenberg NA, Feldman MW (2003) Features of evolution and expansion of modern humans, inferred from genomewide microsatellite markers. *Am. J. Hum. Genet.* **72**, 1171–1186.

CHAPTER TEN

Agricultural expansions

CHAPTER CONTENTS

BOXES

In the previous chapter we saw how modern humans practicing a hunter–gatherer lifestyle moved out of Africa and colonized Eurasia and Australasia within the last 50 KY. Since that time, these regional populations have undergone substantial demographic change and additional migrations. In this chapter we explore the dramatic impact that the shift to agricultural means of food production has had on genetic diversity over the past 10 KY.

It is hard to overestimate the impact of the agricultural revolution on human genetic diversity, and more generally on global biodiversity. The advent of agriculture was a necessary prerequisite for the development of the modern world as we know it. It drove the rampant numerical growth (now approaching one thousand-fold) of our species, the development of complex societies and our conception of the world around us. It made us what we are today. A comprehensive study of the impact of agriculture would be nothing less than the history of the world over the past ten millennia. Here we confine ourselves to the direct impact of agricultural innovation and its spread on genetic diversity in both humans and the species we domesticated.

Farming has been a very successful cultural innovation that spread rapidly over most of the globe. The primary concern of this chapter is whether this flow of culture was mirrored by a flow of genes. In essence, was the spread of agriculture mediated by indigenous peoples learning the techniques of their neighbors (**acculturation**) or by migrations of farmers themselves (**gene flow**) (*Figure 10.1*)? Note that these two alternatives are somewhat analogous to the Multiregional and Out of Africa models for modern human origins. A number of different models for gene flow have been developed, the most popular of which is the **demic diffusion** model of Ammerman and Cavalli-Sforza, which predicts a demographically driven 'wave of advance' of 'farmer' genes (Ammerman and Cavalli-Sforza, 1984). At the same time as considering these models, we must be aware that the histories and interactions of different populations are more complex than any modeler can to take account of, and that there are archaeologists and linguists whose own experience challenges the simplicity that we must, of necessity, introduce.

Firstly we consider the background evidence from other disciplines on the origins, impact and spread of agriculture. We then consider two case studies of human genetic diversity within well-characterized geographical zones of agricultural spread:

▸ out of the Near East into Europe;

▸ out of tropical West Africa into subequatorial Africa.

Finally, we examine the genetic evidence from a selection of domesticates themselves, from several areas of the world, for clues as to the nature of the domestication event(s).

10.1 What is agriculture?

A dictionary tells us that agriculture is:

'. . . the science, art, and business of cultivating soil, producing crops, and raising livestock.'

As well as the inclusive sense of the word outlined above, 'agriculture' is also often used in a narrower sense that specifically refers to the intensive farming of crops and animals in fields, as distinct from a less intensive management of individual plants (**horticulture**) and the herding of animals (**pastoralism**). Agricultural populations tend to live in settlements neighboring their cultivated land. This increased **sedentism** lies in direct contrast to the more mobile life of most hunter–gatherers whose movements are governed more by the availability of seasonal resources.

Although the agricultural revolution has often been presented as a sudden burst of innovation, it is actually an assemblage of innovations which developed much more gradually – the agricultural revolution was a process, not a single event. Many hunter–gatherer groups were adopting more settled lifestyles before the advent of agriculture, and were actively managing the natural resources at their disposal. We can appreciate these distinctions better by dissecting crop cultivation into its component processes:

▸ soil preparation;

▸ selective breeding;

▸ propagation: planting seed (sexual) or taking cuttings (asexual);

▸ crop protection: weeding, use of pesticides and protection from the elements;

▸ harvesting;

▸ storage.

While some of these processes were introduced in quick succession at the dawn of agriculture, others have a much more ancient origin among hunter–gatherer societies and a few are even practiced by nonhuman species. For example, many other species store foodstuffs for less abundant times ahead. Although the adoption of individual practices leads to improved food production, it is the final domestication of plant and animal species that represents the dominant innovation. **Domestication** is defined as the selective breeding of a species to make it more useful to humans, as distinct from the taming of wild animals whose breeding is not manipulated. Many of the modern domesticated animals and plants have undergone great morphological divergence from their ancestral wild species (Section 10.7).

The advent of agriculture is associated with such a dramatic change in tool usage that it is assigned its own cultural period – the **Neolithic** (New Stone Age), and sometimes referred to as the Neolithic revolution. Although originally defined by the use of ground or polished stone implements and weapons, other characteristic features of the Neolithic now include the manufacture of pottery and the appearance of human settlements. The Neolithic appears at different times in different regions, and the reasons for this are explored in the next section. Note that when terms such as Mesolithic and Neolithic were first proposed, the definitions were simple: Mesolithic people were early Holocene hunter–gatherers with flaked stone tools and no

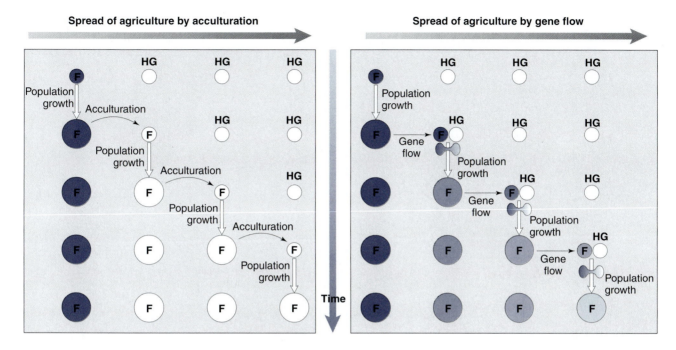

Spread of agriculture by acculturation

Spread of agriculture by gene flow

Figure 10.1: Acculturation and gene flow models for the spread of agriculture.

Shading indicates the genetic ancestry, and the labels F and HG refer to farming and hunter–gatherer economies.

pottery, while Neolithic people were later farmers with polished stone tools and pottery. However, evidence of Mesolithic societies with pottery, and of people with characteristically Neolithic pottery but who were hunter–gatherers, has made these definitions more complex. Confusion is caused when the terms are used to refer to periods of time, to different ways of life, or to types of material culture, but these terms nevertheless persist. When we consider parts of the world far from Europe, where the terms were first used, matters can be even more confusing (see *Box 10.6*).

In Neolithic culture plants were domesticated first, and only later were animals bred for their meat. It is only a long time afterwards that the additional applications of domesticated animals were appreciated. This **'secondary product revolution'** includes the use of hair and hides for clothing and large-bodied mammals for traction and transport (Sherratt, 1981).

In the modern era, agricultural means of food production predominate over hunting and gathering lifestyles. Extant hunter–gatherers tend to reside in marginal areas in which farming is not ecologically feasible (*Figure 10.2*), for example the Inuit north of the Arctic Circle, the Khoisan of the Kalahari Desert and the Aborigines of central Australia. The environmental marginality of many of these populations makes them poor analogs for the prehistoric hunter–gatherers who occupied the richer lands subsequently taken over by farmers.

10.2 Where, when and why did agriculture develop?

10.2.1 Where and when?

From archaeological evidence (*Box 10.1*), the dates of the first appearance of agriculture vary widely in different regions. In addition, the plants and animals domesticated differ among these regions. These two findings suggest multiple, independent origins of farming practices. Dating the prehistoric ranges of these individual farming cultures reveals that they share a common feature: an ancient homeland of agricultural innovation from which farming subsequently expanded. The difficulty in distinguishing between a truly independent origin and a secondary homeland in which new species were domesticated in response to the arrival of agriculture from elsewhere, means that the actual number of truly independent homelands remains a contentious issue (*Figure 10.3*).

The best characterized centers of domestication are the 'Fertile Crescent' in the Near East, Mesoamerica and China. These areas all had native species suitable for domestication (Diamond, 2002). Once the 'agricultural package' had been assembled it could be exported to regions of similar climate but without the suitable native species. Additional domestications of new species often occurred *en route*. Dispersal would have been easier along latitudinal (east to west) axes than along longitudinal (north to south) axes due to the greater similarity of the growing seasons (Diamond, 1998).

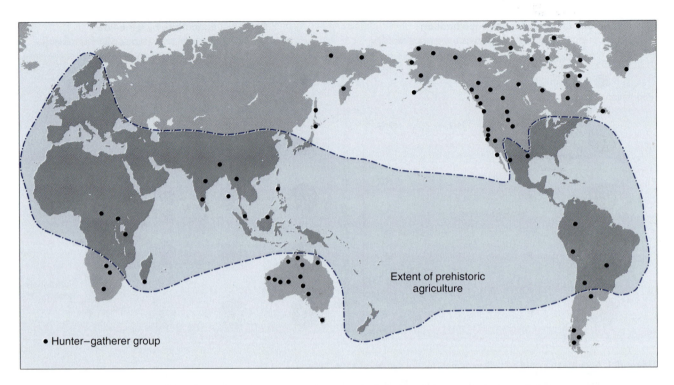

Figure 10.2: Map showing the extent of prehistoric agriculture and the marginality of extant hunter–gatherer societies.

Data from Kelly (1995) and Bellwood (2001).

BOX 10.1 Sources of evidence in the archaeology of agriculture

Examination of archaeological remains reveals direct and indirect evidence of agricultural activity, and ^{14}C-dating (see *Box 9.4*) can be used to estimate when this activity occurred. At least four kinds of evidence are available (Barker, 1985):

▸ *Human bones*: evidence on diet can come from tooth wear, disease, and bone chemistry.

▸ *Animal bones*: patterns of animal remains are distinctively different under domestication compared to wild captures. The fraction of remains made up by a single species increases; animals themselves decrease in size as an environmental and genetic consequence of captivity; and the sex ratio and age profile of animals changes, indicating planned culling rather than more random capture (Meadow, 1989).

▸ *Plant remains*: these are generally less well preserved than animal remains, but can be found in bogs, in frozen soil, or in arid conditions. In semi-arid regions, caves or rock shelters provide some protection from environmental degradation. As well as seeds themselves, which are often recognizably preserved in carbonized form after burning, useful evidence can come from pollen, starch grains, phytoliths (particles of silicon dioxide from vegetable rinds) or parenchyma (the storage material that occurs in tubers and rhizomes). The recognition of a pattern in which tree pollen levels fall dramatically, followed by a rise in grass and cereal pollen, and finally a return to dominance of tree pollen, has been interpreted as the land clearance effects of 'slash and burn' agriculture (*landnam*). However, this picture is not now widely accepted (Barker, 1985). Remains of particular insects have also been taken as evidence of storage of particular seed crops, and analysis of human **coprolites** (Poinar *et al.*, 2001) provides a direct way to see what people were eating.

▸ *Associated cultures*, defined by artifacts (including stone tools, ceramics and rock-art), architecture, burial practices and the layouts of settlements. For example, a distinctive type of pottery (**linear pottery**, or ***linienbandkeramik*; LBK**) is associated with the early European Danubian farming culture, and another type, **Cardial ware**, impressed with the serrated edge of the cardium (cockle) shell, is associated with the spread of farming in Italy, southern France, Spain, Sardinia and Corsica.

A key point when considering archaeological evidence for agricultural activity is that it has been greatly affected by political contexts: for example, there was work during the 1950s and 1960s in Iraq and Iran but little since then; and for decades huge areas of the Middle East, Africa, central/south America and east and southeast Asia have been effectively inaccessible for field research. It is important to distinguish between *evidence of absence* of agriculture in a region, and *absence of evidence*.

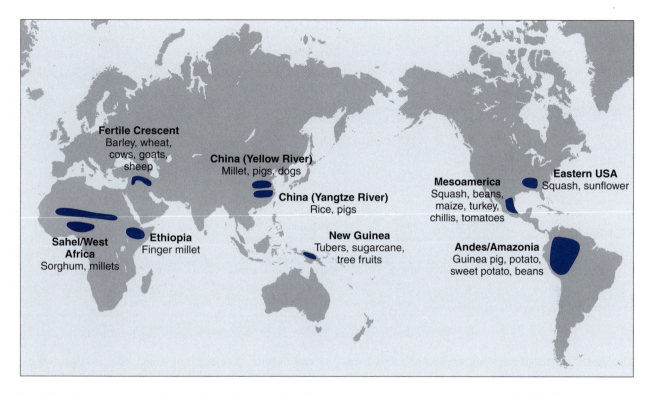

Figure 10.3: Map of centers of agricultural innovation together with some of the plants and animals domesticated within them.

It is worth noting that these past centers of agricultural innovation are not the most productive areas for farming today. Climate change and long term environmental degradation have taken their toll.

Although dates for the first evidence for these different centers of agricultural innovation are spread over six millennia (*Figure 10.4*), when viewed from the perspective of the 100 KY of modern human evolution a question must be asked: why do all these independent origins of agriculture appear within such a defined window?

10.2.2 Why?

The Neolithic revolution required no biological changes of humans, but was a purely cultural phenomenon (although there were numerous biological consequences, which will be discussed in the next section). The traditional view of the Neolithic revolution was that it was a step on the natural progression towards 'civilized man', and that adopting farming was the only sane option for any self-respecting hunter–gatherer. However, the eventual outcome of developing farming practices would have been far from obvious to prehistoric populations. In addition, the view of an inevitable progression does not explain why we see a delay of 30 KY between the completed development of truly modern behavior by 40 KYA and the advent of farming at 10 KYA. Then, as now, individuals would have made complex decisions by appreciating the immediate costs and benefits of crop cultivation and animal husbandry. We should therefore be asking what contemporary issues might have driven the process of developing agriculture, specifically the domestication of plant and animal species.

The end of the last ice age roughly 14 KYA was followed by relative climatic stability (the largest fluctuation being the Younger Dryas cold period which was far smaller than preceding events) in which Mesolithic human populations prospered and grew in size. At the same time many of the animal species preferentially targeted by hunters, especially large-bodied vertebrates, were being driven to extinction by a combination of hunting and other environment changes (*Table 10.1*). These extinctions occurred on all continents during the period 50–10 KYA, but were much more marked in Australia (Roberts *et al.*, 2001) and the Americas (Alroy, 2001; Martin, 1973) than in Africa and Eurasia. The only continent to retain many species of large-bodied wild mammals is Africa, where these species have been exposed to human hunting practices the longest and may have adapted behaviorally. However, temporal gaps exist between many of these extinction events and the onset of agriculture on these continents. This suggests that there is no direct causal relationship between the two, but that as yet unappreciated factors were also important. But why, then, did agriculture only develop in certain areas?

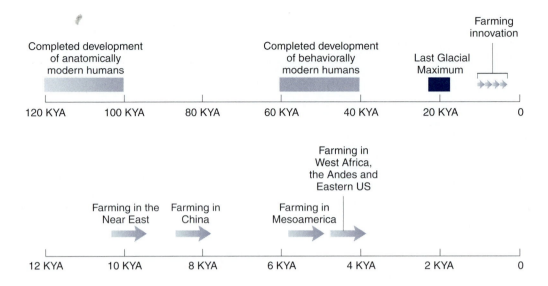

Figure 10.4: Timelines of agricultural development.

The geographical locations of centers of agricultural innovation were largely determined by biogeographic luck (Diamond, 1998). A common archaeological finding is that the most important domesticated species in early farming homelands were cereals and other grasses: wheat and barley in the Near East, millet and rice in China, maize (corn) in the Americas, sorghum in West Africa and sugarcane in New Guinea. The distribution of wild grasses with large seeds is highly irregular, with the highest concentration in Southwest

Asia. Other 'founder' crops and animals that could be domesticated also had limited biogeographic distributions. Agricultural homelands formed in the areas where these distributions overlapped (Diamond, 2002).

Finally, it is worth noting the irreversibility of farming. It supports a larger population than hunter-gathering from the same area of productive land, and such a population could not be sustained if there were a return to the previous way of life. This ratchet-like mechanism ensures that the number of populations practicing agricultural means of food production could only increase over time.

10.2.3 Which domesticates?

Within each agricultural homeland, the diversity of plants and animals that were domesticated was much less than that of the plants and animals consumed by hunter–gatherer societies within the same area. In addition, after the initial burst of domestication relatively few species were added to each agricultural package. Wheat and rice are amongst the earliest domesticates yet still occupy the greatest area of agricultural land of all domesticated plants. This suggests that only certain species were amenable to selective breeding for desirable traits. *Table 10.2* lists some of the desirable characteristics of these species.

As well as having nutritional value worthy of the energetic investment in their propagation, these species also required a number of additional characteristics that facilitate selective breeding. Hermaphroditic plants that are capable of self-fertilization will produce offspring that maintain the desirable attributes of their parents, in contrast to plants that preferentially outcross. Annual plants have shorter generation times, and can therefore undergo more rapid genetic change than perennials. In addition, offspring are more likely to retain desirable characteristics in species in which a desirable

TABLE 10.1: EXTINCTIONS OF LARGE BODIED MAMMALS (> 44 kg) BETWEEN 50 AND 10 KYA.

Species	Continent
Woolly Mammoth	Europe
Giant Irish Elk	Europe
Woolly Rhino	Europe
Saber-tooth Cat	North and South America
American Camel	North America
Giant Beaver	North America
Giant Ground Sloth	North America
American Mastodon	North America
Big-horned Bison	North America
Glyptodon (Giant Armadillo)	South America
Marsupial Lion	Australia
Giant Short-faced Kangaroo	Australia
Giant Wombat	Australia

TABLE 10.2: DESIRABLE CHARACTERISTICS OF SPECIES SUITABLE FOR DOMESTICATION.	
Animals	**Plants**
Diet easily supplied by humans	Cereals with large seeds, or plants with large tubers
Rapid growth rate and reproduction rate (short birth interval)	Annuals (short generation time)
Placid disposition	Selfing hermaphroditic reproduction
Breed in captivity	Easily harvested
Used to dominance hierarchies – easy to manage	Can be stored for long periods
Remain calm when put into enclosures	Monogenic control of crucial traits

trait is controlled by a small number of genes. By contrast, it is easier to control the reproduction of animals, but harder to manage their daily existence.

Francis Galton, as with many other subjects, had something to say on the qualities of animals suitable for domestication (Galton, 1865):

'1. They should be hardy; 2. they should have an inborn liking for man; 3. they should be comfort-loving; 4. they should be found useful to the savages; 5. they should breed freely, 6. they should be easy to tend.

It would appear that every wild animal has had its chance of being domesticated, that those few which fulfilled the above conditions were domesticated long ago, but that the large remainder, who fail sometimes in only one small particular, are destined to perpetual wildness as long as their race continues. As civilization extends they are doomed to be gradually destroyed off the face of the earth as useless consumers of cultivated produce.'

10.3 Outcomes of agriculture

The outcomes of this cultural revolution for the animal and plant species that humans domesticated are considered in Section 10.7. Here, we concentrate on the consequences for humans themselves.

10.3.1 Demographic and disease outcomes

The long-term demographic outcomes of the Neolithic revolution are obvious. While population sizes may have increased slowly during the period 100–10 KYA, today we have a global population hundreds of times greater than that of pre-agricultural hunter–gatherers (*Figure 10.5*). We also enjoy much greater life expectancy in the modern era compared to pre-agricultural societies.

Why did shifting means of food production result in such dramatic population growth? A number of potentially synergistic factors have been proposed to be important (Hassan and Sengel, 1973; Diamond, 2002):

▶ Higher population densities could be supported by increased yields of edible foodstuffs;

▶ Better nutrition results in earlier menarche (first menstrual cycle) for women, resulting in a longer period of fertility for the same life expectancy;

▶ A more stable food supply could result in increased fertility through fewer miscarriages;

▶ Lower mobility resulting from greater settlement allows a shorter interval between births. Previously hunter–gatherers would have had to carry their young children around with them, limiting the number of children that could be brought up at any one time. Modern hunter–gatherer mothers tend to wean their children later than do mothers in agricultural societies;

▶ Availability of milk from animals allowing earlier weaning and closer-spaced pregnancies;

▶ Less need for 'natural' methods of birth control: less infanticide and induced abortion;

▶ A reduction in sex ratio fluctuations could lead to higher population fertility. Smaller hunter–gatherer groups are more prone to such fluctuations, especially given the sex-specific differences in occupational mortality and rates of infanticide.

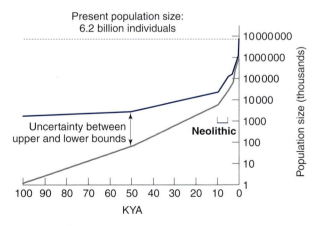

Figure 10.5: Global population size estimates over the past 100 thousand years.

A zone of plausible population sizes between upper and lower bounds is shown on a *logarithmic* scale. Uncertainty in population size estimates is greater in the more distant past.

Despite these apparent advantages, rates of population growth have been much more dramatic in recent times than during the early years of farming. In fact there are a number of indications that farming initially imposed a heavy disease burden on agricultural populations (Cohen and Armelagos, 1984), and indeed the average life expectancy may well have been lower in early farmers than hunter–gatherers!

Dietary changes in early farmers were not beneficial. Skeletal remains demonstrate lower bone porosity among farmers than hunter–gatherers, indicative of anemia. This and other skeletal indicators suggest that malnutrition was common (Cohen and Armelagos, 1984). Comparative studies of Paleolithic and Mesolithic populations from the same region typically show a decline in nutrition over time. While contemporaneous hunter–gatherers adapted to the reduced resources available to them by exploiting a broader spectrum of species in their diet, early farmers, being dependent on far fewer species, were prone to catastrophic crop failure. Even without crop failures, dependence on too few crops can result in nutrient deficiencies.

Early farmers also appear to have experienced a higher prevalence of infectious disease. Skeletal remains of farmers have greater frequencies of lesions resulting from pathogenic infections. It is likely that there is a synergy between malnutrition and infectious disease, although other factors are also important. Many infectious diseases in humans result from pathogens transferring from an animal host; these diseases are known as **zoonoses** (Cockburn, 1971; Barrett et al., 1998). Animal husbandry requires greater daily contact with animals than hunting, and this closer proximity would have facilitated transfer of these diseases. In addition, higher population densities may well have exacerbated the impact of zoonoses. Many pathogens require a high population density before they can be sustained within the population (e.g., smallpox, measles and mumps). Increased human population densities resulting from both population growth and increased **sedentism** could support pathogens previously confined to herd animals. Sedentism also resulted in accumulations of human and animal wastes, which attract vector-borne diseases such as the bubonic plague. Increased pressure on supplies of clean water could have resulted in epidemics of cholera. An increased reliance on food storage during leaner times may have led to increased morbidity though spoilage.

The preponderance of mammals suitable for domestication in Southwest Asia meant that many zoonoses were especially prevalent among populations derived from that agricultural homeland. The eventual development of resistance to these zoonotic pathogens would have later relevance for epidemics resulting from contact between previously isolated populations. From the fifteenth century onwards, the impact of infectious disease was often far more severe on indigenous populations than it was on the European colonists with whom they came into contact (Diamond, 1998; see Section 12.23). This differential susceptibility was not unknown to the protagonists at the time, and one of the earliest forms of bioterrorism is revealed by the reply, in 1763, of British General Amherst to Colonel Bouquet's suggestion that smallpox-infected blankets be given to Native Americans to precipitate an epidemic:

'You will do well to try and inoculate the Indians by means of blanketts (sic), as well as every other method that can serve to extirpate this execrable race.'

Both infectious diseases and dietary changes altered the selective landscape for humans, by, for example, selecting for tolerance to high lactose diets in adults. In Chapters 13 and 14 we explore some of the genetic consequences of these novel selective pressures.

10.3.2 Societal outcomes

Living in an early farming society would have been significantly different from living in a hunting-gathering society. Life was more settled, and was spent in larger populations. The dominant social unit became the household rather than the whole group, which reinforced the concept of private property. Trading networks would have been required with outside communities to maintain access to distant resources. Permanent settlement allows the development of nonportable technology. New tools would have been required for food harvesting, storage and preparation. Although the earliest pottery is found sparsely among a few pre-agricultural societies, clayware is a consistent indicator of the Neolithic transition.

Despite an initial focus on subsistence, it was the ability of agriculture to produce a surplus that transformed human societies. The surplus could be used to support full time craftspeople and a bureaucracy to manage the society. Intensification of agriculture resulted in the new phenomenon of wealth. This varied among households, eventually leading to social stratification. Settlement imposes greater costs on group fission and new mechanisms for conflict resolution were required. Greater dependence on specific tracts of land also led to the potential for conflict between neighboring settlements. The central management of multiple settlements represented the first stages of state formation. The development of trade specialization, structured societies and complex economies eventually precipitated the birth of civilizations. The earliest evidence for writing is largely represented by the administration of economic transactions on clay tablets.

10.4 The language codispersal hypothesis

10.4.1 Recent language expansions

The approximately 6500 languages spoken throughout the world today can be grouped into more than 100 families, each on the basis of a hypothetical shared common ancestral language. The ancestral relationships between language families themselves are highly contentious, and are probably beyond linguistic reconstruction. A few language families contain disproportionately large numbers of languages (*Figure 10.6*); for example, two language families (Austronesian and Niger-Congo) together comprise about 40% of all individual languages. The geographic distribution of these language families is highly skewed: some occupy large continental regions in which few, if any, other languages are present ('spread zones'; *Figure 10.7*), whereas other families inhabit areas which contain a high diversity of unrelated languages with limited distributions (Nichols, 1997).

The existence of a single origin of languages within a family that is distributed over a wide area suggests a recent range expansion of those languages. Few linguists believe that any language family can be more than 10 KY old. By studying which languages within a family share linguistic innovations, the ancestral relationships within a language family can be reconstructed. This process is somewhat analogous to the reconstruction of a phylogenetic tree. Two characteristics of linguistic diversity have been used to try and identify the likely geographic origin of the language family:

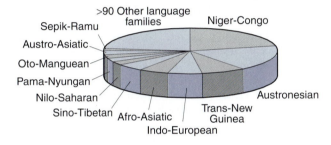

Figure 10.6: A few language families contain disproportionately large numbers of languages.

Data from Ethnologue: http://www.ethnologue.com/

▶ the region of greatest linguistic diversity;

▶ the region in which the deepest-rooting languages are spoken.

The likely homelands of many of the largest language families appear to be situated in and around the centers of agricultural innovation identified above. This has led to the hypothesis that many of these language families moved along with the expansion of agriculture in many different areas of the world (Bellwood, 2001; Renfrew, 1988; *Box 10.2*). This view is supported by the finding that areas of patchwork-like language distributions are found in regions in which there is little archaeological evidence for agricultural expansions.

10.4.2 Testing the hypothesis with linguistic evidence

In principle, if it were possible to date accurately the spread of languages from linguistic evidence itself, it would be easy to test whether linguistic expansions accompany agricultural expansions. However, despite early attempts at linguistic dating, most linguists believe that known differences in rates of change between languages disprove the existence of a 'linguistic clock', analogous to the molecular clock (Section 5.6.4) used in genetic dating methods.

An alternative approach derives from attempts to reconstruct the vocabularies of ancestral languages, known as **proto-languages**. A word present in the same form in all languages within a family indicate that this word was present in the proto-language. By reconstructing a number of different words it should be possible to build up a cultural context for this ancestral language that will allow indirect

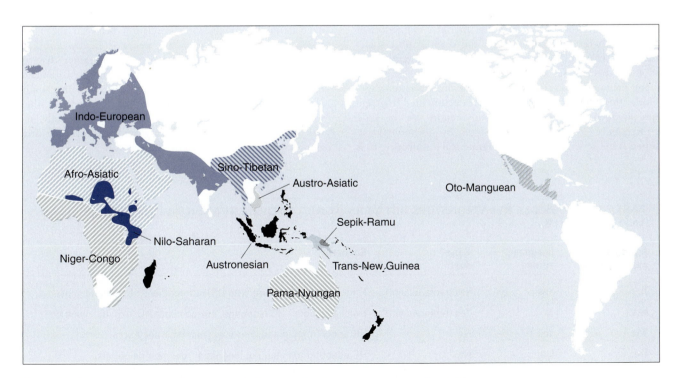

Figure 10.7: Map showing the distributions of major language families that contain more than 100 languages.

dating through archaeological evidence. For example, one of the problems with the theory of agricultural origins for Indo-European languages is that it appears to be possible to reconstruct terms involved in horse-drawn chariots in 'proto-Indo-European'. The first archaeological evidence for these vehicles postdates the spread of agriculture by several millennia, suggesting that the expansion of this language family took place long after the spread of agriculture.

There are alternative explanations for the heterogeneous distribution of language families – why some families have spread far and wide, while others remain localized. Ecological, geographical and economic factors can play important roles in determining linguistic diversity (Nichols, 1997). Cultures adapted to coastal environments or arid plains can rapidly spread their languages over large distances. By contrast, highlands provide refuge for remnant languages and often harbor high linguistic diversity.

There are historically documented language spreads that have been associated neither with agricultural expansions, nor with high levels of gene flow. These include the spread of Romance languages derived from Latin throughout Western Europe, and the adoption of a Turkic language by populations previously speaking Indo-European languages in Anatolia (Turkey).

10.4.3 Genetic implications of language spreads

While cultures may spread without the concomitant spread of genes, languages are less easy to acquire during adulthood than other cultural transformations. It has been argued that the unlikely nature of this **language shift** in an entire population suggests that the languages must have been spread by the farmers themselves, and that the later extinction of hunter–gatherer languages reflects the demographic superiority of farmers (Bellwood, 2001). In such instances we should expect to see clear genetic evidence for gene flow along the same axes as the spread of agriculture and language. Under this framework, the historically documented language spreads identified in the previous section are of little consequence as they were imposed by technologically superior elites with little need for numerical dominance, thus resulting in negligible gene flow. By contrast, during agricultural expansions farmers and hunter–gatherers would have found themselves competing directly for land, and numerical imbalances might have determined the eventual linguistic outcome.

Other considerations suggest that gene flow need not accompany agriculturally-associated language expansions. In the previous section we saw the dramatic societal changes that the advent of agriculture entailed. It can be argued that the all-encompassing nature of these changes precludes the simple adoption of agricultural practices into the pre-existing languages spoken by hunter–gatherers; large elements of the vocabulary would have to be invented wholesale. Under such great social pressures language shift may well have been more frequent than assessments of its infrequent modern occurrence might suggest. These different scenarios are summarized in *Table 10.3*.

The above models for agriculturally associated language expansion with and without concomitant gene flow are overly reliant upon discrete packages of genes labeled 'farmers' and 'hunter–gatherers'. Admixture could have occurred between these two groups during the large timescales over which these language expansions took place (see Chapter 12, and the next section in this chapter). Such admixture is unlikely to have been completely inhibited by the different means of food production of their distant ancestors. These intermediate models incorporating admixture are explored in *Table 10.4*. Such admixture could result in offspring with parents speaking different languages; the adoption by the child of a single language would by necessity uncouple genes and language to a limited degree.

These additional complications indicate that the simple expectation of an absolute correlation between means of food production, language and genes is highly unlikely. Complex models of individual interactions rather than of nonadmixing, clonal, populations are required to generate realistic expectations of the patterns of gene flow that might accompany agricultural expansions, whether or not they were accompanied by the spread of languages. Initially, we will examine whether there is any evidence for gene flow along known axes of agricultural expansion.

TABLE 10.3: POSSIBLE RELATIONSHIPS BETWEEN AGRICULTURE EXPANSION, LANGUAGE SPREAD AND GENE FLOW.

Agricultural expansion	Language spread	Gene flow	Scenario
Yes	Yes	Yes (replacement)	F move in, no admixture with HG, out-compete HG, expand, move on
Yes	No	Yes (replacement)	F move in, adopt HG language, out-compete HG, expand, move on
Yes	Yes	No	HG adopt farming and language from their neighbors
Yes	No	No	HG adopt farming but not language from their neighbors

F, Farmers, HG, hunter–gatherers.

TABLE 10.4: POSSIBLE RELATIONSHIPS BETWEEN AGRICULTURAL EXPANSION, LANGUAGE SPREAD AND GENE FLOW INCORPORATING ADMIXTURE.

Agricultural expansion	Language spread	Gene flow with admixture	Scenario
Yes	Yes	Yes (partial)	F move in, assimilate HG, retain language, expand, move on
Yes	No	Yes (partial)	F move in, assimilate HG, adopt HG language, expand, move on

F, Farmers; HG, hunter–gatherers.

10.5 Out of the Near East into Europe

10.5.1 Archaeological evidence and dates for the European Neolithic

Figure 10.8 shows the distribution of the archaeological sites at which the earliest evidence for agriculture has been found throughout Europe (Ammerman and Cavalli-Sforza, 1984). There is a time gradient: the oldest sites, indicating the origins of agriculture and dating to almost 10 000 YA, are in the 'Fertile Crescent' in the Near East, that is, in modern Syria and Israel. Sites become consistently younger towards the northwest of Europe, with agricultural practices arriving at the Baltic and the British Isles between 5500 and 4200 YA. Although there is debate about interpretation of individual pieces of archaeological evidence, and the small-scale pattern of spread, this overall picture from archaeology has provided a framework for the disputes of genetics.

A traditional view of the arrival and spread of agriculture in Europe can be summarized as follows (Price, 2000b). The crops (wheat, barley, pulses and flax) and animals (cattle, pigs, sheep and goats) of the European Neolithic were originally domesticated in the Near East, and brought to Europe by farmers. Farming communities appeared in the Aegean area and Greece around 9000 YA, and there followed two streams of movement, one into southeast Europe, and the other along the Mediterranean coast. A rapid expansion of farming then occurred, occupying an area from the Netherlands in the west to the Ukraine in the east, and from the Alps to northern Germany and Poland. From the Mediterranean there were expansions into Italy, France and the Iberian peninsula. Regional differentiation of Neolithic groups then took place, including the development of the characteristic **megaliths** (structures built of large stones, including single standing stones) of the west, and dispersal to the extreme west, including Ireland, was completed by 4000 YA. This apparently regular pattern of spread has formed the basis for the genetic studies discussed below.

This view of agricultural spread is by no means universally accepted, however. The idea of a 'package', in which all of the elements (permanent village of rectangular houses, domesticated crops and animals, evidence of religion, ground stone tools and pottery) arrive at once, does not always fit the archaeological evidence. The existence of permanent pre-Neolithic settlements, and evidence in some places of the cultivation of crops or the domestication of animals dating many centuries before the arrival of the Neolithic (Price, 2000b), suggests that the populations already present may have played a considerable role in the process of cultural change (Barker, 1985). While some anthropological evidence from craniometry (the measurement of skull proportions) suggests Neolithic population replacement in the Iberian peninsula (Lalueza Fox *et al.*, 1996), other data showing 'continuity' in dental morphological traits in some areas between the Mesolithic and Neolithic suggests cultural change but little change in the biological composition of the population (Jackes *et al.*, 1997). As was discussed in Section 10.4.2, the inclusion within the Neolithic 'package' of an Indo-European language has also been challenged.

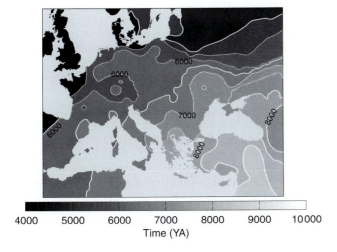

Figure 10.8: Map of the distributions of the earliest archaeological sites in Europe and the Middle East showing evidence of agriculture.

Constructed from data in Ammerman and Cavalli-Sforza (1984).

10.5.2 Expectations for genetic outcomes of different models of expansion

As stated at the outset of this chapter, there are two models to explain the spread of agriculture: in the first model, **acculturation** (also known as **cultural diffusion**), the farmers did not move, but the technology and ideas of agriculture did. The second model involves gene flow, in which the farmers themselves moved, taking agricultural practices with them. While there are many different gene-flow models that can be imagined, the most widely discussed is **demic diffusion,** also known as the **wave of advance.** In this model, spread is stimulated by population growth (itself a result of increasing availability of food) and local migratory activity. What patterns do we expect to find in the genetic diversity of modern European populations if either of these models were true? The answer depends greatly upon the starting conditions – the gene pool carried by the pre-existing Paleolithic hunter–gatherer populations compared

with that of the Neolithic farmers – and this important issue will be addressed below. Let us first assume for simplicity that the dispersed groups of hunter–gatherers have a uniform composition of genetic markers, and that there is no overlap with that of the farmers (see p. 307).

In a purely demic diffusion model without interbreeding, the migration of farmers would have the effect of replacing the gene pool of the indigenous Europeans with that of the Near Eastern immigrants. We would expect to see strong affinities between the genes of all modern European

---

BOX 10.2 Opinion: Language/farming dispersals

When you look at a map of the world's major language families, it is clear that some families (such as the Indo-European family – including English, French, Spanish, Greek, Iranian, Hindi and most of the European languages; or the Sino-Tibetan family, including Chinese) have come to occupy vast areas of the earth's surface. These territories have been called by linguists 'spread zones'. Certain other language families occupy a space much more tightly packed with less-closely related languages whose origins may go back much deeper in time: these are the 'mosaic zone' families, seen for instance in the Caucasus, and in New Guinea and in North Australia.

Thirty years ago I suggested that the distribution of the Indo-European language family was best explained by the dispersal to Europe of the first farmers, spreading out of Anatolia (modern Turkey), first to Greece and then into the Balkans and also along the north Mediterranean coast, carrying with them their Proto-Indo-European language (Renfrew, 1988). The implication was that the Proto-Indo-European homeland must have been in Anatolia, at the time of the first farmers there around 9 KYA.

This suggestion caused a storm of disagreement among linguists at the time: it contradicted the then standard or classic view that the first Indo-Europeans were nomad pastoral horse riders spreading westward from the Steppes north of the Black Sea around 5 KYA. But linguists are now beginning to accept earlier dates for the proto-languages of major language families. Moreover, at the same time, Peter Bellwood, working in Australia, quite independently suggested a farming/language dispersal model for the spread of the Austronesian languages (including the Polynesian languages) in the Pacific. It seems likely also that other 'spread zone' languages were dispersed in this way, for instance that the Niger-Kordofanian languages, including the Bantu languages, accompanied the spread of farming out of west Africa. The underlying model is based upon the pronounced increase in population density that accompanies the transition of hunter–gatherer communities to farming: a demographic process which Ammerman and Cavalli-Sforza have called the 'wave of advance'.

When Cavalli-Sforza and his colleagues came to study the distribution in Europe of classical genetic markers using principal components analysis, they found that their first principal component showed a strong cline from southeast to

northwest, which they suggested was the result of the spread of farming from Anatolia to Europe at the beginning of the Neolithic period. They did not at that time seek to relate that to the spread of the proto-Indo-European language, but many researchers now do. At first the new evidence that became available with DNA studies from mtDNA and then from the Y chromosome, seemed to contradict this view. For it was claimed that many of the haplogroups represented were the result of population events which had taken place already in the Upper Paleolithic period, long before the coming of farming to Europe. But Y chromosome studies have now not only recognized haplogroups whose arrival in Europe should be of early Neolithic date (this is true for mtDNA haplogroups also). They have shown a marked cline in those haplogroups from southeast to northwest. More sophisticated modeling, for instance by Lounès Chikhi, allows one to envisage that each budding-off movement by a new generation of farmers from the ancestral village could be mainly (say 90%) direct descendants of the ancestral farming communities and thus still speaking a variety of the proto-language. But if in each generation, with each 20 or 30 kilometer translocation to the north or west, there was a 10% admixture of the local hunter–gatherer genes from the Mesolithic population of Europe, the 'farming' genes would come to decrease exponentially with distance from Anatolia. This is very much the pattern that is found. So the genetic evidence can be seen to be in harmony with a demic diffusion dispersal process, which may well be the root cause for the spread of the Proto-Indo-European language.

These are early days yet in the application of molecular genetics to historical linguistics, and of course there are no genes automatically associated with specific languages. There is no Indo-European or Bantu or Afroasiatic gene, if we take those terms in a strictly linguist sense. There is hope however that the procedures of archaeogenetics will continue to clarify the processes by which languages spread, and by which language families replace or are replaced. Farming/language dispersal is only one of those processes, but it may be one of the most important, and a key to explaining how the current world map of language diversity was formed (Bellwood and Renfrew, 2003).

Colin Renfrew, McDonald Institute for Archaeological Research, University of Cambridge, UK

populations and those of the populations in the Near East, but little genetic differentiation within Europe. In a purely acculturation model the spread of farming should not be associated with major genetic changes, and we expect a sharp distinction between the genes of most Europeans and those of Near Eastern populations (*Figure 10.1*, left panel). An intermediate model, in which farmers spread and interbred with indigenous peoples (*Figure 10.1*, right panel), should have left its imprint as a **cline**, or gradient, of gene frequencies (Section 6.7) from the Near East towards the northwest. The model espoused by Ammerman and Cavalli-Sforza (1984) is often described as a demic diffusion model, but, as they make clear, it is actually a mixed acculturation–demic diffusion model. The debate that has arisen around the genetic evidence is in fact based not upon either extreme position, which nobody accepts as reasonable, but on degrees of the mixed intermediate model, and the consequences of these mixed models for modern European genetic diversity. However, in the discussion below we abide by convention by referring to Ammerman and Cavalli-Sforza's model as the 'demic diffusion' model.

The early work on the demic diffusion model has been notable because of its use of computer simulations as a means to identify the relative importance of different variables, and to estimate the power of different analytical methods to identify the genetic outcomes of the different models (Ammerman and Cavalli-Sforza, 1984). *Figure 10.9* shows an example of simple simulation of the wave-of-advance, which used a migration rate of 1 km per year – a rate estimated from the overall archaeological picture of agricultural spread in Europe.

So far, we have been considering an unrealistically simplistic situation; in reality, there are many complicating issues in this debate:

▸ The available genetic data refer to modern populations, separated in time by a few hundred generations from the populations in which agriculture was first established; there has been much opportunity for

genetic drift and for subsequent migration events, both small- and large-scale, to cause changes in allele frequencies. For example, the so-called 'barbarian' invasions ~ 1500 YA involved the immigration of peoples from as far afield as Russia and Mongolia, Hungary was invaded by the Magyars from the Steppes around 1150 YA, and Arabs occupied Sicily, Sardinia, southern Italy and Spain from 1300 to 1100 YA. The extent of the genetic impact of such events is not clear.

▸ The genetic composition of the indigenous Europeans before the advent of agriculture is unlikely to have been uniform. *Figure 10.10* illustrates the two other major demographic events in European prehistory, aside from a possible Neolithic demic diffusion from the Near East: these are the original occupation in the Paleolithic, which followed a similar route to agricultural expansion, and the later re-expansion after the last glacial maximum from southerly **glacial refugia** in the peninsulas of Iberia, Italy, and the Balkans. The sum of these population movements is very likely to have produced heterogeneous patterns of gene frequencies among hunter–gatherer populations well before the advent of agriculture, and the movements could themselves have resulted in clinal patterns, for example by repeated founder events during the initial Paleolithic expansion that colonized the continent.

▸ The indigenous Europeans and the incoming farmers are likely to have had alleles in common, since each ultimately shared a common origin. Some alleles will have been at characteristically different frequencies, while some will have been at similar frequencies.

▸ Interactions between residents and incomers are likely to have been more complex than simple indifference, intermarriage or displacement. Factors such as warfare and disease on the one hand, and trade and cooperation on the other, may have played important roles in the development of populations.

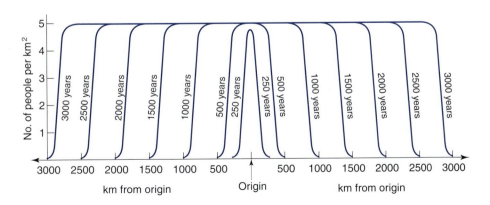

Figure 10.9: Simulating the wave of advance.

Curves showing local population density indicate the position of the wave of advance at 500-year intervals, given a rate of advance of 1 km/year. Redrawn from Ammerman and Cavalli-Sforza(1984).

▶ Where sufficient archaeological evidence exists to address the question, it suggests that the rate of advance of agriculture was far from uniform. Clear evidence for rapid expansion (at least 5 km/year) followed by lengthy periods of stasis, are seen, for example, in the LBK (linear pottery) and Cardial ware areas (Price, 2000a). New radiocarbon dates for the earliest western Mediterranean Cardial ware sites ranging from central Italy to Portugal are barely distinguishable, indicating that the spread was very rapid in this region, perhaps taking only six generations (Zilhão, 2001), and suggesting colonization by use of boats. These kinds of processes are inadequately described by the simple wave of advance model.

Having stated these caveats, we can now ask what the genetic evidence can, or cannot, tell us.

10.5.3 Evidence from 'classical' genetic markers

Extensive data on patterns of variation of European allele frequencies for 'classical' markers (see *Box 3.3*), representing individual loci in the nuclear genome other than the Y chromosome, have been collected and analyzed to test the hypothesis of Neolithic demic diffusion (Cavalli-Sforza *et al.*, 1994; Menozzi *et al.*, 1978). In all, 94 alleles at 34 loci, distributed among 16 different chromosomes were studied (*Table 10.5*), with sample sizes per population varying from a few tens for some loci to almost 3000 for alleles of the ABO blood-group system (see *Box 3.3*).

Patterns for individual loci are heterogeneous, and the pattern of a single locus has to be treated with caution: some might be under natural selection, some might be uninformative (having had no initial difference in frequency between farmers and hunter–gatherers), and some might

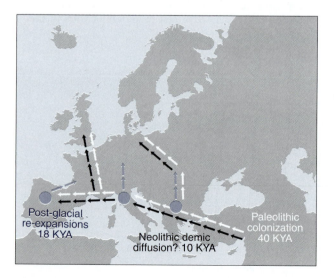

Figure 10.10: Major population movements in European prehistory.

Arrows indicate Paleolithic colonization, postglacial expansion, and Neolithic input from the Middle East. Redrawn from Simoni *et al.* (2000a).

indicate different population movements. Unlike selection, migration has the same effect on all genes, so in order to circumvent these problems and to discern broad underlying patterns, **principal components analysis** (PCA; Section 6.4.4) was used to analyze data from all alleles simultaneously. This is a graphical means of summarizing the maximum amount of the variance in a multivariate data-set with the minimum loss of information – it allows multidimensional data to be represented in a comprehensible, two-dimensional form. A problem with combining data from many alleles was the extreme irregularity of distribution of sample sites for different markers across Europe. To overcome this, the irregular pattern of data points was converted into a common regular grid of 'virtual' data points for each allele by a process known as interpolation (see Section 6.7.1). The complex method for doing this is summarized by Cavalli-Sforza *et al.* (1994).

The output of PCA is usually a plot of one principal component (PC) against another, upon which populations are indicated by points (see *Figure 6.10*). Cavalli-Sforza and colleagues developed a method to present the PCA output in a more useful way, by displaying it in the form of synthetic geographical maps of individual PCs. These should allow common patterns due to single migrations to be abstracted from data on all alleles, and the element of a geographical map is an important **heuristic** device that enables easy recognition of the various patterns and their discussion in terms of migrations and the archaeological record. These geographic maps of PCs have a number of properties that must be considered when they are interpreted:

▶ the maps have 'hills' and 'valleys' (depicted as dark and light areas), but these are interchangeable since the sign given to PC values is arbitrary;

▶ a hill or a valley in a particular area indicates a local population(s) with extreme allele frequencies;

▶ a hill and valley in a given map indicate opposing characteristics.

In a rather rude remark about these maps, MacEachern (2000) likens them to 'Rorschach inkblot' tests, in which the viewer is asked what a particular pattern suggests; the important point of this criticism is that synthetic maps may appear compatible with particular explanations, but that this does not prove that these explanations are correct.

The first four PCs of European allele frequencies together summarize 68% of the total variance in the data (*Table 10.6*); it is the synthetic map of the first PC, summarizing 28% of the variance, which is taken as providing support for the Neolithic demic diffusion hypothesis (*Figure 10.11*). The map has a strong focus in the Near East, and an approximately concentric gradient extending out to the northwestern fringes of Europe that resembles the map of dates of the arrival of agriculture (*Figure 10.8*). There need be no simple relationship between the proportion of variance summarized by a PC and the genetic contribution of migrants who contributed to it; indeed simulation studies (Barbujani *et al.*, 1995; Rendine *et al.*, 1986) suggest that the

TABLE 10.5: 'CLASSICAL' MARKERS USED IN PRINCIPAL COMPONENT ANALYSIS OF EUROPEAN POPULATIONS.

Locus type	Locus name	Chromosome	Alleles analyzed (n)
Blood group	ABO	9q	5
	Duffy	1q	1
	Kell	7q, Xp	1
	Kidd	18q	1
	Lewis	19p	2
	Lutheran	19q	1
	MNS	4q	6
	Rhesus	1p	10
	Secretor	19q	1
Immunological	Complement component 3	19p	1
	HLA (human leukocyte antigen)-A	6p	15
	HLA-B	6p	17
	Immunoglobulin GM1; GM3	14q	5
	Immunoglobulin KM	2p	2
Other proteins	Acid phosphatase 1	2p	2
	Adenosine deaminase	20q	1
	Adenylate kinase 1	9q	1
	α-1-antitrypsin	14q	1
	β-lipoprotein, Ag system	?	1
	β-lipoprotein, Lp system	6q	1
	Cholinesterase 1	3q	1
	Cholinesterase 2	2q	1
	Esterase D	13q	1
	Glucose-6-phosphate dehydrogenase	Xq	2
	Glutamic-pyruvate transaminase	8q	1
	Glycine-rich β-glycoprotein; factor B	6p	4
	Glyoxylase 1	6p	1
	Group-specific component	4q	2
	Haptoglobin	16q	2
	P protein	22q	1
	Phosphoglucomutase 1	1p	1
	Phosphogluconate dehydrogenase	1p	1
	Transferrin	3q	1
Other phenotypes	Phenylthiocarbamide tasting	7q	1
			94

Secretor status is included as an 'honorary' blood group, as the locus governs whether or not the antigens of the ABO system are secreted in body fluids such as saliva. Data taken from Cavalli-Sforza *et al.* (1994).

observed patterns would occur only if the Neolithic proportion was greater than 66% at the time of admixture. It is interesting that, in a later publication, Cavalli-Sforza suggests that in this case the proportion of Neolithic immigrants was 'probably not very far' from the 28% value (Cavalli-Sforza and Minch, 1997). This is an important issue, and will be returned to later.

The genetic data, as represented in the first PC, are compatible with the demic diffusion hypothesis, but why should we reject an alternative origin for this pattern in the initial colonization of Europe in the Paleolithic? Support for a Neolithic origin for the gradient of classical gene frequencies has been adduced from Mantel testing (see Section 6.7.4) of partial correlations between three kinds of distance: genetic, geographical, and temporal – the latter based on archaeological evidence for the times of origin of agriculture in different regions. This question is discussed further below. When geographical distances are held constant, genetic distances for several loci show significant correlations with the agriculturally-based temporal distances (Sokal *et al.*, 1991). The other argument that has been used to support a Neolithic origin for the gene frequency patterns is based on the language codispersal hypothesis (Section 10.4). If Indo-European languages are no more ancient that the Neolithic, as many believe, then the fact that they have spread so widely within (and beyond) Europe suggests a

TABLE 10.6: PERCENTAGES OF TOTAL VARIANCE OF THE FIRST FOUR PRINCIPAL COMPONENTS OF EUROPEAN ALLELE FREQUENCIES.

Principal component	% of total variance
1	28.1
2	22.2
3	10.6
4	7.0
1+2+3+4	67.9

Data taken from Cavalli-Sforza *et al.* (1994).

corresponding spread of agriculturalists (however, see Section 10.4.2, suggesting a later spread of Indo-European). At this stage, it is necessary to note the importance attached to the Basques (see *Box 10.3*).

10.5.4 Evidence from biparentally-inherited nuclear DNA markers

The availability of polymorphic nuclear DNA markers allows the direct analysis of patterns of genetic diversity in Europe, and also the added consideration of molecular distance between alleles. In one study, extensive data on seven highly variable nuclear loci, typed in an average of 18 650 European chromosomes as part of forensic profiling efforts, were analyzed (Chikhi *et al.*, 1998). The loci – four microsatellites, two minisatellites, and the HLA-DQα system, where variability is due to base substitution – were initially developed to distinguish between individuals (see Chapter 15), and have an average of ~16 alleles each. Their consequent large within-population diversity means that

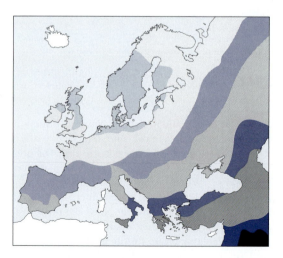

Figure 10.11: Synthetic map of Europe and Western Asia obtained using the first principal component of classical genetic data.

Redrawn from Cavalli-Sforza *et al.* (1994).

inter-population differences are small; however, using spatial autocorrelation analysis (Section 6.7.3), and taking into account the degree of molecular similarity between alleles, significant clinal patterns were nonetheless observed for all seven loci (*Figure 10.12*). Using the microsatellite data and an estimate of the mutation rate, the $\delta\mu^2$ method (Section 6.4.2) was employed to calculate population divergence times for 467 pairs of populations: for only 15 of these pairs were the times greater than 10 000 years, and despite great uncertainty in individual estimates, this was taken as evidence supporting the generation of the observed clines in times more recent than the Paleolithic. Studies of the distributions of some disease alleles, for example the R408W-1.8 **phenylketonuria** allele, also show patterns across Europe that have been taken to reflect demic diffusion (O'Donnell *et al.*, 2002).

10.5.5 Evidence from mtDNA

Much of the fiercest debate on the issue of Neolithic demic diffusion in Europe has been engendered by the patterns of mtDNA diversity and their interpretation. While we have deliberately avoided an over-historical account of human evolutionary genetics in this book, the history of this particular debate is interesting because it illustrates well how philosophical and methodological differences combined with salvoes of barbed comments can divide researchers into bitterly opposed camps. This debate is reflected in the Opinion Boxes by Guido Barbujani (a founder of the camp we can call 'demist'; *Box 10.4*) and Martin Richards (of the 'acculturationist' camp; *Box 10.5*), where these authors present very different views based essentially upon the same data.

In a study that triggered the controversy (Richards *et al.*, 1996), Martin Richards and colleagues used HVS I sequence variation to study the diversity of mtDNAs in 821 individuals belonging to 14 populations, one representing the Middle East (42 individuals, mainly Bedouins), and the rest European. Haplotypic diversity was very high, and many haplotypes were found only once in the sample; to simplify phylogenetic analysis, relationships were considered only between haplotypes present at least twice in the dataset. Conventional tree-based methods were rejected, and instead analysis of the relationships between haplotypes was done using reduced median networks (see Section 6.5.6). This approach allowed five major lineage groups, or clusters, to be identified (*Figure 10.13*) as well as some subclusters within these, and their TMRCAs were estimated based on the mean pairwise sequence difference between their European members. The conclusions of this analysis were at odds with the Neolithic demic diffusion hypothesis in a number of respects, and instead suggested that farming had been a largely culturally transmitted phenomenon in Europe.

▶ All but one of the TMRCAs for clusters or subclusters were dated earlier than 10 KYA, which was taken to indicate that the great majority of mtDNA haplotypes were brought into Europe in the Paleolithic (< 20% in the early Upper Paleolithic, ~35 KYA, and almost 70% in the late Upper Paleolithic, 17.5–23.5 KYA).

BOX 10.3 The Basques: A Paleolithic relic in Europe?

Whichever European genetic study you care to examine, one thing will be clear: the Basques are special. Along with Saami (Lapps), Sardinians, and Icelanders, they are the traditional outliers within the European gene pool. Studies using paleoecological information to partition Europe into subregions (Richards *et al.*, 2000) are happy to free the Basques from the straitjacket of paleoecology, and consider them separately.

While Saami are distinguished by their geographical extremity and lifestyle (reindeer herding), and Sardinians and Icelanders dwell on islands, and are therefore likely to be susceptible to founder effects and drift, the Basques are these days an agricultural community firmly within continental Europe, and are not surrounded by impenetrable geographical barriers (the mountains of the Pyrenees do not separate the Basque country from the French to the north, but rather lie *within* the Basque territory). So, why the special status? The key is linguistic: the Basque language is a remarkable non-Indo-European isolate within the relatively homogenous Indo-European landscape of Europe, and it is this linguistic quirk that has suggested to many that the Basques somehow withstood the introgressions of the Neolithic, and even today preserve some aspects of a Paleolithic population (Chikhi *et al.*, 2002). Mark Kurlansky cites the connection of 'aitzur' (hoe), 'aizkora' (axe), and 'aizto' (knife), with 'haitz', formerly 'aitz', meaning 'stone', as a direct link to the Stone Age (Kurlansky, 1999). This linguistic issue is an interesting one, because there is evidence from place names, coins and inscriptions that other European populations (for example Etruscans in Italy, and probably Picts in Scotland) also preserved non-Indo-European languages over 2000 years ago, but lost them relatively recently. The supremacy of the Roman Empire was responsible for the 'Romanization' of many languages. The Basques dealt well with the arrival of the Romans, and are documented to have served as soldiers for them as far afield as Hungary, the lower Rhine and Hadrian's Wall, in Northern England (Collins, 1986). So, linguistic replacement, often more rapid than replacement of genes, may suggest that there are other European populations that could be considered as Paleolithic relics, aside from the Basques.

There is no doubt that Basques have unusual allele frequencies for some genes: for example, they have the highest known frequency of the Rhesus negative blood group (Mourant *et al.*, 1976), a finding that was made as early as the 1940s. However, the combination of linguistic uniqueness, genetic differentiation, and, critically, a desire for self-determination may have proved to be a self-fortifying cocktail. Once interest in a particular population has been aroused, it becomes an object for special study, and prophecies may become self-fulfilling. We can ask the question: if Galicians or Gascons had preserved non-Indo-European languages, would we now be studying them in similar depth, putting them at the top of our DNA-sample wish-lists, and considering them as Paleolithic relics? It may indeed be true that the Basques have preserved a higher than average frequency of alleles resident in Europe in the Paleolithic, but the almost automatic assumption that 'Basque equals Paleolithic' seems far too simple to be true. Indeed, recent analysis of a young, Iberian-specific Y-chromosomal lineage shared between the Basque's and other populations of the Iberian peninsula suggests that the Basque's can not be considered both Paleolithic, and an isolate (Hurles *et al.*, 1999).

Origins are part of modern politics for many populations, but especially so for the Basques. As Roger Collins has put it: '*what is tantamount to a politicization of normally abstruse and recherché anthropological arguments about the Stone Age is a distinctive feature of the ideological underpinning of modern Basque nationalism, in which the longevity of the people and the continuity of their occupation of their western Pyrenean homelands are arguments of central importance*' (Collins, 1986).

▸ The networks of the clusters associated with the late Upper Paleolithic dates both showed 'star-like' patterns, indicating population expansion (see Section 6.6.2); it was suggested that these expansions could have occurred as a result of the improvement of living conditions after the last glacial maximum.

▸ The two youngest subclusters (together corresponding to the currently designated haplogroup J), were found in western (TMRCA, 12.5 KYA) and central/northern Europe (TMRCA, 6 KYA), and were linked by ancestral haplotypes found only in the Middle East, suggesting that these subclusters originated in the Middle East and were spread during the Neolithic. Their overall frequency in the European sample was only 12%, indicating a rather low Neolithic contribution to modern European populations.

▸ Frequencies of the subclusters associated with the Neolithic varied greatly from population to popu-

lation, but notably, did not follow a clear southeast to northwest gradient.

▸ As a corollary of this mtDNA-based interpretation, the inarguably clinal patterns seen in the classical gene frequency data were explained by events during the Paleolithic.

The publication of this study was soon followed by a vigorous and often acrimonious debate that has continued to the present day:

▸ Initial criticisms (Cavalli-Sforza and Minch, 1997) of Richards *et al.* (1996) included fundamental attacks on the use of mtDNA itself as a marker, on the grounds of the homogenizing effects of **patrilocality** on a maternally-inherited locus (a factor that must still be borne in mind today), possible problems with the mutation rate due to **heteroplasmy** (see Section 3.2.8; not now widely thought to be a major issue), and the failure to use all the observed haplotypes in the analysis.

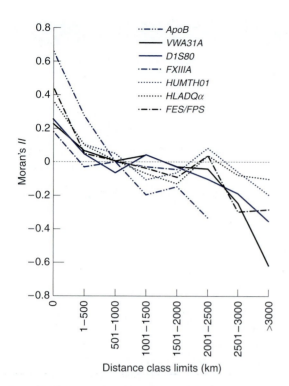

Figure 10.12: Clines of nuclear DNA markers in Europe.

Spatial autocorrelation analysis of molecular variation at seven highly variable nuclear DNA loci shows significant clines in all cases: autocorrelation is positive and significant at short distances, and negative and significant at long distances. Drawn from data in Chikhi *et al.* (1998). The largest class for *ApoB* corresponds to > 2000 km, and the largest class for *FXIIIA* to > 1500 km.

▶ An important philosophical rift also emerged in this early debate: Cavalli-Sforza reanalyzed the mtDNA data using the frequency-based approach that had been used for classical markers, and concluded that the only significant inter-population difference discernible in the Richards *et al.* (1996) dataset was that between the Near Easterners and the rest, and that there was not enough differentiation within Europe to draw any conclusions. The response (Richards *et al.*, 1997) was that a frequency-based approach was the only one available for classical gene frequency data, but was futile for the interpretation of data on a non-recombining molecule such as mtDNA, where a 'genealogical approach' was more useful, defining clusters or clades within a gene tree and then finding and interpreting their patterns among populations (phylogeography).

▶ Another criticism focused on the apparent assumption that the age of a cluster could be equated with the age of a population (Barbujani *et al.*, 1998). It was in this context that the analogy of a future group of European colonists of Mars was made – their MRCA might be in the Paleolithic, but this would not imply anything about the age of the colony (see Section 6.6.2). The

counter-claim (Richards and Sykes, 1998) against this accusation was that this assumption had not really been made, but instead that European founder haplotypes had been identified, and used as a baseline from which founder events associated with haplotype clusters could be identified and dated.

▶ One acknowledged problem was the inadequate size and distribution of the European sample (there had been a bias towards the north and west), and the fact that the Near Eastern sample was small and largely from only one population; this was subsequently addressed in studies of mtDNA variation from both camps.

Combined analysis of HVS I together with coding region RFLPs and one HVS II variant, improved the robustness of the mtDNA phylogeny (Richards *et al.*, 1998) and introduced a cladistic nomenclature (see *Boxes 8.5* and *8.6*). *Figure 10.14* shows the distribution of the various haplogroups within Europe, with data taken from a later study (Richards *et al.*, 2000), and using a regional subdivision system based on **paleoecology** (Gamble, 1999). One haplogroup, H, predominates in all populations, and, at this level of haplotype classification, no obvious patterns emerge apart from clear distinction between Europe and the Near Eastern/Caucasus samples, and the distinction of the Basques from other European populations.

In a study emerging from the 'demist' camp, a sample of 2619 HVS I sequences from Europe and the Near East, including in the latter category Kurds and Israeli-Druze, was analyzed by two spatial autocorrelation methods (Simoni *et al.*, 2000a), one using only information on the deduced frequencies of mtDNA haplogroups, and another considering sequence differences between haplotypes. The conclusion from both methods was that there was very little geographical structure to mtDNA diversity in most of Europe, but that clinal patterns could be observed around the Mediterranean Sea, consistent with their spread during the Neolithic. No real explanation was offered of the different behavior of mtDNA in Europe to other markers (including the Y chromosome: see following section), though reference was again made to the implications for mtDNA phylogeography of patrilocality and consequent high female mobility, and to a role for selection in shaping patterns. No evidence was found of postglacial expansions from south to north. The 'acculturationist' response to this study was dismissive (Torroni *et al.*, 2000), and centered on the apparently inadequate method by which haplogroups had been defined, and numerous errors in the arrangement of data in a key table. The counter-blast (Simoni *et al.*, 2000b) admitted that the published table had been wrong, but pointed out that a correct version had been used as input for the spatial autocorrelation analysis. It was claimed that haplogroup definition had been based on HVS I sequence data alone because 'acculturationist' criteria for haplogroup definition had not been made publicly available. A more general point was made that, while 'acculturationists' seemed to believe that mtDNA had special status, deserving of specific methods of analysis, it ought to be capable of

BOX 10.4 Opinion: Neolithic demic diffusion

Genetic diversity shows a very regular geographic pattern in Europe. Many genes are distributed in broad, essentially parallel gradients, encompassing the area between the eastern Mediterranean and northwestern Europe. Mitochondrial DNA represents the most conspicuous (if partial) exception, in that the distribution of its alleles appears clinal only around the Mediterranean coasts.

Although several loci may well have evolved under selective pressures of some sort, multiple parallel gradients are unlikely to reflect adaptation. By contrast, migration is expected to cause similar spatial structuring across loci, because individuals migrate with all their genes. In particular, directional expansions tend to determine multilocus clines. Therefore, the main feature of the global European diversity probably reflects a large-scale population movement along a southeast–northwest axis. On top of that, many other processes, including reproductive isolation and local gene flow, must have been important at a smaller geographic scale.

When was the current European population structure established? Clines do not carry a date, but two main processes originating from the Near East are documented in the European archaeological record. One is the initial Paleolithic colonization (starting some 35 KYA), and the other is the Neolithic diffusion of farming and animal breeding (starting some 10 KYA). Schematically, if the clines we see now originated in Paleolithic times, the genes of current Europeans are largely derived from those of the first hunting–gathering populations, and cultural transmission is sufficient to explain the Neolithic shift to food production. On the contrary, if the European clines reflect a Neolithic expansion, most ancestors of current Europeans lived out of Europe, in the countries of the eastern Mediterranean, until 10 KYA, and farming would have entered Europe when the farmers' novel ability to produce food prompted population growth and dispersal. It is reasonable to imagine that the current European gene pool results from admixture between Paleolithic settlers and immigrating Neolithic farmers. The question is whether the contribution of Near Eastern immigrants was large or small.

Theoretical calculations, and simulation studies, showed that the hypothesis of a large demographic impact of Neolithic gene flow accounts for the observed continentwide allele-frequency clines significantly better than alternative models. Recent analyses of nuclear DNA markers confirmed these results, although more complex trends emerge on a smaller geographic scale, pointing to migratory inputs from other sources. Conversely, gradients were not apparent in the first mitochondrial studies, in which the ages of most groups of evolutionarily related haplotypes (haplogroups) were estimated between 15 KY and 55 KY. These findings led to the proposal that most European mitochondrial genomes (and presumably most individuals) had a local ancestry in late Paleolithic times, and that the main demographic process shaping current genetic diversity was the postglacial (less than 18 KYA) northward re-expansion from refugia in Southern Europe. Note that this Neolithic gene flow hypothesis differs from that of a major impact of the first Paleolithic colonization. The genetic contribution of the early farmers from the Near East was regarded as minor, and approximately quantified as 22% in a recent Y-chromosome study.

I see two problems with the hypothesis of an important role of postglacial re-expansions in determining current genetic diversity in Europe. First, these re-expansions should have generated largely south–north patterns. However, such patterns have not been described yet, not only for the tens of nuclear loci studied so far, but, despite some claims to the contrary, even for mtDNA. Under that hypothesis one cannot account for the strong structuring of European genetic variation along a southeast–northwest axis. Second, it is not clear why the pre-Neolithic age of haplogroups should support a *local* ancestry of the populations in Paleolithic times. People who move carry with them their genes, no matter how old are the mutations in these genes. When a population is established, some initial polymorphism is generally present, and so the allele genealogies are consistently deeper than the populations' genealogies. Only in one case does the age of a haplogroup approximate the population's age, namely when all founders were genetically identical. However, there is too much diversity in Europe for founder effects to have systematically occurred. In brief, if clines do not carry a date, haplogroups do, but it is the wrong one. The estimated age of a lineage represents the age of a biochemical process, a mutation in somebody's germ-line. The population process that spread the allele over the area where it is found is of course independent, and it occurred necessarily later.

The relative contributions of Paleolithic and Neolithic parental populations to the European gene pool have been recently estimated for the first time by a likelihood-based method, using the same Y-chromosome dataset that was interpreted as suggesting a 22% Neolithic contribution. The estimates vary across regions, and carry with them a large error, which is unsurprising, given the high diversity known to exist within (but not between) most human populations. Still, the most likely overall figure representing the Paleolithic contribution at the continental level is around 30%. These methods will be applied to more loci in the future, and it will be interesting to see if mitochondrial data yield similar results. However, the evidence available so far is at odds with what would be expected if the current European diversity had been shaped by postglacial expansions from refugia in Southern and Central Europe. On the contrary, all the available data, including the ages of mitochondrial haplogroups, seem fully compatible with a substantial contribution of genes from the Near East to the current European gene pool.

Guido Barbujani, Department of Biology, University of Ferrara, Italy

---

BOX 10.5 Opinion: DNA and the Neolithic

Ammerman and Cavalli-Sforza's *The Neolithic Transition and the Genetics of Populations in Europe* was published in 1984, pioneering the idea of collaboration between geneticists and archaeologists. It set out two models for the spread of agriculture into Europe from the Near East: cultural diffusion, without a substantial movement of people, and demic diffusion, in which the new technology was brought by migrating Near Eastern people. Genetics, they claimed – with some subtlety and rigor – could distinguish between these two hypotheses, because if there had been an influx of newcomers, they would have left genetic patterns that could still be detected today. Moreover, they claimed that they had indeed detected such patterns, in the blood groups and enzyme distributions of modern Europeans: these showed a genetic gradient from southeast to northwest that resembled the radiocarbon map for the spread of agriculture into Europe. They did not claim to be able to quantify the influence, however, but attempted to model it using Fisher's 'wave of advance.'

In many ways they were right. Genetics is the most direct way to identify prehistoric migrations, and indeed there are signals in modern Europeans of Neolithic dispersals into Europe. But as their position took hold, it also began to harden. The picture was backed by an assumption that agricultural surpluses necessarily led to dramatic population growth, entailing demographic expansion; and ethnographic comparisons that suggested that European hunter–gatherers had been very thin on the ground. By 1994, in Cavalli-Sforza, Menozzi and Piazza's monumental *History and Geography of Human Genes*, the hapless little bands of European hunter–gatherers were being completely engulfed by *'the much more numerous population of farmers, which could grow to higher population densities.'* The wave of advance had become a tidal wave.

The earliest critiques of this position came not from fellow geneticists, but from archaeologists such as Marek Zvelebil. By the 1990s, the archaeological picture was very different from that painted by geneticists. The radiocarbon map had been updated, and no longer suggested the steady rate of expansion that Ammerman and Cavalli-Sforza had assumed. Moreover, farming and settled life were no longer seen to go hand in hand, damaging the argument for expansion based on surpluses. In fact, detailed archaeological accounts of the

transition in various parts of Europe implied that it had taken place in many different ways, and in some parts of the continent over many thousands of years. Only in southeast and central Europe had farming arrived as the archaeological 'package' identified by Ammerman and Cavalli-Sforza – crops, stock, pottery and novel architecture all arriving at once. On the hunter–gatherer side, it had been realized that comparisons of Mesolithic Europeans with modern Australian Aborigines were misleading; more appropriate comparisons, such as with Pacific northwest coastal communities, implied a much higher population density for the native population, more difficult for the incoming farmers to overwhelm.

This meant that when, in 1996, mitochondrial DNA (mtDNA) came onto the scene, the surprise was as much that about 10–20% of modern lineages could indeed be traced to Neolithic immigrants as that most of the remaining 80–90% were likely to be indigenous. The mtDNA work was much criticized for its approach and (more justifiably) for its rather limited sampling; and doubts were expressed as to mtDNA's suitability as a marker. But subsequent work, with more carefully considered methods and much larger sample sizes, has given much the same picture (Richards *et al.*, 2000). Not only that, but new work from Cavalli-Sforza's lab, by Ornella Semino, Peter Underhill and their colleagues, has suggested that the Y chromosome seems to show a similar pattern to the mitochondrial DNA: again, about 20% Neolithic and 80% pre-Neolithic (Semino *et al.*, 2000). It seems that 'we Europeans' go back a long way.

There has been much discussion of the genetic gradients that Cavalli-Sforza and his colleagues first identified, and what they may mean. About a third of classical markers in Europe show southeast–northwest gradients, and various gradients also show up in the Y chromosome and mitochondrial DNA. They may well have been generated in part by Neolithic expansions, as Cavalli-Sforza suggested, but these may have overlaid earlier expansions and dispersals, and in turn have been overlaid by many more since. The archaeologists have a word for this – a 'palimpsest'. Basically, a lot of things have happened in Europe in the last 50 KY, and teasing one from another is a big job. In the study of demographic history, the truth is rarely pure and never simple.

Martin Richards, Department of Chemical and Biological Sciences, University of Huddersfield, UK

---

meaningful analysis using tried-and-tested population genetic methods.

Work from Richards and colleagues (Richards *et al.*, 2000) has now used larger sample sizes than previous studies, and formalized an explicit procedure for identifying founder haplotypes in Europe, and for dealing with criticisms of earlier work, and of founder analysis in particular (*Box 6.5*):

▶ diversity of founder lineages is corrected for initial diversity in the Near East by subtraction of Near

Eastern diversity; this is done to deal with the 'Martian colonization' problem discussed above;

▶ a means of identifying and eliminating back–migrants from Europe and recurrent mutations in both source and migrant populations is used;

▶ allowance is made for the spread of lineages in more than one migration event;

▶ the problem of whether the modern Near East population accurately represents a source population

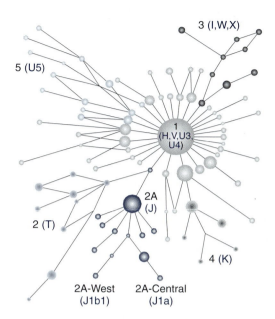

Figure 10.13: Network of European mtDNA HVS I haplotypes.

Circles represent haplotypes, with an area proportional to their frequency; lines represent genetic differences. New nomenclature for the clusters 1–5 is taken from Richards *et al.* (1998), and given in blue. Redrawn from Richards *et al.* (1996).

for European lineages spread during much earlier periods remains, but this is a difficulty common to most genetic studies.

Ages of lineages were again calculated in this study, using a mutation rate of 1 transition per 20 180 years (see Section 3.2.8), and the ρ (rho) statistic (the mean number of mutations to the root of a cluster, and an improvement on mean pairwise differences; see Section 6.6.2). Since most lineages appear to show star-like genealogies, indicating population expansions, errors due to variance in the demographic history were claimed to be minimal. *Figure 10.15* illustrates the ages of the major haplogroups: J and T1 remain the only Neolithic lineages. When allowance is made for multiple dispersals of the common H lineage, the Neolithic component of modern European populations is estimated at around 20%. The major demographic impact is again concluded to be postglacial expansion in the late Upper Paleolithic, a phenomenon linked to the distribution of haplogroup V, and explored more fully in other studies (Torroni *et al.*, 2001).

Despite their early distaste for such analyses, the 'acculturationist' camp (Richards *et al.*, 2002) has more recently applied the frequency-based methods of PC analysis to a large dataset of 3113 European, 208 North Caucasus and 1234 Near Eastern mtDNA haplotypes divided into the main haplogroups, and again using the geographical subdivisions shown in *Figure 10.14*. After exclusion of the African haplogroups, the first PC summarizes 51% of the variation and separates Europe from the Near East. There is a broadly southwest to northeast pattern, as might be expected under the Neolithic demic diffusion hypothesis, but, while central and eastern Mediterranean and southeast European populations cluster with the Near East, the western Mediterranean clusters with central and northern Europe. This is taken to indicate a lack of fit between the mtDNA data and the demic diffusion hypothesis, since the archaeological evidence indicates the major expansion of agriculture to be into central Europe. Despite the earlier claims for the primacy of the postglacial expansion from south to north, and, in agreement with the spatial autocorrelation analysis discussed above (Simoni *et al.*, 2000a), no major PC shows a north-south orientation.

10.5.6 Evidence from the Y chromosome

The first attempt to use the Y chromosome to address the issue of European diversity (Semino *et al.*, 1996) employed the first two polymorphic Y-specific markers to be identified: one was the 12f2 deletion (Casanova *et al.*, 1985), which in current nomenclature defines Y-chromosomal haplogroup J (see *Box 8.8*), and the other was one of the haplotypes defined by the complex pattern of hybridization of the probe 49f to genomic DNA digested with the enzyme *Taq*I (Ngo *et al.*, 1986). This marker, known as 49f/haplotype (ht) 15, is less easy to equate to a currently defined haplogroup, but is largely equivalent to haplogroup R*[xR1a] (Jobling, 1994; Jobling and Tyler-Smith, 2000). European sampling in this study was poor, with no samples at all from the British Isles, Scandinavia, or central/eastern Europe. Within this sampling framework, the overall pattern was a high frequency of R*[xR1a] chromosomes (see *Box 8.6* for an explanation of the nomenclature) in the north and west, reaching a peak in the supposedly 'Paleolithic' Basques (see *Box 10.3*), and a low frequency in the south and west; the frequency of haplogroup J chromosomes was correspondingly high in the south and east, and low in the north and west. Though no statistical measure of clinality was offered in this study, it was claimed that there was a gradient of Y chromosomes in Europe that offered support to the Neolithic demic diffusion hypothesis.

Two later studies (Rosser *et al.*, 2000; Semino *et al.*, 2000) exploited advances in the development of binary markers on the Y chromosome, and used larger sample sizes. The first (Rosser *et al.*, 2000) included over 3500 chromosomes, including samples from central and northern Europe and the British Isles, and typed 11 binary markers, defining six common (each with frequency > 5%) haplogroups, together comprising about 98% of the sample. However, one group of chromosomes ('haplogroup 2'), comprising 22% of the sample, was defined only by an absence of specific derived markers, and this uninformative group therefore contained a collection of disparate lineages. Spatial autocorrelation methods were used to identify statistically significant clines for all of the five remaining major haplogroups, including clines for two haplogroups (J and R*[xR1a, R1b8]; see *Boxes 8.6* and *8.8*), representing 46% of chromosomes in the dataset, with a strong southeast to northwest orientation. Haplogroup J was at high frequency in the south and east, and rare in the west; haplogroup R*(xR1a, R1b8) was at

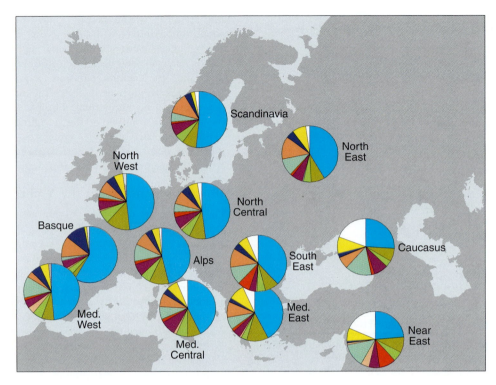

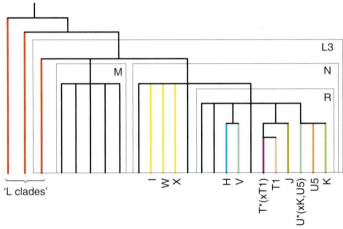

Figure 10.14: Distribution of major mtDNA haplogroups in Europe.

Each pie chart shows the frequencies of major haplogroups in each of 12 subregions defined on the basis of paleoecology by Gamble (1999). Data are from Richards *et al.* (2000) – see http://www.stats.ox.ac.uk/~macaulay/founder2000/. Note the relatively high frequency of the African haplogroup (informally referred to here as 'L clades') in the Near Eastern and south European populations, the marked difference between Near Eastern and Caucasian populations and Europe, and the overall similarity between European populations, with the exception of the Basques, who show markedly higher frequencies of haplogroups H, V and U5 than the rest. Sample sizes in Europe range from 156 (Basques) to 456 (North West) and are 1088 for the Near East. The tree below shows the phylogenetic relationships of the different haplogroups. White sectors within pie charts represent a collection of other minor haplogroups.

low frequency in the east, but reached very high frequencies in the west. In a separate detailed study of Irish Y-chromosomal diversity it was shown to be almost fixed (98.5% frequency) in the west of Ireland (Hill *et al.*, 2000). These clines were claimed to be compatible with an origin in Neolithic demic diffusion. Under this scenario, haplogroup J

chromosomes could be considered as a signal of Neolithic contributions to Europe, and haplogroup R*(xR1a, R1b8) as a Paleolithic substrate, but no attempt was made to equate the haplogroup J proportion to the Neolithic proportion. However, there were other, different, clinal patterns too, suggesting a complex history: haplogroup N3 was at high

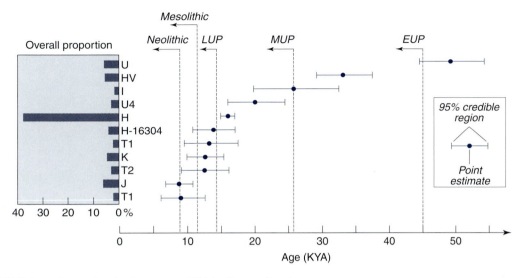

Figure 10.15: Estimated ages of major European mtDNA haplogroup founders.

Ages are taken from Richards *et al.* (2000), and were calculated using the statistic ρ. Proportions of each haplogroup in Europe as a whole are given in the bar chart to the left. EUP, MUP, LUP, Early, Middle and Late Upper Paleolithic.

frequency immediately east of the Baltic but rare elsewhere; haplogroup R1a was at high frequency in central Europe and southern Scandinavia, and chromosomes within haplogroup E showed a pronounced south to north gradient. Principal components analysis showed strong geographic structuring with a southeast–northwest axis. In genetic barrier analysis, the major boundaries were in the center of the European landscape, and did not correspond to obvious geographical barriers. Although not stressed in the published study, the correspondence of these boundaries with the 'suture zones' defined by the contact of many animal and plant species re-expanding from southern refugia after the last glacial maximum (Hewitt, 2000) may suggest a role for postglacial expansion in shaping the modern Y-chromosomal landscape.

The second study (Semino *et al.*, 2000) had a smaller overall sample size (1007 chromosomes) and poorer geographical coverage than that discussed above (Rosser *et al.*, 2000), with an absence of samples from the British Isles, Scandinavia (except the Saami), and much of Central and Eastern Europe (*Figure 10.16*). However, 22 binary markers resolved clades with the tree that were unresolved in the other study, and also defined six major haplogroups. Some markers were shared between the two studies, and encouragingly, where comparisons can be made, the distributions of lineages in these two independent studies show similar patterns, indicating that there are not major problems with sample sizes. Though not tested statistically, a clinal pattern was claimed, and, largely on the grounds of their distribution with respect to a southeast–northwest axis, a number of haplogroups were identified as candidate Paleolithic (haplogroups I and R) and Neolithic [haplogroups E3b, F*(xI,J2,G,H,K), G and J2] lineages. Attempts were made to date particular haplogroups without any consideration of the problem of whether or not the age of a lineage equates simply to the age of an expansion or

population. Explicit identification was made between two Paleolithic lineages and specific archaeological cultures, the Aurignacian (haplogroup R, 35–40 KYA) and the Gravettian (haplogroup I, 22 KYA). By 'eyeballing' the distribution of lineages, the signal of postglacial expansion from the Balkans and the Iberian peninsula was also recognized, in haplogroups R1a and R*(xR1a) respectively. With Cavalli-Sforza as an author, this paper can be considered as emerging from the 'demist' camp; however, having identified 'Neolithic Y chromosomes' in a simple way, the authors then take the proportion of these in Europe as a whole (22%) as a simple measure of the contribution of the Neolithic farmers. The glee of the 'acculturationists' at this easy capitulation can be imagined, and the lack of rigor in this study is in contrast to the earlier work cited at the beginning of this section (Ammerman and Cavalli-Sforza, 1984).

Higher resolution studies of the Mediterranean region have also been published (Malaspina *et al.*, 2001; Scozzari *et al.*, 2001), and use microsatellites to subdivide haplogroup J into two components that have different distributions and independent histories. This finding illustrates the important point that the apparent number of lineages in any study reflects the resolution of markers, and that dating of these lineages may fail to date the population historical events that underlie their modern distributions.

Despite the apparent concurrence on an overall Neolithic proportion of about 20% between mtDNA and Y-chromosome studies outlined above, the debate is far from dead. A reanalysis of the data of Semino *et al.* (2000) produced a proportion of 50% (*Figure 10.17*). This reanalysis (Chikhi *et al.*, 2002) is critical of the *post hoc* identification of Neolithic chromosomes and the simple equation of their proportion with the Neolithic contribution to Europe (Semino *et al.*, 2000). Instead, the question is treated here as

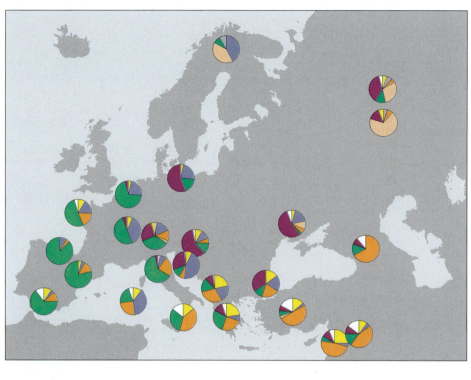

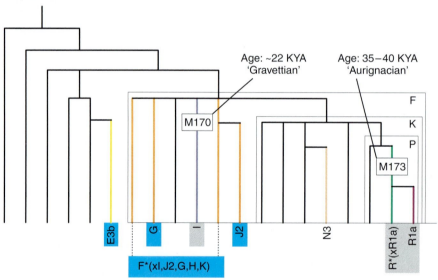

Figure 10.16: Distribution of major Y-chromosomal haplogroups within Europe.

Pie charts show the relative frequencies of different haplogroups, proportional to sector area. Average sample size is 42, and ranges from 16 to 77. Redrawn from Semino *et al.* (2000). The tree below the map shows the phylogenetic relationships and names of the haplogroups, using YCC nomenclature. Some haplogroups were further subdivided in this study, but are not shown here. White sectors within pie charts represent a collection of other minor haplogroups. Clades associated with a Paleolithic contribution are shaded in gray on the tree, and the dates estimated for two key mutations are shown. Clades associated with a Neolithic contribution are shaded in blue.

an admixture problem, with an explicit model of the admixture process, and based upon estimates of the ancestral allele frequencies in all populations. In this genealogical likelihood–based method, there is no attempt to identify 'Neolithic' or 'Paleolithic' chromosomes in modern populations, but instead parental populations are specified. A pooled population from Turkey, Lebanon and Syria was chosen as the descendants of Near Eastern Neolithic farmers, while the Basques were chosen as the descendants of a Paleolithic population (see *Box 10.3*). The admixture model

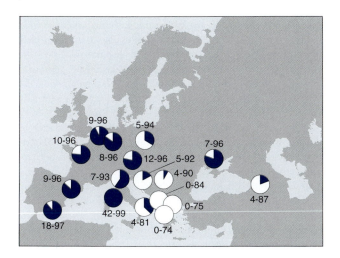

Figure 10.17: Proportions of Paleolithic contributions to 17 modern European populations, based on Y-chromosomal haplogroup data.

Proportions were calculated by a maximum-likelihood admixture estimation method (Chikhi *et al.*, 2002) from the data of Semino *et al.* (2000). Each pie chart shows the modal estimate of Paleolithic contribution as a blue sector. The 90% confidence interval (in percentages) is given next to each chart.

10.5.7 The Neolithic in Europe: Where do we go from here?

The fact that this area of research remains a controversial and difficult one should not surprise us. Europe combines the factors of relatively easy acquisition of samples with an ancient and complex history: modern humans have been around in this part of the world for over 40 KY, and, as Martin Richards points out (*Box 10.5*), we certainly have a **palimpsest** to deal with. If much shorter histories, such as those of the more recently colonized and previously unpopulated Americas and Pacific islands (Chapter 11) are difficult to resolve, we cannot expect simple resolution of the problem of the European Neolithic. How can we reconcile the debates, and what future work can be done to provide a more solid foundation for discussion?

▸ The clines in the protein and nuclear DNA data are real, and require explanation. Arguments that they may represent the result of selection (Fix, 1999; see *Box 6.6*), since most clinal markers in the classical dataset are HLA genes, and possibly involved in resistance to disease after animal domestication, require critical assessment. It is to be hoped that large datasets on nuclear DNA variation (in the form of haplotypes as well as allele frequencies) become available in the near future.

▸ Clines do not come with dates conveniently attached. Apart from the correlation methods discussed above, ancient DNA studies could in principle help here, since the allelic compositions of prehistoric populations might be directly investigated. However, barring unforeseeable advances, these approaches will be confined to mtDNA, and the authenticity of results will be difficult or impossible to verify (see *Boxes 4.5* and *4.7*).

▸ The pattern of mtDNA lineages in Europe is not clinal in those regions predicted by the demic diffusion model. It would be helpful to have better resolution of some major mtDNA haplogroups such as H, and this may emerge as more complete sequences of mtDNAs become available (Richards and Macaulay, 2001). Despite some methodological rapprochement between 'demists' and 'acculturationists', with the former enthusiastically adopting gene genealogies (Semino *et al.*, 2000), and the latter signing up to PC analysis (Richards *et al.*, 2002), there is still little common ground. There is suspicion about the value of founder analysis (Goldstein and Chikhi, 2002), and comparative application of this method to Y-chromosomal data would be interesting and helpful. One view might be that not all loci are expected to show clines after demic diffusion, and that mtDNA is just one of those that doesn't. The fact that it does not does not invalidate the hypothesis. The absence of clines could reflect starting conditions before the Neolithic (no clines would be found if there were not differences between the founding populations), female migration, or the effects of selection − for example, due to environmental temperature (Mishmar *et al.*, 2003).

assumes that the two independent parental populations contribute genes to a third, admixed population at some time in the past. Subsequent mutation is neglected (reasonable for Y-SNPs), the populations are assumed to be isolated from one another after the admixture event, and they diverge because of drift. Using a statistical method (see Chapter 12.3.1) ancestral haplogroup frequencies compatible with observed data are reconstructed, and the probabilities of obtaining the observed frequencies are calculated for different proportions of parental contributions in the past. Probabilities of different values of a parameter estimating the amount of drift (the time since admixture divided by the population size) are also estimated. *Figure 10.17* illustrates the modal (most probable) values for the Paleolithic contribution to 17 populations across Europe. Although confidence intervals are very wide, there is a trend across Europe which was shown to be statistically significant, and, at 50% overall, significantly different from the 22% estimate (Semino *et al.*, 2000). Since the Basques contain an unknown proportion of Y chromosomes introduced in the Neolithic, these Paleolithic proportions are suggested to be probable overestimates. Estimates for the amount of drift increase with distance from the Near East, which is explained as being due to the later population expansion times − larger populations experience less drift. This relatively sophisticated approach to the issue seems attractive, but has its limitations: the implications of back-migration from Europe to the Turkish/Lebanese/Syrian sample, contributions from a third population, gene flow within Europe after admixture, or gene flow from Africa into Southern Europe, are unclear.

▸ The apparent consensus on the proportion of Neolithic lineages in Europe (at around 20%) is an illusion, and the estimation of this proportion is far from trivial (Chikhi *et al.*, 2002). Application of model-based methods and discussions of their limitations are necessary; it seems clear that the abstraction of the proportion of a given lineage, or of the amount of variance summarized by a PC, to the Neolithic proportion, is over-simplistic.

10.6 Out of tropical West Africa into subequatorial Africa

If the history of human habitation within Europe is inconveniently ancient for the investigation of relatively recent population movement, it is even more ancient in Africa. Hence, we might expect the question of African agricultural expansion to be as mired in genetic controversy as the European Neolithic. This is not so, however, and this probably reflects:

▸ a later spread of agriculture, linked to the expansion of a relatively small number of specific populations over the last 3000 years;

▸ a reasonable consensus on the implications of the linguistic evidence;

▸ a paucity of archaeological evidence compared to Europe;

▸ a paucity of genetic data compared to Europe, and, so far, an absence of vicious exchanges.

10.6.1 Prehistoric background, and dates for African agricultural expansion

There is a reasonable consensus that agriculture spread from the Near East into Egypt between 9500 and 7000 YA (though some archaeologists claim that agricultural behaviors may have been developing earlier than this in the Nile Valley and the Sahara). Emmer wheat and barley were grown in the fertile Nile Valley, and farming cultures including domesticated pigs and goats became firmly established by 5500 YA. To the west, evidence of cattle herding in the mountainous regions of the Sahara dates back to 8000 YA, in the form of striking rock paintings found over a wide geographical area (though this kind of evidence is notoriously difficult to date). At this time the Sahara was not the extremely arid environment it is today, but semi-arid grassland, rich in antelope and wild oxen, *Bos primigenius*, the progenitor of modern cattle. Whether the cattle eventually tamed in Northern Africa were descended from local oxen or were brought in from the Near East has been a matter of debate, and is addressed in Section 10.7.3. Cattle herders may have also been using wild cereal grasses and perhaps cultivating other local crops; they hunted with bows and arrows, and used flaked and polished stone adzes (Phillipson, 1993) – cutting tools with a flat heavy blade at right angles to the haft. The term 'Neolithic' is sometimes used in the context of African archaeology, but has a confused history (*Box 10.6*).

Around 5500 YA the Sahara became drier and lakes disappeared. The northern limit of the area in which the tsetse fly was endemic shifted southwards: this fly carries the trypanosome parasite that causes **trypanosomiasis**, fatal to cattle. Under these, and other, 'push' and 'pull' factors, the pastoralists moved south into the savanna regions, probably cultivating sorghum, millet and yams. By 3500 YA, cereal agriculture was widespread throughout the belt of savanna south of the Sahara, with farming communities using the shifting agriculture of woodland soils ('slash and burn'), and using digging sticks, hoes and axes. Soil had to be managed carefully, and required different practices for different places, so this cannot be regarded as an unsophisticated farming culture.

Africa is different from other parts of the Old World in that there was (except in Egypt and some other northern areas), no distinct 'Bronze Age' or 'Copper Age' during which the softer metals were utilized but the technology of smelting iron had yet to be established. Particularly in sub-Saharan Africa, iron was the first metal to be introduced, possibly from northern Africa west of Egypt, or from Egypt itself. Its arrival south of the Sahara around 2700 YA coincided with the rapid spread of farming economies throughout sub-Saharan Africa. This spread, a complex series of population movements that is poorly understood, is known as the **Bantu expansion**, since Bantu languages are

---

BOX 10.6 What is the African Neolithic?

The term 'neolithic' refers to a change in practice of stone tool making. However, by extension, and from a history of studies within Europe, it has come to encompass an association of tools – polished stone axes and barbed arrows – with pottery, and with agriculture and the domestication of animals. V. Gordon Childe's (1936) Neolithic Revolution was analogous to the Industrial Revolution – a fundamental change in the human way of life. A consequence of this view was that 'neolithic' came increasingly to mean 'food-producing'. Many European archaeologists working in Africa tried to shoehorn their findings into the European terminological framework, with confusing results, which do not properly accommodate the complexity of African economic and cultural interactions. Similar problems apply to the imported term 'Iron Age': it proves difficult to identify in the African archaeological record the beginnings and ends of these different cultural phases, and evidence for populations who both used stone tools and knew about iron-working cannot be fitted into traditional Eurocentric models. Rejection of the term 'neolithic' in the African context, has seemed a good idea to some archaeologists (Sinclair *et al.*, 1993), and we do not use it here.

suggested to have been codispersed with agriculture over much of east, central and southern Africa from a homeland in eastern West Africa. The linguistic evidence is the subject of the next section. It is worth noting that some archaeologists regard the term 'Bantu expansion' as unhelpful or misleading, since it assumes that the traits of iron-working, farming and Bantu-speaking were spread at the same time by the same people, and this assumption can be challenged (Okafor, 1993; Vansina, 1995). Though there is evidence that farming practices were established over a considerable area before the arrival of iron, its availability certainly made farming much more efficient; by 1300 YA domesticates originating in equatorial Africa, stock-breeding, iron-working and new forms of social organization were spread widely through the center and south of the continent.

There is much archaeological evidence for the early existence of iron working, for example, in the remains of furnaces at Nok, in Nigeria. However, the details of the origins and pattern of spread of the expansion from this center southwards, and possibly to another center in the east, Urewe, are far from clear (Phillipson, 1993). A simple picture of the probable routes and relevant dates is shown in *Figure 10.18*. Because of their high degree of homogeneity in the archaeological record, the early iron-working communities of eastern and southern Africa are generally considered as a single archaeological complex, once known as the Early Iron Age, but also now known as the Chifumbaze complex (Phillipson, 1993).

It is important to recognize that there are serious practical difficulties in studying the archaeological evidence for agriculture in tropical Africa; the area is covered with trees, which makes fieldwork difficult; the important early crops were tubers, and the methods to study them (see *Box 10.1*) have only recently been introduced; and, last but not least, the political situation in many tropical African countries is a severe hindrance to archaeological work. The apparent simplicity of the archaeology of sub-Saharan Africa may be illusory, and simply a reflection of the relative paucity of evidence that has been obtained so far.

10.6.2 The linguistic evidence

The distribution of language families spoken within Africa is shown in *Figure 10.19a*. The dominance of the Bantu subgroup of the Niger–Kordofanian family in central and southern Africa is striking in comparison with the diversity of subgroups of Niger–Kordofanian and of other families in the equatorial region and parts of the north. The 500 languages within the Bantu group are closely related, with high mutual intelligibility, suggesting a recent common origin, and their distribution coincides well with that of the Chifumbaze complex. This has led to the idea that the modern speakers of Bantu, now numbering over 200 million people (about 28% of Africans) spread over almost 9 million square kilometers, owe their wide distribution to the expansion of iron-working agriculturalists. Linguistically speaking, at least, this expansion has its origin in what is now Cameroon and eastern Nigeria, where linguists believe that the ancestral Bantu language was spoken.

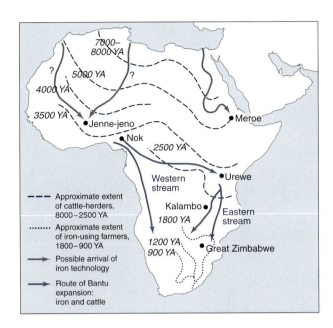

Figure 10.18: The spread of agriculture and iron working in Africa.

Modern Bantu languages can be split into two major groups, spoken in the eastern and western parts of Bantu territory respectively. The boundary between the two follows the eastern edge of the equatorial forest, and the western branch of the Rift Valley, but is less clearly defined further south. The earliest dispersal of Bantu speakers seems to have been that of the western group in the equatorial forest (*Figure 10.19b*), with several distinct stages suggested by **lexicostatistical** studies. The final expansion took place into much of the southern savanna, where subsequent interaction took place between speakers of eastern and western Bantu. The southward expansion of eastern Bantu is less well understood, partly because the languages in this group are very closely related and the events that led to their spread are comparatively recent. There is also much evidence of word borrowing; studies of non-Bantu loanwords suggest that the origin of the eastern group was in the area around Urewe (*Figure 10.18*). The origin of this eastern center itself is somewhat mysterious: it is presumed to have originated in a migration from the western Bantu center, but there is no archaeological or linguistic evidence to support this directly.

Finding specific evidence to link the Bantu languages with the spread of agriculture is not simple, partly because the agricultural practices in the equatorial west were very different from those in the eastern savanna, and so we expect linguistic differences in agricultural terms from these different regions. However, some words (e.g., that for goat) can be traced to a common ancestral form and the geographical distribution in Bantu of many loan words for cattle, sheep and cereal cultivation suggests that the people who spread the farming practices were largely Bantu speakers. A maximum parsimony approach to Bantu language classification leads to

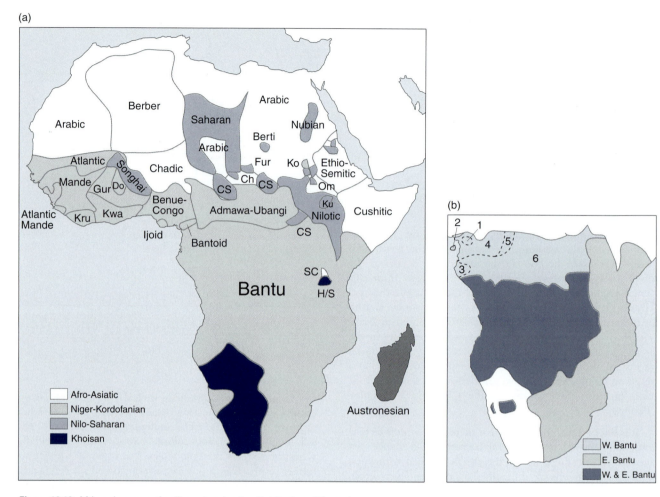

Figure 10.19: African language families, showing the distribution of Bantu languages.

(a) Bantu languages are widespread in central and southern Africa. Not all language families are named here. Do: Dogon; Ch: Chadic; CS: Central Sudanic; H/S: Hadza/Sandawe; Ko: Kordofanian; Ku: Kuliak; Om: Omotic; SC: South Cushitic. Redrawn from Blench (1993). (b) The distribution of Bantu subfamilies. The numbers 1 to 6 refer to the stages of dispersal based on lexicostatistical evidence. Redrawn from Phillipson (1993).

well-supported trees of Bantu languages that are highly consistent with archaeological evidence for the spread of agriculture (Holden, 2001).

10.6.3 The genetic evidence

Principal coordinates analysis of classical gene frequency data from selected African populations (Cavalli-Sforza et al., 1994) shows that Bantu speakers from different parts of Africa tend to cluster together. F_{ST} values between different groups, even as geographically separated as those in the northwest and southeast, are significantly smaller than those among a group of West African populations, for example. On the basis of these genetic distances, very good agreement has been claimed with the archaeological evidence. In synthetic PC maps of Africa the major influence, seen in the first PC, is differentiation between the north and the south; a pattern ascribable to the Bantu expansion does not appear until the fourth PC map, and summarizes only 7% of the total variation.

Despite the suggestion (Cavalli-Sforza et al., 1994) that the Bantu expansion took place 'in a vacuum', since the new territory was only sparsely populated by hunter–gatherers, and thus analogous to the Neolithic in Europe, consideration of the PC maps is claimed also to provide evidence of gene flow from Khoisan to Bantu populations in the south. In areas where early farming settlement was dense, the indigenous peoples may have been displaced rapidly, while in other areas there may have been aloof coexistence (Phillipson, 1993).

Many early studies of African mtDNA diversity focused upon the issue of the origin of modern humans and early events in human prehistory. More recently the issue of the Bantu expansion has begun to be addressed in some detail (Pereira et al., 2001; Salas et al., 2002), though sampling within this vast continent remains rather poor, and definition of what constitutes a population or ethnic group is not straightforward (MacEachern, 2000).

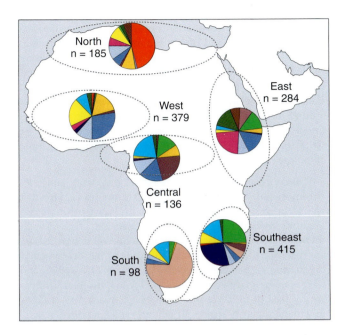

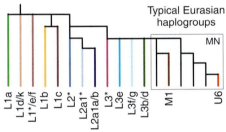

Figure 10.20: Distribution of mtDNA haplogroups in Africa.

Pie charts show the frequencies of different haplogroups represented as sector areas. The tree shows the phylogenetic relationships of the haplogroups, with color-coding corresponding to the pie charts. Data from Salas *et al.* (2002).

Salas *et al.* (2002) classified into haplogroups 307 mtDNAs from 16 different Bantu-speaking population groups from Mozambique, in southeastern Africa, and surrounding areas, then analyzed these data together with previously published data. *Figure 10.20* shows the distributions of mtDNA haplogroups in Africa (Salas *et al.*, 2002). Note that the decision was made to group the samples into six regional categories, south, southeast, central, west, east, and north; this grouping is not explicitly justified, and if not historically meaningful, could act to obscure informative phylogeographic patterns. The southeastern group clusters tightly together in PC analysis (*Figure 10.21*), and AMOVA analysis (Section 6.2.4) shows that almost all (98.8%) of the genetic variation is within-population, rather then between. This shows that these populations as a group have high homogeneity, compatible with a very recent common origin. Despite this homogeneity, high haplogroup diversity has been maintained, and this indicates that substantial numbers of people were probably involved in the dispersals that formed them. This may support the idea that the Bantu

expansion was not a single population movement, but rather a complex series of shorter-range movements.

The great majority of all the African mtDNAs, including all of those in the southeastern group, lie in the 'L clades' of the mtDNA phylogeny (see *Box 8.5*; note that 'L' itself is not a clade). By considering the diversity within these clades, using network construction and founder analysis, a scenario for the origins of mtDNA haplogroups within the southeastern Bantu speakers was constructed (*Figure 10.22*). Five percent of lineages (belonging to L1d) were probably contributed by local Khoisan populations, and the diversity of these lineages suggests that this represents acculturation of a reasonable number of individuals, rather than a founder event. A cluster of lineages representing about a quarter of the mtDNAs are from east Africa, and are thought to originate in a local contribution that was subsequently spread southwards from the eastern center; candidate founders of this cluster are dated to 1900 ± 750 YA and 800 ± 550 YA. From the founder analysis about a half of lineages have a West African origin, with rather uncertain dates around 4000 YA. The remainder (about 28%) are from Central Africa, but there is uncertainty as to whether these were from the western Bantu source region itself, or from local assimilation in the forest zone during southerly movements. This mtDNA study therefore presents a picture that seems to fit well with those archaeological and linguistic studies suggesting a major expansion, albeit not necessarily in a single event, from both western and eastern sources, and suggests a small amount of acculturation of local southern populations. However, the interpretation is restricted by the sparse sampling, and employs the methods of founder analysis that have been criticized in European studies (see Section 10.5.5);

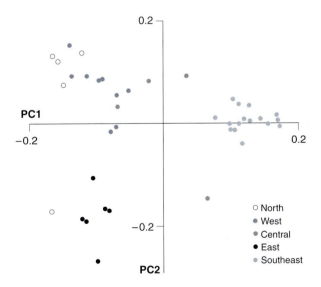

Figure 10.21: Principal components analysis of mtDNA haplogroup frequencies for African populations.

Note the clustering of southeastern populations. Redrawn from Salas *et al.* (2002).

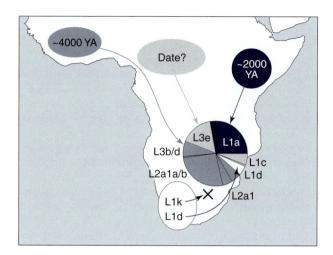

Figure 10.22: Putative origins of mtDNA lineages in the southeastern Bantu.

The pie chart indicates mtDNA haplogroups in the southeastern Bantu speakers, with minor haplogroups unlabeled for clarity. Colors indicate origins in West, Central, East and Southern Africa; haplogroup L1d is supposed to have been contributed by acculturation of local populations, but note that L1k, also present at appreciable frequency in Khoisan, is not found. Approximate dates of origin of some haplogroups are indicated. Adapted from Salas *et al.* (2002).

this suggests that it is unlikely to go unchallenged by other researchers in the future.

It has been known for some time that Y-chromosomal haplogroup diversity within Africa presents an unusual pattern (Section 8.5.2 and *Figure 9.18*). While the diversity of many genetic markers is greatest in Africa – a fact that is often taken as evidence for an African origin for modern humans – the diversity of Y chromosomes measured using binary markers appears to be significantly lower in Africa than it is in either Asia or Europe (*Table 8.1*). The geographical distribution of the major clades within the Y phylogeny shows that the two deepest rooting clades, A and B, are African-specific. However, these are present at only 13% overall in a sample of 608 chromosomes (Cruciani *et al.*, 2002); there is one predominant and derived lineage, haplogroup E3a, that accounts for over 47% of African chromosomes overall, and about 80% of chromosomes in Cameroon and West Africa (Cruciani *et al.*, 2002). The distribution of these and other haplogroups is shown in *Figure 10.23*.

From simple visual inspection of this map, the major distinction is a north–south divide, in agreement with the classical gene frequency analysis; East Africa is also differentiated from the rest. The focus of this study (Cruciani *et al.*, 2002) was on the presence of a lineage (in haplogroup R) of apparently Asian origin in the northern Cameroon sample (see *Box 9.6*), but for the purposes of this discussion we concentrate on what the Y data have to say about the Bantu expansion. Three haplogroups, two within haplogroup

E3b and one a subgroup of E2b, are present throughout sub-Saharan Africa. Network analysis of sparse microsatellite data (two dinucleotide and three tetranucleotide repeat loci) showed low geographical differentiation, suggesting relatively recent origins for these lineages in population expansions. On the basis of these lineages, introgression of Bantu Y-chromosomal lineages into Khoisan is estimated at around 50%; lineages within haplogroup A present at high frequency in Khoisan are also present almost exclusively (at 6% frequency) in southern African Bantu, suggesting some admixture.

Taken together, the classical, mtDNA and Y-chromosomal data present a fairly harmonious picture that seems in reasonable agreement with archaeological and linguistic evidence for the Bantu expansion, even though there are archaeologists and linguists who will disagree. Data are still scanty, however, and it remains to be seen whether the harmony is maintained as sample sizes increase and further complexity is introduced. This may happen if hopes are realized to improve the coverage of African populations in genetic diversity studies in general, and studies of genetic disease in particular (Tishkoff and Williams, 2002).

10.7 Genetic analysis of domesticated animals and plants

Genetic studies of animals and plants have recently made significant advances, so that a serious analysis of the genetic history of domestication can be undertaken. There are three principal reasons for this progress:

▶ technologies for the mapping and sequence analysis of DNA have advanced and become widely available, so that DNA sequence data can readily be obtained from almost any species;

▶ data emerging from the human and mouse genome projects has greatly increased our understanding of the organization and function of mammalian genomes, with a consequent impact on our understanding of these features of the genomes of mammalian domesticates. Likewise, data on the genomes of *Arabidopsis thaliana* (mustard weed – a mere ~ 100 Mb in size) and rice (only four times larger than *Arabidopsis*) provide a framework for understanding plant genome structure and function in general;

▶ ancient DNA techniques have made sequences from preserved animal and plant remains directly available.

These advances make it possible to address questions of the genetic effects of artificial selection, and the number and geographical origin of domestication events, which inform our understanding of the whole process of the development of agriculture.

10.7.1 The impact of selective regimes

Since the beginning of agriculture people have selected desirable traits to improve product quality and manageability of animal and crop species. Modern animal and plant

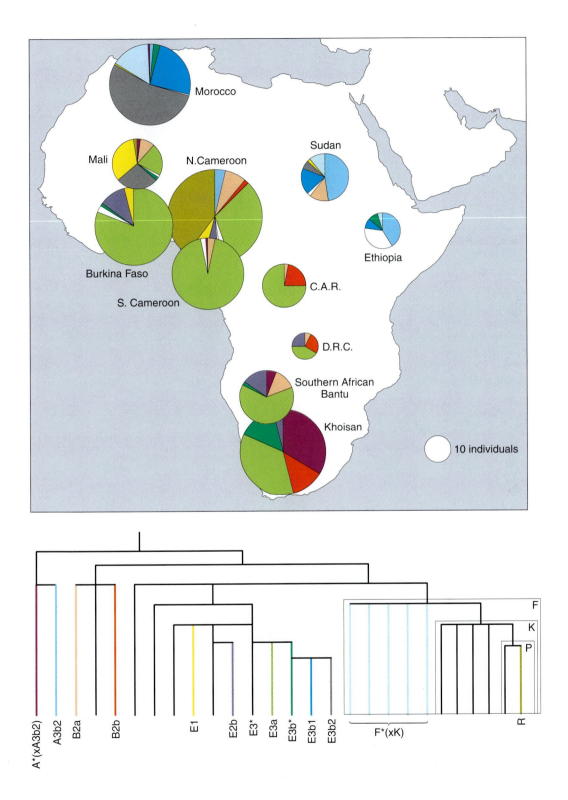

Figure 10.23: Distribution of major Y-chromosomal haplogroups in Africa.

Pie charts represent populations, with area of chart proportional to sample size (the smallest sample is that from the Democratic Republic of Congo [D.R.C.], at 12 chromosomes). Areas of sectors are proportional to haplogroup frequencies. Data are from Cruciani *et al.* (2002) and Underhill *et al.* (2000). C.A.R.: Central African Republic. The tree below shows the phylogenetic relationships of haplogroups and identifies them by color. White sectors within pie charts represent a collection of other minor haplogroups.

domesticates differ significantly in their morphology and biochemistry from the wild species from which they were selectively bred. The propagation of individuals with desirable characteristics can lead to dramatic phenotypic changes, as demonstrated by the different sizes and shapes of pedigree dogs bred for different purposes, or between wild and cultivated sunflowers first domesticated 4000 YA in the Americas (Burke *et al.*, 2002). A chihuaha and a great-dane, or a cultivated and wild sunflower, are **interfertile** and comprise members of a single species, despite their evident differences (*Figure 10.24*). These artificial adaptations become apparent at different times in the prehistory of a domesticated species. In the archaeological record it is easier to identify morphological changes than biochemical alterations.

Plant domesticates share a number of changes resulting from domestication. Domesticated cereals have a suite of three traits in common that distinguishes them from their wild progenitors (Salamini *et al.*, 2002); these are:

▸ tough **rachis** (the 'backbone' of an ear of seeds, in wheat, for example), allowing the ear to be harvested as a whole, rather than the natural process of **shattering**, where the spikelets (each containing a seed – see *Figure 10.25*) are released individually on maturity. Note that nonshattering mutants in the wild will be strongly selected against;

▸ seeds that are freely released during threshing from their **glumes** (leaf-like structures that protect them);

▸ considerably larger seeds. **Polyploidy** (a genome copy number greater than two) frequently accompanies the seed enlargement.

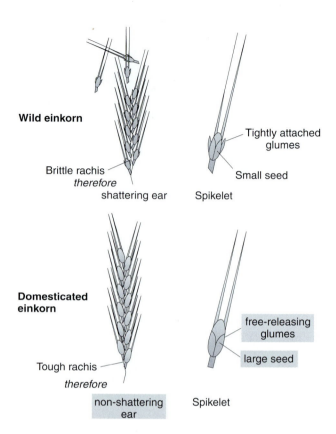

Figure 10.25: The three key domestication traits of cereals.

Adapted from Salamini *et al.* (2002).

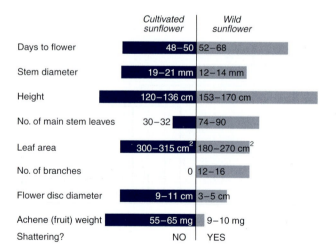

Figure 10.24: Some phenotypic differences between wild and cultivated sunflower.

Schematic illustration of nine of 18 major traits that differ. Lengths of bars represent mean values. Note that time to flowering is not only shorter in cultivated sunflower, but also more predictable. These plants are interfertile, and regarded as the same species. Data from Burke *et al.* (2002).

Analysis of **quantitative trait loci** (**QTLs**; *Boxes 10.7* and *13.2*) of different domesticated cereal species has shown that the genes controlling the critical phenotypic differences are rather few in number. For example, the striking morphological differences between domesticated maize and its wild progenitor, **teosinte**, can be ascribed to as few as five QTLs, each having a major effect. Comparative genomic analysis shows that one of these, controlling shattering behavior, is **orthologous** to loci in sorghum and foxtail millet controlling the same phenotype (Devos and Gale, 2000). Similarly, three QTLs influencing seed size correspond between sorghum, rice and maize (Buckler *et al.*, 2001). Note that such comparative genomic approaches are difficult in the *Triticeae* cereals, including wheat and barley, because of their enormous genome sizes. Modern bread wheat, for example, is a hexaploid species formed from the hybridization of three diploid species, and has a genome rich in repeated sequences and, at about 17 000 Mb, nearly three times the size of the human genome. In noncereal species, too, it is often found that only a few QTLs are responsible for major selected changes. In the eggplant (aubergine), just six loci control many aspects of fruit size, color, and shape and also plant prickliness, and several of these are orthologous to loci having the same effects in other members of the *Solanaceae*: tomato, pepper and potato (Doganlar *et al.*, 2002).

TABLE 10.7: INCREASES IN YIELD OF DOMESTICATED CEREALS COMPARED TO WILD PROGENITORS.

Cereal	Wild yield (tonnes /hectare)	Yield in 1961 (tonnes /hectare)	Yield in 2000 (tonnes /hectare)	Gain over progenitor (-fold)
Maize	0.16	2.5	9.1	55.9
Sorghum	<0.60	2.4	5.9	<9.9
Rice	1.12	4.1	5.6	5.0
Oats	2.93	1.8	4.5	1.5
Barley	0.65	2.4	6.4	9.8
Wheat	0.65	2.4	7.1	11.0
Pearl Millet	<0.55	1.1	1.5	<2.7

Taken from Buckler *et al.* (2001).

After the establishment of the basic three traits described above, selective breeding in cereals has concentrated on increasing yield. *Table 10.7* illustrates the dramatic increases in cereal yield since domestication, and also over the last 40 years; recent increases have a substantial genetic component due to advances in plant breeding. In some plant species, selective breeding has also concentrated on making the edible parts more palatable by removing the chemical and physical defenses that protect them against over-grazing in the wild. Plants belonging to the *Cucurbitaceae* (e.g., pumpkins, melons and cucumbers) have lost their natural bitterness due to down-regulated production of the bitter compound cucurbitacin, and also reduced production of phytoliths (particles of solid silicon dioxide) in their rinds (Piperno *et al.*, 2002). Similarly, cultivated tomatoes (*Lycopersicon esculentum*) have down-regulated production of the unpleasant-smelling 2-phenylethanol and phenylacetaldehyde compared to their wild relatives (*Lycopersicon pennellii*) (Tadmor *et al.*, 2002).

It might be imagined that continuous selection for advantageous traits among cereal species would have led to low genetic diversity. Surprisingly, this is not the case. General estimates for maize, sorghum, rice, oats, wheat and pearl millet indicate that the domesticates have retained on average two-thirds of the nucleotide diversity of their wild progenitors, and cultivated barley has even more diversity than wild. This high diversity is probably a result of out-crossing with other strains during and after domestication, and also to the very large numbers of plants that would have to be maintained in a population while a crop was being used for subsistence (Buckler *et al.*, 2001).

However, when genetic studies are focused upon specific loci controlling important selected traits, reduced diversity is apparent. One major difference between domesticated maize and its progenitor teosinte is that teosinte has long branches with tassels at the tips, while maize has short branches tipped by ears. This difference is controlled by the gene *teosinte branched1* (*tb1*) (Doebley *et al.*, 1997). A study of genetic diversity at this locus (Wang *et al.*, 1999) shows that, while the coding region of the gene shows no evidence of the action of strong recent selection, with maize carrying 39% of the nucleotide diversity of teosinte, a 1.1-kb region containing the promoter carries only 3% of the teosinte diversity (*Figure 10.26*). In a phylogenetic tree of the coding region sequences, maize falls into many different clades,

BOX 10.7 Quantitative trait loci (QTLs)

A **Quantitative Trait Locus (QTL)** is usually defined as a polymorphic genetic locus identified through the statistical analysis of a continuously distributed trait (such as plant height or animal body weight). These traits are typically affected by more than one gene, as well as by the environment. There can also be interactions between QTLs (**epistasis**) that complicate analysis. The identification of QTLs is like any other genetic mapping experiment, in that the segregation of phenotypic traits in the offspring of crosses is examined and correlated to the segregation of polymorphic genetic markers. In plant and animal QTL analysis controlled breeding experiments between chosen samples are set up, and in some cases homozygous inbred lines can be used. The QTL concept has been expanded to include traits that are not continuously distributed, such as schizophrenia or diabetes in humans, since the risk of developing these disorders is usually assumed to reflect a continuously distributed susceptibility function. In these contexts QTLs are really no different from the complex genetic phenotypes discussed in Chapter 14. More information on quantitative genetics, the discipline underlying QTL analysis, is in *Box 13.2.*

consistent with neutrality, but in contrast the promoter region sequences are monophyletic. An **HKA test** for selection (see Section 6.3.4) is highly significant, and thus the promoter region has been under strong selection during domestication. Studies of another maize locus, *c1*, controlling kernel color (Hanson *et al.*, 1996), and *brittle2*, *sugary1* and *amylose extender1* (Whitt *et al.*, 2002), involved in starch production, also show the effects of selection. In rice, low nucleotide diversity at the *waxy* locus has been driven by selection for the glutinous ('sticky') rice trait (Olsen and Purugganan, 2002). Rather than targeting specific loci, a more general way to detect selected genes throughout the genome is to screen more widely using microsatellites to detect **selective sweeps** (see Section 5.4) in the form of a reduced level of diversity compared to neutral expectations. A screen of 501 maize genes using sets of linked microsatellites revealed 15 genes showing good evidence for selection (Vigouroux *et al.*, 2002).

In comparison to plants, morphological changes in domesticated animals are more species-specific. Size changes are more difficult to discern due to the changes in population structure that may result from harvesting at a younger age (Zeder and Hesse, 2000). Behavioral changes were bred into animal stocks, making them less aggressive and more easily managed. Many domesticated animals also experienced changes in skull shape due to artificial selection for retention of juvenile features in adults, a characteristic known as **neoteny**. Most domesticated animals have experienced reduction in brain size and acuity of the senses compared to their wild progenitors – both essential in the wild, but a waste of energy under domestication (Diamond, 1996). Changes in features exploited during the secondary product revolution appear later in prehistory, for example, sheep hair becomes finer and less pigmented. Most animal QTL studies have concentrated not on identifying genes important in

early domestication, but on traits that are thought useful in improving yield and quality of animal products, in particular meat and milk (reviewed by Andersson, 2001).

10.7.2 Identifying the origin of a domesticated plant

Pinpointing the origin of plant domestication events involves two major lines of enquiry:

▶ the identification of seeds of the wild species in early archaeological sites, with seeds of domesticated species succeeding in later strata;

▶ the identification and location of likely progenitor species still existing, and the confirmation of their progenitor status by genetic approaches. These genetic studies have often used **AFLPs** as polymorphic markers (see *Box 4.2*), which give a genomewide estimate of similarity between a domesticated and candidate progenitor species.

A study of 288 AFLPs putatively identified the geographical origin of domesticated **einkorn** (Heun *et al.*, 1997), the first wheat to be successfully domesticated, as the foothills of the Karacadag mountains in southeastern Turkey (*Figure 10.27*). Though this finding is disputed by some, it has been supported by archaeological evidence from nearby sites. Since the progenitor range of einkorn, rye, and some legumes, including lentil, pea and chickpea overlap in this small region of Turkey, it has been proposed that this region was a 'cradle of agriculture' in which many innovations occurred (Lev-Yadun *et al.*, 2000). Recent evidence from AFLP analysis has added **emmer** wheat, the most important crop of the Neolithic, to the list of cereals whose progenitors map to the Karacadag region (Özkan *et al.*, 2002). However, for political reasons, archaeological, botanical and zoological field research addressing agricultural transitions has been minimal for decades in potentially important regions such as Iraq, Iran, Afghanistan, Pakistan, and Kashmir, and it remains possible that the ranges of the wild progenitor species were greater than is often assumed.

If there was a single domestication event in the history of a species, phylogenetic trees constructed from AFLP data are expected to show the domesticated species lying within a single clade. This is the case for barley, for example, illustrated in *Figure 10.28*. Phylogenies based upon sequence data can be used in the same way – for example, the **promoter** region of the maize *tb1* gene, discussed above, forms a single clade, compatible with a single event, and the position of this clade with respect to teosinte from different regions strongly suggests that the origin was in the Balsas river valley of southwestern Mexico (Wang *et al.*, 1999). As well as AFLPs, some work has been done using microsatellites. In a phylogeny constructed from genetic distances based on the proportion of shared alleles at 99 maize microsatellites, maize falls into a **monophyletic** group with strong **bootstrap** support, and teosinte branches basal to the maize clade all come from the Balsas river region (Matsuoka *et al.*, 2002). The microsatellite data can also be used to date

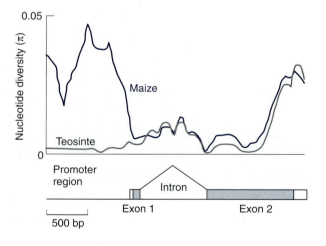

Figure 10.26: The effects of artificial selection on the *teosinte branched1* (*tb1*) gene.

Nucleotide diversity is greatly reduced in the domesticated maize promoter region compared to the same region in its wild progenitor, teosinte. Redrawn from Wang *et al.* (1999).

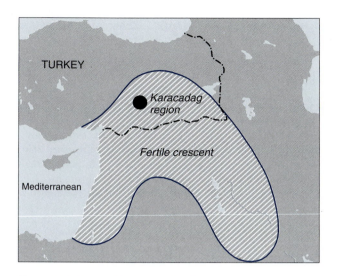

Figure 10.27: The cradle of agriculture?

Map showing the location of the Karacadag region, in southeastern Turkey, proposed as the origin of several plant domestications.

the divergence of the teosinte and maize lineages, and at 9200 YA (with 95% confidence interval 5700–13 100 YA), this is consistent with archaeological evidence. It is not surprising to note that some archaeologists believe that multiple pathways to domestication are more likely, and that the genetic evidence is not entirely conclusive (Jones and Brown, 2000).

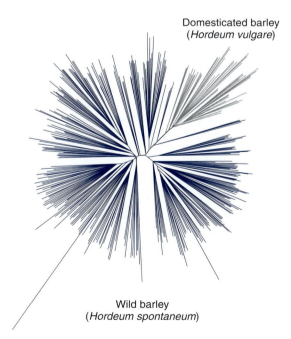

Figure 10.28: Phylogenetic tree showing a single origin for domesticated barley.

Tree based on AFLP data from 400 loci. Redrawn from Salamini *et al.* (2002).

Ancient DNA evidence can be useful when the modern wild descendants of progenitor species have become extinct in a particular region, as is apparently the case for the *japonica* rice progenitor in the region spanning Burma, Thailand and Laos. As has been discussed in Section 4.11, nuclear DNA sequences are almost impossible to recover from ancient sources. Many plant studies therefore rely upon chloroplast DNA (cpDNA) sequences, since, like the mitochondrion, this organelle is present in many copies per cell, thus increasing the probability of DNA survival. Identification of a cpDNA polymorphism differentiating the two modern rice forms, *indica* and *japonica*, and the typing of this same polymorphism in ancient rice DNA samples has been used as a basis to propose that the two forms were domesticated independently (Jones and Brown, 2000).

10.7.3 Identifying the origin of a domesticated animal

The task of finding the place where animal domestication occurred differs from that of finding the origins of plant domestication in a number of respects. Unlike plants, which if left to their own devices and not subject to extreme climate changes will stay where they are, animals are mobile, and liable to be hunted to extinction if not already domesticated. The last **aurochs** (this is the singular noun – the plural is aurochsen), progenitor of modern cattle, is said to have been killed in Poland in 1627 (Clutton-Brock, 1999), and the wild ancestors of the horse, the Old World camel and the goat have also disappeared. Ancient DNA studies clearly can make a contribution to resolving this problem, and recovery of ancient DNA has been most successful in the case of the cattle progenitors.

The vast majority of studies of the origins of animal domestication have concentrated on employing mtDNA. This is largely due to the convenience of studying a simple molecule that has enough sequence conservation to allow the design of universal PCR primers, but enough diversity, particularly in the hypervariable region, to yield a large number of different and informative haplotypes. Also, ancient DNA work is at its most successful when mtDNA is being studied (see Section 4.11). Another advantage accrues to mtDNA when we consider the mating structures of domesticated animal populations: under managed breeding, males have a very much larger number of offspring than do females, and the expectation of this is that past spatial pattern of paternally and biparentally inherited markers are obscured. The maternally inherited mtDNA molecule is therefore expected to show relatively little geographic differentiation, and so, compared to other loci, should show relative continuity with populations of the past (Bradley, 2000). Since all domestic livestock species can interbreed with their wild relatives, introgression is a potential problem in interpreting domestic animal phylogenies; however, most introgression events are thought likely to be from wild males, rather than females, so mtDNA should be relatively insensitive to this problem.

There are two major types of cattle, the European humpless *Bos taurus* and the more arid-adapted Indian zebu *Bos indicus*, named as different species even though they are completely **interfertile**. Archaeological evidence suggests that taurine cattle were domesticated from the aurochs in the Near East 8–10 KYA, and there is some evidence that zebu breeds were developed from taurine breeds later. However, evidence from mtDNA suggests instead that they derive from two independent domestication events. Trees of mtDNA control region haplotypes show 'double-headed broom' topologies (Loftus *et al.*, 1994), with two clusters of branches separated by a long internal branch (*Figure 10.29*). The clusters correspond well to the two cattle taxa. The long branch indicates a predomestication divergence between the taxa of well in excess of 100 KY. Network analysis of haplotypes within the taurine part of the phylogeny (Troy *et al.*, 2001) reveals four clusters with star-like topologies, indicative of population expansions. Haplotypes from Anatolia and the Middle East consist of three star-like clusters, and in Europe one of these clusters predominates, supporting the idea of a Near Eastern origin for modern cattle. African haplotypes also exhibit one cluster predominantly, but haplotypes from within this cluster are rare in the Anatolian and Middle Eastern samples. This finding provides some support for an independent African domestication (see Section 10.6.1). The phylogeny shown in *Figure 10.29* includes six aurochs sequences derived from bone samples from England, and dating from 3720 to 12 400 YA; these cluster together, and lie in a branch neighboring the *Bos taurus* cluster, confirming that *Bos taurus* was domesticated from *Bos primigenius* (the aurochs). The fact that these English sequences form a distinct cluster of their own, well separated from modern taurine sequences, indicates that wild capture throughout Europe is unlikely to have occurred, and again is consistent with a Near Eastern origin. Unfortunately, data from aurochsen from the Near East itself are not yet available.

This picture of a small number of independent, often geographically separated domestications is a common theme for several animals (*Table 10.8*), with the apparent exception of horses, in which, from the evidence of mtDNA, domestication involved a large number of wild captures (Jansen *et al.*, 2002; Vilà *et al.*, 2001).

As yet, nuclear sequence analysis has yet to be widely applied to the problem of animal domestication. One study of cattle has used 15 microsatellites (Hanotte *et al.*, 2002) to address the question of the origins of African cattle. Using **synthetic maps** as a tool, and examining correlations between PCs and distance from different centers under different hypotheses, this study shows a major zebu influx into Africa from the east, interpreted as early imports of these cattle. Some evidence is claimed of support for local domestication of cattle within Africa, and for the spread of cattle to the south during the Bantu expansion being largely via the eastern route.

Summary

▶ The development of agriculture over the last 10 KY was an important cultural innovation that allowed human populations to expand dramatically. A key debate is whether agricultural practices were adopted by existing hunter–gatherers (acculturation), or accompanied by substantial gene flow from farming populations bringing their innovations with them.

▶ Agriculture involved the domestication of animal and plant species, and is recognizable by direct evidence of this, and by changes in associated culture including stone tools and pottery. These changes are sometimes referred to as the 'Neolithic revolution'. For political and practical reasons, some parts of the world have not been well studied for evidence of early agricultural development, and this introduces a bias into our understanding.

▶ Agriculture began independently in different times and places, and involved spread from a number of 'homelands'. Its development was probably triggered by the needs of a growing human population in a period of relative climatic stability, and the extinction of large prey species in most regions.

▶ As well as the eventual population growth enabled by agriculture, there were problems, including initial poor nutrition, and spread of disease from animals to humans.

▶ The geographical relationships of agricultural homelands with centers of large language families suggests that agriculture and language may have dispersed together. However, how far this was so is under debate.

▶ There has been a vigorous debate about the patterns of genetic diversity in Europe, and whether they support demic diffusion of farmers from the Near East (beginning about 10 KYA), or acculturation. The 'wave of advance' model is too simplistic to account for the complex patterns in the archaeological data.

▶ Data on classical gene frequencies, and nuclear DNA including the Y chromosome all show some clinal

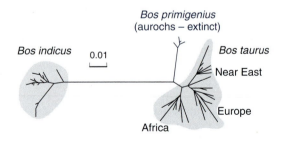

Figure 10.29: Phylogeny of cattle mtDNA control region sequences.

Note the long branch between the *Bos indicus* (Asian) and *Bos taurus* (European and African) clusters, and the clustering of European *Bos primigenius* (aurochs) ancient DNA sequences. The scale bar indicates 1% divergence. Redrawn from Troy *et al.* (2001).

TABLE 10.8: EVIDENCE ON ANIMAL DOMESTICATION EVENTS FROM mtDNA DATA.

Animal	Structure of phylogeny	Interpretation	Reference
Cattle	Two clusters separated by long internal branch	Two independent domestications, one in the Near East and one in Asia	Loftus *et al.*, 1994
Sheep	Two well-separated clusters	Two independent domestications from two mouflon species	Hiendleder *et al.*, 2002
Goat	Three well-separated clusters; weak geographic structure	Three independent domestications; much transport of goats with human migrations	Luikart *et al.*, 2001
Pig	Two clusters for domestic pigs, one Asian and one European	Two independent domestications, one in Europe and one in Asia	Giuffra *et al.*, 2000
Water buffalo	Two clusters separating river buffalo (India, Egypt and the Balkans) and swamp buffalo (S.E. Asia)	Two independent domestications	Reviewed by Bradley, 2000
Horse	17 clusters	Introgression from wild or multiple domestications	Jansen *et al.*, 2002

Some studies used coding region sequences as well as control region sequences, but only the latter are described here.

patterns with foci in the Near East, explicable by demic diffusion. However, dating these clines and correlating them directly with the Neolithic is difficult. mtDNA patterns are not strongly clinal, and this finding has been used to support the acculturation view. Methods for estimating the proportion of the Paleolithic contribution to modern Europeans are not agreed upon.

▸ The spread of iron-working, Bantu languages and agriculture from West Africa into central and southern Africa (the 'Bantu expansion') occurred over the last 3 KY. Patterns of mtDNA and Y-chromosomal diversity seem broadly consistent with the archaeological and linguistic framework, but more data are needed.

▸ Genetic analyses of domesticated animals and plants can locate the origins of agriculture. Selected changes in animals were mostly behavioral, and have received little attention from geneticists. Changes in plants focused on improved ease of harvesting and yield. The same genes have been the target of selection in many plant species, but at the same time a high level of overall genetic diversity has been maintained.

▸ Phylogeographic studies of plant diversity can pinpoint the locations of domestication events. Several species seem to have been domesticated in southeastern Turkey. Many domesticated animals show a pattern consistent with multiple independent domestications, including the Near East for cattle.

Further reading

Diamond J (2002) Evolution, consequences and future of plant and animal domestication. *Nature* **418**, 700–707.

Diamond J, Bellwood P (2003) Farmers and their languages: the first expansions. *Science* **300**, 597–603.

Bellwood P (2001) Early agriculturalist population diasporas? Farming, language and genes. *Ann. Rev. Anthropol.* **30**, 181–207.

Nichols J (1997) Modeling ancient population structures and movement in linguistics. *Ann. Rev. Anthropol.* **26**, 359–384.

Price TD (ed.) (2000) *Europe's First Farmers.* Cambridge University Press, Cambridge.

References

Alroy J (2001) A multispecies overkill simulation of the end-Pleistocene megafaunal mass extinction. *Science* **292**, 1893–1896.

Ammerman AJ, Cavalli-Sforza LL (1984) *Neolithic Transition and the Genetics of Populations in Europe.* Princeton University Press, Princeton, New Jersey.

Andersson L (2001) Genetic dissection of phenotypic diversity in farm animals. *Nature Rev. Genet.* **2**, 130–138.

Barbujani G, Bertorelle G, Chikhi L (1998) Evidence for Paleolithic and Neolithic gene flow in Europe. *Am. J. Hum. Genet.* **62**, 488–491.

Barbujani G, Sokal RR, Oden NL (1995) Indo-European origins: a computer-simulation test of five hypotheses. *Am. J. Phys. Anthropol.* **96**, 109–132.

Barker G (1985) *Prehistoric Farming in Europe.* Cambridge University Press, Cambridge.

Barrett R, Kuzawa CW, McDade T, Armelagos GJ (1998) Emerging and re-emerging infectious diseases: the third epidemiological transition. *Ann. Rev. Anthropol.* **27**, 247–271.

Bellwood P (2001) Early agriculturalist population diasporas? Farming, language and genes. *Ann. Rev. Anthropol.* **30**, 181–207.

Bellwood P, Renfrew C (2003) *Examining the Farming/Language Dispersal Hypothesis.* McDonald Institute for Archaeological Research, Cambridge.

Blench R (1993) Recent developments in African language classification and their implications for prehistory. In: *The Archaeology of Africa: Food, Metals and Towns* (eds T Shaw, P Sinclair, B Andah, A Okpoko). Routledge, London, pp. 126–138.

Bradley DG (2000) Mitochondrial DNA diversity and origins of domestic livestock. In: *Archaeogenetics: DNA and the Population Prehistory of Europe* (eds C Renfrew, K Boyle). Oxbow Books, Oxford, pp. 315–320.

Buckler ES, Thornsberry JM, Kresovich S (2001) Molecular diversity, structure and domestication of grasses. *Genet. Res.* **77**, 213–218.

Burke JM, Tang S, Knapp SJ, Rieseberg LH (2002) Genetic analysis of sunflower domestication. *Genetics* **161**, 1257–1267.

Casanova M, Leroy P, Boucekkine C, *et al.* (1985) A human Y-linked DNA polymorphism and its potential for estimating genetic and evolutionary distance. *Science* **230**, 1403–1406.

Cavalli-Sforza LL, Menozzi P, Piazza A (1994) *The History and Geography of Human Genes.* Princeton University Press, Princeton, New Jersey.

Cavalli-Sforza LL, Minch E (1997) Paleolithic and Neolithic lineages in the European mitochondrial gene pool. *Am. J. Hum. Genet.* **61**, 247–251.

Chikhi L, Destro-Bisol G, Bertorelle G, Pascali V, Barbujani G (1998) Clines of nuclear DNA markers suggest a largely Neolithic ancestry of the European gene pool. *Proc. Natl Acad. Sci. USA* **95**, 9053–9058.

Chikhi L, Nichols RA, Barbujani G, Beaumont MA (2002) Y genetic data support the Neolithic demic diffusion model. *Proc. Natl Acad. Sci. USA* **99**, 11008–11013.

Childe VG (1936) *Man Makes Himself.* Watts, London.

Clutton-Brock J (1999) *A Natural History of Domesticated Mammals.* Cambridge University Press, Cambridge.

Cockburn TA (1971) Infectious diseases in ancient populations. *Curr. Anthropol.* **12**, 45–62.

Cohen MN, Armelagos GJ (1984) *Paleopathology at the Origins of Agriculture.* Academic Press, Orlando.

Collins R (1986) *The Basques.* Blackwell, Oxford.

Cruciani F, Santolamazza P, Shen PD *et al.* (2002) A back migration from Asia to sub-Saharan Africa is supported by high-resolution analysis of human Y-chromosome haplotypes. *Am. J. Hum. Genet.* **70**, 1197–1214.

Devos KM, Gale MD (2000) Genome relationships: the grass model in current research. *Plant Cell* **12**, 637–646.

Diamond J (1996) Competition for brain space. *Nature* **382**, 756–757.

Diamond J (1998) *Guns, Germs and Steel: a Short History of Everybody for the last 13000 years.* Vintage, London.

Diamond J (2002) Evolution, consequences and future of plant and animal domestication. *Nature* 418, 700–707.

Doebley J, Stec A, Hubbard L (1997) The evolution of apical dominance in maize. *Nature* **386**, 485–488.

Doganlar S, Frary A, Daunay MC, Lester RN, Tanksley SD (2002) Conservation of gene function in the Solanaceae as revealed by comparative mapping of domestication traits in eggplant. *Genetics* **161**, 1713–1726.

Fix AG (1999) *Migration and Colonization in Human Microevolution.* Cambridge University Press, Cambridge.

Galton F (1865) The first steps towards the domestication of animals. *Trans. Ethnolog. Soc. Lond.* **3**, 122–138.

Gamble C (1999) *The Palaeolithic Societies of Europe.* Cambridge University Press, Cambridge.

Giuffra E, Kijas JMH, Amarger V, Carlborg Ö, Jeon JT, Andersson L (2000) The origin of the domestic pig: independent domestication and subsequent introgression. *Genetics* **154**, 1785–1791.

Goldstein DB, Chikhi L (2002) Human migrations and population structure: what we know and why it matters. *Ann. Rev. Genomics Hum. Genet.* **3**, 129–152.

Hanotte O, Bradley DG, Ochieng JW, Verjee Y, Hill EW, Rege JEO (2002) African pastoralism: Genetic imprints of origins and migrations. *Science* **296**, 336–339.

Hanson MA, Gaut BS, Stec AO, Fuerstenberg SI, Goodman MM, Coe EH, Doebley J (1996) Evolution of anthocyanin biosynthesis in maize kernels: the role of regulatory and enzymatic loci. *Genetics* **143**, 1395–1407.

Hassan FA, Sengel RA (1973) On mechanisms of population growth during the Neolithic. *Curr. Anthropol.* **14**, 535–542.

Heun M, Schafer-Pregl R, Klawan D, Castagna R, Accerbi M, Borghi B, Salamini F (1997) Site of einkorn wheat domestication identified by DNA fingerprinting. *Science* **278**, 1312–1314.

Hewitt G (2000) The genetic legacy of the Quaternary ice ages. *Nature* **405**, 907–913.

Hiendleder S, Kaupe B, Wassmuth R, Janke A (2002) Molecular analysis of wild and domestic sheep questions current nomenclature and provides evidence for domestication from two different subspecies. *Proc. Roy. Soc. Lond. B* **269**, 893–904.

Hill EW, Jobling MA, Bradley DG (2000) Y chromosomes and Irish origins. *Nature* **404**, 351–352.

Holden CJ (2001) Bantu language trees reflect the spread of farming across sub-Saharan Africa: a maximum-parsimony analysis. *Proc. Roy. Soc. Lond. B* **269**, 793–799.

Hurles ME, Veitia R, Arroyo E *et al.* (1999) Recent male-mediated gene flow over a linguistic barrier in Iberia, suggested by analysis of a Y-chromosomal DNA polymorphism. *Am. J. Hum. Genet.* **65**, 1437–1448.

Jackes M, Lubell D, Meiklehohn C (1997) On physical anthropological aspects of the Mesolithic-Neolithic transition in the Iberian Peninsula. *Curr. Anthropol.* **38**, 839–846.

Jansen T, Forster P, Levine MA *et al.* (2002) Mitochondrial DNA and the origins of the domestic horse. *Proc. Natl Acad. Sci. USA* **99**, 10905–10910.

Jobling MA (1994) A survey of long-range DNA polymorphisms on the human Y chromosome. *Hum. Mol. Genet.* **3**, 107–114.

Jobling MA, Tyler-Smith C (2000) New uses for new haplotypes: the human Y chromosome, disease, and selection. *Trends Genet.* **16**, 356–362.

Jones M, Brown T (2000) Agricultural origins: the evidence of modern and ancient DNA. *Holocene* **10**, 769–776.

Kelly RL (1995) *The Foraging Spectrum: Diversity in Hunter–Gatherer Lifeways.* Smithsonian Institution Press, Washington.

Kurlansky M (1999) *The Basque History of the World.* Jonathan Cape, London.

Lalueza Fox C, Gonzales Martin A, Vives Civit S (1996) Cranial variation in the Iberian Peninsula and the Balearic Islands: inferences about the history of the population. *Am. J. Phys. Anthropol.* **99**, 413–428.

Lev-Yadun S, Gopher A, Abbo S (2000) The cradle of agriculture. *Science* **288**, 1602–1603.

Loftus RT, MacHugh DE, Bradley DG, Sharp PM, Cunningham P (1994) Evidence for two independent domestications of cattle. *Proc. Natl Acad. Sci. USA* **91**, 2757–2761.

Luikart G, Gielly L, Excoffier L, Vigne JD, Bouvet J, Taberlet P (2001) Multiple maternal origins and weak phylogeographic structure in domestic goats. *Proc. Natl Acad. Sci. USA* **98**, 5927–5932.

MacEachern S (2000) Genes, tribes and African history. *Curr. Anthropol.* **41**, 357–384.

Malaspina P, Tsopanomichalou M, Duman T *et al.* (2001) A multistep process for the dispersal of a Y chromosomal lineage in the Mediterranean area. *Ann. Hum. Genet.* **65**, 339–349.

Martin PS (1973) The discovery of America. *Science* **179**, 969–974.

Matsuoka Y, Vigouroux Y, Goodman MM, Sanchez JG, Buckelr E, Doebley J (2002) A single domestication for maize shown by multilocus microsatellite genotyping. *Proc. Natl Acad. Sci. USA* **99**, 6080–6084.

Meadow RH (1989) Osteological evidence for the process of animal domestication. In: *The Walking Larder: Patterns of Domestication, Pastoralism and Predation* (ed. J Clutton-Brock). Unwin Hyman, London, pp. 80–90.

Menozzi P, Piazza A, Cavalli-Sforza LL (1978) Synthetic maps of human gene frequencies in Europeans. *Science* **201**, 786–792.

Mishmar D, Ruiz-Pesini E, Golik P *et al.* (2003) Natural selection shaped regional mtDNA variation in humans. *Proc. Natl Acad. Sci. USA* **100**, 171–176.

Mourant AE, Kopec AC, Domaniewska-Sobczak K (1976) *The Distribution of the Human Blood Groups and Other Polymorphisms.* Oxford University Press, London.

Ngo KY, Vergnaud G, Johnsson C, Lucotte G, Weissenbach J (1986) A DNA probe detecting multiple haplotypes of the human Y chromosome. *Am. J. Hum. Genet.* **38**, 407–418.

Nichols J (1997) Modeling ancient population structures and movement in linguistics. *Ann. Rev. Anthropol.* **26**, 359–384.

O'Donnell KA, O'Neill C, Tighe O, Bertorelle G, Naughten E, Mayne PD, Croke DT (2002) The mutation spectrum of hyperphenylalaninemia in the Republic of Ireland: the population history of the Irish revisited. *Eur. J. Hum. Genet.* **10**, 530–538.

Okafor EE (1993) New evidence for early iron-smelting from southeastern Nigeria. In: *The Archaeology of Africa: Food, Metals and Towns* (eds T Shaw, P Sinclair, B Andah, A Okpoko). Routledge, London, pp. 432–448.

Olsen KM, Purugganan MD (2002) Molecular evidence on the origin and evolution of glutinous rice. *Genetics* **162**, 941–950.

Özkan H, Brandolini A, Schafer-Pregl R, Salamini F (2002) AFLP analysis of a collection of tetraploid wheats indicates the origin of emmer and hard wheat domestication in southeast Turkey. *Mol. Biol. Evol.* **19**, 1797–1801.

Pereira L, Macaulay V, Torroni A, Scozzari R, Prata MJ, Amorim A (2001) Prehistoric and historic traces in the mtDNA of Mozambique: insights into the Bantu expansions and the slave trade. *Ann. Hum. Genet.* **65**, 439–458.

Phillipson DW (1993) *African Archaeology,* 2nd Edn. Cambridge University Press, Cambridge.

Piperno DR, Holst I, Wessel-Beaver L, Andres TC (2002) Evidence for the control of phytolith formation in *Cucurbita* fruits by the hard rind (*Hr*) genetic locus: archaeological and ecological implications. *Proc. Natl Acad. Sci. USA* **99**, 10923–10928.

Poinar HN, Kuch M, Sobolik KD *et al.* (2001) A molecular analysis of dietary diversity for three archaic Native Americans. *Proc. Natl Acad. Sci. USA* **98**, 4317–4322.

Price TD (2000a) *Europe's First Farmers.* Cambridge University Press, Cambridge.

Price TD (2000b) Europe's first farmers: an introduction. In: *Europe's First Farmers* (ed. TD Price). Cambridge University Press, Cambridge, pp. 1–18.

Rendine S, Piazza A, Cavalli-Sforza LL (1986) Simulation and separation by principal components of multiple demic expansions in Europe. *Am. Nature* **128**, 681–706.

Renfrew AC (1988) *Archaeology and Language: the Puzzle of Indo-European Origins*. Cambridge University Press, Cambridge.

Richards M, Corte-Real H, Forster P *et al.* (1996) Paleolithic and neolithic lineages in the European mitochondrial gene pool. *Am. J. Hum. Genet.* **59**, 185–203.

Richards M, Macaulay V (2001) The mitochondrial gene tree comes of age. *Am. J. Hum. Genet.* **68**, 1315–1320.

Richards M, Macaulay V, Hickey E *et al.* (2000) Tracing European founder lineages in the near eastern mtDNA pool. *Am. J. Hum. Genet.* **67**, 1251–1276.

Richards M, Macaulay V, Sykes B, Pettitt P, Hedges R, Forster P, Bandelt HJ (1997) Paleolithic and neolithic lineages in the European mitochondrial gene pool - Reply. *Am. J. Hum. Genet.* **61**, 251–254.

Richards M, Macaulay V, Torroni A, Bandelt HJ (2002) In search of geographical patterns in European mitochondrial DNA. *Am. J. Hum. Genet.* **71**, 1168–1174.

Richards M, Sykes B (1998) Evidence for Paleolithic and Neolithic gene flow in Europe – Reply. *Am. J. Hum. Genet.* **62**, 491–492.

Richards MB, Macaulay VA, Bandelt HJ, Sykes BC (1998) Phylogeography of mitochondrial DNA in western Europe. *Ann. Hum. Genet.* **62**, 241–260.

Roberts RG, Flannery TF, Ayliffe LK *et al.* (2001) New ages for the last Australian megafauna: continent-wide extinction about 46,000 years ago. *Science* **292**, 1888–1892.

Rosser ZH, Zerjal T, Hurles ME *et al.* (2000) Y-chromosomal diversity within Europe is clinal and influenced primarily by geography, rather than by language. *Am. J. Hum. Genet.* **67**, 1526–1543.

Salamini F, Ozkan H, Brandolini A, Schafer-Pregl R, Martin W (2002) Genetics and geography of wild cereal domestication in the Near East. *Nature Rev. Genet.* **3**, 429–441.

Salas A, Richards M, De la Fe T *et al.* (2002) The making of the African mtDNA landscape. *Am. J. Hum. Genet.* **71**, 1082–1111.

Scozzari R, Cruciani F, Pangrazio A, Santolamazza P, Vona G, Moral P, Latini V, *et al.* (2001) Human Y-chromosome variation in the Western Mediterranean area: Implications for the peopling of the region. *Hum. Immunol.* **62**, 871–884.

Semino O, Passarino G, Brega A, Fellous M, Santachiara-Benerecetti AS (1996) A view of the Neolithic demic diffusion in Europe through two Y chromosome-specific markers. *Am. J. Hum. Genet.* **59**, 964–968.

Semino O, Passarino G, Oefner PJ *et al.* (2000) The genetic legacy of Paleolithic *Homo sapiens sapiens* in extant Europeans: a Y chromosome perspective. *Science* **290**, 1155–1159.

Sherratt AG (1981) Plough and pastoralism: aspects of the secondary product revolution. In: *Pattern of the Past: Studies in Memory of David Clarke* (eds I Hodder, G Isaac,

N Hammond). Cambridge University Press, Cambridge, pp. 261–305.

Simoni L, Calafell F, Pettener D, Bertranpetit J, Barbujani G (2000a) Geographic patterns of mtDNA diversity in Europe. *Am. J. Hum. Genet.* **66**, 262–278.

Simoni L, Calafell F, Pettener D, Bertranpetit J, Barbujani G (2000b) Reconstruction of prehistory on the basis of genetic data. *Am. J. Hum. Genet.* **66**, 1177–1179.

Sinclair PJJ, Shaw T, Andah B (1993) Introduction. In: *The Archaeology of Africa: Food, Metals and Towns* (eds T Shaw, P Sinclair, B Andah, A Okpoko). Routledge, London, pp. 1–31.

Sokal RR, Oden NL, Wilson C (1991) Genetic evidence for the spread of agriculture in Europe by demic diffusion. *Nature* **351**, 143–145.

Tadmor Y, Fridman E, Gur A *et al.* (2002) Identification of *malodorous*, a wild species allele affecting tomato aroma that was selected against during domestication. *J. Agric. Food Chem.* **50**, 2005–2009.

Tishkoff SA, Williams SM (2002) Genetic analysis of African populations: human evolution and complex disease. *Nature Rev. Genet.* **3**, 611–621.

Torroni A, Bandelt HJ, Macaulay V *et al.* (2001) A signal, from human mtDNA, of postglacial recolonization in Europe. *Am. J. Hum. Genet.* **69**, 844–852.

Torroni A, Richards M, Macaulay V *et al.* (2000) mtDNA haplogroups and frequency patterns in Europe. *Am. J. Hum. Genet.* **66**, 1173–1177.

Troy CS, MacHugh DE, Bailey JF *et al.* (2001) Genetic evidence for Near Eastern origins of European cattle. *Nature* **410**, 1088–1091.

Underhill PA, Shen P, Lin AA *et al.* (2000) Y chromosome sequence variation and the history of human populations. *Nature Genet.* **26**, 358–361.

Vansina J (1995) New linguistic evidence and 'the Bantu expansion'. *J. Afr. Hist.* **36**, 173–195.

Vigouroux Y, McMullen M, Hittinger CT *et al.* (2002) Identifying genes of agronomic importance in maize by screening microsatellites for evidence of selection during domestication. *Proc. Natl Acad. Sci. USA* **99**, 9650–9655.

Vilà C, Leonard JA, Gotherstrom A *et al.* (2001) Widespread origins of domestic horse lineages. *Science* **291**, 474–477.

Wang RL, Stec A, Hey J, Lukens L, Doebley J (1999) The limits of selection during maize domestication. *Nature* **398**, 236–239.

Whitt SR, Wilson LM, Tenaillon MI, Gaut BS, Buckler ES (2002) Genetic diversity and selection in the maize starch pathway. *Proc. Natl Acad. Sci. USA* **99**, 12959–12962.

Zeder MA, Hesse B (2000) The initial domestication of goats (*Capra hircus*) in the Zagros mountains 10,000 years ago. *Science* **287**, 2254–2257.

Zilhão J (2001) Radiocarbon evidence for maritime pioneer colonization at the origins of farming in west Mediterranean Europe. *Proc. Natl Acad. Sci. USA* **98**, 14180–14185.

CHAPTER ELEVEN

Into new found lands

11.1 Introduction

In Chapter 9 we discussed the evidence for a recent origin of **anatomically modern humans** in Africa. We also saw how, 60–40 KYA, modern humans settled Eurasia and Australia. Since that time, settlement patterns have not been fixed. Clearly not all areas presently occupied by humans were settled during this primary exodus from Africa. These 'new found lands' include the Americas, islands in the Pacific and elsewhere, and the extreme northerly latitudes of Eurasia. In contrast, some areas previously settled by humans are now no longer inhabited. As we shall see, this is primarily due to rising sea levels submerging low-lying **continental shelves.**

In this chapter we will explore the evidence, both genetic and nongenetic, for the settlement of these new lands. We will be asking when it was accomplished and by whom. We will investigate the role of changing sea levels in influencing human dispersals, and then consider two case studies that have captured the imagination of generations of prehistorians:

▶ the settlement of the Americas;

▶ the **colonization** of the Pacific.

These new found lands, and the act of reaching them would have exerted novel selective pressures on migrating populations. In addition, the **founder effects** that inevitably accompanied such migrations will have also shaped modern genetic diversity. We shall also encounter the evidence for these two processes, as well as their implications, during this chapter.

11.1.1 Sea level changes since the out-of-Africa migration

In Section 9.5.1 we saw that over the past 100 KY the global climate fluctuated considerably and was appreciably cooler than the past 10 KY of stable warmth. We also saw that during this colder period, far more of the world's water was tied up in huge **ice sheets** at extreme latitudes, and that sea levels were therefore much lower than they are at present. Sea levels determine both the amount of available land, and the accessibility of that land. Hence, we shall explore in more detail the nature of sea level change.

The proportion of the world's water tied up in ice sheets is only one (**eustatic**) factor that determines sea levels (*Figure 11.1*). Two other important factors are described below.

▶ Regional (**isostatic**) changes to the Earth's crust that can either alter the volume of an oceanic basin, or the height of a landmass. The Earth's crust lies on top of the liquid interior of our planet in a set of tiled tectonic plates. Consequently the weight of water in oceanic basins and ice caps on landmasses causes the crust to be depressed. This effect is most clearly shown in the 'rebound' of land previously covered by the ice caps. For example, Chesapeake Bay in northeast USA is still rising by as much as 3–4 mm per year as a result of the removal of this weight.

▶ The thermal expansion of water. At higher temperatures water becomes less dense and the same mass of water occupies a larger volume.

Figure 11.2 shows the approximate sea levels over the past 40 KY (Lambeck *et al.*, 2002). The lowest sea levels existed during the last glacial maximum (LGM), ~ 20 KYA. Sea level rises since the LGM have not been smooth, since sudden thaws of the ice sheets released huge volumes of water that resulted in rapid sea level rises known as **meltwater pulses**. These can be observed to occur at the same time in sea level records from around the world (e.g., radiocarbon dated remains of submerged corals or mangrove swamps, which were previously close to sea level). Two meltwater pulses are thought to have bracketed the cold period, known as the Younger Dryas, that occurred between 13 and 11.5 KYA. Rising sea levels finally tailed off, about 6 KYA, after having risen approximately 120 meters over roughly 10 KY.

To put these changes into perspective, it has been predicted if all remaining glaciers (> 160 000), ice caps (70) and ice sheets (two) were to melt, present sea levels would only rise by about 70 meters (IPCC, 2001).

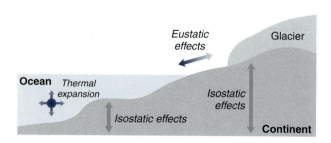

Figure 11.1: Three factors influencing sea levels.

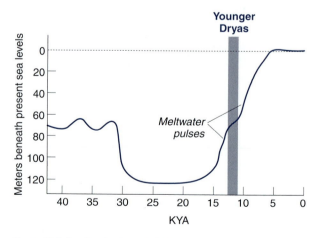

Figure 11.2: Sea levels since 40 KYA.

The Younger Dryas cold period interrupts the two meltwater pulses at the end of the last ice age. Redrawn from IPCC (2001) and Lambeck *et al.* (2002).

The global average sea levels shown in *Figure 11.2* can be used to predict the approximate positions of coastlines at any time in the past 40 KY from an accurate map of modern **seafloor topography** (Smith and Sandwell, 1997). More precise reconstructions would require that regional isostatic effects are taken into account, but these are only known accurately for a few locations worldwide. *Figure 11.3* shows the approximate extent of additional exposed land during the LGM, when sea levels were 120 meters lower than today, and at the end of the Younger Dryas ~ 11.5 KYA, when sea levels were about 50 meters lower. Rising sea levels would have had the greatest effect on continental shelves with shallow gradients, where a small change in sea level could cause a major change in landmass.

It is clear that four areas of the world were especially prone to large changes in landmass as a result of rising sea levels (Peltier, 1994), they are:

▸ the English Channel and the North Sea which, at the time of the LGM and until about 7–8 KYA, formed a land bridge between Britain and the European continent;

▸ island Southeast Asia, which during the LGM formed a large landmass known as **Sunda**;

▸ the Bering Straits between Alaska and Eastern Siberia, which formed a land bridge between the Old and New Worlds, sometimes known as **Beringia**;

▸ the **Sahul** landmass, which consisted of Australia, Tasmania and New Guinea.

These dramatic changes in sea levels would have precipitated rapid changes in settlement patterns and perhaps population crashes (Oppenheimer, 1998); many **hunter–gatherers** occupied coastal niches. Despite the apparently recent advent of malaria, which may have forced tropical populations to higher altitudes, modern populations are still disproportionately skewed towards coastal regions (Cohen and Small, 1998).

Recent evidence shows that about 14 KYA, sea levels may have risen by as much 16 meters per century (Hanebuth *et al.*, 2000). By comparison, in our admittedly more densely populated modern world, a sea level rise of only 1 meter over the next century (at the upper end of recent predictions) would result in the likely relocation of 140 million people (10% of the population) in China and Bangladesh alone (IPCC, 1995).

11.1.2 What drives new colonizations of uninhabited lands?

The agricultural dispersals described in Chapter 10 were greatly facilitated by a period of unprecedented climatic stability, and almost exclusively occupied lands previously settled by hunter–gatherers. By contrast, since the out-of-Africa migrations, changing environments have facilitated the settlement of previously uninhabited lands by a number of means.

▸ Retreating ice sheets reveal land bridges leading to uninhabited landmasses; however, in some cases sea level rises mean that these land bridges are only passable during a limited time window (e.g., Beringian land bridge).

▸ Warmer climates allow flora and fauna to follow retreating ice sheets into more extreme latitudes. Humans could then have extended their range without needing to change their diet. In theory, this could allow the mixing of populations that were previously confined to geographically restricted southerly **glacial refugia**. However, there are differences of opinion as to whether these refugia were indeed isolated from one another. It has been proposed that some environments could have been buffered against climate change (Tzedakis *et al.*, 2002), and others have suggested that certain topographical features in more northerly latitudes might form hospitable microclimates resulting in the survival of small 'cryptic refugia' (Stewart and Lister, 2001).

Thus far we have primarily considered external factors that determine when and where humans might migrate. Clearly, however, humans occupy a far broader range of environments than any other mammalian species by virtue of our behavioral adaptability and capacity for technological innovation. These cultural changes can facilitate the settlement of previously uninhabited lands by:

▸ allowing previously inaccessible lands to be reached, perhaps by use of a new means of transport or navigation (e.g., the double-hulled ocean-going canoes of Polynesians);

▸ enabling previously inhospitable conditions to be tolerated, through changes in diet, lifestyle and construction of shelters (e.g., the construction of stone houses on treeless islands).

11.2 Peopling of the Americas

The origins of the Americans have attracted enormous attention and controversy, focusing particularly on two questions:

▸ when did humans first enter the Americas and where did they come from?

▸ how many major migrations were there?

Despite the controversy about the time of first entry, which is discussed below, it is agreed that the Americas were the last of the inhabited continents to be colonized by humans, and that the time was after 40 KYA; possibly much later than this. A glance at a map of the world reveals the proximity of eastern Siberia and western Alaska, and this is generally thought to be the direction of entry. Two possible routes have been proposed: by land or along the coast (Section 11.2.1 and *Box 11.1*). Alternatives, such as travel across the Northern Atlantic, have been suggested, but are much less

Coastlines with sea level −120 m

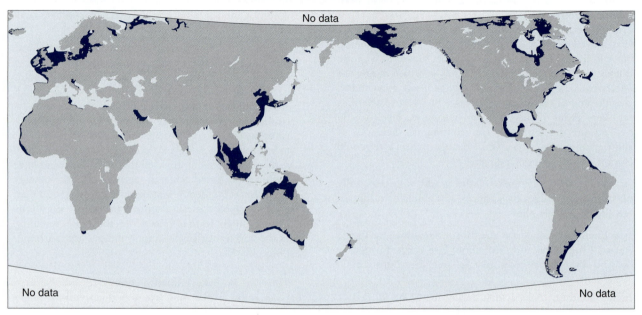

Coastlines with sea level −50m

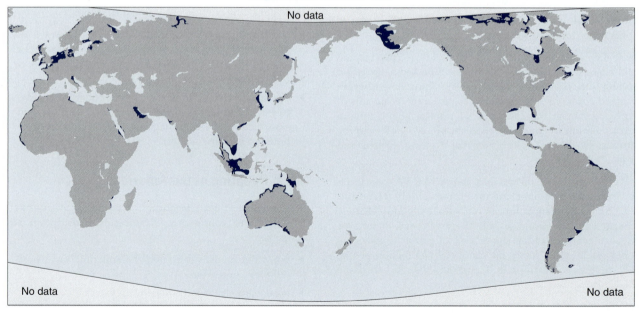

Figure 11.3: Approximate coastlines during the LGM and at the end of the Younger Dryas.

Present day continents are shown in gray; dark blue denotes the extra landmass at lower sea levels. Bathymetric (sea depth) data from Smith and Sandwell (1997).

plausible. Sources of evidence for the peopling of the Americas, as for other parts of the world (Chapter 9), are information about the environmental conditions, linguistics, paleontology and archaeology, and genetics; moreover, the development and first applications of several methods have been driven by the desire to answer questions about human entry into the Americas.

> **BOX 11.1 What was the route of entry into North America?**
>
> Two routes have been proposed (see *Figure 11.5*):
>
> ▶ by land across the Beringian land bridge and down the ice-free corridor;
>
> ▶ by sea along the coast of Siberia, Beringia and North America.
>
> The question is important because sections of the land route were only passable during limited periods, and so constrain the possible dates of entry. In contrast, the coastal route does not impose such obvious constraints. The Beringian land bridge was most extensive when most ice was present in ice caps, but this consequently impeded movement further into North America; the simultaneous presence of both the land bridge and the ice-free corridor was thus limited to the time before 36 cal KYA and a short period near the end of the ice age (*Figure 11.4*). People could have moved into Alaska in between these dates, and could thus have been present in North America, but, according to this model, would not have been able to penetrate far inland at this time. Such a land route would have been compatible with a 'Clovis-first' model, but not with a human presence a few thousand years earlier. Moreover, there is no direct evidence for use of the ice-free corridor by early humans.
>
> Analysis of the northwest coastal environment is complicated by the sea level changes that have taken place: the late Pleistocene coastline is now under water and difficult to investigate for archaeological sites. Nevertheless, human remains from near the coast dated to 11.0 cal KYA (Section 11.2.2), and stone tools recovered from under water off the coast of British Columbia from deposits 10.6 cal KY old (Josenhans *et al.*, 1997), demonstrate human presence in the early postglacial period. Earlier use of this route during the glacial period seems possible and is an important focus for future research.

11.2.1 The environment

Climate during the period of interest was dominated by the Ice Age, with the full glacial period or LGM around 18–21 calibrated KYA (15–18 radiocarbon KYA; for a discussion of the difference between calibrated and radiocarbon years, see *Boxes 9.4* and *11.2*. Dates in this chapter will be calibrated unless otherwise stated, but at the risk of boring the reader, we will still emphasize this by specifying 'cal KYA'). In addition to the low temperature, and its consequences for plant and animal life, there were significant effects on sea level (*Figure 11.3*). North America was joined by land to Asia by the **Beringian land bridge** situated between Siberia and Alaska where the Bering Straits are now found. This land bridge was present between about 65 and 36 cal KYA, and again between 30 and perhaps as recently as 13 cal KYA (*Figure 11.4*), and reached a maximum width of about 2000 km (~ 1200 miles). The land in this region was not covered by an ice cap, but was probably tundra, dominated by grass, sedge and *Artemisia* (a genus that includes sagebrush) and inhabited by a rich fauna including mammoths, bison, horses and camels. It would thus have been possible for humans adapted to such an environment to walk from Asia to North America. There is, however, no certainty that this land route was used by the ancestors of modern Native Americans: an alternative would have been to travel along the coast using boats (see *Figure 11.5* and *Box 11.1*). On land, for much of the **Pleistocene**, any travelers would then have encountered a major obstacle: an ice sheet covering a large part of North America, which would have formed an impenetrable barrier. At times, it was divided into eastern (Laurentide) and western (Cordilleran) sections by an ice-free corridor stretching from present-day Yukon, through Canada, to Montana. While the environment in this corridor would still have been inhospitable, it constitutes one possible route into the rest of the Americas. It is thought that this corridor was open before the LGM at 55–30 cal KYA and again after 13.5 cal KYA (Burenhult, 1993; Mandryk *et al.*, 2001).

11.2.2 Fossil and archaeological evidence

Fossils

Human fossils provide unequivocal evidence for the presence of humans in the Americas, and those associated with reliable dates give valuable information about the timing of the early peopling of these regions and the morphology of the colonists. Their study also raises ethical and, in some countries, legal issues, which can be complicated by the difficulty of establishing whose ancestors any fossil is most likely to represent. The number of early fossils is small, but they often attract much attention, so that some are now well known. Examples (*Figure 11.6*) include:

▶ 'Luzia', possibly the earliest, named after the famous *Australopithecus afarensis* fossil 'Lucy' (Section 8.2.2). Luzia's skull and some additional bones were found in 1975 in the Lapa Vermelha rock shelter in Minas Gerais, Brazil, but the date of 13.5 cal KYA was obtained more recently from associated material, so there is some doubt about how closely this corresponds to the date of the bones, which contained insufficient collagen for direct dating.

▶ 'Buhl Woman' was found in a quarry near Buhl, Idaho in 1989 with artifacts including an obsidian biface and a bone needle, suggesting a burial. She was judged to be a healthy 17–21-year-old and the cause of death was unclear. Dating of bone collagen gave ~ 12.9 cal KYA. She was reburied in 1991, after only limited study, in accordance with the Native American Graves Protection and Repatriation Act.

▶ 'Prince of Wales Island Man', discovered in 1996, is represented by a lower jaw, vertebrae and pelvis found with a stone point in a cave near the coast of Prince of Wales Island, Alaska. The remains were studied by scientists in association with the Tlingit and Haida people, and a fragment of the jaw was dated to ~ 11 cal

BOX 11.2 Radiocarbon and calibrated dates

We emphasize again that dates for fossils or archaeological remains are often given in 'radiocarbon years' rather than 'calibrated' or 'calendar years' (*Box 9.4*). The difference for the time period discussed here can be as much as 2 KY, so that a radiocarbon date of 11 KYA corresponds to a calibrated date of 13 KYA (see *Figure*). Note also that it is often unclear whether dates encountered in the literature are calibrated or not.

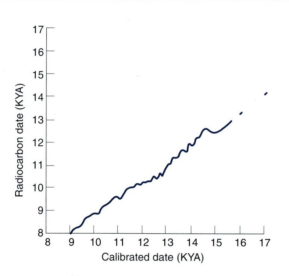

Calibration curve for radiocarbon dating.

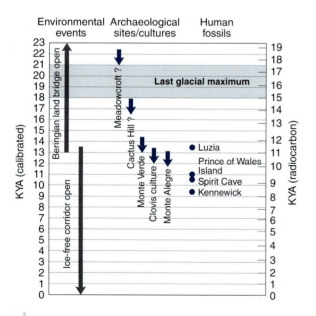

Figure 11.4: Chronology of events and sites relevant to the peopling of the Americas.

Geological and climatic events determined the environmental background. Selected archaeological sites and human fossils are included. Note the different radiocarbon (right) and calibrated (left) timescales.

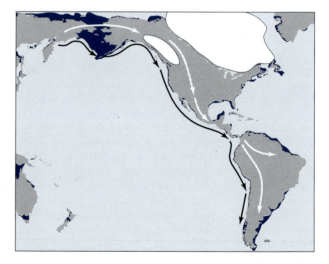

Figure 11.5: Suggested routes into the Americas.

Colonists could have traveled by land (white arrows) across the Beringian land bridge into Alaska and then down the ice-free corridor, although these two routes were not usually available at the same time. Alternatively, they could have traveled along the coast (black arrows), using land or sea. Scenarios in which travel is inland for some of the distance and coastal for the rest are also possible. Gray, present coastline; dark blue, coastline during LGM; white, ice caps.

KYA. Such an early date so close to the coast provides support for the use of coastal routes for early travel in the Americas, and perhaps for the initial entry.

▶ 'Spirit Cave Man' is represented by a burial of a short man aged 40–45 years with signs of dental abscesses and several injuries, partially mummified by the dry conditions and thus providing unusual insights into ancient perishable materials. He was found in 1940 in a dry rockshelter in Nevada and was wearing moccasins and wrapped in a skin robe and two tule (bulrush) mats. His bones were also dated recently to 10.6 cal KYA. Examination of his stomach contents showed that his last meal included fish.

▶ 'Kennewick Man' (*Figure 11.7*) was discovered in 1996 eroding from the banks of the Columbia River in Washington State and has been the subject of considerable controversy and legal action, which have led to his wide notoriety. Initially classified as a modern skeleton, it was only after a stone projectile point was discovered in the right pelvis bone that ^{14}C dating and DNA analysis were initiated. The radiocarbon date was ~ 8.4 KYA, corresponding to 9.3–9.5 cal KYA. The skeleton was almost complete, and was that of a 40–55-year-old man who had suffered numerous injuries in life, including fractures of six ribs and atrophy of the left arm. Controversy focused on his 'European' skeletal morphology: was he truly ancestral to modern Native Americans? Attempts to amplify DNA by several labs have so far been unsuccessful (http://www.cr.nps.gov/aad/kennewick/index.htm). Differences between the early inhabitants of the Americas and modern Native Americans are discussed in *Box 11.3*.

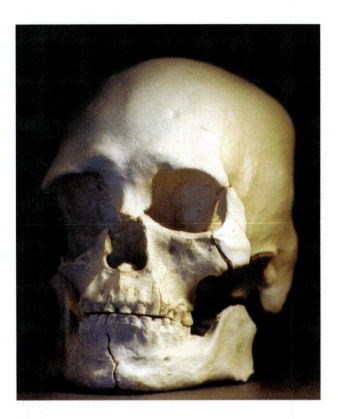

Figure 11.7: Kennewick Man.

A near-complete skeleton of a 40–55-year-old man dating to 9.5–9.3 KYA. Considerable controversy has accompanied this find, including disagreements about the morphological similarities to modern populations. Attempts to analyze DNA have so far been unsuccessful. Reproduced with permission from Associated Press.

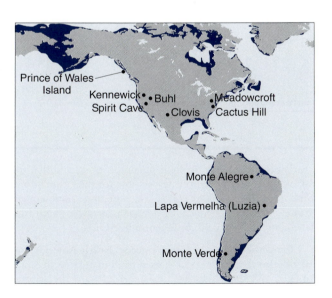

Figure 11.6: Locations of selected fossil and archaeological sites.

Surprisingly, the earliest widely accepted site (Monte Verde) and human fossil (Luzia) are both in South America. Gray, present coastline; dark blue, coastline at LGM.

Archaeological remains

The Clovis culture is universally accepted as clear evidence for humans in the Americas by about 13.5 cal KYA. There have been many claims for 'pre-Clovis' remains, and these have often been highly controversial, although some pre-Clovis sites are now widely accepted. We will therefore start by considering Clovis.

Clovis and the Paleoindians

By 14 cal KYA, the North American ice sheet was receding and parts of central North America supported an extensive megafauna including mammoths and bison. Soon after, the Clovis cultural complex appeared in the archaeological record. It is named after the town of Clovis in New Mexico, near which one of the first sites to be investigated, Blackwater Draw, is located (*Figure 11.6*). It is characterized by **Clovis points** (*Figure 11.8*), fluted (grooved) projectile points, often finely made and beautiful objects with straight sides and the flutes removed from their bases, constructed from a variety of stone types. The earliest Clovis remains date from around ~ 13.5 cal KYA, and the culture appears to have spread over much of the nonglaciated part of North America within a few hundred years, but was soon replaced by other

BOX 11.3 Did the first colonists go extinct?

The early (> 8 KYA) Paleoindian skeletal material is often morphologically distinct from modern Native Americans: an observation reflected in the media by descriptions of Kennewick Man as 'European' in appearance, or Luzia as 'Australian'. Attempts have been made to summarize the differences by describing the early individuals as having long (measured front to back) skulls with narrow faces, prominent noses and inconspicuous cheekbones, in contrast to modern Native Americans who have more rounded skulls, flatter broader faces, smaller noses and more widely flaring cheekbones. However, there are so few reliably dated ancient crania, and they are so diverse, that it may be unreasonable to regard them as a single group. Morphometric comparisons (see *Figure*) of modern population samples with ancient fossils (Jantz and Owsley, 2001) illustrate the diversity of the four oldest crania used in that study (which included Spirit Cave), and, for three of them, their difference from most modern Native American populations: they show more similarity to Inuit (a surprising observation whose significance is unclear), Polynesians or Ainu. There is evidence that some of these morphologically diverse populations survived into historical times, for example, in the isolated Baja California Peninsula in Mexico, where skulls dating between 2.8 and 0.3 KYA resembled Paleoindians (González-José *et al.*, 2003).

It thus appears that there are real differences in morphology, but what do these tell us? Cranial morphology is influenced by environmental as well as genetic factors (see Section 13.2.1). Lifestyle and diet differed significantly between the late Pleistocene/early Holocene and more recent times, even before European contact. It would thus be premature to conclude that the initial colonists were genetically distinct: ancient DNA analysis is needed for this, and would be of considerable scientific interest. It would be even more rash to deduce that there were different origins or migrations for these first settlers. More data are needed.

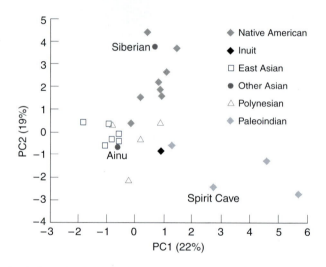

Morphometric analysis comparing individual early North American crania with modern population samples.

Data from Jantz and Owsley (2001).

styles of points. These include Folsom (~ 12.9–12 cal KYA), distinguished by their smaller size and longer flutes.

The Clovis people were undoubtedly big game hunters and their points are found associated with, or occasionally embedded in, the bones of large animals. Mammoths were a common prey species, but **mastodons** and smaller animals were also hunted. The Folsom people are particularly associated with bison. However, it is likely that all these **Paleoindians** ate a wide variety of foods that could have included small game, fish, shellfish and plants. These may be less well preserved in archaeological sites, so the relative importance of the different resources can be difficult to assess. It has been suggested that the wide and rapid spread of the

Clovis people was due to their pursuit of prey, and that the Paleoindians may have been responsible for the extinction of some prey species (*Box 11.4*), necessitating a subsequent change in lifestyle. Abundant evidence is available about later developments, but will not be considered here.

Clovis points are not found outside North America, but Paleoindian remains are also known from many sites in South America and these sites are characterized by a diversity of artifacts including fishtail, willow-leaf and triangular stemmed points. Some are contemporaries of Clovis or Folsom, including those from the cave 'Caverna da Pedra Pintada' ('Cave of the Painted Rock') near Monte Alegre in Amazonian Brazil. This site contains rock paintings and

Figure 11.8: Clovis and related points.

After ~ 13.5 KYA, finely made points were manufactured in North America. They include (left to right) the Clovis, Folsom, Scottsbluff and Hell Gap cultures. Reproduced with permission from Blackwell Science Ltd.

yielded biological remains indicating the use of Brazil nuts, fish, turtles and mussels, as well as triangular points made of quartz and chalcedony that could have been used for hunting larger animals (Roosevelt *et al.*, 1996). Radiocarbon dates from single plant specimens suggested that the initial occupation occurred from ~ 11.2 to 10.5 radiocarbon KYA (~ 13.2–12.5 cal KYA), only ~ 300 years after the earliest Clovis dates. Thus by ~ 13 cal KYA, there were distinct Paleoindian populations with varied lifestyles in both North and South America.

Pre-Clovis sites

The abundance of Clovis and other Paleoindian remains, contrasted with the difficulty of reliably identifying pre-Clovis sites, led to the 'Clovis-first' hypothesis: the idea that the Clovis people were the first humans in the Americas. Work over the last few decades, however, has persuaded the majority of investigators that earlier sites do exist. The most widely accepted of these is Monte Verde in southern Chile, a wet site in an upland bog containing stone implements and also organic remains such as charcoal and wooden tools, and even chewed leaves, and also bones with soft tissue adhering (Meltzer, 1997). The earliest dates obtained were around 14.5 cal KYA, a thousand years earlier than Clovis. In North America, the Meadowcroft rock shelter in Pennsylvania has yielded some 20 000 stone flakes, 1 million animal remains and 1.4 million plant remains. The **stratigraphy** dates to > 30 cal KYA, with the oldest signs of human occupation at 22–23 cal KYA, but critics suggest that contamination of the dated material with natural carbon may produce artifactually old dates. Other claims for early sites include Cactus Hill, Virginia (~ 18 cal KYA), but again critics are not convinced that the site is free from disturbance that might associate the remains signaling human occupation with the earlier material used for dating.

Unresolved issues

More work is required to validate these and other pre-Clovis sites, and major uncertainty still surrounds the question of when humans first entered the Americas. Was it a short time before Clovis, or much earlier? The main reason why a very early date (before 20 cal KYA) appears unlikely is that

---

BOX 11.4 Why did the megafauna go extinct?

At ~ 14 cal KYA, some 35 genera of large mammals became extinct, including mammoths, mastodons, camels, horses, giant ground sloths, bears and saber-toothed cats. Suggested explanations include:

▶ climate change;

▶ disease, perhaps introduced by humans;

▶ 'overkill' by human hunters.

It has not been easy to distinguish between these possibilities because climatic warming and human entry (or expansion in numbers to become visible in the archaeological record) occurred at about the same time. However, there have been many fluctuations in the climate over the last few million years which did not lead to mass extinctions and there is no good reason why the Pleistocene–Holocene change should have been different, except for the presence of humans. Disease rarely kills all members of a species, as illustrated by the ability of Australian rabbits to survive myxomatosis, and would be unlikely to affect so many different species. In contrast, highly skilled hunters, encountering naïve prey, could rapidly exterminate a large proportion of the species, according to the 'blitzkrieg' overkill hypothesis. Modeling by Alroy (2001) examined the effects of varying parameters such as dispersal rate and competition among prey species, human population growth rate and hunting ability. Most simulations led to a major mass extinction. Thus, as a result of both qualitative and quantitative considerations, it seems likely that humans caused the extinction of the megafauna. Indeed, wherever modern humans have encountered naïve animals, including in Australia, Madagascar and New Zealand, mass extinction has followed. Only the African megafauna, which evolved in contact with hominids for millions of years, avoided mass extinction: they probably adapted to the gradually increasing hunting skills.

In Alroy's scenarios, extinction follows rapidly after the appearance of humans: the median time in these models from the introduction of 100 humans to the extinction of the prey was 1229 years. If this work is a reliable guide to the consequences of hunters entering a new territory, mass extinction would itself provide a recognizable marker in the fossil record for the appearance of modern humans.

humans entering such a favorable environment would be expected to increase rapidly in numbers and leave obvious traces. Few, if any, such traces have been found, although it is of course possible that future discoveries will reveal more. However, the acceptance of even a single pre-Clovis site, Monte Verde, has major implications. It would have taken a significant amount of time for humans to travel the 10 000 miles from Alaska to Chile (*Figure 11.6*), since people adapted to a cold Arctic environment would have had to move through temperate and tropical regions. How long would this have required? If it was a few thousand years, the ice-free corridor might not have been open. Did these people travel by a coastal route? These questions remain unanswered.

Another issue is the origin of the Clovis point technology. What were its precursors? According to a Clovis-first scenario, we might expect to find these in Siberia a short time before 13.5 cal KYA. However, the Upper Paleolithic technology in Siberia used blades, not flakes, and was thus significantly different. Some have commented on the similarities between Clovis and Solutrean points from southwestern France and the north of Spain (~ 19–24 cal KYA), even proposing contact across the Atlantic. One view is that both Clovis and Solutrean could share a common origin in the earlier Streletskayan culture of the Don region of Russia (Pearson, 1997). However, it is not clear that the similarities represent more than the convergent development of finely crafted points. With the likelihood of a period of several thousand years of pre-Clovis occupation of the Americas, it seems more plausible to propose that the Clovis technology developed in the New World; a critical evaluation of pre-Clovis tools is needed. It is tempting to link the appearance of Clovis points with a local adaptation to big game hunting and a large increase in the human presence, at least as it affects their visibility in the archaeological record.

11.2.3 Linguistic evidence: The three-migration hypothesis

In 1986, Greenberg, Turner and Zegura published a proposal that has been at the center of much of the subsequent debate: in a synthesis of linguistic, dental and genetic evidence, they suggested that three separate migrations had contributed to the settlement of the Americas (Greenberg *et al.*, 1986). Languages were classified into three families:

▶ **Eskimo-Aleut** (the name 'Inuit' is used for the people, but 'Eskimo' for the language), spoken in the far north, as well as in Greenland and parts of Siberia. This family is widely recognized and accepted;

▶ **Na-Dene**, spoken in parts of North America. This family has deeper divisions than Eskimo-Aleut, and Greenberg *et al.* acknowledge that some would question the affiliation between Haida and the other Na-Dene languages, but again it is widely recognized;

▶ **Amerind**, a vast family containing the remaining indigenous languages spoken in North America, and all endemic languages from Central and South America. In contrast to the two previous families, the grouping

of these diverse languages into a single family, with the implication of a common descent from a hypothetical proto-Amerind, is highly controversial among linguists. Some would place them in 70 to 80 different families, with around 80 languages remaining as unclassified isolates (Renfrew, 2000). Suggested examples of unifying Amerind characteristics are **grammatical** elements like personal pronouns prefixed with *n-* in the first person and *m-* in the second person, and **lexical** elements like the root of the word describing children, which has three grades depending on the sex: *t'ina 'son, brother', *t'ana 'child, sibling', *t'una 'daughter, sister' (Renfrew, 2000 p. 169). However, others have pointed out that these features are not universal in Amerind languages and suggested that Greenberg's entire pattern of Amerind correspondences may be due to chance resemblances: many of these would be expected since he examined around 900 languages. It is worth noting that, despite the skepticism of some linguists, this family has proven popular among geneticists, perhaps because of its simplicity. We follow this practice here, because whether or not Amerind languages all descend from a single common proto-language, this grouping represents the residue once the other two better-supported language families have been removed.

Greenberg *et al.* then hypothesize that '*the three linguistic stocks represent separate migrations*', and consider the ability of **glottochronology** (a controversial method of dating the divergence time of two languages by assuming a clocklike rate of change in a basic vocabulary) to provide divergence times for these language families. They favored times of > 11 KY for Amerind, around 9 KY for Na-Dene, and 4 KY for Eskimo-Aleut.

11.2.4 Genetic evidence

Greenberg *et al.* reviewed the genetic evidence available to them, mainly from classical markers, and concluded that it was compatible with their three-migration hypothesis, but provided only '*supplementary rather than primary*' support. Considerably more genetic data are now available and can be used to ask:

▶ which old world populations are most similar to Native Americans?

▶ do Native Americans show reduced genetic diversity?

▶ are there distinct genetic subgroups within Native Americans?

▶ what insights into migration times are provided?

A problem in interpreting the results of genetic analysis in Native American populations, particularly those from North America, is that of pervasive admixture with other populations, including Europeans and Africans, over the last few hundred years (Chakraborty, 1986; see Chapter 12). This admixture has important consequences: it increases genetic diversity, and it introduces foreign lineages that could lead to incorrect conclusions if their origins are not identified. In

addition, the definition of population membership is perhaps less straightforward than in other parts of the world (see *Box 1.2* and Section 9.2.4).

Overall, **classical marker** analyses suggested that Native American populations were most similar to North Asian populations (*Figure 9.14*), consistent with entry into the Americas through Beringia. A more extensive analysis of classical marker data was presented by Cavalli-Sforza and colleagues (Cavalli-Sforza *et al.*, 1994) who calculated **pairwise** F_{ST} values between 20 American populations or groups of populations, and three Siberian populations. Their data (*Figure 11.9*) showed the distinction between the American and Siberian populations, and revealed two clusters within America: one containing the Arctic populations, and the other the remaining North American and all the South American populations. There was thus some correlation with language, since all Eskimo-Aleut-speakers fell into the Arctic cluster and all Amerind-speakers into the second cluster. However, the Na-Dene-speakers did not form a separate third cluster, but resembled their geographical neighbors: the northern Na-Dene fell into the Arctic cluster and the southern Na-Dene into the Amerind cluster. The authors concluded that '*genetic analysis fully confirms the division of American natives into three major clusters*', but this required the assumption of considerable **admixture** between the Na-Dene and their neighbors, and/or similar source populations for Eskimo-Aleut and Na-Dene. An alternative conclusion would be that there are two genetic groups and the Na-Dene do not form a single genetic unit.

A large body of evidence shows that genetic diversity is lower in the Americas than in other continents. For example, a recent study used 377 autosomal microsatellites to characterize 1056 individuals from the HGDP diversity panel

(Section 9.2.3) and found that **heterozygosity** was, on average, lowest in the Americas (Rosenberg *et al.*, 2002; *Figure 11.10*). Although only five American populations were sampled, the large number of loci investigated makes this a robust conclusion. Individual loci show **stochastic** variation, and *CD4* haplotype diversity in the Americas was low, but not quite as low as in the Pacific (*Figure 8.8*). The reduced diversity suggests that a bottleneck, but not an extreme bottleneck, occurred during or after the peopling of the Americas.

mtDNA phylogeography

When it became possible to carry out molecular analysis of mtDNA variation, one of the first applications was to investigate diversity within Native Americans. Most (> 95%) mtDNAs fall into one of four distinct haplogroups; since these were the first haplogroups to be identified, they were designated A, B, C and D. They were initially defined by RFLPs, but these haplogroups are associated with characteristic coding region variants (*Figure 11.11*; the fifth haplogroup is discussed below). The distributions and times to the most recent common ancestor (TMRCAs) of the haplogroups have been of considerable interest. It must be remembered that a TMRCA does not date a migration, but the TMRCA of a lineage shared between two locales must predate a migration between them. While substantial variation is found between individual population samples because of different contributions of the founding lineages, subsequent genetic drift and admixture, and error due to small sample size, a general pattern can still be discerned. Eskimo-Aleut populations mostly have high frequencies of A and some have a significant proportion of D, Na-Dene show

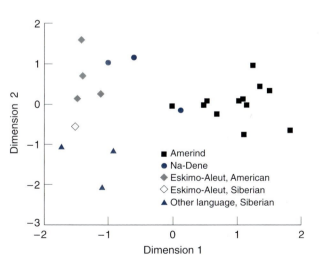

Figure 11.9: Multi-Dimensional Scaling analysis of population pairwise F_{ST} values calculated from classical marker frequencies.

Redrawn from the data in Table 6.9.1 of Cavalli-Sforza *et al.* (1994). For more information on Multi-Dimensional Scaling see Section 6.4.4.

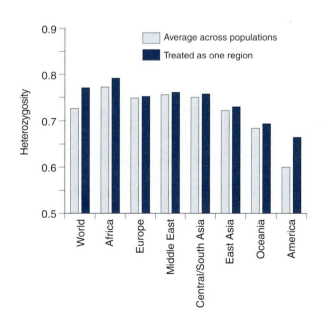

Figure 11.10: Heterozygosities of 377 microsatellites in different continental populations.

Based on data from Rosenberg *et al.* (2002).

mainly A, and Amerind have a more even distribution of all four haplogroups (*Figure 11.12*). Outside the Americas, haplogroups A, C and D are common in Siberia and much of eastern Asia, while haplogroup B is found in more southern parts of east Asia, but is rare in Siberia.

TMRCAs for these haplogroups have been calculated from several sample sets using either coding region or HVS data (*Table 11.1*). Although some studies have found lower levels of variation (and hence a younger TMRCA) in haplogroup B, most have found rather similar diversities and TMRCAs for the four haplogroups: for example, a recent study using coding region sequence estimated TMRCAs of between 15 and 28 cal KYA for all haplogroups.

In a variant of this approach, Forster *et al.* identified founder sequences (*Box 6.5*) by comparing 472 American mtDNA control region sequences with 309 Asian sequences and used the statistic ρ (Section 6.6.2) to measure the diversity that had accumulated within each founding lineage, averaging across the haplogroups within each population (Forster *et al.*, 1996). These ρ values then needed to be calibrated in order to convert them into times measured in years, and this was done by assuming that the diversity within haplogroup A2 in Siberians, Eskimos and Na-Dene had developed since the Younger Dryas 13.3 KYA, when Arctic regions were likely to have been depopulated. ρ values for North, Central and South Amerinds were between 1.4 and 2 times higher than for the A2 average, and so the Amerind founding event was placed at 16–23 KYA. Note that this is

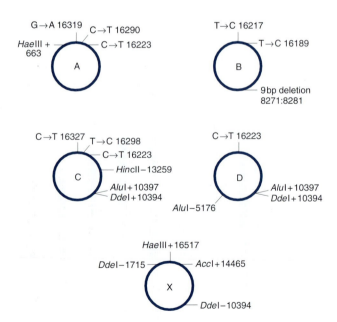

Figure 11.11: Polymorphisms used to identify the five American mtDNA haplogroups A, B, C, D and X.

These include restriction enzyme site cleavage (indicated as + or −), the 9-bp deletion, and base substitutions. Note that some variants are shared by more than one haplogroup: the complete haplotype defines the haplogroup.

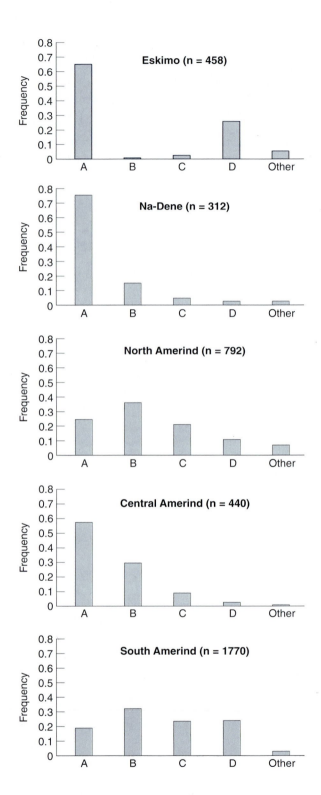

Figure 11.12: Frequencies of the main mtDNA haplogroups A, B, C and D in American populations grouped according to language.

Amerind speakers are subdivided according to geography. Data from Table II of Salzano (2002).

TABLE 11.1: SELECTED TMRCAS OR EXPANSION TIMES FOR mtDNA HAPLOGROUPS A, B, C AND D.

TMRCA [KY (95% CI)]					
A	**B**	**C**	**D**	**Data used and calibration**	**Reference**
26–34	12–15	33–44	18–24	RFLP, 2.9% or 2.2%/MY	Torroni *et al.*, 1994
28 (24–32)	27 (23–31)	28 (23–32)	23 (27–37)	HVS I, 15%/MY	Bonatto and Salzano, 1997
21 (16–25)	18 (15–28)	22 (17–26)	24 (19–28)	8.8-kb coding region	Silva *et al.*, 2002

the range of three point estimates, not a 95% confidence interval.

Haplogroup X

A fifth haplogroup, haplogroup X, has subsequently been identified as another probable founding lineage. It is present only at low frequency in the Americas (~ 3%), where it has been detected in North American Amerind and Na-Dene (but not Eskimo-Aleut) populations, but not thus far in Central or Southern Americans (Brown *et al.*, 1998). Its geographical distribution in the rest of the world also differs from that of haplogroups A–D in that it has been found in Europe at low frequency, but was initially not detected in Asia. More recently, it was found in **Altaians** from southern Siberia (Derenko *et al.*, 2001), but the overall frequency in Asia according to these two studies is < 0.4%. Moreover, most (21/22) American haplogroup X mtDNAs were distinct from both Asian and European haplogroup X sequences, showing an A at position 16213 and a G at 200. Estimates of the TMRCA of the American subgroup calculated using the ρ measure (see Section 6.6.2) ranged from 31 ± 4 to 36 ± 7 KYA (Brown *et al.*, 1998). Although there is considerable uncertainty associated with these dates and their interpretation (was there one haplogroup X founder in the Americas, or many?), the time and geographical specificity suggest a prehistoric, rather than historic, origin, a conclusion supported by the finding of haplogroup X sequences in pre-Columbian skeletons (e.g., Stone and Stoneking, 1998). The geographical origin of the American haplogroup X lineage remains uncertain: the divergence from European haplogroup X sequences makes an origin in Europe unlikely and it seems more probable that the lineage originated in Asia, but it also differs from the few known Asian haplogroup X mtDNAs. Perhaps the American-specific sublineage arose in America; alternatively, it could have arisen in Asia but drifted below the detection threshold or to extinction there.

The few Native American mtDNAs that fall outside these five haplogroups are interpreted as arising from recent admixture; it remains possible that additional indigenous lineages will be found, but, in view of the large sample sizes that have already been examined, any such lineages are likely to be very rare.

Interpretation of the mtDNA data

While the distribution of mtDNAs in the Americas and potential source regions are reasonably well established, the insights they provide into the number and timing of the migrations remain contentious. The starting assumption is that the modern population data provide useful insights into ancient populations from the same area, but this assumption may not be reliable if there is extensive drift, admixture or movement of people. A second factor is that information about molecular lineages provides only indirect information about population origins: the TMRCA of a lineage, for example, does not itself tell us whether the lineage arose in Asia or America.

A glance at *Figure 11.12* shows that the four major haplogroups are found within pooled samples derived from all language groups, although at significantly different frequencies: haplogroup B is very rare in Eskimo-Aleut speakers, for example. Some have emphasized the similarities between the lineages present in the different population groups, while others have emphasized the frequency differences. Thus Torroni and colleagues have used the predominance of haplogroup A in the Na-Dene, and its lower level of variation than in the Amerinds, to argue for a separate and later Na-Dene migration, suggesting entry between 22 and 29 cal KYA for the Amerinds (Torroni *et al.*, 1994). Furthermore, their finding of a lower level of variation within Amerind haplogroup B, together with its different Asian distribution, led them to suggest that two separate migrations may have contributed to the Amerinds: an earlier one introducing haplogroups A, C and D, and a later one introducing haplogroup B. This scenario would thus have involved three migrations: the first two at 34–26 and 15–12 cal KYA, and the third Na-Dene migration at 10–7 cal KYA. The number of migrations is the same as that proposed by Greenberg *et al.*, but the groups involved are not. In contrast, Merriwether, Kolman and others (Kolman *et al.*, 1996; Merriwether *et al.*, 1996) have emphasized the presence of all four haplogroups A–D in populations from North, Central and South America, including all three linguistic groups, together with the scarcity of Asian populations carrying these four haplogroups and the large number of haplogroups in most Asian populations, and argued for a single migration. Multiple migrations from populations containing many haplogroups would be unlikely to sample only the same four each time. Additional evidence comes from the observation of a subset of haplogroup A carrying a C to T transition at position 16 111 that is widespread in the Americas, but rare in Asia. If there were independent migrations, the chance of them carrying the same rare haplogroup A variant each time would be small. A possible Asian source country, where all four haplogroups are found, is Mongolia (*Figure 11.13*) (Kolman *et al.*, 1996; Merriwether *et al.*, 1996). Most recent papers (e.g., Silva *et al*, 2002) have favored a single migration, although subsequent **population differentiation** within the Americas is required to explain the differences in the present distribution of mtDNA lineages, such as the high frequency of A in the north.

Y phylogeography

Early work demonstrated the predominance of a single Y lineage in Native Americans, originally recognized by a combination of a complex alphoid DNA **heteroduplex** pattern and a microsatellite allele (Pena *et al.*, 1995), but subsequently defined more simply by a SNP (the T allele at *DYS199*, also known as M3; Underhill *et al.*, 1996), and now designated haplogroup Q3. Q3 chromosomes are present in all three linguistic groups in the Americas, but are very rare elsewhere, being found only in a few Siberian populations who live near the Bering Strait and could have acquired their Q3 chromosomes by back-migration. This observation does not fit easily with the three-migration hypothesis and, like some interpretations of the mtDNA data, has led to suggestions of a single major migration. The time and place of origin of this lineage are thus of some interest, and are discussed below.

While this lineage is frequent in the Americas, making up 58% of the Y chromosomes according to the combined results of two substantial surveys including both North and South American populations (Karafet *et al.*, 1999; Lell *et al.*, 2002), additional lineages are also present. Do these represent recent admixture, which is likely to be predominantly male-mediated and thus contribute more to the pool of Y chromosomes than to the mtDNAs (Section 12.4), or other indigenous lineages? It is assumed that admixture will be mainly of European (colonist) or African (slave) origin, so lineages found in Native Americans which are frequent in these external populations are ascribed to admixture, while those that are rare or absent from these known sources are candidates for further indigenous lineages. According to these criteria, additional founding lineages are likely to include one or more within each of haplogroups P (which is further subdivided into haplogroups Q and R: *Figure 11.14*), and C (*RPS4Y₇₁₁* T). The status of other lineages remains uncertain, but some of them, e.g., N3 (Tat C), within haplogroup N (*Figure 11.14*) may represent further rare founders.

As with the mtDNA lineages, we can now ask what are the most likely geographical sources for the founding Y lineages. Since Q3 seems to have an origin within Beringia or the Americas, this question comes down to finding sources for the precursor to Q3, any additional subdivisions of P, and C. For haplogroup C, this is relatively simple. Haplogroup C itself is very common in much of East Asia, but when additional information about microsatellite subtypes is taken into account, the most likely source is from the region of Lake Baikal (Karafet *et al.*, 1999; Lell *et al.*, 2002). For the lineages within haplogroup P, however, the question is more complex. This section of the phylogeny is now relatively well resolved (*Figure 11.14*), but, unfortunately, detailed information about the geographical distributions of most of the lineages is not yet available. Early studies using a small number of markers identified central southern Siberia as the most likely source of the lineage that gave rise to Q3 (Karafet *et al.*, 1999; Santos *et al.*, 1999; *Figure 11.13*), but subsequently the American haplogroup P chromosomes have been subdivided into two classes using the marker M173 (Lell *et al.*, 2002; *Figure 11.14*). One class (called M45a, and probably corresponding to Q*) again has a likely origin in central southern Siberia, but the second (M45b, now designated R1; *Figure 11.14*) is suggested to have originated in eastern Siberia. While this region is further east than the Lake Baikal origin suggested for the C lineage, the sparse geographical sampling means that the two lineages may well share the same source. Thus the evidence from the Y chromosome could be interpreted as suggesting two migrations: one from central southern Siberia contributing the major Q* and Q3 lineages that are found throughout the Americas, and a second from eastern Siberia contributing the C and R1 lineages that are restricted, in the Americas, to Northern and Central populations (*Figure 11.13*). This interpretation has, however, been challenged. Tarazona-Santos and Santos (2002) have pointed out that lineage C is found in central Siberia, and that R1, the most frequent haplogroup in many western European populations, could represent recent admixture. Thus they suggest that the Y data can most simply be explained by a single migration from central Siberia plus recent European admixture. All such analyses assume that lineage distributions in modern Siberian populations are representative of those in the same area prior to the settlement of the Americas. Yet there is likely to have been much subsequent change due to migration, admixture, drift and other causes. Indeed, as climate has improved since the first migration from Asia to the Americas, be it ~ 14 KYA or > 20 KYA, it is possible that subsequent repopulation of northern Asian latitudes has obscured the ancestral gene pool of Native Americans. There is some evidence for this

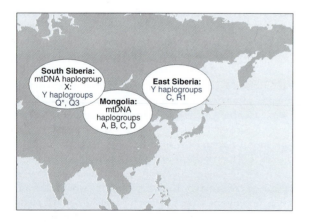

Figure 11.13: Suggested Asian source regions for American mtDNA and Y-chromosomal lineages.

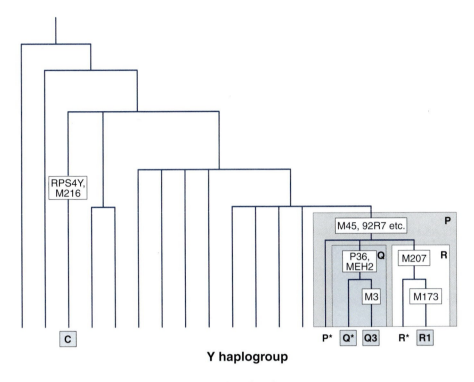

Y haplogroup

Figure 11.14: Y-chromosomal phylogeny relevant to the peopling of the Americas.

Indigenous Native American haplogroups are boxed, although lineage N may represent a further rare founder lineage. Note that additional markers can subdivide some of these haplogroups further.

repopulation scenario from mtDNA studies (Forster *et al.*, 2001). If this has happened, it will be very difficult to identify a source region.

The next question is what can be deduced about the timing of the migrations from the Y data. The wider geographical distribution and higher frequency of the Q*/Q3 lineages in the Americas suggests that they or their precursors were carried by the earlier (or only) migration, with the C and R1 lineages entering simultaneously, a little later or much later. If Q* lineages are found in Siberia and the Americas, but Q3 only in the Americas, the dates of the two mutations defining these lineages bracket the migration. Estimates of these dates have great uncertainty, but taken at face value would place the migration after ~ 20 KYA but

before ~ 10 KYA (*Table 11.2*): consistent with the archaeological data, but doing little to refine it.

Unfortunately, no useful information has yet been provided about the timing of the proposed second migration, since no SNPs define the American-specific C and R1 lineages, and no TMRCAs have been calculated from the microsatellite variation. In principle, such information could be obtained.

Conclusions from the genetic data
The genetic data do not support the hypothesis of three migrations corresponding to the three linguistic groups of Greenberg *et al.* (1986). Most analyses reveal the similarity of American populations to one another and their difference

TABLE 11.2: SELECTED ESTIMATES OF TIMES FOR THE Y-CHROMOSOMAL LINEAGE Q

Event	Data	Method	Time, KYA ± SD or (95% CI)	Reference
P36/MEH2 mutation	SNP	Coalescent-based	17.7 ± 4.8	Hammer and Zegura, 2002
M3 mutation	SNP	Coalescent-based	7.6 ± 5	Karafet *et al.*, 1999
M3 TMRCA	Microsatellite	ASD	11.5 (9.4–13.8)	Ruiz-Linares *et al.*, 1999
M3 TMRCA	Microsatellite	Variance	9.3	Ruiz-Linares *et al.*, 1999

from the rest of the world. Within the Americas, northern populations tend to differ from more southerly ones. This is seen with classical marker frequencies, mtDNA haplogroup frequencies with haplogroup X reported only from the north, and Y haplogroup frequencies with C identified only in the north. Although these patterns might result from a single entry followed by loss of diversity because of drift during subsequent southwards movements, it is tempting to explain the observations by two migrations. However, it is not clear whether all the distinct properties listed for the northern populations have a common origin and thus could all be explained by the same second migration. Nevertheless, an initial migration at the time suggested by the archaeological data, ~ 20–15 KYA, establishing the genetic pattern seen throughout most of the New World, with a second migration at an unknown but later time modifying the pattern seen in North America, would provide a coherent explanation for our current observations.

11.3 Peopling of the Pacific

The origins of Pacific islanders have intrigued researchers from many disciplines since the voyages of Captain Cook in the late eighteenth century. Cook himself was struck by the mutual intelligibility of languages spoken on islands separated by thousands of kilometers, and by the navigational sophistication necessary to voyage between them.

As we shall see, the likely origins of Pacific islanders to the west, in Island Southeast Asia, require that we examine in more detail the changing environment over the past 40 KY. Alfred Wallace (the co-discoverer of natural selection) was one of the first to notice the sharp faunal differences that exist within this region. Consequently, the division between the typically Asian ecology in the west and the distinctly different ecology in the east is known as **Wallace's line**. This line closely corresponds with the Sunda landmass in existence during the LGM, described earlier in this chapter (see *Figure 11.15*). It is even possible to reconstruct a major drainage system on the Sunda continent, known as the Molengraaff river, of which many of the major Indonesian rivers on different islands would have been tributaries (Pelejero *et al.*, 1999). The islands that reside between Sunda and Sahul landmasses are known collectively as **Wallacea**.

The land bridges that joined these islands into their respective landmasses would have been present throughout the period of out-of-Africa colonization events and would only have been submerged about 8000 years ago. Now we shall turn to the fossil and archaeological evidence for first settlement of the Pacific.

11.3.1 Fossil and archaeological evidence

The islands of the Pacific are classified into **'Near Oceania'**, first settled approximately 30 KYA, and **'Remote Oceania'**, first settled within the last 3500 years (Green, 1991b). Near Oceania includes the island of New Guinea and islands lying off its northeast coast. New Guinea formed part of the Sahul landmass when it was first settled, probably by about 45 KYA. The evidence for this settlement is

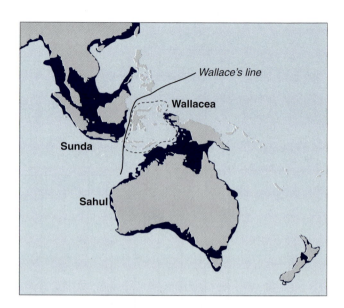

Figure 11.15: Sunda, Sahul and the Wallace line.

The islands of Wallacea lie between the two great landmasses of the last ice age

discussed in Section 9.5. However, the further reaches of Near Oceania, which includes some islands in what are now the Solomon Islands, would have required substantial voyages of 50–100 km. The earliest evidence for human occupation of the Solomon Islands, the majority of which may have been combined into a single island at the time, dates to 29 KYA. There are earlier finds on New Ireland that date to about 35 KYA, which support this early migration eastward from the Sahul continent. These earliest Pacific islanders would have been hunter–gatherers, dependent on local wild resources. The islands of Remote Oceania have substantially fewer floral and faunal resources (in terms of genera and species) than those in Near Oceania.

There is no evidence for human occupation further to the east, where distances between islands increase markedly (> 350 km), until some 25 KY later. One factor that has been emphasized in determining the limit between Near and Remote Oceania is the 'intervisibility' of voyages towards new lands; in other words, the destination can be seen from the island of origin, albeit from a high point on that island in some cases. Alternatively, the destination may be seen from the water, before the island of origin has disappeared from view. It has been estimated that all of the sea voyages accomplished during the settlement of Near Oceania would have been towards visible destinations, whereas this is certainly not true of voyages in Remote Oceania. Voyaging continued among the populations of Near Oceania as shown by trade in **obsidian** from New Britain to other islands from 12 KYA onwards.

The earliest settlements of islands further to the east, in Remote Oceania, are all associated with the same cultural package, known as **Lapita** (Kirch, 1999). This package included the means for agricultural food production, characteristic pottery and tools, and superior voyaging and navigational technologies. The Lapita complex appears to have its origins 3.5 KYA in the islands just off the north coast of New Guinea, in New Britain and the Bismarck archipelago. From here, there is a rapid eastward spread of this culture, taking only 500–600 years to reach Fiji, Samoa and Tonga (*Figure 11.16*).

Pacific islands have traditionally been divided into three groups: **Polynesia** ('many islands') is the area within a triangle with points at Hawaii, New Zealand (Aotearoa) and Easter Island (Rapanui); **Melanesia** ('black islands') is the area to the west of Polynesia, up to and including New Guinea (see *Box 11.5*); and **Micronesia** ('small islands') are the low-lying coral atolls to the north of Melanesia. After the initial spread of the Lapita culture to the western fringe of Polynesia there was a time lag of about 1000 years before a distinctive, Polynesian, culture developed and spread to previously uninhabited islands, first into central Polynesia and then finally dispersed to the periphery – to Hawaii, **Aotearoa** and **Rapanui** – by about 800 YA (see *Figure 11.16*). There is also evidence of back-migration of Polynesian cultures to some of the islands to the west, from whence they came. The timing and origins of first migrations to Micronesia are less well characterized. The western fringe of Micronesia appears to have been settled directly from

Island Southeast Asia around 3 KYA, whereas most of Micronesia was settled slightly later, probably from islands to the south.

After their initial settlement, and despite the great distances between them, the island groups of Polynesia did not become isolated from one another. Archaeological evidence for long distance voyaging comes from the frequent identification of stone tools in one **archipelago** with isotope frequencies diagnostic of rocks from other archipelagos (Weisler and Kirch, 1996).

As with the settlement of the Americas, the archaeological record details the destructive ecological impact of these new arrivals on their various islands. Indeed, dramatic ecological changes are commonly used on Pacific islands as one of the first markers of human settlement. These impacts result from a number of human practices, including:

▶ habitat modification (e.g., land clearance);

▶ introduction of **commensal** and domesticated animals and plants that compete for resources with indigenous species (e.g., rats);

▶ hunting to extinction (e.g., of flightless birds in New Zealand).

These impacts could be very rapid. The absence of ground-dwelling mammals in New Zealand resulted in several bird

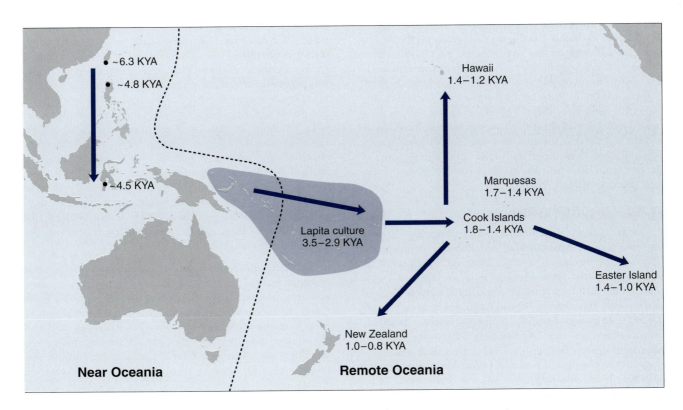

Figure 11.16: Near Oceania, Remote Oceania and the Lapita culture.
The dashed line divides the islands of the Pacific Ocean into Near and Remote Oceania. Labels indicate the earliest dates for agriculture in different locations, and arrows indicate the direction of movement inferred from archaeological dates. The shaded area represents the region occupied by the Lapita culture which spans the boundary between Near and Remote Oceania.

BOX 11.5 Is the category 'Melanesian' meaningless?

The traditional tripartite classification of the islands of the Pacific into Polynesia, Melanesia, Micronesia was based partially on dramatic changes in human phenotypes. Melanesians share darker skin and 'frizzy' hair, whereas Polynesians and Micronesians have a more **'mongoloid'** appearance. As discussed in Section 13.3, skin color can be an unreliable marker of population prehistory due to its interaction with environmental factors. In addition, only Polynesia appears likely to be a coherent grouping with regard to the archaeologicar and linguistic evidence. All languages within this region belong to a single subfamily, 'Polynesian', to the exclusion of other Oceanic languages. By contrast, Melanesia contains islands colonized 30 KYA and others colonized 3 KYA, and peoples speaking many different language families. Micronesia contains islands colonized from the west and islands colonized from the south, and although all are Austronesian speakers, they speak languages belonging to different subfamilies; Oceanic languages are only spoken in central and eastern Micronesia. For these reasons we avoid describing genetic lineages, languages or cultures as having a 'Melanesian origin'. Peoples residing within the geographical area 'Melanesia' are so heterogeneous as to make this description highly ambiguous and capable of almost any interpretation (Green, 1991b). This classification has also focused attention on Polynesians to the detriment of other inhabitants of Remote Oceania.

---

species evolving to efficiently exploit this ecological niche. Loss of the ability to fly was one common consequence among many of these species. This is best exemplified by the 11 species of Moa that existed on New Zealand at the time of the first Polynesian settlements. Some of these birds were over 2 meters tall. Within a few hundred years, all of these species had been hunted to extinction, over an area of 270 000 square kilometers.

A fully developed agricultural package utilizing diverse domesticated species, pottery and complex navigational skills does not tend to develop in isolation on small islands. This raises the question: what are the origins of the Lapita culture? Agriculture in Island Southeast Asia predates the Lapita culture, and appears to arrive from the North, from continental Southeast Asia. The earliest **Neolithic** remains in this region have produced a southward gradient of dates, with the oldest in Taiwan. Elements of the Lapita culture (e.g., the distinctive pottery) clearly owe ancestry to these older Neolithic assemblages. However, **horticulture** has an ancient independent ancestry in New Guinea (see Chapter 10) and other elements of the Lapita culture (e.g., some of the tree crops) appear to have been derived from these more local practices (Green, 1991a).

11.3.2 Linguistic evidence

Languages spoken on Pacific islands are traditionally classified into two groups: **Austronesian** and **Papuan**. Austronesian languages all belong to a single language family and are widely distributed from Madagascar in the west to Easter Island in the east; from Taiwan in the north to New Zealand in the south. By contrast, Papuan languages belong to many different language families, lumped together primarily because they are not Austronesian. As with Amerind languages, Papuan languages represent a residual group that may or may not share a common **proto-language**. Papuan languages have a much more restricted distribution, being confined to New Guinea and a few nearby islands, from Timor in the west to the Santa Cruz group of the Eastern Solomon Islands in the east. The eastern limit of Papuan languages fits closely, but not exactly, with the boundary of Near Oceania, which lies between the Santa Cruz group and

the rest of the Solomon Islands to the west (see *Figure 11.17*). Papuan languages predominate in New Guinea, where Austronesian languages are restricted to coastal areas. The vast diversity of Papuan languages spoken in the New Guinean highlands makes this the most linguistically diverse area on the planet.

The fact that all Austronesian languages belong to a single family, and therefore have a recent common origin, is of itself evidence for a recent spread of these languages over their wide geographical range. By examining the linguistic diversity within Austronesian languages, we can gain further information as to the nature of that spread. Investigating which Austronesian languages share unique innovations (**synapomorphies**) allows subfamilies of related languages to be defined, and a branching pattern similar to a **phylogenetic** tree to be reconstructed. From this evidence it has become apparent that Taiwan contains by far the most diverse set of Austronesian languages found in a single locale, with nine out of the 10 major Austronesian subfamilies confined to this island (Blust, 1999). All remaining Austronesian languages belong to a single subfamily, Malayo-Polynesian. This subfamily can itself be divided into smaller groups of related subfamilies (Pawley, 1999). The branching structure and geographic distribution of these subfamilies is shown in *Figure 11.17*. It can be seen that almost all of the languages of Remote Oceania belong to the Oceanic subgroup of Austronesian languages.

The close fit between the archaeological and linguistic evidence has led to attempts to meld the evidence into a coherent 'archaeo-linguistic' perspective (Pawley and Ross, 1993). It has been suggested that the original inhabitants of Near Oceania spoke non-Austronesian languages, and that the Austronesian languages arrived with farming, which was ultimately from Southeast China, but came via Taiwan. The congruence, in Taiwan, of the greatest diversity of Austronesian languages, together with the oldest farming remains in Island Southeast Asia, supports this scenario, as does the similarity between the limits of Papuan languages and the settlement of Near Oceania. Furthermore, reconstructing elements of the vocabulary that must have been present in the ancestral 'proto-Austronesian' language

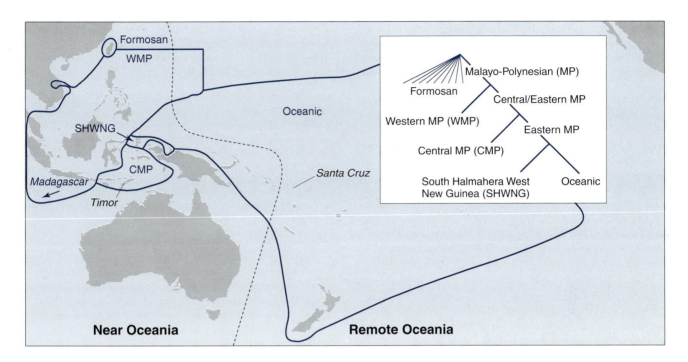

Figure 11.17: Tree of Austronesian languages and their geographical distribution.

The map shows the distribution of different Austronesian language groupings in blue, relative to boundary between Near and Remote Oceania. The inset shows the ancestral relationships between these groupings. Timor and Santa Cruz (black labels) indicate the western and eastern limits of Papuan languages.

reveals that its speakers were indeed farmers and not hunter–gatherers. It has also been suggested that the makers of Lapita pottery would have spoken languages belonging to the Oceanic subgroup of Austronesian languages. This is supported by the reconstruction in the 'proto-Oceanic' language of linguistic terms relating to oceanic voyaging.

The degree to which linguistic and archaeological evidence correlate is impressive. Archaeological evidence for pauses during the dispersal of agriculture correlates well with the linguistic tree of Austronesian languages (Blust, 1999). It might be expected that during these pauses, languages had time to accumulate innovations that enable their modern derivatives to be distinguished as members of a coherent subfamily. This is indeed the case: applying phylogenetic methods to linguistic data reveals that high **bootstrap** values (see Section 6.5.5) are found for language subfamilies that derive from the geographical location of the archaeological pause (Gray and Jordan, 2000). Likewise, archaeological evidence for rapid spread from a particular region is matched by an inability to determine a coherent branching structure for the language subfamilies originating there. For example, the Oceanic subfamily of Austronesian languages, which is associated with the rapid Lapita dispersal, contains a number of subgroups among which a tree-like branching structure is difficult, if not impossible, to determine.

11.3.3 Models for the origins of Pacific Islanders

The ultimate origin of Pacific Islanders, like all other modern humans, is in Africa. There is little doubt that the ancestors of Pacific Islanders would also have had to pass through continental Asia to reach the Pacific; however, here the different models for Pacific origins diverge. Thor Heyerdahl suggested that the first colonists in the Pacific would have been from the east, somewhere in South America. He backed up his hypothesis by taking the unusual step of sailing a vessel similar to that made by Native Americans from the Americas to Polynesia, to show that it could be done (Heyerdahl, 1950). Drifting with the prevailing westerly winds, Heyerdahl argued that colonization of the Pacific from the west would have to have been accomplished in the teeth of these winds. He also put forward as evidence the likely American origins of some important Pacific crops, and similarities between stone facing designs on monuments in Polynesia and South America. The American origin of sweet potato (*Ipomoea batates)* has since been confirmed (Yen, 1974); however, in the face of overwhelming evidence of Asian origins for Pacific languages and cultures, more recent interpretations of this connection have suggested limited prehistoric trading contacts between Polynesians and Native Americans (Green, 2000).

All other models envisage a movement into the Pacific from the west; however, there is huge diversity among them. As we have seen above, there is good evidence for two major cultural movements into the islands southeast of the Asian continent, the first associated with first settlement of the region by modern humans, and the second with the spread of agriculture. If we are to distinguish between the relative contributions to Pacific islanders of ancestral peoples whose migrations may have accompanied these different cultural expansions, it is necessary to distinguish between *proximate* and *ultimate* origins. The *ultimate* origins, say 45 KYA, of genetic lineages from either source population, would be in continental Asia. However, from 29 to 6 KYA, the *proximate* origins of genetic lineages derived from either original settlers or farmers, would be in Near Oceania or continental Asia respectively. Thus we must consider not only the geographical origin of a lineage, but also its time of origin.

Essentially there are two models that are polar opposites, between which lie further models in what archaeologist Peter Bellwood calls 'the continuum of reality'. Metaphors have unfortunately run riot in this arena of prehistory. At one end of the continuum, the 'Express Train' model (also known as the 'Out of Taiwan' model) has Austronesian speakers arrive from the north and, with negligible admixture with indigenous Near Oceanians, disperse into the Pacific (Diamond, 1988). Under this model we should expect proximate origins for Pacific lineages in Taiwan and continental Asia. At the other end of the continuum, the 'Entangled Bank' model sees the movement into Remote Oceania as being the culmination of 40 KY of interactions, in a 'voyaging corridor', among indigenous Near Oceanians (Terrell, 1988; Terrell *et al.*, 1997). By contrast this would suggest genetic lineages in Remote Oceania have a proximate origin in Near Oceania. Between these opposing views are a number of intermediate models that suggest significant genetic contributions from both populations, these include the 'Slow Train' and 'Triple I' (Intrusion, Integration and Innovation) models (Green, 1991a). These models are summarized in *Figure 11.18*.

There is a final model that emphasizes the role of the break up of the Sunda continent in driving population dispersals (Oppenheimer, 1998). This 'Slow Boat' model proposes that the bulk of genetic lineages in Remote Oceania trace to Wallacea (Oppenheimer and Richards, 2001b); see *Box 11.6*.

A few comments are needed to put these different models for the origins of human settlement of Remote Oceania into context:

▸ One model may not fit all the evidence. Whereas a certain model may be more appropriate for the spread of languages, another may fit better to the spread of cultures and genetic lineages. Indeed, many of these models originate in different disciplines, and their original proponents may not have intended them to have any validity beyond that discipline.

▸ These models are not complete: they only describe the contributions of the first settlers of Remote Oceania, and take no account of postsettlement gene flow, which also plays a role in shaping modern genetic diversity. At least some secondary movements of non-Austronesian speakers into Remote Oceania must have taken place, if only to account for the Papuan languages on the Santa Cruz islands.

These two important characteristics of models – that one size does not necessarily fit all, and that they lack sufficient complexity – are common to many different debates for the origins of individual peoples.

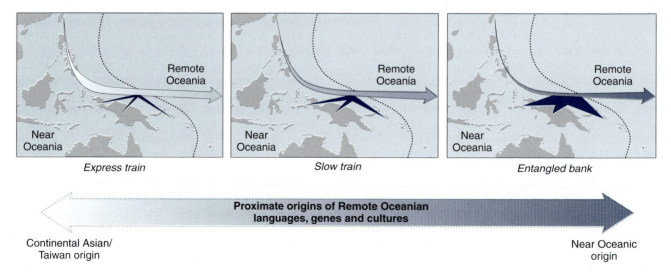

Figure 11.18: Spectrum of alternative models for the origins of the inhabitants of Remote Oceania.

The width of the arrows conveys the relative ancestral contributions of Island Southeast Asian and Near Oceanian populations to the inhabitants of Remote Oceania. The dashed line is the boundary between Near and Remote Oceania. The spectrum beneath represents the continuum of admixture proportions encapsulated within each model.

BOX 11.6 Opinion: Genetic substrates for Austronesian language expansions in the Pacific: Out of Taiwan or Wallacea?

The origins of people in the small islands of the South Pacific, and the Austronesian languages that they mostly speak, has been hotly argued for more than 200 years. For 25 years, the predominant orthodoxy for the spread of Austronesian languages has been the 'Express Train from Taiwan to Polynesia'. This model proposed that the ancestors of the Polynesians and, by association all Austronesian speakers, were early rice farmers who dispersed south from an Austronesian-speaking homeland in Taiwan, through Island Southeast Asia, carrying domesticates such as chickens, pigs and dogs and replacing a hypothetical indigenous 'Australoid' hunter–gatherer population, and then on east, out into the Pacific (see *Figure*). The two cultural markers proposed to date these expansions after Taiwan were red-slipped pottery arriving in the northern Moluccas, perhaps from the Philippines by 3.5 KYA, and a different pottery style, Lapita, appearing with no antecedents in the Bismarck Islands of northeastern Melanesia 3.5 KYA and spreading rapidly to Samoa in the central Pacific, as first evidence of occupation of Polynesia, by 3.2 KYA.

Archaeological and genetic evidence, however, undermines this late expanding Chinese rice-farmer model, suggesting instead that maritime expansions to the Pacific started earlier carrying, not rice, but other intrusive domesticates and genes – both originating, not from Taiwan, but more locally from within Island Southeast Asia. There is no evidence for rice growing in the Philippines or Wallacea at the right time to fuel the expansion of the Express Train. The red-slipped pottery was not associated with rice but instead with traces of root crops, staple food items still used today by all Pacific Islanders. These staples, in turn, were not domesticated in Taiwan or the Philippines, rather in Island Southeast Asia, Wallacea or even Melanesia. There is no evidence that any of the three animal domesticates, dogs, pigs and chickens, came from Taiwan or the Philippines. Dogs and pigs arrived in New Guinea before 5 KYA and both were domesticated locally in Wallacea.

Deliberate Pacific maritime trading and expansion started long before Lapita pottery. Shell tools and betel nut, imports associated with Austronesian-speaking cultures, appeared on the North Coast of New Guinea and the sailors had extended to the Southern Solomons by 6 KYA. **Obsidian** had moved between Melanesian islands long before that. This new evidence implies that the Polynesians were not the first but simply the latest maritime expansion, and that they derived not from China or Taiwan, but from Eastern Indonesia, somewhere between Wallace's line and the island of New Guinea.

Several genetic marker systems point to a primarily Island Southeast Asian ancestry for Polynesians. The 'Polynesian motif', a unique suite of four single nucleotide polymorphisms in mtDNA (see Section 11.3.4), identifies an Oceanic subgroup within a widespread East Asian mtDNA cluster, known as haplogroup B, and characterized by an intergenic 9-base pair

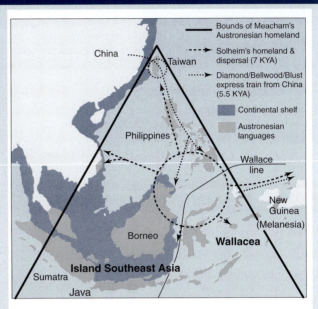

Map showing the two main alternative archaeological views of Austronesian origins.

The oldest view represented by Meacham (solid triangle) and Solheim (dashed line and circle) argues an Island Southeast Asian homeland ($>$ 7 KYA). This is supported by an ancient mitochondrial sequence haplotype, the 'Polynesian motif', found only to the east of Wallace's line. Bellwood's & Diamond's view of a recent rapid migration out of China (5–6 KYA), spreading to replace all the older populations of Indonesia after 4 KYA is shown as a dotted line. Reproduced with permission from Oppenheimer (2001b).

deletion. This Oceanic subgroup is also the main variant throughout the lowland populations of coastal Melanesia, and the bio-geographic zone of Wallacea. Most importantly, it is almost absent to the west of Wallace's Line. It is not found in the Philippines, Taiwan or China – all key stations along the Express Train route. Instead, in these regions we find its immediate ancestor with only three of the four polymorphisms.

The possibility that the final mutation at nucleotide 16 247 occurred '*en route* in the Express Train', is rendered less likely by a study of the diversity accumulated by the Polynesian motif in Wallacea and Melanesia to estimate its age using the molecular clock. In other words, the motif originated *before* an Express Train carrying Taiwanese farmers could have arrived in Wallacea. This seems to break the train ride somewhere around Wallacea (see *Figure*). This finding suggests that Wallacea, a buffer zone between Island Southeast Asia and Melanesia, might have harbored an ancient, indigenous population (of ultimately Asian origin) from which the Polynesian colonists emerged. Study of Y-chromosome variation in the region supports a similar conclusion, and indeed earlier autosomal studies and physical anthropology also suggest ancient differentiation between mainland Asia, Taiwan, Island Southeast Asia, and Melanesia. It is difficult to reconcile this evidence with the Express Train out-of-Taiwan; it seems more consistent with very ancient Austronesian origins *within* tropical Island Southeast Asia.

Stephen Oppenheimer
Green College, Oxford University, UK

11.3.4 Genetic evidence

While in this chapter our focus is on evidence for initial settlement, because we draw our inferences from modern genetic diversity, we must tease apart the conflated influence of postsettlement gene flow from the signal of the first migrations. We must also take account of time-depth in order to distinguish between proximate and ultimate origins. For these reasons we might not expect classical markers to be highly informative.

Classical markers

When analyzed using classical markers, Australian and New Guinean populations exhibit few similarities, suggesting that later region-specific migrations (i.e., to New Guinea but not Australia, or vice versa) have played an important role in shaping modern diversity. An east–west gradient of gene frequencies across New Guinea has been taken as evidence of the predominance of longitudinal migrations across this island: however, these patterns may derive from multiple migrations, and remain undated (Cavalli-Sforza *et al.*, 1994).

The analysis of classical markers among Pacific islanders has contributed relatively little to our understanding of the first settlement of Remote Oceania. While there is little power in these data to distinguish between the different Asian models for Remote Oceanic ancestry outlined above, weak Asian affinities of Pacific populations suggested that a predominantly South American origin was unlikely. In regions of Near Oceania where neighboring populations speak Austronesian and Papuan languages there is little correlation between linguistic and genetic relationships that cannot be accounted for by geography (see the description of the Mantel test in Section 6.7.4). This lack of correlation has been attributed to the impact of admixture between populations speaking languages from the different families since they first came into contact over 3 KYA, a process that may be exacerbated by the small sizes of the island populations involved.

A tree of Near and Remote Oceanic populations constructed from genetic distances between them (see Section 6.5.3) shows little correspondence with the traditional (Polynesian, Melanesian and Micronesian) classification of these islands (Cavalli-Sforza *et al.*, 1994; see *Figure 11.19*). The authors attributed this to the influence of high levels of postsettlement gene flow from Near Oceania to Remote Oceania obscuring the pre-existing, and more distinct, patterns of genetic diversity. However there is some clustering of Eastern Polynesian populations, perhaps indicating that these populations were less prone to postsettlement gene flow by virtue of their isolation.

It is worth noting that population differentiation apparent within trees of populations is generally attributed to the operation of **genetic drift**. The successive founder effects implicit within the colonization of the Pacific would lead to variable effective population sizes and as a consequence different rates of divergence between different islands. These imbalances might be expected to cause such trees to assume unusual topologies. This is indeed the case in *Figure 11.19*, where the Eastern Polynesian populations appear as an outgroup to all the other populations in Near and Remote Oceania, rather than being a subclade within other populations of Remote Oceania as might be expected. Selection might be expected to affect the diversity of individual loci, but is unlikely to significantly perturb a multi-locus tree such as that shown in *Figure 11.19*. One potentially strong selection pressure could have derived from the arduous Oceanic voyages that may have favored those colonists with greater metabolic efficiency (see *Box 11.7*).

Globin gene mutations

One set of autosomal markers has been particularly informative in the Pacific. These markers are deletions of the α-globin genes that result in **α-thalassemia**. These mutations tend to be geographically specific, which makes them useful markers for prehistoric migrations. Region-specific markers are often only present at low frequencies due to a recent mutational origin; however, globin deletions are present at appreciable frequencies because they protect against malaria. Malaria is prevalent in coastal regions from New Guinea to Vanuatu, but absent from more easterly locations in Remote Oceania (Hill and Serjeantson, 1989).

The α-globin gene lies on chromosome 16, where there are two copies that have arisen though an ancient **gene duplication** (see *Box 2.5*). The high frequencies in the Pacific of two different deletions removing a single α-globin gene means that in some locations individuals with the normal copy number of four are actually quite rare, and many more have two or three copies. Unlike more severe forms of thalassemia, the loss of only one or two copies of α-globin cannot be detected by protein electrophoresis in adults; such deletions must be detected at the DNA level.

Recurrent 4.2- and 3.7-kb deletions ($\alpha^{3.7}$ and $\alpha^{4.2}$) that result in the removal of a single α-globin gene result from **nonallelic homologous recombination** between **paralogous** sequences (see Section 3.5) and are found worldwide among tropical and subtropical populations. The $\alpha^{3.7}$ deletions can be subdivided by the precise positioning of the deletion breakpoints in one of three blocks of homology (I, II or III in *Figure 11.20*). In addition, both $\alpha^{3.7}$ and $\alpha^{4.2}$ deletions can be classified according to the **RFLP** haplotype

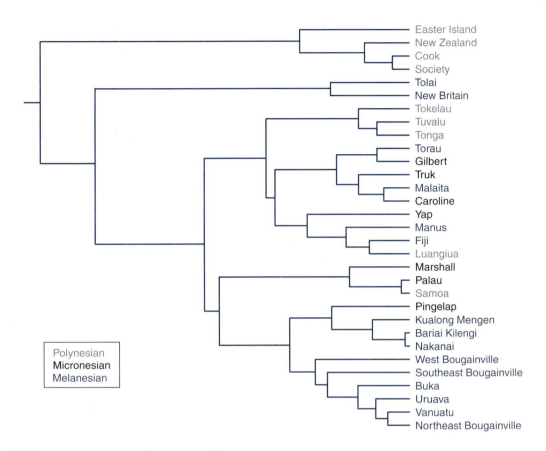

Figure 11.19: Tree of near and remote Oceanic populations.

A neighbor-joining tree of Pacific populations constructed from classical marker frequencies. Redrawn from Cavalli-Sforza *et al.* (1994).

on which the deletion occurs. This enables chromosomes carrying the same deletion to be attributed to unique mutational classes (Hill and Serjeantson, 1989).

In Remote Oceania, the two major α-globin deletions are an $\alpha^{3.7}$ deletion of the III subtype ($\alpha^{3.7III}$) and an $\alpha^{4.2}$ deletion on an RFLP haplotype designated IIIa. Whereas the $\alpha^{3.7III}$

deletion predominates all over Remote Oceania, the $\alpha^{4.2}$ deletion is only present at appreciable frequencies on the western fringes and at low frequency in Tonga (see *Figure 11.21*). Neither of these particular deletions has been found further west than the Wallace line. Although deletions of the same size are found at high frequencies throughout Island Southeast Asia, these occur on different RFLP haplotypes

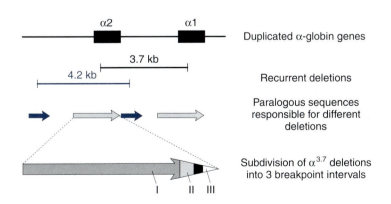

Figure 11.20: Repeats sponsoring single α-globin deletions.

The two size classes of α-globin deletion are shown relative to the two sets of repeated sequences that underlie the deletion events.

BOX 11.7 Opinion: Thrifty genes in Polynesia?

Linguists, archaeologists, and geneticists disagree about the origin of the Polynesians. However, most scholars agree that about 3.5 KYA pre-Polynesian ancestors carrying the Lapita cultural complex arrived in the Bismarck Archipelago, interbred with existing populations, and shortly thereafter achieved the sailing technology for substantial open ocean voyaging. Within a few hundred years these peoples surmounted one of the last great impediments to human settlement: the 500 miles of ocean between Vanuatu and Fiji, sailing against prevailing winds and currents. Both cultural and biological adaptations were required to make this crossing possible. The cultural innovations likely included improvements in sailing technology and navigational expertise and enhanced food preservation techniques. The biological adaptations probably included the robust body build that facilitated paddling the vessels and metabolic adaptations to dietary and cold stress. All of these biological adaptations may be the result of selection for a 'thrifty phenotype'.

The Lapita sailing vessels had a large triangular sail that precluded sailing too close to the wind. The canoes sat low to the water, making paddling possible when winds were too low or from the wrong direction. Skeletal remains from Lapita sites and measurements on modern Polynesians depict a people of substantial stature with broad shoulders and hips, and robust, long limbs. The broad body, long limbs, and sizeable muscle mass are particularly well suited to the biomechanics of paddling – a voyaging strategy that only would have come into play in situations critical to survival. Thus, this body build would have had a strong selective advantage.

Houghton (1996) has modeled the severe cold stress that early Pacific voyagers would have experienced. Maximum cold stress is achieved overnight when moderately low temperatures, high wind chill, and wet clothes and skin combine to produce substantial cold stress. Overnight voyages would have been very rare prior to the 'break out' from the Solomon Islands. The same bodies that are well suited to paddling canoes also have a favorable low surface area to body mass ratio, excellent for conservation of body heat in cold stress (see Section 13.2.1). Again, a large, robust body build would provide an advantage.

Traditionally the **thrifty genotype** argument (see Section 13.4.4) has been used to explain adaptation to periodic famine. In the case of the Pacific it has been invoked as an adaptive response to caloric restriction associated with voyaging and settlement of the islands (Bindon and Baker, 1997). Metabolic efficiency in storage of excess calories is achieved through over-secretion of insulin which increases fat tissue formation and the accumulation of an energy store. This would also increase subcutaneous fat tissue which acts as an insulation against cold stress. It has been suggested, however, that a population as well adapted to a marine

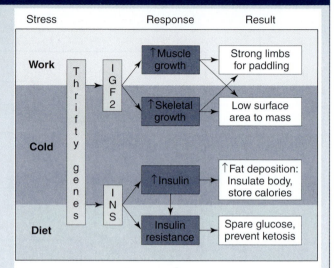

Consequences of a thrifty genotype.

environment as the Lapita people were, may not have suffered from extreme caloric deprivation during voyaging and settlement. Even so, their diet would have been drastically altered: lower carbohydrate intake and an increase in protein intake. They would have eaten through their supply of the poi-like fermented crops (taro, breadfruit, banana) that they were carrying on their voyage and then had to wait for new crops to grow before regaining their normal carbohydrate intake. Several people have argued that the thrifty genotype provides a metabolic adaptation to just such a high-protein low-carbohydrate diet through the metabolic shifts involved in hyperinsulinemia and insulin resistance (see *Figure*).

Recent research into thrifty genes has provided some clues that the cold- and work-adapted body build and the metabolic shift to accommodate dietary stress may be related. These adaptations may be the result of mutations in the region of the insulin gene (*INS*), like the variable number tandem repeat (VNTR) polymorphism near *INS* that modulates transcription of both the *INS* gene and the nearby *Insulin-like Growth Factor 2* (*IGF2*) gene. Increasing transcription of *INS* could generate high blood insulin levels (hyperinsulinemia) and decrease sensitivity to insulin binding in peripheral cells (insulin resistance). Meanwhile, high levels of IGF2 stimulate muscular and skeletal growth predisposing to a large, robust body. I do not mean to imply that this particular VNTR polymorphism represents the thrifty gene, but it points to a possible area for exploration and integrates the biological adaptations found in modern-day Polynesians that appear to result from their voyaging history.

Jim Bindon

Department of Anthropology, University of Alabama, USA

and appear to be independent events. In addition, at least three independent double α-globin deletions are found in Island Southeast Asia at high frequencies, but these are not observed in either Near or Remote Oceania (see *Table 11.3*).

An additional base substitution, called Hb J[Tongariki], has taken place on a chromosome carrying the α[3.7III] deletion, which has a more localized distribution across the boundary of Near and Remote Oceania in coastal New Guinea, New Britain, the Solomon Islands and Vanuatu.

Among chromosomes without any α-globin deletions, RFLP haplotypes give further indications of population ancestry. Interestingly, Polynesians have haplotype Ia, which is frequent in Eurasians but very rare in Near Oceania; in contrast they also have haplotypes IIIa and IVa, which predominate in Near Oceania, but are infrequent elsewhere.

The inference most commonly drawn from the geographical distributions of α-globin deletions given above is that the inhabitants of Remote Oceania have dual genetic

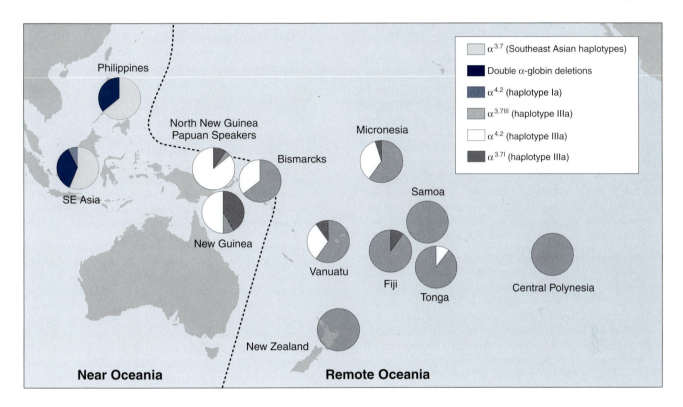

Figure 11.21: Map of α-globin deletion frequencies.

Redrawn from Oppenheimer and Richards (2001a) with extra data from O'Shaughnessy *et al.* (1990).

TABLE 11.3: DELETIONS OF α-GLOBIN GENES IN ISLAND SOUTHEAST ASIA AND THE PACIFIC.

α-Globin deletions	Haplotype	Distribution
SEA double α deletion (~ 20 kb)	–	Southeast Asia
THAI double α deletion (34–38 kb)	–	Southeast Asia
Fil. double α deletion (30–34 kb)	–	Southeast Asia
α[3.7I]	Ia and IIa	Southeast Asia
α[3.7I]	IIIa	Near and Remote Oceania
α[3.7II]	–	Southeast Asia
α[3.7III]	IIIa	Near and Remote Oceania
α[4.2]	1a	Southeast Asia
α[4.2]	IIIa and IVa	Near and Remote Oceania

ancestry from Island Southeast Asia and Near Oceania. The analysis of mutations in other globin genes also supports this interpretation. The $\alpha^{3.7III}$ and $\alpha^{4.2}$ deletions both arose on a IIIa RFLP haplotype in Near Oceania, so why are their distributions in Oceania so different? There are two possible explanations:

▶ only the $\alpha^{3.7III}$ deletion was picked up during the initial Lapita dispersal, and the movement of the $\alpha^{4.2}$ deletion into Remote Oceania resulted from later migrations;

▶ the two deletions were dispersed by the same migration, but the cumulative impact of sequential founder effects skewed their relative frequencies.

Attempts to resolve this issue have tended to focus upon the current distribution of these two deletions within Near Oceania. The $\alpha^{4.2}$ deletion predominates among populations speaking Papuan languages, whereas the $\alpha^{3.7III}$ deletion is more frequent in the Bismarck Archipelago off the New Guinean coast, where the Lapita homeland is thought to be. This differentiation in present-day Near Oceania has led most people to favor the first of the two options detailed above. The limited distributions of the Hb J^{Tongariki} mutation and a less frequent Pacific-specific $\alpha^{3.71}$ deletion on a IIIA haplotype, which straddle the divide between Near and Remote Oceania, also support the existence of later migrations into the western fringes of Remote Oceania.

Mitochondrial DNA (mtDNA)

Mitochondrial DNA diversity within Remote Oceania has been almost exclusively investigated by assaying sequence diversity in hypervariable segment I (HVS I, see Section 3.2.8) of the control region. Despite this segment only representing a small portion of the whole 16.5-kb circular molecule, almost all sequences within Remote Oceania can be apportioned into one of several well-defined lineages. These lineages belong to both of the 'Out of Africa' haplogroups (M and N), and can be related together by a phylogeny, shown in *Figure 11.22*.

All of these lineages can be found in populations to the west of Remote Oceania. To examine the origins of individual lineages in more detail, it is necessary to consider the wider geographical distributions of these lineages among modern populations in Island Southeast Asia, Near Oceania and Remote Oceania (see *Figure 11.23*).

Lineage diversity in Remote Oceania decreases from west to east, until a single lineage becomes almost fixed in some Eastern Polynesian islands. This single lineage is defined by four characteristic mutations at 16 189, 16 217, 16 247 and 16 261 within HVS I, and has become predominant in all Polynesian populations. The universality of this lineage has led to it being dubbed the **'Polynesian motif'** (Redd *et al.*, 1995); however, we can see that the overall distribution of this motif is somewhat broader than this name might suggest. The Polynesian motif has been generated by a sequence of mutations within haplogroup B, which is itself defined by a mutation at 16 189 and a 9-bp deletion elsewhere in the mitochondrial genome (see *Figure 11.22*). This allows us to identify intermediate haplotypes ancestral to the Polynesian motif, which share some but not all of its mutations.

Descendants with these ancestral haplotypes are also found in modern populations. The geographical distribution of these haplotypes stretches back into continental Asia, but they are not found in highland New Guinea, giving perhaps the clearest indication of the ultimately Asian origin of these mtDNAs. Before we examine the origins of the Polynesian motif in more detail, we shall first consider the other lineages found in Remote Oceania. There are two explanations that may account for the different degrees of penetration of these lineages into Remote Oceania:

▶ founder effects resulting from sequential colonization events. These processes ultimately result in the near-fixation of the Polynesian motif in the east;

▶ postsettlement gene flow from the west. This is a more likely explanation for lineages found only on the western fringes of Remote Oceania.

Several other lineages found in Remote Oceania are also found in Island Southeast Asian, but not highland New Guinean, populations. By contrast, two of the lineages found in Remote Oceania (P and Q) are also found at high frequency among highland New Guinean populations, but not among Island Southeast Asian populations. While haplogroup Q can be found throughout Remote Oceania, haplogroup P is confined to the western fringes.

The clear distinctions between the lineage distributions found in Island Southeast Asia and Near Oceania greatly facilitate discussions of lineage origins. However, no modern population should be considered a 'fossilized' remnant of ancient diversity. Despite this fact, isolated populations are less prone to recent admixture and may have retained lineages derived from more ancient migrations. For these reasons, a contrast is often drawn between highland and coastal New Guinean populations. The assumption is that the more isolated highland populations are more likely to have retained lineages derived from the earliest settlement of Sahul. Support for this assumption comes from the observation that the lineages found in coastal New Guinean populations at higher frequencies than in the highlands are also found in other Southeast Asian populations. In addition, the diversity within the P and Q lineages – found at highest frequencies in highland populations – has been used to date the origin of these lineages to more than 15 KYA (Forster *et al.*, 2001). Thus they have an ancient origin, yet appear to have been confined to New Guinea and its neighboring islands prior to the settlement of Remote Oceania.

From analysis of the geographical distributions of mtDNA lineages in Remote Oceania, it appears that there are at least two proximate origins for lineages found in this region, in Near Oceania and Island Southeast Asia. Could it be that there was only one origin for the lineages carried by the initial settlers, and that the other contribution to modern populations results from postsettlement gene flow? If this were true we might expect to see that all lineages from one proximate origin were confined to the western fringes of Remote Oceania. This appears not to be the case: haplogroup Q and haplogroup B4a are both found throughout Remote Oceania, and derive from different

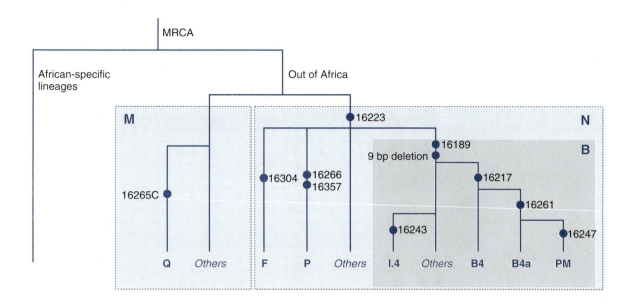

Figure 11.22: Phylogeny of mtDNA lineages in Remote Oceania.

Labeled circles indicate some of the mutations in HVS I that characterize the different lineages; all are transitions except the 16265 transversion. PM, Polynesian motif.

origins. Thus it appears likely that both proximate origins of modern mtDNA lineages were represented among the initial settlers of Remote Oceania.

The limited distribution of the I.4 lineage – shared between Borneo, the Philippines, and the Marianas islands in Remote Oceania – may support the idea that the settlement of western Micronesia was achieved from a different source population than the rest of Micronesia, in agreement with the linguistic evidence.

Now we turn our attention to the origins of the predominant mtDNA lineage in Remote Oceania – the Polynesian motif. The geographical distribution of this lineage does not fit into either of the two patterns identified above: it is found neither in highland New Guinea, nor throughout Island Southeast Asia, but in the latter region is almost exclusively confined to Wallacea. The haplotype on which the final mutation at 16 247 arose, B4a, represents the ancestral lineage to the Polynesian motif, and is spread throughout Island Southeast Asia. The distribution of the ancestral lineage therefore gives little insight into the place of origin of the Polynesian motif. However, it is in Wallacea that the Polynesian motif exhibits the greatest diversity, and consequently it is likely to have originated there. Given that the Polynesian motif was certainly present among the initial settlers of Remote Oceania, there are two possible explanations for its unusual distribution:

▶ The 16 247 mutation arose in Wallacea just prior to the movement into Remote Oceania, on an ancestral B4a haplotype brought from the north. Under this scenario the dominance of the Polynesian motif in Remote Oceania would accord with the 'express train' model.

▶ The 16 247 mutation arose in Wallacea, well before the dispersal into Remote Oceania, in an indigenous pre-Neolithic population, and only much later was incorporated into the pool of mtDNA lineages of mixed origins that dispersed into Remote Oceania. Under this scenario, the dominance of the Polynesian motif would fit better with the 'slow train', or 'entangled bank' model.

The primary distinction between the two models for the origin of the Polynesian motif outlined above is the age of the mutation. It is therefore of interest to date the origin of the Polynesian motif in Wallacea. This is complicated by the rarity of the lineage in this region, and by the uncertainties in the mtDNA mutation rate. However, using the ρ dating method described in Section 6.6.2, the Polynesian motif has been dated to roughly 17 KYA, well before the express train left the proverbial station. For the reasons outlined above, the 95% credible range of these estimates is wide (5.5–34 KYA; Richards *et al.*, 1998). This estimate assumes that the mutation rate of the mtDNA molecule calculated from evolutionary data is more accurate than the much faster rate calculated from pedigree data (that would give a much younger age estimate; see Section 3.2.8 and *Box 3.6*). This assumption is supported by the dating of the first appearance of the Polynesian motif in the Cook Islands, where the age estimate agrees closely with the archaeological dates for first settlement only when the evolutionary mutation rate is used.

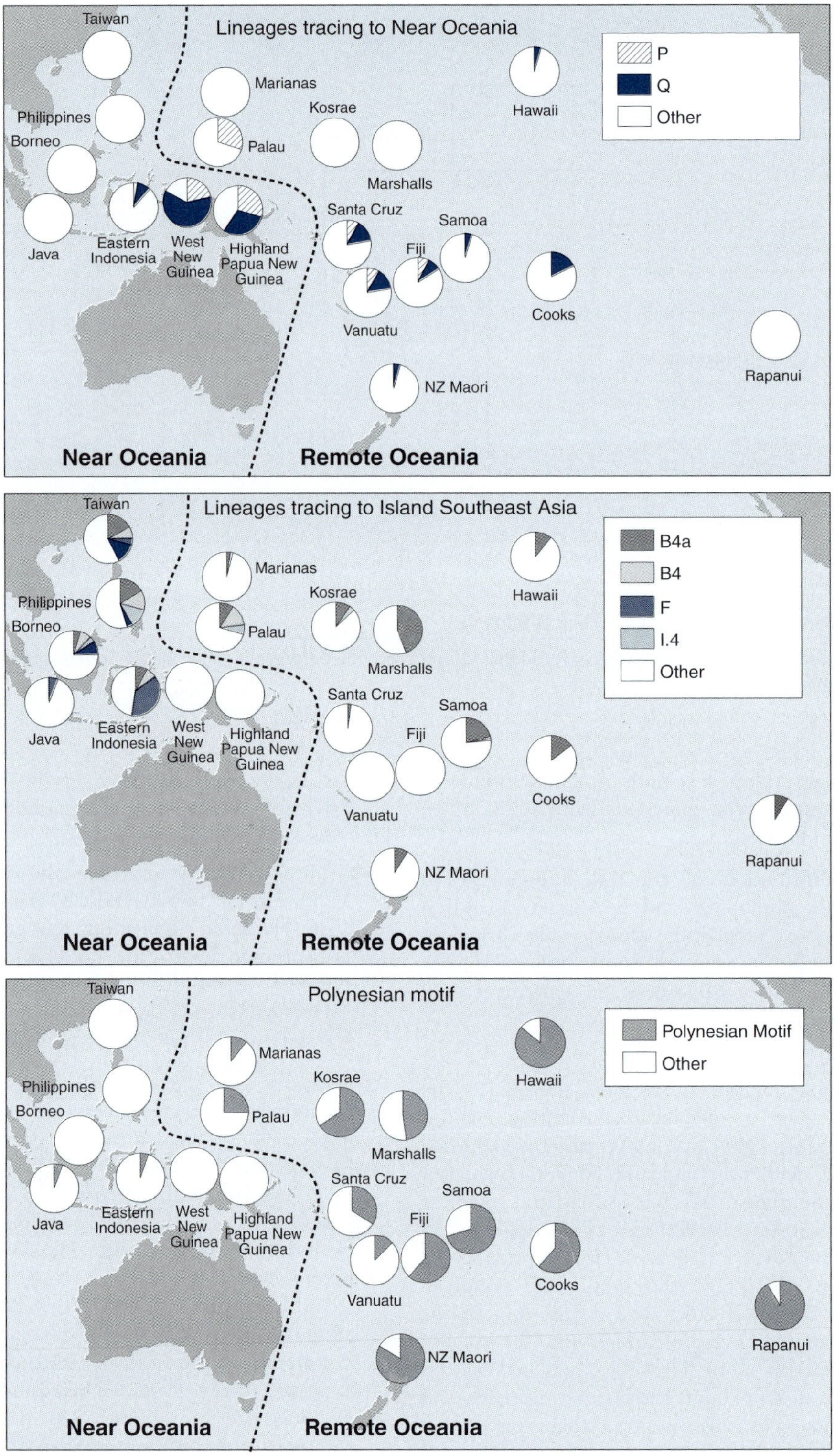

Figure 11.23: Geographical distribution of the major mtDNA lineages in Southeast Asia and the Pacific.

Data from Redd *et al.*, 1995; Lum and Cann, 1998; Melton *et al.*, 1998; Redd and Stoneking, 1999; Lum and Cann, 2000; Friedlaender *et al.*, 2002; Tommaseo-Ponzetta *et al.*, 2002.

It is instructive to compare the clinal distribution of the Polynesian motif with the various genetic lineages associated with the spread of farming into Europe from Southwest Asia (see Section 10.5). In the former case, the frequency of the lineage increases in islands further away from the origin of the migration, whereas in Europe, lineage frequencies decline with increasing distance from the source. Thus the direction of a **cline** is not necessarily a reliable indicator of the place of origin of the migration which generated it. It appears that the nature of the colonization process, and whether migration is into empty or populated territory, determine how migrations generate clines of lineage frequencies. The cumulative founder events necessitated by the discontinuous landmasses of Remote Oceania were instrumental in establishing the observed clinal distribution of the Polynesian motif.

It has been possible to investigate these founder events in more detail by estimating the size of the founding population from genetic data. Forward simulations have been run, using different founding population sizes, to find the size that best recreates the modern genetic diversity. The colonization of New Zealand has been simulated *in silico* by taking random samples of eastern Polynesian mtDNA sequences, allowing the population to grow over time, and then randomly selecting a set of mtDNAs from the artificial population and comparing their genetic diversity to an identically sized set of real mtDNA sequences. Despite the numerous assumptions required by simulations of this nature the founding population was estimated to contain between 50 and 100 women, a figure which agrees strikingly with those given in the **oral histories** handed down over generations (Murray-McIntosh *et al.*, 1998).

Y chromosome

Y-chromosomal diversity within Remote Oceania has been investigated using SNP, microsatellite and minisatellite variability. **Monophyletic** lineages have been defined primarily on the basis of SNP haplotypes, but novel structures of the minisatellite MSY1 have also proved informative (Hurles *et al.*, 2002). Microsatellite diversity within these lineages can then be used to make inferences on their spatial and temporal origins.

As in America, the male bias of European groups throughout Remote Oceania (sailors, whalers, traders, missionaries) over the past 250 years means that, in addition to postsettlement gene flow, European admixture may complicate the signal of initial settlement present in the paternal lineages of modern Oceanic populations. However, in practice the high geographic differentiation of the Y chromosome means that European admixture can be easily identified and discounted before further analysis (Hurles *et al.*, 1998). A number of studies show that European admixture varies greatly between different islands of Remote Oceania, with perhaps the greatest influence in the Cook Islands where approximately one third of all Y chromosomes appear to belong to three diagnostically European lineages. The relative abundance of lineages reflecting admixture can be used to pinpoint their likely origin within Europe, assuming that they all come from the same region. These considerations support an origin for these lineages in northwest Europe, in agreement with the historical evidence.

Having discounted any European admixture, the remaining indigenous paternal lineages of Remote Oceania can be related by the phylogeny shown in *Figure 11.24*.

The distribution of these paternal lineages is shown in *Figure 11.25*. The pattern of lineage distribution is strikingly reminiscent of the mtDNA lineage pattern. The lineage distributions of Near Oceania and Island Southeast Asia are quite distinct and lineages found in Remote Oceania can be traced to both source areas. Again, the predominant lineage in Remote Oceania, haplogroup C, can be traced as far west as Wallacea within Island Southeast Asia.

Although haplogroup C chromosomes are absent from northern Island Southeast Asia, they can be found at appreciable frequencies in continental East Asia, the Americas and Australia. The haplogroup C chromosomes found in Remote Oceania all belong to a monophyletic sublineage of this wider haplogroup. This sublineage can be defined by multiple polymorphisms: a specific deletion within the microsatellite *DYS390* (Kayser *et al.*, 2000), a unique allele structure at the minisatellite MSY1 (Hurles *et al.*, 2002) and an additional SNP, M38 (Underhill *et al.*, 2001), although the order in which these mutations arose is presently unknown. This 'Oceanic motif' is found only in Remote Oceania, Near Oceania and Wallacea, and only in Wallacea is it found together with other haplogroup C chromosomes that do not belong to this lineage. It is also in Wallacea that the greatest diversity within this Oceanic motif is found, pinpointing this region as the likely place of origin.

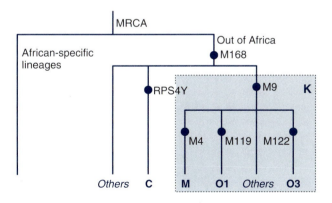

Figure 11.24: Phylogeny of Y-chromosomal lineages in Remote Oceania.

Labeled circles indicate some of the mutations that characterize the different lineages. Lineages resulting from European admixture are not included.

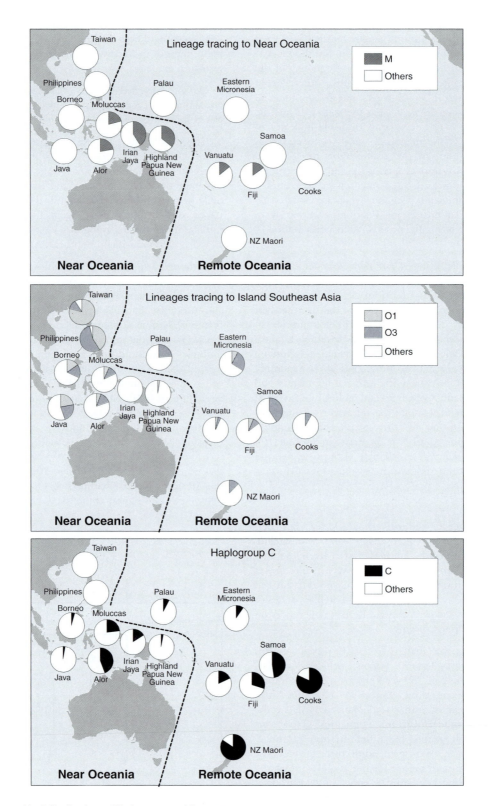

Figure 11.25: Geographical distributions of Y-chromosomal lineages.

Data from Su *et al.*, 2000; Capelli *et al.*, 2001; Kayser *et al.*, 2001; Underhill *et al.*, 2001.

Clearly the ultimate origin of chromosomes carrying the Oceanic motif is in continental East Asia. As with its mtDNA counterpart – the Polynesian motif – estimating the age of the mutations forming the Oceanic motif is required to determine the proximate origins of this lineage. In other words, could the mutation have plausibly taken place in Wallacea during the spread of farming throughout the region, or was it an earlier event? All estimates for the age of the MRCA of the extant Y chromosomes belonging to this lineage predate the spread of farming. For example, the MRCA of chromosomes carrying the *DYS390* deletion described above has been dated using a coalescent-based method (see Section 6.6.2) to ~ 11 KYA with the 95% confidence interval 5–31 KYA (Kayser *et al.*, 2001).

In conclusion, the predominant maternal and paternal lineages in Remote Oceania appear to have an origin in Wallacea that predates the spread of farming into this region. However, in addition to these lineages, there are contributions from Island Southeast Asian and New Guinean sources, indicating that the initial settlers of the remote islands of the Pacific were an admixed population with at least three sources of genetic input.

11.3.5 Evidence from other species

In recent years, studies of the biological impact of the last ice age have been greatly informed by noting common patterns of genetic diversity among several extant floral and faunal species (Hewitt, 2000). Each species provides independent evidence on a common evolutionary process, in this case, the recolonization of previously inhospitable lands. The term **'comparative phylogeography'** has been coined to describe this search for similar geographical distributions of lineages from different species that have undergone the same evolutionary process.

When humans settle previously uninhabited lands they bring with them (both intentionally and unintentionally) a panoply of other species, including infectious microbes, domesticated plants and animals, and nondomesticated species that co-exist with humans (known as **commensals**). Thus the geographical patterns of genetic diversity in these species have been shaped, at least in part, by a common process – human migration. Can a similar comparative approach to the one described above allow the analysis of genetic diversity in nonhuman species to shed light on human migrations?

In principle, such an approach to studying human migration has several advantages over using human samples. In particular, ancient DNA studies are often facilitated by a relative abundance of samples (especially from domesticated species) and the absence of problems caused by human DNA contamination (see Section 4.11).

The potential for gaining insight into human migration from the genetics of nonhuman species has been explored in the colonization of the Pacific. Several domesticates have been studied (e.g., the Polynesian food rat, *Rattus exulans*, and pigs), as has a commensal lizard (*Lipinia noctua*) that was presumably a stowaway on ocean-going canoes, and an infrequently **pathogenic** virus (human polyomavirus JC).

Despite the relative paucity of studies on individual species at present, there is support from analysis of all the species mentioned above for colonization of the Pacific from the West. Analysis of whole genome sequences (5.1 kb) of the JC virus in Remote Oceania suggest that the settlers carried at least two distinct viral subtypes, one with East Asian origins and the other from New Guinea (Yanagihara *et al.*, 2002). In addition, mtDNA phylogenies of *Lipinia noctua* exhibit a clear signal of the rapid transport into Remote Oceania of this native New Guinean lizard (Austin, 1999). More detailed studies of mtDNA diversity of the Polynesian food rat in Remote Oceania suggest that substantial levels of inter-island voyaging opposed the isolation of previously settled islands (Matisoo-Smith *et al.*, 1998). Also these data show that islands in Remote Oceania now uninhabited (Kermadecs) were formerly settled by humans either full time or on a seasonal basis.

In conclusion, genetic studies of nonhuman species support the rapid colonization of Remote Oceania from the west by an admixed population. In addition, initial settlement was followed by high levels of inter-island voyaging that indicate that rather than being a barrier to gene-flow, the large Oceanic distances between islands could more properly be considered as well-trodden highways.

Summary

▸ Rising sea level has both driven human migration by submerging previously settled lands, and restricted it by inundating land bridges.

▸ Technological ability and environmental opportunity both govern when and from where new found lands could be settled in prehistory.

▸ The number of migrations leading to the peopling of the Americas, their origins, routes and timing are all contentious.

▸ Evidence from human fossils and archaeology suggests that humans were present in the Americas by at least 14.5 cal KYA and soon spread over both continents, probably exterminating most of the megafauna.

▸ Modern Native American classical marker frequencies suggest a Siberian origin, and DNA analysis shows that they carry a restricted set of mtDNA and Y-chromosomal lineages which can be traced back to sources in southern and possibly eastern Siberia.

▸ The genetic evidence suggests that most Native American diversity can be accounted for by one major migration 20–15 KYA, but that a second, later, minor migration may have contributed to the North American gene pool.

▸ In the Pacific, genetic evidence supports the archaeological and linguistic evidence for the origins of Pacific islanders from the west.

▶ The origins of the mtDNA and Y-chromosomal lineages that predominate on recently settled Pacific islands can be placed in both time and space, and suggest that the peoples who first settled these remote islands were an admixed population with contributions, in order of decreasing input, from indigenous Wallaceans, dispersing farmers and indigenous New Guineans.

▶ Migrations into new found lands can produce clinal patterns of lineage frequencies that oppose the direction of movement as a result of cumulative founder events.

▶ Analyzing the genetic diversity of pathogens, commensals and domesticated species that owe their geographic distribution to human migration can itself be informative about patterns of human migration.

Further reading

Burenhult G (ed) (1993) *The First Humans: Human Origins and History to 10,000 BC.* Harper, San Francisco.

Hewitt G (2000) The genetic legacy of the Quaternary ice ages. *Nature* **405**, 907–913.

Hurles ME, Matisoo-Smith E, Gray RD, Penny D (2003) Untangling Oceanic settlement: the edge of the knowable. *Trends Ecol. Evol.* **18**, 531–540.

Renfrew C (ed) (2000) *America Past, America Present: Genes and Languages in the Americas and Beyond.* McDonald Institute, Cambridge

Electronic references

Environment during the last 130 KY:
http://members.cox.net/quaternary/

Summary of American fossil and archaeological sites:
http://www.uwm.edu/~trinrud/evidence_1.htm

Beringia:
http://www.ngdc.noaa.gov/paleo/parcs/atlas/beringia/index.html

References

Alroy J (2001) A multispecies overkill simulation of the end-Pleistocene megafaunal mass extinction. *Science* **292**, 1893–1896.

Austin CC (1999) Lizards took express train to Polynesia. *Nature* **397**, 113–114.

Bindon JR, Baker PT (1997) Bergmann's rule and the thrifty genotype. *Am. J. Phys. Anthropol.* **104**, 201–210.

Blust R (1999) Subgrouping, circularity and extinction: some issues in Austronesian comparative linguistics. *Symp. Ser. Inst. Ling. Acad. Sin.* **1**, 31–94.

Bonatto SL, Salzano FM (1997) Diversity and age of the four major mtDNA haplogroups, and their implications for the peopling of the New World. *Am. J. Hum. Genet.* **61**, 1413–1423.

Brown MD, Hosseini SH, Torroni A et al. (1998) mtDNA haplogroup X: an ancient link between Europe/Western Asia and North America? *Am. J. Hum. Genet.* **63**, 1852–1861.

Capelli C, Wilson JF, Richards M et al. (2001) A predominantly indigenous paternal heritage for the Austronesian-speaking peoples of insular Southeast Asia and Oceania. *Am. J. Hum. Genet.* **68**, 432–443.

Cavalli-Sforza LL, Menozzi P, Piazza A (1994) *The History and Geography of Human Genes.* Princeton University Press, Princeton, NJ.

Chakraborty R (1986) Gene admixture in human populations: models and predictions. *Yearbk. Phys. Anthropol.* **29**, 1–43.

Cohen JE, Small C (1998) Hypsographic demography: the distribution of human population by altitude. *Proc. Natl Acad. Sci. USA* **95**, 14009–14014.

Derenko MV, Grzybowski T, Malyarchuk BA et al. (2001) The presence of mitochondrial haplogroup X in Altaians from South Siberia. *Am. J. Hum. Genet.* **69**, 237–241.

Diamond JM (1988) Express train to Polynesia. *Nature* **336**, 307–308.

Forster P, Harding R, Torroni A et al. (1996) Origin and evolution of Native American mtDNA variation: a reappraisal. *Am. J. Hum. Genet.* **59**, 935–945.

Forster P, Torroni A, Renfrew C, Rohl A (2001) Phylogenetic star contraction applied to Asian and Papuan mtDNA evolution. *Mol. Biol. Evol.* **18**, 1864–1881.

Friedlaender JS, Gentz F, Green K, Merriwether DA (2002) A cautionary tale on ancient migration detection: mitochondrial DNA variation in Santa Cruz islands, Solomon islands. *Hum. Biol.* **74**, 453–471.

González-José R, González-Martin A, Hernández M et al. (2003) Craniometric evidence for Paleoamerican survival in Baja California. *Nature* **425**, 62–65.

Gray RD, Jordan FM (2000) Language trees support the express-train sequence of Austronesian expansion. *Nature* **405**, 1052–1055.

Green RC (1991a) The Lapita cultural complex: current evidence and proposed models. *Indo-Pacific Prehist. Assoc. Bull.* **11**, 295–305.

Green RC (1991b) *Near and Remote Oceania – Disestablishing 'Melanesia' in culture history. Man and a half: Essays on Pacific Anthropology and Ethnobiology in Honour of Ralph Bulmer.* A. Pawley, Auckland, The Polynesia Society.

Green RC (2000) A range of disciplines support a dual origin for the bottle gourd in the Pacific. *J. Polynesian Soc.* **109**, 191–197.

Greenberg JH, Turner II CG, Zegura SL (1986) The settlement of the Americas: a comparison of the linguistic, dental and genetic evidence. *Curr. Anthropol.* **27**, 477–497.

Hammer MF, Zegura SL (2002) The human Y chromosome haplogroup tree: nomenclature and phylogeny of its major divisions. *Annu. Rev. Anthropol.* **31**, 303–321.

Hanebuth T, Stattegger K, Grootes PM (2000) Rapid flooding of the Sunda shelf: A late-glacial sea-level record. *Science* **288**, 1033–1035.

Hewitt G (2000) The genetic legacy of the Quaternary ice ages. *Nature* **405**, 907–913.

Heyerdahl T (1950) *Kontiki: Across the Pacific by Raft*. Rand McNally, Chicago, IL.

Hill AVS, Serjeantson SW (1989) *The Colonization of the Pacific: a Genetic Trail*. Clarendon, Oxford.

Houghton P (1996) *People of the Great Ocean: Aspects of Human Biology of the Early Pacific*. Cambridge University Press, Cambridge.

Hurles ME, Irven C, Nicholson J *et al.* (1998) European Y-chromosomal lineages in Polynesians: A contrast to the population structure revealed by mtDNA. *Am. J. Hum. Genet.* **63**, 1793–1806.

Hurles ME, Nicholson J, Bosch E *et al.* (2002) Y chromosomal evidence for the origins of oceanic-speaking peoples. *Genetics* **160**, 289–303.

IPCC (1995) *Climate Change 1995: Impacts, Adaptations and Mitigation of Climate Change*. Intergovernmental Panel on Climate Change, Geneva.

IPCC (2001) *Climate Change 2001: The Scientific Basis*. Intergovernmental Panel on Climate Change, Geneva.

Jantz RL, Owsley DW (2001) Variation among early North American crania. *Am. J. Phys. Anthropol.* **114**, 146–155.

Josenhans H, Fedje D, Pienitz R *et al.* (1997) Early humans and rapidly changing Holocene sea levels in the Queen Charlotte Islands–Hecate Strait, British Colombia, Canada. *Science* **277**, 71–74.

Karafet TM, Zegura SL, Posukh O *et al.* (1999) Ancestral Asian source(s) of New World Y-chromosome founder haplotypes. *Am. J. Hum. Genet.* **64**, 817–831.

Kayser M, Braver S, Weiss G, Underhill PA, Rower L, Schiefenhovel W, Stoneking M (2000) Melanesian origin of Polynesian Y chromosomes. *Curr. Biol.* **10**, 1237–1246.

Kayser M, Brauer S, Weiss G, Schiefenhovel W, Underhill P, Stoneking M (2001) Independent histories of human Y chromosomes from Melanesia and Australia. *Am. J. Hum. Genet.* **68**, 173–190.

Kirch PV (1999) *The Lapita Peoples*. Blackwell, Oxford.

Kolman CJ, Sambuughin N, Bermingham E (1996) Mitochondrial DNA analysis of Mongolian populations and implications for the origin of New World founders. *Genetics* **142**, 1321–1334.

Lambeck K, Esat TM, Potter E-K (2002) Links between climate and sea levels for the past three million years. *Nature* **419**, 199–206.

Lell JT, Sukernik RI, Starikovskaya YB *et al.* (2002) The dual origin and Siberian affinities of Native American Y chromosomes. *Am. J. Hum. Genet.* **70**, 192–206.

Lum JK, Cann RL (1998) mtDNA and language support a common origin of Micronesians and Polynesians in Island Southeast Asia. *Am. J. Phys. Anthropol.* **105**, 109–119.

Lum JK, Cann RL (2000) mtDNA lineage analyses: origins and migrations of Micronesians and Polynesians. *Am. J. Phys. Anthropol.* **113**, 151–168.

Mandryk CAS, Josenhans H, Fedje DW, Matthewes RS (2001) Late Quarternary paleoenvironment of Northwestern North America: implications for inland versus coastal migration routes. *Quaternary Sci. Rev.* **20**, 301–314.

Matisoo-Smith E, Roberts RM, Irwin GJ, Allen JS, Penny D, Lambert DM (1998) Patterns of prehistoric human mobility in Polynesia indicated by mtDNA from the Pacific rat. *Proc. Natl Acad. Sci. USA* **95**, 15145–15150.

Melton T, Clifford S, Martinson J, Batzer M, Stoneking M (1998) Genetic evidence for the Proto-Austronesian Homeland in Asia: mtDNA and nuclear DNA variation in Taiwanese aboriginal tribes. *Am. J. Hum. Genet.* **63**, 1807–1823.

Meltzer DA (1997) Monte Verde and the Pleistocene peopling of the Americas. *Science* **276**, 754–755.

Merriwether DA, Hall WW, Vahlne A *et al.* (1996) mtDNA variation indicates Mongolia may have been the source for the founding population for the New World. *Am. J. Hum. Genet.* **59**, 204–212.

Murray-McIntosh RP, Scrimshaw BJ, Hatfield PJ *et al.* (1998) Testing migration patterns and estimating founding population size in Polynesia by using human mtDNA sequences. *Proc. Natl Acad. Sci. USA* **95**, 9047–9052.

Oppenheimer SJ, Richards M (2001a) Fast trains, slow boats, and the ancestry of the Polynesian islanders. *Science Prog.* **84**, 157–181.

Oppenheimer SJ, Richards M (2001b) Slow boat to Melanesia? *Nature* **410**, 166–167.

Oppenheimer SJ (1998) *Eden in the East: the Drowned Continent of Southeast Asia*. Weidenfeld & Nicolson, London.

O'Shaughnessy DF, Hill AVS, Bowden DK, Weatherall DJ, Clegg JB (1990) Globin genes in Micronesia: origins and affinities of Pacific island peoples. *Am. J. Hum. Genet.* **46**, 144–155.

Pawley A (1999) Chasing rainbows: implications of the rapid dispersal of Austronesian languages for sub-grouping and reconstruction. In: *Selected papers from the Eighth International Conference on Austronesian Linguistics* (eds. Zeitoun E. and Li PJ-K). *Academia Sinica*, Taipei pp. 95–138.

Pawley A, Ross M (1993) Austronesian historical linguistics and culture history. *Annu. Rev. Anthropol.* **22**, 425–459.

Pearson GA (1997) Further thoughts on Clovis Old World origins. *Curr. Res. Pleistocene* **14**, 74–76.

Pelejero C, Kienast M, Wang L *et al.* (1999) The flooding of Sundaland during the last glaciation: imprints in hemipelagic sediments from the southern South China Sea. *Earth Planet. Sci. Lett.* **171**, 661–671.

Peltier WR (1994) Ice Age paleotopography. *Science* **265**, 195–201.

Pena SD, Santos FR, Bianchi NO *et al.* (1995) A major founder Y-chromosome haplotype in Amerindians. *Nature Genet.* **11**, 15–16.

Redd AJ, Stoneking M (1999) Peopling of Sahul: mtDNA variation in Aboriginal Australian and Papua New Guinean populations. *Am. J. Hum. Genet.* **65**, 808–828.

Redd AJ, Takezaki N, Sherry ST *et al.* (1995) Evolutionary history of the COII/tRNA lys intergenic 9 base pair deletion in human mitochondrial DNAs from the Pacific. *Mol. Biol. Evol.* **12**, 604–615.

Richards M, Oppenheimer S, Sykes B (1998) MtDNA suggests Polynesian origins in eastern Indonesia. *Am. J. Hum. Genet.* **63**, 1234–1236.

Roosevelt AC, Lima da Costa M, Lopes Machado C *et al.* (1996) Paleoindian cave dwellers in the Amazon: the peopling of the Americas. *Science* **272**, 373–384.

Rosenberg NA, Pritchard JK, Weber JL *et al.* (2002) Genetic structure of human populations. *Science* **298**, 2381–2385.

Ruiz-Linares A, Ortiz-Barrientos D, Figueroa M *et al.* (1999) Microsatellites provide evidence for Y chromosome diversity among the founders of the New World. *Proc. Natl Acad. Sci. USA* **96**, 6312–6317.

Salzano FM (2002) Molecular variability in Amerindians: widespread but uneven information. *An. Acad. Bras. Cienc.* **74**, 223–263.

Santos FR, Pandya A, Tyler-Smith C *et al.* (1999) The central Siberian origin for Native American Y chromosomes. *Am. J. Hum. Genet.* **64**, 619–628.

Silva WA, Jr., Bonatto SL, Holanda AJ *et al.* (2002) Mitochondrial genome diversity of Native Americans supports a single early entry of founder populations into America. *Am. J. Hum. Genet.* **71**, 187–192.

Smith WHF, Sandwell DT (1997) Global sea floor topography from satellite altimetry and ship depth soundings. *Science* **277**, 1956–1962.

Stewart JR, Lister AM (2001) Cryptic northern refugia and the origin of modern biota. *Trends Ecol. Evol.* **16**, 608–613.

Stone AC, Stoneking M (1998) mtDNA analysis of a prehistoric Oneota population: implications for the peopling of the New World. *Am. J. Hum. Genet.* **62**, 1152–1170.

Su B, Jin L, Underhill P *et al.* (2000) Polynesian origins: Insights from the Y chromosome. *Proc. Natl Acad. Sci. USA* 97, 8225–8228.

Tarazona-Santos E, Santos FR (2002) The peopling of the Americas: a second major migration? *Am. J. Hum. Genet.* **70**, 1377–1380.

Terrell J, Hunt TL, Gosden C (1997) The dimensions of social life in the Pacific: human diversity and the myth of the primitive isolate. *Curr. Anthropol.* **38**, 155–195.

Terrell JE (1988) History as a family tree, history as an entangled bank: Constructing images and interpretations of prehistory in the South Pacific. *Antiquity* **62**, 642–657.

Tommaseo-Ponzetta M, Attimonelli M, De Robertis M *et al.* (2002) Mitochondrial DNA variability of West New Guinea populations. *Am. J. Phys. Anthropol.* **117**, 49–67.

Torroni A, Neel JV, Barrantes R *et al.* (1994) Mitochondrial DNA 'clock' for the Amerinds and its implications for timing their entry into North America. *Proc. Natl Acad. Sci. USA* **91**, 1158–1162.

Tzedakis PC, Lawson IT, Frogley MR *et al.* (2002) Buffered tree population changes in a quaternary refugium: evolutionary implications. *Science* **297**, 2044–2047.

Underhill PA, Jin L, Zemans R *et al.* (1996) A pre-Columbian Y chromosome-specific transition and its implications for human evolutionary history. *Proc. Natl Acad. Sci. USA* **93**, 196–200.

Underhill PA, Passarino G, Lin AA *et al.* (2001) Maori origins, Y-chromosome haplotypes and implications for human history in the Pacific. *Hum. Mutat.* **17**, 271–280.

Weisler MI, Kirch PV (1996) Interisland and interarchipelago transfer of stone tools in prehistoric Polynesia. *Proc. Natl Acad. Sci. USA* **93**, 1381–1385.

Yanagihara R, Nerurkar VR, Scheirich I *et al.* (2002) JC virus genotypes in the western Pacific suggest Asian mainland relationships and virus association with early population movements. *Hum. Biol.* **74**, 473–488.

Yen DE (1974) *The Sweet Potato and Oceania*. Bernice P. Bishop Museum Bulletin, Honolulu.

CHAPTER TWELVE

What happens when populations meet?

CHAPTER CONTENTS

BOXES

12.1 Introduction

12.1.1 What is genetic admixture?

Populations are not discrete entities: they can readily exchange their constituent parts – individuals. This characteristic lies in direct contrast to the discrete nature of both organisms and species, the two neighboring levels to populations in the hierarchy of life. Populations are able to mix, generating hybrid populations. This process of mixing is known as **admixture**, and is a common genetic consequence of the meeting of populations.

Neighboring populations frequently exchange individuals by an ongoing process of bi-directional migration. However, a third, hybrid population does not usually result from this kind of exchange. The term 'admixture' is often reserved for the formation of a hybrid population from the mixing of ancestral populations that have previously been in relative isolation from one another. The range expansion of one population into a region inhabited by a previously isolated population is one such scenario. Thus, admixture can be thought of as being initiated at a specific point in time, when the populations first came into contact.

As with most studies described in this book, we are almost exclusively limited to examining modern genetic diversity. When we examine modern populations, we detect not simply the proportions of admixture established when the populations first met, but the summation of cumulative gene flow from when they first met to the present day. Thus the consequences of admixture and gene flow may be difficult to distinguish. Of course, the imprint of past admixture in modern populations has also been modified by the **drift**, selection and mutation processes that shape all genetic diversity.

Many different issues of population prehistory can be viewed as questions about admixture. All that is required is that alternative ancestral populations can be differentiated from one another in either time or space. For example, the relative contributions to the modern gene pool of several

migrations to the same location can be thought of as admixture between the source populations for each migration. It is in this framework that the relative contributions of **Paleolithic hunter–gatherers** and **Neolithic** farmers to modern European diversity has been considered (see *Figure 12.1*). Whilst the two ancestral populations spread from largely similar geographical origins in the Near East, they are separated in time by thousands of years. This specific example is explored in greater depth in Section 10.5.

The processes of isolation and range expansion that result in subsequent admixture can be driven by environmental changes. During the recent **ice ages**, the environment in more northerly latitudes became uninhabitable. Humans and other plant and animal species found refuge in pockets of more hospitable climate, known as **'glacial refugia'**. These refugia were often isolated from one another. For example, it is known that three major European glacial refugia have been:

▶ the Iberian peninsula;

▶ Italy;

▶ the Balkans and Greece.

After the end of the last ice age some 14 KYA, many species started the long process of re-colonizing the more northerly latitudes from these refugia. During this process, many previously isolated populations were brought back into contact with one another, the genetic consequences of which can be analyzed through a consideration of admixture.

More recent historical events that can be studied through an appreciation of admixture processes are episodes of enforced migration. These episodes have often been motivated by the creation of a subjugated labor force – in other words, a slave trade. Whilst the eighteenth century Atlantic slave trade has received most attention, slavery was widespread throughout the Ancient world including the Egyptian, Greek and Roman empires, among Arabs, in Iceland and in the Pacific.

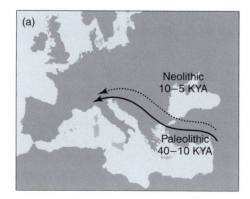
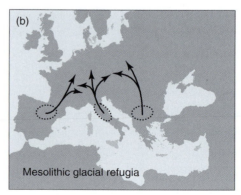

Figure 12.1: Maps showing potential past admixture in Europe.

(a) Neolithic and Paleolithic peoples migrated into Europe at different times, although by similar routes, and both are thought to have contributed to the modern European gene pool. (b) Peoples from different **glacial refugia** (circled), may have been isolated from one another during the ice age, and as conditions improved northward migration would have presented opportunities for admixture.

Historically, some of the first studies of genetic admixture at the molecular level were those that analyzed the frequencies of different blood group protein alleles in African–Americans, comparing them to European–Americans and Africans (Glass and Li, 1953). The aim has been primarily to quantify European admixture among African–Americans. Both marital records and a supposed lightening of African–American skin color provided external evidence of admixture. A number of studies of different blood groups in different US populations were published throughout the 1950s and 60s (e.g. Reed, 1969). They demonstrated that the extent of European admixture varied considerably among the different regional populations of African–Americans over the 10 generations or so since the major period of African slavery. The proportion of European genes within different African–American populations was shown to vary from ~ 4% to ~ 30%, with southern populations having consistently lower levels of European admixture. Although the molecule of more recent investigations has been DNA rather than protein, these conclusions have stood the test of time.

Often studies of prehistoric admixture events are initiated when evidence from non-genetic sources indicates that admixture might have occurred. This is because, unsurprisingly, the meeting of previously isolated populations has impacts beyond the realm of genetics (see *Box 12.1*). Nevertheless, it begs the question: can genetic admixture be recognized in the absence of corroborative historical or prehistorical information? This question will be addressed later on in this chapter.

12.1.2. The genetic imprint of admixture

The process of admixture shapes genetic diversity in a number of different ways. In this chapter we will explore how seeking these different imprints in modern genetic diversity can lead to the inference that admixture occurred some time in the past. Our ability to detect admixture depends in part on how differentiated the source populations were from one another. As we shall see, the more different the ancestral populations were, the easier it is to detect admixture.

While many studies of genetic admixture have been of a single genetic locus, other studies have looked at multiple loci in each individual. This raises the additional complication of admixture within a single genome: some alleles may have their ancestry in one parental population

---

BOX 12.1 The ever-changing terminology of people with mixed ancestry

Societies undergoing admixture have often sought to classify individuals on the basis of their proportions of admixture. Many of these largely historical terms introduced fine gradations of admixture, but the number of terms required to cover all possible proportions doubles each generation, and it quickly becomes impractical to maintain a word for each fraction. As a result many of these words have all but died out, whilst others have been retained in common usage as general terms for people of mixed ancestry.

Term	Parents	Proportion
Mulatto	Black and White	1/2 Black
Quadroon	Mulatto and White	1/4 Black
Octoroon	Quadroon and White	1/8 Black
Mustifee	Octoroon and White	1/16 Black
Mustifino	Mustifee and White	1/32 Black
Cascos	Mulatto and Mulatto	1/2 Black
Sambo	Mulatto and Black	3/4 Black
Mango	Sambo and Black	7/8 Black
Metisse	White and Native American	General
Mestizo	White and Native American	General
Griffe	Black and Native American	General
Hapa	Asian/Polynesian and White	General

As with all terminologies associated with contentious societal issues, many of these words have been considered offensive at some point in time. It could be argued that the rapid turnover of names for individuals of mixed ancestry is driven by society's need to find neutral words free from negative connotations. However, as the societal inequalities remain, these new terms attract derogatory associations and fresh terms need to be invented at regular intervals.

while other alleles have their ancestry in another. This is an inevitable consequence of sexual reproduction and diploidy. In fact, it is highly unlikely that any individual in an admixed population will be able to trace all their genes to a single source population (see *Figure 12.2*); different genomes within an admixed population are likely to exhibit differing amounts of admixture. Thus an estimate of population admixture can only be an average of the admixture among the individual genomes within it.

The complex genetic ancestry of a genome often contrasts markedly with the simplicity of an individual's perceived identity. Consequently, public dissemination of the results from admixture studies needs to be undertaken responsibly and with due care for their potential impact. It is worth remembering that ancestral populations themselves are likely to be the result of earlier admixture.

In this chapter we will consider the wide variety of statistical methods that have been brought to bear on this important issue in population prehistory. Then we will examine some case studies that illustrate the power of these methods and demonstrate the diverse outcomes of population encounters in the past. But first, we shall consider the impact of admixture on other features of the populations involved.

12.2 **The impact of admixture**

Genetic admixture is not the only consequence of the meeting of populations. Such events often impact significantly upon the cultural lives of the populations involved. Thus many episodes of prehistoric admixture may well be detected using other records of prehistory, although it should be remembered that there need not necessarily be an archaeological or linguistic correlate for every genetic episode, and *vice versa*.

In the first few generations of admixture, individuals that descend primarily from one of the ancestral populations are often easily identifiable through their language or appearance. Individuals apportioned to the different ancestries are rarely on an equal footing in the nascent society. For example, the status of African slaves, European settlers and Native Americans within the Americas was far from equal. Genetics cannot be divorced from these sociological considerations as they directly influence the nature of the admixture. Rather, integrating genetic evidence with other prehistorical and historical records affords a richer appreciation of population encounters in prehistory.

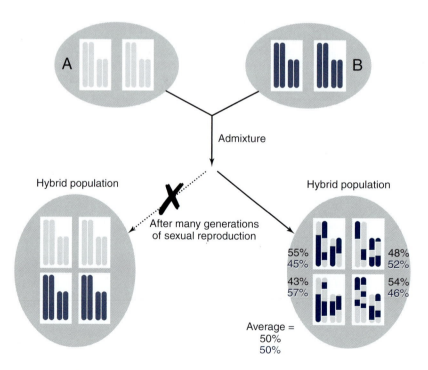

Figure 12.2: Admixture within individual genomes.

Diploid genomes comprising two autosomes are shown schematically in ancestral and hybrid populations. There are two individuals (white rectangles) in each ancestral population (gray ovals), and four individuals in each hybrid population. Admixture does not result in a population in which individuals can trace the ancestry of their entire genome to one of two ancestral populations, but rather a population in which all individuals have genomes of mixed ancestry.

12.2.1 Linguistic evidence for admixture

Populations that have been isolated from one another will accumulate linguistic differences relatively rapidly, and perhaps even speak different, mutually unintelligible, languages. What kinds of linguistic changes in a hybrid population might we expect to see as a result of their admixture? Bilingualism is one outcome, and seems to involve little change to either language, but as we shall see, this itself can be a cause of language change.

The first point to appreciate is that a language is not like a genome. Whereas it is perfectly possible to assemble a fully functioning hybrid genome from several ancestral genomes with no associated costs, the same is not generally true of languages, because languages must maintain a certain level of coherence to function adequately. Linguists call native languages that represent true hybrids of parental languages **'creoles'**. A number of well-known modern creoles derive from European languages (French, Spanish, Portuguese and English) and indigenous languages brought together by the actions of European colonial powers over the past few hundred years. For example the Cajun language of Louisiana is a creole derived from French and languages spoken by African slaves. Creolization is considered to be a relatively rare process in language evolution.

Much more common than the development of creoles is the limited incorporation of certain features of one language into a much more dominant substrate from another language. These linguistic borrowings can affect different aspects of the language. The simplest example is the incorporation of outside words into a language. For example, among Polynesian languages, the term for the sweet potato (*'kumara'* in New Zealand Maori) clearly derives from the word *'kumar'* from the Quechuan languages spoken in South America. However, word borrowing into a vocabulary is not the only possible linguistic borrowing. Elements of structure can also be borrowed. For example, the order of subject, verb and object within a sentence are often different between languages. Some **Austronesian** languages spoken in areas with neighboring **Papuan** speakers have adopted the 'verb last' order of these Papuan languages (subject-object-verb – 'he it hit'), as opposed to the more common 'verb medial' organization (subject-verb-object – 'he hit it') found among closely related Austronesian languages. These types of structural change result from what linguist Malcolm Ross calls:

> '. . . the natural pressure to relieve the bilingual speakers' mental burden by expressing meanings in parallel ways in both languages.'

There are thus a wide variety of possible linguistic consequences of contact between populations speaking different languages, and which particular consequence follows in which situation is – as with the genetic outcomes – largely determined by the social context of this contact.

Spoken languages are not the only linguistic source of evidence for admixture: surnames and place names (**toponymy**) are both capable of revealing the hybrid nature of a population. For example, English towns have names derived from Celtic, French (Norman) and Scandinavian languages as well as from Anglo-Saxon. It is worth noting, however, that evidence of past contact from sources such as place names does not imply that significant genetic admixture will be found in the current inhabitants. The Caribbean island populations of today, for example, may contain little genetic input from the original inhabitants.

12.2.2 Archaeological evidence for admixture

The answers that archaeology provides to the question 'what happens when populations meet?' are primarily cultural in nature. The temporal and spatial distribution of archaeological sites can be used to demonstrate contact between cultures, and subsequent cultural change. However, such evidence only establishes the *potential* for genetic admixture. Before genetic admixture can be inferred, it must be assumed that:

▸ the movement of artifacts is mirrored by the movement of people. In other words, artifacts are not being distributed by a set of sequential trading exchanges;

▸ populations with different **material cultures** are also different genetically;

▸ offspring with parents from different cultures are tolerated.

The cultural changes that accompany the meeting of peoples cover a range of outcomes too vast to be covered here in any detail. New cultures can be adopted wholesale, or elements of individual cultures can be combined together in a process of integration. The integration of two cultural traditions may be accompanied by genetic admixture, or it may not. Similarly, a wholesale replacement of cultural practices may or may not be associated with a similar replacement of genes. With these caveats in mind it is worth noting that the spatio-temporal spread of an archaeological culture does indicate the geographical location of likely ancestral populations. In addition, the precision of archaeological dating provides good estimates for the time-scale of potential admixture processes.

Approaches based on **physical anthropology** have been adopted to seek phenotypic changes associated with genetic admixture in skeletal remains. Such work is often contentious, as human populations can rarely be well differentiated on skeletal evidence alone. In addition, alterations in cultural practices, for example in diet, may cause significant morphological changes through **developmental plasticity**, rather than any change of genes (Kaplan, 1954; see Section 13.2).

12.2.3 The biological impact of admixture

Our focus in this chapter is on identifying past admixture. As throughout this book, this requires a focus on noncoding or neutral genetic markers in an attempt to exclude selection processes as explanations for modern patterns of genetic diversity. Nonetheless, all genetic admixture will lead to a variety of phenotypic effects. Any quantitative trait that is genetically encoded and well differentiated between populations will be altered in admixed populations. Obvious physical examples include skin color, hair color and stature (Chapter 13). In the past, in the absence of genetic data, these phenotypic data have been used to calculate admixture proportions. However, without knowledge of the genetic interactions underlying such traits such estimates can be highly variable.

In societies where surnames follow clear lines of inheritance, they have often been used for population genetic analyses. Admixture studies are no exception. Patterns of surname introgression have been clearly shown to be correlated with levels of admixture in a number of different populations (reviewed in Chakraborty, 1986). These conclusions have subsequently been reinforced by genetic typing. Nevertheless, surname analysis has been dubbed the 'poor man's population genetics', and is of real use only where genetic data are unavailable, and when admixture has occurred within the time-frame of surname usage, which varies greatly from population to population, and may be very recent. However, if records are detailed enough, surname analysis can reveal how admixture processes may have changed over time.

Disease **prevalences** are often clearly different between ancestral populations (see Chapter 14 for a discussion). An obvious medical consequence of admixture is that the hybrid population is expected to have disease prevalences for **Mendelian** disorders that are intermediate between those of the ancestral populations. When the most frequent diseases differ between the populations, this can lead to an overall lowering of the disease burden through a reduction in the probability of having two parents carrying the same deleterious **recessive** allele (*Table 12.1*).

Given the variation in degree of admixture among individuals in an admixed population, it may be found that the proportion of admixture can be correlated with susceptibility to certain diseases more prevalent in one or other of the ancestral populations. It has been proposed that the prevalence of type 2 diabetes (OMIM 125853) in different Native American populations is positively correlated with the proportion of Native American genes, irrespective of whether this proportion is assessed phenotypically, genealogically or genetically (Chakraborty, 1986). More recently, it has been shown that nondiabetic Pima Indians have twice as much European admixture on average than diabetic members of the same population (Williams *et al.*, 2000). However, it is unlikely that assessing overall levels of individual admixture will have major predictive value to individuals.

For polygenic diseases, the possibility remains that each ancestral population contains individuals with co-adapted combinations of alleles that will be disrupted by admixture, resulting in a higher burden of disease in the hybrid population than in either ancestral population. However, no example of this **outbreeding depression** has yet been demonstrated in humans.

There remains another mechanism by which admixture can result in an increased disease burden. Human populations often harbor different populations of pathogens to which they have previously developed resistance. The release of these pathogens into an 'immunologically naïve' population could result in a substantial increase in infectious disease. The resulting selective pressures would result in a substantial bias towards contributions from the resistant ancestral population in the admixed population. In early generations this bias would extend towards all genomic loci irrespective of their linkage to the locus conferring disease resistance, although in later generations this bias would be confined to linked loci. This type of episode is exemplified by the population crashes witnessed in Polynesia and the Americas on first contact with Europeans bearing novel pathogens, such as those causing smallpox and measles.

Dramatic demographic changes need not be wrought solely by infectious disease. Economic factors too can play an important role. Competition for resources, potentially resulting in conflict, is a frequent consequence of population encounters. The cohabitation of large parts of the Old World by modern and archaic forms of *Homo sapiens*, the most famous examples of which are **Neanderthal** and Cro-Magnon man in Europe, is one scenario in which economic factors,

TABLE 12.1: ADMIXTURE CAN REDUCE THE DISEASE BURDEN OF RECESSIVE SINGLE GENE DISORDERS.

Population	Carrier frequency of allele A	Carrier frequency of allele B	Incidence of disease A	Incidence of disease B	Total disease incidence
A	1/10	0	1/400	0	1/400
B	0	1/15	0	1/900	1/900
1 : 1 admixture of A and B	1/20	1/30	1/1600	1/3600	~ 1/1100 (13/14400)

Only a quarter of children with carrier parents will be affected by a recessive disease.

and conflict, have been invoked. Some combination of these three factors: disease, environmental and economic, may explain the relatively rapid demise of the Neanderthals.

12.3 Detecting admixture

12.3.1 Methods based on allele frequencies

Allele frequency-based methods were among the first admixture detection analyses to be developed, as a result of their applicability to protein data. Only subsequently have they been applied to DNA data. The simplest scenario occurs when no alleles are shared between the ancestral populations. Each allele in the hybrid population can then be unambiguously assigned to an ancestral population, and the proportion of admixture calculated by simply counting up the number of alleles assigned to each population. However, an absolute lack of allele sharing between ancestral populations is relatively rare; more often, alleles are found within many populations at differing frequencies. In principle, it is easy to estimate the proportion of admixture in a hybrid population formed from two ancestral populations (see *Figure 12.3*).

For any given allele, if we know its frequency in the ancestral populations A and B (p_A and p_B) and in the hybrid population (p_H), we can estimate the proportion (M) that ancestral population A contributed to the admixed population by rearranging the equation (Bernstein, 1931);

$$p_H = Mp_A + (1-M)p_B$$

to give:

$$M = (p_H-p_B)/(p_A-p_B)$$

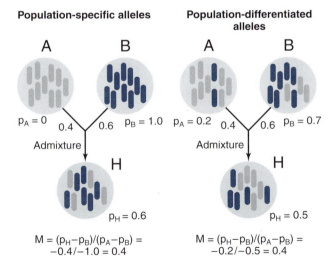

Population-specific alleles

Population-differentiated alleles

A   B   A   B

$p_A = 0$ 0.4   0.6 $p_B = 1.0$ $p_A = 0.2$ 0.4   0.6 $p_B = 0.7$

Admixture   Admixture

H   H

$p_H = 0.6$ $p_H = 0.5$

$M = (p_H-p_B)/(p_A-p_B) =$
$-0.4/-1.0 = 0.4$

$M = (p_H-p_B)/(p_A-p_B) =$
$-0.2/-0.5 = 0.4$

Figure 12.3: Calculating admixture proportions (M) when alleles are population-specific, and when they are present in both populations but at different frequencies.

p_A, p_B and p_H are the frequencies of an allele in the two parental populations A and B, and the hybrid population, H, respectively. The admixture proportions (M) are calculated using the equation described in the text, for two different scenarios.

Obviously, this approach requires the unambiguous identification of the ancestral populations. Calculating precise allele frequencies in these ancestral populations requires a sample size beyond the reach of most ancient DNA investigations. Thus we must try and identity modern populations that faithfully represent the genetic diversity of the ancestral populations. Choosing alternative modern analogs for ancestral populations can have a dramatic effect on the resulting admixture estimates, and considerable care must be taken when specifying these populations.

It should be clear that because we estimate the true allele frequency within a population by taking a sample from it, we need to take account of sampling errors when assessing our confidence in estimates of admixture proportions. However, even if we assume there has been no subsequent migration between the ancestral and hybrid population after the admixture event, there are a number of additional potential complicating factors:

▸ If more than one locus is being studied, how should they be averaged?

▸ What has been the effect of genetic drift upon allele frequencies in all three populations since admixture?

▸ What if we misidentify the ancestral populations?

▸ What if there were more than two ancestral populations?

▸ What if an allele in the admixed population is not found in either ancestral population?

These complications have led to the development of a whole series of different admixture estimation procedures. These can be classified on the basis of the type of data used, the assumed model for admixture and whether they seek to estimate population or individual admixture. *Figure 12.4* illlustrates a number of different admixture models.

Given a set of multi-locus allele frequencies in ancestral and hybrid populations, what is the best way of getting a single estimate of admixture proportions from these data? Admixture estimates can be calculated for each allele, or locus, individually and then averaged. There are a number of different ways of averaging this information across loci.

The equation for M given above suggests that estimates of admixture proportions from different alleles should be related linearly. In other words plotting (p_H-p_A) against (p_A-p_B) for different alleles should give a straight line of gradient M. *Figure 12.5* shows this property for an idealized admixture situation where all alleles give the same estimate of M.

The plot in *Figure 12.5* immediately suggests one method of averaging information from different estimates, namely to plot the least-squares regression line between the points and take its gradient as the multi-locus estimate of M (Roberts and Hiorns, 1962). This estimator of M is often known as **m_R**.

The above method assumes that the allele frequencies are known without error; it does not take account of the different levels of precision associated with each individual

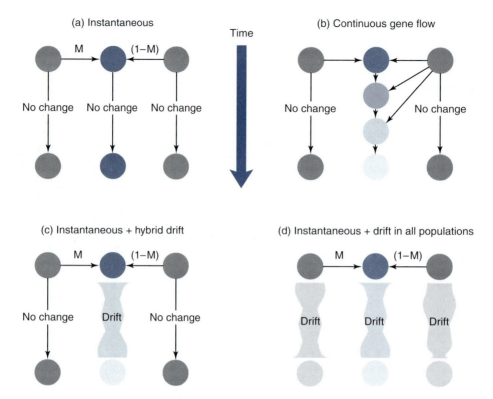

Figure 12.4: Different admixture models of varying complexity.

(a) Instantaneous admixture, (b) cumulative effect of gene flow across many generations, (c) instantaneous admixture allowing for drift in the hybrid population and (d) instantaneous admixture allowing for drift in all three populations. Gray circles, parental populations; blue circles, hybrid populations.

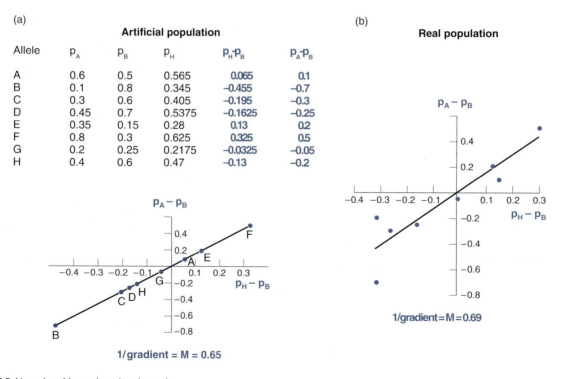

(a)

Artificial population

Allele	p_A	p_B	p_H	$p_H\text{-}p_B$	$p_A\text{-}p_B$
A	0.6	0.5	0.565	0.065	0.1
B	0.1	0.8	0.345	−0.455	−0.7
C	0.3	0.6	0.405	−0.195	−0.3
D	0.45	0.7	0.5375	−0.1625	−0.25
E	0.35	0.15	0.28	0.13	0.2
F	0.8	0.3	0.625	0.325	0.5
G	0.2	0.25	0.2175	−0.0325	−0.05
H	0.4	0.6	0.47	−0.13	−0.2

(b)

Real population

Figure 12.5: Linearity of (p_H–p_A) against (p_A–p_B).

(a) An idealized case of admixture in which eight alleles (A–H) give exactly the same value for M, and (b) a more realistic set of variant estimates of M from different alleles, M is estimated by fitting a best fit line through the points.

estimate. This can be taken into account by averaging the different estimates weighted according to their precision, as assessed by their **variances**. These variances depend on the size of the samples. The weighting factor commonly used is the inverse of the variance of the estimate. In other words, the higher the variance of the estimate, the less we are sure that it is accurate so the lower the weight we give it (Cavalli-Sforza and Bodmer, 1971). This weighted average approach takes into consideration sampling effects on all allele frequency estimates.

One factor not considered up to now is that admixture estimates from alleles at the same locus are not independent of one another. This is because allele frequencies at a single locus must sum to one. If we consider a locus at which there are only two alleles, if one allele gives a reliable estimate of M, then the other allele at the same locus is likely to agree with it. Alternatively, if one allele gives a deviant admixture estimate, the other is likely to be similarly deviant. This problem is known statistically as 'constrained variables', but has been surmounted by a weighted average of maximum likelihood estimates of admixture proportions at each locus (Long and Smouse, 1983). This estimator for admixture is often known as the weighted least-squares (**WLS**) approach.

Admixture should affect allele frequencies at multiple loci to an equal degree, since all depend upon the same parameter, M. However, this is only true for neutrally evolving loci. Selection can bias the frequency of alleles in the admixed population. To see why this might be the case we need to appreciate that the admixed population often inhabits an environment which exerts different selective pressures from that inhabited by either ancestral population. A change in selective pressures will cause a change in allele frequencies at those loci irrespective of any admixture event. Consider an allele that was previously maintained in one of the ancestral populations by **balancing selection** that might now be present in an admixed population in which it has no **heterozygote advantage**. For example, the sickle-cell anemia (OMIM 603903) allele (Hb^S) in heterozygotes protects against malaria in Africa, but in the admixed population of African–Americans in the USA, where malaria is absent, is simply deleterious in homozygotes. Consequently, the frequency of Hb^S in African–Americans is much closer to that in European–Americans than we might otherwise expect. Admixture proportions calculated solely on the basis of this allele give much higher estimates for the contribution of European genes to African–Americans than do other, neutral, loci.

This finding can be turned on its head to provide a means for identifying selection acting upon specific alleles. Heterogeneity among admixture estimates derived from the frequencies of different alleles can be used to pinpoint those alleles whose admixture estimates deviate significantly from those of most other alleles (see *Figure 12.6*).

However, selection is not the only evolutionary process that can distort allele frequencies and the admixture estimates derived from them. **Genetic drift** also influences these estimates. Distinguishing between systematic biases in

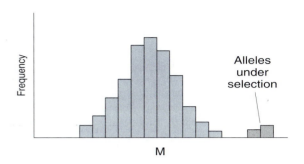

Figure 12.6: Deviant estimates of M reveal outlier alleles under selection.

A change of selective environments between ancestral and hybrid populations can lead to selection pressures on some alleles resulting in estimates of M that are outliers when compared to neutral alleles.

admixture estimates resulting from selection, on the one hand, and random biases introduced by sampling effects and genetic drift, on the other, is far from easy.

Jeffrey Long has developed a method that seeks to disentangle these different sources of bias, by apportioning the variation in admixture estimates produced by different alleles into that caused by sampling, and that caused by evolutionary processes such as drift and selection (Long, 1991). He has shown for a number of well-known case studies that variation among different admixture estimates is due more to the action of genetic drift than to sampling effects. He has also shown that once variation due to drift has been accounted for, it is very difficult to detect the imprint of selection in the heterogeneity of admixture estimates. It should be noted that this method only takes account of drift in the admixed population and not in the ancestral populations.

An alternative approach to calculating admixture proportions from allele frequency data is to transform these data into a measure of **genetic distance** between the hybrid and each ancestral population. The smaller the genetic distance, the greater has been the contribution of that ancestral population to the hybrid. It turns out that measures of similarity are more reliable than measures of distance so more recent applications of this approach have considered gene identities rather than genetic distances. As in the equation relating allele frequencies above, gene identities between populations can be simply related to the proportion of admixture, M, by the equation:

$$I_{AH} = MI_{AA} + (1-M)I_{AB}$$

Where I_{AH} is the gene identity between the hybrid population (H) and one of the parental populations (A). This method has been further developed to take into account any number of ancestral populations (Chakraborty, 1986).

The application of **coalescent theory** is revolutionizing standard population genetic analyses (see *Box 6.4* for more details) and admixture studies are no exception. Recently, a method based on coalescent simulations has been devised which enables drift in hybrid *and* ancestral populations to be incorporated into a **likelihood**-based estimation of admixture (Chikhi *et al.*, 2001). This method, named LEA, has the additional advantage of allowing the simultaneous estimation of the admixture proportions and the time since the admixture event. The method has been recently applied to the issue of the relative Paleolithic and Neolithic contributions to modern European populations (Chikhi *et al.*, 2002). Their findings are discussed in greater depth in Section 10.5.

12.3.2 Methods taking molecular information into account

The interrogation of genetic diversity at the DNA level allows much greater resolution of variation than that at the protein level. This variation typically comes in the form of DNA sequences or haplotypes of linked microsatellite alleles; here we use the term haplotype to include both types of data. We are much more likely to find haplotypes in the hybrid population that have no matches in any ancestral population if we use DNA data than if we use data from protein markers. In the equation given above both p_A and p_B are zero and so this **'private' haplotype** is ignored in subsequent calculations. How can we take account of these haplotypes when estimating admixture proportions?

A novel haplotype may be not present in the ancestral populations for one of two reasons:

▶ it has been lost from an ancestral population through drift, or sampling;

▶ it has been derived from an **ancestral haplotype** by mutation since the admixture event.

Thus we must decide from which ancestral population the novel haplotype is most likely to have derived. This requires a consideration of which are the most likely ancestral haplotypes, and of their frequency within each ancestral population. The **principle of parsimony** dictates that the most likely ancestral haplotypes are those that are the fewest number of mutational steps away from the novel haplotype. Consequently, these methods must take account of molecular distances between haplotypes, thus incorporating mutation into the admixture model.

One method for inferring likely ancestral haplotypes uses phylogenetic networks of haplotypes present in ancestral and hybrid populations to identify those haplotypes closest phylogenetically to a 'private' haplotype. The frequency of these inferred ancestral haplotypes in the ancestral populations is then used to calculate a frequency-based admixture estimator known as $\mathbf{m}_\rho$ (Helgason *et al.*, 2000). This estimator uses a phylogenetic distance, rather than simply the observed number of mutational steps between haplotypes, and so is reminiscent of the molecular distance, ρ, used for genetic dating purposes (described in more detail in Section 6.6.2). This estimator has been applied to Y-chromosomal microsatellite data to demonstrate that both Scandinavian and Irish Gaelic males contributed to the

colonists of Iceland, with Scandinavians providing the lion's share (see *Figure 12.7*).

An alternative approach to incorporating mutation processes into admixture studies is to adopt a coalescent-based analysis. Molecular diversity can be related to **coalescent time**: the greater the molecular diversity, the older the coalescent time. In this manner, the mean coalescence times of pairs of haplotypes drawn from an ancestral population, and of pairs of haplotypes drawn one from the hybrid population and one from the ancestral population can be calculated, and related to the admixture proportions (Bertorelle and Excoffier, 1998). The admixture estimate obtained is known as $\mathbf{m}_Y$. This approach requires that a matrix of molecular distances between haplotypes be specified. Consequently, this is applicable to many different types of molecular data, including sequence data and microsatellite haplotypes, and can incorporate observed molecular distances or phylogenetic distances.

Although initially formulated for a pair of ancestral populations, this coalescent-based approach has been expanded to situations with more than two possible ancestral populations (Dupanloup and Bertorelle, 2001). This latter method has been used to address the relative contributions of Europeans, North Africans and sub-Saharan Africans to the modern day population of the Canary Islands situated off the coast of northwest Africa (see *Figure 12.8*). From mtDNA sequences, it appears that the major female contribution was made by the North African Berbers who are known to be the first settlers of the islands, with only minor contributions from Spanish and Guinean populations.

Different demographies have been simulated to compare the performance of admixture **estimators** that include molecular distance information with those that do not. No

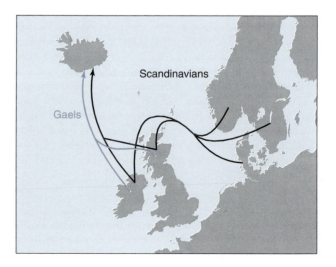

Figure 12.7: Populations contributing to Icelandic genetic diversity.

Scandinavian populations colonized Iceland via settlements in Gaelic-speaking regions of Ireland and Scotland, leading to two major contributions to the Icelandic gene pool.

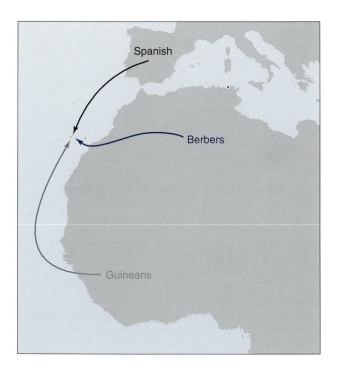

Figure 12.8: Genetic contributions to the inhabitants of the Canary Islands.

Since the initial colonization of the Canary Islands from northwest Africa, Spanish and Guinean populations have also contributed to the modern Canarian gene pool.

single estimator is the best in all scenarios. As expected, the molecular estimators perform better when mutational processes have been dominant over genetic drift in determining differentiation of the populations. These include situations where the ancestral populations have been isolated from one another over long periods of time and the genetic loci being studied have high values of θ (high mutation rate, low drift).

No single estimator of admixture takes account of all of the complexities listed at the start of Section 12.3.1. For example, the coalescent-based molecular estimators considered above do not allow for the ancestral and hybrid populations to experience different levels of genetic drift (Bertorelle and Excoffier, 1998). By contrast the methods which do allow more complex patterns of genetic drift to be modeled are not capable of taking account of mutational processes (Chikhi *et al.*, 2001). Thus the appropriate admixture estimator must be chosen on the basis of whether genetic drift or mutation is likely to have been the dominant mechanism causing population differentiation. No doubt future coalescent-based approaches to admixture estimation will incorporate more complex population models.

However, all methods discussed above will produce estimates of M even if the ancestral populations have been grossly misidentified, and so great care will always be necessary to identify them correctly. These assumptions

about population history often require the support of historical, archaeological and linguistic evidence.

12.3.3 Estimates of admixture proportions in individuals

The calculation of admixture proportions within a hybrid population does not mean that a number of individuals within that population trace all their ancestry from one ancestral population whereas others derive from a different ancestral population. In reality, after the passage of only a few generations, most or all individuals belonging to the hybrid population carry haplotypes that derive from more than one ancestral population. Clearly there will be variation in the amount of admixture within individual genomes. Substructure within the admixed population, perhaps resulting from **assortative mating**, will reveal itself in unusually large variation in individual admixture estimates.

Estimates of population admixture can be considered to be averages of admixture within individuals (remember *Figure 12.2*). Levels of population admixture can be estimated from individual admixture estimates, but not *vice versa*. How can we investigate this finer-grained admixture?

An appreciation of genomewide admixture can only be gained by inferring the ancestry of multiple unlinked loci within that genome, and thus requires substantial typing effort. Some alleles may not be particularly well differentiated between different ancestral populations, and thus may not be particularly informative. To reduce the typing load many studies have focused on what are called **'population-specific alleles'** (PSAs). These are alleles for which the ancestral populations exhibit considerable ($> 45\%$) differences in frequency (Shriver *et al.*, 1997; this work is discussed in a forensic context in Section 15.2.2). Analysis of these markers has the additional advantage of giving admixture estimates with lower variance than those derived from less well-differentiated alleles. The high degree of population differentiation exhibited by the nonrecombining portion of the Y chromosome makes it a valuable source of such markers. However, this entire 30-Mb chromosomal region represents but a single locus, with a single evolutionary history that may not be representative of the rest of the genome.

There are potential problems in focusing solely on PSAs. Because of the low degree of **genetic differentiation** among modern humans most alleles do not show high population specificity (see Section 9.3 for more details). There are at least two reasons why some alleles do show appreciable population specificity:

▶ they result from relative recent mutations that have not had sufficient time to disperse to any great degree;

▶ they result from the action of selection, which influences allele frequency differently in different selective environments.

In the first instance, the recent nature of PSAs means that they are generally only found at low frequency, and so require large sample numbers for reliable detection. In the second instance, selection, rather than levels of admixture, may have played a significant role in determining the allele frequency in the hybrid population. Consequently, estimates of admixture based on such an allele will be biased. Remember the lower than expected frequency of the sickle cell anemia allele in admixed African–Americans. Similarly, alleles at the Duffy blood group locus (*FY*; OMIM 110700) exhibit high degrees of population specificity and recent evidence suggests that this locus has undergone selection as a result of its role in determining susceptibility to malaria caused by the parasite *Plasmodium vivax* (Hamblin *et al.*, 2002) – see Section 9.4.3. Nevertheless, the typing considerations outlined above have resulted in a near-exclusive focus on PSAs when calculating individual admixture. The availability of high-density microsatellite and binary polymorphisms in the human genome provides a huge potential resource to identify many more PSAs than have been used so far (Collins-Schramm *et al.*, 2002; see Section 15.2.2).

The actual calculation of individual admixture levels is statistically complex, because information has to be incorporated from many alleles at many loci, and interested readers can consult the references for more details. One commonly adopted method finds the value of M for each individual by considering the likelihood of observing the genotypes across all loci within that individual's genome (MacLean and Workman, 1973). More recently a **Bayesian**-based method has been developed that can take account of uncertainties in the ancestral population frequencies (McKeigue *et al.*, 2000).

Earlier we posed the important question as to whether or not it was possible to detect admixture in the absence of any prior information from external sources. In Section 12.5 we will see one way in which admixture leaves an unmistakable imprint in the pattern of genetic diversity within populations. However, a consideration of individual genetic diversity can also reveal 'cryptic' **population structures** that were not previously known to exist.

Clustering methods can take multi-locus genotypes of individuals from several populations and apportion them into well-resolved clusters that are clearly differentiated from one another. One fundamental problem is deciding upon the most likely number of clusters. Recently a model-based clustering method (STRUCTURE) has been devised that determines the most likely number of clusters, the frequency of any given allele in each cluster and the proportion of each individual's genome that owes ancestry to each cluster (Pritchard *et al.*, 2000) – *Box 9.3*. Thus, under the assumption that all loci are in **Hardy–Weinberg equilibrium** in all populations, STRUCTURE is capable of calculating individual admixture, where the ancestral populations are genetic clusters and not populations imposed by the sampling process, nor presupposed on external evidence (for example, people speaking the same language). Simulated evolution of a pair of populations evolving in the absence of admixture has shown that the two clusters formed correspond to the populations themselves and that individuals

owe all their ancestry to a single cluster. By contrast, when admixture is included within the simulation, again two clusters are formed but many individuals find some ancestry in both, see *Figure 12.9*.

The identification of cryptic population structure is important for other applications. For example, if undetected it can cause spurious associations when hunting disease genes (see Chapter 14) and unreliable match probabilities in forensic situations (see Chapter 15). *Table 12.2* lists software for estimating admixture proportions.

12.4 Example 1: Sex-biased admixture

12.4.1 What is sex-biased admixture?

So far we have considered that all neutrally evolving loci should give the same estimate of admixture proportions. However, there is a situation in which this scenario does not hold true. Unequal contributions of the different sexes of an ancestral population to an admixed population result in a phenomenon known as **'sex-biased admixture'**. Males or females may contribute disproportionate amounts of admixture. In the extreme case, admixture may be restricted to one sex only, so-called **'sex-specific admixture'**. Sex-biased admixture can result from a sex-bias in the make-up of one of the ancestral populations. There are clear examples of this from recent colonial admixture. The European explorers, traders and missionaries who have traveled the world over the past 500 years have been predominantly male. As a consequence, admixed populations resulting from these contacts are likely to exhibit male-biased admixture. We saw

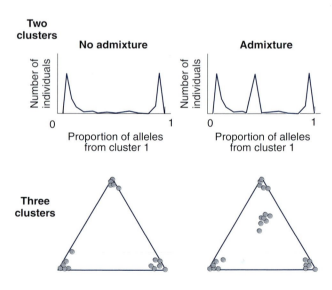

Figure 12.9: Clustering identifies 'cryptic' admixture.

Schematic graphical representations of how much ancestry each individual traces from each of two or three ancestral populations (clusters). In the absence of admixture the majority of alleles in each individual can be traced to a single cluster. Admixture can be identified when groups of individuals are found to fall between clusters.

TABLE 12.2: SOFTWARE FOR ADMIXTURE ESTIMATION.

Method	Software	URL
LEA	LEA	http://www.rubic.rdg.ac.uk/~mab/software.html
m_Y (pair of ancestral populations) m_R	ADMIX 1.0	http://web.unife.it/progetti/genetica/Giorgio/giorgio_soft.html
m_Y (multiple ancestral populations)	ADMIX2.0	http://web.unife.it/progetti/genetica/Isabelle/admix2_0.html
Gene identity	ADMIX95	http://www.genetica.fmed.edu.uy/software.htm
Bayesian individual admixture	ADMIXMAP	http://www.lshtm.ac.uk/eu/genetics/admix.html
Model-based clustering	STRUCTURE	http://pritch.bsd.uchicago.edu/

an unambiguous example of this in Chapter 11, where the paternally-inherited Y chromosomes of Cook Islanders in the Pacific are one third European, but no European admixture appears among their maternally-inherited mitochondrial DNA (Hurles *et al.*, 1998).

Sex-biased admixture is not restricted to the scenario outlined above; it may also result from admixture between ancestral populations, neither of which is sex-biased (see *Figure 12.10*). As was emphasized in previous sections, we cannot divorce genetic admixture from the wider social

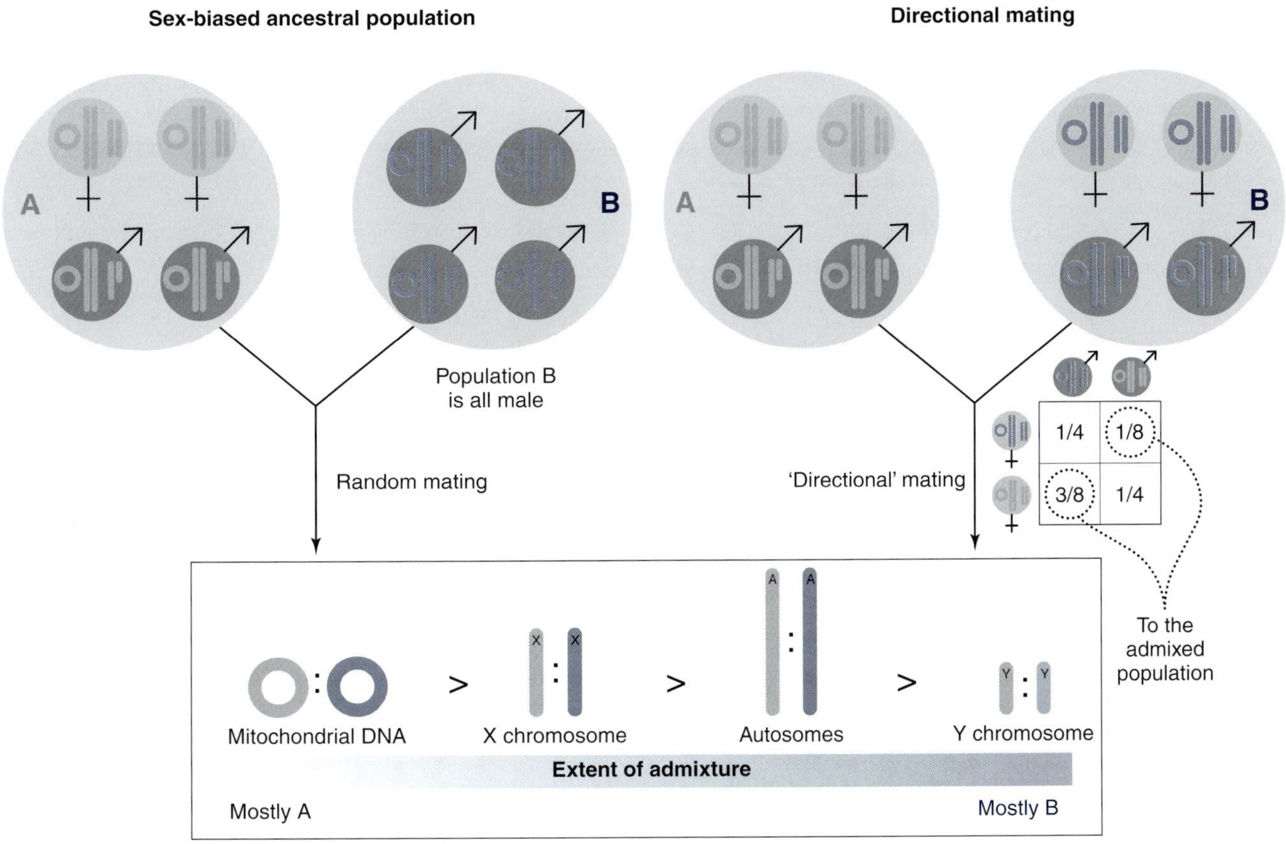

Figure 12.10: Scenarios under which sex-biased admixture may occur.

Two different scenarios that lead to sex-biased admixture are shown. Each individual is represented by a pair of autosomes, a pair of sex chromosomes and a mitochondrial genome (circle) colored according to their population affiliation. In each scenario two ancestral populations (A and B) contribute to the admixed population. In the first scenario Population B consists only of males and so contributes no mitochondrial genomes, many Y chromosomes, and intermediate levels of autosomes and X chromosomes to the admixed population. In the second scenario, mating of males from population B with females from population A represents three-quarters of the mixed matings that contribute to the admixed population, and outweighs matings of females from population B with males from population A, leading to a sex-bias in the contributions to the admixed population.

context in which it occurs. Ancestral populations rarely have equal status when they encounter one another for the first time. The colonial situation illustrates this clearly. In addition, human populations rarely exhibit random mating especially across perceived racial or socioeconomic boundaries. Such boundaries are often more permeable to one sex than another, an imbalance that is sometimes dependent on whether the individual is mating 'above' or 'beneath' themselves in status terms. For example, in the Indian caste system, it is easier for a woman to marry a man from a higher caste than *vice versa*. This feature is also apparent in Western societies. In England, the frequency of marriages between white females and African–Caribbean males is greater than that between African–Caribbean females and white males (Model and Fisher, 2002). This phenomenon is also known as 'directional mating'. The social treatment of mixed marriages is an important factor: how are unions across status boundaries incorporated into the population? Mixed unions are often stigmatized and so are incorporated into the lower status group, or are ostracized from all groups. All these factors potentially skew the contributions of males and females to all admixed populations. Thus sex-biased admixture may be a relatively common feature of admixture, but how can it be detected?

12.4.2 Detecting sex-biased admixture

Sex-biased admixture will cause admixture estimates from loci with different patterns of inheritance to differ markedly. Thus, admixture estimates are not pooled from all loci, but are compared between loci with different patterns of inheritance. There are four different patterns of inheritance in the human genome:

▶ exclusively maternal inheritance – mitochondrial DNA;

▶ exclusively paternal inheritance – Y chromosome;

▶ 2 : 1 female-biased inheritance – X chromosome;

▶ equal, biparental inheritance – autosomes.

If females contributed more than males to an admixed population, estimates of admixture will be lie on a gradient:

mitochondrial DNA > X chromosome > autosomes > Y chromosome

By contrast, if males contribute a greater proportion, the gradient is reversed:

Y chromosome > autosomes > X chromosome > mitochondrial DNA

In principle comparisons between any two of these types of loci can reveal sex-biased admixture. However, for largely historical and technical reasons, the most common comparisons made in practice have been between either mtDNA and autosomal estimates, or mtDNA and Y-chromosomal estimates. Given the often substantial variance around admixture estimates described above, it makes sense to compare loci at which the greatest differences in admixture estimates should be expected. Only the largest sex biases are going to be apparent from significantly different

admixture estimates between X-chromosomal and autosomal loci. Because mtDNA and Y-chromosomal markers have only recently been developed (compared to **classical protein markers**), the detection of sex-biased admixture at the genetic level has become possible only in the last decade. It is important to take drift into account when comparing mtDNA and Y-chromosomal admixture estimates, which, due to the small effective population sizes of these loci, are prone to **stochastic** fluctuations.

As with all admixture studies, an additional factor to be considered when contrasting admixture estimates from different loci is the allele frequency difference between ancestral populations; the larger this difference, the more precise the admixture estimates. We saw in Chapter 9 that the level of population differentiation differs between loci with different inheritance patterns, with Y-chromosomal markers exhibiting by far the highest levels of differentiation. This means that, on average, Y-chromosomal markers will exhibit the greatest difference in frequency between ancestral populations. This makes the inclusion of these markers particularly attractive for studying sex-biased admixture.

Care must always be taken when comparing loci with different patterns of inheritance because the markers being analyzed often have different mutation dynamics. Any differences between them may result from the mutation-rate differences rather than the inheritance differences. For example, comparisons are commonly made between mtDNA sequences and autosomal microsatellites (e.g., Seielstad *et al.*, 1998), which by virtue of dissimilar mutation dynamics may be more unreliable than comparisons of more similar loci. However, this is less of a problem for admixture studies than it is for other analyses.

12.4.3 Sex-biased admixture resulting from directional mating

A more complex pattern of admixture than that described in the previous section is revealed when examining the origins of South American populations. Three major ancestral populations have contributed to modern genetic diversity on this continent: Native Americans, European colonists, and African slaves. The timing of these contributions is rather more complex than the admixture models outlined earlier in this chapter. Contributions occurred at different times and were from populations of different sizes. If we consider Brazil in isolation we find that at the time of the Portuguese 'discovery' in 1500 the Native American population was thought to number 2.4 million. Some half a million European settlers, predominantly male Portuguese, had arrived by 1808. The next two centuries saw more extensive settlement from a wider variety of source populations. Roughly 6 million settlers arrived during this latter period, of which 70% came from Portugal and Italy (other sources included Spain, Germany, Syria, Lebanon, and Japan). Meanwhile, in the 300 years between the mid-sixteenth and nineteenth centuries, some 4 million African slaves were imported into the country. These movements are summarized in *Figure 12.11*.

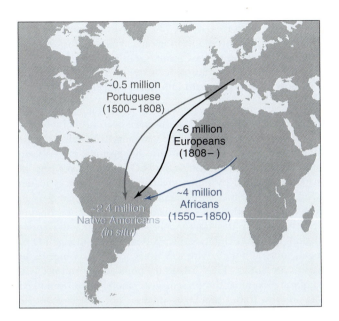

Figure 12.11: Map of genetic sources of Brazilian populations.

The different contributions to the modern Brazilian gene pool are described in the text.

The present day populations of Brazil, and the rest of South America, are far from homogenous. There is substantial **population structure**, with groups tracing predominant ancestry to different source populations, each with a very different socioeconomic status. A predominantly white middle class tends to occupy a position within society 'above' groups with more apparent ancestry from Native Americans or Africans, or both. However, levels of segregation differ greatly between different South American

countries. Genetic studies have been conducted on populations from all these groups within South American societies. These studies reveal a striking picture of sex-biased admixture in all groups, with directional mating between European males and Native American and African females. Having said that, the male-biased demography of the earliest settlers is also likely to have played a role in establishing the sex-biased admixture among modern populations.

Table 12.3 gives admixture estimates from Y-chromosomal, mitochondrial and autosomal loci in three populations: an Afro-Uruguayan population claiming predominantly African ancestry, a countrywide Brazilian 'white' population and a local Colombian population. Admixture estimates in these three populations were obtained by the gene identity method, allele counting, and the weighted least-squares approach, respectively.

A number of conclusions can be drawn from these results:

▸ as expected, autosomal values for admixture lie between mitochondrial and Y-chromosomal estimates;

▸ all three population samples contain genes from all three source populations;

▸ European ancestry is found at higher frequency among Y chromosomes than among mtDNAs;

▸ African and Native American ancestry is found at higher frequency among mtDNAs than among Y chromosomes.

Thus all the South American groups exhibit the same pattern of directional mating, despite their different socioeconomic status within society. Studies of African–American populations from the USA also demonstrate that European admixture is biased towards males. More surprisingly, low levels of female-biased Native American admixture have been identified in these populations, revealing them to have

TABLE 12.3: ADMIXTURE ESTIMATES IN SOUTH AMERICAN POPULATIONS FOR LOCI WITH DIFFERENT INHERITANCE PATTERNS.

Population/*locus*	% African	% European	% Native American
Afro-Uruguayan[a]			
Y-chromosomal	30	64	6
Autosomal	47	38	15
Mitochondrial	52	19	29
Brazilian Whites[b]			
Y-chromosomal	3	97	0
Mitochondrial	28	39	33
Colombians[c]			
Y-chromosomal	5	94	1
Mitochondrial	0–10	0–10	90

Data from [a]Sans *et al.*, 2002; [b]Alves-Silva *et al.*, 2000 and Carvalho-Silva *et al.*, 2001, [c]Carvajal-Carmona *et al.*, 2000.

three ancestral populations, rather than simply being a European–African mix (Parra *et al.*, 2001). The events of the past 500 years have clearly had a much more detrimental effect on the frequency of Native American paternal lineages than it has had on the frequency of their maternal lineages.

12.5 Example 2: Admixture and linkage disequilibrium

12.5.1 How does admixture generate linkage disequilibrium?

In Chapters 3 and 5 we encountered the phenomenon of **linkage disequilibrium** (LD), whereby two alleles at different loci tend to be co-inherited more often than we might otherwise expect. In theory, admixture events should generate LD between all loci at which differences in allele frequency exist between the two ancestral populations (see *Figure 12.12*). However, LD between unlinked loci (self-contradictory though this may sound) dissipates rapidly over a few generations as a result of **chromosomal segregation**. LD at physically linked loci decays more slowly due to recombination events. As a result, recently admixed populations should exhibit LD over greater genetic distances than non-admixed populations. This makes them of potential use for mapping disease genes, as discussed in Chapter 14.

A number of factors affect the extent of LD exhibited by an admixed population. These include:

▶ the time since admixture;

▶ the admixture dynamics, e.g., instantaneous or continuous gene flow;

▶ the relative contributions of different ancestral populations;

▶ the allele frequency differences between ancestral populations;

▶ the pattern of recombination in the human genome.

The presence of longer than expected lengths of LD represents an opportunity to identify 'cryptic' admixture events, for which historical evidence is lacking or contentious. These theoretical predictions about the relationship between LD and admixture have been backed up by computer simulations (Pfaff *et al.*, 2001). However, very little is known about the pattern of recombination in the human genome, and so it is of interest to see if empirical evidence can be found for extended LD in admixed human populations. One confounding factor is that length of detectable LD appears to be highly variable between different parts of the genome, making comparisons between different loci impossible.

12.5.2 Admixture and LD in the Lemba of southern Africa

The Lemba of southern Africa speak **Bantu** languages but claim Jewish ancestry (see *Figure 12.13*). Evidence, primarily from Y-chromosomal diversity, has confirmed that they are an admixed population with substantial contributions from sub-Saharan African and Semitic, potentially Judaic, populations (Thomas *et al.*, 2000). Mitochondrial DNA lineages within this population are typically African and appear to show no sign of admixture, suggesting a potential sex-bias in the admixture process (see *Box 12.2*). What do we know of the extent of LD in this population?

Comparisons of the extent of LD in the Lemba with that in the likely ancestral populations have been made by typing 67 microsatellite loci spaced at approximately regular intervals along the entirety of the X chromosome (Wilson and Goldstein, 2000). Long-range LD is found over pairs of markers separated by much greater genetic distances in the Lemba than in either ancestral population (see *Figure 12.14*). An artificial admixed population created by mixing genotypes from the Bantu and Ashkenazi ancestral populations *in silico* exhibits similar patterns of long range LD to the Lemba, although with 1.5 times more marker pairs in significant LD. This artificial population mimics an admixture event in the present generation, and so unlinked loci are also found in LD, a situation not observed in the Lemba. This suggests that admixture event in the Lemba occurred a sufficient number of generations ago to dissipate LD between unlinked markers but not so long ago that long-range LD between linked markers was also lost.

12.6 Example 3: Transnational isolates

12.6.1 What are transnational isolates?

Isolated populations are those that by virtue of their geography, history and/or culture have experienced little gene flow with surrounding populations. The term 'isolation' is a relative one; no threshold of per-generation gene flow has been set that defines a population isolate. Rather, a population is isolated when its surrounding populations more readily exchange genes with one another than with the isolate. This isolation can be revealed by unusual allele frequencies within the population compared to surrounding populations, and is often associated with linguistic and geographical boundaries. For example, Basques (see *Box 10.3*) and Finns are typically regarded as population isolates within Europe. As a result of their isolation, population isolates often have unusually high frequencies of typically rare genetic diseases. The Finnish genetic disease heritage is discussed in greater detail in Chapter 14.

Population isolates are commonly restricted to defined geographical regions. Luba Kalaydjieva has coined the term **'transnational isolates'** to refer to groups which, despite a more widespread geographical distribution, remain isolated, largely through the social practice of **endogamy**. While she used the word to describe the peculiar genetics of European Roma, here it is extended to other groups. As with traditionally defined population isolates, the genetic coherence of transnational isolates is readily apparent in their common disease heritage.

The paradox between the genetic coherence and geographical dispersal of these transnational isolates could result from one of two processes:

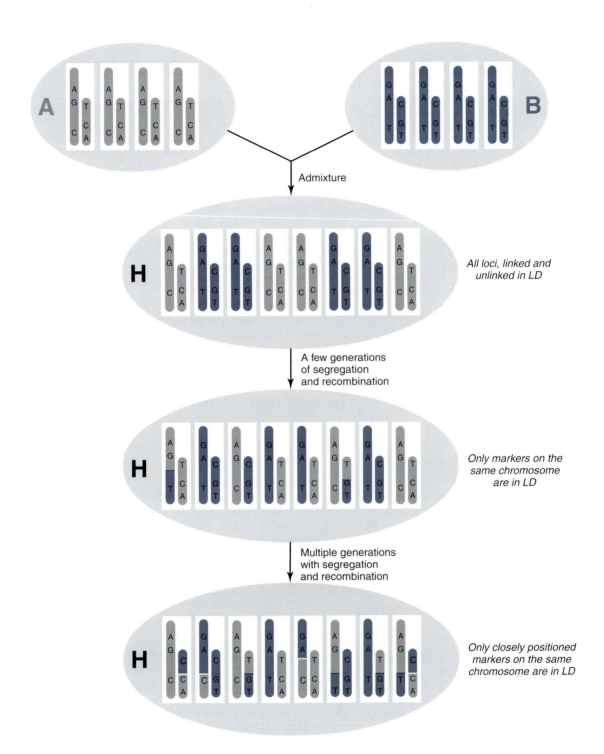

Figure 12.12: Admixture generates linkage disequilibrium (LD).

Decline in extent of LD over the generations after admixture. Two ancestral populations (A and B; gray ovals) contribute four haploid gametes (white rectangles) each represented by two chromosomes to a hybrid population (H). Three alleles are shown on each chromosome, and the chromosomal segment on which they lie is shaded according to the ancestral population from which it came. Immediately after admixture any alleles with frequency differences between the populations, irrespective of their genomic position, are in LD. After a few generations of chromosomal segregation only alleles on the same chromosome are in LD. As the number of generations increases, recombination events break down this LD until only closely positioned alleles on the same chromosome are in LD.

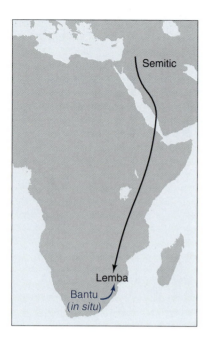

Figure 12.13: Population ancestry of the Lemba.

The Lemba claim Jewish ancestry, and genetic evidence of this is apparent only in their paternal lineages (see *Box 12.2*), along with substantial Bantu ancestry.

▶ coherence is actively maintained through mating over large distances, but within the group;

▶ coherence results from a recent migration from common origin, but is decaying over time.

As we shall see, the latter process is the more frequent explanation.

12.6.2 European Roma (Gypsies) as a transnational isolate

European Roma, often called Gypsies, represent a population of about 8 million spread over the entire European continent. They are found in highest concentrations in Southeastern Europe and the Iberian peninsula. Historical records indicate they entered Europe about 1000 YA, gradually spreading across the continent from the Southeast (see *Figure 12.15*). The Roma speak a variety of Romani dialects although some have adopted languages of surrounding populations. The linguistic affinities of Romani languages clearly indicate an origin somewhere on the Indian subcontinent.

The social structure of the Roma is orientated around small, endogamous groups, often associated with a specific trade and religion (see *Box 12.3*). However, these religions differ greatly between groups. Islam, Roman Catholicism, Protestantism and the Eastern Orthodox church are all represented among European Roma.

A wide variety of **Mendelian** disorders are found among the Roma. These disorders are characterized by homogeneity, with single mutations underlying most cases. These mutations tend to exist on a single **haplotypic background**, thus indicating a single common and recent origin. Some of these disorders are common in other European populations; others are specific to the Roma. Some of the Roma-specific mutations are found only in certain groups whereas others are spread across all European Roma (Kalaydjieva *et al.*, 2001). Thus, while there is an obvious **founder effect** resulting from a common origin of European Roma, there is also substantial heterogeneity between groups resulting in considerable internal diversity. Genetic distances between Roma groups both within and between countries are typically larger than those between the surrounding European populations. Thus individual Roma groups can be thought of as isolates within a larger isolate. Three processes could cause this population differentiation:

▶ high levels of drift due to endogamous practices in small populations;

▶ different levels of admixture between Roma groups and the surrounding populations;

▶ original substructure in the ancestral Roma population, maintained over time.

As a consequence of these issues, population genetic studies have tended to focus on three aspects of Roma diversity:

▶ similarities between the different Roma groups;

▶ admixture between Roma groups and the surrounding European populations;

▶ genetic affiliations between Roma and potential source populations on the Indian subcontinent.

Genetic evidence from classical, Y-chromosomal and mtDNA markers have shown that the Roma share alleles and lineages with populations on the Indian subcontinent that are not found in other European populations. Nevertheless, it has not been possible to identify a single likely ancestral population. This inability severely hampers our ability to perform the kinds of quantitative admixture analysis examined previously in this chapter.

However, by observing the frequencies of European-specific Y-chromosomal lineages among different Roma populations some conclusions can be drawn about the nature of the admixture (Gresham *et al.*, 2001). The two lineages shown in *Figure 12.16* are not common among Indian subcontinental populations, and thus the sum of their frequencies represents a minimum estimate for European admixture. Two conclusions can be drawn. Firstly, the degree of admixture is highly variable between different populations, ranging from 0 to ~60% within Bulgarian Roma populations. Secondly, the admixed lineages reflect the lineage distributions within surrounding populations. It can therefore be inferred that multiple independent admixture events have occurred in the different populations, and that admixture has played a significant role in population differentiation among the different Roma groups.

BOX 12.2 Opinion: The Lemba: the 'Black Jews' of Southern Africa

The Lemba are a Bantu-speaking South African tribe with an unusual oral history. They claim descent from a group of male Jews who came 'from Sena in the north by boat', but the location of Sena is not known. Sometimes it is suggested that Sena was in Yemen but at other times Egypt, Ethiopia or ancient Israel are proposed. It is related that half of the original group were lost at sea, while the remainder, once they reached the coast, settled and married local women.

Europeans have known about the Lemba at least since the early eighteenth century when they were mentioned in a Dutch report. Like Jews they practice circumcision and do not eat pork. Nevertheless it has been pointed out that Muslims as well as Jews avoid pork and many African tribes have circumcision rituals.

The Lemba's origin story is well suited to genetic investigation, an approach that the Lemba themselves have favored. The Lemba's oral tradition predicts that Lemba mtDNA will be like that of other South African Bantu speaking groups and the Y chromosomes more like those of people living in the Middle East.

This is exactly the pattern that Trefor Jenkins, a South African geneticist, and his colleagues found when they undertook an early study. The particular polymorphisms they analyzed, however, did not permit them to distinguish a Jewish from a more general Semitic paternal contribution.

In a later study of the origins of the paternally inherited Jewish priesthood (Cohanim), Thomas et al. (1998) identified a particular Y chromosome type present at moderately high frequencies in Jewish communities and in almost 50% of Jewish priests but, based on later research, either absent or rare in non-Jewish Middle Eastern populations. This chromosome type, defined by five single nucleotide polymorphisms, an *Alu* insertion and six microsatellites,

which they called the Cohen Modal Haplotype, was suggested as a 'signature haplotype' of the ancient Hebrew population.

In 1999 Lemba DNA was analyzed, along with that of Jews, Bantu speakers and Yemeni (Thomas et al., 2000). About two thirds of the Lemba Y chromosomes were of the 'Semitic type' and one third of the 'South African Bantu-speaking type'. In addition the geneticists discovered that about 9% of the total number matched the Cohen Modal Haplotype. But that was not all. The Lemba are divided into many clans, the oldest of which is the Buba, and which for some ritual purposes is also the most important. It was among members of this Buba clan that the Cohen Modal Haplotype was most common (over 50%).

This finding does not prove that the Lemba oral history is correct, but it does provide persuasive support. Further support for the oral history concerning the Buba comes from two documents, the text of a funeral oration and a privately published book, both of which predate the 1999 study. In the former, a Lemba elder states that *the Senas left Judea under the leadership of Buba and settled in Yemen where they built their city of Sena*. In the latter, Professor Mathivha, also a Lemba, writes *the Bhuba lineage came down from Judea as the leading lineage of the Basena when they left Judea in their early migration to the Yemen where they settled and built the city of Sena. They ruled over all the lineages in good manner*.

Neil Bradman[1], Mark G. Thomas[1], Michael E. Weale[1] and David B. Goldstein[2].

[1]**The Centre for Genetic Anthropology, University College London, UK**

[2]**Department of Biology (Galton Laboratory), University College London, UK**

DISTRIBUTION OF Y CHROMOSOMES FROM FOUR POPULATIONS AMONG FOUR GROUPS DEFINED BY SIX POLYMORPHISMS.

Haplogroup	Frequency (n) in							
	AI		Y		L		B	
1 YAP−GACCT	0.650	(39)	0.735	(36)	0.654	(89)	0.169	(13)
2 YAP−GACTT	0.183	(11)	0.163	(8)	0.015	(2)	0.000	(0)
3 YAP+AACCT	0.167	(10)	0.061	(3)	0.029	(4)	0.026	(2)
4 YAP+AGCGT	0.000	(0)	0.041	(2)	0.302	(41)	0.805	(62)
Total	1.000	(60)	1.000	(49)	1.000	(136)	1.000	(77)

Six polymorphisms (in the order YAP, SRY 4064, sY81, SRY+465, 92R7, Tat) define four different haplogroups. The Lemba have a haplogroup distribution somewhat between that of the Ashkenazic Jews and the Bantu speakers. AI, Ashkenazic Israelites; Y, Yemeni; L, Lemba; B, Bantu speakers; A, adenine; C, cytosine; G, guanine; T, thymine. Modified from Thomas et al. (2000).

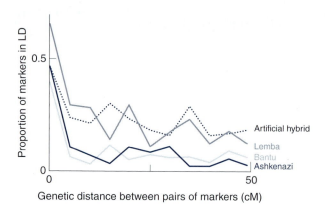

Figure 12.14: Extent of LD in Lemba, Bantu, Ashkenazi Jewish and an artificial hybrid population.

The admixed Lemba exhibit LD between X-chromosomal microsatellites over longer genetic distances than non-admixed Bantu and Ashkenazi populations. An artificial hybrid population generated *in silico* from the Bantu and Ashkenazi populations has similar levels of LD to the Lemba. Data from Wilson and Goldstein (2000).

12.6.3 The Jewish people as a transnational isolate

A common religion, language and traditions unite the Jewish people. Historical and linguistic evidence attests to their Bronze Age origins in the Middle East. The ~ 14 million modern Jews reside mostly in the USA (~ 6 million) and Israel (~ 5 million). Despite the maternal inheritance of 'Jewishness' prescribed by religious law, the practice of endogamy ensures a degree of both paternal and maternal genetic continuity; the Jewish religion does not seek converts with the same enthusiasm as some other faiths. Jews are

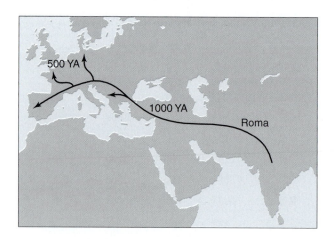

Figure 12.15: The Roma diaspora

The recent Indian origins of the Roma are supported by historical, linguistic and genetic evidence. Approximate dates derive from historical records.

generally classified into three groups on the basis of their ancestral migrations (see *Figure 12.17*):

▶ Ashkenazi Jews had migrated from the Middle East into Central Europe by around 1100 YA and subsequently shifted within Northern Europe often attempting to avoid persecution. During the nineteenth and twentieth centuries many Ashkenazi Jews left Europe for the Americas, Australia and South Africa, and as a consequence they now make up ~ 90% of the US Jewish population.

▶ Sephardic Jews had resided in the Iberian peninsula for centuries prior to being persecuted by the Spanish Inquisition during the fifteenth century. This led to their dispersal to mainly Mediterranean countries (Italy, Balkans, North Africa, Turkey and Lebanon), Syria and the Americas.

▶ Oriental (Middle-Eastern) Jews remained in the Levant (lands on the Eastern edge of the Mediterranean Sea) and surrounding countries (Iran, Iraq and the Arabian Peninsula).

The Mendelian disorders of Jewish people have been investigated in great depth. A plethora of genetic diseases have been identified which typically result from one or two common founder mutations. As with the Roma, some of these mutations are common to many Jewish groups whereas others are specific to certain populations. For example, Tay-Sachs disease (OMIM 272800) is prominent only among Ashkenazi Jews where the mutation responsible reaches a heterozygote carrier frequency of 1/25, whereas the carrier frequency of alleles responsible for familial Mediterranean fever (OMIM 249100) is between 1/10 and 1/5 in populations from all three major Jewish groups. The likelihood that an Ashkenazi Jew is a carrier for one of the eight most common disease alleles is 1/4. It has been suggested that either the common mutations derive from mutational events at different times during the Jewish diaspora, or that genetic drift has resulted in the loss of the disease alleles from some populations, perhaps as a result of founder effects. The dating of different disease alleles to different times supports the former explanation (Ostrer, 2001). However, as an additional complication, some disease alleles appear to have been recent introductions via admixture, as they are common in surrounding populations, but not in other Jewish groups. The utility of population isolates in identifying disease-related genes is explored in greater detail in Section 14.3.

Many alleles of classical markers and haplotypes of molecular markers have been identified that are shared by all three Jewish populations but not their surrounding populations. This provides considerable support for the common origin of these groups and the partial maintenance of their genetic integrity through endogamous practices over the past 2 KY. One specific Y-chromosomal haplotype that appears to have been co-inherited with the paternally-inherited Cohanim priesthood is discussed in Section 15.3.3 (Thomas *et al.*, 1998).

BOX 12.3 Opinion: The Roma (Gypsies): disregard for social history, self-identity, and tradition can interfere with understanding genetic disease

Origins

The Roma arrived in Byzantium (roughly corresponding to modern day Turkey) in the eleventh to twelfth century and by the fifteenth century had spread across Europe. The eighteenth century linguistic hypothesis of their Indian origins has been supported by genetic studies of classical markers, showing a closer similarity to Indians than to other Europeans. Strong evidence that the European Roma originated in the Indian subcontinent was provided by the identification of an Asian-specific haplogroup shared by 44.8% of Romani Y chromosomes and an Asian-specific maternal lineage shared by 26% of mtDNA samples (Gresham *et al.*, 2001). Limited Y STR haplotype and mtDNA sequence diversity within these haplogroups was consistent with a profound bottleneck event and a small number of related founders. Additional paternal and maternal lineages were the product of early admixture on the route to Europe, or of multiple admixture events within Europe.

Who is a Gypsy today?

The term 'Gypsy', a misnomer derived from an early legend about Egyptian origins, is being replaced by 'Rom, Roma', meaning 'Man, People'. The Roma defy the conventional definition of population studies: they are spread across many countries with no nation-state of their own, speak different languages, belong to many religions and, as a result of a long history of discrimination and persecution, may be reluctant to declare Romani ethnic identity. Official data on population size (including the reported 8–10 million Roma in Europe) rely on arbitrary criteria that differ widely depending on the type of authority and should therefore be regarded as only a rough approximation. The criteria used by the wider community are as arbitrary, based on appearances (skin color in the first place), and traditional stereotypes about patterns of behavior. Genetic studies are conducted by 'outsiders', and relationships with the subjects involved will inevitably be influenced by the wider social context and the historical experience of the Roma in the specific country. The sensitive issues generally involved in genetic research are of even greater importance in this case, not least because of the role played by genetics and eugenics as the 'scientific' basis of the Holocaust. Trust, mutual respect and understanding are mandatory and, once established, reveal an extremely friendly and supportive community, where information on ethnic identity and extended family history is readily and reliably provided.

How many populations?

The social organization of the Roma resembles the endogamous professional *jatis* of India. Romani groups, with **ethnonyms** usually derived from traditional trades, are defined by self-identity, customs and traditions, historical migrations, language, etc. Groups residing in close proximity can be widely divergent and strictly endogamous, but at the same time group identity and intermarriage can cross national boundaries. Studies of the genetics of Romani groups (Gresham *et al.*, 2001; Kalaydjieva *et al.*, 2001) document their common origins and the recent nature of the population fissions, occurring at different times after the arrival in Europe. Genetic affinities parallel most closely the history of early migrations within Europe, classifying the Roma into Balkan settlers, Vlax groups (long residing in present-day Romania and Moldova) and early migrants into Western Europe. The history of the Roma, like that of the Jews, abounds in repressive legislation and persecution. The external social pressures, leading to the initial splits into small groups, have been complemented by an internal restitution of the ancient social traditions of India, consolidating the group divisions. These have served to further reduce effective population size, enhancing the effects of genetic drift. Today, the Roma of Europe are best described as a fragmented founder population comprised of multiple groups, whose genetic structure is the compound product of a string of bottlenecks and founder effects, strong drift and differential admixture. Genetic diversity (especially male) is strikingly limited in some groups, whereas in others it is similar to most European populations.

Genetic disorders

Mendelian disorders in the Roma can be caused by 'private' founder mutations, causing novel or known disorders, as well as by mutations 'imported' by admixture (Kalaydjieva *et al.*, 2000). The epidemiology and molecular characteristics of founder mutations have been shaped by the population history of the Roma. Ancient mutations, whose origins pre-date the exodus from India, are characterized by: (i) wide spread across Romani groups; (ii) dramatic differences in frequency resulting from secondary founder effects and drift; (iii) a diversity of related disease haplotypes inherited from the ancestral population; (iv) recent independent haplotype evolution resulting in a unique profile of each endogamous Romani group. 'Private' mutations of recent origin are confined to a single or a small number of Romani groups (geographic spread can still be very wide) and display limited haplotype diversity. The characteristics of 'imported' mutations are related to those in the surrounding populations and depend on the number and sources of admixture events. Knowledge of the social organization and population history of the Roma is clearly relevant to public health interventions and medical genetic research. Arbitrary sampling of 'Gypsies' will predictably result in incomplete and distorted (possibly extremely so) data on carrier rates, impacting on public health strategies, and in a limited representation of the diversity that is so valuable in the mapping and cloning of disease genes. Detailed knowledge of population structure will be crucial in studies of genetically complex disorders.

Luba Kalaydjieva

Western Australian Institute for Medical Research and Centre for Medical Research, University of Western Australia, Perth, Australia

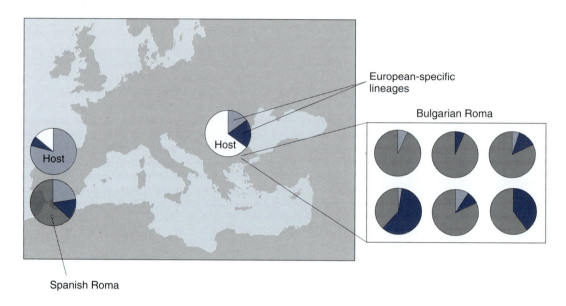

Figure 12.16: Frequencies of two European-specific Y-chromosomal lineages among different Roma groups.

The light blue European-specific lineage is found at higher frequency in Spain than Bulgaria, and is found at higher frequencies among Spanish Roma than Bulgarian Roma, wherein the dark blue European-specific lineage predominates. Black and white shading indicates uninformative lineages in the Roma and European populations respectively. Data taken from Gresham *et al.* (2001).

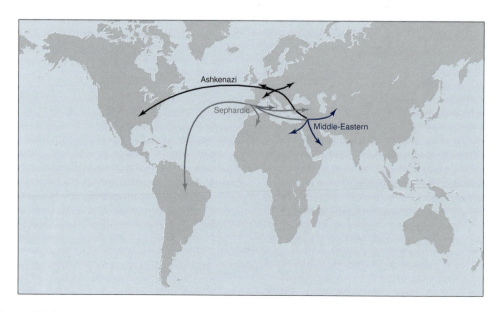

Figure 12.17: The Jewish diaspora.

The tripartite division of Jewish peoples based on their history of migrations is discussed in the text.

For calculating admixture estimates, the well-defined initial origins of Jewish populations means that one ancestral population, that of the Middle East, can be assigned with relative confidence (although subsequent migrations to the region could have altered allele frequencies). However, given the migratory history associated with both Ashkenazi and Sephardic Jews, other genetic contributions could come from a number of other ancestral 'host' populations. The maternal inheritance of 'Jewishness' suggests that admixture might be less in maternal lineages than paternal lineages, and thus it is of *a priori* interest to compare admixture estimates for Y-chromosomal and mtDNA markers. A comparison of genetic diversity among diverse Jewish and non-Jewish 'host' populations showed that whereas Y-chromosomal diversity was not generally less diverse than that in the host population, mtDNA diversity was frequently significantly less in Jewish populations (Thomas *et al.*, 2002; see *Table 12.4*). There are two potential explanations for the observations:

▶ founder effects and other causes of genetic drift were stronger in maternal lineages;

▶ admixture has been consistently male-biased, resulting in the introduction of paternal lineage diversity.

We can distinguish between these two possibilities on the basis of genetic distances (Livshits *et al.*, 1991). Each of the Jewish populations studied is associated with a different host population. If the former hypothesis is correct, we should expect that Jewish populations will show high maternal genetic distances (calculated from mtDNA haplotypes) as drift acts separately in each population to differentiate them. Alternatively, if the latter hypothesis is true and paternal admixture is prevalent, Y-chromosomal genetic distances should be larger as they incorporate lineages from different populations, whereas maternal lineages derive from a common origin, see *Table 12.5*.

As we can see, the former prediction is correct, in that mtDNA genetic distances are far greater than Y-chromosomal distances among Jewish populations. As we saw in Section 9.3, this is an unusual finding; globally the Y-chromosome shows the greatest genetic differentiation of all loci. Thus it appears that extreme maternal founder effects during the Jewish **diaspora** have shaped the present pattern of diversity.

While the hypothesis of sex-biased admixture remains unproven for Jewish populations, evidence for admixture with host populations has been identified in other studies. Recently, Hammer and colleagues adopted a standard approach to detecting admixture among different Jewish groups (Hammer *et al.*, 2000). Applying a coalescent-based

TABLE 12.4: GENETIC DIVERSITY AMONG DIVERSE JEWISH POPULATIONS AND THEIR HOST, NON-JEWISH, POPULATIONS.

Jewish population	Host population	Genetic diversity, Jewish<Host?	
		Y-chromosomal	Mitochondrial
Ashkenazi	German	No	Yes
Moroccan	Berber	Yes	Yes
Iraqi	Syrian	No	Yes
Georgian	Georgian	Yes	Yes
Bukharan	Uzbekistani	Yes	Yes
Yemeni	Yemeni	No	Yes
Ethiopian	Ethiopian	No	Yes
Indian	Hindu	No	Yes

Data taken from Thomas *et al.* (2002).

TABLE 12.5: ADMIXTURE ESTIMATES AMONG THREE JEWISH GROUPS.

Jewish population	Host population	Admixture, M.
Ashkenazi	German, Austrian, Russian	0.130 ± 0.099
Roman	Italians	0.289 ± 0.183
Lemba	South African East Bantu speakers	0.465 ± 0.123

admixture estimation method (Bertorelle and Excoffier, 1998) to paternal lineage frequencies with the greatest differential between ancestral populations, they demonstrated significant variation in admixture rates among three different Jewish groups (see *Table 12.5*).

Summary

▶ Genetic admixture is the process by which a hybrid population is formed from contributions by two or more parental, or ancestral, populations. Many important issues in recent human evolution can be considered to be questions of admixture.

▶ Past admixture events can be identified in the historical, linguistic and archaeological records as well as through their impact on patterns of genetic diversity.

▶ The genetic contribution of an ancestral population to the hybrid population can be estimated by considering the frequencies of a given allele in the hybrid population and both ancestral populations. Admixture estimates require the correct identification of the ancestral populations and are most accurate when an allele is present at very different frequencies in the two ancestral populations.

▶ A number of different methods for estimating admixture proportions have been devised to best combine information from multiple alleles into a single estimate, and that take into account some of the confounding factors, such as genetic drift in both ancestral and hybrid populations since the admixture event.

▶ In an admixed population, individual genomes are themselves admixed to a greater or lesser degree. By typing many markers in the same individual, levels of individual admixture can be estimated and compared within a population.

▶ Under a number of different admixture scenarios, the contributions of males and females from a given ancestral population may not be equal. This sex-biased admixture reveals itself in discrepant admixture estimates from loci with different patterns of inheritance (i.e., mtDNA, Y chromosomes, X chromosomes and autosomes). Sex-biased admixture is commonly observed in the formation of admixed populations as a result of the Atlantic slave trade.

▶ Admixture results in elevated levels of linkage disequilibrium, which decays over the time since the admixture event, but can be detected in the Bantu-speaking Lemba population in Southern Africa that claims Jewish paternal ancestry.

▶ Admixture cannot be divorced from its social context. Social practices such as endogamy restrict admixture with other populations, and the relative socioeconomic status of the different ancestral populations can lead to directional mating between males and females of the two groups.

▶ Transnational isolates (e.g., the Roma and the Jews) are populations who maintain genetic coherence over vast geographical distances as a result of recent dispersal from a common origin and endogamous mating practices that restrict admixture.

Further reading

Chakraborty R (1986) Gene admixture in human populations: models and predictions. *Yearbook Physical Anthropol.* **29**, 1–43.

References

Alves-Silva J, da Silva Santos M, Guimaraes PE, Ferreira AC, Bandelt HJ, Pena SD, Prado VF (2000) The ancestry of Brazilian mtDNA lineages. *Am. J. Hum. Genet.* **67**, 444–461.

Bernstein F (1931) In: *Comitato Italiano per o Studio dei Problemi Della Populazione.* Instituto Poligrafico dello Stato, Roma, pp. 227–243.

Bertorelle G, Excoffier L (1998) Inferring admixture proportions from molecular data. *Mol. Biol. Evol.* **15**, 1298–1311.

Carvajal-Carmona LG, Soto ID, Pineda N et al. (2000) Strong Amerind/white sex bias and a possible Sephardic contribution among the founders of a population in northwest Colombia. *Am. J. Hum. Genet.* **67**, 1287–1295.

Carvalho-Silva DR, Santos FR, Rocha J, Pena SD (2001) The phylogeography of Brazilian Y-chromosome lineages. *Am. J. Hum. Genet.* **68**, 281–286.

Cavalli Sforza LL, Bodmer WF (1971) *The Genetics of Human Populations.* W.H. Freeman, San Francisco.

Chakraborty R (1986) Gene admixture in human populations: models and predictions. *Yearbook Physical Anthropol.* **29**, 1–43.

Chikhi L, Bruford MW, Beaumont MA (2001) Estimation of admixture proportions: a likelihood-based approach using Markov chain Monte Carlo. *Genetics* **158**, 1347–1362.

Chikhi L, Nichols RA, Barbujani G, Beaumont MA (2002) Y genetic data support the Neolithic demic diffusion model. *Proc. Natl Acad. Sci. USA* **99**, 11008–11013.

Collins-Schramm HE, Phillips CM, Operario DJ et al. (2002) Ethnic-difference markers for use in mapping by admixture linkage disequilibrium. *Am. J. Hum. Genet.* **70**, 737–750.

Dupanloup I, Bertorelle G (2001) Inferring admixture proportions from molecular data: extension to any number of parental populations. *Mol. Biol. Evol.* **18**, 672–675.

Glass B, Li CC (1953) The dynamics of racial intermixture – an analysis based on the American Negro. *Am. J. Hum. Genet.* **5**, 1–19.

Gresham D, Morar, B, Underhill P *et al.* (2001) Origins and divergence of the Roma (Gypsies). *Am. J. Hum. Genet.* **69**, 1314–1331.

Hamblin MT, Thompson EE, Di Rienzo A (2002) Complex signatures of natural selection at the Duffy blood group locus. *Am. J. Hum. Genet.* **70**, 369–383.

Hammer MF, Redd AJ, Wood ET *et al.* (2000) Jewish and Middle Eastern non-Jewish populations share a common pool of Y-chromosome biallelic haplotypes. *Proc. Natl Acad. Sci. USA* **97**, 6769–6774.

Helgason A, Siguroardottir S, Nicholson J *et al.* (2000) Estimating Scandinavian and Gaelic ancestry in the male settlers of Iceland. *Am. J. Hum. Genet.* **67**, 697–717.

Hurles ME, Irven C, Nicholson J *et al.* (1998) European Y-chromosomal lineages in Polynesians: A contrast to the population structure revealed by mtDNA. *Am. J. Hum. Genet.* **63**, 1793–1806.

Kalaydjieva L, Gresham D, Calafell F (2001) Genetic studies of the Roma (Gypsies): a review. *BMC Med. Genet.* **2**, 5.

Kalaydjieva L, Gresham D, Gooding R (2000) N-myc downstream-regulated gene 1 is mutated in hereditary motor and sensory neuropathy-Lom. *Am. J. Hum. Genet.* **67**, 47–58.

Kaplan B (1954) Environment and human plasticity. *Am. Anthropologist* **56**, 781–799.

Livshits G, Sokal RR, Kobyliansky E (1991) Genetic affinities of Jewish populations. *Am. J. Hum. Genet.* **49**, 131–146.

Long JC, Smouse PE (1983) Intertribal gene flow between the Ye'cuana and Yanomama: genetic analysis of an admixed village. *Am. J. Phys. Anthropol.* **61**, 411–422.

Long JC (1991) The genetic structure of admixed populations. *Genetics* **127**, 417–428.

MacLean CJ, Workman PL (1973) Genetic studies on hybrid populations. I. Individual estimates of ancestry and their relation to quantitative traits. *Ann. Hum. Genet.* **36**, 341–351.

McKeigue PM, Carpenter JR, Parra EJ, Shriver MD (2000) Estimation of admixture and detection of linkage in admixed populations by a Bayesian approach: application to African-American populations. *Ann. Hum. Genet.* **64**, 171–186.

Model S, Fisher G (2002) Unions between blacks and whites: England and the US compared. *Ethnic and Racial Studies* **25**, 728–754.

Ostrer H (2001) A genetic profile of contemporary Jewish populations. *Nature Rev. Genet.* **2**, 891–898.

Parra EJ, Kittles RA, Argyropoulos G *et al.* (2001) Ancestral proportions and admixture dynamics in geographically defined African Americans living in South Carolina. *Am. J. Phys. Anthropol.* **114**, 18–29.

Pfaff CL, Parra EJ, Bonilla C *et al.* (2001) Population structure in admixed populations: effect of admixture dynamics on the pattern of linkage disequilibrium. *Am. J. Hum. Genet.* **68**, 198–207.

Pritchard JK, Stephens M, Donnelly P (2000) Inference of population structure using multilocus genotype data. *Genetics* **155**, 945–959.

Reed TE (1969) Caucasian genes in American Negroes. *Science* **165**, 762–768.

Roberts DF, Hiorns RW (1962) The dynamics of racial admixture. *Am. J. Hum. Genet.* **14**, 261–277.

Sans M, Weimer TA, Franco MH *et al.* (2002) Unequal contributions of male and female gene pools from parental populations in the African descendants of the city of Melo, Uruguay. *Am. J. Phys. Anthropol.* **118**, 33–44.

Seielstad MT, Minch E, Cavalli-Sforza LL (1998) Genetic evidence for a higher female migration rate in humans. *Nature Genet.* **20**, 278–280.

Shriver MD, Smith MW, Jin L, Marcini A, Akey JM, Deka R, Ferrell RE (1997) Ethnic-affiliation estimation by use of population-specific DNA markers. *Am. J. Hum. Genet.* **60**, 957–964.

Thomas MG, Skorecki K, Ben-Ami H, Parfitt T, Bradman N, Goldstein DB (1998) Origins of Old Testament priests. *Nature* **384**, 138–140.

Thomas MG, Parfitt T, Weiss DA *et al.* (2000) Y chromosomes traveling south: The Cohen modal haplotype and the origins of the Lemba – the "black Jews of Southern Africa". *Am. J. Hum. Genet.* **66**, 674–686.

Thomas MG, Weale ME, Jones AL *et al.* (2002) Founding mothers of Jewish communities: geographically separated Jewish groups were independently founded by very few female ancestors. *Am. J. Hum. Genet.* **70**, 1411–1420.

Williams RC, Long JC, Hanson RL, Sievers ML, Knowler WC (2000) Individual estimates of European genetic admixture associated with lower body-mass index, plasma glucose, and prevalence of type 2 diabetes in Pima Indians. *Am. J. Hum. Genet.* **66**, 527–538.

Wilson JF, Goldstein DB (2000) Consistent long-range linkage disequilibrium generated by admixture in a Bantu-Semitic hybrid population. *Am. J. Hum. Genet.* **67**, 926–935.

SECTION SIX

How is an evolutionary perspective useful?

Having an evolutionary perspective on human genetic diversity not only illuminates the origins of our species and the way that early humans spread across the world, but it has practical applications too. It gives us an essential framework for understanding normal and pathogenic phenotypic variation among people, and is important in making reasonable deductions about individuals from their DNA.

CHAPTER 13   *Understanding the past and future of phenotypic variation*

This chapter discusses 'normal' phenotypic variation among people, including the complex phenotypes of morphology, pigmentation and dietary differences. We ask to what extent these can be understood in terms of adaptation, drift and sexual selection. An evolutionary perspective leads to some predictions about the future of our species, and these are explored in terms of the changes in population size, reproductive technologies, and new challenges from disease and the environment.

CHAPTER 14   *Health implications of our evolutionary heritage*

Many aspects of our health are influenced by our genome and the variants it contains, and since our genome is the product of our evolutionary history, evolutionary genetics can have important implications for diagnosing, understanding and treating medical conditions. In this chapter we examine genetic influences on the distributions and frequencies of diseases, the use of evolutionary information in identifying disease genes, and the importance of evolutionary information for medical treatments, including pharmacogenetics, which is concerned with the genetic basis for different responses to drugs.

CHAPTER 15   *Identity and identification*

Except for the genomes of identical twins, each copy of the human genome is different, and exploiting these differences allows the identification of individuals from their DNA, in forensic analysis. This final chapter explains how this is done, and also examines the use of DNA in determining family relationships, phenotypic characteristics and membership of broader groups such as genealogies. These studies must be placed in a broader framework of the history and structure of human populations if their interpretations are to be reasonable.

CHAPTER THIRTEEN

Understanding the past and future of phenotypic variation

CHAPTER CONTENTS

BOXES

13.1 Introduction: Traits, phenotypes and diseases

So far, the emphasis of this book has been upon genetic variation that is thought to be selectively neutral, and to have little or no influence on our phenotype. Here, and in the following chapter, we change our focus, and begin an exploration of the phenotypic diversity of modern human populations from an evolutionary genetic perspective. The material already presented in this book provides a framework for this exploration, since human phenotypic variation is influenced by many aspects of our evolutionary past:

▸ the way that our species originated, and the effect of this upon our genetic diversity;

▸ the early differentiation and spread of human populations;

▸ the initial colonization of new environments and accommodation to climatic change;

▸ the development of agriculture, recent demographic expansions, change in diet, and close contact with animal species;

▸ the recent admixture of populations with different histories, and recent migrations.

This chapter examines the genetic basis of 'normal' phenotypic variation, while Chapter 14 focuses upon variation influencing genetic or infectious disease. However, this is not always a simple distinction to make. For example, everyone would agree that cystic fibrosis is certainly a disease, while having red hair and fair skin is more likely to be considered as normal variation, and referred to as a 'trait'. But having red hair and fair skin increases the risk of developing the serious cancer melanoma (Box *et al.*, 2001), through a genetic mechanism. In this chapter we will discuss the genetic bases of phenotypic variation among 'normal' healthy people, but will point out how these may influence the risk of pathogenic consequences. Similarly, in Chapter 14 we will examine phenotypes that are usually detrimental, but can be neutral or advantageous under some circumstances. Some of this 'normal' variation represents adaptations to the environment, such as climate and diet, and at the same time its expression can be environmentally influenced. This environmental influence does not remain constant, and there are examples of adaptations that may have been beneficial in our ancestors that are detrimental today in modern populations (an example, the 'thrifty genotype', is discussed in Section 13.4.4). It is important to bear in mind that both 'normal' and pathogenic genetic variation reflects the interaction of our genes with the wider selective world around us.

The geographic patterns of normal and pathogenic variation that we observe in modern human populations are the outcomes of our evolutionary history; as we have seen in earlier chapters, the extent and distribution of human genetic diversity is very different from that of our closest animal relatives, the great apes. We are a young species and have undergone a rapid dispersal followed by explosive population growth in the Neolithic, and, only very recently, extensive remixing by inter-continental migrations. The consequence of this history is relatively low genetic diversity, a preponderance of 'young' mutations, and little differentiation among populations in different places. Unlike the great apes, we have moved into a wide range of very different environments, with different climates, altitudes, food sources and pathogens. Selective regimes are therefore very diverse for humans.

In attempting to understand phenotypic variation in terms of our evolutionary history, uncertainty about particular questions has an important influence on our choice of explanations. For example, the debate over the extent of gene flow between pre-Neolithic continental groups (discussed in Chapter 8) leads to uncertainty over whether those of our phenotypic characteristics that do show strong geographical differentiation are of ancient or more recent origin. If there had been little gene flow, an ancient origin is more likely, but if gene flow had been greater, this would favor a more recent origin (see Section 13.3.5).

In Sections 13.5–13.7 of this chapter, we move from a consideration of the influence of our evolutionary past, to a consideration of our evolutionary future. As our species continues to grow in number and appears to become more and more dominated by cultural rather than biological influences, have we become insulated from the forces of natural selection? Ultimately, can we use genetics to predict anything about the unity, or indeed the survival, of our species?

13.2 Known variation in human phenotypes

13.2.1 What is known about human phenotypic variation?

Before we begin a discussion of phenotypic variation we need to distinguish between a number of different underlying mechanisms. First, there are physiological mechanisms representing short-term and sometimes rapid reversible responses to environmental change, known as **acclimatization**. Examples are tanning, where pigmentation increases as a response to increased sun exposure, and the elevated production of hemoglobin in response to reduced oxygen concentrations at high altitude. Although there are genetic influences acting on the efficiency of these responses, we will not consider acclimatization further here. Second, there is **developmental plasticity**, in which the environment influences the long-term development of an individual. Examples are the possible influence of changed environment on the proportions of the skull (see this Section, below); the possible influence of poor nutrition on aspects of metabolism, hormone secretion and response, and pancreatic cell development and function (see Section 13.4.4); and some adaptations to high altitude. Again, these mechanisms are expected to contain genetic components, though they are poorly understood. Finally, there are genetic mechanisms directly responsible for phenotypic variation, which are the primary subject of this chapter. The distribution of alleles underlying these phenotypes is governed by the principles outlined in Chapter 5: genetic drift, founder effects, gene flow between populations, and adaptation through natural selection. Until recently, studies of the basis of human phenotypic variation were almost exclusively the preserve of biological anthropologists. Major efforts went into describing the patterns of variation among human populations in pigmentation, morphology, physiology, behavior and life history. Hypotheses have been advanced for the adaptive (or nonadaptive) significance of many different phenotypes. Some of these will be described below, and later we will focus on what recent evidence is available from genetic studies.

As has been pointed out already, humans inhabit an unusually broad range of environments. Adaptive responses to similar environments are likely to be similar, and result in similar phenotypes: an often-used example is dark skin in Africans, South Indians, Australian Aborigines and Melanesians. Because these physical characteristics are likely to be responsive to the environment (see Section 13.3), the apparent similarities they present are unlikely to be representative of the rest of the genome, and we might expect them to be poor indicators of population affinity. However, arguments persist that particular shared phenotypes indicate some relatedness; this is connected to the issue, discussed above, of uncertainty about the extent of gene flow between populations before the Neolithic.

What kinds of 'normal' phenotypic variation have formed a basis for study? One of the most obvious kinds of difference between populations is pigmentation of skin, hair and eyes; another striking difference is in digestion, in particular the degree to which lactose (a sugar from milk) is tolerated in adult life. Because genetics has contributed relatively heavily to these two areas, they will form the basis of two subsequent sections of this chapter. Before doing this, we will consider briefly some other differences, most of which have been little investigated from a genetic point of view.

Morphology and temperature adaptation

Body size and proportions vary greatly among human populations, and this is generally interpreted as an adaptive response to different thermal environments. Studies of other mammals, and also of birds, have led to the recognition of a general relationship between climate and morphology – body size increases with distance from the equator. Physical anthropologists refer to two rules that describe the relationships between body morphology and climate; both are essentially renderings of the same simple physical principle. **Bergmann's rule** states that body size increases as climate becomes colder, because as mass increases, surface area does not increase proportionately. Heat is lost at the surface, so increased body mass means better heat retention. **Allen's rule** tells us that shorter appendages (limbs) are favored in colder climates because they have relatively high ratios of mass to surface area, and therefore retain heat. Conversely, longer appendages are favored in hotter climates. According to these principles, people are best adapted to hot climates when they have a linear body shape with long limbs, and best adapted to cold climates when they are stocky and

short-limbed. Studies on many populations generally confirm this to be the case, with East African pastoralists and Inuit from the Arctic often being used as typical examples (Ruff, 2002) (*Figure 13.1*). These morphological differences are under genetic control, and taking an East African child to the Arctic would not endow him or her with the stocky Inuit build in adulthood. However, not all populations conform to the rules. The small body size of African Pygmies is sometimes claimed to be an adaptation to the hot climate of central Africa, but there is no direct evidence to support this. It is sometimes claimed to be an adaptation to moving through dense vegetation, and may also be a result of sexual selection (see Section 13.3.5).

Physiological adaptations to altitude

As has already been mentioned, people from low altitudes can become acclimatized to high altitudes through physiological mechanisms. However, there are about 25 million people who are permanently settled in regions over 3000 meters in altitude (in the Andes, Ethiopia and the Himalayas). In some of these people there is evidence of adaptation to high altitude, presumably with a genetic basis. For example, some populations of Tibet do not show the reduced birth-weight characteristic of other high altitude populations or recent immigrants, and this may reflect an adaptive difference in maternal blood-flow to the placenta during pregnancy (Jurmain *et al.*, 2000).

Facial features

Differences in facial features between populations have long been used as part of the traditional descriptions of races (Howells, 1992). Degree of flatness of the face, shapes of ears, nose and lips, and the distribution and texture of hair (color is considered below), vary among populations. Adaptive

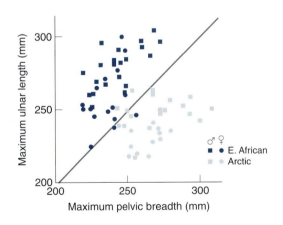

Figure 13.1: Differing body proportions in East African and Arctic populations.

Maximum length of ulna (the larger of the two bones of the forearm) plotted against bi-iliac (maximum pelvic) breadth for modern East Africans and modern samples from Alaskan Inupiat and Aleut. East Africans generally have long arms and narrow pelvises, while most of the Arctic peoples have short arms and broad pelvises. Adapted from Ruff (2002).

explanations have been put forward for some of these differences. Nose shape shows some association with climate, and it has been suggested that a narrow nose may warm and moisten air before it reaches the lungs more efficiently than a broad one. The internal skin-fold of the eyelid (**epicanthal fold**) seen in many Asian populations has been explained as a protection from cold air and snow glare (Diamond, 1991). However, these are 'Just So' stories, with little or no direct evidence, and again the features do not show geographical distributions compatible with simple adaptive explanations.

Tooth morphology and cranial proportions

Complex tooth crown shape is prevalent in some Asian and Native American populations, and a particular upper incisor morphology ('shoveled' incisors) is found in similar populations, as well as in the Khoisan. The shape of the cranium has been the focus of special interest from anthropologists for many years. **Cephalic index** (**CI**, also called cranial index) is the ratio of the breadth to the length of the skull multiplied by 100, and provides a single, if crude, figure to describe the proportions of a skull. Individuals with CIs below 75 have long, narrow heads and are termed **dolichocephalic**, while those with CIs over 80 have broad heads and are **brachycephalic**. Those with CIs between 75 and 80 are **mesocephalic**. Many data were gathered on the CIs within different populations. However, highly influential work by Franz Boas, an American anthropologist, indicated that the CI of children born in the USA of European immigrant parents changed compared to their European born siblings (Boas, 1910). In other words, inherited aspects of the index were not as important as early environmental influence, and this suggested that cranial proportion was a developmentally plastic trait, and therefore not a reliable phenotype to use as a classificatory tool, either for modern or for ancient samples. Boas's conclusions were an important part of the refutation of scientific racism. Despite 90 years of acceptance, this view is now under challenge because of reanalyses of Boas's original data. A recent study (Sparks and Jantz, 2002) has used modern statistical methods and models from quantitative genetics, and finds that there is high heritability of CI, and no evidence to support Boas's claims at all. In complete contrast, another statistical reanalysis (Gravlee *et al.*, 2003) supports the main conclusions of Boas's study. This debate has just begun, and seems certain to be vigorous.

Behavioral differences

Sociobiology, and its outgrowth, evolutionary psychology, are concerned with finding explanations in our primate past and our adaptive human past for complex modern behaviors and traits. These are contentious fields, in which the unit of investigation tends to be the human species as a whole. Discovering meaningful behavioral differences *between* populations that may have adaptive (and therefore underlying genetic) causes is not trivial. Human behavioral ecologists have studied important aspects of behavior such as inter-birth interval, marriage practice, and foraging strategy (Laland and Brown, 2002). However, while a phenotype like pigmentation may be easy to study and quantify (though not so easy to understand – see Section 13.3) in *any* population,

these behavioral studies are difficult and are focused on a small number of 'traditional' populations. Modern industrialized populations are considered less suitable material, as if they are somehow insulated from the evolutionary forces that might have shaped these behaviors originally (see Section 13.5). It does not take an anthropologist to realize that there are differences in (for example) mating strategies between populations. Humans are supremely flexible in their responses to different physical and social environments, and our chief adaptations to the environment are socially transmitted cultural ones (see *Box 13.6*) – for example, without the wearing of warm clothes Inuit people could certainly not survive in the Arctic. Even if we do assume that behavioral adaptations have an underlying genetic component, there is still potential cause for confusion. Some of the behaviors that are observed could be:

▸ ancient adaptations, with underlying genetic causes, and maintained because of continuity of the selective environment;

▸ adaptive to the current environment, but without the same basis in ancient adaptation;

▸ nonadaptive traits arising as evolutionary by-products, or as the result of forces such as sexual selection (see Section 13.3.5).

Studies of differences between populations in intelligence, as measured through IQ tests, have a very checkered history, and there is serious doubt over whether IQ testing is a universally and fairly applicable cross-cultural assessment of intelligence. However, as genes are identified whose products play specific roles in neurotransmitter function, and are associated with specific behavioral phenotypes, haplotype diversity studies will be performed in different populations and adaptive interpretations will be put forward. An example is given in *Box 13.1*, which discusses the diversity of haplotypes at the *DRD4* gene (Ding *et al.*, 2002), one allele of which has been associated with novelty-seeking behavior in some studies. There are several other examples of candidate genes whose products may influence behavioral disorders, including monoamine oxidase A (*MAOA*), which metabolizes biogenic amines such as the neurotransmitter serotonin; tryptophan hyroxylase (*TPH*), the rate-limiting enzyme for serotonin biosynthesis, and the serotonin transporter (*SLC6A4*), important in serotonin uptake at synapses.

13.2.2 How do we uncover genotypes underlying phenotypes?

As should be clear from the brief discussion of the debate surrounding Boas's research, anthropologists want to be able to distinguish between an innate (genetic) basis for a characteristic, and an environmental one. This teasing apart of genes and environment is a central part of the study of any complex trait, and is addressed by **quantitative genetics** (see *Box 13.2*). Boas asked about the environmental effect on cephalic index by comparing the offspring of immigrants born in the USA with those born in the country of origin;

BOX 13.1 *DRD4* haplotype diversity: Selection at a locus affecting behavior?

Neurotransmitters, such as dopamine, are small molecules that mediate communication between neurons and target cells. They are released by the neuron at the synapse and interact with a neurotransmitter receptor (NR) in the membrane of a postsynaptic cell. Opening of ion channels in the target cell alters its electrical potential and thus an electrical signal is transmitted from the excited nerve. There are many different kinds of NR, and polymorphisms in their genes are promising candidates for mutations modulating behavior.

Independently confirmed association studies (Faraone *et al.*, 2001) have shown that one allele of the *DRD4* gene, encoding the D4 dopamine receptor, is significantly associated with attention-deficit hyperactivity disorder (ADHD) in childhood, a condition involving long-term pervasive inattention, hyperactivity and impulsiveness. Intriguingly, a number of reports also linked this allele with novelty-seeking behavior in adults (Benjamin *et al.*, 1996; Ebstein *et al.*, 1996). This personality trait is diagnosed by a specific questionnaire; people who score highly on the novelty-seeking scale are characterized as impulsive, exploratory, fickle, excitable, quick-tempered and extravagant, while low-scorers are reflective, rigid, loyal, stoical, slow-tempered and frugal. However, importantly, recent **meta-analyses** (statistical analysis of the results from several individual studies designed to integrate the findings) have failed to confirm this association with novelty-seeking (Schinka *et al.*, 2002).

Allelic diversity in *DRD4* is largely due to variation in the number of 48-bp repeat units at a VNTR within exon 3 of the gene, encoding variable 16-amino acid repeats within an intracellular loop of the receptor. Variant alleles contain between two and 11 repeats (denoted 2R to 11R). The allele implicated in ADHD has seven repeats.

Frequencies of the different alleles vary considerably from population to population, with the 7R allele being present at 48% in Native Americans, but below 20% in other groups, and at only ~ 2% in east and south Asians (Chang *et al.*, 1996). **Resequencing** of 600 *DRD4* alleles in a global sample (Ding *et al.*, 2002) has shown that the origin of the 2R-6R alleles can be explained by simple one-step recombination/mutation events. However, the 7R allele is not related to the other common alleles in a simple way, but differs from them by more than six events. A high level of LD around the 7R allele suggests that it is a young mutation. These considerations have led to the suggestion that this allele originated recently (30–50 KYA) in a rare mutational event, and has risen to high frequency by positive selection; however, the repetitive nature of the variable region makes it difficult to apply standard selection tests.

It has been speculated that positive selection for *DRD4* allele 7R through an associated phenotype of novelty-seeking and perseverance occurred during the major expansion of modern humans into new and difficult territories (Ding *et al.*, 2002). The presence of the 7R allele at appreciable, but markedly different, frequencies in different populations has been explained as a result of **frequency-dependent selection** operating at different levels in different populations. Perhaps 7R-bearing men have risky 'show-off' behaviors that exhibit their good genes (Harpending and Cochran, 2002), and have been favored in societies (like those of South America) where most of the agricultural work is done by women. In contrast, men lacking 7R may have been favored in more intensively agricultural societies such as that of East Asia. Such speculations may be entertaining, but are potentially dangerous if they reinforce racist thinking. It is essential to stress that there is currently no convincing link between *DRD4* alleles and normal behavioral variation, and that this study therefore provides no evidence for genetically influenced behavioral differences between populations. More work on the molecular basis for any difference in 'normal' behavior, and more rigorous examination of the role of selection in shaping *DRD4* allele frequencies, is needed.

such studies are still of use. However, a more widely used approach to separating the effects of genes and environment is twin studies, the value of which was initially pointed out by Francis Galton. Identical (**monozygotic, MZ**) twins have exactly the same genotype, so their many phenotypic similarities might be assumed to have a genetic basis. However, they also shared a uterus, and their post–natal environment too. Fraternal (**dizygotic, DZ**) share their environments, but genetically are no more similar to each other than are ordinary siblings. A comparison of the two kinds of twins should therefore identify genetic components responsible for any greater degree of resemblance between MZ than between DZ twins. Unfortunately, the assumption that the extent of environmental sharing is identical for the two twin types is not entirely valid: MZ twins are always of the same sex, while on average half of DZ twins are brother-sister pairs, and MZ twins may be treated more similarly because they look more alike. Studies of separated MZ twins

are also problematic because sample sizes are very small, and because of early environmental sharing. Notwithstanding these problems, once a genetic basis for a phenotype has been confirmed, there are different approaches that can be taken to identify the genes responsible.

As more becomes known about the physiological or biochemical roles of particular gene products, this knowledge allows a **candidate gene** approach to be taken to understanding the genetic basis of a particular complex phenotype. For example, this approach has identified over 200 candidate genes that may affect obesity, and many others that might in principle underlie variation in pigmentation (see Section 13.3.3). However, in both of these cases attempts to associate allelic variation at candidate genes with variation in phenotypes has not proved very fruitful. The candidate gene approach has helped to delineate the biochemical pathways involved in the phenotypes, but has

---

BOX 13.2 Quantitative genetics and complex traits

The discipline of quantitative genetics (Falconer and Mackay, 1996) is fundamental to the identification of quantitative trait loci (**QTLs**; see *Box 10.7*) contributing to complex traits. It aims to decompose the total variance in a phenotype (V_P) into its genetic (V_G) and environmental (V_E) variance components, such that:

$$V_P = V_G + V_E$$

The genetic and environmental components can be further decomposed. Genetic variance components can be:

▶ additive (V_A) (due to homozygous alleles);

▶ dominant (V_D) (due to heterozygous alleles);

▶ epistatic (due to interactions between genes).

Environmental variance components can be due to specific, identified elements in the environment, or to unidentified factors.

The proportion of variance in a trait that is explained by genetic factors is known as its **heritability**. Heritability is subdivided into **broad sense heritability** (the proportion of genetic variance that can be attributed to all the genetic effects listed above, i.e., V_G/V_P), and **narrow sense heritability** (only that proportion of genetic variance attributable to additive effects, i.e., V_A/V_P). If a trait is shown to be heritable, an attempt can be made to identify genes (QTLs) responsible for the differences between individuals. This is done by linkage analysis (Strachan and Read, 1999).

The ability to identify QTLs depends on sample size, the nature of the units (e.g., pedigrees or sib-pairs) being studied, and the strength of the effect of the QTL itself. QTLs accounting for 10–15% of variance in a trait should be identifiable in humans, given a good study design (Rogers *et al.*, 1999). However, there is difficulty in interpreting the results of linkage analysis. Traditional linkage methods are based on a specified genetic model that incorporates a defined mode of inheritance, **penetrance** (probability of manifesting the trait given possession of the allele), and frequency of the underlying allele. In some cases this model can be successfully applied to a trait – an example is the persistence of intestinal lactase into adulthood (see Section 13.4.3). However, most complex traits are governed by multiple loci for which such parameters are not known, as well as by environmental influences, so the application of these traditional models is inappropriate. More sophisticated models have been developed (Rogers *et al.*, 1999).

---

failed to yield many genes having significant effects in human populations. One reason may be that many candidates have been identified as a result of the phenotypes of naturally occurring mutants or gene 'knockouts' in mice, and the evolutionary distance between humans and mice may be sufficiently large for functional divergence to have occurred. A further reason is that some human genes have been chosen as candidates for affecting normal variation in a trait because mutations in these genes can give rise to specific abnormal human phenotypes. The connection between normal and abnormal variation is often not straightforward. An example of this is the limited success of using variation in genes underlying hypopigmentation phenotypes such as albinism to explain normal pigmentation variation (see Section 13.3).

An alternative, and less biased, approach to the problem is a **genome scan** for loci involved in a phenotype, in which genome variation is surveyed using a large number of anonymous polymorphic markers spread throughout the genome. Once an interval has been identified within which variation is associated with the phenotype it can sometimes be refined further (resulting in a shorter interval containing fewer genes) using targeted linkage approaches, but often will still contain many genes. Identification of the individual gene underlying the phenotype can be attempted by systematic surveys of variation within all genes, in the hope of identifying functional variants. However, sometimes a **'positional candidate'** approach is adopted, where fewer genes in the interval are chosen for initial investigation on the basis of their perceived possible roles in the phenotype of interest

13.2.3 What have we discovered about genotypes underlying phenotypes?

Despite the fact that biological anthropologists have gathered huge amounts of descriptive data on morphological, physiological and behavioral variation, and formulated theories to explain its adaptive significance, almost nothing has been discovered about the molecular genetic basis of differences in these phenotypes between populations. One thing seems clear – most phenotypes have a multigenic basis – none of the most frequently quantified traits exhibit **Mendelian** inheritance. QTL analysis could be used to identify individual genes contributing to a phenotype, and although more difficult to carry out in humans than in experimental animals, should be applicable to a number of these complex phenotypes. For example, its application to the phenotype of obesity (Comuzzie, 2002) has identified several QTLs, including a locus influencing levels of the hormone leptin on chromosome 2, a finding that has been replicated in several studies. This QTL accounts for 47% of the variation in serum leptin levels in Mexican–Americans, and 56% of its variation in African–Americans, and contains the candidate gene *POMC*, encoding the prohormone pro-opiomelanocortin, a potential regulator of appetite. This study demonstrates the feasibility of identifying QTLs underlying anthropologically interesting traits, and it is to be hoped that more research is directed towards this in the future (Rogers *et al.*, 1999). If QTLs for complex human traits are not sought or identified, selective hypotheses regarding these traits remain unfalsifiable – more 'Just So' stories.

13.3 Adaptation to climate: The evolutionary genetics of pigmentation

Skin color is one of the most obvious ways in which humans differ, and has been widely used in attempts to define races (Section 9.2). It shows a highly nonrandom geographical distribution (*Figure 13.2*), particularly in the indigenous populations of the Old World. We find populations with the darkest skin colors in the tropics, and those with lighter pigmentation in more northerly latitudes. A large number of adaptive explanations have been put forward for this variation in pigmentation. However, as with many apparently adaptive phenotypes in humans there are populations who do not fit into this neat adaptive pattern, and other explanations must be borne in mind.

If we consider pigmentation as a quantitative trait, its proportion of variation between populations is clearly high. One estimate gives it an F_{ST} of 0.6 (Relethford, 1992), as opposed to 0.15 for neutral autosomal markers (see Section 9.3). The apportionment of diversity in other commonly observed phenotypes behaves in a similar way. As Harpending and Rogers (2000) have pointed out, Nei and Roychoudhury (1982), in a key early study on the genetics

of human races, stated that their results on neutral markers did not apply to '*those genes which control morphological characters such as pigmentation and facial structure*'. This reinforces the point that the easily observed differences between populations are probably the result of adaptation or other selective forces, and therefore are a poor reflection of the relationships between populations. It undermines any attempt to construct racial groups on traditional lines of observable phenotypic differences.

13.3.1 Melanin, melanocytes and melanosomes

The most important pigment influencing skin color is **melanin** (Sturm *et al.*, 1998). This granular substance is produced by specialized cells called **melanocytes** that lie at the boundary between the dermis and epidermis. Melanin is concentrated in vesicles known as melanosomes, and these are transported into keratinocytes. Melanin is responsible not only for the color of skin, but also that of hair and (to a large extent) eyes. Hair follicles and surrounding keratinocytes receive melanosomes from melanocytes in a similar way to skin pigmentation, while in the iris the melanosomes are retained within the melanocytes themselves. For an interesting discussion of the mystical properties ascribed to melanin by some people, see Jones (1996).

The number of melanocytes is about the same in different individuals, and variation in skin pigmentation results from differences in the number, size and distribution of melanosomes within the keratinocytes. Dark skin contains

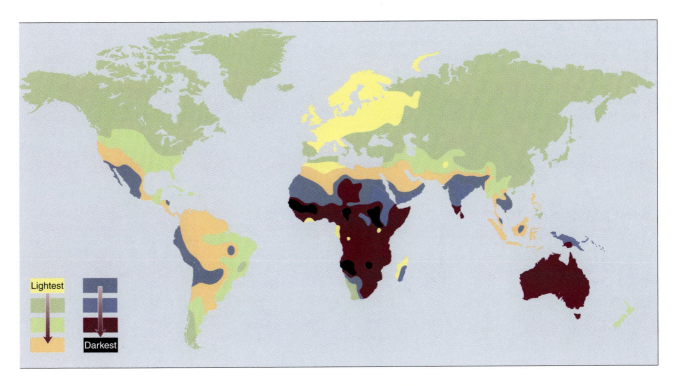

Figure 13.2: Distribution of different skin colors in indigenous populations of the world.

Redrawn from Jurmain *et al.* (2000), after Biasutti (1959).

many large, very dark, melanosomes, while lighter skin contains smaller and less dense melanosomes (*Figure 13.3*); these differences exist at birth, though subsequent exposure to the sun can alter melanosome size and clustering, and result in changed pigmentation (tanning).

As well as the size and distribution of melanosomes, the type of melanin within them also influences pigmentation. The exact chemical structures of different classes of melanin polymer are complex and difficult to define, but it is clear that black/brown pigments are formed by synthesis of eumelanin, while red/yellow pigments result from synthesis of the sulfur-containing pheomelanin. A melanocyte can synthesize both types of melanin.

13.3.2 An adaptive explanation for skin color variation

Anthropologists have addressed the question of the probable ancestral state of human skin color (Jablonski and Chaplin, 2000). It is thought likely that the skin of the earliest hominids was similar to that of the chimpanzee – white, but covered with dark hair. Indeed, the hair-covered skin of most primates is white, suggesting that this is the primate ancestral state, but exposed skin in all primates is pigmented to some extent, indicating that the potential for melanogenesis is also probably ancestral. **Bipedalism** and later brain expansion in early human evolution is thought to have required the evolution of a sensitive whole-body cooling mechanism capable of regulating brain temperature precisely. This involved the loss of hair, and a concomitant increase in number of sweat glands. Bare skin is at risk from damage from UV radiation (UVR), in particular UV-B (wavelength 290–315 nm), the most energetic form of UVR that normally reaches the earth's surface. In order to protect the skin from UVR damage, skin melanization was strongly favored. The immediate ancestors of modern humans, therefore, are generally agreed to have had dark skins, and the issue of skin color variation therefore becomes a question of explaining the different degrees of loss of pigmentation.

Melanin has a twofold role in protecting us from UVR. First, it reduces the amount of radiation entering the deeper layers of the epidermis by absorbing or scattering it. Second, by virtue of its chemical nature, it acts as a filter, absorbing chemical by-products of UVR damage that would otherwise be toxic or carcinogenic. What are the potential effects of UVR exposure that might have acted as selective pressures upon melanization?

▶ *Short-term UVR exposure causes sunburn:* almost everyone with light skin knows the unpleasant after-effects of too much sun. Apart from discomfort and lowering of the pain threshold, more importantly there is damage to sweat glands and suppression of sweating which disrupts thermoregulation. Severe sunburn can be fatal. Light skin is therefore likely to have been strongly disadvantageous in UVR-exposed foraging societies.

▶ *Long-term UVR exposure causes cancers*: degenerative changes in the dermis and epidermis due to UVR can eventually lead to basal cell and squamous cell carcinomas, and also the often-fatal melanomas. However, these cancers are usually manifested after reproductive age, and are therefore not thought to be a strong selective force.

▶ *UVR causes nutrient photodegradation in the skin:* important nutrients such as flavins, carotenoids, tocopherol and folate are sensitive to photo-degradation, and it has been suggested that this negative effect of UVR has formed part of the selective basis for skin pigmentation. The importance of folate has been particularly stressed (Jablonski and Chaplin, 2000). This vitamin is essential in the synthesis of nucleotides, the building blocks of DNA, and its deficiency has been shown to be responsible for neural tube defects (NTDs) such as spina bifida, as well as infertility through impaired spermatogenesis. The connection between UVR, folate and NTDs is supported by the reduction in folate levels caused by

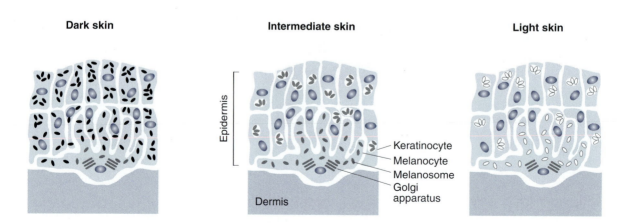

Figure 13.3: Different skin colors are determined by melanosome type and distribution.

In dark skin, melanosomes accumulate eumelanin and remain as single particles, while in lighter skins they contain increasing amounts of pheomelanin and cluster in membrane-bound organelles.

exposure to simulated sunlight, and reports of NTDs in the offspring of mothers who underwent UVR treatment on sun-beds during the early weeks of pregnancy. However, in a recent controlled randomized study, no evidence was found for a reduction of serum folate levels by UV-A (315–400 nm wavelength) exposure, at least (Gambichler *et al.*, 2001). The adaptive effect of folate photodegradation would be to favor dark skins in regions exposed to high levels of UVR, in order to maintain normal fertility and development.

▶ *UVR is essential for vitamin D synthesis:* while the effects of UVR are almost universally harmful, it does have one benefit: the synthesis of vitamin D. This vitamin plays an essential role in the mineralization and normal growth of bone during infancy and childhood. It is present in a number of foodstuffs, including liver, fish oils, egg yolk and milk products, and the proportion of the diet that these foods comprise varies from population to population. For most populations, however, the majority of the vitamin D requirement comes from the action of UVR on a steroid precursor in the skin, 7-dehydrocholesterol. In its active form, vitamin D acts as a hormone to regulate calcium absorption from the intestine and to regulate levels of calcium and phosphate in the bones. Insufficient amounts of vitamin D result in **rickets** in childhood, and **osteomalacia** in adulthood, conditions in which bones are soft, and become distorted, particularly in the weight-bearing parts of the skeleton, the legs and pelvis. Rickets was described in 1645 by Daniel Webster:

'. . . *the whole bony structure is as flexible as softened wax, so that the flaccid and enervated legs can hardly support the superposed weight of the body; hence the tibia, giving way beneath the overpowering weight of the frame, bend inwards . . . and the back, by reason of the bending of the spine, sticks out in a hump in the lumbar region . . . the patients in their weakness cannot (in the most severe stages of the disease) bear to sit upright, much less stand . . .'* (quoted in Stryer, 1988).

Pelvic deformities are a particular problem for women during childbirth, since narrowing of the birth canal can lead to the death of both mother and baby. A demand for adequate synthesis of vitamin D in low sunlight climates may therefore have favored reduction in skin pigmentation. Evidence to support this idea comes from the higher incidence of rickets in African–American inhabitants of the northern USA than in light-skinned European–Americans living in the same place, and similar cases among Ethiopians living in Israel and people from the Indian subcontinent living in the UK. While vitamin D is toxic in excess, there is no evidence that over-exposure to UVR leads to synthesis to toxic levels, since photodegradation of the vitamin results in a steady-state level. Cultural adaptations to vitamin D deficiencies are dietary: in the Arctic they involve the traditional consumption of plentiful oily fish, and in many countries a daily spoon of cod-liver oil or vitamin supplementation of milk and other foods.

Figure 13.4 illustrates how these various factors could interact to influence skin pigmentation. In order to evaluate the different hypotheses, it is necessary to have reliable data on pigmentation in different populations as well as accurate information on the amount of UVR exposure that these populations experience. Skin color is measured using a device such as a reflectometer, which assesses how much light of a particular wavelength is reflected from the skin, usually in a region that is little exposed to the sun (e.g., the underarm area). There is evidence for inter-observer differences in these measurements, as well as difficulties in comparing different measuring devices across datasets. Traditionally, estimating levels of UVR has also been difficult because of the need to take into account atmospheric absorption and scattering of UVR (which depends on the angle of solar elevation), as well as absorption by the ozone layer, clouds, dust and organic compounds. This problem has now been circumvented by the availability of spectrometric readings taken from satellites orbiting the earth. These readings give direct measurements of the actual amount of UVR reaching any point on the earth's surface. Jablonski and Chaplin (2000) compare these UVR data with extensive skin reflectance data from different populations, and interpret their results in terms of the potential for UV-induced vitamin D synthesis in different regions and in different skin colors. The UV dose considered in this study is the minimum-erythremal dose (UVMED), which is the amount of UVR required to produce a barely perceptible reddening of lightly pigmented skin. *Figure 13.5* shows the three latitudinal regions defined in this way; the correlation observed between skin color distribution and predicted vitamin D synthetic potential is taken to support the vitamin D hypothesis. Although the amount of data analyzed in this study is impressive, it does not seem to represent a truly objective test of opposing hypotheses. Cultural adaptations are complicating factors that are difficult to take into account: dietary practices can cope with a lack of vitamin D (or, indeed, folate), and shade-seeking or the wearing of clothes can provide protection from the damaging effects of UVR. Also, there is little paleoanthropological evidence for rickets or ostelomalacia in ancient populations. It seems likely that a need for vitamin D synthesis and photo-protection have both played a role in the distribution of skin color, together with cultural adaptations, and, possibly, sexual selection (see Section 13.3.5).

13.3.3 Genes and gene products in human pigmentation

The biochemical pathways underlying melanin synthesis have been elucidated with the help of mouse coat color mutations. Almost 100 genes have been identified that affect mouse coat color – many of their phenotypes were once highly prized by collectors of 'fancy mice'. Now over 40 of the genes underlying these traits have been cloned (Oetting and Bennett, 2003), and many of them have corresponding human phenotypes; *Table 13.1* lists a selection. Some of these phenotypes are primarily in pigmentation: examples are the different forms of oculocutaneous albinism (*OCA*), in which

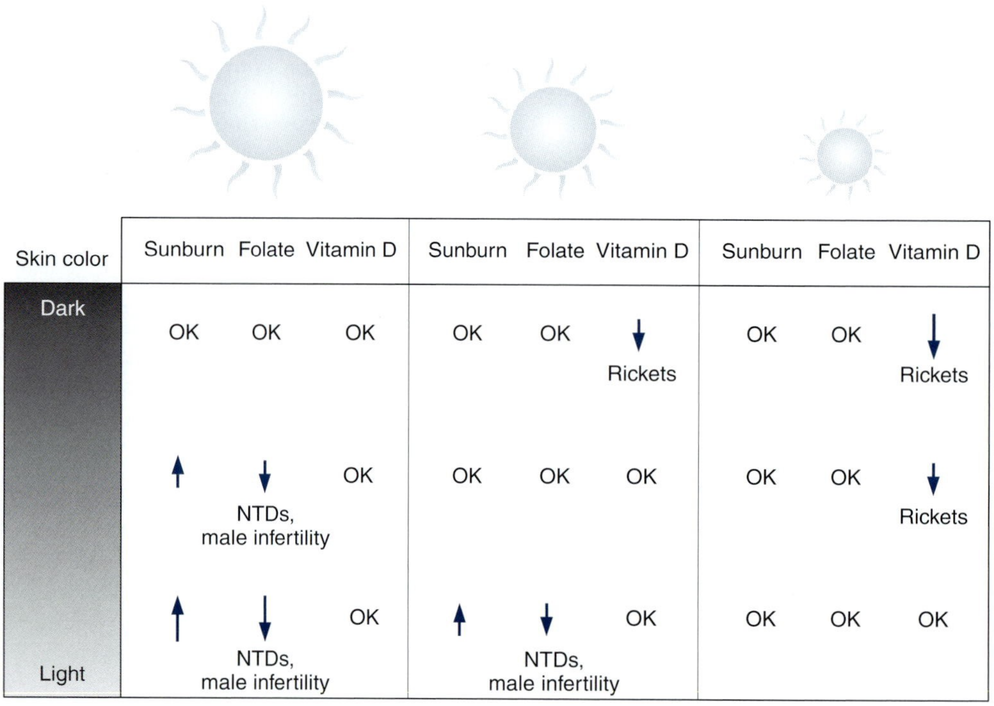

Skin color	Sunburn	Folate	Vitamin D	Sunburn	Folate	Vitamin D	Sunburn	Folate	Vitamin D
Dark	OK	OK	OK	OK	OK	↓ Rickets	OK	OK	↓ Rickets
	↑	↓ NTDs, male infertility	OK	OK	OK	OK	OK	OK	↓ Rickets
Light	↑	↓ NTDs, male infertility	OK	↑	↓ NTDs, male infertility	OK	OK	OK	OK

Figure 13.4: Balancing the harmful and beneficial effects of UVR through skin color adaptation.

The harmful effects of sunburn and folate photodegradation are balanced against the benefit of vitamin D synthesis through UVR, and together these form an adaptive explanation for the distribution of skin color variation. Note that cultural adaptations (protection of skin by clothing and shade, and dietary folate and vitamin D) are also important – see text.

there is lack of pigment. Others have wider developmental or physiological effects that include a pigmentation abnormality: an example is Waardenburg syndrome Type 1 (OMIM 193500), in which a frontal white blaze of hair, variably pigmented irises, white eye lashes and white patches of skin are accompanied by deafness. The association between pigmentation and deafness here results from another function of melanocytes: they are needed within the cochlea for the development of normal hearing. The gene defect underlying this phenotype is in the *PAX3* gene, which is required for the correct migration of melanocytes from the neural crest, within which they originate. *Figure 13.6* shows the subcellular location and function of some of the gene products affecting pigmentation.

While there are now many human genes known whose mutant phenotype includes unusual pigmentation, only a minority of these is known to affect 'normal' variation in skin color. Attempts to implicate particular genes include association studies with particular phenotypes, such as red hair, and also experiments to measure the activities or amounts of particular proteins in cultured melanocytes derived from skin of different colors.

Chief among the genes known to affect normal variation is the melanocortin 1 receptor gene (*MC1R*). Mutations in

this gene affect not only human pigmentation, but also that of wild animals, mice, and many domestic animals, including cattle, horses, sheep, pigs, dogs and chickens (Andersson, 2001). The gene product lies in the cell membrane of the melanocyte (*Figure 13.6*), and is the receptor for α-melanocyte stimulating factor (α-MSH). Receptor stimulation leads to elevation of intracellular concentration of cyclic AMP (cAMP), a signaling molecule, thus inducing changes in protein activity through phosphorylation, altered gene expression, and ultimately the generation of a mature eumelanogenic melanosome. In the absence of a signal via MC1R, the eumelanosome cannot form, and instead the immature pheomelanosome persists, with a consequent phenotype of reduced pigmentation.

The *MC1R* gene is highly polymorphic, with over 30 variant alleles involving amino acid substitutions reported to date: three of these have been shown to be associated with red hair, fair skin and freckling (Bastiaens *et al.*, 2001; Box *et al.*, 1997; Valverde *et al.*, 1995), and a fourth has been associated with fair/blonde hair (Box *et al.*, 1997). These variants map to the intracellular part of the protein, and the first three have been shown to reduce the ability of MC1R to stimulate increases in the concentration of the intracellular messenger molecule cAMP. The mode of inheritance of red hair has been thought to be autosomal recessive, and the

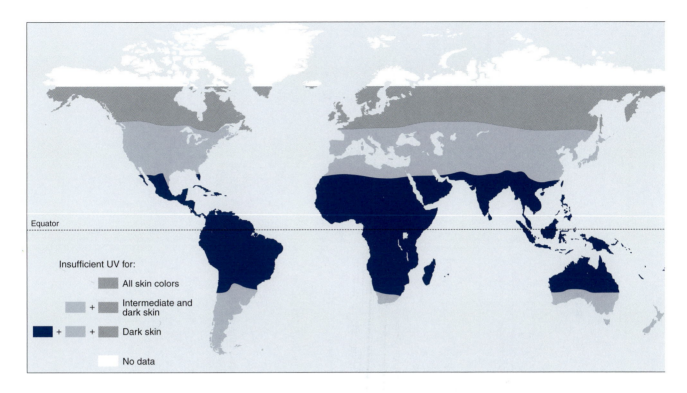

Figure 13.5: Estimated areas in which UVR is insufficient for vitamin D synthesis in different skin colors.

In northern Europe, for example, there is insufficient UVR for the adequate synthesis of vitamin D. The dose of UV considered is the average annual UV minimal erythremal dose, i.e., the amount of UVR causing a barely perceptible reddening of the skin. Redrawn from Jablonski and Chaplin (2000).

inheritance of the variant alleles is broadly consistent with this (Flanagan *et al.*, 2000); this apparent simplicity of inheritance of a pigmentation phenotype has attracted the interest of forensic scientists (see Section 15.2.3). However, heterozygotes show a difference in their ability to tan compared to wild-type homozygotes, so some gene dosage effect is also likely. As well as hair color, there is a general association of the number of *MC1R* variants carried and skin color (*Figure 13.7*), though sharing of identical *MC1R* haplotypes among people with very different pigmentation phenotypes indicates that other loci must be involved (Harding *et al.*, 2000; Rana *et al.*, 1999). Individuals with red hair and fair skin are at increased risk of the cancer cutaneous malignant melanoma, and this is a particular problem for people of north European ancestry who now live in sunnier climes such as Australia. This increased risk is not purely a secondary consequence of the pigmentation phenotype, however. Carrying particular *MC1R* variants also increases melanoma risk in people whose darker skin color might otherwise be considered to be protective against this cancer (Palmer *et al.*, 2000).

Apart from *MC1R*, the identification of genes influencing 'normal' pigmentation, without accompanying pathological features, has been difficult. There is some suggestive

evidence, and a model for pigmentation control based on the suspect loci is presented in *Box 13.3*.

▶ Mutations in the gene *ASIP*, encoding an antagonist of MC1R stimulation, have been reported to be associated with brown hair and eyes (Kanetsky *et al.*, 2002), although this association has not yet been independently confirmed.

▶ A high European frequency of a nonsynonymous substitution in the *MATP* gene, encoding a melanosome-specific transporter protein (Newton *et al.*, 2001), suggests that it may play a role in normal pigmentation, though this idea has yet to be formally tested.

▶ The P protein (*Figure 13.6*) is involved in eumelanogenesis through its role in reduction of acidity within the melanosome. A study of a Tibetan population (Akey *et al.*, 2001) shows no association between skin color and two SNPs in the *P* gene, or three SNPs in the *MC1R* gene. However, an **epistatic** model allowing for interaction between genes did show a statistically significant effect upon skin color of the two genes acting in concert. These quantitative genetic models of pigmentation seem to be the best way forward for untangling this complex phenotype.

TABLE 13.1: EXAMPLES OF HUMAN GENES ASSOCIATED WITH PIGMENTATION PHENOTYPES.

Gene	Protein/function	Mutant human phenotype	OMIM no.	Mouse coat color or other phenotype
Melanocyte function				
TYR	Tyrosinase – oxidation of tyrosine, dopa	Oculocutaneous albinism (OCA) type 1	203100	albino
OCA2	P-protein, regulating pH of melanosome	OCA2	203200	pinkeyed dilute
TYRP1	Oxidation of DHICA, stabilization of tyrosinase	OCA3	203290	brown
MATP	membrane-associated transporter protein	OCA4	606574	underwhite
ASIP	Agouti signal protein – pheomelanogenic stimulation	Association with dark hair and brown eyes	—	agouti
MC1R	Melanocyte-stimulating hormone (MSH) receptor	Red hair or blonde/brown hair	155555	extension
POMC	Pro-opiomelano-cortin (from which MSH and adreno-corticotrophic hormone are produced) – pheomelanogenesis	Red hair with severe early-onset obesity and adrenal insufficiency	—	pomc1
OA1	G-protein coupled receptor	Ocular albinism type 1	300500	oa1
Melanosome transport/uptake by keratinocyte				
MYO5A	Myosin Va motor protein	Griscelli syndrome (includes partial albinism)	214450	dilute
RAB27A	RAS family protein	Griscelli syndrome	214450	ashen
Developmental				
KIT	c-kit receptor – melanoblast migration	Piebaldism	172800	dominant
PAX3	Transcription factor in neural tube development	Waardenburg syndrome type 1	193500	splotch
MITF	Transcription factor	Waardenburg syndrome type 2	193510	microphthalmia

Adapted from Sturm et al. (2001) with additional information from the Albinism Database, which maintains a complete list (http://www.cbc.umn.edu/tad/genes.htm; Oetting and Bennett, 2003).

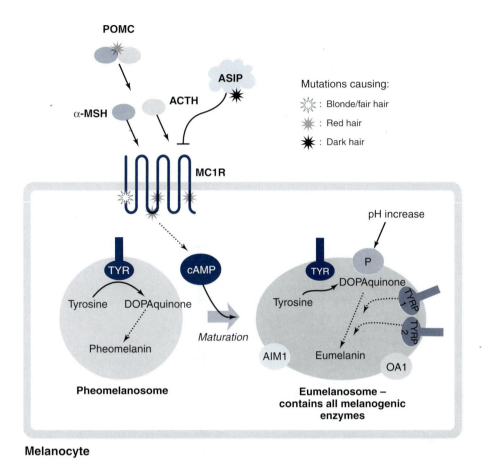

Figure 13.6: The control of pigment type switching in the melanocyte.

The POMC precursor is cleaved to give α-MSH and ACTH (adrenocorticotrophic hormone), which stimulate the MC1R protein. The ASIP protein is an antagonist of stimulation. Stimulated MC1R elevates intracellular cAMP, which acts to stimulate melanosome maturation. The mature eumelanosome contains all the enzymes necessary for synthesis of the dark pigment eumelanin. In the absence of MC1R stimulation, tyrosine is converted to the red/yellow pheomelanin in the pheomelanosome. Key mutations affecting normal pigmentation are shown by stars. Only selected proteins are indicated, and the pathway of conversion of tyrosine is not shown in detail. Adapted from Sturm *et al.* (1998; 2001).

13.3.4 Geographical pattern of genetic variation in human pigmentation genes

If selection has been acting on human pigmentation genes, then the signal of that selection may be discernible in patterns of haplotype diversity. The only gene that has been studied from this perspective is *MC1R*. A study of *MC1R* variation based on **resequencing** of the gene (Harding *et al.*, 2000) revealed a striking difference in the distribution of haplotypes between African and Eurasian populations (*Figure 13.8*). There are only five African haplotypes in the sample, and a complete absence of nonsynonymous base substitutions. In Eurasia, by contrast, there are 13 haplotypes, with 10 nonsynonymous mutations and three synonymous ones. Comparing these findings with the known number of synonymous and nonsynonymous changes between the

human consensus sequence and the chimpanzee sequence shows that the high degree of amino acid sequence conservation in Africa is highly unlikely to have arisen by chance. It probably reflects strong functional constraint in Africa, where any diversion from eumelanin production is strongly deleterious. Despite the vitamin D hypothesis, discussed above, which could provide a mechanism for selection, the pattern in Eurasia seems compatible with low selective constraint, rather than selective enhancement of diversity. This is supported by **HKA selection tests** (see Section 6.3.4) comparing *MC1R* and the β-globin gene between human populations and between humans and chimpanzees. On the face of it, this represents a challenge to the vitamin D hypothesis, but it must be remembered, however, that the power of these tests to detect selection is

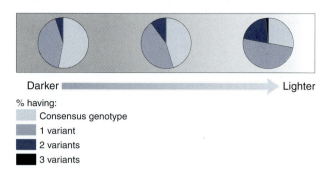

Darker ➤ Lighter

% having:

Consensus genotype
1 variant
2 variants
3 variants

Figure 13.7: *MC1R* genotypes in different skin colors in a population from Australia.

The percentage of the consensus *MC1R* genotype, and those of individuals carrying 1, 2, or 3 variants in their *MC1R* genes (a gene can contain more than one variant), are shown in three different skin color groups (total sample size 637). Lighter skin is associated with a higher proportion of variant alleles. The variants considered here are three nonsynonymous variants previously associated with red hair, one associated with blonde/fair hair, and one further variant with no reported pigmentation association. The population sample in this study was from Queensland, Australia, and presumably of mixed ancestry, but with a substantial European component. Adapted from Palmer *et al.* (2000).

limited. One result of strong selection in Africa and no selection outside Africa is that the *MC1R* gene shows a high among-population variance, of 29% compared to the typical 10–15% (see Section 9.3).

Sequence analysis of a 6.6-kb region including the *MC1R* promoter (Makova *et al.*, 2001) reveals a very high degree of nucleotide diversity; unlike the coding region, but like many other regions of the genome, there is greater diversity among Africans than among non-Africans. No attempt has yet been made to associate variants in this potentially important region with pigmentation phenotypes.

13.3.5 Sexual selection in human phenotypic variation?

An adaptive model for human skin color variation is described above, and, as has been mentioned at various points in this chapter, there have been adaptive explanations proposed for many other differences between human populations. However, two major problems remain. The first is that, even when an adaptive explanation seems reasonable and persuasive, there are often populations that are presently in the 'wrong' place environmentally to fit into the adaptive mold. The second is that some adaptive explanations are far from persuasive in the first place. An alternative to adaptation through natural selection is to turn instead to the peacock's tail as a model. This spectacular ornament is nothing to do with natural selection, and everything to do with sexual selection. Females choose mates on the basis of the quality of their tails, which leads to an increasing degree of tail ornamentation, to the extent that this can actually become

maladaptive, endangering the survival of its bearer. It has been suggested that a model of sexual selection might be applicable to phenotypic variation among human populations (Darwin, 1871; Diamond, 1991; Harpending and Rogers, 2000). Some of Charles Darwin's views on this are summarized in *Box 13.4*.

If this were the case then adaptive explanations would be inadequate, which is what we observe. African Pygmies have been suggested as a population in which a trait (small stature) could have been maintained by sexual selection even when neutral genetic marker differentiation between Pygmies and surrounding populations had been homogenized. Harpending and Rogers (2000) speculate that visible differences traditionally regarded as 'racial' reflect, in some sense, ancient population differences that have survived because of sexual selection. As a consequence, traits under sexual selection will be poor indicators of population relationships, which are more faithfully represented by genetic diversity at neutrally evolving loci. This resurrection of the idea of ancestral racial phenotypes is interesting, but it is worth noting that structured sexual selection could lead to dramatic modern phenotypic differences even in the absence of any ancient differences. These issues will be illuminated by an examination of the long-term stability and influence of sexual selection, as well as the hoped-for identification of genes underlying relevant quantitative traits in humans, and perhaps molecular studies aimed specifically at detecting the imprint of sexual selection. Note that the amount of credence given to ideas of ancient population differences depends on the chosen model of early human evolution, and specifically the amount of gene flow between early populations (see Section 13.1).

13.4 Adaptation to diet: Lactase persistence

13.4.1 Features of lactase persistence

In most of the world's population the ability to digest lactose, the major sugar of milk (*Figure 13.9*), declines rapidly after weaning, because of falling levels of the enzyme lactase in the small intestine. In these people, ingestion of more than a small quantity of milk (which contains lactose at a concentration of 4–8%) in adult life causes abdominal pain, flatus and diarrhea, since lactose causes osmotic transport of water into the small intestine, and is fermented in the colon by gut bacteria (Swagerty *et al.*, 2002). This permanent reduction in lactase levels is common to all mammals, and may be important in encouraging young animals to be weaned towards an adult diet. In lactose-intolerant people, milk products that are soured or otherwise treated, such as cheeses and yogurts, cause few problems because they contain relatively low levels of lactose, or even bacteria that themselves secrete lactases (e.g., *Lactobacillus acidophilus*). This is a cultural adaptation to the trait of lactose intolerance.

However, substantial numbers of people worldwide can continue to drink fresh milk without any problems, because lactase activity persists into adulthood (**lactase persistence**). In such adults, prolonged avoidance of lactose does not result

BOX 13.3 Opinion: Color in our genes – natural selection operating through climatic, dietary or immunological pressures?

There is no doubt that visual impressions of body form and color are important in the interactions within and between human communities. Differences in these human color traits can be graded between individuals of darkest to those of lightest pigmentation in a continuum. In the case of skin, hair and eye color such differences are mainly produced by one pigment known as melanin which has a variety of functions that may be under natural selection. By incorporating different chemical subunits the melanin polymer can vary from black/brown to red/yellow, which can account for some of the color qualities, with melanin particle size, shape, density and distribution contributing to the degree of opacity. The goal of understanding the differences of human pigmentation, so often reduced to the single index of color, is really one of much higher complexity in identifying and understanding the nature of the genes regulating melanin formation. Population studies have now identified functionally significant polymorphic differences within the melanocortin-1 receptor (*MC1R*) regulating melanin type switching, and two genes responsible for different forms of oculocutaneous albinism (*OCA*), the P-protein (*OCA2*) and a membrane-associated transporter protein (*OCA4*). Together these three genes may provide an explanation for how human pigmentation changes have evolved and allow a true molecular understanding of the dynamics of color genes in different societies.

The degree of pigmentation of the small group of ancestral humans originating in Africa and who radiated through the rest of the world was presumably uniform but of unknown pigmentation state. The presence of full strength (wild-type) alleles for *MC1R*, *OCA2* and *OCA4* genes that act in a dominant fashion in modern African populations argues for the fact that black hair, brown eyes and dark skin color must be considered the primordial state for humans before transcontinental migration. The selective pressure to retain a darkly pigmented epidermis has been proposed to include the ultraviolet (UV) light absorbing properties of melanin which would protect the skin from sunburn and eventual skin cancer as well as eye damage. Excessive sunburn and skin peeling would most likely limit foraging and hunting. Although skin cancer and premature blindness would be unlikely to occur during the reproductive years, this could affect the longevity of the grandmother and reduce the reproductive success of her daughters or survival of grandchildren. It has now been recognized that protection of the sweat glands from UV-damage is also necessary to ensure

UV-light →

Vitamin-D ... **Melanin**

	MC1R OCA2 OCA4	MC1R OCA2 oca4	MC1R oca2 OCA4	MC1R oca2 oca4	mc1r OCA2 OCA4	mc1r OCA2 oca4	mc1r oca2 OCA4	mc1r oca2 oca4
Africa MC1R OCA2 OCA4	MC1R/MC1R OCA2/OCA2 OCA4/OCA4							
MC1R OCA2 oca4	MC1R/MC1R OCA2/OCA2 OCA4/oca4	MC1R/MC1R OCA2/OCA2 oca4/oca4						
MC1R oca2 OCA4	MC1R/MC1R OCA2/oca2 OCA4/OCA4	MC1R/MC1R OCA2/oca2 OCA4/oca4	MC1R/MC1R oca2/oca2 OCA4/OCA4					
Asia MC1R oca2 oca4	MC1R/MC1R OCA2/oca2 OCA4/oca4	MC1R/MC1R OCA2/oca2 oca4/oca4	MC1R/MC1R oca2/oca2 OCA4/oca4	MC1R/MC1R oca2/oca2 oca4/oca4				
mc1r OCA2 OCA4	MC1R/mc1r OCA2/OCA2 OCA4/OCA4	MC1R/mc1r OCA2/OCA2 OCA4/oca4	MC1R/mc1r OCA2/oca2 OCA4/OCA4	MC1R/mc1r OCA2/oca2 OCA4/oca4	mc1r/mc1r OCA2/OCA2 OCA4/OCA4			
mc1r OCA2 oca4	MC1R/mc1r OCA2/OCA2 OCA4/oca4	MC1R/mc1r OCA2/OCA2 oca4/oca4	MC1R/mc1r OCA2/oca2 OCA4/oca4	MC1R/mc1r OCA2/oca2 oca4/oca4	mc1r/mc1r OCA2/OCA2 OCA4/oca4	mc1r/mc1r OCA2/OCA2 oca4/oca4		
mc1r oca2 OCA4	MC1R/mc1r OCA2/oca2 OCA4/OCA4	MC1R/mc1r OCA2/oca2 OCA4/oca4	MC1R/mc1r oca2/oca2 OCA4/OCA4	MC1R/mc1r oca2/oca2 OCA4/oca4	mc1r/mc1r OCA2/oca2 OCA4/OCA4	mc1r/mc1r OCA2/oca2 OCA4/oca4	mc1r/mc1r oca2/oca2 OCA4/OCA4	
Europe mc1r oca2 oca4	MC1R/mc1r OCA2/oca2 OCA4/oca4	MC1R/mc1r OCA2/oca2 oca4/oca4	MC1R/mc1r oca2/oca2 OCA4/oca4	MC1R/mc1r oca2/oca2 oca4/oca4	mc1r/mc1r OCA2/oca2 OCA4/oca4	mc1r/mc1r OCA2/oca2 oca4/oca4	mc1r/mc1r oca2/oca2 OCA4/oca4	mc1r/mc1r oca2/oca2 oca4/oca4

Diet ↓ Vitamin-D

Melanin ↑

thermoregulation and that photodegradation of folate essential for embryonic health must be avoided, which would again select for dark skin in high incident UV-climates. Lastly, the presence of high levels of melanin in the skin may be involved in protection from fungal infections in steamy tropical climates providing for an additional immunological selection for pigmentation.

In contrast to the extraordinary conservation of the wild-type *MC1R* genotype in Africa, indicative of a strong selective pressure to retain dark skin, there is extensive allelic polymorphism within this locus in light-skinned European populations, with over 30 *MC1R* alleles so far reported (Sturm *et al.*, 2001). There also appear to be some *MC1R* polymorphic alleles of higher frequency in the Asian community. There must have been disparate selective pressures acting upon human pigmentation during migrations into Europe and Asia with the associated climatic changes, and a later adaptation to an agrarian economy must also be considered. With lower incident UV-light it is possible that the strong positive selection for darker complexions in Africa was reduced allowing for lighter skin color in Northern European populations, with Asia perhaps somewhere in the middle. More likely was a reciprocal positive selection for lower levels of epidermal melanin in Europeans acting through the

evolution of multiple lower strength *MC1R* alleles. In addition there are major differences in the allele strength or frequency of the two *OCA* genes in human populations, with weaker *OCA2* alleles associated with skin lightening and blue eyes seen in Europeans and strong *OCA2* alleles responsible for brown eyes and darker skin. The finding of an exclusive Leu374Phe amino acid change within the *OCA4* allele of most Europeans must also be considered functionally significant and indicative of an early selective pressure for depigmentation.

To have so many genetic changes in multiple loci occurring in the European population argues for a strong positive selection for lighter skin color in glacial and temperate zones. The vitamin D hypothesis is that a lack of sunlight in Europe combined with a change of diet to one consisting predominantly of grains (further reducing the vitamin D intake), caused rickets. This would result in selection for lighter skin color to increase the UV-induced synthesis of this vitamin in the skin during pregnancy and lactation. Other selective pressures for a lower epidermal melanin content may have included susceptibility to frost bite and also that skin color is a visible sign of kinship.

Richard A. Sturm, Institute for Molecular Bioscience, University of Queensland, Australia

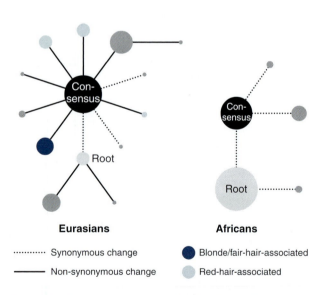

Eurasians   **Africans**

·········· Synonymous change   ● Blonde/fair-hair-associated

——— Non-synonymous change   ● Red-hair-associated

Figure 13.8: Gene trees connecting *MC1R* haplotypes in Eurasians and Africans.

Circles represent haplotypes, with area proportional to frequency. Each line is a single base substitution, and dashes indicate synonymous changes. Note that the African haplotypes include no nonsynonymous changes, and that the root haplotype is most common in Africa. The gene trees show no evidence of historical recombination events. Haplotypes previously shown to be associated with particular hair colors are indicated. Adapted from Harding *et al.* (2000).

in lactose intolerance: the trait is an innate one, representing a genetic adaptation. Note that the term initially used to describe this phenomenon, the disease state 'lactose intolerance', reflects a **Eurocentric** bias: the tolerant state found in Europeans, or people of European descent, was by default defined as the 'wild-type' state with which the phenotype in other populations was contrasted. The intolerant state is the 'wild-type' or ancestral state, and it is the Europeans who are 'mutant' or derived.

13.4.2 Geographical distribution of lactase persistence, and adaptive hypotheses

The geographical distribution of lactase persistence is highly nonuniform (*Figure 13.10*): it is highly prevalent (> 70% population frequency) in Europeans and in certain African pastoralists, such as the Beja of Sudan. It is at intermediate (30–70%) frequency in the Middle East, around the Mediterranean, and in south and central Asia, and low in native Americans and Pacific islanders, as well as in much of sub-Saharan Africa and southeast Asia. The phenotype thus seems to be most frequent among people who have a history of drinking fresh milk, and least frequent among people who have no such history. Lactase persistence has therefore long been regarded as a probable adaptation to dietary change brought about by the development of agriculture and animal domestication in the Near East (see Section 10.5). Before discussing this and other hypotheses in more detail, it is important to note that data are very patchy for some regions of the world. Tests used to determine the lactase persistence phenotype are outlined in *Box 13.5*.

BOX 13.4 Darwin's 'The Descent of Man and Selection in Relation to Sex'

In *The Descent of Man*, published in 1871, Darwin turns his attention from a general consideration of the relationships and origins of species to the origin and diversity of one species, *Homo sapiens*. In a series of chapters in the middle of the book he establishes the importance of sexual selection in evolution by describing examples among animals, and in the last few chapters turns his attention to humans.

'During many years it has seemed to me highly probable that sexual selection has played an important part in differentiating the races of man; but in my Origin of Species I contented myself by merely alluding to this belief. . . . let us see how far the men are attracted by the appearance of their women, and what are their ideas of beauty.'

Darwin gives many examples, drawn from the experiences of travelers and early anthropologists. These examples demonstrate his simple yet powerful observation that physical traits deemed beautiful (e.g., hairiness, facial features and stature) vary hugely among different populations, yet are always found at high frequency within the population that favors them. From our perspective at the beginning of the twenty-first century, many of these anecdotal comments might seem offensive; in particular, white male anthropologists in the past have come in for justified criticism because of their attention to sexual characteristics of the women of 'native' populations. Here are a few of the examples used by Darwin:

'Pallas, who visited the northern parts of the Chinese empire, says, "those women are preferred who have the Mandschu form; that is to say, a broad face, high cheek-bones, very broad noses, and enormous ears"; and Vogt remarks that the obliquity of the eye, which is proper to the Chinese and Japanese, is exaggerated in their pictures for the purpose, as it "seems, of exhibiting its beauty, as contrasted with the eye of the red-haired barbarians." It is well known, as Huc repeatedly remarks, that the Chinese of the interior think Europeans hideous, with their white skins and prominent noses. The nose is far from being too prominent, according to our ideas, in the natives of Ceylon; yet 'the Chinese in the seventh century, accustomed to the flat features of the Mongol races, were surprised at the prominent noses of the Cingalese; and Thsang described them as having "the beak of a bird, with the body of a man."

It is well known that with many Hottentot women the posterior part of the body projects in a wonderful manner; they are steatopygous; and Sir Andrew Smith is certain that this peculiarity is greatly admired by the men. . . . some of the women in various negro tribes have the same peculiarity; and, according to Burton, the Somal men are said to choose their wives by ranging them in a line, and by picking her out who projects farthest a tergo. Nothing can be more hateful to a negro than the opposite form.

With mammals the general rule appears to be that characters of all kinds are inherited equally by the males and females; we might therefore expect that with mankind any characters gained by the females or by the males through sexual selection would commonly be transferred to the offspring of both sexes. If any change has thus been effected, it is almost certain that the different races would be differently modified, as each has its own standard of beauty.'

Darwin surmises that the strength of sexual selection among developed nations may be diminished:

'Civilised men are largely attracted by the mental charms of women, by their wealth, and especially by their social position; for men rarely marry into a much lower rank. The men who succeed in obtaining the more beautiful women will not have a better chance of leaving a long line of descendants than other men with plainer wives . . .'

However, social stratification may mean that there are subpopulations in which sexual selection is still an active force:

'Many persons are convinced, as it appears to me with justice, that our aristocracy, including under this term all wealthy families in which primogeniture has long prevailed, from having chosen during many generations from all classes the more beautiful women as their wives, have become handsomer, according to the European standard, than the middle classes . . .'

Thus, Darwin had a strong belief that sexual selection was a major force in shaping human phenotypes, and in particular the phenotypic differences between different human populations.

The adaptive hypotheses make the reasonable assumption (given the absence of lactase persistence in other mammals) that nonpersistence is the ancestral state. At least four selective explanations have been proposed for the variation in lactase persistence, three of them based on the increase in frequency of persistence alleles in populations practicing dairying (Hollox and Swallow, 2002).

▶ *Food value:* the lactose component of fresh milk provides a valuable extra source of nutrition.

▶ *Water content:* milk provides important additional fluid in particularly arid regions. For some desert nomads

milk may be the only source of water at some times of year, and diarrhea as a consequence of lactose intolerance would be strongly disadvantageous.

▶ *Improved calcium absorption:* this hypothesis (Flatz and Rotthauwe, 1973) was put forward to explain the rapid establishment of a high frequency of the phenotype in northern Europe, even though mixed farming meant lower reliance upon milk than elsewhere. As was discussed in Section 13.3.2, populations in northern latitudes may be susceptible to rickets and osteomalacia because of low sunlight and consequent low vitamin D

β-1,4-galactosidic linkage

Lactose

Galactose　　　　**Glucose**

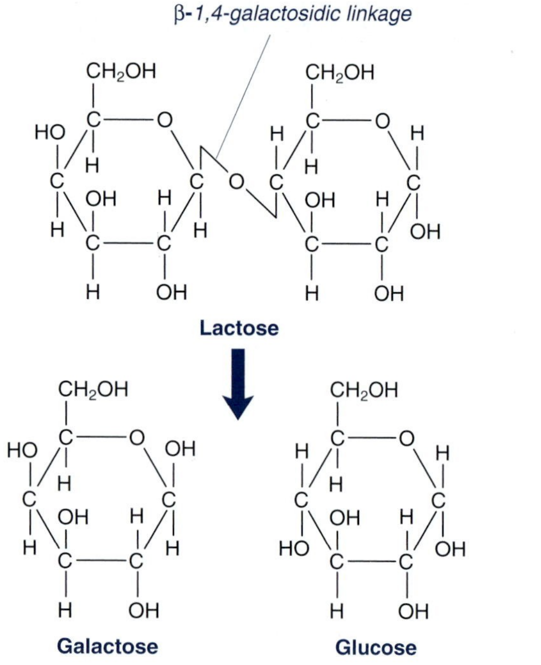

Figure 13.9: Cleavage of the disaccharide lactose by the enzyme lactase.

levels. Calcium can help to prevent rickets, probably by reducing the breakdown of vitamin D in the liver (Thacher *et al.*, 1999). Milk contains calcium, and lactose promotes its absorption; therefore lactase persistent individuals should be able to absorb more calcium than nonpersistent individuals.

▸ *Protection against malaria:* this fourth hypothesis (not widely supported) is based on the observation that individuals who have mild flavin deficiency are somewhat protected against *Plasmodium falciparum* malaria. Milk is rich in riboflavin, so people who cannot drink it may be less susceptible to malarial disease. Lactase nonpersistence is more prevalent than persistence in malarial regions.

All of these hypotheses seem plausible explanations for the distribution of lactase persistence. It is worth noting that some people have taken the view that ingestion of fresh milk in large quantities is such a recent innovation that selection cannot be plausibly invoked – these commentators prefer drift. Nonetheless, the patchy distribution of the trait within Africa and its apparent association with **pastoralism** is persuasive. How can we choose between the different hypotheses? We could try to understand the distribution by examining the correlations between pastoralism, aridity and amount of sunshine on the one hand, and lactase persistence on the other (setting aside the malarial hypothesis). The difficulty with this is that, in effect, it regards each population with high lactase persistence as an independent occurrence of the trait, and ignores population coancestry. An attempt has

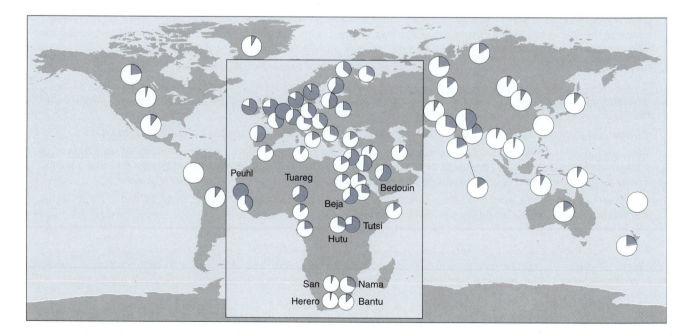

Figure 13.10: Frequency of the lactase-persistence (*LCT*P*) allele in different populations.

The blue sector of each pie chart represents the *LCT*P* allele frequency, as deduced from the frequency of the lactase-persistence phenotype. Note the high frequency of the allele in northern Europe and in some African populations, and the sharp differences between some neighboring groups in Africa. Drawn from data given in Swallow and Hollox (2000); some populations were pooled, and some were omitted because of very small sample sizes, recent assumed admixture, or (in central Europe), for clarity.

BOX 13.5 Lactose-tolerance testing

Asking someone if they are lactose intolerant is not a reliable way to diagnose this trait, since people often attribute symptoms to lactose that actually have other causes (Suarez *et al.*, 1995). Lactose-tolerance tests (Swallow and Hollox, 2000) include:

▶ taking a careful case history, supported by experimental administration of a 50-g dose of lactose and observation of symptoms. Note that this is equivalent to drinking about a liter of milk at one time, and is therefore a greater dose of lactose than most people normally experience;

▶ monitoring the concentration of hydrogen (a by-product of bacterial digestion of lactose in the colon) in the breath after lactose administration;

▶ measuring blood glucose level after lactose administration – a large increase is expected in lactose-tolerant subjects;

▶ administering a dose of ethanol, which inhibits the conversion of galactose to glucose in the liver, and then measuring blood or urinary galactose level after lactose administration.

Lactase levels can also be assayed directly in surgical biopsy material from the small intestine (a precise, but not routine method).

---

been made to surmount this problem by taking population relationships into account using a phylogenetic approach (Holden and Mace, 1997). Both a genetic phylogeny (based on Cavalli-Sforza's classical gene frequency data) and a cultural phylogeny (based on Ruhlen's linguistic classifications) were used (but note, as Cavalli-Sforza does, that they are strongly correlated). Within the framework of these phylogenies, it was asked which of the three variables (pastoralism, aridity, and amount of sunshine) explained the greatest amount of variance in lactase persistence levels. The answer was pastoralism, supporting the hypothesis that the trait is an adaptation to dairying, but not the other hypotheses. Furthermore, a **maximum likelihood** approach to the possible pathways of change between the four possible com-

binations of lactase persistence/nonpersistence and dairying/nondairying indicated that dairying probably originated before lactase persistence. This linkage of the genetic trait of lactase persistence to the cultural trait of milk-production is an excellent example of **gene–culture coevolution** (Feldman and Cavalli-Sforza, 1989) (see *Box 13.6*).

13.4.3 The lactase gene and the distribution of lactase haplotypes

Family and twin studies have shown that the mode of inheritance of lactase persistence is consistent with an underlying single dominant mutation with high **penetrance** (see *Box 3.1*). Therefore in principle it should be easier to identify the genetic factors in this case than it is in the case

---

BOX 13.6 Gene–culture coevolutionary theory

We inherit two kinds of information from our ancestors, genetic and cultural, and both genetic and cultural adaptations are important in human evolution. The study of gene–culture coevolution (Laland and Brown, 2002) investigates the interaction of these two spheres and their effects on the evolutionary process. Culture is an evolving set of beliefs, ideas, values and knowledge that is learned and can be socially transmitted between individuals. Sometimes, whether or not culture is adopted by an individual will depend on their genetic constitution (the example of lactase-persistence is described below). On the other hand, selective pressures acting upon genes can be modified by culture.

The quantitative study of gene–culture coevolution was originated by Feldman and Cavalli-Sforza (1976). Their models incorporated both the differential transmission of genes from one generation to the next, and the diffusion of cultural information, allowing the evolution of the two to be mutually dependent. While gene transmission is an exclusively 'vertical' process (from parents to children), cultural transmission too can be 'vertical', but also 'oblique' (e.g., from teachers to children) or 'horizontal' (among siblings or friends within the same generation). Oblique or horizontal transmission is faster than vertical transmission, which is necessarily measured in generations.

Gene–culture coevolutionary theory can be applied to the question of whether selection pressures following the adoption of dairying led to the spread of lactase persistence (Feldman and Cavalli-Sforza, 1989). This analysis has shown that the persistence allele (see Section 13.4.3) will increase to high frequencies within ~ 300 generations (the interval between the origin of dairying and the present) only if there is strong vertical cultural transmission of milk consumption. If a significant proportion of the offspring of milk consumers did not themselves drink milk, it would require an unrealistically high selective advantage to milk drinking for the allele frequency to rise. The implication of this is that differences between cultures in the strength of cultural transmission might have had a large effect on the current distribution of the persistence trait. In this example the complicating effect of culture means that the predictions of traditional genetic models could arrive at the wrong answer.

of more complex traits. Lactose intolerant individuals are homozygous for an autosomal recessive allele (*LCT*R*/*LCT*R*; where 'R' stands for 'restriction'), and lactase persistent individuals are either heterozygous (*LCT*P*/*LCT*R*) or homozygous (*LCT*P*/*LCT*P*) for a dominant allele preventing the normal decline of lactase activity. Lactase activity assays in adult intestine samples show a trimodal (three peaks) pattern of activity, reflecting the three different genotypes. Nonpersistence is accompanied by reduced levels of lactase mRNA, indicating that variation in levels is due to differences in transcription or mRNA stability. Because *LCT*P*/*LCT*R* heterozygotes show elevation of mRNA derived from one lactase allele only (Wang *et al.*, 1995), the effect must be in *cis* – in other words, the regulatory difference lies on the same chromosome as the upregulated gene copy. A mutation in or near the lactase locus is therefore the likely cause of the change in levels.

The lactase gene spans about 70 kb of genomic DNA on chromosome 2q21 and contains 17 exons. Efforts to identify the molecular basis for persistence/nonpersistence initially concentrated on the promoter region, since animal models suggested that this region was responsible for downregulation after weaning: 1 kb of pig lactase promoter is sufficient to control downregulation in a transgenic mouse model (Troelsen *et al.*, 1994). Sequence analysis identified several polymorphisms in the promoter region of the human gene, as well as nonsynonymous base substitutions within exons. However, phenotype–genotype correlations show that none of these is responsible for lactase persistence. Lactase haplotype studies (Hollox *et al.*, 2000b) identify four globally common haplotypes, one of which, haplotype A, is common only in northern European populations, where there is a high frequency of lactase persistence (*Figure 13.11*). This haplotype is therefore likely to be linked to the causative variant.

Recently, LD studies and haplotype analysis in Finnish pedigrees segregating lactase persistence/nonpersistence localized the variant responsible for persistence to an interval of 47 kb 5′ to the gene. Resequencing of this interval in four lactase persistent and three nonpersistent individuals identified two candidate sequence variants associating with persistence (Enattah *et al.*, 2002). C/T-13910 lies about 14 kb upstream of the start codon of the lactase gene, and G/A-22018 about 8 kb further upstream (*Figure 13.12*); the three sequenced nonpersistent individuals were homozygous for the C and G alleles respectively. Presence of the T allele at

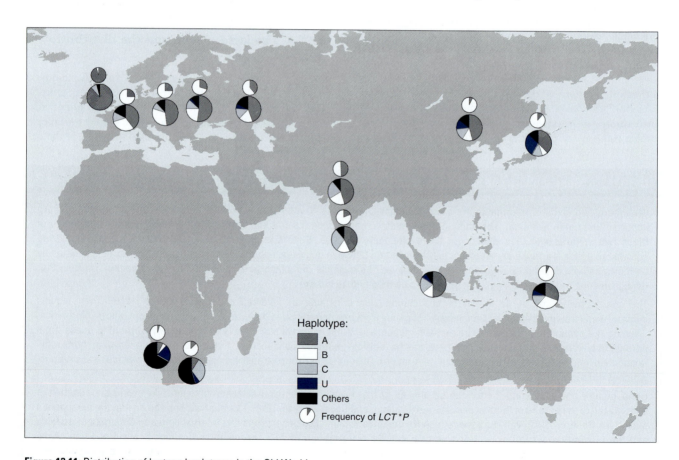

Figure 13.11: Distribution of lactase haplotypes in the Old World.

Frequency of the A haplotype fits quite well with *LCT*P* frequency in Europe and Africa. However, the fit is poor in Asia and Oceania. The blue sector of each small pie chart represents estimated *LCT*P* allele frequency; there are no data for the Singapore sample. Adapted from Hollox *et al.* (2000a).

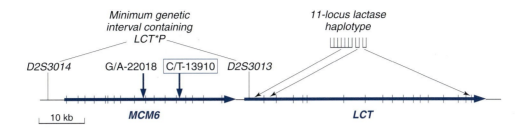

Figure 13.12: The causative variant of lactase persistence lies in an upstream gene.

The genetic interval containing the element responsible for lactase persistence (*LCT*P*) lies between microsatellites *D2S3013* and *D2S3014*. The best candidate for the critical variant site is C/T-13910, where the T allele is completely associated with dominant lactase persistence. Variant sites defining the 11-locus lactase haplotype are shown. Genes are indicated by thick blue lines, with the arrow showing the direction of transcription. Vertical lines within genes indicate exons. Adapted from Hollox *et al.* (2000b) and Enattah *et al.* (2002).

C/T-13910 is completely associated with lactase persistence both in pedigrees, and in a sample of 236 individuals from four different populations (Finns, French, European–Americans, and African–Americans); association is not complete for the G/A-22018 variant, since seven out of 236 individuals with lactose nonpersistence carried A, rather than G, alleles. C/T-13910 is therefore an excellent candidate as the causative variant, though G/A-22018 may also play a role. Studies of these variants in global population samples with known lactase status will be of great interest. The T allele at C/T-13910 disrupts a consensus binding site for the transcription factor AP-2, and this may explain the change in developmental regulation of lactase expression, though proof of this idea will require careful experiments in appropriate intestinal cell lines. Interestingly, the two candidate variants lie not in intergenic DNA, but within introns of an adjacent gene, *MCM6*, the human homolog of a yeast gene involved in the cell cycle. This emphasizes the complexity of the regulation of gene expression, and illustrates the point that apparently neutral DNA variants in introns may play unexpected functional roles many kilobases away.

13.4.4 Are there other consequences of dietary change?

The agricultural revolution was accompanied by major changes in diet for many human populations, and it would be strange if the ability to digest lactose were the only example of a dietary adaptation. There are other possible adaptations to increased carbohydrate and other elements in the diet.

▶ The domestication of plants such as wheat is likely to have led to a large increase in dietary maltose, the main disaccharide produced by digestion of starch by amylases. There is little evidence of dietary problems associated with variation in maltose digestion; however, studies of haplotype diversity around the gene for maltase-glucoamylase, responsible for maltose digestion, would be worthwhile.

▶ Sucrose is likely to have formed a large part of the diet only in relatively recent times. Deficiency of sucrase-isomaltase, the enzyme responsible for sucrose digestion, is not uncommon and reaches 10% frequency in Inuit of Greenland (McNair *et al.*, 1972), a population whose calories traditionally come not from sugar, but from animal products. Again, worldwide studies and haplotype analysis of the sucrase-isomaltase gene would be of interest.

▶ It has been suggested (somewhat speculatively) that the ability to tolerate large amounts of the wheat protein gluten in the diet has been selected for, and that cases of gluten-intolerance are due to the residuum of 'hunter–gatherer genes' in Europe (Greco, 1997).

▶ No hunter–gatherers studied over the last century have been able to make alcoholic drinks, and so drinking alcohol (more properly, ethanol) is generally regarded as an innovation of the Neolithic (Eaton *et al.*, 1988). After ingestion, alcohol is converted by the enzyme alcohol dehydrogenase (ADH) into the toxic compound acetaldehyde, which itself is normally quickly converted into acetic acid by the enzyme aldehyde dehydrogenase I (ALDH I). Variation in the activity of ADH may contribute to 'alcohol sensitivity', in which people drinking alcohol become flushed and rapidly intoxicated, but ALDH I deficiency is the best understood cause (Stinson, 1992). The deficiency allele has an apparent autosomal recessive mode of inheritance, but the molecular basis is not known. The frequency of the trait is high (25–50%) in East Asia and in indigenous peoples of South America, but low or absent in Africa and Europe. The most commonly suggested explanation for this interesting distribution is drift, since there is no convincing adaptive hypothesis. Nonetheless, it has been suggested that selection may be acting through possible roles of ADH and ALDH I in neurotransmitter and hormone metabolism, or through the inhibition of growth of intestinal parasites by elevated acetaldehyde (reviewed by Stinson, 1992).

Our early human ancestors were hunter–gatherers, and only in the relatively recent past have we become agriculturalists; this transition was accompanied by major changes in diet for most human populations (reviewed by Eaton and Eaton, 1999). As well as adaptations to current diets, therefore, we need to consider traits that were previously beneficial, but are now harmful in our changed nutritional environment. In 1962 James Neel suggested that the disease **diabetes mellitus** was the result of '*a thrifty genotype rendered detrimental by "progress"*' (Neel, 1962). His hypothesis was that the rapid release of the hormone insulin in response to elevated blood-sugar levels (hyperglycemia) was advantageous to our ancestors, allowing them to build up fat deposits in times of plenty. However, in an environment where there is continuously plentiful or even overabundant food, this rapid response is detrimental – overproduction of insulin leads to insulin resistance, subsequent high levels of blood glucose, and the set of debilitating symptoms constituting diabetes. Combined with the relative physical inactivity of many people in the developed world, it also leads to obesity. Neel later updated and refined his hypothesis (Neel, 1982) so that it applied specifically to the complex disorder noninsulin dependent diabetes mellitus (NIDDM; also known as type 2 diabetes; OMIM 125853). The thrifty genotype hypothesis has been widely used to explain the high incidences of NIDDM among westernized Native Americans, Australian Aborigines and Pacific Islanders (see *Box 11.7*); on the island of Nauru in Micronesia, for example, more than 30% of people over the age of 15 years have diabetes. It is argued that, because these regions of the world were settled by small numbers of people under difficult circumstances, the thrifty genotype was strongly favored. However, it seems likely that food shortages were an ever-present threat for our ancestors, and unlikely that this threat suddenly disappeared with the advent of agriculture; indeed, paleopathological evidence suggests that nutrition among early farmers was often poorer than among hunter–gatherers (see Section 10.3.1), and so the thrifty genotype could have been favorable in a wider group of populations until very recently. Note that the definition of the thrifty genotype and its association with disease is another example of **Eurocentrism** – this condition may well have been ancestral, so we should perhaps be focusing our attention on the 'nonthrifty genotype' of Europeans (Allen and Cheer, 1996).

Neel's thrifty genotype hypothesis has not gone unchallenged: it has been proposed that poor nutrition *in utero* and in early infancy may be a major cause of increased risk of NIDDM. In this **'thrifty phenotype'** hypothesis (Hales and Barker, 1992), the fetus adapts to maternal malnutrition by itself becoming nutritionally thrifty, resulting in decreased growth, hormonal and metabolic adaptations, and altered growth and function in the cells of the pancreas responsible for insulin secretion. A transition to over-nutrition later in life makes this developmental adaptation disadvantageous, as it predisposes to NIDDM through reduced secretion of insulin, or insulin resistance.

The notion that a thrifty genotype may be associated with disease after a change of environment has also been put forward to explain the high prevalence of hypertension (high blood pressure) among African–Americans (reviewed by Stinson, 1992). The African ancestors of these people had a low salt diet but were prone to salt loss through sweating in a hot climate. The thrifty adaptation was increased sodium retention, which in their US descendants, with a diet high in salt, became detrimental. This interesting idea has yet to be rigorously tested, though a gene has been isolated that might be considered as a candidate (reviewed by Siffert, 2001). *GNB3* encodes a subunit of a G protein, an intracellular molecule that is an important component in cell signaling – the complex cascade of molecular events triggered by the cell receiving a specific signal from its extra-cellular environment. G proteins are transiently activated by a G protein-coupled receptor, a process that involves the binding of GTP, hence the protein's name. Blood cells from hypertensive individuals showed elevated activity of a cell membrane ion transport system, the Na^+/H^+ exchanger. This activity was linked to enhanced G protein activation, and specifically to a common synonymous base substitution (C to T at nucleotide 825) in exon 10 of the *GNB3* gene (Siffert *et al.,* 1998). Through an unknown mechanism, this polymorphism, or another variant linked closely to it, causes a novel splicing event that gives rise to a truncated but functional protein showing enhanced activity. A number of studies in different populations, including Europeans and people of African descent, have shown the 825T allele to be associated with hypertension. Globally, 825T is most frequent (60–80% allele frequency) in sub-Saharan Africans, African–Americans and Australian Aborigines; it is at 40–60% in East Asians, and about 30% in Europeans, which might support an adaptive explanation of sodium retention. However, the mechanism by which elevated G protein activation leads to hypertension is still poorly understood. It may be that association with obesity is more important, and that hypertension in patients is a secondary effect, since there is an association between the 825T allele and obesity in individuals with normal blood pressure.

There is a wide perception that most modern humans are eating the 'wrong' diet – we should be returning to the 'natural' diet of our Paleolithic hunter–gatherer ancestors, rejecting the products of the agricultural Neolithic, and thus avoiding the diseases of civilization, such as coronary heart disease, diabetes and cancers. Analysis of the diets of 229 modern hunter–gatherer populations (Cordain *et al.*, 2000) indicates that they obtain 19–35% of their energy from protein, 22–40% from carbohydrate, and 28–58% from fat. The equivalent figures for US adults are 15.5% from protein, 49% from carbohydrate, 34% from fat (and 3% from alcohol). Thus, the US diet's energy from protein and carbohydrate is outside the range of the hunter–gatherer populations. A paleo-diet called 'NeanderThin' ('Eat like a caveman to achieve a lean, strong, healthy body') is followed by some neo-Paleolithics, and includes a lot of meat, and no 'Neolithic' products. An alternative view (Milton, 2000) takes into account the fact that modern-day hunter–gatherers are largely free of the 'diseases of civilization' regardless of whether they eat largely animal or plant foods, or even if, like the Yanomamo of South America, they rely on domesticated plant foods taken from a single species. Also, a consideration of our great ape relatives and the likely progenitor species of chimpanzees and humans suggests that our distant ancestors had a strongly plant-based diet. While it is certainly true that many of us nowadays do have very unhealthy diets, humans are successful omnivores, and it will take careful study rather than supposition to disentangle the evolutionary web of diet, adaptation and disease.

13.5 Have we stopped evolving?

The human population is clearly not in a state of equilibrium. Present patterns of migration and population growth are far removed from those that have predominated over much of human prehistory. But while the human species is undoubtedly in a state of flux, *have we stopped evolving?* The motivation for this question derives from the common observation that humans have come to occupy such a broad range of environments and respond to environmental changes by evolving in a predominantly cultural rather than biological fashion. While we may yet uncover genetic changes that have aided their colonization of isolated islands, Polynesians have prospered in the South Pacific by virtue of their innovations in long distance navigation and vessel construction, and ability to support large populations in limited environments through fishing and agriculture. Likewise, despite their dutiful adherence to Bergmann and Allen's rules, Inuit would not survive long in the Arctic without the cultural acquisitions of clothing, shelter and fire.

As we shall see, this apparently simple question about the future of human evolution encompasses a diversity of more specific lines of enquiry – different attempts to answer this question have addressed different expectations, and include:

▸ have human allele frequencies been frozen in time?

▸ are humans still undergoing natural selection?

▸ can we expect any changes to the human phenotype?

▸ will humans speciate into two (or more) species some time in the future?

▸ when, if at all, will humans become extinct?

The predominance of cultural evolution does not in itself answer these questions. It is also worth bearing in mind Fisher's comment that: '*Evolution is not natural selection*'.

The issues raised by the set of questions listed above cover a huge time-scale, from the next generation to the end of life on Earth. Some of these queries address the **microevolutionary** pressures – mutation, drift and selection – operating on modern humans, while other queries focus on the **macroevolutionary** future – whether speciation is likely or indeed inevitable. All speculations about the future are just that – speculation. We can have more confidence in the accuracy of predictions about the near future than those on our more distant future. In this section we confine ourselves almost exclusively to microevolutionary processes in the near future. The further forward we try to predict, the more we come into the realm of other disciplines, most notably geology and astronomy. Whilst the catastrophic death throes of our own sun in a few billions of years time exercises the minds of some, this time-span exceeds more than a hundred-fold the average lifespan of a primate species. More pressing considerations abound; the inter-galactic cart should not be put before the Earth-bound horse.

13.5.1 Impact of changes in human demography and population structure

Over the past few centuries, modern sanitation, medicine and improved nutrition have led to dramatic reductions in infant mortality and increases in life expectancy. This has resulted in an exponential increase in population size: our present census population size of over 6 billion contrasts sharply with our long-term **effective population size** of only about 10 000 individuals (see Section 5.3.1). The human population is predicted to keep growing in size for some time yet. Although alternative models of future population growth differ significantly in their projections (UN, 1999), most agree that the rate of population increase is finally slowing, perhaps for the first time since the advent of agriculture some 10 000 years ago. Irrespective of which population forecast turns out to be most accurate, the effective population size of humans is likely to increase substantially over future generations.

An increase in effective population size would be expected to substantially reduce the impact of genetic drift on allele frequencies. However, once we factor in the vastly different growth rates experienced in different continental regions, we can see that this expectation underestimates the complexity of our heavily structured populations. According to one recent model, whereas European populations represented 13.4% of the world's population in 2000, by 2100 they will represent only 7.2% (Lutz *et al.*, 2001). Over the same time span, sub-Saharan African populations are expected to increase from 10.0% to 17.8% of the world's population. Coupled with allele frequency differences between different regions of the world, these changes can be expected to have a significant effect on the species-wide frequencies of different alleles (*Figure 13.13*).

This difference in growth rates results predominantly from the dramatic decline in birth rates in developed countries. Over time regional differences in birth rates are expected to lessen, as birth rates in less developed countries also decline significantly (UN, 2001). Consequently, this demographically driven change in species-wide allele frequencies will not persist far into the future, but nonetheless represents an important mechanism of allele frequency change in the short term.

Another important impact of increasing effective population size is a significant shift in the balance between selection and drift. As we saw in Chapter 5, in larger populations, as genetic drift lessens, smaller selection pressures become able to influence allele frequencies. Thus, in theory, an increasing effective population size should allow selection pressures of lesser magnitude that were previously overridden by stochastic allele frequency changes to make their presence

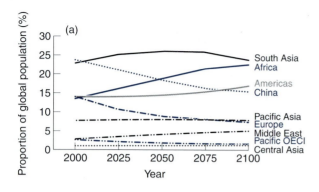

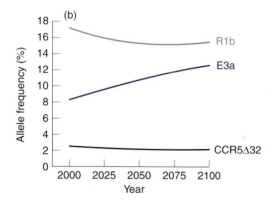

Figure 13.13: Impact of predicted changes in regional population size.

(a) Predicted changes in regional population size.
Data from Lutz *et al.* (Lutz *et al.*, 2001).
(b) Demographically driven changes in allele frequencies, in the absence of drift, mutation and selection.
Data from Hammer *et al.* (2001) and Dean *et al.* (2002). R1b and E3a refer to Y-chromosomal lineages within the phylogeny shown in *Box 8.8*. CCR5Δ32 is the 32-bp deletion that confers reduced risk to HIV infection (see *Figure 14.14*).

felt. This feature of future human populations may interact synergistically with the different selective environments discussed in the previous sections of this chapter.

Effective population size also has a role in determining how long it takes to fix a new allele. It has recently been argued that if the effective population size of humans approached the current census size of 6 billion, it would take more time to fix a new neutral allele by drift than the time available until the end of the solar system (Klein and Takahata, 2002)! However, positive selection decreases the time to fixation massively: a new allele with a 1% selective advantage can be fixed in the same population in about 10 000 years. Moreover, it is highly unlikely that the effective population size will ever get that high. Other species with large populations often have effective population sizes several orders of magnitude beneath their census size.

Effective population size also determines the amount of neutral genetic diversity within a species. As genetic drift

lessens, the balance between mutation introducing new alleles and drift removing them will be adjusted (see Chapter 6). Thus, if the population remains at a large and constant size for a long period of time, we can expect much higher levels of species-wide genetic diversity in the future. Potential ramifications of this additional diversity are considered in Section 13.7.2.

Population size is not the only facet of population structure that has changed significantly in the recent past, and that can be expected to change over subsequent centuries. Individual mobility has also increased massively. Distances between the birthplaces of wives and husbands have increased, and are increasing, breaking down prior population substructure. As discussed in Chapter 5, even small levels of migration over multiple generations can homogenize previously differentiated populations. Over time, phenotypic and genotypic differences between regional populations can be expected to lessen as a result of this increased mobility. A greater proportion of global genetic diversity will be found within any single population. Measures of apportionment of diversity, such as F_{ST}, will tend to zero if our species approaches **panmixis**.

13.5.2 Will the mutation rate change?

Both endogenous and exogenous mutagenic factors are likely to determine whether mutation rates remain constant in the near future.

Endogenous mutagenic factors include: replication errors, reactive by-products of metabolism and DNA repair efficiency. Exogenous mutagenic factors include: chemical mutagens, ultraviolet (UV) radiation and ionizing radiation.

First we shall consider whether the incidence of replication errors per year is likely to be altered by future demographic changes, and then we will go on to consider exogenous and endogenous physico-chemical mutagenesis.

On average, humans are waiting longer before having their first child and are able to have children later than in previous generations. The result of this is an ever-increasing length of the average generation. In Chapter 3, we saw that as generation times increase, the number of male germ-line replications per generation increases while the number in females stays static (Hurst and Ellegren, 1998). Assuming that the ratio of male and female mutation rates (known as the **alpha factor**) is driven by the relative number of replications, this ratio will increase as humans reproduce later in life.

Although increases in generation time will result in increases in the number of replication errors per generation for males, what is the effect on the annual replication error rate? As females have a fixed number of replications per generation, a longer generation time means a lower annual replication error rate in the female germ line. However, the opposite is true of the male germ line. In males, a fixed number of germ-line replications occurs prior to puberty (~ 35 years) and then the replication rate accelerates to a much faster annual rate (~ 23 per year). As generation times get longer, the proportion of the generation spent during the high rate replication phase, post-puberty, increases. Consequently, the annual germ-line replication rate in males increases with longer generation times.

Assuming that the future will most likely entail increased generation times over the species as a whole, what will be the overall impact on annual replication error rates of the opposing directions of female and male germ-line replication error rates described above? The autosomal replication error rate is simply an average of the two sex-specific rates. We can see in *Figure 13.14* that the increase in the male annual replication error rate outweighs the decrease in females, and as a consequence autosomal replication error rates per year increase with longer generation times.

In contrast to the autosomes, the sex chromosomes do not pass equally through the male and female germ lines. Y chromosomes pass solely through the male germ line, whereas X chromosomes pass through the female germ line twice as often as the male germ line. As a result, increasing generation times will have different effects on the annual replication error rate of these chromosomes than on autosomes. *Figure 13.15* shows that the Y-chromosomal annual replication error rate will increase more rapidly than that of the autosomes, whereas the replication error rate of the X chromosome will also increase, but at a reduced rate compared to the autosomes.

Thus far, we have assumed that germ-line replications of males and females are equally mutagenic. Interesting conclusions can result if we drop this assumption, which, as the following reasoning shows, may not be tenable. For example, for a generation time of 25 years, the number of male germ-line replications exceeds the number of female replications by a factor of ~ 14. However, most estimates of the alpha factor are considerably lower than this (typically 1.5–5.0; see *Box 5.8*). If replication errors are the sole source of mutation, this discrepancy could only be explained if replication is less mutagenic in the male germ line than in the

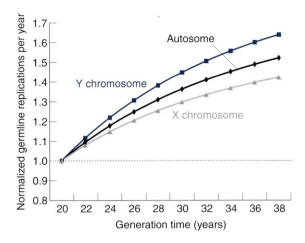

Figure 13.15: Generation time increases have disproportionate effects on loci with different patterns of inheritance.

The numbers of replications per year have been normalized to facilitate comparison.

female germ line. If we downweight the number of male germ-line replications accordingly, the female germ line starts playing a larger role in determining the annual mutation rate and longer generation times should inflate autosomal mutation rates less dramatically. Again this will have disproportionate effects on loci with different inheritance patterns. *Figure 13.16* shows that, under a set of plausible assumptions, the annual mutation rate of the X-chromosome can actually decrease with longer generation times. This raises the interesting prospect that increases in human generation time might result in both increases *and* decreases in annual mutation rates in different parts of the genome.

An alternative explanation of the mismatch between the alpha factor and the excess of male replications outlined

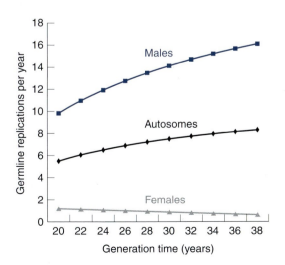

Figure 13.14: Generation time increases drive higher annual autosomal mutation rates.

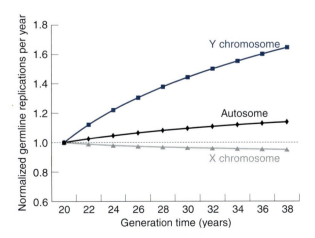

Figure 13.16: Lower mutagenicity of male germ-line replications can lead to longer generation times, resulting in lower X-chromosomal mutation rates.

Assuming that generation time = 25 years, alpha factor = 2.

above is that replication errors are not the sole source of germ-line mutations. We can estimate the contribution (c) of the other possible endogenous and exogenous sources of mutation by assuming no sex differences in mutation dynamics and rearranging the equation below:

$$\alpha = (c + 14r)/(c + r)$$

where r is the number of female germ-line replications, to give:

$$c = (14r - \alpha r)/(1 - \alpha)$$

Using this equation we can see that if the alpha factor equals 5 then other mutagenic factors are contributing approximately twice as many mutations than are replication errors. However, if the alpha factor is only 1.5, then these other factors contribute 23 times as many mutations!

One potential source of additional mutations is error-prone DNA repair processes. It is difficult to hypothesize how the efficiency of DNA repair might change in future human populations. If repair efficiencies are age related, an aging population might exhibit lower levels of DNA repair. It would be important to see whether impairment of DNA repair manifests itself exclusively postreproductively.

By contrast, it is easier to explore the possible changes to the physico-chemical challenges to the genome in the future:

▶ The number of synthetic chemicals in the environment is increasing; however, their mutagenic potential is assessed prior to their release and levels are regulated, at least in principle. The monitoring of predetermined safe levels by environmental agencies has thus far ensured that global chemical mutagenicity has not altered significantly, although local cases have been documented. There is, however, a possibility that improved methods for the bio-monitoring of environmental mutagens might reveal hitherto unrecognized global contributions to the mutational process.

▶ Nuclear weapons testing and radiation leaks have increased levels of background radiation, albeit minimally. The global increase in radiation exposure as a result of our nuclear activities is far outweighed by regional variation in levels of natural radioactivity. However, species have never (prior to nuclear weapons and nuclear accidents) been exposed to such high-energy ionizing radiation (IR) before, and there may well be unpredictable consequences in the way that cells behave. Mutation rate at human minisatellites has been shown to be elevated by radiation exposure (Dubrova *et al.*, 1996). Recent work in mice has suggested that increased mutation rates as a result of exposure to high-energy IR are not restricted to the exposed generation, but can be inherited by future generations (Barber *et al.*, 2002), themselves unexposed.

▶ Cosmic radiation with mutagenic capabilities is relatively constant in magnitude. Increased UV radiation has reached extreme latitudes through the ozone holes formed by **anthropogenic** degradation, resulting in an elevated prevalence of skin cancer, especially in Australasia. However, the low penetration of UV radiation means that it does not affect germ-line mutation rates. There is reason to hope that the ozone holes will recover, provided legislative action is effectively applied worldwide.

▶ The future impact of endogenous mutagenic chemicals is more difficult to assess. While oxidative metabolism generates mutagenic by-products, the increasing calorific intake of modern populations may not be matched by increased metabolism.

In summary, there is no good reason to expect that nuclear mutation rates will increase in the near future as a result of physico-chemical assaults, although recent evidence suggests mtDNA might be more susceptible to environmental influences. In one study, an Indian population exposed to high levels of natural radioactivity exhibited an elevated rate of base substitution in the mtDNA control region (Forster *et al.*, 2002).

13.5.3 Alterations to selective environments

For natural selection to drive evolutionary change, variation at a locus must be correlated with reproductive success and be genetically heritable. Reproductive success is the over-representation of an individual's genes in successive generations, and may result from increases in either survival to reproductive age (differential mortality) or the production of offspring (differential fertility). This selective environment has changed enormously in the past century, most notably in developed countries. Many individuals who would not previously have had the opportunity to reproduce, are now living longer and having children. As a consequence, greater levels of deleterious variation at genetic loci can now be passed on to future generations. It was the fear of ever-decreasing genetic fitness that drove the eugenics movement in the early twentieth century. It is a little discussed fact that R.A. Fisher spent much of the latter half of his seminal population genetics text (Fisher, 1930) arguing that greater reproduction of lower levels of society caused the eventual failure of historical empires. Others regard the mitigation of natural selection as a defining feature of civilization (see *Box 13.4*, for Darwin's view on sexual selection). Irrespective of the ethical implications, we can examine the degree to which natural selection has become attenuated in humans. We shall consider changes in differential mortality and fertility in turn.

Differential mortality

The proportion of all babies surviving to reproductive age is increasing; in the developed world the probability that a baby will survive to reproductive age is well over 90%. The majority of deaths in the developed world result from post-reproductive disorders such as cardiovascular disease, cancers and diabetes. Infectious disease, which once accounted for the majority of human mortality, has been relegated to a more peripheral role in these countries. The strength of selection on post-reproductive disorders is presumed to be small, as by their definition they do not impact directly upon reproductive success. Nevertheless, future changes in human lifestyles may cause a reduction in age of onset of such disorders, illustrated by the rise in childhood obesity (Rocchini, 2002). U.S. Surgeon General David Satcher has commented that: *'Childhood obesity is at epidemic levels in the United States'.*

Childhood obesity is closely correlated with rises in both early-onset diabetes and cardiovascular disease. Such social changes could bring these disorders into the realm of natural selection by resulting in pre-reproductive mortality.

Infectious disease plays a far more dominant role in mortality in the developing world (*Figure 13.17*). Tuberculosis (TB) kills roughly 1.5 million people annually, and every year about 1% of the world's population is newly infected, with about 2 billion people carrying infection in total (WHO, 1999). Malaria causes 300 million acute illnesses annually and kills over a million people (WHO, 1999). AIDS infects some 40 million and killed about 3 million people in 2001 (UNAIDS, 2002). The annual AIDS death rate is predicted to rise substantially over the next two decades. Interestingly, the mortality of modern infectious diseases like AIDS is geographically non-random as is seen for earlier diseases.

The spread of infectious disease has been greatly enhanced by increases in mobility and especially intercontinental air travel. Every year there are unexpected outbreaks of emerging and re-emerging infectious diseases. In addition, novel pathogens are recognized on an almost annual basis. Irrespective of whether humans are evolving or not, pathogens will continue to evolve, adapting to new hosts and circumventing both immune and chemical defenses. Six billion large-bodied mammals with limited genetic diversity and high mobility represents a host environment of almost unparalleled riches for a new pathogen.

Undoubtedly, much of our response to these new challenges will be cultural in nature, involving the development of new vaccines, cures and treatments as well as greater education and preventative action. However, it is unlikely that infectious disease will be eradicated from all regions of the world (the World Health Organization reports that smallpox is the only disease to be completely eradicated, although polio and Guinea worm are both on their way out). Epidemics of new diseases will still arise suddenly, and within affected populations, differential mortality is likely to result from different genotypes. The result may be a selective sweep around haplotypes conferring resistance. The potential for infectious disease to cause selective sweeps has been recently suggested to be the cause of the surprisingly low MHC diversity within chimpanzees (de Groot *et al.*, 2002). Proteins encoded by MHC alleles have different specificities for peptides derived from different pathogens. It has been hypothesized that this selective sweep in chimpanzee MHC diversity might relate to the development of protection against the chimpanzee form of human immunodeficiency virus (HIV), simian immunodeficiency virus (SIV). Certainly, modern chimpanzees appear to be resistant to developing an immunodeficiency syndrome from SIV infection. In addition, many of the common chimpanzee MHC class I alleles target conserved HIV protein motifs **(epitopes)**, and these same motifs are recognized by the specific class I alleles that are present in humans with resistance to the development of full blown AIDS, known as long-term nonprogressors (de Groot *et al.*, 2002).

We already know of specific alleles that either confer predisposition or resistance to mortality and morbidity from infectious diseases. Some, like the sickle cell anemia allele, Hb^S, and some thalassemias have risen to high frequency due to past (and possibly present, in some remote areas) selection pressures, in this case from malaria. Others, like the CCR5-Δ32 allele that confers resistance to HIV infection, may have arisen either through neutral processes or as a result of selection resulting from unknown ancient diseases. Still others, such as the prion protein gene variant that, in homozygous form, confers susceptibility to variant Creutzfeldt-Jakob disease (CJD) – a 'new' disease – may be under novel selection because of novel environmental effects (see *Box 13.7*).

It is worth noting that the eradication of certain infectious diseases of longstanding impact on humans has important implications for the evolution of some protecting balanced polymorphisms. These polymorphisms are presently maintained in the population by a balance of negative selective pressures on both homozygotes; the heterozygote has a selective advantage over either homozygote. The removal of negative selection on one of these homozygotes, by for example the eradication of malaria, might result in directional selection pressures eradicating the allele that

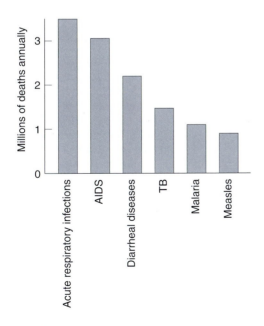

Figure 13.17: Six diseases account for 90% of deaths from infectious disease (WHO, 1999).

BOX 13.7 vCJD in the UK: A new opportunity for genetic selection?

The prion diseases, or transmissible spongiform encephalopathies (TSEs), are a remarkable group of fatal neurodegenerative diseases that affect humans and animals. The etiological agent has been shown to be an abnormal **isoform** of a host-encoded protein in the brain that is apparently derived from its normal counterpart by a conformational change. Crucially, this pathogenic conversion can be triggered by the presence of the abnormal isoform in a quasi-catalytic fashion.

Some of these diseases [such as familial Creutzfeldt-Jakob disease (CJD; OMIM 123400) and Gerstmann-Sträussler syndrome (OMIM 137440)] are inherited, and caused by dominant mutations in the prion protein gene (*PRNP*); others are sporadic, with a uniform worldwide incidence of around 1 in 10^6. The subset that has caused most alarm over recent years, however, is the acquired TSEs. The first of these to be described was kuru, a disease caused by eating the brain of an infected person, and observed in the Fore people of Papua New Guinea, who practiced ritual cannibalism as part of the process of mourning for their dead. **Iatrogenic** CJD results from medical intervention, and recognized routes of infection include:

▶ treatment with human growth hormone or gonadotrophin derived from pituitary glands taken from infected people after death;

▶ grafting of the cornea of the eye, or the *dura mater* (the tough outer covering of the brain, used to provide protection after brain and head trauma surgeries);

▶ inadequately sterilized neurosurgical instruments.

Of most concern has been the emergence in 1996 of variant CJD (vCJD), affecting unusually young people, and with characteristic and distinct clinicopathological features. vCJD is believed to be caused by the infectious agent of bovine spongiform encephalopathy (BSE), also known as 'mad cow disease', and transmitted to humans as a novel **zoonosis** by the eating of infected beef. Exposure to BSE-infected meat in the UK was widespread in the late 1980s and early 1990s, and exposure also occurred in several other European countries on a smaller scale.

The number of cases of vCJD reported in the UK by the end of 2002 was 129, with a few confirmed cases in Ireland and France. Much effort has gone into epidemiological predictions of the likely scale of future vCJD; however, because of the small current number of cases, the long incubation period, and other factors, considerable uncertainty remains, with the 95% confidence interval for future cases being from 10 to 7000 deaths (Ghani *et al.*, 2003). These figures are nonetheless substantially smaller than the upper limits of earlier predictions, which were in the hundreds of thousands.

The *PRNP* gene encodes a protein of 253 amino acids, and contains a common nonsynonymous variant, in which an A/G transition results in a methionine /valine polymorphism at amino acid 129. Studies of cases of iatrogenic and sporadic CJD as well as kuru showed that Met-129 homozygotes were significantly over-represented among affected individuals; more strikingly, all cases of vCJD that have been analyzed are Met-129 homozygotes (Andrews *et al.*, 2003; Collinge *et al.*, 1996). An accurate estimate of the Met-129 allele frequency is not available, but is slightly over 60%; Met-129 homozygotes in both the UK and the French populations exist at a frequency of about 40% (Beaudry *et al.*, 2002; Collinge *et al.*, 1991). vCJD is acting as a novel selective force against Met-129 homozygotes, but, fortunately for us, the strength of this selection seems likely to be weak. Nevertheless, it illustrates one way in which new forms of selection can still arise and affect the human population.

---

previously conferred resistance to infection (Dean *et al.*, 2002). (*Figure 13.18*)

Differential fertility

The phenomenon of differential fertility is easy to demonstrate, but showing that it has any genetic basis is far more difficult. A proof of its **heritability** is not sufficient, because there is much potential for social as well as biological inheritance (see *Box 5.2*). Given these difficulties, it is difficult to ascertain the degree to which natural selection is presently acting on differential fertility.

Differential fertility is a common feature of social inequalities, although it appears that a shift in direction has occurred in recent times. Ethnographic studies of indigenous societies often show that high status affords better access to resources, more wives and more offspring. By contrast, in the developed world it is the lower socioeconomic sections of society that consistently out-reproduce the higher groups. However, as population growth lessens throughout the world, the magnitude of differential fertility should diminish.

Nevertheless, a recent long-term study of twins suggests that there is an appreciable genetic basis underlying differential fertility (Kirk *et al.*, 2001). The researchers compared the family sizes of monozygotic and dizygotic twins and found that about 40% of the variance in reproductive success could be attributed to genetic factors, even after taking into account differences in education background and religious affiliation amongst the women. These environmental factors do play a significant role: Roman Catholic women had 1.4 times more reproductive success than non-religious women did, and those educated the least also had 1.4 times more reproductive success than those with a university education. The latter statistic is the kind that disturbs eugenicists.

Such a high degree of genetic heritability could not be sustained over the long term, as beneficial alleles would quickly become fixed. The authors therefore suggested that selection might be acting on different traits than in previous generations. This provides a mechanism linking cultural and biological evolution. A constantly changing cultural world

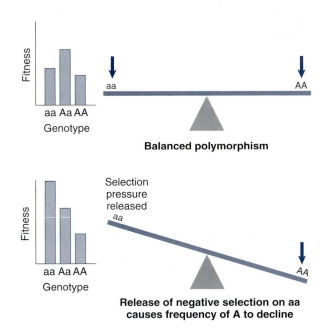

Figure 13.18: Balancing selection.

The removal of one of the selective pressures reduces the fitness of the heterozygote relative to one of the homozygotes, resulting in the selective loss of the deleterious allele.

allows previously unimportant genetically-encoded traits to determine reproductive fitness, thus maintaining a high level of genetic heritability. For example, it might be that in pre-industrial societies, reproductive biology (e.g., wide birth canal) was the most important determinant of reproductive fitness, whereas in modern developed societies, behavioral traits (e.g., desire for a large family) may have become more important. Over the long term, if the reproductive culture remains relatively stable, the heritability of differential fitness is likely to decline as a result of the spread of the beneficial alleles. In the meantime, natural selection on differential fertility might be driving dramatic changes in specific alleles responsible for greater reproductive fitness.

13.5.4 The future of phenotypic diversity

Finally, we should remember the easily recognized phenotypic differences between humans discussed in the earlier parts of this chapter – physical characteristics such as morphology and facial features, pigmentation and dietary tolerances and intolerances. If we accept a role for natural or sexual selection in the establishment of some of these phenotypes, then what is their likely future?

These phenotypes were established over long periods when population sizes were small and migration between different groups was much less than today, and when our ability to influence our own immediate environment through technology was more limited. Modern technologies mean that natural selection seems unlikely to remain a major factor influencing the frequency of many of these traits, and their geographical specificity will be reduced by migration

and admixture. The trait of lactase persistence, with its dominant mode of inheritance, has already been spread widely by admixture with European populations.

By contrast, sexual selection may well continue to play a role in shaping phenotypic traits. Despite increased modern migration it is probable that pockets of small endogamous groups will remain, in which particular attractive traits will be maintained far into the future. In addition, within some populations conceptions of beauty are changing. For example, whereas for many societies in the past darker skin tones were deemed unattractive and indicative of manual labor in the fields, more recently in some developed societies tanned skin is considered attractive and has become associated with foreign holidays and disposable wealth (leading to growth in use of sun-beds). Furthermore, global media have dispersed Western concepts of beauty to societies where until recently different traits were considered attractive. This globalization of desirable traits can be expected to have significant impact in populations in which the frequency of those traits has previously been low. Sexual selection may also become more influential in any population in which many parents adopt assisted reproductive technologies that allow them to choose gamete donors on the basis of their 'attractiveness' (see Section 13.6.2).

13.6 What new factors will influence human evolution?

Are there future factors that might influence human evolution, which cannot be understood through a simple extrapolation from our present circumstances? Here, two such possibilities are considered:

▶ environmental factors: climate change and sustainable development;

▶ technological factors: assisted reproduction and gene therapy.

13.6.1 Environmental factors

As we have seen in preceding chapters, environmental changes in the past have been instrumental in facilitating the spread of our species and in determining our patterns of land use. The past 10 KY have seen a far more stable global climate than in any period of similar length in the previous 100 KY. This has allowed us to develop complex societies. The next century will see major changes in how humans interact with the global environment. Climate change appears to be inevitable, although the outcomes for individual regions are likely to be highly diverse, including increases and decreases in temperatures, and greater frequency and magnitude of extreme events such as droughts and floods (IPCC, 2001). These changes will have different impacts on individual countries which themselves vary in their abilities to adapt culturally. This is predicted to lead to greater adverse impacts on developing countries (IPCC, 2001). Sea level rises resulting from melting glaciers and ice sheets will change settlement patterns in flood-prone areas, resulting in the migration of huge numbers of people. Climate change will also affect crop yields in many areas of the world, which again will drive extensive migration. The aftermath of an extreme event is often associated with disease epidemics, for example floods and subsequent cholera outbreaks. These epidemics may increase mortality rates from infectious disease, which have been declining for the past century. The geographical distributions of certain infectious diseases are already changing and show clear dependence on climatic oscillations (Patz, 2002). The movement of infectious disease into new areas inhabited by populations without partial genetic resistance afforded by long-term

exposure can also be expected to increase infectious disease mortality rates.

In addition to climate change, the need for adaptation to a sustainable human economy (see *Box 13.8*) will impose unknown demands on human adaptability. Whether these demands can be satisfied purely by cultural evolution alone remains to be seen. However, as the need for action becomes more acute, the probability increases that some form of biological adaptation, mediated through catastrophic mortality, will be forced upon us. The example of a dramatic population crash in the wake of the complete deforestation

---

BOX 13.8: The oversized ecological footprint of *Homo sapiens*

The consumption of environmental resources by the 6 billion human inhabitants of planet Earth is presently unsustainable. A consideration of our species' 'ecological footprint' is perhaps the most effective method for demonstrating this (Wackernagel *et al.*, 2002). An ecological footprint is '*the area of biologically productive land and water that is required to produce the resources consumed and wastes generated by humanity*'. This 'footprint' can be compared to the amount of biologically productive land available on our planet. Seven major activities determine our land requirements:

▸ growing crops;

▸ grazing animals;

▸ harvesting timber;

▸ marine and freshwater fishing;

▸ accommodation of residential and industrial facilities;

▸ burning fossil fuels;

▸ waste disposal.

From this list it can be appreciated that an individual's ecological footprint is dependent on their lifestyle. Thus footprints can be calculated for the average inhabitant of any given country to compare the equality of land usage and predict future ecological footprints on the basis of demography and development. Calculations based primarily on the first six of the activities listed above reveal that humanity's ecological footprint already exceeds the total biologically productive land area available on our planet, and has done so for over two decades (Wackernagel *et al.*, 2002). Average individual ecological footprints are predicted to increase substantially in the near future (WWF, 2002). Allied to the growing human population, our ecological footprint will expand even faster. If technological advances in the efficient use of resources are not implemented within the next 20 years, average human welfare is predicted to start to decline rapidly before 2030 (WWF, 2002).

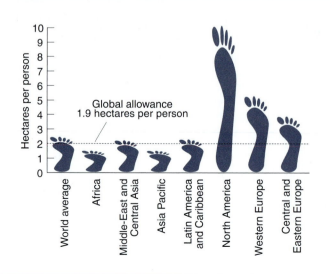

of Easter Island provides a stark reminder of what can happen (Bahn, 1992).

13.6.2 Technological factors

Assisted reproduction and **gene therapy** both provide the opportunity to manipulate the genotype of succeeding generations. For example, genetic typing of embryos fertilized *in vitro* prior to implantation gives parents the opportunity to choose the alleles carried by their children. For the first time in evolution, the frequencies of alleles in the next generation are not determined by the random union of parental gametes.

The proportion of live births resulting from normal, unassisted, reproduction massively exceeds that from assisted alternatives (see *Box 13.9*). Unless these alternative methods conceive a much higher proportion of all children than they do at present, any impact upon allele frequencies is likely to be minimal. In addition, the majority of assisted reproduction does not require gamete donors and does not entail genetic testing of embryos prior to implantation.

However, these methods can lead to some males being over-represented in the next generation. Sperm donors represent a tiny sample of the general population. In some countries, lax regulation has led to a very few sperm donors fathering large numbers of children. In 1992, Dr. Cecil Jacobson was convicted for fraud in which he replaced sperm from donors with his own, and may have fathered up to 75 children. There are other potential problems. For example, in 1998, a man whose sperm was used to conceive 18 children was subsequently diagnosed with autosomal dominant cerebellar ataxia (OMIM 117210), which causes gradual shriveling of the cerebellum. These children have a 50% chance of having inherited the disease. The introduction of sperm catalogs with individual donors ranked by their desirable properties only increases opportunities for skewed representation of sperm donors in the next generation, and represents a novel and potent form of sexual selection. Recent regulations in some countries restrict the number of children born from a single donor. In the UK, the Human Fertilization and Embryology Authority limits the number of children per donor to 10.

Assisted reproductive technologies (ART) raise the unusual prospect that genetic defects causing infertility can now be inherited. For example, the technique of Intra-cytoplasmic Sperm Injection (ICSI) can be used to allow fathers to pass on Y-chromosomal microdeletions that remove crucial spermatogenic genes to their sons (Page *et al.*, 1999). It has been suggested that should ICSI be taken up by a substantial portion of males whose infertility derives from genetic causes, then the prevalence of male infertility can increase rapidly within the population (Faddy *et al.*, 2001), thus resulting in an increased reliance on ART to maintain the human population.

In addition, it has been suggested that ART is not without associated risks to the health of the resultant offspring. Recent comparisons of birth defects among infants conceived naturally and through ART concluded that both ICSI and IVF double the risk of a major birth defect from ~ 4% to ~ 9% (Hansen *et al.*, 2002).

Gene therapy is the treatment of disease through the modification of the patient's genetic material (Somia and Verma, 2000). There are various strategies by which this may be achieved (see *Box 13.10*). By contrast, **genetic enhancement** describes the use of genetic techniques to engineer the introduction of desirable characteristics into an individual. Some argue that there is substantial gray area between the two definitions. Only modification of germ-line cells introduces heritable changes into the genome. The present ethical climate allows somatic gene therapy but forbids germ-line gene therapy and all forms of genetic enhancement. Whether these ethical distinctions should change in the near future is the subject of furious debate. If gene therapy (or enhancement) is to contribute to human evolution, modifications must be heritable. Therefore, if germ-line modification continues to be deemed ethically

---

BOX 13.9 Assisted reproduction technologies (ART)

Infertility affects 10–20% of all couples, with causal factors divided roughly equally between the sexes. A number of different strategies have been developed to allow these couples to increase their chance of reproducing.

▶ **Artificial insemination** (AI). Injection of sperm into the uterus of a woman during or just preceding ovulation. The sperm may come from the woman's partner or from a sperm donor.

▶ **In vitro fertilization** (IVF) involves the fertilization of an egg outside of the body, with the resultant embryo being implanted back into the mother. Although eggs and embryos can be donated, donation is not common due to the invasive nature of the procedure for harvesting eggs.

▶ **Intra-cytoplasmic sperm injection** (ICSI) is a variant of IVF that specifically addresses male infertility by injecting into the egg sperm that would be incapable of entering unaided. The rise of ICSI in the past decade has been mirrored by a decline in donor insemination (DI) as fathers prefer to have genetic input into their offspring.

In the UK in 1999, just over 1% of annual births resulted from DI, IVF and ICSI combined (HFEA, 2000a; HFEA, 2000b). Of these, 16% used donor sperm, eggs or embryos. In the USA, for 1998, the frequency of IVF and ICSI among live births was 0.7% (CDC, 1999; CDC, 2001). About 9% of these used gamete donors. Because of the nature of ART, multiple births (twins, triplets etc.) are common. In 1997 it was estimated that over 40% of higher-order births (triplet or greater) in the USA resulted from ART.

improper, there will be no impact on future allele frequencies. In principle, however, germ-line modification allied to parental choice would see both the reduction in frequency of deleterious alleles, and increased frequencies of desirable alleles. However, desirable traits for genetic enhancement are likely to vary among different populations and in the same population over time.

While changes to the ethical environment are *necessary* to allow gene therapy to impact upon human evolution, they are not *sufficient*. In addition, the relative uptake of such technologies must be considered. Therapies resulting from biotechnological advances tend to be very expensive and consequently rarely achieve widespread application. Few individual patients can afford them and state-funded health services are similarly restricted by cost. Globally, the total number of individuals receiving gene therapy for all diseases up to 2002 lies in the hundreds. General application of such technologies would therefore require massive reductions in costs before any impact upon allele frequencies could be expected. These practical limitations raise the possibility of a two-tier society based on genetic enhancement, which many people regard as an ethically unattractive outcome. These limitations also suggest that genetic modification is unlikely to have appreciable impact on allele frequencies in the short term, although a cumulative effect on global allele frequencies might be noticed after hundreds of generations. However, it is worth noting that those sections of society most able to afford such enhancements are also those that reproduce least.

In addition to its therapeutic potential, gene therapy can also cause deleterious changes. The imprecise nature of most human genome modification (see *Box 13.10*) effectively constitutes a mutagenic process. The addition of a new gene

into a genome generally involves quasi-random integration into the host genome. This integration can cause unforeseen pathogenic consequences if it occurs within, or adjacent to, a coding sequence. Recent evidence suggests that a case of a leukemia-like disorder in a boy undergoing somatic gene therapy for Severe Combined Immunodeficiency (SCID) results from the integration of the functional gene next to a gene with oncogenic potential (Check, 2002). The promoter of the introduced gene apparently changes the normal expression patterns of the neighboring gene. Large-scale genetic modification at multiple loci may entail a considerable mutagenic cost.

While germ-line modification as envisaged above has not been conducted on humans, there is one form of assisted reproduction that does result in heritable changes to the genome of a resultant child. It has been found that the transfer of cytoplasm from healthy donor oocytes to developmentally compromised oocytes can increase the chances of successful IVF. Whilst no nuclear DNA transfer occurs during this procedure, mitochondria from the donor are delivered into the recipient oocyte. These children were found to be heteroplasmic for mtDNA markers, demonstrating the inheritance of both mitochondrial genomes present in the oocyte (Barritt *et al.*, 2001). Effectively, these children have genetic contributions from three individuals. It should be possible for the mtDNA from the donor oocyte to be passed on to future generations.

While ooplasmic transfer is regarded as the most minor form of germ-line modification, it still raises the ethical issue of the inability of future modified offspring to give informed consent prior to the procedure. At present, this issue is typically regarded as being of minor importance relative to the safety risks associated with germ-line modification; however, as these risks are reduced with improved

BOX 13.10 Gene therapeutic approaches

Different strategies for gene therapy can be classified according to multiple criteria, which include the mode of DNA delivery, the type of cell being modified and the type of modification. Here we use the latter criterion, because we are interested in the possible mutagenic potential of gene therapy.

▶ **Gene addition** is the most common strategy currently employed. A functional copy of the affected gene is integrated at random into the host genome. This compensates for the lack of a necessary activity from the endogenous gene, but there is little if any control over the site of integration, which may determine the expression levels of the introduced gene. The gene is typically introduced along with its own promoter in order to ensure expression. Gene addition also provides an opportunity to introduce synthetic genes with novel activities that, for example, might turn off the expression of a mutated gene that has acquired a deleterious function.

▶ **Gene replacement** uses homologous recombination to replace a large section of a mutated gene. Because it relies on the recombinational ability of the host cell, the efficiency of replacement is low and restricted to certain cell types. The new genetic material occupies the same location in the genome as the pre-existing gene, so it is not necessary to introduce **promoter** sequences, and only a portion of the gene need be replaced.

▶ **Gene repair** employs the endogenous mismatch repair activity of a cell to precisely repair the mutated position in a gene. A synthetic **oligonucleotide** is introduced into the cell that is homologous to the target gene apart from a single mismatch with the relevant base. This forms a **heteroduplex** that is detected by the cell and repaired. The efficiency of gene repair is also low, but seems to be applicable to a wider variety of cell types than gene replacement. Like gene replacement, gene repair can remove a mutated gene that has gained a deleterious function, as well as generating a gene with the wild-type function.

technologies tested on other organisms, debate on this conundrum should come to the fore.

13.7 Will *Homo sapiens* speciate?

Speciation tends to occur over longer time-scales than we have considered thus far, making any predictions more speculative. In addition, the genetic basis for speciation is poorly understood; it may involve a single gene or hundreds of genes (see Chapter 7). Many genetic changes can be associated with speciation events, so it is difficult to differentiate between the few that played a causal role and the majority that are likely to be consequences.

The process of speciation can be classified into two forms: the splitting of a single ancestral species into two non-interbreeding species (**cladogenesis**), or the accumulation of changes in a single species (**anagenesis**). These two alternatives are illustrated in *Figure 13.19*.

Cladogenesis requires the reproductive isolation of populations that subsequently diverge to form different species. The increased mobility of humans would seem to preclude the formation of geographical barriers between populations, but this is not the only potential cause of reproductive isolation. A number of other mechanisms can cause populations to cease interbreeding:

▸ behavioral isolation – populations do not respond to courtship signals;

▸ temporal isolation – populations breed at different times;

▸ mechanical isolation – barriers to the physical process of mating;

▸ gametic isolation – gametes cannot fuse;

▸ reduced hybrid viability;

▸ reduced hybrid fertility.

Among the above factors, perhaps the most likely candidate is behavioral isolation driven by cultural changes. As

mentioned before, some people believe that germ-line genetic enhancement could lead to a two-tier society, in which mating would be discouraged between the tiers. Others have suggested that modern humans and **Neanderthals** may have been **interfertile** (as are **bonobos** and chimpanzees, who share a more ancient common ancestor than humans and Neanderthals), but were prevented from extensive mating by cultural barriers. Thus the likelihood of human cladogenesis is highly dependent on long-term cultural evolution, which is impossible to predict over the time-scales required for speciation to occur.

Rather than adopting an organismal approach to this question of human speciation, we can predict likely future genetic changes in humans and use what little knowledge we have of the genetic basis for speciation to ask whether they are likely to result in anagenesis or cladogenesis. Two such situations are considered here.

13.7.1 Degeneration of the Y chromosome

One form of genetic change thought to drive speciation is the evolution of a new mechanism for sex-determination. The human sex-chromosomal mechanism for sex determination (discussed in Chapter 2) appears to be relatively stable. This system is shared among mammalian species, and only a handful out of over 4000 known mammalian species have developed novel sex-determining mechanisms. The human X and Y chromosomes arose from a pair of autosomes, and so were initially the same size and contained the same number of genes (see *Box 8.8*). Since then the vast majority of Y-chromosomal genes have accumulated nonfunctional mutations or become lost altogether. It is estimated that the human Y chromosome has retained only four of its original ~ 1500 genes over the 170 million years since the mammalian sex-chromosomal system first arose (Graves, 2001). This process of degeneration has been accompanied by the acquisition of new genes, which have typically adopted male-specific functions. Given the small number of functional genes (~ 40) presently on the human Y chromosome, gene loss has heavily outweighed gene acquisition. However, it is difficult to state with any certainty that the disappearance of the Y chromosome as a sex-determining system is inevitable. It is even more difficult to estimate when this might occur, although a recent speculative estimate suggests a 5–10 MY time frame (Graves, 2002).

13.7.2 Increased neutral polymorphism

In an earlier section of this chapter we saw that the human effective population size would increase substantially in future generations, and that this should result in significantly increased neutral polymorphism. At present, we see one polymorphic site every 1250 bp on average (see Section 3.2.7). But why should increased polymorphism have an impact on speciation?

There is much evidence to suggest that homologous recombination is necessary to allow the correct segregation of chromosomes at meiosis. Studies on yeast have shown that an

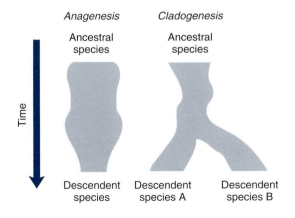

Figure 13.19: Cladogenesis and anagenesis.

The two commonly conceived forms of speciation.

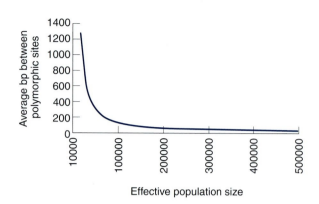

Figure 13.20: Reduction in size of interval between polymorphic sites as effective population size increases.

assessment of homology is required before homologous recombination can proceed during meiosis. This assessment is mediated by the cell's **mismatch repair** system. 'Knocking out' genes within this system allows less well-matched molecules to recombine (Hunter *et al.*, 1996). As a consequence, DNA segments that were previously too diverged to recombine are subsequently capable of undergoing meiotic recombination. Other studies have shown that a minimum length of absolute identity is required before homologous recombination can proceed (Waldman and Liskay, 1988).

If the human effective population size exceeds even 0.02% of the current census size of ~ 6 billion, at a newly established **mutation drift equilibrium** we should expect to see a polymorphic site every 10 bp (see *Figure 13.20*). This length is far below current estimates of the minimal length of sequence identity needed to sponsor homologous recombination and the result would be repression of recombination during meiosis (although it is unclear what effect islands of homology resulting from regions of selective constraint would have on this process). This repression would in turn result in frequent chromosomal **aneuploidy**, leading to a dramatic reduction in human fertility. As this situation started to develop, the offspring of closely related individuals would be better able to produce mature gametes and would therefore have greater fertility than offspring with more distantly related parents (**heterozygote disadvantage**). This might force the human population into isolated breeding groups, with barriers of dramatically reduced hybrid fertility leading to eventual speciation.

Elevated levels of other forms of polymorphism might also interfere with hybrid fertility. For example, increased levels of structural polymorphism, such as balanced translocations, would also lead to heterozygote disadvantage, due to a high frequency of recombination resulting in unbalanced rearrangements in which genes are lost or gained.

It is difficult to predict how rapidly the human effective population size will rise to levels at which chromosomal **aneuploidy** results in dramatic reductions in fertility.

Nevertheless, if we assume that the population remains stable at 6 billion, and that the harmonic mean of the historical census population sizes over a 100-KY period approximates the effective population size (see Chapter 5), we can forecast that within a million years, the average distance between polymorphic sites will be reduced to about 100 bp.

This seemingly inevitable barrier to the long-term integrity of a large species like *Homo sapiens* may not be generally applicable to other species, where environmental change, sexual selection for courtship behavior and other processes may fragment larger populations before polymorphism has a chance to accrue to levels required to cause hybrid sterility (Coyne and Orr, 1998).

Summary

▶ There is much variation in morphology, facial features and behavior among humans. The adaptive significance of much of this is hard to prove, and the genetic basis has not yet been properly explored.

▶ Variation in pigmentation has been widely studied, and many hypotheses have been put forward to explain it. It probably results from a balance between the opposing needs for UV exposure for vitamin D synthesis, and protection from UV damage, sometimes supplemented or replaced by sexual selection.

▶ Different pigmentation phenotypes result from differences in the size, composition and distribution of melanosomes, which contain the pigment melanin. Genes underlying the pigmentation pathways have been identified using mouse coat color mutations, and various human disorders involving altered pigmentation.

▶ The clearest example of a gene influencing normal pigmentation is *MC1R*, in which mutations can cause red hair and fair skin. Patterns of *MC1R* haplotype diversity show low diversity in Africa, consistent with selection for the function of melanin production in dark skin, and high diversity elsewhere, consistent with a lack of selective constraint.

▶ Persistence of intestinal lactase into adulthood is most readily explicable as an adaptation to milk-drinking; if this is so, it represents an example of 'gene-culture coevolution' that has developed since the beginning of agriculture 10 KYA.

▶ Diet has changed in the transition from hunter–gathering to agriculture, and more recently into a state of overnutrition in many developed societies. Some of the health problems associated with this transition could reflect alleles of adaptive value to our ancestors that are disadvantageous in our new nutritional environment.

▶ In the future, new mutations are likely to arise at rates not too different from those experienced in the past. If anything, these annual mutation rates will be slightly higher as a result of demographic changes and ever-present environmental mutagens.

▸ Opportunities for natural selection to act on genetic diversity have been greatly attenuated compared to previous generations. However, rapid evolution of novel pathogens will most likely preclude the complete eradication of prereproductive mortality from infectious diseases.

▸ Cultural evolution and biological evolution need not be mutually exclusive – changes in society can reveal new targets for biological evolution, by causing earlier onset of previously postreproductive disorders, and by changing the genetic basis of differential fertility.

▸ Gene therapy and assisted reproduction technologies have the potential to skew allele frequencies, but, given their restriction to those that can afford them, seem unlikely to cause a major perturbation in the foreseeable future.

▸ Ongoing population growth and greater effective population sizes will lead to much higher levels of genetic diversity in humans, which might cause widespread infertility as a result of impaired meiotic recombination. This may drive the reproductive isolation of small breeding groups that would eventually lead to speciation.

Further reading

Coyne JA, Orr HA (1998) The evolutionary genetics of speciation. *Phil. Trans. R. Soc. Lond. B Biol. Sci.* **353**, 287–305.

Electronic references

World population trends – United Nations: http://www.un.org/popin/wdtrends.htm

Climate change – Intergovernmental panel on climate change: http://www.ipcc.ch/about/about.htm

Infectious diseases – World Health Organization: http://www.who.int/health_topics/infectious_diseases/en/

Implications of human germ-line engineering: http://research.mednet.ucla.edu/pmts/Germline/default.htm

Assisted reproductive technologies: http://www.nature.com/fertility/

References

Akey JM, Wang H, Xiong M, Wu H, Liu W, Shriver MD, Jin L (2001) Interaction between the melanocortin-1 receptor and P genes contributes to inter-individual variation in skin pigmentation phenotypes in a Tibetan population. *Hum. Genet.* **108**, 516–520.

Allen JS, Cheer SM (1996) The non-thrifty genotype. *Curr. Anthropol.* **37**, 831–842.

Andersson L (2001) Genetic dissection of phenotypic diversity in farm animals. *Nature Rev. Genet.* **2**, 130–138.

Andrews NJ, Farrington CP, Ward HJT, et al. (2003) Deaths from variant Creutzfeldt-Jakob disease in the UK. *Lancet* **361**, 751–752.

Bahn P, Flenley J (1992) *Easter Island, Earth Island*. Thames and Hudson, New York.

Barber R, Plumb MA, Boulton E, Roux I, Dubrova YE (2002) Elevated mutation rates in the germ line of first- and second-generation offspring of irradiated male mice. *Proc. Natl Acad. Sci. USA* **99**, 6877–6882.

Barritt JA, Brenner CA, Malter HE, Cohen J (2001) Mitochondria in human offspring derived from ooplasmic transplantation. *Hum. Reprod.* **16**, 513–516.

Bastiaens M, ter Huurne J, Gruis N, Bergman W, Westendorp R, Vermeer B-J, Bavinck J-NB (2001) The melanocortin-1 receptor gene is the major freckle gene. *Hum. Mol. Genet.* **10**, 1701–1708.

Beaudry P, Parchi P, Peoc'h K, et al. (2002) A French cluster of Creutzfeldt-Jakob disease: a molecular analysis. *Eur. J. Neurol.* **9**, 457–462.

Benjamin J, Li L, Patterson C, Greenberg BD, Murphy DL, Hamer DH (1996) Population and familial association between the D4 dopamine receptor gene and measures of Novelty Seeking. *Nature Genet.* **12**, 81–84.

Biasutti R (1959) Razze e popoli della terra, Torino.

Boas F (1910) Changes in bodily form of descendants of immigrants. United States Immigration Commission, Senate Document No. 208, 61st Congress. Government Printing Office, Washington, D.C.

Box NF, Duffy DL, Chen W, Stark M, Martin NG, Sturm RA, Hayward NK (2001) MC1R genotype modifies risk of melanoma in families segregating CDKN2A mutations. *Am. J. Hum. Genet.* **69**, 765–773.

Box NF, Wyeth JR, O'Gorman LE, Martin NG, Sturm RA (1997) Characterization of melanocyte stimulating hormone receptor variant alleles in twins with red hair. *Hum. Mol. Genet.* **11**, 1891–1897.

CDC (1999) Assisted reproductive technology success rates. Centers for Disease Control and Prevention, Atlanta, GA.

CDC (2001) National Vital Statistics Report: Births, Marriages, Divorces, and Deaths: Provisional Data for 2001. Centers for disease control and prevention, Atlanta, GA.

Chang FM, Kidd JR, Livak KJ, Pakstis AJ, Kidd KK (1996) The world-wide distribution of allele frequencies at the human dopamine D4 receptor locus. *Hum. Genet.* **98**, 91–101.

Check E (2002) Regulators split on gene therapy as patient shows signs of cancer. *Nature* **419**, 545–546.

Collinge J, Palmer MS, Dryden AJ (1991) Genetic predisposition to iatrogenic Creutzfeldt-Jakob disease. *Lancet* **337**, 1441–1442.

Collinge J, Sidle KCL, Meads J, Ironside J, Hill AF (1996) Molecular analysis of prion strain variation and the aetiology of 'new variant' CJD. *Nature* **383**, 685–690.

Comuzzie AG (2002) The emerging pattern of the genetic contribution to human obesity. *Best Pract. Res. Clin. Endocrinol. Metab.* **16**, 611–621.

Cordain L, Miller JB, Eaton SB, Mann N, Holt SHA, Speth JD (2000) Plant-animal subsistence ratios and macronutrient energy estimations in worldwide hunter–gatherer diets. *Am. J. Clin. Nutr.* **71**, 682–692.

Coyne JA, Orr HA (1998) The evolutionary genetics of speciation. *Phil. Trans. R. Soc. Lond. B Biol. Sci.* **353**, 287–305.

Darwin C (1871) *The Descent of Man and Selection in Relation to Sex.* John Murray, London.

de Groot NG, Otting N, Doxiadis GM, et al. (2002) Evidence for an ancient selective sweep in the MHC class I gene repertoire of chimpanzees. *Proc. Natl Acad. Sci. USA* **99**, 11748–11753.

Dean M, Carrington M, O'Brien SJ (2002) Balanced polymorphism selected by genetic versus infectious human disease. *Annu. Rev. Genomics Hum. Genet.* **3**, 263–292.

Diamond J (1991) The rise and fall of the third chimpanzee. Vintage, London.

Diamond J (2003) The double puzzle of diabetes. *Nature* **423**, 599–602.

Ding Y, Chi H-C, Grady DL, et al. (2002) Evidence of positive selection acting at the human dopamine receptor D4 gene locus. *Proc. Natl Acad. Sci. USA* **99**, 309–314.

Dubrova YE, Nesterov VN, Krouchinsky NG, Ostapenko VA, Neumann R, Neil DL, Jeffreys AJ (1996) Human minisatellite mutation rate after the Chernobyl accident. *Nature* **380**, 683–686.

Eaton SB, Eaton SB, III (1999) The evolutionary context of chronic degenerative diseases. In: *Evolution in Health and Disease* (ed. SC Stearns). Oxford University Press, Oxford, pp. 251–259.

Eaton SB, Konner M, Shostak M (1988) Stone Agers in the fast lane: chronic degenerative diseases in evolutionary perspective. *Am. J. Med.* **84**, 739–749.

Ebstein RP, Novick O, Umansky R, et al. (1996) Dopamine D4 receptor (D4DR) exon III polymorphism associated with the human personality trait of Novelty Seeking. *Nature Genet.* **12**, 78–80.

Enattah NS, Sahi T, Savilahti E, Terwilliger JD, Peltonen L, Järvelä I (2002) Identification of a variant associated with adult-type hypolactasia. *Nature Genet.* **30**, 233–237.

Faddy MJ, Silber SJ, Gosden RG (2001) Intracytoplasmic sperm injection and infertility. *Nature Genet.* **29**, 131.

Falconer DS, Mackay TFC (1996) Introduction to Quantitative Genetics, 4th Edn. Addison-Wesley Publishing Co, Boston, MA.

Faraone SV, Doyle AE, Mick E, Biederman J (2001) Meta-analysis of the association between the 7-repeat allele of the dopamine D(4) receptor gene and attention deficit hyperactivity disorder. *Am. J. Psychiat.* **158**, 1052–1057.

Feldman MW, Cavalli-Sforza LL (1976) Cultural and biological processes: selection for a trait under complex transmission. *Theor. Pop. Biol.* **9**, 238–259.

Feldman MW, Cavalli-Sforza LL (1989) On the theory of evolution under genetic and cultural transmission with application to the lactose absorption problem. In: *Mathematical Evolutionary Theory* (ed. MW Feldman). Princeton University Press, Princeton, NJ.

Fisher RA (1930) *The Genetical Theory of Natural Selection.* Clarendon Press, Oxford.

Flanagan N, Healy E, Ray A, et al. (2000) Pleiotropic effects of the melanocortin 1 receptor (MC1R) gene on human pigmentation. *Hum. Mol. Genet.* **9**, 2531–2537.

Flatz G, Rotthauwe HW (1973) Lactose nutrition and natural selection. *Lancet* **2**, 76–77.

Forster L, Forster P, Lutz-Bonengel S, Willkomm H, Brinkmann B (2002) Natural radioactivity and human mitochondrial DNA mutations. *Proc. Natl Acad. Sci. USA* **99**, 13950–13954.

Gambichler T, Bader A, Sauermann K, Altmeyer P, Hoffmann K (2001) Serum folate levels after UVA exposure: a two-group parallel randomised controlled trial. *BMC Dermatol.* **1**, 8.

Ghani AC, Ferguson NM, Donnelly CA, Anderson RM (2003) Factors determining the pattern of the variant Creutzfeldt-Jakob disease (vCJD) epidemic in the UK. *Proc. Roy. Soc. Lond. B*, **270**, 689–698

Graves JA (2001) From brain determination to testis determination: evolution of the mammalian sex-determining gene. *Reprod. Fertil. Dev.* **13**, 665–672.

Graves JAM (2002) The rise and fall of SRY. *Trends Genet.* **18**, 259–264.

Gravlee CC, Bernard HR, Leonard WR (2003) Heredity, environment, and cranial form: a reanalysis of Boas's immigrant data. *Am. Anthropol.* **105**, 125–138.

Greco L (1997) From the Neolithic revolution to gluten intolerance: benefits and problems associated with the cultivation of wheat. *J. Pediatr. Gastroenterol. Nutr.* **4**, S14–S17.

Hales CN, Barker DJP (1992) Type 2 (non-insulin dependent) diabetes mellitus: the thrifty phenotype hypothesis. *Diabetologia* **35**, 595–601.

Hammer MF, Karafet TM, Redd AJ, Jarjanazi H, Santachiara-Benerecetti S, Soodyall H, Zegura SL (2001) Hierarchical patterns of global human Y-chromosome diversity. *Mol. Biol. Evol.* **18**, 1189–1203.

Hansen M, Kurinczuk JJ, Bower C, Webb S (2002) The risk of major birth defects after intracytoplasmic sperm injection and in vitro fertilization. *N. Engl. J. Med.* **346**, 725–730.

Harding RM, Healy E, Ray AJ, *et al.* (2000) Evidence for variable selective pressures at MC1R. *Am. J. Hum. Genet.* **66**, 1351–1361.

Harpending H, Cochran G (2002) In our genes. *Proc. Natl Acad. Sci. USA* **99**, 10–12.

Harpending H, Rogers A (2000) Genetic perspectives on human origins and differentiation. *Ann. Rev. Genomics Hum. Genet.* **1**, 361–385.

HFEA (2000a) *The Patients' Guide to DI.* Human Fertilisation and Embryology Authority, London.

HFEA (2000b) *The Patients' Guide to IVF Clinics.* Human Fertilisation and Embryology Authority, London.

Holden C, Mace R (1997) Phylogenetic analysis of the evolution of lactose digestion in adults. *Hum. Biol.* **69**, 605–628.

Hollox EJ, Poulter M, Swallow DM (2000a) Lactase haplotype diversity in the Old World. In: *Archaeogenetics: DNA and the Population Prehistory of Europe* (eds C Renfrew, K Boyle). Oxbow Books, Oxford, pp 305–308.

Hollox EJ, Poulter M, Zvarik M, *et al.* (2000b) Lactase haplotype diversity in the Old World. *Am. J. Hum. Genet.* **68**, 160–172.

Hollox EJ, Swallow DM (2002) Lactase deficiency – biological and medical aspects of the adult human lactase polymorphism. In: *Genetic Basis of Common Diseases* (eds RA King, JI Rotter, AG Motulsky). Oxford University Press, Oxford.

Howells WW (1992) The dispersion of modern humans. In: *The Cambridge Encyclopedia of Human Evolution* (eds S Jones, R Martin, D Pilbeam). Cambridge University Press, Cambridge, pp. 389–401.

Hunter N, Chambers SR, Louis EJ, Borts RH (1996) The mismatch repair system contributes to meiotic sterility in an interspecific yeast hybrid. *EMBO J.* **15**, 1726–1733.

Hurst LD, Ellegren H (1998) Sex biases in the mutation rate. *Trends Genet.* **14**, 446–452.

IPCC (2001) *Climate Change 2001: Impacts, Adaptation and Vulnerability (Technical Summary).* Intergovernmental panel on climate change, Geneva.

Jablonski NG, Chaplin G (2000) The evolution of skin coloration. *J. Hum. Evol.* **39**, 57–106.

Jones S (1996) *In the Blood: God, Genes and Destiny.* Flamingo, London.

Jurmain R, Nelson H, Kilgore L, Trevathan W (2000) *Introduction to Physical Anthropology.* Wadsworth/Thomson Learning, Belmont, CA.

Kanetsky PA, Swoyer J, Panossian S, Holmes R, Guerry D, Rebbeck TR (2002) A polymorphism in the agouti signaling protein gene is associated with human pigmentation. *Am. J. Hum. Genet.* **70**, 770–775.

Kirk KM, Blomberg SP, Duffy DL, Heath AC, Owens IP, Martin NG (2001) Natural selection and quantitative genetics of life-history traits in Western women: a twin study. *Evolution* **55**, 423–435.

Klein J, Takahata N (2002) *Where do we Come From?* Springer, Berlin.

Laland KN, Brown GR (2002) *Sense and Nonsense: Evolutionary Perspectives on Human Behaviour.* Oxford University Press, Oxford.

Lutz W, Sanderson W, Scherbov S (2001) The end of world population growth. *Nature* **412**, 543–545.

Makova KD, Ramsay M, Jenkins T, Li W-H (2001) Human DNA sequence variation in a 6.6-kb region containing the melanocortin 1 receptor promoter. *Genetics* **158**, 1253–1268.

McNair A, Gudman Hoyer E, Jarnum S, Orrild L (1972) Sucrose maladsorption in Greenland. *Br. Med. J.* **2**, 19–21.

Milton K (2000) Hunter-gatherer diets – a different perspective? *Am. J. Clin. Nutr.* **71**, 665–667.

Neel JV (1962) Diabetes mellitus: a thrifty genotype rendered detrimental by "progress"? *Am. J. Hum. Genet.* **14**, 353–362.

Neel JV (1982) The thrifty genotype revisited. In: *The Genetics of Diabetes Mellitus* (eds J Kobberling, R Tattersall). Academic Press, London, pp. 283–293.

Nei M, Roychoudhury A (1982) Genetic relationship and evolution of human races. *Evol. Biol.* **14**, 1–59.

Newton JM, Cohen-Barak O, Hagiwara N, Gardner JM, Davisson MT, King RA, Brilliant MH (2001) Mutations in the human orthologue of the mouse underwhite gene (uw) underlie a new form of oculocutaneous albinism, OCA4. *Am. J. Hum. Genet.* **69**, 981–988.

Oetting WS, Bennett D (2003) *Mouse Coat Color Genes.* Vol. 2003. The Albinism Database. http://www.cbc.umn.edu/tad/genes.htm

Page DC, Silber S, Brown LG (1999) Men with infertility caused by AZFc deletion can produce sons by intracytoplasmic sperm injection, but are likely to transmit the deletion and infertility. *Hum. Reprod.* **14**, 1722–1726.

Palmer JS, Duffy DL, Box NF, *et al.* (2000) Melanocortin-1 receptor polymorphisms and risk of melanoma: is the association explained solely by pigmentation phenotype? *Am. J. Hum. Genet.* **66**, 176–186.

Patz JA (2002) A human disease indicator for the effects of recent global climate change. *Proc. Natl Acad. Sci. USA* **99**, 12506–12508.

Rana NK, Hewett-Emmett D, Jin L, *et al.* (1999) High polymorphism at the human melanocortin 1 receptor locus. *Genetics* **151**, 1547–1557.

Relethford JH (1992) Cross-cultural analysis of migration rates: effects of geographic distance and population size. *Am. J. Phys. Anthropol.* **89**, 459–466.

Rocchini AP (2002) Childhood obesity and a diabetes epidemic. *N. Engl. J. Med.* **346**, 854–855.

Rogers J, Mahaney MC, Almasy L, Comuzzie AG, Blangero J (1999) Quantitative trait linkage mapping in anthropology. *Yearbk. Phys. Anthropol.* **42**, 127–151.

Ruff C (2002) Variation in human body size and shape. *Ann. Rev. Anthropol.* **31**, 211–232.

Schinka JA, Letsch EA, Crawford FC (2002) DRD4 and Novelty Seeking: results of meta-analyses. *Am. J. Med. Genet.* **114**, 643–648.

Siffert W (2001) Molecular genetics of G proteins and atherosclerosis risk. *Basic Res. Cardiol.* **96**, 606–611.

Siffert W, Rosskopf D, Siffert G, et al. (1998) Association of a human G-protein b3 subunit variant with hypertension. *Nature Genet.* **18**, 45–48.

Somia N, Verma IM (2000) Gene therapy: trials and tribulations. *Nature Rev. Genet.* **1**, 91–99.

Sparks CS, Jantz RL (2002) A reassessment of human cranial plasticity: Boas revisited. *Proc. Natl Acad. Sci. USA* **99**, 14636–14639.

Stinson S (1992) Nutritional adaptation. *Ann. Rev. Anthropol.* **21**, 143–170.

Strachan T, Read AP (2004) *Human Molecular Genetics 3.* Garland Science, New York.

Stryer L (1988) *Biochemistry.* W.H. Freeman & Co., New York.

Sturm RA, Box NF, Ramsay M (1998) Human pigmentation genetics: the difference is only skin deep. *BioEssays* **20**, 712–721.

Sturm RA, Teasdale RD, Box NF (2001) Human pigmentation genes: identification, structure and consequences of polymorphic variation. *Gene* **277**, 49–62.

Suarez FL, Savaiano DA, Levitt MD (1995) A comparison of symptoms after the consumption of milk or lactose-hydrolyzed milk by people with self-reported lactose intolerance. *New Engl. J. Med.* **333**, 1–4.

Swagerty DL, Walling AD, Klein RM (2002) Lactose intolerance. *Am. Fam. Physician* **65**, 1845–1850.

Swallow DM, Hollox EJ (2000) The genetic polymorphism of intestinal lactase activity in adult humans. In: *Metabolic and Molecular Basis of Inherited Disease* (eds CR Scriver, AL Beaudet, WS Sly, D Valle). McGraw-Hill, New York, pp. 1651–1663.

Thacher TD, Fischer PR, Pettifor JM, Lawson JO, Isichei CO, Reading JC, Chan GM (1999) A comparison of calcium, vitamin D, or both for nutritional rickets in Nigerian children. *N. Engl. J. Med.* **341**, 563–568.

Troelsen JH, Mehlum A, Olsen J, et al. (1994) 1 kb of the lactase-phlorizin hydrolase promoter directs post-weaning decline and small intestinal-specific expression in transgenic mice. *FEBS Lett.* **342**, 291–296.

UN (1999) *Long-range World Population Projections: Based on the 1998 Revision.* United Nations, New York

UN (2001) *World Population Prospects: The 2000 Revision.* United Nations, New York

UNAIDS (2002) *Report on the Global HIV/AIDS Epidemic.* Joint United Nations programme on HIV/AIDS, Barcelona. http://www.unaids.org

Valverde P, Healy E, Jackson I, Rees JL, Thody AJ (1995) Variants of the melanocyte-stimulating hormone receptor gene are associated with red hair and fair skin in humans. *Nature Genet.* **11**, 328–330.

Wackernagel M, Schulz NB, Deumling D, et al. (2002) Tracking the ecological overshoot of the human economy. *Proc. Natl Acad. Sci. USA* **99**, 9266–9271.

Waldman AS, Liskay RM (1988) Dependence of intrachromosomal recombination in mammalian cells on uninterrupted homology. *Mol. Cell. Biol.* **8**, 5350–5357.

Wang Y, Harvey CB, Pratt WS, et al. (1995) The lactase persistence/non-persistence polymorphism is controlled by a cis-acting element. *Hum. Mol. Genet.* **4**, 657–662.

WHO (1999) *WHO Report on Infectious Diseases: Removing Obstacles to Healthy Development.* World Health Organization, Geneva

WWF (2002) *Living Planet Report 2001.* World Wide Fund for Nature, Gland, Switzerland

CHAPTER FOURTEEN

Health implications of our evolutionary heritage

CHAPTER CONTENTS

BOXES

14.1 Introduction

Many aspects of our health are influenced by our genotype, and since our genotype is the product of our evolutionary history, evolutionary genetics can have important implications for diagnosing, understanding and treating medical conditions. In this chapter we will examine:

▸ genetic influences on disease **incidence** or **prevalence** (incidence is the number of people who develop a disease during a particular period, e.g., their lifetime, and is distinguished from prevalence, which is the number who have it at any one time. In the UK, for example, the lifetime incidence of cancer is over 30%, but the prevalence is around 2%). There are many environmental, social, political, cultural and other reasons why the frequency of a disease varies from one population to another, and among the latter are genetic factors;

▸ the use of evolutionary information in identifying disease genes;

▸ the importance of evolutionary information for medical treatments.

We have seen in Chapter 13 that there is no clear distinction between 'normal' genetic variation and variation associated with disease: the phenotypic consequence of a DNA variant depends on the environment. Nevertheless, it is useful to draw a distinction between variants that are disadvantageous in most environments and those that are usually neutral. For example, phenylketonuria (PKU; OMIM 261600) develops when individuals with insufficient activity of the enzyme phenylalanine hydroxylase live on a diet containing more than minimal levels of the amino acid phenylalanine: the phenylalanine is not metabolized and accumulates, leading to mental retardation and other symptoms. Fortunately, the deficiency can be diagnosed soon after birth, and affected individuals can live a normal life as long as they eat a low-phenylalanine diet. Nevertheless, most foods contain abundant phenylalanine, so this variant is disadvantageous in most environments, and phenylalanine hydroxylase deficiency is classified as detrimental. In this chapter, we are concerned with variants that affect our health. These are most commonly disadvantageous, but some are, or have been in the past, advantageous. This subject falls within the field of medical genetics, which is covered extensively in many standard works. Here, we are not in the main concerned with many of the usual topics of medical genetics, and will not provide a comprehensive coverage of the field, but will instead focus on four questions:

▸ why do genetic diseases vary in their incidence between populations?

▸ how can we use evolutionary information to identify genes causing disease?

▸ what are the effects of past genetic diseases on modern patterns of genetic variation?

▸ how can we use evolutionary knowledge in medicine?

These questions are not independent: indeed, the first and third are intimately related. Note that 'genetic disease' is used here to mean a disease influenced by the genotype: this influence may be large or small and, as we saw above, will depend on the environment.

14.2 Geographical distributions of genetic diseases

14.2.1 Distribution of simple genetic diseases

Just as there is no complete division between normal genetic variation and disease-associated variation, so there is no clear distinction between 'simple' and 'complex' genetic diseases. The complexity of even the well-understood simple disease PKU considered above has been emphasized (Scriver and Waters, 1999). Nevertheless, it remains a helpful division, and one that will be used here: simple genetic diseases are those where individuals carrying a disease allele usually manifest the disease phenotype, despite variation due to **stochasticity** in the development of the phenotype, environmental effects and the effects of other genes. Complex genetic diseases are those where the carrier may well not develop the disease; the most useful distinguishing criterion is the inheritance pattern in families: if Mendelian inheritance can be discerned, the disease is simple.

What distribution of frequency and geographical spread would we expect for genetic disease? Diseases are, by definition, disadvantageous, and genes leading to them will be selected against in the population. In the most extreme case, that of a fully-penetrant dominant disease or condition that prevents reproduction of affected individuals (e.g., because they die in childhood or are infertile) all mutations will produce affected individuals, who will then invariably fail to transmit the mutation. Therefore, all cases of the disease will be due to independent *de novo* mutations, and the incidence of the disease will equal the mutation rate (this is a special case of mutation-selection balance, see Section 5.6.1). This incidence will be low, and mutations will probably occur with equal frequency in different populations, so the disease will be rare and have a relatively uniform geographical distribution.

If, however, the phenotype is milder and individuals carrying the mutant allele reproduce, other factors including the strength of the selection and **random genetic drift** come into play, and the resulting incidence and distribution of the disease will be influenced by population processes, which include structure and history (e.g., founder events). Nonetheless, the default expectation remains that the most disadvantageous individual mutations would not spread far, so diseases would be rare, found at similar frequencies in different populations, and originate from many different mutations. Many genetic diseases do show this pattern: hemophilia B (OMIM 306900), for example, is rare (1109 patients known in the UK, which has a population of around 60 million, giving a disease frequency of less than 0.002%), found at similar frequencies in other countries, and diverse.

A study of 424 independent UK families succeeded in identifying mutations in the Factor IX gene in 412 of the families; of these mutations, 167 were family-specific, and distinct haplotypic backgrounds suggested that some of the shared mutations had arisen more than once, such that at least 302 different mutations were represented (Green *et al.*, 1999). This pattern is common (see Online Mendelian Inheritance in Man, http://www.ncbi.nlm.nih.gov/omim/), but a few exceptional Mendelian disorders are more frequent than would be expected. For these, we therefore need to ask:

▶ why are they found at high frequency?

▶ does the frequency vary between populations, and if so, why?

Factors influencing the frequency of diseases in individual populations include:

▶ mutation rate;

▶ mode of inheritance (dominant or recessive, autosomal or X-linked);

▶ selection: what is the extent of selection against the allele? Is it even advantageous under some circumstances? One of the factors influencing selection is whether the disease develops early or late in life. Late-acting (especially postreproductive) genes are considered unlikely to have experienced strong selection;

▶ migration, including recent population movements. For example, thalassemia was virtually unknown in Melbourne and Sydney (Australia) before 1948, but has now been introduced as a result of recent immigration from Mediterranean countries;

▶ past demography. For example, we saw in Section 5.6 that those populations with smaller effective population sizes (as a result of their specific demography histories) are less able to eliminate deleterious mutations through purifying selection than larger populations. This is because genetic drift is predominant in determining the future frequencies of weakly deleterious alleles.

Selected examples of the incidences of genetic diseases are given in *Table 14.1*. In each class, some of the most frequent diseases have been chosen. The incidences of the complex diseases listed are nonetheless far higher than the simple diseases; indeed the equations simple = rare, and complex = common are sometimes made. We will now discuss some of the most frequent simple genetic diseases, and refer to their 'high' incidence, but this is *relative* to other simple diseases, not complex diseases. It must be remembered that these examples represent unusual Mendelian diseases which, despite being frequently discussed in medical genetics journals (partly due to their elevated disease burden), are not representative of most monogenic diseases.

Duchenne muscular dystrophy (DMD)

Duchenne muscular dystrophy (DMD) has an incidence of around 30 per 100 000 boys, but shows little variation in frequency between populations. The high incidence of the disease is explained by a combination of two factors:

▶ the large size of the gene. *DMD* is the largest gene known in the genome (Section 2.2), and provides many targets for mutation, resulting in a high mutation rate. Unusually for such a common disorder, about one-third of cases are due to new (*de novo*) mutations;

▶ the mode of inheritance of the disease: DMD is recessive, and the gene is located on the X chromosome. This means that all males who inherit a mutant copy of the gene manifest the disease. Females with one mutant copy are unaffected but can transmit it, so half of their sons will inherit the disease. *DMD* mutations, approximately one-third of which are deletions, duplications or insertions which can be large (http://www.dmd.nl/dmd_all.html), arise on a wide variety of haplotypic backgrounds, so there is little geographical variation in the incidence of the disease.

Other diseases, such as sickle cell anemia, cystic fibrosis (CF) and Huntington disease (HD), show considerable differences in incidence between populations as well as high frequencies relative to other simple genetic diseases. For sickle cell anemia and CF, this can probably be explained by **balancing selection**. In the case of sickle cell anemia, the preferential survival of individuals heterozygous for the Hb^s allele because of their resistance to the most lethal form of malaria, *falciparum* malaria, has led to an increase in frequency of a small number of Hb^s mutations in populations where malaria has been endemic (Section 5.4.1 and *Figure 5.10*). Individuals homozygous for the Hb^s allele, however, show the damaging and lethal sickle cell anemia. They are found in populations currently or previously exposed to malaria, or their descendants. The high incidence of the thalassemias can be explained in a similar way by selection of heterozygotes (**heterozygote advantage)** for resistance to malaria, with the disease manifested in homozygotes.

Cystic fibrosis (CF)

CF is found at high frequency in Europeans, where it is the most common severe autosomal recessive disease. Affected individuals carry two inactive copies of the *CFTR* (cystic fibrosis transmembrane conductance regulator) gene. One allele, *ΔF508*, a deletion of 3 bp which removes the phenylalanine at amino acid 508 and leads to a misfolded, nonfunctional protein, accounts for about two-thirds of CF chromosomes, although a total of over 1000 mutations is known, most of which are very rare (http://www.genet.sickkids.on.ca/cftr/). It has been proposed that the high frequency of the disease is due to a selective advantage of the heterozygote (Romeo *et al.*, 1989). Evidence for selection is less clear than for sickle cell anemia, but heterozygotes may be more resistant to diseases such as cholera, typhoid fever or bronchial asthma. The CFTR protein, as suggested by its name, is located in the membrane of several cell types, including the epithelial cells lining the small intestine, where it regulates chloride ion transport, which is in turn linked to fluid secretion. Untreated cholera leads to excessive chloride ion and fluid secretion that can be fatal. In a mouse model, animals carrying one copy of the *CFTR* gene secreted only half the amount of fluid and chloride ions in response to

TABLE 14.1: SELECTED EXAMPLES OF SIMPLE (S) AND COMPLEX (C) GENETIC DISEASES.

Disease	OMIM		Phenotype		Inheritance	Population	Incidence per 100 000
Duchenne muscular dystrophy, DMD	310200	S	A degenerative disorder affecting the muscles, leading to wheelchair dependency (median age 10 years) and death (median age 17 years)	X-linked	Recessive	World	~ 30 (affects males only)
Sickle cell anemia	603903	S	Red blood cells are sickle-shaped instead of donut-shaped; blocking of blood vessels leads to severe pain and damage to organs. Median survival 42 years (men), 48 years (women)	Autosomal	Recessive	African–Americans European–Americans	270 2
Cystic fibrosis, CF	219700	S	Affects the lungs, small intestine, sweat glands and pancreas; increased mucus secretion; infertility is common especially in males. Median survival 31 years	Autosomal	Recessive	Europeans Asia (Hawaii)	40 ~1
Huntington disease, HD	143100	S	Degenerative brain disorder with symptoms including involuntary twitching (chorea) and dementia; usually develops between the ages of 35 and 44 and has a median duration of 16 years before death	Autosomal	Dominant	Europeans Finns Chinese South African Bantu-speakers	5–10 0.5 0.2–0.4 < 0.1
Diabetes, type 2	125853	C	High blood sugar; major cause of adult blindness; two- to fourfold increase in cardiovascular mortality			Developed countries Pima Indians	10 000–20 000 > 60 000
Cancer (lifetime risk)	Many	C	Diverse set of diseases characterized by improperly controlled cell growth			UK	~ 30 000
Schizophrenia	181500	C	A psychosis (disorder of thought and sense of self); onset is usually before the age of 25 years			World	~ 1000

cholera toxin compared with animals carrying two copies (Gabriel et al., 1994), so it seems reasonable to think that the same would happen in humans. Measurement of intestinal chloride secretion in humans in response to prostaglandin stimulation (considered to be a model for cholera toxin exposure), however, revealed no difference between unaffected individuals and CF heterozygotes (Högenauer et al., 2000), so some doubt remains about whether this scenario provides an adequate explanation in humans. An alternative, or additional, possibility comes from work demonstrating that CFTR is used by *Salmonella typhi*, the bacterium that causes typhoid fever, as a receptor to enter epithelial cells (Pier et al., 1998). Cultured mouse cells take up little *S. typhi*, but will do so if they express wild-type human CFTR. If, however, they express *ΔF508* CFTR, they take up only the basal level. Thus human CF heterozygotes might be partially protected against typhoid fever by reduced uptake of the bacterium. Statistical analysis of the pattern of variation around the *CFTR* locus can also be used to assess whether *ΔF508* heterozygotes are likely to have had a selective advantage, and some studies have indicated such an effect, although this conclusion depends greatly on the assumptions made (Slatkin and Bertorelle, 2001; Wiuf, 2001). Cholera itself was first reported in Europe in 1832, while, as will be discussed below, the *ΔF508* mutation is much older than this, so cholera is unlikely to have been the selective agent, but typhoid fever, or another diarrhea-causing pathogen could have selected for CF heterozygotes. The time and place of origin of the *ΔF508* mutation has thus been a focus of interest, and some of the dating methods described in Section 6.6.3 were developed in order to answer this question. Information about the frequency of the deletion, the extent of linkage disequilibrium, and the variability of closely-linked microsatellites can all be used. *Table 14.2* summarizes some of the published estimates.

A large dataset of variation at three microsatellites within the *CFTR* gene (*Figure 14.1A*) was assembled from 1738 *ΔF508* chromosomes by Morral and co-workers (Morral et al., 1994). Although 54 different haplotypes were observed, only four were present above a frequency of 2% in their sample, and most of the haplotypes fell into two main clusters (*Figure 14.1B*). After excluding a few rare divergent haplotypes as likely to have arisen by recombination, Morral et al. assumed a single origin for the remaining chromosomes

and calculated a TMRCA of > 52 KY based on the average distance from the inferred root and an estimated mutation rate. The mutation rate and estimated number of mutations were criticized by Kaplan et al., who argued that if a higher mutation rate (taking into account the three-locus haplotype rather than a single locus) and the genealogical relationship between the chromosomes were taken into account, the data suggested a considerably more recent TMRCA (Kaplan et al., 1994). This is because not all derived mutations are independent; some have been inherited from common ancestors, meaning that fewer mutations than previously estimated have actually occurred. Most subsequent analyses (Slatkin and Bertorelle, 2001; Wiuf, 2001) have also supported a more recent date (*Table 14.2*). One conclusion from these studies could be that our quantitative understanding of the *ΔF508* selective advantage (if any), demography and mutation rate are so poor that, even with the most sophisticated modeling, the geographical distribution of the chromosomes probably provides the best guide to their age. Their widespread distribution within Europe, but rarity outside populations of European origin, implies a date after the out-of-Africa expansion ~ 50 KYA, but sufficiently long ago to spread through much of Europe, which would require many thousands of years. A best guess might therefore be around 10 KYA (*Table 14.2*). Thus, despite some uncertainties, a past selective advantage for heterozygotes acting in Europe but not elsewhere provides the best explanation for the high but localized incidence of the disease.

Huntington disease (HD)

The explanation for the high incidence and differential distribution of HD is more complex. It is most frequent in populations with a Western European origin, where, except in the Finns, it is around 10 to 100 times more frequent than in populations from other parts of the world. Before its molecular basis was elucidated, it was therefore expected to have a low mutation rate and originate from a single founder. Subsequently, however, this explanation has proved inadequate. The phenotype is due to the expansion of a $(CAG)_n$ array within the first exon of the huntingtin protein which codes for a polyglutamine tract in the protein. The length of the array is polymorphic in unaffected individuals, with a mean of 18 copies and a range from ~10 to 35 copies, but has pathogenic consequences in striatal cells (a subset of neurons) in adults when it exceeds 35 copies. Pedigree

TABLE 14.2: ESTIMATES OF THE TIME OF ORIGIN OF THE *CFTR ΔF508* MUTATION.

Time (KYA), 20 years/generation	Time (generations)	Basis	Reference
4	200	LD pattern	Serre *et al.*, 1990
> 52	> 2627	Variation at 3 microsatellites	Morral *et al.*, 1994
18	> 900	Reanalysis of Morral *et al.* data	Kaplan *et al.*, 1994
10	> 500	Variation at 3 microsatellites	Slatkin and Bertorelle, 2001
12	> 580	Coalescent analysis of Morral *et al.* data	Wiuf, 2001

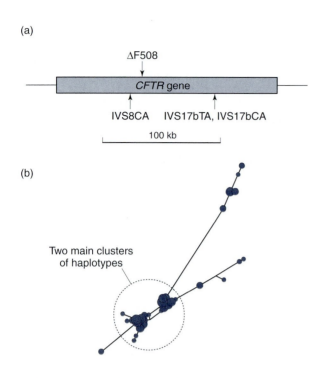

Figure 14.1: Variation within the *CFTR* gene.

(a) Genomic structure of the locus, showing the *CFTR* gene (blue box, representing both exons and introns), location of the common *ΔF508* mutation and three intronic microsatellites (IVS8CA, IVS17bTA and IVS17bCA) used to assess variation. (b) Network of microsatellite haplotypes associated with the *ΔF508* mutation. Most haplotypes fall within two main clusters (circled); because of the large number of haplotypes, not all connecting lines can be seen, but their diversity is thought to reflect mutational events occurring on this lineage. The more diverse haplotypes lying on long branches are likely to arise by recombination of the *ΔF508* mutation with other haplotypes. Redrawn from the data of Morral *et al.*, 1994.

studies showed that a significant proportion of families (~ 3%) manifest the disease as a result of new mutations, so the mutation rate is not low and there is no single founder. But if there are many recurrent mutations creating long disease-related alleles, why should they occur in Europeans so much more often than in other populations? This question has been addressed by examining the evolutionary history of the mutant chromosomes (Squitieri *et al.*, 1994). Most (~ 78%) HD chromosomes belonged to only two of the six common haplotypes, huntingtin haplotype 1 and huntingtin haplotype 3, defined by three flanking markers (*Figure 14.2*). Together, these two haplotypes made up only ~ 39% of the chromosomes in unaffected individuals from the same families (*Figure 14.2*). These haplotypes differ only in the allelic state of the Δ2642 polymorphism (a deletion of GAG from a run of four consecutive GAG codons leading to a loss of a glutamic acid residue) ~ 150 kb from the $(CAG)_n$ expansion, and could have arisen from a distant common progenitor by recombination. Huntingtin haplotype 1 and huntingtin haplotype 3 chromosomes in normal individuals

had mean $(CAG)_n$ lengths of 19.1 ± 2.4 and 22.5 ± 4.4 repeats, compared with a range of 16.5 ± 2.1 to 18.3 ± 2.1 for the other common haplotypes in the unaffected individuals. Further surveys have shown that the Δ2642 deletion allele (allele with three repeats in *Figure 14.2*) is associated with both HD (where it is about fivefold over-represented) and larger $(CAG)_n$ arrays: 23.1 ± 5.0 for alleles with three repeat units compared with 16.9 ± 2.7 for alleles with four repeat units (Almqvist *et al.*, 1995). Furthermore, in the 645 individuals surveyed, the three-repeat unit allele was only found in Western European populations, and not in the African, Chinese or Japanese samples. While a functional effect of one or both of the nearby $(CCG)_n$ and Δ2642 polymorphisms on $(CAG)_n$ expansion has been considered, it is simpler to interpret them as neutral markers of a haplotype $(CCG)_n = 7$, $Δ2642 = 3$ carrying a large $(CAG)_n$ array which predisposes to further expansion leading to the disease phenotype. Thus the immediate cause of the high incidence of HD in Europeans is the high mutation rate of the longer $(CAG)_n$ array and the dominant manifestation of the disease at an age after reproduction, which allows affected individuals to pass on the disease gene, whereas the underlying cause (and the explanation for population specificity) is the drift to high frequency of a predisposing haplotype (carrying what is sometimes known as a 'premutation') in a single geographical region.

The Jewish disease heritage

Over 40 Mendelian disorders have been identified that are present at markedly higher frequencies in Jewish populations than in surrounding populations (reviewed by Ostrer, 2001). This concentration of normally rare disorders within a

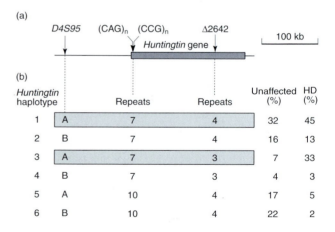

Figure 14.2: Variation in and around the huntingtin gene.

(a) Genomic structure showing the huntingtin gene (blue box, representing both exons and introns), the location of the disease-causing expansion $(CAG)_n$, and three flanking markers (*D4S95*, $(CCG)_n$ and Δ2642), which are probably neutral. (b) Haplotypes found in a sample of unaffected and HD chromosomes. Each haplotype is given a number (1–6) and consists of the allelic states at the *D4S95*, $(CCG)_n$ and Δ2642 markers, which are shown above their respective column of alleles. The blue boxes indicate huntingtin haplotype 1 and huntingtin haplotype 3, which are both over-represented in HD patients. Data from Squitieri *et al.* (1994).

population is known as a 'genetic disease heritage', and, because of the unlinked nature of the causal loci underlying these disorders, must result from some evolutionary process that affects many genomic regions simultaneously. Unlike selection or locus-specific mutation rates, genetic drift affects all alleles within a population, and past demography that resulted in a low effective population size is a frequent cause of a recognizable genetic disease heritage. These demographic processes include founder events, population bottlenecks and long-term endogamy. In the case of Jewish populations, their disease heritage relates in part to their complex history of diaspora (see Section 12.6.3). It is often found that populations with distinct genetic disease heritages also harbor unusual frequencies of neutral alleles as well, further supporting the causal role of genetic drift. Furthermore, there appears to be an inverse relationship between disease frequency and allelic heterogeneity. In a population where a particular Mendelian disorder is common as a result of genetic drift, the disorder is often caused by fewer mutations than in populations in which the disease is less frequent. This feature of populations with distinctive disease heritages can be exploited by gene mappers searching for susceptibility alleles to complex disorders as well as Mendelian disorders (Section 14.3.2).

We thus have a good understanding of the distribution of many simple genetic diseases, and the factors that have led to the high incidence of a few of them in some populations. Evolutionary forces, including drift, mutation and selection, have been important influences in several of these cases. The examples discussed in most detail, CF and HD, are present at their highest frequencies in people with European ancestry. This reflects **ascertainment bias**: disproportionate medical and scientific attention have been directed at these populations. It is likely that similar evolutionary forces have led to other simple genetic diseases reaching high frequencies in other populations, and we will learn more about these as

studies of diseases prevalent in other parts of the world progress.

14.2.2 Distribution of complex genetic diseases

There is also a strong ascertainment bias in our information about the distribution of complex genetic diseases, again towards those that are frequent in developed countries. Reduction in mortality due to infectious disease and malnutrition allows other causes of mortality and morbidity to come to the fore. Indeed, most of the leading causes of death in developed countries can be considered to have a genetic component (*Figure 14.3*): in the USA, the top three were heart disease, cancer and cerebrovascular disease, while diabetes and Alzheimer disease were sixth and ninth, respectively. Complex diseases are thus much more common than simple ones (*Table 14.1*). Disease results in morbidity as well as mortality, and complex diseases that are receiving attention from geneticists include cancer, cardiovascular disease, hypertension, diabetes, asthma, obesity and mental disorders.

Data on the incidence or prevalence of these conditions in different parts of the world are limited, and may be complicated by the difficulty of defining a disease phenotype in a consistent way in diverse environments and cultures. Unlike the simple diseases discussed in the previous section, the genetic variants underlying these susceptibilities are generally poorly understood, although the identification of these genes is one of the most active areas of current research. Thus it is at present impossible to discuss the geographical distribution of variants contributing to these diseases in the way that we did for simple genetic diseases. We will, however, consider the frequencies of two examples of complex disorders: schizophrenia and diabetes.

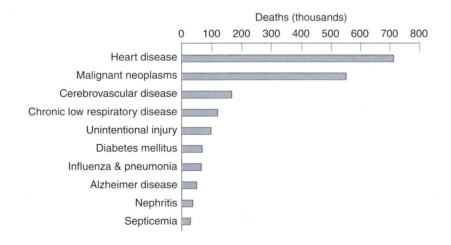

Figure 14.3: Leading causes of death in the USA in 2000.

Categories and numbers of deaths (in thousands) are from the Centers for Disease Control and Prevention (CDC) website: http://webapp.cdc.gov/sasweb/ncipc/leadcaus10.html

Schizophrenia

Schizophrenia is characterized by symptoms that include disorganized thinking, hallucinations, loss of goal-directed behaviors and deterioration of social functions. It is one of the top 10 global causes of life years lost to disability, and accounts for a significant proportion of healthcare costs: 5.4% of state inpatient funding in the UK, for example (Schultz and Andreasen, 1999). It is generally considered to have a rather uniform geographical distribution, with a prevalence of about 1% (*Table 14.1*). Its incidence in eight countries is summarized in *Figure 14.4A*. Apart from Russia, the figures from diverse populations that include Nigeria (Africa), Denmark (Europe) and Japan (Asia) vary by less than twofold. The low incidence in Russia may be partly dependent on the definition used: with a broader definition, the same study found Russia to have the second-highest incidence (Jablensky, 2000). While schizophrenia remains poorly understood, it has a clear genetic component, with the **relative risk** predicted from an affected family member being 45–50 for a monozygotic twin, 14 for a dizygotic twin, and 9–12 for a nontwin sibling (Jablensky, 2000). The genetic influence and relatively uniform distribution have led to the hypothesis that it is '*the price* Homo sapiens *pays for language*' (Crow, 1997; and *Box 7.11*).

Diabetes

Diabetes mellitus is a condition in which there is excessive glucose in the blood, and is classified into type 1 or insulin-dependent, and type 2 or noninsulin-dependent; the latter is the more common. The worldwide prevalence of diabetes was estimated at around 4% of the adult population in 2000, but this figure is increasing rapidly and is projected to reach 5.4% by the year 2025. In addition, the time of onset is also changing: type 2 diabetes was formerly known as adult-onset diabetes, but this name is no longer appropriate because it is now also appearing in the young (see Section 13.5.3). In contrast to schizophrenia, very extensive data on the worldwide prevalence are available, and have been summarized by King *et al.* (1998). Data for the same countries as previously given for schizophrenia are shown in *Figure 14.4B*. Differences of almost 10-fold in prevalence rates are seen. The rate in Nigeria is low, and this is typical of sub-Saharan African countries, where the prevalence averaged across 49 countries was 1.1%. In contrast, prevalences in China were 2.2%, the rest of Asia (excluding India and Japan) 3.2%, Latin America 6.0%, the Middle East 6.5% and the USA 7.6%. Much of this variation is thought to be environmental and related to diet and lifestyle, as is also the rapid increase in recent times. A further illustration of the likely importance of the environment is provided by the figure of 6.9% in Japan: more than twice the average for the rest of Asia. It seems possible that the prevalence in most countries would rise to the 'Western' level if a Western lifestyle was adopted. Nevertheless, the broad figures hide much local variation between populations, and type 2 diabetes is one of the few complex genetic disorders where some information on genetic susceptibilities is available (Section 14.3.3).

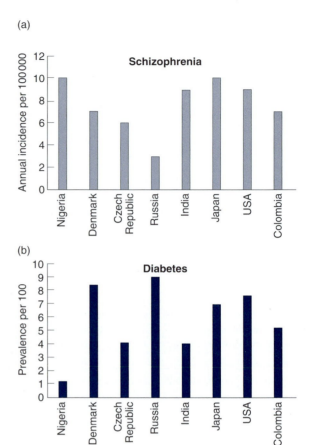

Figure 14.4: Distribution of complex genetic disorders.

(a) Annual incidence of schizophrenia for both sexes in eight countries, using a restrictive definition. Data from Jablensky (2000). (b) Prevalence of diabetes in the same eight countries extrapolated to the year 2000. Data from King *et al.* (1998). The two diseases cannot be compared with one another, but comparisons can be made between countries for the same disease.

14.2.3 The human genetic disease heritage

We saw earlier that particular populations have distinctive 'genetic disease heritages' because of their past demography. We can now consider whether the past demography of our species as a whole has been important in shaping the global distribution and frequency of Mendelian and complex genetic diseases. In previous chapters we have discovered that the human past is characterized by a recent common origin in Africa, followed by a rapid settlement of most of the habitable land surface of the planet, and subsequently by rampant population growth. What are the consequences of these processes for the frequencies of genetic disorders and the alleles that underlie them?

Alleles underlying simple genetic disorders tend to be young and infrequent as they are rapidly removed from the population, and therefore the more ancient processes described above tend to be of little consequence. However, the recent population explosion that has led to an excess of rare alleles at most loci relative to neutral expectations will have contributed to the diversity of disease alleles underlying these disorders in most populations. As detailed above, locus-specific selection pressures or mutation dynamics can cause individual alleles to depart from this general trend.

The typically low incidence of most simple genetic disorders means that they impose relatively low selective costs on the population; the historically small effective population size of humans may have precluded the selective elimination of alleles that could have been expunged in a species experiencing less genetic drift. This may have contributed to the wide range of Mendelian disorders found in humans.

The demography of the more distant past outlined in preceding chapters has had more of an effect on alleles underlying complex disorders than on those underlying Mendelian disorders, because the former can persist in the population for longer. In the absence of region-specific selective pressures, ancient (pre-out-of-Africa migration) susceptibility alleles should reflect the low levels of population differentiation seen in humans, and should be found in most populations at similar frequencies. Younger (post-out-of-Africa) susceptibility alleles may be numerous (again due to recent rapid population growth), but individually are likely to be rare, not having had sufficient time to reach high local frequencies. There are likely to be very few continent-specific, continent-wide alleles, whether deleterious or neutral. We currently know little about the underlying susceptibility alleles (Section 14.3.3), but they may well show similar even distribution patterns to neutral alleles (Section 9.3), with some exceptions because of selection.

At first sight, it seems that even if this prediction is true of susceptibility alleles, it is not fulfilled by the complex disorders themselves. Some of these disorders (e.g., obesity) have distributions that are approximately continent-wide and continent-specific. This, however, can readily be explained by environmental factors: the diet and lifestyle that allow obesity (and many other complex disorders) to develop are more often available on some continents than others. Over the range of human habitation, many disease-related environmental factors (e.g., malarial endemicity or calorific intake) are more variable than are human genotypes. As a consequence, complex diseases with a substantial environmental component often exhibit far greater differentiation than do diseases with lower contributions from environmental factors.

14.3 Identifying genes relevant to disease

14.3.1 Approaches to identifying disease susceptibility genes

If susceptibility to a disease has some genetic basis, a search for the relevant gene(s) can be undertaken. The best approach depends principally on the strength of the genetic susceptibility, revealed by whether or not the disease shows Mendelian inheritance. For Mendelian disorders, it is now a routine matter to use **linkage analysis** in pedigrees to identify the approximate physical location of the gene on a chromosome (often to within a few megabases); there are then several ways in which the responsible gene within this interval can be identified. This approach often requires little or no input from evolutionary genetics, and so lies largely outside the scope of this book. Evolutionary considerations, however, can influence the choice of the most suitable population, and these are discussed in the next section (14.3.2). If a disease does not show Mendelian inheritance, it will be more difficult to identify the relevant genes. However, as we have seen (*Table 14.1*), complex genetic diseases are much more common than simple ones, so it is important to find the relevant genes. The search for such susceptibility genes is therefore now a major focus for medical genetics; indeed, it provided one of the underlying motivations for the human genome sequencing and SNP mapping projects. Success has so far been limited, but evolutionary considerations have had a significant input into the formulation of the underlying models and the design of studies.

Linkage or association for complex disorders?
Although Mendelian inheritance in large pedigrees has little power to locate genes influencing susceptibility to complex disorders, other forms of linkage analysis can be applied; alternatively, **association studies** can be used. An influential paper entitled 'The future of genetic studies of complex human diseases' (Risch and Merikangas, 1996) compared the two. In this theoretical study, Risch and Merikangas compared the power of a linkage approach, the use of **affected sib pairs** (ASP; sib = sibling = brother or sister), with an association approach, the **transmission disequilibrium test** (TDT; both are discussed below and illustrated in *Figure 14.5*) to successfully map a complex disease gene. They needed to define the quantitative effect of a susceptibility allele, the genotypic relative risk (GRR), as the increased chance that an individual with a particular genotype would develop the disease. They considered GRR values of 1.5 and 2.0, among others. An individual can carry zero, one or two copies of the susceptibility allele, and the risk associated with two copies was multiplicative, i.e., individuals homozygous for the susceptibility allele have GRRs of 1.5^2 ($= 2.25$) or 2.0^2 ($= 4.0$), respectively. They then calculated the number of families needed using linkage or association.

▸ The linkage study, using ASP (*Figure 14.5*), considered a genome screen with 500 microsatellites. Risch and Merikangas required a $> 95\%$ chance of no false positives and an 80% power of detecting linkage, and calculated the number of families needed (*Table 14.3*).

▸ The association study, using the TDT design (*Figure 14.5*), was based on a genome screen with 1 000 000 SNPs and again required a $> 95\%$ chance of no false positives and an 80% power of detecting linkage.

The number of families needed depended strongly on the relative risk and frequency of the susceptibility allele, but was

TABLE 14.3: COMPARISON OF THE POWER OF LINKAGE AND ASSOCIATION STUDIES TO IDENTIFY A COMPLEX DISEASE SUSCEPTIBILITY GENE.

Genotypic relative risk	Frequency of disease allele	Number of families needed	
		Linkage[a]	Association[b]
2.0	0.01	296 710	5823
	0.1	5382	695
	0.5	2498	340
1.5	0.01	4 620 807	19 320
	0.1	67 816	2218
	0.5	17 997	949

[a]Each family consisted of two parents plus two affected children. [b]Each family consisted of two parents and one affected child. Numbers from Risch and Merikangas (1996).

always less for the association study than for linkage (*Table 14.3*). A major reason for this is that TDT uses more information than ASP: it takes into account *which* allele is transmitted, while ASP only takes into account that *an* allele is shared identical by descent (IBD), without considering which allele it is. While these conclusions were derived using only a single design of linkage and association study, they have proved generally applicable.

An alternative form of association study is to use unrelated cases and controls; here care must be taken to determine whether any association discovered is due to true association with the disease or **population structure**, also referred to as population stratification. In an **admixed** population, a marker and disease that were both present at a higher frequency in one parental population than the other may show association. This can be measured using other markers and corrected for, or supplemented/replaced by the TDT, which is not affected by stratification since the non-transmitted chromosome in the same families provides the control. In the example above of an association study, Risch and Merikangas assumed that the causal variant would be tested directly. If, instead, nearby variants are used, it is a **linkage disequilibrium** (LD) study, but the terms 'association mapping' and 'LD mapping' are sometimes used loosely.

Thus, when a disorder does not show Mendelian inheritance, the would-be gene mapper has a number of options:

▶ subdivide the phenotype: perhaps some forms of the disorder will show simple inheritance and allow mapping by pedigree studies. For example, early-onset Alzheimer disease is a simple (and rare) genetic disease, while late-onset Alzheimer disease is complex (and common); see Section 14.3.3. Classification of the disease phenotype according to age of onset allowed several genes relevant to the early-onset form to be identified, e.g., *APP*, *PSEN1* and *PSEN2*;

▶ find a population derived from a small number of founders where genetic and environmental heterogeneity may be lower; this approach is not always successful, as will be seen in the search for bipolar affective disorder susceptibility genes (Section 14.3.2);

▶ carry out an association study using, for example, the TDT, case/control comparisons or mapping by admixture linkage disequilibrium (MALD) – see Section 12.5.1. This can examine:

  ▸ candidate genes. The scale and cost of the study are reduced, but biological insight is needed to suggest suitable candidates, and this is not always available;

  ▸ the whole genome. A formidable undertaking, which will be discussed further in Section 14.3.3.

14.3.2 Identifying genes underlying Mendelian disorders

One of the great successes of human genetics over the last two decades or so has been the identification of the genes responsible for many Mendelian disorders (*Box 3.1*). Even when the underlying molecular cause of the disease was unknown, genetic linkage analysis in pedigrees segregating the disease could be used to identify a critical region of the genome, defined by flanking recombination events, within which the gene must lie. Then chromosomal rearrangements, deletions, testing of candidate genes or simply a systematic evaluation of all genes within the critical region for mutations that would inactivate or alter the gene would reliably lead to the identification of the relevant gene. This procedure could be complicated by factors such as the degree of penetrance (does an individual carrying the mutated gene always exhibit the phenotype?) or heterogeneity (do mutations in more than one gene result in the same phenotype?), but has nevertheless led to the successful identification of a large number of important genes, with the immediate possibility of offering diagnosis and the more distant prospect that improved

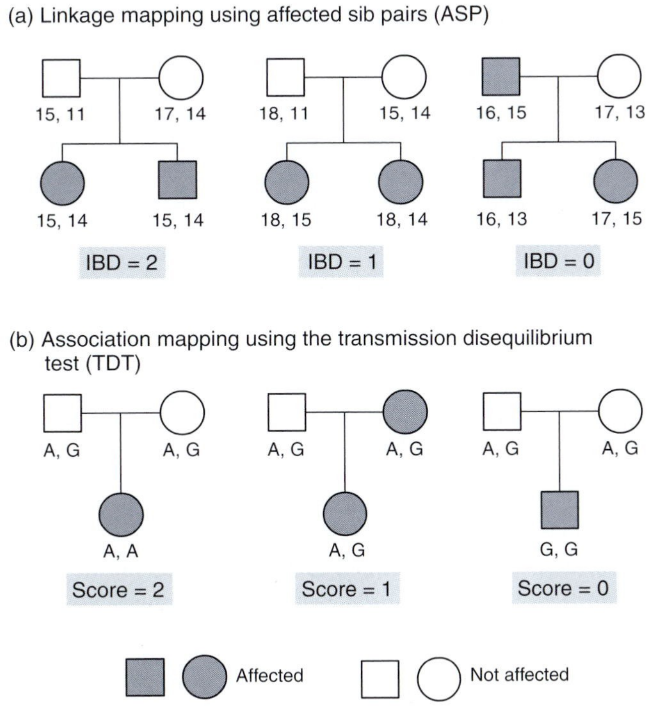

(a) Linkage mapping using affected sib pairs (ASP)

(b) Association mapping using the transmission disequilibrium test (TDT)

Affected   Not affected

Figure 14.5: Comparison of linkage and association mapping of a complex disorder.

(a) The linkage study uses families consisting of two affected children (blue; an affected sib pair) and their parents (who may or may not be affected). All individuals are typed with a genomewide set of microsatellite markers. Here, the results for one locus in three families are shown. The number of alleles at each locus that are shared identical by descent by the affected sibs is counted: this number can be 2, 1 or 0. Note that, in the family on the left affected sibs both typed as 17, 15; 17, 11; or 14, 11 would also score 2; also, not all families will be informative for every locus. The expected null distribution of 2, 1, 0 scores is 0.25, 0.5, 0.25 and each locus is tested for a departure from this distribution. (b) The association study uses an affected child and his or her parents. All individuals are typed with a genomewide set of SNPs and the transmission of each allele (here, the A allele) to the affected child is scored. The null expectation is that the distribution of transmissions, nontransmissions will be 0.5, 0.5 and each SNP is tested for departure from this distribution.

Note that other designs of these tests are possible: here, we illustrate the forms discussed by Risch and Merikangas (1996). The power of these tests is summarized in *Table 14.3*.

treatment may result from an understanding of the target and pathway affected.

Nevertheless, an understanding of population history can be important in the selection of the individuals to be studied and the strategy used. The best example of this is the use of isolated populations, particularly those established from a small number of founders, such as the Finns. Because of genetic drift in a small population, the genetic disease spectrum may be very different from that in surrounding populations, as described above in Section 14.2.1. In the Finns, as we have seen, diseases such as CF are much rarer than in most Europeans, but other genetic diseases that may be very rare in the rest of the world, are present at high frequency. This results in the 'Finnish disease heritage' (*Box 14.1*). These 'Finnish' diseases are often derived from a single founder, which reduces the genetic heterogeneity, and this assumption allows the use of linkage disequilibrium (LD) analyses in the search for the gene responsible. The Roma

also have a distinctive history and set of genetic disorders, and studies of this population are discussed in *Box 12.3*.

The cloning of the diastrophic dysplasia (DTD; OMIM 222600) gene provides an example of the usefulness of LD in defining the position of a disease gene. This autosomal recessive disease, which affects the skeleton and joints, is very rare in almost all populations, but is one of the most common recessive disorders in Finland, with a carrier frequency estimated at 1–2%. Initially, linkage analysis in pedigrees was used to localize the gene to the long arm of chromosome 5, close to the marker *CSF1R*. The interval defined in this way, however, was probably 1–2 Mb in size (which could be expected to contain 10–20 genes on the basis of the genomewide average gene density). LD analysis showed that ~ 95% of Finnish DTD chromosomes were associated with a *CSF1R* haplotype which was present on only 3% of normal Finnish chromosomes. Assuming that all these DTD alleles shared a single origin, and that the 5% of DTD chromosomes

BOX 14.1 Genetic, linguistic and archaeological puzzles of the Finns

The Finns have attracted wide interest from geneticists during the last few decades. The major trigger for this interest was a dramatically different spectrum of inherited, mostly recessive, diseases in Finland, when compared to the neighboring European populations. Many recessive diseases that are common in European populations, such as PKU and CF, are rare or absent in Finland. These features of the disease spectrum reflect the impact of a founder effect and long-lasting isolation, which have molded the Finnish gene pool for thousands of years. The clinical and molecular characterization of more than 30 diseases which constitute the 'Finnish disease heritage' is almost completed; in addition, the population frequencies of the mutations have been established, making the Finns one of the best genetically characterized populations in the world. Typically, one mutation is responsible for 75–98% of disease alleles, allowing efficient and accurate carrier testing. One Finn out of five is a carrier of one of the 'Finnish mutations'. The disease mutations are often embedded in wide chromosomal regions exhibiting linkage disequilibrium (LD) over intervals of up to 13 cM in size.

Within Finland, the diseases showing the shortest LD-intervals surrounding the disease alleles often occur throughout the entire country, while others, with longer LD intervals, are found in geographically more restricted areas (see *Figure*). This reflects the effect of multiple population bottlenecks during the peopling of this relatively large country: small founder populations from the southwestern coastline expanded towards the northeastern wilderness after the sixteenth century. The regional variation in the frequency of disease alleles can be observed also for more common disease alleles, like Factor V Leiden (OMIM 227400) or hemochromatosis (C282Y; OMIM 235200), raising hopes that the genetic background of common diseases could be also more homogeneous and more tractable to genetic analysis in this population.

The isolation of the Finnish population has not been due only to its remote geographical position and sparse population. In addition, the language has been a barrier to immigration and, consequently, to admixture. The Finnish language belongs to a small group of Uralic languages, which differ drastically from most languages spoken in Europe, which belong to the Indo-European family. The best known among the Uralic-speaking populations are the Finns (~ 5.2 million), the Saami (~ 50 000–80 000), the Estonians (~ 1.5 million) and the Hungarians (~ 12 million), but there are several smaller regions in Russia and Siberia where Uralic languages are also spoken.

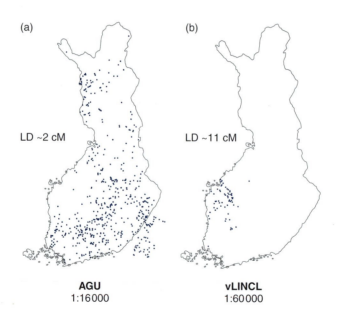

Geographical distribution of genetic diseases in Finland.

(a) AGU, Aspartylglucosaminuria (OMIM 208400), a lysosomal disease caused by deficiency of the enzyme N-aspartyl-beta-glucosaminidase. Note the wide distribution in Finland, relatively high prevalence, and short region of LD. (b) vLINCL, infantile neuronal ceroid lipofuscinosis (OMIM 256730), a neurodegenerative disease in children. Note the restricted distribution, lower prevalence, and longer region of LD.

Although archaeological data provide signs of human activity before 9 KYA, and the arrival of agriculture can be placed at about 4–3.3 KYA, indisputable evidence for permanent settlement of the southern and western coasts of Finland dates back only about 2 KY. However, it is difficult, if not impossible, to deduce who those people were, and what language they spoke. This paucity of human remains, the distinct language, and the unique genetic disease heritage have attracted molecular anthropologists to the Finns for population history studies. Early studies in the 1970s and 1980s, using classical markers, placed the Finns within the generally homogenous genetic landscape of Europe. In a genetic tree reconstructed from allele frequencies from European populations, the Finns are usually located near to their geographical neighbors. Recent studies using DNA variation have shown that the Finns do share the majority of their maternally inherited mtDNA lineages with other European populations, despite the language difference. However, most of those studies have so far been limited to the HVS I and II regions. As the technology will soon allow analysis of full mtDNA sequences, new light may be shed on the maternal history of the Finns and their neighbors. The picture drawn from the paternally inherited Y chromosome is slightly more puzzling. As expected, Y-chromosomal diversity is lower than in the neighboring populations. Also, local communities in Finland show large differences in Y-chromosomal diversity, suggesting few male founders in parts of the country. An interesting finding has been the detection of at least two distinct Y haplogroups in the Finnish male samples. One of those is frequently found also in other European populations, but the other (haplogroup N; *Box 8.5*) most probably has arrived here from the east.

Although several intriguing questions still remain unanswered concerning the timing and number of colonizing waves reaching Finland, and the subsequent population demography of the Finns, the existing studies have clearly shown the power of molecular tools in 'genetic excavation' of disease alleles, internal migration events, and finally, the origins of the Finns.

Antti Sajantila, Department of Forensic Medicine, University of Helsinki, Finland; Leena Peltonen, Department of Molecular Medicine, National Public Health Institute and Department of Medical Genetics, University of Helsinki, Finland.

carrying different haplotypes had arisen by recombination, Hästbacka *et al.* predicted that the DTD gene should lie about 60 kb proximal to *CSF1R*. Although there was some confusion about the orientation of this region on the chromosome, a sulfate transporter gene (designated *DTDST*) was indeed present ~ 70 kb proximal to *CSF1R* (Hästbacka *et al.*, 1994). It was identified as the gene responsible for the disease by the absence of its transcript from most Finnish patients, and the presence of diverse mutations that would inactivate the gene in the rare affected individuals from elsewhere. The precise correspondence between the predicted and confirmed locations of the gene was fortuitous, but does illustrate the value of an understanding of population history and using LD analysis in a suitable population.

There can, however, be perils in relying too heavily on LD as an indicator for the identification of a gene. In 1987, *Nature* published a six-page article announcing the cloning of a 'strong candidate' for the CF gene (Estivill *et al.*, 1987). The authors had identified an expressed sequence that showed LD with CF, and calculated from the measured LD that it must lie within 10 kb of CF. Subsequent studies, however, revealed that this candidate was *WNT2*, a secreted protein irrelevant to CF, and that the true CF gene lay over 130 kb away. LD may extend for long distances and must be supplemented by mutational information and/or functional studies if it is to lead reliably to the identification of disease genes.

Susceptibility genes for mental disorders: a cautionary tale
Mental disorders are medically important, as we have seen for schizophrenia. They are also of great scientific and evolutionary interest. Many mental disorders have a genetic component, so it is not surprising that numerous attempts have been made over many years to identify the genes

responsible. These efforts have been impressive for their determination, but less so for their success.

Early investigations used segregation in large families in an attempt to trace relevant genes. Recognizing the probable genetic complexity of bipolar affective disorders (manic depression), Egeland *et al.* (1987) first identified a small isolated population, the Old Order Amish in Pennsylvania, USA, who comprise ~ 15 000 people descended from ~ 30 progenitor couples, according to some estimates. In such a population, genetic influences might be less heterogeneous than in a large outbred population, and in a single affected family there might be only one disease gene segregating. Thus a disease that is complex in the world population might behave as a simple Mendelian disorder in a family like this. They therefore carried out linkage analysis on 81 individuals from a single large pedigree consisting of 19 affected and 62 unaffected individuals (Egeland *et al.*, 1987). Good evidence of linkage (a LOD score greater than 3 or 4, indicating odds of linkage to the disease of greater than 1000 : 1 or 10 000 : 1 respectively, depending on the penetrance assumed) was found to the marker *HRAS1* on chromosome 11p, with multipoint linkage analysis yielding a LOD score of ~ 5, indicating odds of 100 000 : 1 in favor of linkage (see *Box 3.5* for an explanation of LOD scores). Unfortunately, further work by the same team of investigators revealed that this conclusion was spurious (Kelsoe *et al.*, 1989). Crucial changes were the development of manic depression in two previously unaffected members of the kindred, and the discovery of two additional branches of the pedigree that increased the sample size to 120. Reanalysis produced a LOD score of about −9 with *HRAS1*, implying odds of greater than a billion to one *against* linkage. The initial strong evidence for linkage probably reflected the multiple tests that had been carried out, using many markers and models of inheritance, and a bias towards interest in 'positive' findings led to its publication.

Early searches for schizophrenia susceptibility genes ran into similar problems, with strong linkage to 5q11–13 (LOD score 6.49) failing to be replicated in independent samples and later being retracted, leading to the subject being dubbed a 'graveyard for molecular geneticists'. In a recent summary of the field, Baron (2001) accepted linkage to 14 regions as significant or suggestive (*Figure 14.6*). Such analyses can be augmented by testing LD to SNPs in candidate genes or elsewhere within the identified regions, and four genes have so far been pinpointed in this way (Cloninger, 2002; *Figure 14.6*). For example, Stefansson *et al.* (2002) confirmed that there was evidence suggestive of linkage to 8p (LOD score 3.48 using the best model) in a sample of 105 patients from 33 families in Iceland – a relatively small and, according to some, homogeneous, population. They then examined LD between the disease and haplotypes in their families, and identified a shared region of about 600 kb associated with the disease. Two genes were found in this region, including neuregulin 1. Extensive analysis of neuregulin 1 SNPs, including 15 nonsynonymous SNPs, in a total of 478 patients and 394 controls identified a core haplotype associated with the disease that spanned 290 kb and included the $5'$ neuregulin 1 exon, but no convincing pathogenic mutation could be found and so this gene's role in schizophrenia was not conclusively demonstrated. However, some indirect support for its involvement was provided by its biology, since mice with only one active copy of neuregulin 1 exhibited abnormal behavior considered to resemble mouse models of schizophrenia, and this could be partially reversed by treatment with clozapine, an antipsychotic drug used to treat human schizophrenia. Moreover, the four genes identified so far may all share a common mechanism and influence glutaminergic signaling through the N-methyl-D-aspartate (NMDA) pathway (Cloninger, 2002). While these studies are promising, they illustrate more than anything the great difficulty in finding genes that significantly influence susceptibility to mental disorders, where a fundamental limitation may be the absence of a clear diagnosis.

14.3.3 Searching for genes underlying susceptibility to complex disorders

If LD can be used to refine the approximate locations of genes obtained by pedigree analysis, can the initial study be omitted and the entire genome scanned by LD? This possibility is particularly attractive for complex disorders, where no pedigrees showing Mendelian segregation are available, and alternative approaches using linkage have low power (Section 14.3.1). A recent major focus has been on SNPs rather than microsatellites due to the greater density and stability of the former class of marker. Significant LD would indicate a region of the genome likely to be involved in the disease, so nearby genes could be investigated further using genetic and biochemical approaches. In practice, there are substantial difficulties. The density of SNPs required has been one of the contentious points, and is discussed further below, but it is universally agreed that a large number of

markers would need to be typed in a large number of individuals, and the technology needed to work on this scale is only now becoming available (Section 4.5.3). This, however, is not the major problem: few doubt that the necessary tools can be developed in the near future, although the cost remains uncertain. The more fundamental difficulty is whether the '**allelic architecture**' – the number and frequency of susceptibility alleles – of genes involved in complex disease is such that LD-based methods will lead to success. Modeling suggests that if there is one or a few susceptibility alleles at each locus, the approach will be successful, but that if there is a high degree of allelic heterogeneity, it will not. We will first consider the optimistic scenario.

The common disease/common variant hypothesis
The **common disease/common variant** (CD/CV) hypothesis, as suggested by its name, proposes that common diseases will usually be influenced by one or a small number of susceptibility alleles at each locus. It is often associated with a paper by Lander (1996), but was developed in more detail in a subsequent publication (Reich and Lander, 2001). Very little empirical evidence on the susceptibility alleles for common diseases is available (see below), so the hypothesis is based on theoretical expectations. Reich and Lander's argument involves the following steps:

▶ In the past, the human population was small in size and had reached mutation–drift equilibrium (see Section 5.6.1). Assuming that it was also panmictic and of constant size, the approximate diversity (D) of disease alleles, or its reciprocal (n, the effective number of alleles) is given by:

$$D_{\text{disease}} \approx \frac{1}{1 + 4N_e\mu(1 - f_0)}$$

where N_e is effective population size, μ is the mutation rate for a nondisease allele mutating into a disease allele, and f_0 is the combined equilibrium frequency of all the disease alleles at a locus;

▶ taking $N_e = 10^4$ and the mutation rate per gene per generation as 3.2×10^{-6}, Reich and Lander compare the number of alleles expected for a rare disease ($f_0 = 0.001$) and a common disease ($f_0 = 0.2$). For both, the expression $4N_e\mu(1-f_0)$ is small and the effective number of alleles, n, is close to 1: about 1.1, with a single disease allele accounting for ~ 90% of the total (*Figure 14.7*);

▶ the human population is now large; they consider the value $N_e = 6 \times 10^9$. This is the current census size, not the current N_e, but it may represent a future N_e, and their calculations do not depend critically on this value. The expected number of alleles for a rare disease and a common disease at equilibrium are again similar to one another, but because of the larger N_e, the expression $4N_e\mu(1-f_0)$ is large, and n is about 77 000 for the rare class and 61 000 for the common class. This should not be a surprise: large populations contain more diversity

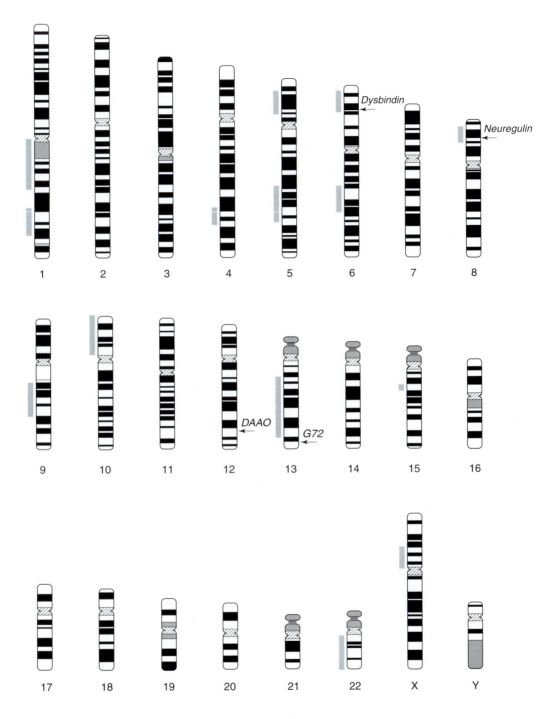

Figure 14.6: Locations of genes implicated in susceptibility to schizophrenia.

Blue bars: regions identified by linkage (Baron, 2001). Arrows: candidate genes (Cloninger, 2002).

than small ones; it also predicts that the allelic spectrum of a common disease will be very diverse *at mutation–drift equilibrium*;

▶ the human population, however, is not at equilibrium because it has expanded in the recent past, and time is required to reach equilibrium. There is a crucial

difference between the two types of disease in the rate at which equilibrium is attained after an expansion. Initially, if the expansion was instantaneous, n would still be ~ 1.1 for both. The subsequent increase in diversity will, however, be rapid for the rare class, but slow for the common class. This is because, although the mutation rate is assumed to be the same for both, a new mutation in a rare disease adds significant diversity to the small pool of disease alleles, while a new mutation in a common disease adds little to the large pool of disease alleles that already exists. In addition, there is stronger selection against the rare disease, removing some of the alleles that are already present, while selection against the common disease is weaker and the original disease allele decreases less rapidly (*Figure 14.7*).

Thus, for a population that expanded some 50 KYA, *n* is barely above 1 for the common disease, but it is approaching its equilibrium value for the rare disease (*Figure 14.8*). Therefore rare diseases will have many alleles (as is often observed, Section 14.2.1), but common diseases are predicted to have far fewer alleles at each locus. Reich and Lander adopt an effective number of alleles of 10 or fewer as a criterion for a simple allelic spectrum.

This model is, of course, a gross oversimplification, but the main conclusions remained the same even if population growth was more gradual or less extensive, or if the ancestral population fluctuated in size. Nevertheless, the real demography is much more complex and the importance of additional variables needs to be explored. In another study, it was argued that the mutation rate should be in the range 2.5×10^{-6} to 1.3×10^{-4} (as much as 40 times higher than that used by Reich and Lander), and that the upper part of this range is the most relevant because genes with higher mutation rates will be more polymorphic, and so more likely to contribute variants that influence disease (Pritchard and Cox, 2002). This would lead to greater allelic heterogeneity. The strength of any selection acting on variants contributing to common disease is also unknown: are they really under negative selection, or have they been in the main neutral?

The alternative is the genetic heterogeneity model: that many rare alleles at numerous loci contribute to common disease. While the CD/CV hypothesis focuses primarily on the number of alleles expected at each locus, some subdivide the models further according to the number of loci influencing the disease (Smith and Lusis, 2002):

▸ few loci, few alleles;
▸ few loci, many alleles;
▸ many loci, few alleles;
▸ many loci, many alleles.

The uncertainties underlying predictions based on theory are very large. Many of the issues are empirical. What answers can the data provide?

Empirical evidence about the complexity of susceptibility alleles for common diseases Empirical evidence about the nature of the variants affecting common disease is, unfortunately, limited by the small number of examples available, and also by possible ascertainment bias: it may be easier to identify the relevant genes when a single variant has a major effect than when many variants have minor effects. Nevertheless, we will consider a few of the examples available.

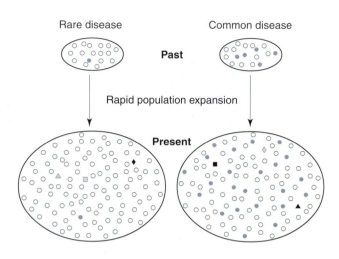

Figure 14.7: The common disease/common variant (CD/CV) hypothesis.

In a small population in the past (top) at mutation–drift equilibrium, there is likely to be only one disease allele present at a chosen locus (blue circle). After rapid population expansion in the present (bottom), a rare disease allele experiences strong negative selection and is rapidly diluted by new disease alleles appearing at the same locus (other shaded symbols), so a complex allelic spectrum is found. In contrast, a common disease allele experiences weaker negative selection and new disease alleles contribute only a small proportion of the total, so a simple allelic spectrum is found. Eventually, this will become more complex, but there has been insufficient time for this in the current human population. Open circles: unaffected alleles. Blue circles: disease alleles.

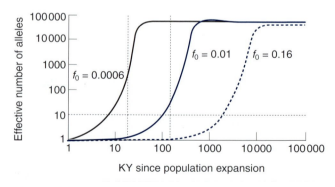

Figure 14.8: Increase in allelic complexity over time.

As the population expands, the number of disease-causing alleles at a disease locus increases. However, this occurs at different rates for common and rare diseases. A rare disease ($f_0 = 0.0006$) has a large effective number of alleles after 10 KY, while a common disease ($f_0 = 0.01$ or 0.16) still has a small effective number of alleles even after 100 KY. Vertical dotted lines: likely time range of human expansion (18–150 KYA). Horizontal dotted line: cutoff for simple allelic spectrum, which has an effective number of alleles < 10. From Reich and Lander, 2001.

Alzheimer disease (AD) is the most common form of dementia, being rare before the age of 65 years but affecting about 10% of people over 65 and almost 50% of those over 85, and costing around $100 billion a year in the USA. Before modern sanitation, nutrition and health care increased longevity, it was rare, but now it is a major, and increasing, health problem in many countries. Linkage analysis and association studies both identified a variant of the apolipoprotein E (*APOE*) gene, *APOE*E4*, defined by arginine instead of cysteine residues at two amino acid positions, 112 and 158, as a significant risk factor. For example, *APOE*E4* was present in 31% of controls but 64% of sporadic late onset cases in one survey, and dosage of this allele was also important: homozygosity for *APOE*E4* was virtually sufficient to cause AD by the age of 80 (Corder *et al.*, 1993). In a compilation of data from 43 populations (> 30 000 individuals), the mean population frequency of *APOE*E4* was 17% (Corbo and Scacchi, 1999): thus it can certainly be considered a common variant affecting susceptibility to a common disease. The *APOE*E4* allele is ancestral (showing the same amino acids at the two variant positions as chimpanzee), and has been suggested to contribute to the 'thrifty genotype' (see Section 13.4.4 and *Box 11.7*) through its involvement in lipid metabolism. Resequencing of a ~ 5.5-kb region in 96 individuals (192 chromosomes) from four populations revealed a fairly normal level of variation ($\pi = 5 \times 10^{-4}$; see *Table 6.1*), although many variant sites exhibited **homoplasy** (Fullerton *et al.*, 2000). Little evidence for departure from neutrality was found, and coalescent-based dating analysis (see Section 6.6.2), which required exclusion of the sites showing homoplasy and assumed neutrality, suggested an age of 220 KYA (120–440 KYA) for the amino acid change leading from the *APOE*E4* allele to the other alleles: a relatively recent time. Whether or not the increase in frequency of derived *APOE* alleles has been due to selection, it has spared the modern human population from an even higher incidence of AD.

A second example is provided by type 2 diabetes susceptibility. One variant associated with susceptibility is the Pro12Ala substitution in the peroxisome proliferator-activated receptor-γ (PPARγ) protein, a receptor involved in adipocyte differentiation and thus discovered by testing candidate genes. The Ala allele was common (~ 85%) and was associated with a protective affect against diabetes leading to a relative risk of ~ 0.8 (Altshuler *et al.*, 2000). While this may seem small, the authors point out that if this estimate were reliable and the Ala allele was fixed in the population, the prevalence of type 2 diabetes would be 25% lower. A second locus associated with type 2 diabetes susceptibility in Mexican Americans is in the cysteine protease gene calpain-10 (*CAPN10*) (Horikawa *et al.*, 2000). This gene was not previously considered a candidate because it did not have a known functional link to diabetes, but a genotype defined by two intronic SNPs and an indel was associated with susceptibility and proposed to explain 14% of the type 2 diabetes risk in Mexican Americans and 4% of that in Europeans. The frequencies of the variants defining this risk genotype have been measured in world populations, and the rare alleles were present at 17%, 29% and 47% in a sample of more than 1100 individuals (Fullerton *et al.*, 2002). Biological understanding of the basis of this association remains incomplete – how is expression of the gene altered in carriers of the susceptibility haplotype, and how does this lead to manifestation of the disease? – but the variants identified are common.

The third example that will be considered is Crohn disease (OMIM 266600), a chronic inflammatory disorder of the gastrointestinal tract with a prevalence estimated at ~ 0.1% in Western populations. Positional cloning has identified *CARD15* (Caspase recruitment domain protein 15, also known as *NOD2*, nucleotide-binding oligomerization domain protein 2) as a susceptibility gene (Hugot *et al.*, 2001; Ogura *et al.*, 2001). Its involvement is supported by its biochemical function: the C-terminal region of the protein interacts with lipopolysaccharides from bacteria and activates NF-κB (nuclear factor κB) as part of an immune response pathway; this defect in innate immunity could then lead to the inflammation and tissue damage characteristic of Crohn disease. Several coding variants have been identified that change the sequence of this region of the protein (Arg702Trp and Gly908Arg), or truncate it (Leu1007fsinsC; fsins = frameshift insertion). These variants share the properties that they are defective in activating NF-κB and present at increased frequency in Crohn disease patients. Their frequencies in normal individuals from several populations have been summarized (Bonen and Cho, 2003) and examples are shown in *Figure 14.9*. They were present at < 6% in unaffected individuals from populations of European origin, and were entirely absent from a Japanese sample of more than 900 individuals (including both patients and controls; Crohn disease has an incidence of about 5 per million in Japan). They would therefore not be classified as common variants according to Reich and Lander's criterion.

Thus the limited empirical evidence on the allelic architecture of susceptibility genes for common disease is equivocal: common variants are known at some genes, but at others there appears to be many rare variants. It is unclear whether AD or Crohn disease provides the more typical example. Additional information is likely to become available in the near future and it will then be possible to form a more complete view.

How many markers are needed? The number of markers needed to carry out whole-genome LD mapping is an important determinant of the practicality and likely success of such projects. It depends on the extent of LD in the genome, which is only just beginning to be understood, but is already known to vary between chromosomal regions and populations. A landmark paper was published in 1999 by Kruglyak, who used simulations based on simplified assumptions about population history, mutational processes and recombination rates to predict the extent of LD expected. The population was assumed to have a constant size with an N_e of 10 000 until 100 KYA (5000 generations ago), and then to expand exponentially to a size of 5×10^9.

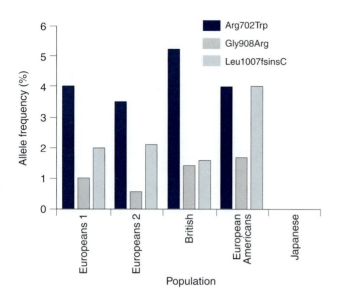

Figure 14.9: Complex allelic spectrum at the *CARD15/NOD2* gene, a susceptibility locus for Crohn disease.

All of the susceptibility alleles are present at low frequency (< 6%) in all the populations tested, and were entirely absent from the Japanese sample. Data from Bonen and Cho, 2003.

DNA variants were assumed to be neutral, to arise as unique events and not to undergo gene conversion. The recombination rate was set at 1% per megabase. LD between a common variant with 50% frequency and a nearby marker was then calculated (*Figure 14.10*). Levels of LD were low, and Kruglyak predicted that '*useful levels of LD are unlikely to extend beyond an average distance of approximately 3 kb in the general population. This result implies that roughly 500 000 SNPs will be required*' (Kruglyak, 1999).

Alternative models can give quite different predictions about the extent of LD, and Pritchard and Przeworski have presented simulations carried out using a number of different demographic scenarios (Pritchard and Przeworski, 2001). Their implementation of Kruglyak's model of extreme population growth produced very few marker pairs more than 10 kb apart in LD, as he had found. In contrast, the standard neutral model (also known as the Wright–Fisher model; Section 5.6) with a constant population size ($N_e = 10\,000$) predicted higher levels of LD, with a few marker pairs in LD across the entire ~ 1-Mb region used in the simulations, while models that incorporated population structure (Section 5.5) could lead to high levels of LD among many marker pairs. The human population has obviously grown in size, but it is also subdivided. It is thus difficult to know what the level of LD would be under more realistic scenarios: all the available models are over-simplified. Empirical data on LD is therefore important.

One relevant study has examined the region around the AD locus where, as we saw, the *APOE*E4* allele is a genetic susceptibility factor for the disease (Martin *et al.*, 2000). SNPs at known distances from the causal mutation were tested in

case–control association studies, and the results were highly variable (*Figure 14.11*). A SNP 16 kb away showed a very strong association, while one 8 kb away showed no significant association. An examination of 19 randomly chosen genomic regions found that, on average, LD extended 60 kb from a chosen common allele in a population of European origin, but less than 5 kb in an African population, that of the Yoruba from Nigeria (Reich *et al.*, 2001). Even within Europeans, there was great variation between loci, with one showing high LD for > 155 kb. Why is LD so variable? When it is expressed in terms of physical distance (as opposed to genetic distance), differences in recombination rate between different genomic regions can have a major effect, and we have seen that this varies at scales from that of the whole chromosome down to individual hotspots 1–2 kb in size (Section 3.7). Drift, selection and other forces can also have differential effects on different genomic regions within the same population (*Box 5.7*), while demography, as shown by Pritchard and Przewowski, has a major influence. Thus while there may be regions of the genome and populations where LD is high, the empirical results suggest that whole genome LD mapping studies based on **pairwise** LD between a disease and randomly chosen markers will require a very high density of markers and will thus remain impractical for some time to come.

There may, however, be an alternative approach. Haplotypes, as we have seen many times, provide more information than single markers. In all regions of the genome except mtDNA and the Y chromosome outside the pseudoautosomal regions, haplotypes are shuffled by recombination. What haplotype structure does this produce?

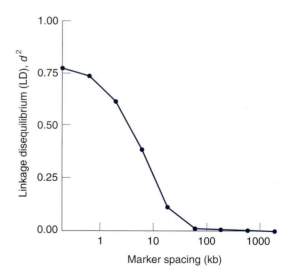

Figure 14.10: Kruglyak's prediction of linkage disequilibrium (LD).

LD was measured by d^2 between a variant and a marker [d^2 is similar to other measures of LD (*Box 3.11*); it is the squared difference between the frequency of the associated marker allele on variant and normal chromosomes]. Note that LD at distances > 10 kb is very low. From Kruglyak (1999).

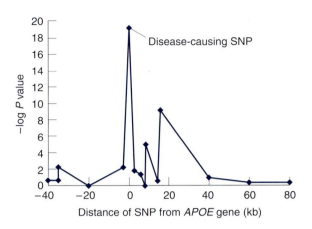

Figure 14.11: Association of SNPs with Alzheimer disease (AD).

P values for association of SNPs with AD in cases and controls are shown relative to their distance from the disease-causing SNP in the *APOE4* gene. Note the logarithmic scale of the *P* values. Data from Martin *et al.* (2000).

Haplotype blocks The idea that the genome is organized into **haplotype blocks** is currently a topic of great interest. It has two bases:

▸ SNPs have been typed in family and population samples, haplotypes assembled or deduced (see Section 4.10) and pairwise LD measures calculated (see *Box 3.11*). A number of criteria can then be applied to identify regions of high LD designated 'haplotype blocks';

▸ the distribution of recombination events has been mapped in great detail in part of the MHC locus by typing sperm DNA, revealing a highly **punctate** pattern of hotspots and cold regions (Section 3.7; Jeffreys *et al.*, 2001).

Note that these lines of evidence are not necessarily connected: recombination hotspots are likely to result in haplotype blocks, but blocks may arise in the absence of punctate recombination. Analyses of large sets of data have revealed that:

▸ there are several ways in which blocks can be identified from haplotype data, and they can identify different blocks; even the SNP chosen as the starting point can affect the location of the block boundaries;

▸ block size depends on the density of SNPs used: the fewer the SNPs, the larger the blocks. Other SNP characteristics, such as their rare allele frequency and how they were ascertained, may also be important.

Thus, while blocks undoubtedly exist, in the sense that methods can identify them, the mechanisms that produce them are unclear. Modeling shows that blocks can result from stochastic and population processes, and that hotspots and cold regions are not necessary. For example, coalescent simulations (Phillips *et al.*, 2003) could match the observed

block frequency spectrum on chromosome 19 well using a model that incorporated uniform recombination rates. This is an active area of research; some implications for genomewide association studies are independent of the mechanism of origin of blocks, but others will differ according to whether they are generated largely by stochastic and population processes, or by variable recombination rates.

▸ Since haplotype diversity within blocks is limited, with > 90% of chromosomes having one of five or fewer haplotypes within each block, many SNPs will be redundant: LD between them is so high that typing the whole set will provide no more information than typing one. Thus a minimal number of informative markers can be used to identify the common haplotypes in each block: these are known as **tag SNPs**. This can provide a considerable saving of genotyping effort and expense, which should be available whatever the origin of blocks.

▸ A disease susceptibility locus would be localized to a block, but there might be little information to localize it to a particular gene, if there were more than one gene within the block.

▸ Block structure may differ between populations. If blocks result largely from population processes, there may be little correspondence between the positions and sizes of blocks in different populations. However, if blocks result largely from recombination hotspots, there should be more correspondence between populations. A study looking at haplotype blocks in populations of African, Asian and European origin found correlations between block boundaries across populations, but also showed that block sizes differ markedly (Gabriel *et al.*, 2002). In contrast, a study of the MHC region in which hotspots were detected by sperm DNA analysis showed that the same blocks of LD are observed in three populations (British, Saami and Zimbabweans) having very different demographic histories (Kauppi *et al.*, 2003). This issue has important implications for association studies: tag SNPs identified in one population will not be useful in another if they lie in different blocks in the second population. Similarly, it may not be helpful to combine individuals from several populations if a susceptibility allele is embedded in a different block structure in the different populations.

A large-scale project, the HapMap (Haplotype Map) project, has now been initiated to determine the haplotype block structure of the genome (*Box 14.2*), with results due by October 2005.

Prospects for mapping common disease susceptibility genes There have been more papers published discussing the prospects for mapping common disease susceptibility genes than there have been reports of successfully finding them. Whole-genome case–control association studies remain in the future, but studies using family information to provide an approximate location, followed by evaluation of candidate genes, have been going on for several years. While there have

---

BOX 14.2 The HapMap project

At the end of October 2002, a project to determine the haplotype structure of the human genome was announced. It is a large-scale international endeavor modeled in several respects on the human genome sequencing project, estimated to take 3 years and cost around $100 million. The plan is to sample a total of 200–400 people from four populations: Yorubas (Nigeria, Africa), Han (China), Japanese, and European–Americans. According to their website, the participants expect that 65–85% of the genome is organized into haplotype blocks of 10 kb or larger, and aim to identify 300 000–600 000 carefully selected tag SNPs that will allow this block structure to be characterized. Subsequent association studies can then use only these tag SNPs to scan most of the genome.

Further information is available from the website: http://genome.gov/page.cfm?pageID=10001688

---

been some successes, why have these been so few, and what are the prospects for better success in the future?

Allelic architecture, as we have discussed, provides one possible explanation for the low success rate: if there are large numbers of low-frequency susceptibility alleles at most loci, these will be difficult to detect. It is also possible that larger population samples are needed, and this requirement is being addressed by the establishment of some large-scale projects, involving up to a million subjects (*Box 14.3*). A meta-analysis of published association studies concluded, optimistically, that '*a sizable fraction (but under half) of reported associations have strong evidence of replication*' (Lohmueller *et al.*, 2003).

One further explanation for the difficulties in identifying susceptibility genes could be that there are additional complexities in their basis. A possible example was provided by the work of Gratacòs *et al.* (2001), on panic disorder, a condition that can affect 1.6–3% of the population. These researchers identified a region of chromosome 15q24–26 (DUP25) that was duplicated in 7% of control individuals and 97% of unrelated patients with both panic disorder and agoraphobia, although the frequency of DUP25 in affected individuals depended on which aspects of the phenotype were considered. Up to this point, the findings fit the CD/CV model, but further aspects do not. The sequences included in the duplicated region, which was always mosaic, differed between individuals and even between different cells from the same individual; furthermore, the inheritance of DUP25 was non-Mendelian. This paper was notable for its lengthy and thorough review, taking almost 3 years between submission and acceptance, and if substantiated would indicate a novel form of genetic influence on a common

---

BOX 14.3 Large-scale projects to map genetic disease susceptibility loci: deCODE and its successors

The health and commercial implications of identifying susceptibility genes for common diseases, coupled with the difficulties encountered so far, have led to the establishment of some large-scale projects in this area (see *Table*).

The first of these was deCODE, a commercial company in Iceland which attracted attention and concern because of the whole-population scale of its work and, particularly, its assumption of the 'presumed consent' of participants, a nebulous concept allowing an individual's details to be included in a database unless they opt out (Annas, 2000; Gulcher and Stefansson, 2000). Other studies are relying on the more traditional informed consent, and together these initiatives provide the promise that one limitation of previous work, at least, the limited numbers of participants, will be overcome. Their progress will be followed with great interest.

EXAMPLES OF ESTABLISHED AND PROPOSED LARGE-SCALE PROJECTS AIMED AT MAPPING SUSCEPTIBILITY GENES FOR COMMON DISEASES.

Country	Project	Participants	Start	Further information
Iceland	deCODE	270 000	1999	http://www.decode.com/
		Whole population[a]		
Estonia	Estonian Genome Project	1 000 000	2003	http://www.geenivaramu.ee/
		~ 70% of population		
UK	UK Biobank	500 000; age 45–69 years	2004	http://www.biobank.ac.uk/
		< 1% of population		

[a]Individuals can opt out.

disease. Unfortunately, reports that the bizarre cytogenetic findings could not be replicated, even in positive control cell lines, have now appeared from highly experienced cytogeneticists (Tabiner *et al.*, 2003) and the status of the original observations is unclear. Is this just another peculiar episode in the search for genes affecting mental disorders? The debate continues. A less controversial complication in the genetics of susceptibility to common disease could be modification of the disease phenotype by **epigenetic** factors: inherited differences that do not involve changes to the DNA sequence (Dennis, 2003). One example of these is DNA methylation (Section 3.2.5), and large-scale studies of methylation patterns have been initiated by the Human Epigenome Consortium (Beck *et al.*, 1999). The significance of any of these factors in susceptibility to common disease is far from clear, but, if any were large, the role of evolutionary history could be less important.

14.4 The imprint of prehistoric infectious disease on patterns of modern genetic diversity

Infectious diseases are likely to have been major causes of mortality for much of human evolution, and, over time, changes in the environment, human demography (e.g., increasing population densities; see Chapter 10) and host–disease interactions have significantly altered the disease spectrum. Disease mortality and thus reproductive success has probably been influenced by an individual's genotype. Consequently, some aspects of modern patterns of diversity have been determined by prehistoric diseases. The clearest examples are provided by malaria, which even now affects 500 million people each year and kills some two million. The disease can be caused by infection by *Plasmodium* parasites belonging to any of four species: *P. falciparum, P. vivax, P. malariae* and *P. ovale*, but *falciparum* malaria is the most lethal. We have encountered the selective pressure that malaria has applied to human populations several times (*Table 14.4*): the heterozygous advantage of the sickle cell trait and unbalanced globin levels have led to large increases in frequency of the alleles underlying these variants, reaching an equilibrium level maintained by the disadvantage to the corresponding homozygotes. Where the homozygous state is not disadvantageous, as in the case of the Duffy O allele and resistance to *vivax* malaria, near-fixation of the variant has occurred in exposed populations (Section 9.4.3). Several other examples of genetic variants associated with malaria resistance have been identified, including at the *G6PD* and *CD40L* (*TNFSF5*) loci. Studies of the malaria parasite population genetics, and that of their vector hosts, are also important avenues of research, since they may be co-evolving with humans (Joy *et al.*, 2003). Here, we will consider the effect that malarial selection for *G6PD* deficiency has had on the gene and its surrounding sequences, and then a probable example of the result of selection by an unknown disease at the *CCR5* locus.

14.4.1 Selection at the *G6PD* locus

Glucose-6-phosphate dehydrogenase (G6PD) is a **housekeeping** enzyme (found in all cells), the gene for which is located on the X chromosome which catalyzes the reaction of glucose-6-phosphate with NADP to form 6-phosphogluconate and NADPH in a pathway leading eventually to the production of ribose. The production of NADPH is of particular significance because this reduced (i.e., NADPH instead of NADP) form is necessary to avoid oxidative damage in cells, and G6PD is the only enzyme in red blood cells that can produce it. Complete deficiency of G6PD is unknown, and is probably lethal, but reduced enzyme activity is associated with both resistance to *falciparum* malaria and a number of pathological conditions affecting some 400 million people worldwide, or around 6% of the human population. The resistance to malaria has been demonstrated in two ways:

▶ epidemiological studies have shown that in heterozygous females and hemizygous males, the risk of severe malaria is reduced by about 50% (*Table 14.5*; female homozygotes were not evaluated in this study because they were too rare);

▶ *in vitro* studies have shown that *P. falciparum* grows poorly in G6PD-deficient red blood cells, at least until it adapts by producing its own G6PD.

It is thought that *Plasmodium* depletes the red blood cell of NADPH, leading to peroxide-induced hemolysis and thus loss of a few cells, but curtailment of the parasite's development. Unfortunately, there are other mechanisms that can also increase oxidation in red blood cells, and these may lead to a general hemolytic anemia. They include infection by other agents, which may have been the most important factor in prehistoric times, and exposure to some foods such as broad beans (*Vicia fava*) which leads to an acute hemolytic anemia called favism, and drugs such as primaquine. These disadvantages of G6PD deficiency explain why it is only encountered at significant frequency in association with past or present exposure to malaria. Nevertheless, if the selective advantage of homozygous G6PD-deficient females were similar to that of heterozygotes, deficiency alleles would be expected to rapidly reach fixation in exposed populations. Although they are found at frequencies of up to 25% in some populations, these fall short of fixation, suggesting either that homozygous females are at a disadvantage, or that the selective pressure

TABLE 14.4: THE INFLUENCE OF SELECTION DUE TO MALARIA ON THE HUMAN GENOME.

Subject	Section
Sickle cell trait, disease	5.4.1
	14.2.1
Duffy blood group	9.4.3
G6PD	14.3.1

varies over time or space. Thus, although some aspects of the selection remain to be elucidated, *G6PD* provides one of the clearest examples of selection in the human genome, and we will assume for the rest of this section that deficiency alleles have been selected because they confer resistance to malaria.

We will now examine variation in and around the *G6PD* gene and ask what traces of selection it reveals. The ancestral *G6PD* allele, judged by comparison with ape sequences, is B (*Figure 14.12*). Several hundred G6PD deficiency alleles have been identified, but few of them are common. In the Mediterranean region, the Middle East and India, the Med allele (Ser188Phe) with ~ 3% of B enzyme activity is present at a frequency of ~ 2–20% in different populations, while in sub-Saharan Africa, the A and A– alleles are found. The A allele (Asn126Asp) can represent up to 40% of a population and has ~ 85% of B activity, but does not protect against severe malaria and should not be considered in the same way as other deficiency alleles; it remains unclear whether it conferred a selective advantage in the past. The most studied deficiency allele in sub-Saharan Africa is A– (Val68Met on an A background), which has ~ 12% of B activity, and can be present at a frequency of up to 25%. The patterns of variation associated with these alleles have been investigated in several ways; here, we will concentrate mainly on those associated with the A– and A alleles in Africa.

Two studies have investigated the sequence variation within the *G6PD* gene, re-sequencing about 5 kb in 47 (Saunders *et al.*, 2002) and 216 (Verrelli *et al.*, 2002) males. Haplotypes could be determined directly from the single copy of the X chromosome, and little evidence for recombination or recurrent mutation was found, so the haplotypes formed a simple network with a single branch carrying the A and A– alleles (*Figure 14.12*). Diversity of the A and A– alleles was reduced compared to the B alleles as measured by π (*Table 14.6*); indeed 16 out of the 17 A-alleles examined in one of the studies had the same sequence, and the seventeenth differed at a single position (Verrelli *et al.*, 2002). Similarly, haplotype diversity measured using three microsatellites within 20 kb of the gene (*Figure 14.12*) was also reduced: 0.96 ± 0.02 for African B alleles, 0.91 ± 0.04 for African A alleles, and 0.72 ± 0.08 for African A-alleles (Tishkoff *et al.*, 2001).

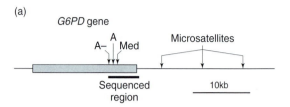

(a)

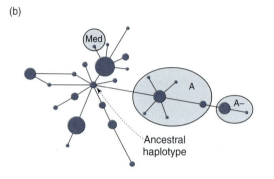

(b)

Figure 14.12: Variation of the *G6PD* gene.

(a) Structure of the gene showing location of the variants and the resequenced region. The gene (including both exons and introns) is represented by a blue box. A, A– and Med are alleles defined by coding-region variants; A– and Med have greatly reduced enzyme activity. (b) Network of haplotypes defined by sequence variants. Haplotypes are shown as circles with an area proportional to their frequency; mutations are shown as lines representing one, two or three mutations according to their length. Most haplotypes belong to the ancestral, B allele, class, and the ancestral haplotype is indicated. The derived A, A– and Med alleles are indicated. Redrawn from data in Verelli *et al.* (2002).

Low diversity accompanied by relatively high frequency is expected if there has been recent positive selection for an allele (see Section 6.3.5), but could also arise by chance. Do the diversity measurements associated with the A– alleles provide evidence for recent positive selection? Tishkoff *et al.* (2001) carried out coalescent simulations in which the expected distributions of statistics such as the number of

TABLE 14.5: G6PD DEFICIENCY PROTECTS AGAINST SEVERE *FALCIPARUM* MALARIA.

	Controls			Severe malaria cases		
	n	**G6PD A– (n)**	**G6PD A– (%)**	**n**	**G6PD A– (n)**	**G6PD A– (%)**
Female heterozygotes	325	64	19.7	388	42	10.8
Male hemizygotes	388	42	10.9	396	17	4.3

Severe malaria cases from Gambia and Kenya were genotyped for the G6PD deficiency allele (A–) and the frequency was compared with that in controls from the same populations. For both female heterozygotes and male hemizygotes, the frequency of the *G6PD* A– allele is reduced to about half in cases, demonstrating its protective effect. Data from Ruwende *et al.*, 1995.

microsatellite alleles and their variance, given the observed frequency of the A− allele, were determined. They concluded that the observed variability was often significantly less than expected, and thus forces other than drift (i.e., selection) must have contributed to the rapid expansion of this allele.

Times of origin or spread of the deficiency alleles are thus of interest since they can be compared with information from other sources (e.g., archaeology), either to evaluate the hypothesis of malarial selection, or (if this is assumed to be true) to assess the reliability of the genetic calculations. Estimates of the TMRCAs of subsets of the chromosomes have been obtained in several ways (*Table 14.7*). Verrelli *et al.* (2002) used coalescent-based dating to calculate that the TMRCA for the entire African sample was 620 (480–760) KYA, a time that is well within the range expected for a neutral X-linked segment of DNA. The coalescence time for A alleles was also ancient (316 ± 244 KYA), but that for the A− alleles was more recent (45 ± 20 KYA). Saunders *et al.* (2002) also estimated the age of the A− allele from their data: coalescent-based dating gave 11 ± 9 KYA, while an analysis of the LD pattern (discussed further below) suggested 1.5–21 KYA. Their dating method assumes neutrality and is thus likely to overestimate the age of the A− allele. As usual, there are very large uncertainties associated with these times, but those for the A− allele are clearly recent, and thus consistent with malarial selection and with the idea that meaningful (although imprecise) estimates are obtained from genetic data. The ancient date for the A allele suggests that it is neutral, or at least not involved in resistance to malaria.

These investigations suggest that malarial selection has markedly influenced the pattern of variation within the *G6PD* gene and at flanking sequences up to 18 kb away, but how far along the chromosome can the influence be detected? An analysis of a ~ 450 kb region spanning the *G6PD* gene has been carried out (Sabeti *et al.*, 2002). This study first identified core haplotypes defined by 11 SNPs in a 14-kb region near the gene, including the SNP defining the A− allele. SNPs at known distances from the core haplotype were then added, and the LD between the core haplotype and each SNP was calculated, defining the measure 'extended haplotype homozygosity' (EHH; see Section 6.3.5). EHH is the probability that two randomly chosen chromosomes carrying a particular core haplotype are identical, as shown by homozygosity at all SNPs. EHH could be visualized as a bifurcating tree, where a haplotype with unusually low variation should be seen as a long thick branch. The haplotype associated with the A− allele did indeed stand out in this analysis (*Figure 14.13a*): its EHH value at each distance from the gene was the highest of any of the nine haplotypes examined (*Figure 14.13b*). To assess the significance of this finding, simulations of the results expected under neutral conditions were carried out, and the probability of obtaining such a result under neutrality was estimated at < 0.0006 to < 0.0008 according to the demographic model used. It thus seems that recent selection for malarial resistance has influenced diversity over a large region of DNA covering > 400 kb in some populations.

14.4.2 Selection at the *CCR5* locus

AIDS is the most devastating new infectious disease to affect the human population in recent times, with 42 million people reported to be infected by HIV at the end of 2002, approaching 1% of the world population. It is most prevalent in parts of Africa, with adult HIV infection levels of over 30% in Botswana (38.8%), Lesotho (31%), Swaziland (33.4%) and Zimbabwe (33.7%) (http://hivinsite.ucsf.edu/InSite.jsp?page =Country). Genetic susceptibility to HIV infection and progression to AIDS vary, with a few individuals showing significant resistance. One of the first resistance loci to be identified was the chemokine receptor *CCR5*, a cell surface molecule which acts as a cofactor for HIV entry into T cells (Dean *et al.*, 1996; Samson *et al.*, 1996). Dean *et al.* (1996) for example, tested 156 at-risk individuals for association between 170 candidate loci and HIV antibody status, and found one significant association: with a partial deletion mutant of *CCR5*, *CCR5Δ32*, which removes 32 bp from within the gene introducing a frameshift and resulting in a nonfunctional protein. Study of a larger cohort of 1955 individuals at risk of HIV infection revealed that while about 69% of *CCR5* +/+ or heterozygous individuals were HIV positive, indicating high levels of exposure of the group, no (0/17) homozygous *CCR5Δ32/ CCR5Δ32* individuals were positive: a highly significant difference (*Figure 14.14*). While heterozygotes did not show a different level of infection, they did take 2–3 years longer to progress from seroconversion (becoming antibody-positive) to AIDS. Thus the *CCR5Δ32* allele provides strong protection against AIDS. No obvious pathology is associated with a homozygous lack of the *CCR5* gene product.

It was apparent from these first findings that *CCR5Δ32* was present at remarkably high frequencies in some populations of European origin, and at much lower

---

TABLE 14.6: DIVERSITY MEASUREMENTS OF A 5.2-KB SEGMENT OF DNA WITHIN THE *G6PD* GENE.

Sample	n	Diversity ($\pi \times 10^4$)
Global B	165	5.5
A	32	2.0
A−	17	0.3
African B	112	5.1
A	32	2.0
A−	16	0.0
Non-African (all B)	53	3.0

Measurements show the value of π (average number of pairwise differences per site), based on silent-site SNPs (from Verrelli *et al.*, 2002).

TABLE 14.7: ESTIMATES OF THE AGE OF THE *G6PD* A– ALLELE.

Age (95% CI), ± SD or range (KY)	Basis	Selection	Generation time (years)	Reference
6 (4–12)	Microsatellite diversity	Yes		Tishkoff *et al.*, 2001
11 ± 9	Sequence diversity	No	25	Saunders *et al.*, 2002
1–21	LD decay	No	25	Saunders *et al.*, 2002
45 ± 20	Sequence diversity	No	20	Verrelli *et al.*, 2002

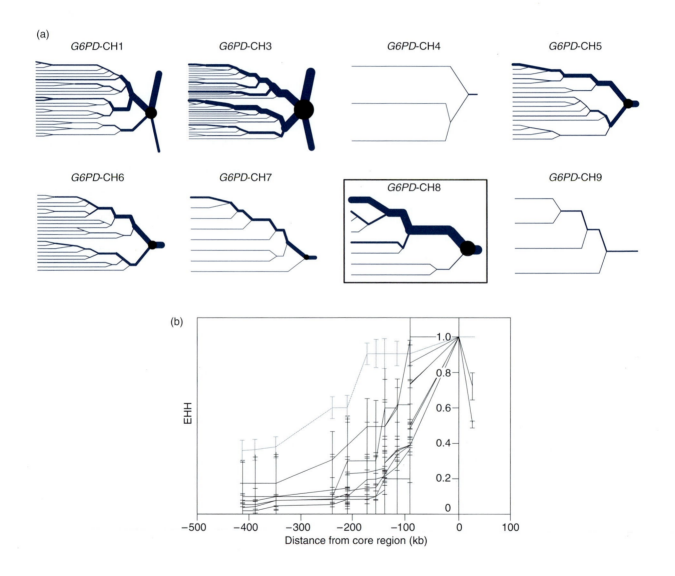

Figure 14.13: Haplotype structure of a ~ 400-kb region around the *G6PD* gene.

(a) Core haplotypes CH1–9 were defined by 11 SNPs within a 15 kb-segment of DNA containing the *G6PD* gene and the branching diagrams represent the extension of these haplotypes into the flanking sequences. Each bifurcation corresponds to two alternative haplotypes, so that thick branches represent common haplotypes. CH8 (boxed) contains the A– allele of *G6PD* conferring resistance to malaria and is associated with a thick branch over the entire > 400-kb region, representing an extended haplotype that has increased in frequency because of selection. (b) Extended haplotype homozygosity (EHH) of the core haplotypes at each distance from the core region. CH8 is shown in blue. From Sabeti *et al.* (2002).

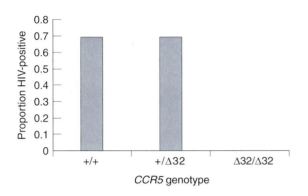

Figure 14.14: Protective effect of the *CCR5Δ32* allele against HIV infection.

The proportion of a sample of at-risk individuals showing infection by HIV is greatly reduced if they carry the *CCR5Δ32/ CCR5Δ32* genotype. From data of Dean *et al.* (1996).

frequencies in other populations. Large-scale studies of its distribution have now been carried out, approaching the scale used for classical markers, and a recent review summarized worldwide results from over 26 000 individuals (Dean *et al.*, 2002; *Figure 14.15*). The highest allele frequencies (~ 0.15) were found in northwest Asia and northern Europe, while on other continents (excluding recent population movements) the frequency was usually < 0.01 and could largely, or entirely, reflect recent European admixture. Although there has been controversy about the origins of HIV and AIDS, it is agreed that they have only appeared in the human population in the last few decades. The high frequency and wide distribution of *CCR5Δ32* alone are sufficient to exclude AIDS as the selective agent,

and more detailed studies of the origin of this allele have suggested that the mutation must have arisen several hundred years ago, or more. Stephens *et al.* (1998) estimated the time of origin of the *CCR5Δ32* allele from the extent of LD with two flanking microsatellites. Significant LD was found between the two microsatellites and *CCR5Δ32*, suggesting a single origin for the deletion, with ~ 85% of deletion chromosomes sharing a single haplotype. This was also the most common haplotype in nondeletion chromosomes but only represented ~ 36% of these. *CCR5Δ32* haplotypes which differed from the consensus haplotype were considered to have arisen by mutation and/or recombination, and a combined estimate for the rate of these was obtained using general measurements of microsatellite mutation rates and a translation of the authors' **radiation-hybrid distances** into recombination distances. These considerations suggested that it would have taken about 27.5 (11–75) generations or around 700 (300–1900) years for the ~ 15% of variant haplotypes to develop. This estimate is highly dependent on the value of the combined (mutation + recombination) rate used, which is uncertain; indeed, the map order of these loci suggested by the current DNA sequence differs from the order used in the calculations of Stephens *et al.* (1998). In addition, the estimation of local recombination rates from regional chromosomal averages is extremely sensitive to perturbations caused by recombination rate heterogeneity that may be a general feature of the recombination process (see Section 3.7). Nevertheless, this age estimate, and the geographical distribution – wide within Europe and Western Asia but globally restricted – suggest an origin within the last few thousand years, but before a few hundred years ago.

The question of whether or not selection has acted to increase the frequency of the *CCR5Δ32* allele has been

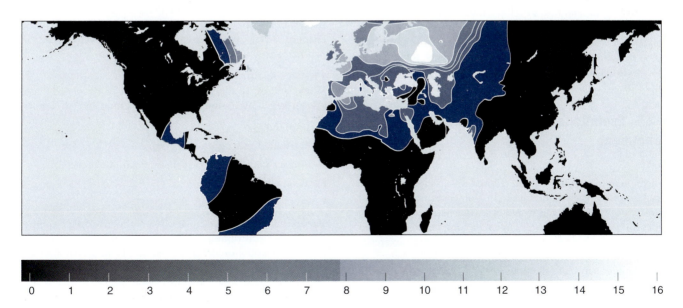

Figure 14.15: Worldwide distribution of the *CCR5Δ32* allele.

Land colors represent the interpolated frequencies of the *CCR5Δ32* allele, ranging from 0 (black), through shades of blue to > 14% (white), according to the scale at the bottom. From data compiled by Dean *et al.* (2002).

addressed more formally by considering the relationship between the frequencies of the *Δ32* allele and the linked microsatellite alleles (Slatkin and Bertorelle, 2001) (see Section 6.3.5). Using the haplotype data and mutation and recombination rate estimates of Stephens *et al.* (1998), Slatkin and Bertorelle (2001) calculated the chance of finding the *CCR5Δ32* haplotype at the observed high frequency with such a low degree of variation, for each microsatellite separately under neutrality. These probabilities were very low: 3.2×10^{-9} or 2.2×10^{-12}. Changing the parameter values, which are only known very approximately, by a factor of 10 still led to a convincing rejection of neutrality, strongly suggesting that the *Δ32* allele has been selected in European populations, with a selection coefficient of > 0.2, which would be extremely high. The identity of the selective agent is still unknown, but might be found using information from the current allele distribution (*Figure 14.15*) and historical records or analysis of human remains for pathogens using ancient DNA techniques.

14.4.3 The importance of infectious disease in shaping genome diversity

Past selection has thus had a substantial influence on some allele frequencies in modern populations, and this can extend several hundred kilobases into the flanking sequences. The change in allele frequencies can also have important health implications. One lesson to note is that the same allele of the same gene can be a target of selection at different times by different agents: *CCR5Δ32* by an unknown organism hundreds of years ago and HIV today. Attempts to explain the patterns we find by a single episode of selection may sometimes be over-simplistic.

An exciting challenge for the future is to recognize these patterns in the DNA without prior knowledge of the selective agent, and thus develop a more complete picture of the importance of selection in influencing genome variation and the forces that have shaped modern humans. Targets of past selection also provide potential targets for medical intervention, and so may be of interest to the pharmaceutical industries.

14.5 Evolutionary history and medical treatment

One goal of some medical geneticists is 'personalized medicine': treatments that are tailored to the specific genetic make-up of each individual. In some future scenarios, everyone would be genotyped at a large number of loci to identify their genetic susceptibility to complex diseases, and would then modify their lifestyle using diet, exercise, medication and other means to reduce their chance of developing any of the diseases to which they were predisposed. This prospect, especially in its last parts, is far from becoming a reality, and it is unclear to what extent it is desirable. There are ethical considerations: some genetic screening tests, like that of newborns for PKU are generally, if not universally, considered desirable, but not all are seen

in the same way. PKU testing is acceptable because the consequences of genetically susceptible individuals failing to modify their diet are well understood, severe and inevitable, and modification makes a large difference. The advantages of testing when the development of the disease is uncertain and cannot readily be treated or altered are less obvious: who wants to know that they have a 70% chance of developing Alzheimer disease by a certain age, and can possibly reduce this to 60%? This subject will be increasingly debated as more susceptibility genes are identified.

In one area, however, there are clear benefits from determining genetic variation and applying the information to medical treatment: in individual variation in response to drug treatment. The response, both therapeutic and adverse, to the same dose of a drug can vary enormously between different people, and some of this variation is genetically determined. This in itself makes the subject of interest to evolutionary geneticists, but there is an additional factor: response often differs between populations. An understanding of evolutionary genetics can contribute to understanding the current distribution of this variation, and this field is now receiving much attention. Furthermore, this variation may well reflect selection from past environmental agents such as foods (see Section 13.4.4), and may thus be a rich source of insights into our evolutionary past, an area that has not received the attention it deserves.

14.5.1 Pharmacogenetics

Pharmacogenetics is the study of individual genetic variation in response to drugs, and is related to pharmacogenomics, which takes a more global view of the effects of drugs on patterns of gene expression; here, we will mainly be concerned with the former. '*If it were not for the great variability among individuals medicine might as well be a science and not an art*' wrote Sir William Osler in 1892, a view that is still held today; but some optimists think that medicine may be changed from an art to a science by developments in pharmacogenetics (Roses, 2000).

Processes affecting drug response
When a drug is administered, its fate can be divided into several stages (Weinshilboum, 2003). It must be:

▸ absorbed;

▸ distributed to its site of action;

▸ interact with its targets such as enzymes or receptors;

▸ metabolized;

▸ excreted.

Variation in any of these processes can influence drug response; absorption, distribution and excretion are sometimes collectively known as drug 'disposition'. Examples of variation in genes involved in transport, such as ABCB1 (ATP binding cassette B1; also known as MDR1, multidrug resistance 1), and drug targets such as the β_2-adrenoreceptor, variation of which affects response to β_2-**agonists**, are known (Evans and McLeod, 2003). However, variation in metabolism has received most attention and will be discussed here.

Cytochromes P450 (CYPs)

The cytochrome P450 (*CYP*) superfamily of genes is large and widely distributed among living organisms: humans are estimated to have 57 *CYP* genes and 33 pseudogenes belonging to 18 different families (Nebert and Russell, 2002). While this is a large number, it should be borne in mind that the compact genome of the plant *Arabidopsis thaliana* has some 249 *CYP* genes, about 1% of its total gene complement. These gene products metabolize a wide variety of substrates (*Table 14.8*), and appear to be subject to rapid evolutionary turnover; the mouse genome exhibits a surprisingly different repertoire of *CYP* genes to that of humans (Waterston *et al.*, 2002).

Modifications to the substrates include oxidative, peroxidative and reductive changes. The main families involved in metabolizing foreign (exogenous) chemicals in humans are *CYP1*, *CYP2* and *CYP3*, with some involvement of *CYP4*. Between them, they contain 35 genes: nearly two-thirds of the total in the superfamily. An indication of their relevance to genetic medicine is provided by the proportion of drugs metabolized by some members (*Table 14.9*), together with the finding that several of them are highly polymorphic. Among them, the *CYP2* family metabolizes over half of all frequently prescribed drugs, and has received considerable attention. We will consider one example from this family, *CYP2D6*.

CYP2D6 *CYP2D6* metabolizes around a fifth of drugs, and is highly variable in its enzyme activity between individuals. Examples of its importance (Rogers *et al.*, 2002) include the following:

▸ Codeine is a prodrug that must be metabolized into an active form, morphine, to produce pain relief. *CYP2D6* catalyzes this process, and consequently patients with low activity find that codeine is not an effective analgesic and, in demanding more, may be suspected of drug-seeking behavior.

▸ Tricyclic antidepressants (e.g., nortriptyline) have a narrow therapeutic range, so administration of a

TABLE 14.8: SUBSTRATES METABOLIZED BY MEMBERS OF THE CYTOCHROME P450 FAMILY.

Endogenous substrates	Exogenous substrates
Fatty acids	Drugs
Eicosanoids	Environmental pollutants
Sterols and steroids	Natural plant products
Bile acids	
Vitamin D$_3$ derivatives	
Retinoids	
Uroporphyrinogens	

Data from Nebert and Russell, 2002.

TABLE 14.9: PROPORTION OF COMMERCIALLY AVAILABLE DRUGS METABOLIZED IN PART BY SELECTED CYTOCHROME P450 FAMILY MEMBERS.

Enzyme	Percentage of drugs metabolized
CYP1A1/2	11
CYP2A6	3
CYP2B6	3
CYP2C8/9	16
CYP2C19	8
CYP2D6	19
CYP2E1	4
CYP3A4/5	36

Data from Rogers *et al.* (2002).

standard dose to ultra-rapid metabolizers (see *Figure 14.16* for an explanation of the terminology used for metabolizers) may have little therapeutic effect, while poor metabolizers may experience adverse effects such as abnormally rapid heartbeat and fatigue.

▸ Anti-arrythmic agents that regulate heart contraction (e.g., propafenone) can be insufficiently metabolized by poor metabolizers, leading to side effects including visual blurring, nausea and abnormal skin sensations.

These and two further properties make this gene particularly interesting to the evolutionary geneticist:

▸ There is little post-transcriptional regulation, so an individual's enzyme activity depends in a simple way on the structure of their *CYP2D6* genes.

▸ *CYP2D6* variation is strongly structured by geography, and thus evolutionary history (see below).

Much is known about the genetic basis of *CYP2D6* variability. This includes variation in copy number between zero and at least 13 per individual, amplified in a tandem array (Johansson *et al.*, 1993). One consequence of this copy number variation is that the sequence of the gene was missing from the original version of the public human genome sequence because the bacterial artificial chromosome (BAC) covering the region was derived from a chromosome lacking the gene: there is no such thing as *the* human genome sequence or *the* number of human genes. The low degree of post-transcriptional control of *CYP2D6* means that different copy numbers of the gene lead to very different enzyme activities, and consequently different levels of drug *in vivo* after a standard dose (*Figure 14.17*; Dalén *et al.*, 1998).

In addition to variation in copy number, there are many variants within the gene which affect enzyme activity. These are given an allele number *CYP2D6*1*, *CYP2D6*2*, etc. according to an agreed nomenclature (http://www.imm.ki.

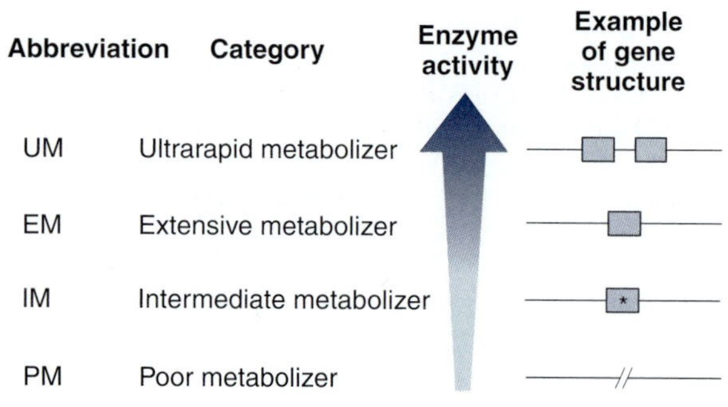

Abbreviation	Category	Enzyme activity	Example of gene structure
UM	Ultrarapid metabolizer		
EM	Extensive metabolizer		
IM	Intermediate metabolizer		
PM	Poor metabolizer		

Figure 14.16: *CYP2D6* phenotypes and examples of the genotypes associated with them.

The 'metabolizer' status reflects the CYP2D6 enzyme activity. 'Normal' individuals are categorized as 'extensive metabolizers', although the use of the term 'normal' is questionable. Much genetic variation is present within each category; UMs can have up to 13 copies of the gene, for example.

se/CYPalleles) and the data available on their frequencies in different populations have recently been summarized (Bradford, 2002). When alleles are grouped into broad functional classes according to the level of enzyme activity, and continentwide classes of populations are considered, substantial differences are seen, with Europeans standing out from Africans and Asians (*Figure 14.18A*): they have more inactive and 'normal' activity alleles, and less low activity alleles than the other populations. When single alleles are considered within individual populations, even larger differences are seen. For example, the functional allele *CYP2D6*1* was common in all the populations examined (Bradford, 2002; *Figure 14.18B*), varying less than threefold in frequency (23–57%); but an allele with reduced activity, *CYP2D6*10*, made up 50% of the Chinese sample and was common in several other East Asian populations, although it

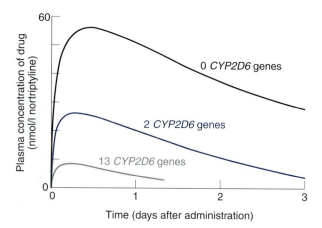

Figure 14.17: Effect of polymorphism in *CYP2D6* copy number on drug metabolism.

was rare (generally below 5%) in the African and European samples. Similarly, the non-functional allele *CYP2D6*4* was present at intermediate frequencies (commonly 10–20%) in European samples but was found at less than 10% in African samples and was absent from some Asian samples (*Figure 14.18B*). These variations are far wider than those found for neutral markers (Section 9.3) and are likely to reflect past episodes of differential selection.

Prospects Candidate genes that may be involved in modulating an individual's response to a drug are readily identified from studies of the biochemistry of the drug and many are already known. Variant proteins can be tested for alterations to biochemical function, or associations between genetic polymorphisms and different drug responses can be investigated. Thus association studies are much easier to carry out in the pharmacogenetic field than in investigations of susceptibility to common diseases, and have been more fruitful. If genetic typing is available to the clinician, the pharmacogenetic information can be put directly to use by modifying the choice or dose of drug used. It thus seems likely that pharmacogenetics will have a more immediate impact on medicine than studies of the genetics of common diseases; an evolutionary perspective can provide a sound background for it.

14.5.2 An evolutionary perspective in medicine

So far in this chapter, we have considered how differences between individuals, arising through neutral processes or selection, contribute to our health and disease. In addition, there are features of our evolutionary past with health implications that arose before the appearance of modern humans and are thus shared by all people. For example:

▶ **bipedalism** required large-scale changes to the skeleton and muscles, and lower back pain in humans is commonly explained as a side effect of our upright posture and locomotion;

▶ our large brain but narrow birth canal makes childbirth difficult. Some statistics show > 500 000 maternal deaths per year worldwide, with enormous variations between regions: a lifetime risk of one in 4000 in developed countries, but one in 54 in South Asia and one in 13 in sub-Saharan Africa (http://www.childinfo.org/eddb/mat_mortal/).

There is no obvious way in which evolutionary genetics can contribute to solving these problems, but there are other ways in which a general evolutionary perspective is important in medicine. This approach is described in *Box 14.4*.

Summary

▶ Genetic diseases, considered broadly as disorders that are influenced by an individual's genotype, can be divided into 'simple' diseases which show Mendelian inheritance in families, and 'complex' diseases which do not show such an inheritance pattern.

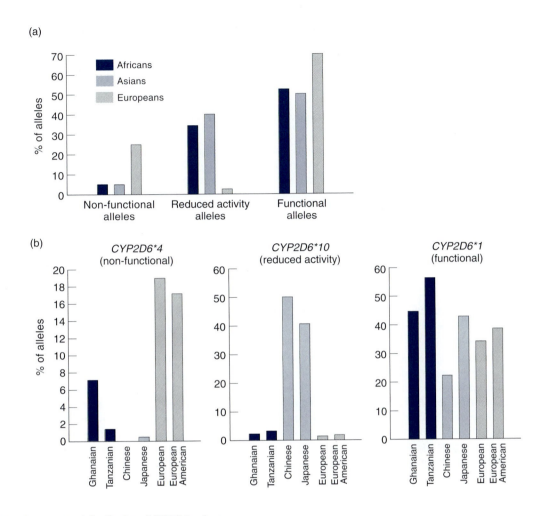

Figure 14.18: Geographical distribution of *CYP2D6* variants.

(a) Frequencies of alleles grouped into broad classes according to their enzyme activity in Africa, Asia and Europe. Note the distinct profile of Europeans. (b) Frequencies of selected alleles in six individual populations; note that the scales differ. All data are from the compilation of Bradford (2002)

▸ Simple genetic diseases are, by default, expected to be rare, evenly distributed in different geographical regions, and originate from many independent mutations. This is true for many of them.

▸ Exceptions to this pattern can, in part, be explained by a high mutation rate, selection, or late onset. In addition, drift of a susceptibility haplotype carrying a 'premutation' to high frequency in Europeans but not other populations is needed to explain the relatively high incidence and European distribution of Huntington disease. The high frequency and European specificity of cystic fibrosis can be accounted for by a selective advantage such as resistance to typhoid fever shown by the heterozygotes, with the disease manifested by homozygotes.

▸ Complex genetic diseases can be much more common than simple ones because selection against the disease alleles is weaker. Many of the major sources of morbidity and mortality in developed countries fall into this category. Some, such as schizophrenia, show a relatively uniform geographical distribution, but many, like type 2 diabetes, show large differences in prevalence rates. These geographical differences may owe their origin to either environmental and/or genetic heterogeneity among different populations.

▸ Many genes underlying Mendelian disorders have been identified by positional cloning; in some cases, this has been aided by knowledge of the evolutionary history of the population and choice of small isolated populations such as the Finns or Roma.

▸ It has been much more difficult to identify genes underlying susceptibility to complex diseases, despite enormous efforts motivated by their medical importance. Some successes have been achieved by combining family studies, linkage disequilibrium and biological information.

BOX 14.4 Darwinian medicine

Evolution is the foundation for biology, and biology is the basis for medicine, so you might well expect that doctors routinely try to understand disease in evolutionary perspective. But they don't. Doctors study development of the individual in extraordinary detail – how a cluster of a few cells develops into a mature phenotype. However, they don't study how our species developed from its ancestors, even though an evolutionary approach is equally essential to understand crucial questions such as why we have wisdom teeth, why we are left with an appendix that causes only trouble, why childbirth is along a route fraught with pain and danger, and why, if lucky, we get old and die.

Systematically applying evolutionary biology to the problems of medicine is surprisingly new. I think this is largely because few physicians have ever had a chance to learn about evolution. Even some who have had a course called 'evolutionary biology' have mainly learned population genetics or phylogeny, with little attention to the wonderful fine points about how natural selection works, why bodies are the way they are, the ubiquity of constraints and trade-offs, and how to study adaptation. In addition, two simple conceptual difficulties pose further obstacles. First, most physicians don't recognize the distinction between proximate and evolutionary explanations. It is beyond me how anyone can try to understand the body without knowing that a full explanation needs both a proximate explanation of how things work and also a complementary evolutionary explanation of how they got that way. Confusion on this point leads to endless unnecessary arguments. The other difficulty is that disease is abnormal and is not shaped by natural selection, so it is a bit hard to see how natural selection could be relevant. However, a slight change of focus makes evolution essential. The new question is why natural selection has left us vulnerable to disease. If lizards can re-grow a lost tail, why can't we re-grow a lost finger? If cell control mechanisms can keep each of several billion cells from dividing out of control, why do they sometimes fail, causing cancer? If fever can cause seizures, why is it still with us? And if pain is so unnecessary that we can use drugs safely to block it, why is it there in the first place? Finding strategies to address such questions has been the focus of most of my work on Darwinian medicine.

When I met George Williams in the mid-1980s, I had spent years trying to understand why selection did not eliminate genes that caused aging. Rates of aging are under genetic control. If they decrease fitness, they should be selected against. It seemed implausible that they could all be mutations whose effects were manifest so late in the life-span that they were not selected against. When I brought this up to Richard Alexander and other evolutionary biologists at Michigan, they laughed and asked what I thought of the 1957 paper by Williams on pleiotropy and senescence. This was in the days before electronic databases. I was not the only one who missed it. The paper, neglected by most aging researchers, suggested that some genes that cause aging are selected for because they give benefits early in life that more than outweigh the costs they exact later in life. The other possible explanation is that genetic variations that influence aging have little effect on fitness because life spans tend to be short in the wild and few individuals are alive to reap any fitness benefits from variations that extend life. After mulling this over for a few months, I found a way to use life table data to calculate the effect of senescence on fitness in wild populations of many species. The results showed enormous variation, from birds, where mortality rates are so high that there was no indication that senescence influenced fitness at all, to large mammals, where senescence had such a spectacular effect that it was implausible to suggest that the genes that caused it were there only because they had not been selected against. By the time I finished this work, I was looking for an evolutionary biologist who might want to collaborate to think through the problem of disease. By happy good fortune, George Williams had simultaneously been looking for a doctor who wanted to collaborate on the same problem. Little did we know when we first met that we would eventually debate every point in our book, and rewrite each others sentences time after time.

The field has grown rapidly thanks to many classes on the subject for undergraduates and much interest among researchers. Medical schools are another matter. Deans of medical education are overwhelmed by the task of somehow trying to keep a rein on the demands for classroom hours from each department, while trying to include the crucial 15 000 facts that a medical student is thought to need. There are no Departments of Evolutionary Biology in medical schools, and few physicians have taken an evolutionary biology course, so there is no lobby to bring this material into the curriculum. Furthermore, it does not readily bring in large research grants like molecular biology does. Nonetheless, more and more young physicians are realizing there is a big gap and beginning to ask why. Better yet, some are realizing that Darwinian medicine offers a solid framework for making sense of all the apparently disorganized facts in the medical curriculum. Things would develop much more quickly if an evolutionary approach provided a real breakthrough for understanding a major disease. My own research is focused on trying to understand how natural selection shaped the mechanisms that regulate mood, and how this knowledge can help us understand, prevent, and treat depression. But even if there are no specific breakthroughs, an evolutionary approach will gradually prove its utility in medicine, as it has in every other field based on biology.

Randolph M. Nesse, M.D., The University of Michigan, USA.

▶ There is great interest in scanning the entire genome for susceptibility loci. This represents a formidable technical challenge because of the scale of data generation required, but its success also depends on the frequency and number of susceptibility alleles at each locus (the 'allelic architecture'). According to the common disease/common variant hypothesis, susceptibility alleles should be common, but theoretical considerations do not lead to unequivocal predictions.

▶ Empirical evidence about susceptibility alleles is limited. There are examples of common variants underlying common diseases, such as *APOE*E4* and Alzheimer disease, but other common diseases such as Crohn disease seem to be associated with many rare variants at *CARD15/NOD2*.

▶ The number of markers needed for whole-genome association studies is unclear, but current work on the haplotype block structure should clarify this question.

▶ Prehistoric infectious diseases have left detectable imprints on modern patterns of genetic diversity. At the *G6PD* locus, reduced enzyme activity is associated with both hemolytic anemia and resistance to malaria; variants such as the A− allele have been selected in Africa, which has had a detectable impact on patterns of genetic diversity over hundreds of kilobases of surrounding DNA. A deficiency allele at the *CCR5* locus (*CCR5Δ32*) appears to have been recently selected in Europe, and now confers strong resistance to HIV infection.

▶ Considerable genetic variation is found between individuals in the response to drugs, which can lead to large differences in both therapeutic and adverse reactions if a standard dose is used.

▶ Many processes are involved in determining drug response, but variations in drug metabolism have received most attention. For example, the *CYP2D6* gene can be present at between zero and 13 copies per individual, and alleles vary greatly in their activity. This variation has a geographical structure and large differences are found between populations, perhaps as a result of past selection. These genetic differences provide an opportunity to optimize medical treatment for each individual according to their genotype.

Further reading

Malcolm S, Goodship J (eds) (2001) *Genotype to Phenotype.* BIOS Scientific Publishers Limited, Oxford.

Nesse RM, Williams GC (1994) *Why We Get Sick: The New Science of Darwinian Medicine.* Vintage, New York.

Strachan T, Read AP (2003) *Human Molecular Genetics 3.* BIOS Scientific Publishers Limited, Oxford.

Electronic references

Online Mendelian Inheritance in Man:
http://www.ncbi.nlm.nih.gov/omim/

The Frequency of Inherited Disorders Database:
http://archive.uwcm.ac.uk/uwcm/mg/fidd/index.html

References

Almqvist E, Spence N, Nichol K *et al.* (1995) Ancestral differences in the distribution of the D2642 glutamic acid polymorphism is associated with varying CAG repeat lengths on normal chromosomes: insights into the genetic evolution of Huntington disease. *Hum. Mol. Genet.* **4**, 207–214.

Altshuler D, Hirschhorn JN, Klannemark M *et al.* (2000) The common PPARg Pro12Ala polymorphism is associated with decreased risk of type 2 diabetes. *Nature Genet.* **26**, 76–80.

Annas GJ (2000) Rules for research on human genetic variation – lessons from Iceland. *N. Engl. J. Med.* **342**, 1830–1833.

Baron M (2001) Genetics of schizophrenia and the new millennium: progress and pitfalls. *Am. J. Hum. Genet.* **68**, 299–312.

Beck S, Olek A, Walter J (1999) From genomics to epigenomics: a loftier view of life. *Nature Biotechnol.* **17**, 1144.

Bonen DK, Cho JH (2003) The genetics of inflammatory bowel disease. *Gastroenterology* **124**, 521–536.

Bradford LD (2002) CYP2D6 allele frequency in European Caucasians, Asians, Africans and their descendants. *Pharmacogenomics* **3**, 229–243.

Cloninger CR (2002) The discovery of susceptibility genes for mental disorders. *Proc. Natl Acad. Sci. USA* **99**, 13365–13367.

Corbo RM, Scacchi R (1999) Apolipoprotein E (APOE) allele distribution in the world. Is APOE*4 a 'thrifty' allele? *Ann. Hum. Genet.* **63**, 301–310.

Corder EH, Saunders AM, Strittmatter WJ *et al.* (1993) Gene dose of apolipoprotein E type 4 allele and the risk of Alzheimer's disease in late onset families. *Science* **261**, 921–923.

Crow TJ (1997) Is schizophrenia the price that *Homo sapiens* pays for language? *Schizophr. Res.* **28**, 127–141.

Dalén P, Dahl ML, Ruiz ML, Nordin J, Bertilsson L (1998) 10-Hydroxylation of nortriptyline in white persons with 0, 1, 2, 3, and 13 functional CYP2D6 genes. *Clin. Pharmacol. Ther.* 63, 444–52.

Dean M, Carrington M, Winkler C *et al.* (1996) Genetic restriction of HIV-1 infection and progression to AIDS by a deletion allele of the CKR5 structural gene. *Science* **273**, 1856–1862.

Dean M, Carrington M, O'Brien SJ (2002) Balanced polymorphism selected by genetic versus infectious human disease. *Annu. Rev. Genomics Hum. Genet.* **3**, 263–292.

Dennis C (2003) Altered states. *Nature* **421**, 686–688.

Egeland JA, Gerhard DS, Pauls DL *et al.* (1987) Bipolar affective disorders linked to DNA markers on chromosome 11. *Nature* **325**, 783–787.

Estivill X, Farrall M, Scambler PJ *et al.* (1987) A candidate for the cystic fibrosis locus isolated by selection for methylation-free islands. *Nature* **326**, 840–845.

Evans WE, McLeod HL (2003) Pharmacogenomics – drug disposition, drug targets, and side effects. *N. Engl. J. Med.* **348**, 538–549.

Fullerton SM, Clark AG, Weiss KM *et al.* (2000) Apolipoprotein E variation at the sequence haplotype level: implications for the origin and maintenance of a major human polymorphism. *Am. J. Hum. Genet.* **67**, 881–900.

Fullerton SM, Bartoszewicz A, Ybazeta G *et al.* (2002) Geographic and haplotype structure of candidate type 2 diabetes susceptibility variants at the calpain-10 locus. *Am. J. Hum. Genet.* **70**, 1096–1106.

Gabriel SB, Schaffner SF, Nguyen H *et al.* (2002) The structure of haplotype blocks in the human genome. *Science* **296**, 2225–2229.

Gabriel SE, Brigman KN, Koller BH, Boucher RC, Stutts MJ (1994) Cystic fibrosis heterozygote resistance to cholera toxin in the cystic fibrosis mouse model. *Science* **266**, 107–109.

Gratacòs M, Nadal M, Martín-Santos R *et al.* (2001) A polymorphic genomic duplication on human chromosome 15 is a susceptibility factor for panic and phobic disorders. *Cell* **106**, 367–379.

Green PM, Saad S, Lewis CM, Giannelli F (1999) Mutation rates in humans. I. Overall and sex-specific rates obtained from a population study of hemophilia B. *Am. J. Hum. Genet.* **65**, 1572–1579.

Gulcher JR, Stefansson K (2000) The Icelandic Healthcare Database and informed consent. *N. Engl. J. Med.* **342**, 1827–1830.

Hästbacka J, de la Chapelle A, Mahtani MM *et al.* (1994) The Diastrophic Dysplasia gene encodes a novel sulfate transporter: positional cloning by fine-structure linkage disequilibrium mapping. *Cell* **78**, 1073–1087.

Högenauer C, Santa Ana CA, Porter JL *et al.* (2000) Active intestinal chloride secretion in human carriers of cystic fibrosis mutations: an evaluation of the hypothesis that heterozygotes have subnormal active intestinal chloride secretion. *Am. J. Hum. Genet.* **67**, 1422–1427.

Horikawa Y, Oda N, Cox NJ *et al.* (2000) Genetic variation in the gene encoding calpain-10 is associated with type 2 diabetes mellitus. *Nature Genet.* **26**, 163–175.

Hugot JP, Chamaillard M, Zouali H *et al.* (2001) Association of NOD2 leucine-rich repeat variants with susceptibility to Crohn's disease. *Nature* **411**, 599–603.

Jablensky A (2000) Prevalence and incidence of schizophrenia spectrum disorders: implications for prevention. *Aust. N. Z. J. Psychiatry* **34** (Suppl.), S26–S34.

Jeffreys AJ, Kauppi L, Neumann R (2001) Intensely punctate meiotic recombination in the class II region of the major histocompatibility complex. *Nature Genet.* **29**, 217–222.

Johansson I, Lundqvist E, Bertilsson L, Dahl M-L, Sjöqvist F, Ingelman-Sundberg M (1993) Inherited amplification of an active gene in the cytochrome P450 CYP2D locus as a cause of ultrarapid metabolism of debrisoquine. *Proc. Natl Acad. Sci. USA* **90**, 11825–11829.

Joy DA, Feng X, Mu J *et al.* (2003) Early origin and recent expansion of *Plasmodium falciparum*. *Science* **300**, 318–321.

Kaplan NL, Lewis PO, Weir BS (1994) Age of the DF508 cystic fibrosis mutation. *Nature Genet.* **8**, 216.

Kauppi L, Sajantila A, Jeffreys AJ (2003) Recombination hotspots rather than population history dominate linkage disequilibrium in the MHC class II region. *Hum. Mol. Genet.* **12**, 33–40.

Kelsoe JR, Ginns EI, Egeland JA *et al.* (1989) Re-evaluation of the linkage relationship between chromosome 11p loci and the gene for bipolar affective disorder in the Old Order Amish. *Nature* **342**, 238–243.

King H, Aubert RE, Herman WH (1998) Global burden of diabetes, 1995–2025: prevalence, numerical estimates, and projections. *Diabetes Care* **21**, 1414–1431.

Kruglyak L (1999) Prospects for whole-genome linkage disequilibrium mapping of common disease genes. *Nature Genet.* **22**, 139–144.

Lander ES (1996) The new genomics: global views of biology. *Science* **274**, 536–539.

Lohmueller KE, Pearce CL, Pike M, Lander ES, Hirschhorn JN (2003) Meta-analysis of genetic association studies supports a contribution of common variants to susceptibility to common disease. *Nature Genet.* **33**, 177–182.

Martin ER, Lai EH, Gilbert JR *et al.* (2000) SNPing away at complex diseases: analysis of single-nucleotide polymorphisms around APOE in Alzheimer disease. *Am. J. Hum. Genet.* **67**, 383–394.

Morral N, Bertranpetit J, Estivill X *et al.* (1994) The origin of the major cystic fibrosis mutation (DF508) in European populations. *Nature Genet.* **7**, 169–175.

Nebert DW, Russell DW (2002) Clinical importance of the cytochromes P450. *Lancet* **360**, 1155–1162.

Ogura Y, Bonen DK, Inohara N *et al.* (2001) A frameshift mutation in NOD2 associated with susceptibility to Crohn's disease. *Nature* **411**, 603–606.

Ostrer H (2001) A genetic profile of contemporary Jewish populations. *Nature Rev. Genet.* **2**, 891–898.

Phillips MS, Lawrence R, Sachidanandam R *et al.* (2003) Chromosome-wide distribution of haplotype blocks and the role of recombination hot spots. *Nature Genet.* **33**, 382–387.

Pier GB, Grout M, Zaidi T *et al.* (1998) *Salmonella typhi* uses CFTR to enter intestinal epithelial cells. *Nature* **393**, 79–82.

Pritchard JK, Przeworski M (2001) Linkage disequilibrium in humans: models and data. *Am. J. Hum. Genet.* **69**, 1–14.

Pritchard JK, Cox NJ (2002) The allelic architecture of human disease genes: common disease–common variant . . . or not? *Hum. Mol. Genet.* **11**, 2417–2423.

Reich DE, Cargill M, Bolk S *et al.* (2001) Linkage disequilibrium in the human genome. *Nature* **411**, 199–204.

Reich DE, Lander ES (2001) On the allelic spectrum of human disease. *Trends Genet.* **17**, 502–510.

Risch N, Merikangas K (1996) The future of genetic studies of complex human diseases. *Science* **273**, 1516–1517.

Rogers JF, Nafziger AN, Bertino JS (2002) Pharmacogenetics affects dosing, efficacy, and toxicity of cytochrome P450-metabolized drugs. *Am. J. Med.* **113**, 746–750.

Romeo G, Devoto M, Galietta LJ (1989) Why is the cystic fibrosis gene so frequent? *Hum. Genet.* **84**, 1–5.

Roses AD (2000) Pharmacogenetics and the practice of medicine. *Nature* **405**, 857–865.

Ruwende C, Khoo SC, Snow RW *et al.* (1995) Natural selection of hemi- and heterozygotes for G6PD deficiency in Africa by resistance to severe malaria. *Nature* **376**, 246–249.

Sabeti PC, Reich DE, Higgins JM *et al.* (2002) Detecting recent positive selection in the human genome from haplotype structure. *Nature* **419**, 832–837.

Samson M, Libert F, Doranz BJ *et al.* (1996) Resistance to HIV-1 infection in caucasian individuals bearing mutant alleles of the CCR-5 chemokine receptor gene. *Nature* **382**, 722–725.

Saunders MA, Hammer MF, Nachman MW (2002) Nucleotide variability at G6pd and the signature of malarial selection in humans. *Genetics* **162**, 1849–1861.

Schultz SK, Andreasen NC (1999) Schizophrenia. *Lancet* **353**, 1425–1430.

Scriver CR, Waters PJ (1999) Monogenic traits are not simple: lessons from phenylketonuria. *Trends Genet.* **15**, 267–272.

Serre JL, Simon-Bouy B, Mornet E *et al.* (1990) Studies of RFLP closely linked to the cystic fibrosis locus throughout Europe lead to new considerations in populations genetics. *Hum. Genet.* **84**, 449–454.

Slatkin M, Bertorelle G (2001) The use of intraallelic variability for testing neutrality and estimating population growth rate. *Genetics* **158**, 865–874.

Smith DJ, Lusis AJ (2002) The allelic structure of common disease. *Hum. Mol. Genet.* **11**, 2455–2461.

Squitieri F, Andrew SE, Goldberg YP *et al.* (1994) DNA haplotype analysis of Huntington disease reveals clues to the origins and mechanisms of CAG expansion and reasons for geographic variations of prevalence. *Hum. Mol. Genet.* **3**, 2103–2114.

Stefansson H, Sigurdsson E, Steinthorsdottir V *et al.* (2002) Neuregulin 1 and susceptibility to schizophrenia. *Am. J. Hum. Genet.* **71**, 877–892.

Stephens JC, Reich DE, Goldstein DB *et al.* (1998) Dating the origin of the CCR5D32 AIDS-resistance allele by the coalescence of haplotypes. *Am. J. Hum. Genet.* **62**, 1507–1515.

Tabiner M, Youings S, Dennis N *et al.* (2003) Failure to find DUP25 in patients with anxiety disorders, in control individuals, or in previously reported positive control cell lines. *Am. J. Hum. Genet.* **72**, 535–538.

Tishkoff SA, Varkonyi R, Cahinhinan N *et al.* (2001) Haplotype diversity and linkage disequilibrium at human G6PD: recent origin of alleles that confer malarial resistance. *Science* **293**, 455–462.

Verrelli BC, McDonald JH, Argyropoulos G *et al.* (2002) Evidence for balancing selection from nucleotide sequence analyses of human G6PD. *Am. J. Hum. Genet.* **71**, 1112–1128.

Waterston RH, Lindblad-Toh K, Birney E *et al.* (2002) Initial sequencing and comparative analysis of the mouse genome. *Nature* **420**, 520–562.

Weinshilboum R (2003) Inheritance and drug response. *N. Engl. J. Med.* **348**, 529–537.

Wiuf C (2001) Do DF508 heterozygotes have a selective advantage? *Genet. Res.* **78**, 41–47.

CHAPTER FIFTEEN

Identity and identification

In the previous chapters of this book genetic variation has been discussed within the framework of the gene pool, and changes in allele frequencies within populations. We have considered human beings as *groups* of individuals:

▶ the species as a whole, and its origins and relationship to other species;

▶ populations – from those of whole continents to those of subregions or islands.

In this, the final chapter, we exchange our telescope for a microscope, and focus on the individual instead of the population. We ask:

▶ how can we best use genetic variation for individual identification of any one of the 6.2 billion human beings?

▶ what can we determine about an individual from their DNA? This includes physical characteristics such as sex and pigmentation, and other categorizations such as population of origin;

▶ how can we decide from genetic variation if two individuals are close relatives, such as parent and child?

▶ and, widening the field of view, how can we tell if two individuals share membership of a larger grouping such as a genealogy, a clan, or some other genetically defined group? This question shades into the more general issues of population history that we have addressed earlier in this book.

Many of these questions are of great social interest, and, in the form of forensic DNA analysis and paternity testing, are key issues in law enforcement and social policy. Questions about the genetics of an individual cannot be asked in isolation, however, because the answers to the questions – such as how certain we can be that two DNA samples derive from the same person – depend on the degree and structure of genetic variation in the population to which that person belongs. Thus, studies of individual variation must be set in the context of population variation, and appropriate statistical methods must be applied to evaluate different hypotheses. These statistical issues may be complex, but at the same time they must often be presented to lay-people in criminal cases in a way that is both comprehensible, and not misleading. Aside from statistical issues, there are practical issues of data quality and error rates in DNA analysis, and serious ethical issues related to the custodianship and legitimate use of databases containing DNA data on individuals.

15.1 Individual identification

The diversity of the human genome, and the mutation processes that underlie this diversity, have been discussed in Chapter 3. On average, we expect to find a base substitutional difference (SNP) between two copies of the genome about every 1250 bp (see Section 3.2.7). This means that there will be a few million of these differences between the diploid genome of one individual and that of another. There are fewer microsatellites and minisatellites than there

are base substitutions, but many of them have many more alleles and much higher **heterozygosities**, so among the few tens of thousands of potentially useful loci we expect the majority to reveal differences between two individuals taken at random. There is thus an enormous number of differences between DNA carried by one person and that carried by another, and if we could access this variation readily it would be a straightforward matter to individually identify someone from a DNA sample, or to show conclusively that two samples came from the same person. In reality, however, there are practical constraints that influence the kinds and numbers of polymorphic markers that are chosen for analysis. The situation in a forensic laboratory is different from that in a research context:

▶ methods for individual identification evolve slowly, because their wide acceptance depends on the steady accumulation of confidence based upon their successful use in the courtroom (through legal precedent) and effective standards in testing (validation). Large databases are expensive and can be slow to set up, so changes in established systems for the detection of variation are undesirable;

▶ DNA samples are often degraded or present in trace amounts, so methods must be particularly sensitive – there is an analogy with ancient DNA studies here, including the concern about possible contamination;

▶ errors in typing can have serious consequences, so all methods and procedures must be reliable and accurate (see *Box 15.1*);

▶ many different laboratories need to analyze the same markers to the same standards, so methods must be widely applicable and robust; novel, specialized and technically difficult methods are not easily accepted, no matter what wealth of variability they may offer;

▶ questions about methods and statistics are often raised in the adversarial and conservative environment of the courtroom, rather than the less charged realm of the scientific seminar. This does not aid a productive and objective debate about the issues.

At this stage we must point to an important distinction between different uses of DNA-based evidence. The discussion above is dealing primarily with evidence that will come to court, be evaluated by a jury and, in many countries, picked over by defense and prosecuting counsel. However, in principle, evidence used only for investigative purposes need never come to court. For example, a match between the DNA of a suspect and a sample from a crime scene can lead to the investigation of that suspect and the confirmation of their guilt or innocence by other, non-DNA based means (e.g., a confession, a blood-stained monkey wrench, or a cast iron alibi). The kind of DNA evidence used in investigations can therefore be more novel and specialized than the points made above might suggest, though in practice DNA evidence more often than not does enter the courtroom. In addition, the possibility must always be considered that a crime scene sample, even if it gives a perfect DNA fingerprint or profile, does not necessarily relate to the

BOX 15.1 Error in forensic analysis

All types of error in forensic casework are potentially very serious, and the primary DNA data, its recording and interpretation must therefore be of the highest quality. Errors are normally categorized as type 1 and type 2:

▶ a **type 1 error** is a false inclusion – an incorrect statement that a person's DNA matches that left at a crime scene by a perpetrator, for example. This could lead to a miscarriage of justice in the conviction of an innocent person. An example is given in Section 15.1.3;

▶ a **type 2 error** is a false exclusion – an incorrect statement that a person's DNA does not match a crime-scene sample, for instance. This could lead to a guilty person being judged innocent, and therefore free to commit further offenses. This kind of error is overwhelmingly the most likely outcome of technical errors in DNA profiling, or administrative errors in entering data. Errors in the database therefore tend to favor the suspect in a criminal case.

These are the same as the type 1 and 2 errors used in general statistics, where a type 1 error is the incorrect rejection of a null hypothesis, in this case that the two samples come from different people.

---

perpetrator of the crime. It only demonstrates that a person has been, at some point in time, at that specific place.

15.1.1 Multilocus minisatellite fingerprinting and single-locus profiling

The technique of multilocus minisatellite fingerprinting was developed by Alec Jeffreys in 1984 (Jeffreys *et al.*, 1985c), and rapidly became established as a highly effective means of distinguishing between DNA samples from different individuals, and reliably establishing family relationships. Prior to the use of DNA in these fields, classical polymorphisms such as blood groups and protein markers (see *Box 3.3*) were the only available tools.

In the DNA fingerprinting method (*Figure 15.1*), radiolabeled DNA **probes** (short, cloned DNA sequences) containing minisatellite sequences are hybridized to DNA that has been digested with a restriction enzyme, separated by agarose gel electrophoresis, and immobilized on a membrane by Southern blotting. The probe hybridizes to a set of minisatellites in the genomic DNA contained in restriction fragments whose sizes differ because of variation in the numbers of repeat units (Section 3.3.2). Excess probe is washed from the membrane at 'low stringency' (a fairly high salt concentration), which allows hybrids that are somewhat mismatched to persist. In this way a collection of minisatellite sequences containing common sequence motifs is detected. Exposure to X-ray film

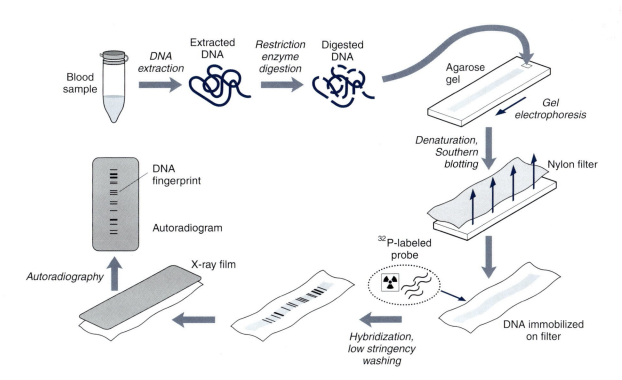

Figure 15.1: Procedure for generating a DNA fingerprint.

(autoradiography) then allows these variable fragments to be visualized, and their sizes compared between individuals.

The most widely used probes, called 33.6 and 33.15, detect independent sets of ~ 17 variable DNA fragments per individual ranging from 3.5 to > 20 kb in size. It was shown that the patterns these probes reveal, called 'DNA fingerprints', are highly individual-specific, with an original estimate of the probability of two unrelated individuals sharing DNA fingerprints for probe 33.15, for example, of 3×10^{-11} (Jeffreys *et al.*, 1985c). The only individuals expected (and demonstrated) to share DNA fingerprints are monozygotic twins.

In multilocus methods the allelic correspondences between bands in one DNA fingerprint and another are not known, and nor is it known *a priori* whether or not any two bands represent loci that are closely linked on a chromosome, and therefore possibly in linkage disequilibrium (LD) with each other (see Section 3.7). If two loci are in LD, then they are not independently informative on ancestry, which could lead to statistical errors. The method also requires at least 250 ng of DNA. To overcome some of these limitations, **single-locus profiling** was developed, and overtook DNA fingerprinting. Here, a single hypervariable locus is detected by a specific probe (single-locus probe; SLP), using high stringency hybridization (washing off excess probe under low-salt conditions). After hybridizing with one probe, a filter bearing DNA could be 'stripped' and reprobed a number of times; typically, four single-locus probes were used, yielding eight alleles per individual. This method was viable with only 10 ng of genomic DNA.

These DNA methods were first applied in a criminal case in 1986 in the investigation of a double rape and murder in Enderby (Leicestershire, UK). First, the two crimes, occurring 3 years apart, were shown to have been committed by the same man by finding identical profiles in semen samples taken from both victims (a **match**). This connection of one crime scene to another is of enormous importance to the police. Second, a prime suspect who had confessed was proven to be innocent because his DNA profile was different from that found in the semen samples (an **exclusion**). A **mass screen** was then organized: ~ 5000 blood samples were taken from men living in the locality, and after a prescreen with 'classical' markers including the ABO blood group, DNA was extracted from 500 blood samples that were not excluded. The true killer, Colin Pitchfork, was clearly persuaded of the power of DNA methods, because he evaded the mass screen by persuading a colleague to give a blood sample on his behalf. Shortly afterwards, the colleague talked about this exploit to friends who tipped off the police; Pitchfork was arrested, his single-locus profile and DNA fingerprint were shown to match the samples from the crime scene, and he pleaded guilty.

Using these methods, exclusions are straightforward – if two DNA fingerprints or SLP profiles are different, they cannot have come from the same person. The convincing establishment of the innocence of a suspect is one of the great benefits of these DNA-based techniques. However, matches

are more difficult – there are important statistical issues, which will be considered in Section 15.1.3, and technical issues, which will be considered in this section.

Are two banding patterns truly identical, or merely similar? For multilocus DNA fingerprinting, little statistical significance was placed on any one band, and the significance of a match gained its power from the large number of bands involved. For SLP profiling, on the other hand, high statistical significance was placed on matching a small number of bands. At the molecular level, the difference between minisatellite alleles is discrete, and can be measured in whole numbers of repeat units (though to complicate matters, these can themselves vary in length; see Section 3.3.2). However, the unit size is small (usually < 50 bp), so in practice the variation detected by Southern blotting is quasi-continuous – individual alleles often cannot be distinguished. Additional uncertainty is introduced if different DNA samples do not run in an identical way on an agarose gel, because of a nonuniform electric field across the gel, or because of problems with DNA samples. DNAs at different concentrations, or prepared by different methods from different starting materials, may not behave similarly, and this may cause uncertainty about whether bands match. Therefore, rather than asking whether a band was of exactly the same size as another, size 'bins' were defined, and it was asked whether two bands fell into the same bin (*Figure 15.2*). Two different kinds of bin, 'fixed bins' (defined relative to a DNA size marker only), and 'floating bins' (defined relative to the observed DNA fragment size, and adjusted for the resolving power of the detection system) were employed. Choice of bin size was important – too small a bin size might mean that a fragment was assigned to the wrong bin, and too large a bin would overestimate the population frequency and increase the probability of a chance match (though at the same time reducing the match probability). Because of these technical issues, several of which were discussed by Lander (1989), standards were established to which forensic laboratories were required to adhere.

These technical problems were widely debated, and overcome. The use of single-locus profiling has been validated through extensive experiments and forensic casework, and for many years provided a robust and valuable system for individual identification.

15.1.2 Microsatellite profiling

DNA profiling has now entered the PCR era, with the universal use of microsatellites [usually referred to in the forensic community by their alternative name, STRs (short tandem repeats)]. This has several advantages over earlier methods:

▶ PCR allows the determination of a profile from very small (< 1 ng) amounts of DNA, though at the same time this sensitivity introduces potential problems of DNA contamination (see *Box 15.2*);

▶ small amplicon sizes allow profiling from degraded DNA samples;

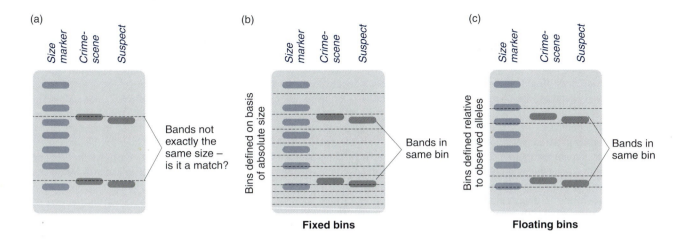

Figure 15.2: The use of size 'bins' in single-locus profiling.

(a) The two alleles in a crime-scene sample and in a suspect are not exactly the same size in profiling, because of running conditions or DNA quality issues. To avoid a Type 2 error (falsely excluding the suspect), size bins are defined, and we ask if the alleles in both samples fall into the same bins. (b) Fixed size bins are defined relative to the size markers, and applied uniformly to all single-locus probes; each bin size represents ± 3 standard deviations of the fragment size (on empirical evidence, ~ ± 1.5–3%). (c) Floating size bins are defined relative to the observed alleles, again using a range such as ± 3 standard deviations. In calculating match probabilities, the population frequencies used must be those of the bins used in the casework.

▸ semi-automated analysis using fluorescent primers on a sequencing platform allows high throughput profiling and the generation of large databases;

▸ discrete allele lengths (and hence repeat numbers) for the microsatellite loci can be measured, which obviates the need for binning and facilitates comparisons between profiles. Fragments are sized with reference to 'allelic ladders' – sets of PCR products corresponding to different alleles across the known size range. Allele length variability is less than that for the minisatellites using in SLP profiling, but several microsatellites can be multiplexed (simultaneously amplified) to give highly variable profiles;

▸ the microsatellites are often stated to be selectively neutral, though active support for this view has not been presented. In fact, it seems highly likely that some of the loci used in profiling are associated with diseases. Many microsatellites chosen initially were discovered because they were close to genes that were being studied. Examples are *HUMVWA*, close to the von Willebrand factor gene, mutations in which cause the blood clotting disorder von Willebrand disease (OMIM 193400), and *HUMTH01*, close to the insulin gene, variation in which is associated with differing susceptibility to type 1 **diabetes** (OMIM 222100).

Markers are usually chosen from different chromosomes to avoid allele association problems (*Figure 15.3*), with high heterozygosity (each above 0.7), PCR product size between 100 and 340 bp, and 'clean' and reliable PCR amplification. Multiplexes containing several markers have been optimized, and the one currently in use in the UK amplifies 10 loci; the system used by the FBI uses 13 loci, and some commercially available multiplexes contain even more (*Figure 15.4*). Typically, for extraction and typing, DNAs are bar-coded to aid tracking of samples. DNA quantification and PCR set-up are automated and gels loaded manually. In the UK Forensic Science Service, electropherograms (examples are given in *Figure 15.4*) were originally read independently by two operators, but now the second operator has been replaced by quality control software. Ten percent of samples are routinely retyped 'blind' as a quality control.

15.1.3 What does the match probability mean?

The evidential weight placed on two DNA profiles matching is measured by a 'match probability' – the chance that a profile chosen at random from the population will match that from the crime scene. This raises a number of important questions, some of which are covered in *Box 15.3*.

▸ *What precision of defining a profile is appropriate?* If there is over-optimism about this precision, unrealistically low matching probabilities could result. One example is the spectacular figure of 1 in 738 000 000 000 000 for a four-locus SLP match that was presented by a commercial forensic laboratory (Lander, 1989). How is this figure, whose denominator is vastly greater than the entire population of the world, supposed to be interpreted by a jury? Large microsatellite multiplexes now typically provide matching probabilities of 1 in 10^{17}–10^{18}.

▸ *Which population should be chosen to form the database to which the profile is compared?* The normal practice was to use the 'product rule': the frequency of a genotype was obtained by multiplying together the population

---

BOX 15.2 The analogy between forensic and ancient DNA analysis

With the introduction of highly sensitive PCR-based profiling, forensic scientists have gained access to DNA profiles from a wide range of samples that were previously inaccessible: single hairs, cigarette butts, saliva on postage stamps, fingerprints (the conventional kind), fragments of bone, etc. Section 15.3.2 describes successful DNA analysis of the skeletal remains of the Russian Czar Nicholas II and his family. Analysis of such DNA can be compared to ancient DNA studies (**aDNA**; see Section 4.11; reviewed by Capelli *et al.*, 2003): template is present in very small quantities, is often extensively degraded, and may be chemically modified so that it is difficult to amplify. Another analogy is the ever-present danger of contamination by modern DNA, for which reason personnel involved in taking samples from the crime scene and processing them have their profiles analyzed so that they can be recognized if they occur in the samples. Purpose-built facilities for analysis are similar to aDNA laboratories, with dedicated equipment, filtered air, and protective clothing. There are differences as well as similarities: unlike aDNA studies, there is no opportunity to gain evidence of authenticity from the phylogenetic relationships between samples, because all are modern. Also, the forensic scientist usually wants to amplify nuclear microsatellites from the sample, while the molecular anthropologist is usually targeting mtDNA sequence variation.

The sensitivity of these 'low copy number' forensic DNA methods is impressive. Touched surfaces (such as a telephone handset, door-handle or keys) can yield full microsatellite profiles (van Oorschot and Jones, 1997), and profiles can also be obtained from single buccal cells by increasing PCR cycle number (Findlay *et al.*, 1997). Problems associated with profiles from such samples (< 100 pg) are allele 'drop-out', where expected peaks on an **electopherogram** fail to appear, artifact formation giving spurious bands, and unavoidable low-level laboratory contamination. Guidelines for such work include duplicate detection of every allele before reporting the profile (Gill *et al.*, 2000). The use of such profiles in criminal casework needs effective monitoring, because DNA can be transferred easily from one individual to another by contact, such as a handshake.

---

frequencies of its component alleles. However, if there is substantial LD or population structure, then some alleles or combinations of alleles may be present at higher frequency than expected in particular subpopulations. Choice of an inappropriate subpopulation could then lead to an artificially reduced match probability. This issue caused a major debate among population geneticists and forensic scientists, and led to a 3-year study by the US National Research Council (reviewed by Lander and Budowle, 1994). The recommendations of this study were that a conservative approach (the 'ceiling principle') should be adopted, where the maximum allele frequency for a given marker among a set of 15–20 populations should be taken, even if all other populations showed much lower frequencies. Note that this is regarded by some as a breach of the basic principles of population genetics; for example, the allele frequencies sum to > 1! Furthermore, a mandatory minimum frequency of 5% was set for each allele. The aim was to provide a system that both prosecution and defense could agree was in the favor of defendants, and thus to nullify problems associated with claims of extremely low matching probabilities. In reality, the effect of the ceiling principle was to increase the probability of a chance match by only two orders of magnitude (from 10^{-8}–10^{-9} to 10^{-6}–10^{-7} – still low enough for a jury to regard as a convincingly low value, though nonetheless the cause of much hostility from forensic scientists). The vigorous debate on this subject was carried out in the context of single-locus profiling; it is interesting that comparatively little analogous debate has been evident on the use of microsatellite profiling.

▶ *Was the crime actually committed by a close relative of the suspect?* Relatives are expected to share alleles with the suspect, and are also much more likely than people taken at random to share an environment, and therefore an opportunity to commit the crime. These possibilities can be incorporated within the statistical interpretation of the evidence (see *Box 15.3*).

▶ *If the DNA evidence matches a suspect to a crime scene but there is clear exonerating evidence of other kinds, which should be believed?* Such low match probabilities have often been given that the DNA evidence may seem overwhelming. For example, a man with advanced Parkinson disease who could not even dress himself unaided was arrested in the UK in 1999 and charged with a burglary committed over 300 km from his home. The arrest was on the basis of a DNA match for six SLPs, with a match probability of 1 in 34 million, but despite the man's condition and an alibi. He was released when further SLP typing showed that he no longer matched the crime scene. This was an example of a false 'cold hit' to a database; the issue of such databases is discussed in further detail below.

Despite the high discriminatory power of DNA evidence, as is explored in *Box 15.3* and by Evett and Weir (1998), the way that it is dealt with in court is often distorting. The 'prosecutor's fallacy' is a commonly cited example. Suppose a DNA match at a crime scene has a probability of occurrence in a population of only 1 in a million. The prosecutor has found that the accused has this match and says the odds are a million to one in favor of the suspect being guilty. It ought to be impossible for the jury to acquit against such apparent odds. Unfortunately, however, there were ten

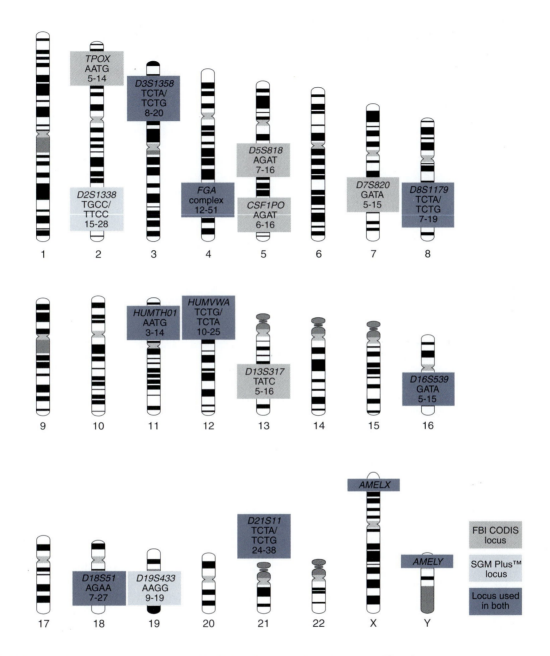

Figure 15.3: Microsatellites commonly used in multiplex DNA profiling, and their chromosomal locations.

Loci forming the US FBI's Combined DNA Index System (CODIS) and the SGM (Second Generation Multiplex) plus system used by the UK Forensic Science Service are shown. Note that, surprisingly, two of the CODIS loci are on the same arm of the same chromosome (see Section 15.1.1), though separated by a large physical and genetic distance. The repeat sequence given for each locus is the predominant repeat; some contain additional or variant repeat types. Information from STRBase (http://www.cstl.nist.gov/biotech/strbase/index.htm).

million people in the city on the night of the crime, so the odds are that 10 people among the population of the city on the night of the crime matched the DNA sample. If the DNA match is the only evidence available, then the odds are 10 to 1 in favor of the suspect being innocent! What the prosecutor should have said was: 'The probability of a match

if the crime-scene sample came from someone other than the suspect, is one in a million'. The discussion should be about the comparing the likelihood of observing the evidence given that the defendant is guilty to the likelihood of observing the evidence given that someone else is guilty.

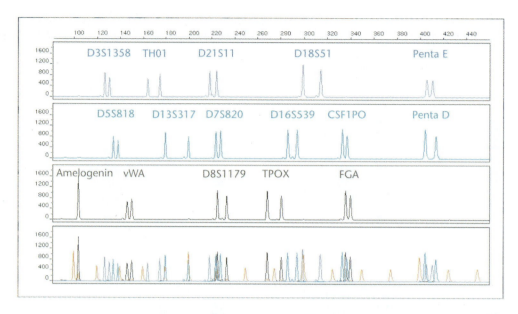

Figure 15.4: A high resolution microsatellite multiplex for DNA profiling.

Result of amplifying a single 0.5 ng genomic DNA sample using the PowerPlex® 16 System from Promega and detected using a capillary electrophoresis system, the ABI PRISM® 310 Genetic Analyzer. The top panel displays the loci *D3S1358, TH01, D21S11, D18S51* and *Penta E* labeled with the dye fluorescein. The second panel displays the loci *D5S818, D13S317, D7S820, D16S539, CSF1PO* and *Penta D* labeled with JOE. The third panel displays the loci Amelogenin (sex test), *vWA, D8S1179, TPOX,* and *FGA* labeled with TMR. The last panel displays all 16 loci and the Internal Lane Standard 600 (ILS600). Figure kindly sponsored by Promega Corporation.

15.1.4 Forensic DNA profiling databases

As well as matching the profile of a crime-scene sample to a suspect, it may be possible to match it to a database of profiles from individuals and thus to identify a suspect in a 'cold hit'. In the USA, databases are run on a state-by-state basis, and the rules specifying types of offense in which DNA sampling for a database is appropriate vary from state to state. In England and Wales, Acts of Parliament passed at the inception of the UK database allowed the police to take nonintimate samples (i.e., **buccal** swabs and hair samples with roots) without consent (sometimes referred to as 'informed nonconsent') from people in police custody who have been charged with a 'recordable offense' (an offense subject to a term of imprisonment). Profiles derived from these samples are stored on the National DNA Database (NDNADB) of England and Wales (the largest in the world; reviewed by Werrett, 1997). Because this database is a particularly large and well established example, we concentrate on it in this section. However, other countries are compiling databases too, and procedures are likely to vary from place to place.

Profiles in the NDNADB can be used to check against profiles derived from other samples taken in connection with other offenses. Profiles are removed from the database if the case against an individual has been dropped, or they have been acquitted or died, and those retained long term on the NDNADB should thus relate only to individuals who have been cautioned or convicted. When profiles are removed, the associated samples are destroyed. The size and performance of the NDNADB are impressive. By the summer of 2002 it contained ~ 1 500 000 profiles (~ 3% of the population), and had provided 180 000 matches of crime scene to suspect, and 18 000 matches of crime scene to crime scene (Forensic Science Service, 2002). Because of the sensitivity of the PCR-based profiling and the size of the database, 'old' cases can now be readdressed provided that crime-scene samples have not been discarded. A recent example was the conviction of a man in 2002 for the killing of a woman in 1991; a microsatellite profile was obtained from the inside of a plastic sleeve in which a swab had been stored, and matched a profile on the database.

Forensic DNA databases are certain to increase in size. The stated projection of the England and Wales National DNA Database is to reach 5 million samples (Werrett, 1997), which is ~ 10% of the total population, ~ 20% of the male population, and more than a third of that age group of men (aged 10–50 years; Home Office, 2002) who are likely offenders. A police estimate of the number of 'active criminals' in the UK is 3 million. Changes in the law since the database's inception allowed nonintimate samples to be taken without consent from people who had been previously convicted of burglary, or violent or sexual offenses and who were still in custody. The Criminal Justice and Police Bill of 2001 makes further major changes:

BOX 15.3 Opinion: Interpreting DNA evidence

DNA profiles can provide strong evidence for the identity of a defendant with the source of a crime stain, or in favor of a relationship such as paternity. Making full and fair use of this powerful evidence has not been without difficulty and controversy, particularly during the first decade after the introduction of DNA profiling in the late 1980s.

Scientists must take some share of the blame for the confusion that reigned in courts. In their enthusiasm to exploit the new technology, some scientists understated its weaknesses. There was also a widespread misunderstanding of basic ideas about measuring evidential weight. This is in part because the scientific paradigm of hypothesis testing isn't appropriate in a court of law: jurors must consider the particular case at hand, and not, for example, the error rates in a conceptual sequence of multiple trials that underpin the notion of a p-value.

Weight-of-evidence in the courtroom

A more suitable approach for the courtroom measures the weight of evidence by how much it changes the probability of guilt. If E is the DNA evidence, D is the defendant, and $G = X$ means 'X is guilty', then:

$$P(G = D|E) = \left(1 + \sum_{X \neq D} \frac{P(E|G = X)\, P(G = X)}{P(E|G = D)\, P(G = D)}\right)^{-1} \quad (1)$$

This equation describes the effect that the DNA evidence has on the probability that the defendant is guilty, $P(G = D|E)$, by considering the likelihood of observing the evidence if the defendant was guilty $[P = (E|G=D)]$ compared to its likelihood if an alternative culprit is guilty $[P(E|G = X)]$. Before hearing the DNA evidence the jury already has some idea from other evidence as to whether the defendant is guilty $[P(G = D)]$, or someone else is $[P(G = X)]$, and this must be taken into account. Because X can refer to many possible culprits besides the defendant, the likelihood of observing the data given their guilt must be calculated for each one and these probabilities are combined together in equation (1) (see Balding, 2000).

Scientists cannot directly use equation (1) to compute a 'correct' answer, because it is up to the juror to assign $P(G = X)|P(G = D)$. However, the scientist can assist the juror by advising on values for the likelihood ratio $P(E|G = X)/P(E|G = D)$ for various possible culprits, X (e.g., a brother, a cousin, an unrelated man). Under assumptions that are usually reasonable, the likelihood ratio simplifies to a *match probability*: the probability that X would also have the same profile as D and the crime scene DNA sample.

The match probability

Scientists have often made misleading statements suggesting that a small match probability alone amounts to convincing evidence of guilt. The match probability is even sometimes used as if it were the probability of innocence, an error known as the 'Prosecutor's Fallacy'. To illustrate why this is misleading, suppose that there are N alternative possible culprits all unrelated to D, and that a juror initially regards all these $N + 1$ individuals as equally under suspicion. Then:

$$P(G = D|E) = \frac{1}{1 + \sum_{X \neq D} r} = \frac{1}{1 + Nr} \quad (2)$$

in which r is the match probability (the same for each X because they are all unrelated to D). If N is large, the juror can obtain a substantial probability of innocence even when r is very small.

It is important to realize that there can be no 'objective' value for the match probability: probabilities measure our lack of knowledge. Note also that the match probability involves two individuals, X and D, and hence four alleles at each locus. A key unknown is the exact ancestral relationships among these four alleles. The match probability is *not* simply the profile frequency in the population: this is appropriate only when X and D are known to have no recent shared ancestry. More generally, use of a profile frequency estimate in place of the match probability is likely to be unfair to the defendant. In practice, match probabilities are based on allele frequencies in a forensic database drawn from a large group (e.g., '**Caucasians**'), together with population genetics theory to justify an assumption of independence both over loci and over the alleles at a locus. This assumption is never strictly valid, but may provide an acceptable approximation provided that the relationship between X and D is accounted for at each locus.

Close relatives and population structure

If we modify the above example so that the alternative possible culprits consist of a brother of D and $N-1$ men unrelated to D, then the probability of guilt becomes:

$$P(G|E) = \frac{1}{1 + b + (N - 1)r} \quad (3)$$

where b denotes the match probability for the brother. Typically b exceeds r by almost an order of magnitude for each locus, and so b can dominate the denominator even if there is no particular evidence inculpating the brother. This contrasts with misleading statements suggesting that brothers can be ignored if there is no specific evidence against them.

Similarly, suppose now that the database consists of mixed 'Caucasians' but the crime occurred in Cornwall and D is Cornish. The general level of shared ancestry of Cornish people implies that a higher match probability, say r', is appropriate for Cornish alternative possible suspects than for other 'Caucasians'. It is difficult to quantify this effect precisely, but population genetics theory offers models in which a parameter, called θ or F_{ST}, can be defined, estimated in actual populations, and used to calculate r'.

Database searches
Thanks to technological improvements and better training about weight of evidence, the presentation of DNA evidence today is much improved over the situation a decade ago. However, there remains scope for errors and misunderstandings.

A continuing misunderstanding concerns the weight of evidence of a DNA profile match obtained via a database search. Scientists, instinctively wary of 'hypothesis trawls', often regard the search as seriously weakening the DNA evidence. This is not the case. The key insight is that, unlike in typical scientific settings, we know in advance that one hypothesis of the form '*X* is the culprit', is true, and all the others are false. A database search excludes many of the false hypotheses, thus increasing the probability that the matching individual is the true culprit.

David Balding, Department of Epidemiology and Public Health, Imperial College, London UK

▶ profiles can be retained, even if the donor is acquitted or charges dropped. There have been high profile cases in which rape and murder suspects matched database profiles, but the suspects have been acquitted: the DNA evidence was ruled inadmissible in court because the profiles should no longer have been on the database after previous acquittals;

▶ profiles taken from volunteers in 'mass-screens' can also be retained, provided that the volunteers consent in writing. Thus the database will contain profiles from people who have never been accused of a recordable offense.

Databases are seen by many as an absolute good. For any criminal whose profile is in these databases, the risk to them of being detected in further criminal activity is high. If this acts as a deterrent to further offending, this is a real benefit, though since many serious crimes are committed on impulse the deterrence may not be strong, and the databases' primary benefit will be the speedy apprehension of culprits. However, there are also potential drawbacks. No database is perfect – all will contain some errors, and the issue is what the error rate is, and what its consequences may be. False exclusions (Type 2 errors) are the most obvious outcome. A review of a small South Australian database suggested an error rate in interpretation of profiles of 5–10% (Dyer, 2002), which is clearly unacceptably high. The error rates of larger databases are not discussed, but probably should be; any bland assurance that there are no errors would be a political, rather than a scientific, claim.

Alec Jeffreys, the originator of DNA fingerprinting, called in September 2002 for the creation of a database storing the profiles of the entire UK population. He has two objections to the current NDNADB and the way that it is operated. First, if the database includes profiles from suspects, these suspects are highly unlikely to be a random subset of the population. Because of the nonrandom geographical and social distribution of crime, and the police's response to this, particular groups in society are likely to be over-represented among suspects, and therefore the database is discriminatory. Second, the database contains private genetic information

about the phenotypes of individuals that should not be in the hands of the police. The example of possible associations between microsatellites used in profiling and disease states has already been given. Other examples are potential information about sex-reversal syndromes (see Section 15.2.1), and other phenotypic characteristics (see Section 15.2). For instance, analysis of genes involved in cranio-facial development, for the prediction of facial features, may well lead to the identification of variants associated with facial dysmorphologies. Jeffreys' view is that both of these problems would be solved by the creation of a new universal database held by an independent body that would consider police requests for access. It would be for investigative purposes only, and if it led to the identification of a suspect, a fresh DNA-based investigation would begin, using an independent set of DNA markers. James Watson, co-discoverer of the structure of DNA, called in February 2003 for a global DNA database that would help to fight global crime and terrorism. Civil liberties debates on these issues are set to continue.

15.1.5 Use of Y-chromosomal and mtDNA markers in individual identification

DNA profiles based on a set of autosomal microsatellites owe their enormous variability to three processes: mutation, which generates new alleles, and independent chromosomal assortment and recombination, which together reassort them each generation. Because of this, the only person expected to share a person's profile is their identical (MZ) twin. Since the frequency of such twins is rare (~ 1 in 300 births, with some geographical variation) and their existence usually known about, this is not a practical problem. Now consider DNA profiles based on nonrecombining segments of DNA, the Y chromosome and mtDNA. First, genetic variability in these systems is due only to mutation (including intrachromosomal recombination), and this greatly reduces haplotype variability. Second, all members of a man's close **patriline** (his father, brother, son, paternal uncle and so on) are expected to share his Y-chromosomal haplotype, and all members of a person's **matriline** are expected to share their mtDNA haplotype. Obviously, some patrilineal or matrilineal relatives will differ from a sample donor due to mutation, and the more distantly related they are, the more likely this is. Nonetheless, we expect that a sizeable number of relatives will be indistinguishable from the donor of a sample using this kind of DNA evidence.

An additional problem for both the Y chromosome and mtDNA is the possible extent of geographic substructuring. We have already addressed the problem of population structure in autosomal profiling, but it is potentially much more severe for these nonrecombining pieces of DNA, each of which is a single genetic locus. Past mating practices and drift (see *Boxes 5.6, 8.5* and *8.8*) are strong potential forces that could lead to local high frequencies of Y-chromosomal or mtDNA haplotypes. An example of a high frequency Y haplotype defined using SNPs and 16 microsatellites is described in Section 15.3.3.

These factors are not a problem for exclusions. For example, if a suspect does not share a Y-chromosomal profile with a crime-scene sample it excludes him, regardless of how common his patrilineal haplotype is. However, this sharing of haplotypes within patrilines or matrilines might seem to preclude the use of the Y chromosome or mtDNA in matching samples to suspects. This is not entirely so: they have both come to play useful, if specialized roles, which are explored below. Because of the problems discussed above, evidence is more likely to be used for investigative purposes than to come to court, but the information that a Y-chromosomal or mtDNA haplotyping failed to exclude a suspect is nonetheless useful.

Y chromosomes in individual identification

The first consideration for Y-chromosomal profiling (reviewed by Jobling *et al.*, 1997) is that a very high proportion of offenders, and an overwhelming proportion of violent offenders, are male. If they leave a sample of body fluid or tissue at a crime scene, it will therefore contain a Y chromosome. In rape cases an assailant-specific profile for autosomal markers can usually be obtained from a vaginal swab by the method of differential lysis, which allows sperm cells to be selectively enriched (Gill *et al.*, 1985). However, in cases where the rapist is azoospermic and has no sperm in his semen (Betz *et al.*, 2001; Sibille *et al.*, 2002) or where there are other body fluid mixtures ('mixed stains'), such as blood mixed with blood, a Y-chromosomal microsatellite profile provides a sensitive means to gain specific information about the assailant. In multiple rapes it may be possible to gain information about the likely number of assailants.

Y chromosome analysis has been taken up with enthusiasm by some forensic services. A set of nine Y-chromosomal microsatellites is widely used, and forms the basis for three online forensic databases covering Europe, the USA and Asia (Roewer *et al.*, 2001; Kayser *et al.*, 2002) that allow a match to be sought. Note that these databases are derived from volunteers chosen by the contributing laboratories. If 'unrelated' individuals are chosen, there is a danger that common haplotypes may be under-represented. Haplotype frequencies can also be estimated through these databases; the 'product rule' referred to above is clearly not appropriate for Y-chromosomal microsatellites, because the alleles are not independently assorting. Frequencies can be determined simply by counting; if a haplotype is absent from the database, the conservative (albeit mathematically unrealistic) assumption that it would be the next one added can be made. Alternative Bayesian methods of estimating haplotype frequencies have also been implemented (Krawczak, 2001). Note that laboratories contributing data to the database must pass a quality control exercise, and that other published data are not incorporated. In the European database, for example, the **'virtual heterozygosity'** (probability of two haplotypes taken at random from the population differing) of the haplotype is high (0.9976), and the most common nine-locus haplotype comprises 2.87% of the 12 675 haplotypes (however, see the point above about local population structuring). Alternative microsatellite systems to the standard loci have been proposed (Butler *et al.*, 2002; Redd *et al.*, 2002), and there are likely to be well over a hundred useful microsatellites on the chromosome that have yet to be exploited. In principle, typing of a large proportion of these would reduce the problem of patrilineal haplotype sharing, because microsatellite mutations would distinguish all but very close relatives, but currently would be expensive and impractical.

mtDNA in individual identification

The major virtue of mtDNA for forensic analysis (reviewed by Holland and Parsons, 1999) is the same as that for the analysis of ancient samples (see Section 4.11). Because mtDNA has a far higher copy number per cell than nuclear DNA, it has a correspondingly greater chance of survival, and is therefore useful in the analysis of forensic samples that contain little DNA (e.g., hair shafts), are old, or have received severe environmental insult. Also, because of its matrilineal inheritance, a matrilineal relative can be used as reference material for matching. A forensic database of mtDNA sequences (containing 341 bp of HVS I and 267 bp of HVS II) has been established (Budowle *et al.*, 1999; Monson *et al.*, 2002), containing an initial 1393 sequences. As with Y-chromosomal haplotypes, the population frequency of a mtDNA sequence cannot be estimated by multiplication of the frequencies of individual variant sites. The normal practice is to do this simply by counting the number of instances in the database: the random match probability in the European–American component of the database is 0.52%, though again this does not take into account possible local population structure.

In samples that have undergone extreme treatments such as burning, mtDNA may be the only recoverable DNA suitable for analysis. For the victims of accidents, warfare or disasters, therefore, this kind of analysis may be the only hope for identification, and is sometimes very successful. In such a case, matrilineal relatives provide a reference source that allows matches to be made. The identification of Michael J. Blassie, interred for 14 years in Arlington Cemetery as the Vietnam Unknown Soldier, was an example of the success of this approach. Even when there are large numbers of identifications to be made, such as the victims of air-crashes, the population of potential sources of the mtDNA sequences at the crash-scene is 'closed' – i.e., confined to an identified set of individuals, and hopes are high that full identification, at least to the level of the matriline, can be made.

15.2 What can we know about Mr or Ms X?

In this section we describe attempts to deduce information about aspects of phenotype from DNA. In practice, the police want this information when they are faced with a crime but no suspects – any information whatever may advance their investigation. Note that, if a comprehensive DNA database of all citizens of a country were available, these phenotypic deduction methods would be unnecessary.

15.2.1 Sex testing

The simplest and most common genetic distinction between human beings is their sex. Since men have a Y chromosome and women do not, it ought to be easy to tell whether a DNA sample is from a male or a female. A sex test would not only be useful in forensic and archaeological contexts, but also in prenatal testing in cases of a suspected sex-linked disorder, and in gender verification in sport. There is a history of occasional athletic and skiing champions in women's events subsequently turning out to have had unfair advantages conferred upon them by a Y chromosome (reviewed by Ferguson-Smith and Ferris, 1991).

A simple PCR-based sex test that amplifies a Y-specific sequence is not a good test, however. If a PCR product is obtained, the sample is from a male, but if not, there are three possible explanations: the sample could be from a female, the DNA could be too degraded to yield a product, or the PCR could have failed for a technical reason. It is therefore important to have an assay that incorporates an internal control to show that the PCR is working properly. An independent pair of primers could be used to co-amplify a second locus, but this would not be ideal because absence of the Y-chromosomal product could then still be explained by a problem with the Y-specific primers. Ideally, the Y-specific and control sequences should be co-amplified by the same primer pair.

Extensive XY-homology offers a natural solution to this problem. This homology derives from the common origin of the sex chromosomes as an ancestral autosomal pair (see *Box 8.8*), and from more recent transpositions of material from the X to the Y chromosome. Many PCR primer pairs designed to amplify a Y-chromosomal sequence will also amplify an X-chromosomal sequence, and all that is necessary is to design an assay in which the X- and Y-specific products can be readily distinguished.

The most widely used assay (Sullivan *et al.*, 1993) amplifies part of intron 1 of the XY-homologous **amelogenin** gene (*AMELY/AMELX*; *Figure 15.5*). A single pair of PCR primers produces a product of 112 bp from the Y chromosome and 106 bp from the X chromosome. A male will therefore yield both products, while a female will yield only the smaller. The 6-bp difference in size is easily resolved on a polyacrylamide gel or capillary sequencing apparatus,

and this assay included in various commercially available PCR mutliplex kits that are widely used for DNA profiling (see Section 15.1.2 and *Figure 15.4*).

Although this test is quite reliable and universally applied, there are rare instances where it will give the wrong answer (and, as discussed in Section 15.1.4, invade genetic privacy).

Sex reversal
46,XX males and 46,XY females have a sex-chromosomal constitution that is not concordant with their phenotypic sex. There are many underlying causes for these conditions. An example of XY femaleness is androgen insensitivity syndrome (AIS; OMIM 300068), an X-linked recessive disorder occurring in about 1 in 20 000 46,XY births, in which affected individuals have female external genitalia and breast development, no uterus, and undescended testes. In this syndrome the testis produces androgens, but a mutation in the androgen receptor gene means that the hormonal signal cannot be acted upon by target tissues. An amelogenin sex test here would suggest a male, though the phenotype is female (*Figure 15.5*). XX maleness (OMIM 278850) has an overall frequency at birth of about 1 in 20 000, and is most commonly due to a translocation of a small portion of the Y chromosome, including the testis-determining gene *SRY*, to the X chromosome. Men carrying these translocation

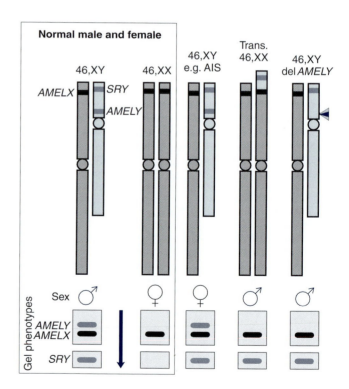

Figure 15.5: DNA-based sex testing.

The amelogenin sex test distinguishes normal males and females, but does not correctly sex 46,XY females such as those with androgen insensitivity syndrome (AIS), translocation 46,XX males, or males carrying *AMELY* deletions. Testing of the presence of *SRY* is helpful in the latter two cases.

chromosomes are infertile because they do not carry long-arm genes necessary for spermatogenesis. Because the translocated portion of the Y chromosome does not include the *AMELY* locus (*Figure 15.5*), the sex test suggests a female, though the phenotype is male. Other sex-reversal syndromes will also give contradictory results.

Deletions of the AMELY locus in normal males
Some normal, fertile males carry interstitial deletions of part of the short arm of the Y chromosome, removing the *AMELY* locus (Santos *et al.*, 1998). Such males would be wrongly diagnosed as females using the standard sex test (*Figure 15.5*). The incidence of *AMELY* deletions is low; for example, it was reported as 0.02% in ~ 29 000 Austrian males (Steinlechner *et al.*, 2002). However, it may reach higher frequencies in some populations through drift: 5 of 270 Indian males (1.85%) showed deletions (Thangaraj *et al.*, 2002).

To surmount some of these problems, a number of additions or alternatives to the basic amelogenin test have been proposed (*Table 15.1*). While these may be useful in special cases, the amelogenin test is now so well established that it seems unlikely to be supplanted.

15.2.2 Prediction of population of origin

The reliable determination of the population of origin of the depositor of a crime-scene sample would be very useful information for law enforcement agencies. As has been discussed in Section 9.3, most (~ 85%) genetic variation is found within, rather than between, human populations. However, significant variation still lies between populations, and this might potentially be used in attempts to predict population of origin.

The microsatellite markers chosen for individual identification were selected on the basis of their high heterogeneity in several population samples. This arises from their high mutation rates, and leads to low interpopulation variance (F_{ST}). They should therefore be poorly suited to predicting the population of origin. However, since a substantial investment of time and money has gone into establishing large databases, and since the marker systems are standardized, forensic geneticists are interested to know how well these markers perform at prediction. An example is a test of the performance of six-locus profiles in classifying individuals into one of five 'ethnic groups' (Lowe *et al.*, 2001). The method used here was to measure allele frequencies in five group-specific databases, and then to simulate 10 000 profiles for each group based on these frequencies. The probability of each real individual profile originating from a particular group was estimated, and these overall probabilities could then be ranked. For example, 56% of profiles known to be from **'Caucasians'** were assigned as such, while 44% were wrongly assigned to other ethnic groups (*Table 15.2*). Thus, a large proportion of profiles is misclassified, and this could have a serious effect on the course of an investigation. In practice, the potential benefit of prediction is assessed by considering the likely reduction in the number of suspect investigations that it would bring about before the actual perpetrator was reached. This may be substantial in some cases; targeting one particular group in the absence of other evidence, however, could lead to accusations of prejudice.

A problem in judging the effectiveness of studies such as this is the way in which population of origin was defined, and this identifies another distinction between studies of 'indigenous' populations by anthropologists where individuals known to be closely related are usually excluded, and the urban, often admixed populations encountered by forensic scientists. In the study discussed above, population of origin was decided in a subjective way by the police, based on five simple categories (*Table 15.2*), and it is not clear, for example, how a person would be classified who had one parent from each of two different ethnic groups. Most anthropological studies use self-definition, and also ask about parents and grandparents of subjects. The problems associated with geneticists' definitions of social groups have been discussed by Foster and Sharp (2002).

A desire to make predictions about population of origin has been stimulated not only by forensic scientists, but also by those who are interested in admixture (see Section 12.3) for epidemiological purposes, or for mapping disease genes using LD. Alleles that exist in one population but not in another have been called 'private' markers (Neel, 1973), or **'population-specific alleles'** (PSAs; Shriver *et al.*, 1997). By screening many loci, alleles at binary or multiallelic autosomal markers that show large (> 50%) frequency differences ($\delta > 0.5$) between four population groups of the USA (African–Americans, European–Americans, Hispanics and Native Americans) have been identified (*Figure 15.6*). These were proposed as useful markers for 'ethnic affiliation estimation' (Shriver *et al.*, 1997), a term that was subsequently dropped in favor of 'estimation of biological ancestry', in recognition of the strong cultural component of ethnicity. These markers perform well, 10 selected loci in effect deciding correctly whether a sample is African–American or European–American in origin in 999 cases out of 1000. This performance is considered implausibly high by some commentators, and the point has been made that categorizing the subjects by interviewing them is unlikely to achieve the same level of confidence (Brenner, 1998). Such markers, now known as **ancestry informative markers** (**AIMs**), have been used to estimate European admixture proportions in various African–American subpopulations (Parra *et al.*, 1998, 2001).

If the aim of isolating these markers is to carry out gene mapping by admixture LD, then there is a need for a larger number of markers spread across the genome. One study (Collins-Schramm *et al.*, 2002) has screened microsatellites and indel polymorphisms to find markers showing $\delta > 0.3$ in comparisons of European–Americans with Mexican–, African– and Native Americans. For the Mexican–American sample, screening of more than 600 markers identified 151 useful candidates, while screening more than 400 African–Americans identified 97. Thus about a quarter of markers turned out to be useful; this study regressed by using the term 'ethnic difference markers'.

TABLE 15.1: DNA-BASED SEX TESTS.

Locus	Features	Advantages	Disadvantages	Reference
AMELY/AMELX	6-bp difference between coamplified Y- and X-products	Small products, so works well with degraded DNA. Standardized method	Sex-reversal and Y-chromosomal deletions – see text	Sullivan et al., 1993
SRY	Sex-determining gene; can be used in combination with AMELY/AMELX	Correctly identifies translocation XX males; not subject to deletion in males	Not very useful on its own. Some XY females carry SRY; Some XX males lack it	Santos et al., 1998
ZFY/ZFX	XY-homologous gene pair with 1151-bp X-product and 729-bp Y-product	Identifies translocation XX males, since Y locus is usually included in translocated segment; can be used for primate sexing	Large PCR products unsuitable for degraded DNA	Wilson and Erlandsson, 1998
DXYS156	XY-homologous microsatellite; Y-copy contains single base insertion, allowing presence of Y allele to be distinguished	Microsatellite variability contributes to DNA profile; ethnic specificity of alleles can give place-of-origin information	Test complicated by allelic variation; requires high resolution (1 bp) analysis	Cali et al., 2002
AluSTYa/AluSTXa	Two assays targeting XY-homologous loci: one Y-specific and one X-specific fixed Alu insertion	Two independent loci unlikely to be affected by same deletion event; simple gel-based assay	Fragment sizes large (200–900 bp); rare individuals may not be fixed for the insertion(s)	Hedges et al., 2003

TABLE 15.2: ASSIGNING A DNA PROFILE TO AN 'ETHNIC GROUP'.

	Predicted group (%)				
True group	**C**	**A-C**	**I**	**SEA**	**ME**
Caucasian (C)	**56**	9	15	9	11
Afro-Caribbean (A-C)	9	**67**	5	8	11
Indian sub-continent (I)	17	8	**43**	18	13
Southeast Asian (SEA)	6	6	13	**66**	9
Middle Eastern (ME)	18	16	19	18	**30**

The percentage of profiles correctly (bold) and incorrectly predicted, for six-locus SLP profiling (from Lowe *et al.*, 2001). Classifications are those used by the police in England and Wales.

Recent studies (Rosenberg *et al.*, 2002; Bamshad *et al.*, 2003) have successfully used the model-based clustering algorithm STRUCTURE (see *Box 9.3*) to identify subgroups that have distinctive allele frequencies; this work is discussed more fully in Section 9.3.3. These studies suggests that conventional ways of defining geographical origins do have some genetic meaning, and might offer hope to forensic scientists. However, it is important to remember that the populations in these collections were chosen to be nonadmixed regional populations (the Rosenberg *et al.* study used the CEPH human diversity panel; Cann *et al.*, 2002). If these methods were applied to urban samples from New York or London they might not perform so well. There may also be practical difficulties in typing as many as 100 markers in the limited samples often available to forensic scientists

(although there are promising technical developments that may overcome this problem – see *Box 4.1*), and in funding such large-scale analyses.

A number of studies have used the Y chromosome and mtDNA to investigate population of origin in admixed groups. These loci show strong geographic differentiation (see *Boxes 8.5* and *8.8*) because their small effective population size leads to strong genetic drift. Mating practices may also contribute to interpopulation differences (see *Box 5.6*). Studies in populations from South America (Bravi *et al.*, 1997; Alves-Silva *et al.*, 2000; Carvajal-Carmona *et al.*, 2000; Carvalho-Silva *et al.*, 2001), Polynesia (Sykes *et al.*, 1995; Hurles *et al.*, 1998) and Greenland (Saillard *et al.*, 2000; Bosch *et al.*, 2003), for example, have identified lineages of different continental origins (the result of sex-biased

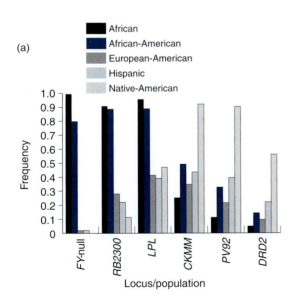

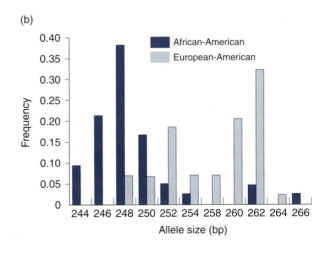

Figure 15.6: Markers showing large allele frequency differences between populations of the USA.

(a) Binary markers; note that one of these, showing high frequency in Africans/African–Americans and low frequency elsewhere, is the FY-null allele conferring resistance to *Plasmodium vivax* malaria (see Section 9.4.3). Some of these markers may owe their large interpopulation frequency differences to selection, as well as drift. (b) Alleles of a dinucleotide microsatellite, *D7S657*. Redrawn from Shriver *et al.* (1997).

admixture; see Section 12.4). However, the susceptibility of the Y chromosome and mtDNA to drift means that quantifying admixture using these loci may be unreliable.

15.2.3 Prediction of phenotypic features

As was discussed in Chapter 13, many human phenotypic features (for example, stature, facial features and pigmentation) have a strong genetic component. If we understood their genetic basis we might hope to be able to make reliable predictions about the phenotype of an individual from a DNA sample. However, there are two problems. First, as we have already seen, these quantitative traits are complex and multigenic, and very little is currently understood about the genes that underlie them. There is no guarantee that, even with knowledge of these genes, a simple predictive test would emerge. Second, the environment has a part to play in these phenotypes. For instance, stature and morphology are influenced by diet, and pigmentation is influenced by climate and habits, as well as by the application of hair dye and novelty contact lenses.

Indirect information about phenotype can be sought by trying to predict population of origin (see previous section). If there is a strong prediction from a DNA profile that an individual is of African, or East Asian, or Northern European descent, such information is likely to influence our expectations of the appearance of the person who deposited the DNA sample. However, as has been discussed, this indirect information may be misleading, particularly in admixed populations.

Direct inference about phenotype would be more useful. The only phenotypic trait that has so far been addressed by forensic scientists is red hair (Section 13.3.3), and its correlation with polymorphism in the melanocortin 1 receptor (*MC1R*) gene. The frequency of a number of *MC1R* variants has been examined in a group of people who self-described their hair color (*Table 15.3*). The results suggest that we can be confident that someone who is homozygous for the consensus sequence will not have red hair (in agreement with Palmer *et al.*, 2000). Also, if someone is homozygous or compound heterozygous for a 'red hair-associated' variant, they have a > 90% chance of having red hair. This test therefore seems to be a potentially useful investigative tool in some populations; if there is a large pool of suspects, and the DNA data show homozygosity or compound heterozygosity suggesting red hair, this would allow prioritization of the red headed among the suspects. In a UK population of European ancestry this would represent only about 5% of the sample (Sunderland, 1956). Note, however, that there are two individuals in the forensic study who do not have red hair, but nonetheless are compound heterozygotes; these people reported having had red hair in their youth, which illustrates that hair color changes with age.

While red hair currently provides the main example, apart from sex, of a forensically useful phenotype that can be predicted from the genotype, this area of research seems likely to expand considerably in the future.

TABLE 15.3: PREDICTIVE POWER OF THE *MC1R* GENOTYPE FOR HAIR COLOR.

MC1R genotype	Red hair	Other color
Wild-type/wild-type	0	35
Wild-type/red variant	4	33
Red variant 1/red variant 2	32	2[a]
Red variant 1/red variant 1	14	0

'Red variant' means a variant thought to be associated with red hair, as defined in this study, and includes the three variants described in Section 13.3.3. Other variants were also found, but are not included in this table. Adapted from Grimes *et al.*, 2001.

[a]Two individuals with blonde and light hair reported as having red hair when young.

15.3 Deducing family and genealogical relationships

15.3.1 General technical and statistical issues

Typing any marker in a mother–father–child trio can give information about the biological parentage of the child. Indeed, nonpaternity can often be revealed as a by-product of testing within a family for a genetic disorder, and this is an important ethical issue for genetic counselors (reviewed by Lucassen and Parker, 2001; and see *Box 15.4*). As soon as DNA fingerprinting was developed, it was shown by family studies that the inheritance of the many bands in a fingerprint was **Mendelian**, and therefore that close family relationships could be verified to near-certainty by DNA analysis (Jeffreys *et al.*, 1985b). An early case in which this was important was the verification that a Ghanaian boy, seeking to re-enter the UK, was indeed the son of the UK-resident woman claiming him as her child (Jeffreys *et al.*, 1985a). This is a maternity test, which is atypical; usually (except in the rare cases of inadvertent switches in hospital nurseries, one of which is described by Pena and Chakraborty, 1994) the mother is known and undisputed, and the question is about the father (*Figure 15.7*). As with forensic applications, technology in **paternity testing** has evolved from fingerprinting, through SLP profiling, to microsatellite profiling. Paternity testing is big business, with about 250 000 tests performed each year in the USA, and 15 000 in the UK, for example.

The technical and statistical issues surrounding paternity testing are similar to those in profiling for other purposes. A paternity case usually compares these two hypotheses:

▶ H_1: the alleged father is indeed the father; and
▶ H_0: the father is some unknown, unrelated man.

Genotypes are determined for child, alleged father, and mother, and the likelihood ratio X/Y is calculated, where X and Y are:

X = P(observing child's genotype given adults' genotypes and assuming H_1), which can be written $P(E|H_1)$

Y = P(observing child's genotype given adults' genotypes and assuming H_0), or $P(E|H_0)$

X/Y, often known as the **paternity index**, is thus a measure of how much more readily the observed results are explained by a true relationship than by coincidence. A simple example using a single marker is illustrated in *Figure 15.7A*, where the probability of true paternity increases with the rarity of a paternal allele. As long as the various systems tested are statistically independent, paternity index for a set of markers is simply the product of the values for the constituent systems. Thus, when large numbers of microsatellites are used, probabilities of true paternity can become extremely large – of the order of a million-fold more likely than false paternity. Often paternity casework requires a prior probability of true paternity to be stated, and by convention

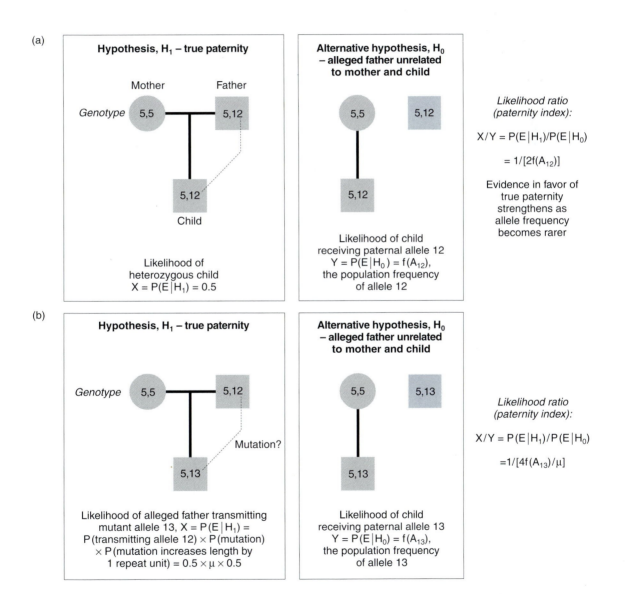

Figure 15.7: DNA-based paternity testing.

Genotypes at a single autosomal microsatellite are shown, with numbers indicating repeat units.
(a) Absence of mutation. Likelihoods for other allele combinations in the trio can be calculated. (b) Taking into account the possibility of single-step mutation in a paternal microsatellite allele. The assumption here is that a single-step increase or decrease is equally likely, and that other larger scale mutations can be neglected (as discussed in Section 3.3.1, this assumption is not very realistic). Formulae are available to take these rarer mutations into account. Note that allele-specific mutation rates are not available, so the average rate has to be used.

BOX 15.4 Nonpaternity rates in human populations

Commonly, if you ask a human geneticist what the nonpaternity rate is, they will come back with a quick answer of 'about 10%'. When you ask them how they know this, they are often not sure, and the figure has been described as an 'urban myth' by some (reviewed by Macintyre and Sooman, 1991). A more extravagant oft-used source is a remark made at a symposium in 1972 that an immunological blood-group study in southeast England had to be stopped because the nonpaternity rate was so high – 30%. Sometimes the rate quoted is that found when typing markers in a family segregating a genetic disorder of some kind, and if so, this could represent an ascertainment bias, since the presence of the disorder in the family itself means that they are not typical of the general population. Similarly, overall rates certainly cannot be estimated from nonpaternity casework (where the anecdotal rate is about 30%), because the sample does *not* represent the general population.

Systematic studies of nonpaternity rates are few and far between. Early experience of cystic fibrosis screening found seven nonpaternities out of 521 families (1.35%) (Brock and Shrimpton, 1991). One study in Switzerland (Sasse *et al.*, 1994) looked at 1607 children and their parents and found 11 exclusions, corresponding to a rate of < 1%. However, it may be that rates vary between populations: that in a Mexican population was estimated at ~ 12% (Cerda-Flores *et al.*, 1999). It is certainly true that there is variation in the degree of acceptability of nonpaternity in different cultures. For example, in the 'customary adoption' practice ('*Kupai Omasker*') of the people of the Torres Strait Islands, a high proportion of children are not the offspring of the adults with whom they live (Ban, 1994).

Paternity often has social and financial implications, so fathers unwilling to support their children sometimes request a paternity test; testing without the consent of both partners is currently available from some commercial companies. The social implications of such genetic information are currently subjects of debate.

---

this is 0.5 – the alleged father is equally likely to be the true father or not.

Because sexual relationships, even more than crimes, often occur among a fairly small group of people including family members, the possibility that the true father is actually a close relative of the suspected father (e.g., a brother) needs to be entertained. Disentangling the offspring of different possible incestuous relationships is even more complex, and computer programs are often used in these cases. Usually, an exclusion of a parent or child is more straightforward; however, there is a complicating factor here, too. A person who gives a sample to a database leaves the identical DNA at a crime scene. A parent who gives a sample to a paternity-testing lab, however, will not necessarily have passed on exactly the same DNA to their child: there could have been a mutation (*Figure 15.7B*). The mutation rate of the microsatellites used for forensic and paternity studies is low: a study of nine-locus genotypes in 10 844 parent–child transfers in paternity casework (Brinkmann *et al.*, 1998) found 23 microsatellite mutations. Mutation rates for the loci were between zero and 0.7% per locus per generation. What is more, most true mutations (22 out of 23) involve a single-step mutation rather than a multi-step change, and this property helps to distinguish them from mismatches due to nonpaternity. These factors can be taken into account when calculating the paternity index (*Figure 15.7B*). Note that, if mutation rate for several microsatellites were simultaneously elevated, for example by a mutation in a component of the mismatch repair pathway (see *Figure 3.19*), then the simple modeling of mutation in paternity analysis would be far from adequate.

Deduction of undisputed mother–father–child relationships using DNA is not difficult. In accidents and disasters, such as the August 1996 Spitzbergen air-crash (Olaisen *et al.*, 1997), possible profiles of missing children can be reconstructed from parents and reliably matched against

profiles from human remains. This allows the remains of victims to be returned to their families, which is of great importance to them. The skeletal remains of a murder victim were identified 8 years after her death by comparison with the profiles of her presumptive parents (Hagelberg *et al.*, 1991). As relationships become more distant, however, the amount of shared DNA becomes less, and the degree of certainty also lessens. For a 10-locus profile, a parent contributes 10 bands to a child; first cousins, on the other hand, share on average only 2.5 bands by descent, which is likely to provide no support for this relationship, given that individuals can often share some bands by chance.

For a male child whose paternity is in question, but where the potential father cannot be traced or is dead, Y-chromosomal markers can be used (*Figure 15.8*). This is called 'deficiency paternity testing', and relies on matching the Y chromosome of the child with that of a potential grandfather or paternal uncle, for example (reviewed by Jobling *et al.*, 1997). In the event of a match, the paternity is supported, with paternity index equal to the inverse of the frequency of the allele or haplotype (*Figure 15.8a*), but in the event of an exclusion, ambiguity remains, since it is impossible to determine in which generation the nonpaternity took place (*Figure 15.8b*). Mutation at Y-specific microsatellites also needs to be taken into account (Rolf *et al.*, 2001), and the caveats stated above (Section 15.1.5) about population structure are again important.

Maternal relationships within families can also be investigated using mtDNA. As has already been mentioned, victims of disasters can be identified, in effect, by maternity testing through mtDNA. Because of their uniparental inheritance, Y chromosomes and mtDNA can provide a way to link individuals reliably across a large number of generations provided unbroken lines of patrilineal or

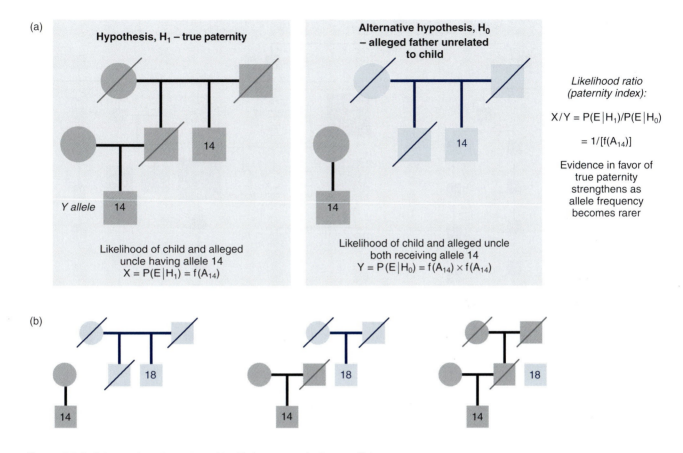

Figure 15.8: Deficiency paternity testing with a Y-chromosomal microsatellite.

(a) Evaluating the likelihood ratio (paternity index) in the case of a shared microsatellite allele between a son and alleged paternal uncle.
(b) In the event of an exclusion that cannot easily be explained by a mutation (in this case, a four-step difference), there remains ambiguity about the nonpaternity event. Typing of further male-line relatives may resolve this.

matrilineal descent exist. Two examples illustrate this nicely: the identification of the remains of the Romanovs using mtDNA; and the investigation of the Thomas Jefferson paternity case using Y chromosomes.

15.3.2 Genealogical studies using the Y chromosome and mtDNA

The Jefferson case is illustrated in *Figure 15.9*. The question to be addressed was whether Thomas Jefferson fathered any of the children of one of his slaves, Sally Hemings. Thomas's wife Martha predeceased him by 44 years after bearing him two daughters. For two of Sally's sons, Tom Woodson (claimed as a son of Sally in the oral history of his descendants) and Eston Hemings Jefferson, modern male-line descendants could be found. Because Thomas did not have any legitimate sons of his own, modern male-line descendants of his uncle, Field Jefferson, were used as a source of the 'Jefferson' Y chromosome. As a defense against the accusation that Sally's children resembled Thomas, his own legitimate descendants had claimed that the true father was one or both of Thomas's nephews Samuel and Peter

Carr, the sons of his sister. To test this hypothesis, modern male-line descendants of the Carrs were also traced. Using Y-specific binary markers, microsatellites and a minisatellite, detailed Y-chromosomal haplotypes were constructed (Foster *et al.*, 1998). Four of the five Field Jefferson descendants shared the same Y-chromosomal haplotype (the fifth differed by one repeat unit at one locus, most likely representing a mutation), and the Carr descendants shared a different haplotype. Therefore it could be asked which, if either, of these haplotypes Sally Hemings's descendants carried. The descendant of her last child, Eston, but not those of Tom Woodson, matched the Field Jefferson descendants, which was consistent with Thomas being Eston's father. The frequency of the Jefferson haplotype in the general population was difficult to estimate, but low: for example, the microsatellite haplotype was not observed in over 1200 other individuals analyzed. Of course, as was pointed out at the time of this study (Abbey, 1999), any one of Thomas's contemporary male-line relatives, including his brother Randolph, could have been the true father, and it is therefore impossible to offer formal proof of paternity. However, historians have shown that Thomas and Sally were together at Monticello (Jefferson's home in Virginia) nine

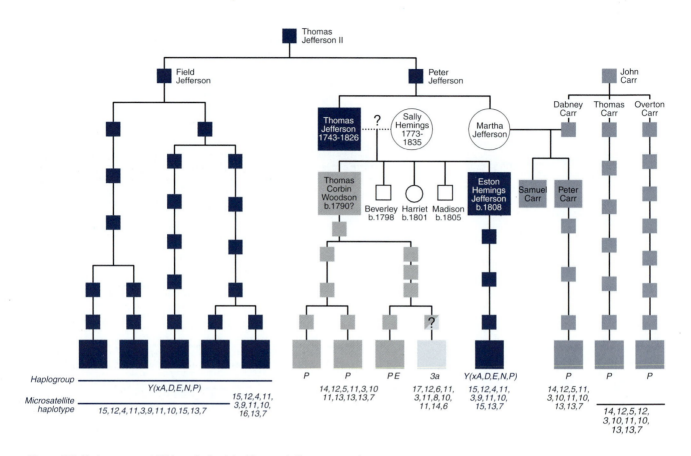

Figure 15.9: Y-chromosomal DNA analysis of the Thomas Jefferson paternity case.

Four of the Field Jefferson descendants and Eston Hemings, Jefferson's descendant, share an identical Y chromosome, distinct from those in the Woodson and Carr descendants, and this is consistent with Thomas Jefferson's fathering Eston Hemings Jefferson, Sally Hemings's last child. Note that there is nonpaternity among the Woodson descendants, and a microsatellite mutation in one of the Field Jefferson descendants and the descendant of Peter Carr. Y haplogroup nomenclature is according to the YCC (see *Box 8.8*), and the microsatellite haplotype is the number of repeat units of each of 11 Y-specific loci. The MSY1 minisatellite was also typed (not shown). For details of markers, see Foster *et al.* (1998).

months before the birth of each of Sally's children (reviewed by Gordon-Reed, 1997), which at least 'fails to exclude' Thomas.

This kind of genealogical study should be generally applicable to the tracing of linkages between patrilineal branches of family trees. Indeed, extending lines of patrilineal descent into the past, we can ask whether all men who share a patrilineal surname share a Y-chromosomal haplotype (reviewed by Jobling, 2001). This would be true if there were a single founder for the surname, no name changes, illegitimacy or unrecorded adoption, and no mutation since foundation. While the first of these conditions might seem reasonable for some rare surnames, the others do not. A study of the English surname Sykes (Sykes and Irven, 2000) showed that 21/48 unrelated men of this name, sampled from three counties of England, share an identical haplotype as defined by four Y-chromosomal microsatellites. The other 27 belong to a range of different haplotypes, and this pattern was interpreted as a single origin for the surname, followed by illegitimacies introducing different haplotypes into the

surname. However, a number of other explanations seem compatible with the observed pattern, including a larger number of founders (men who originally took up the name Sykes) followed by drift so that one founding haplotype reached high frequency while others became rare or even went extinct. It should be possible to throw light on the range of plausible possibilities by simulation. In any case, the hope in some forensic circles that surname might be predictable from Y haplotype certainly seems over-optimistic, but there is great interest among individuals in using Y typing as an aid to tracing their own ancestry.

mtDNA was famously applied to the question of establishing the identity of human remains reputed to be those of Czar Nicholas II of Russia and his family (*Figure 15.10*), killed by the Bolsheviks in 1918 (Gill *et al.*, 1994; Ivanov *et al.*, 1996). Nine skeletons were discovered in a communal grave in Ekaterinburg, Russia, in 1991. These had been tentatively identified as the remains of the Czar, the Czarina, three of their daughters, three servants and the family's doctor. Sexing using the amelogenin test agreed with

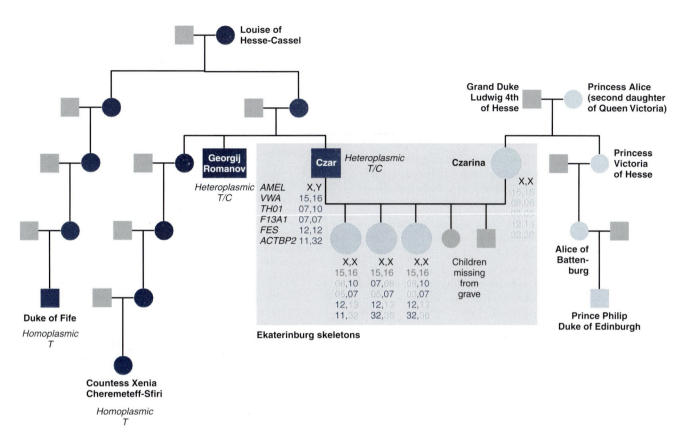

Figure 15.10: Confirmation of the identity of the Romanov skeletons by DNA analysis.

Ausotomal microsatellite data confirm the family relationships of the five skeletons, and exclude the other four adults as parents (not shown). Paternal alleles are shown in dark blue, maternal in pale blue, and uninformative alleles in gray. Modern matrilineal relatives confirm the identity of the Czar and Czarina by mtDNA analysis. The Czarina's maternal lineage is shown by pale blue filled symbols, and that of the Czar in dark blue. The Czar's exhumed brother, Georgij Romanov, shares with him a heteroplasmic position in mtDNA that is homoplasmic in two living relatives. Drawn from information in Gill *et al.* (1994) and Ivanov *et al.* (1996).

the skeletal analysis, and microsatellite profiling using five loci was consistent with the family relationship of five of the skeletons. Sequencing of mtDNA HVS I and II showed that the Czarina and the three daughters had the same sequence, confirming the family relationship. A modern matrilineal relative of the Czarina, HRH Prince Philip, the Duke of Edinburgh, had the same sequence also, supporting her identity. mtDNA analysis of the Czar's remains, however, produced a surprise: his sequence was identical to those of two living matrilineal relatives at all bases except one. Here, while the relatives possessed a T, the Czar had a mixture of T and C – a **heteroplasmy** (Gill *et al.*, 1994). The remains of the Grand Duke of Russia Georgij Romanov, brother of the Czar, were later exhumed and he was also shown to carry the heteroplasmy (Ivanov *et al.*, 1996). This adds considerable weight to the conclusion that the Ekaterinburg remains included those of Czar Nicholas II, and also illustrates that mtDNA heteroplasmy can resolve within four generations to apparent homoplasmy.

15.3.3 Tracing modern diasporas

Genealogical studies such as those described above can be extended further back in time to encompass a lineage or set of lineages associated with a particular group. These studies are potentially useful in tracing the descendants of historical diasporas, such as that of the Jews, or that of people of African descent whose ancestors were taken as slaves to the Americas. Identification of a particular mtDNA or Y-chromosomal haplotype at high frequency in certain population groups can sometimes be taken as a 'population marker', and commercial companies are quick to offer typing of these markers, to somehow legitimize a customer's membership of a group. It is important to remember that each of us has very many ancestors, and that this matrilineal or patrilineal tracing is considering only a single ancestor out of many (see Section 1.2). Also, there are no 'ideal' markers that are present in all members of a population or group, and not present in others.

A particular Y-chromosomal microsatellite haplotype cluster (now known to lie in haplogroup J) has been shown to be at high frequencies in Jewish groups, and in members of the Cohanim priestly lineage in particular (Thomas *et al.*, 1998; Thomas *et al.*, 2002). Priesthood status is patrilinearly inherited, and many men who claim Cohanim status have surname Cohen, or related names such as Kahn or Kane. The microsatellite diversity within the Cohen haplotype cluster was used to estimate a coalescence time for these Y chromosomes (see Section 6.6). This age, 2650 years (95% confidence interval, 2100–3250 years), required some additional assumptions but is consistent with the Temple period in Jewish history, when the Cohanim were established following the exodus from Egypt.

A study of the diversity of Y-chromosomal haplotypes in Asia has revealed the previously unsuspected spread of a lineage that has a persuasive historical explanation (Zerjal *et al.*, 2003). A sublineage within haplogroup C*(xC3c), defined at high resolution with 16 microsatellites, was found in 16 populations throughout much of Asia, from the Pacific to the Caspian Sea, and made up ~8% of the Y chromosomes sampled from the region. The coalescence time of this star-like cluster of haplotypes was calculated at ~1000 years (95% confidence interval, 590–1300 years), and comparative diversity considerations suggested Mongolia as the source. The age, place of origin, and modern distribution of these chromosomes are consistent with their representing the Y-chromosomal lineage of Genghis Khan, his immediate male relatives and their descendants. This fearsome emperor, who lived between c.1162 and 1227, established a long-lasting male dynasty that ruled large parts of Asia for many generations. A Y chromosome that may have been carried by him is now carried by about 0.5% of the world's men, representing a spectacular founder effect that is expected to have an impact upon allele frequencies of loci in other chromosomes.

Summary

▶ There is a vast number of differences between any one copy of the human genome and another (except in identical twins), providing the potential for individual identification based on DNA.

▶ Multilocus fingerprinting, based on the detection of a set of hypervariable autosomal minisatellite loci, was the first DNA-based method used to identify individuals for forensic purposes. It was supplanted first by single-locus probe methods, and then by the PCR-based multiplex detection of a set of autosomal microsatellites (profiling).

▶ The match probability is the chance that a profile chosen at random from the population will match that from a crime scene. There have been problems in courtroom use of DNA evidence because of arguments about population structure, the exclusion of relatives of suspects, and confused statements regarding the evidential weight of the match probability.

▶ Very large DNA databases have been established containing DNA profiles of offenders and some nonoffenders. These are powerful investigative tools.

▶ Y-chromosomal and mtDNA haplotypes are shared by members of a patriline or matriline, respectively, but have specialized uses in forensic analysis.

▶ Prediction of sex from a DNA sample is almost always reliable. Prediction of red hair is also possible, but the genetic basis of other features is not yet well enough understood to allow prediction.

▶ Prediction of population of origin may be reliable if markers are carefully chosen and populations are not recently admixed.

▶ DNA profiles can also be used for establishing family relationships among DNA samples, such as paternity testing. Comparisons must take into account the possibility of mutation, as well as the population frequency of alleles.

▶ Y-chromosomal markers can be used to test patrilineal relatives in 'deficiency' cases, where the alleged father of a son is not available.

▶ Y-chromosomal and mtDNA haplotypes can be used to follow lineages in genealogical cases, and to trace population diasporas in historical times.

Further reading

Krawczak M, Schmidtke J (1998) *DNA Fingerprinting*, 2nd edn. BIOS Scientific Publishers Ltd., Oxford.

Butler JM (2001) Fo*rensic DNA Typing: Biology and Technology Behind STR Markers*. Academic Press, London.

Electronic references

Short Tandem Repeat DNA Internet DataBase:
http://www.cstl.nist.gov/biotech/strbase/index.htm

Y-STR Haplotype Reference DataBase:
http://ystr.org/europe/ and links therein

FBI mtDNA Population Database:
http://www.fbi.gov/hq/lab/fsc/backissu/april2002/miller1.htm

Thomas Jefferson:
http://www.monticello.org/plantation/hemings_resource.html

References

Abbey DM (1999) The Thomas Jefferson paternity case. *Nature* **397**, 32.

Alves-Silva J, Santos MD, Guimarães PEM, Ferreira ACS, Bandelt H-J, Pena SDJ, Prado VF (2000) The ancestry of Brazilian mtDNA lineages. *Am. J. Hum. Genet.* **67**, 444–461.

Balding DJ (2000) In: *Statistical Science in the Courtroom* (JL Gastwirth, ed.). Springer, New York, pp. 51–70.

Bamshad MJ, Wooding S, Watkins WS, Ostler CT, Batzer MA, Jorde LB (2003) Human population structure and the inference of group membership. *Am. J. Hum. Genet.* **72**, 578–589.

Ban P (1994) Customary 'Adoption' in Torres Strait Islands: towards legal recognition. *Abor. Law Bull.* **3**, No. 66.

Betz A, Bassler G, Dietl G, Steil X, Weyermann G, Pflug W (2001) DYS STR analysis with epithelial cells in a rape case. *Forens. Sci. Int.* **118**, 126–130.

Bosch E, Calafell F, Rosser ZH, Nørby S, Lynnerup N, Hurles ME, Jobling MA (2003) High level of male-biased Scandinavian admixture in Greenlandic Inuit shown by Y-chromosomal analysis. *Hum. Genet.* **112**, 353–363.

Bravi CM, Sans M, Bailliet G *et al.* (1997) Characterization of mitochondrial DNA and Y-chromosome haplotypes in a Uruguayan population of African ancestry. *Hum. Biol.* **69**, 641–652.

Brenner CH (1998) Difficulties in the estimation of ethnic affiliation. *Am. J. Hum. Genet.* **62**, 1558–1560.

Brinkmann B, Klintschar M, Neuhuber F, Hühne J, Rolf B (1998) Mutation rate in human microsatellites: influence of the structure and length of the tandem repeat. *Am. J. Hum. Genet.* **62**, 1408–1415.

Brock DJH, Shrimpton AE (1991) Non-paternity and prenatal genetic screening. *Lancet* **338**, 1151.

Budowle B, Wilson MR, DiZinno JA, Stauffer C, Fasano MA, Holland MM, Monson KL (1999) Mitochondrial DNA regions HVI and HVII population data. *Forens. Sci. Int.* **103**, 23–35.

Butler JM, Schoske R, Vallone PM, Kline MC, Redd AJ, Hammer MF (2002) A novel multiplex for simultaneous amplification of 20 Y chromosome STR markers. *Forens. Sci. Int.* **129**, 10–24.

Cali F, Forster P, Kersting C, Mirisola MG, D'Anna R, De Leo G, Romano V (2002) DXYS156: a multi-purpose short tandem repeat locus for determination of sex, paternal and maternal geographic origins and DNA fingerprinting. *Int. J. Legal. Med.* **116**, 133–138.

Cann HM, de Toma C, Cazes L *et al.* (2002) A human genome diversity cell line panel. *Science* **296**, 261–262.

Capelli C, Tschentscher F, Pascali VL (2003) "Ancient" protocols for the crime scene? Similarities and differences between forensic genetics and ancient DNA analysis. *Forens. Sci. Int.* **131**, 59–64.

Carvajal-Carmona LG, Soto ID, Pineda N *et al.* (2000) Strong Amerind/white sex bias and a possible Sephardic contribution among the founders of a population in northwest Colombia. *Am. J. Hum. Genet.* **67**, 1287–1295.

Carvalho-Silva DR, Santos FR, Rocha J, Pena SDJ (2001) The phylogeography of Brazilian Y-chromosome lineages. *Am. J. Hum. Genet.* **68**, 281–286.

Cerda-Flores RM, Barton SA, Marty-Gonzalez LF, Rivas F, Chakraborty R (1999) Estimation of nonpaternity in the Mexican population of Nuevo Leon: A validation study with blood group markers. *Am. J. Phys. Anthropol.* **109**, 281–293.

Collins-Schramm HE, Phillips CM, Operario DJ *et al.* (2002) Ethnic-difference markers for use in mapping by admixture linkage disequilibrium. *Am. J. Hum. Genet.* **70**, 737–750.

Dyer R (2002) New South Wales Standing Committee on Law and Justice: Review of the Crimes (Forensic Procedures) Act 2000, Sydney, NSW.

Evett IW, Weir BS (1998) *Interpreting DNA Evidence: Statistical Genetics for Forensic Scientists.* Sinauer, Sunderland, MA.

Ferguson-Smith MA, Ferris EA (1991) Gender verification in sport: the need for change? *Br. J. Sport Med.* **25**, 17–20.

Findlay I, Frazier R, Taylor A, Urquhart A (1997) Single cell DNA fingerprinting for forensic applications. *Nature* **389**, 555–556.

Forensic Science Service (2002) Annual Report and Accounts (http://www.forensic.gov.uk/forensic/corporate/annual_rep/ann_report_01–02/report.pdf). Her Majesty's Stationery Office.

Foster EA, Jobling MA, Taylor PG *et al.* (1998) Jefferson fathered slave's last child. *Nature* **396**, 27–28.

Foster MW, Sharp RR (2002) Race, ethnicity, and genomics: social classifications as proxies of biological heterogeneity. *Genome Res.* **12**, 844–850.

Gill P, Jeffreys AJ, Werrett DJ (1985) Forensic application of DNA 'fingerprints'. *Nature* **318**, 577–579.

Gill P, Ivanov PL, Kimpton C *et al.* (1994) Identification of the remains of the Romanov family by DNA analysis. *Nature Genet.* **6**, 130–135.

Gill P, Whitaker J, Flaxman C, Brown N, Buckleton J (2000) An investigation of the rigor of interpretation rules for STRs derived from less than 100 pg of DNA. *Forens. Sci. Int.* **112**, 17–40.

Gordon-Reed A (1997) *Thomas Jefferson and Sally Hemings: an American Controversy.* University Press of Virginia, Charlottesville, VA.

Grimes EA, Noake PJ, Dixon L, Urquhart A (2001) Sequence polymorphism in the human melanocortin 1 receptor gene as an indicator of the red hair phenotype. *Forens. Sci. Int.* **122**, 124–129.

Hagelberg E, Gray IC, Jeffreys AJ (1991) Identification of the skeletal remains of a murder victim by DNA analysis. *Nature* **352**, 427–429.

Hedges DJ, Walker JA, Callinan PA, Shewale JG, Sinha SK, Batzer MA (2003) Mobile element-based assay for human gender determination. *Anal. Biochem.* **312**, 77–79.

Holland MM, Parsons TJ (1999) Mitochondrial DNA sequence analysis – validation and use for forensic casework. *Forens. Sci. Rev.* **11**, 21–49.

Home Office (2002) *Criminal Statistics England and Wales 2001.* The Stationery Office, London.

Hurles ME, Irven C, Nicholson J et al. (1998) European Y-chromosomal lineages in Polynesia: a contrast to the population structure revealed by mitochondrial DNA. *Am. J. Hum. Genet.* **63**, 1793–1806.

Ivanov PL, Wadhams MJ, Roby RK, Holland MM, Weedn VW, Parsons TJ (1996) Mitochondrial DNA sequence heteroplasmy in the Grand Duke of Russia Georgij Romanov establishes the authenticity of the remains of Tsar Nicholas II. *Nature Genet.* **12**, 417–420.

Jeffreys AJ, Brookfield JFY, Semeonoff R (1985a) Positive identification of an immigration test-case using human DNA fingerprints. *Nature* **317**, 818–819.

Jeffreys AJ, Wilson V, Thein SL (1985b) Hypervariable 'minisatellite' regions in human DNA. *Nature* **314**, 67–73.

Jeffreys AJ, Wilson V, Thein SL (1985c) Individual-specific 'fingerprints' of human DNA. *Nature* **316**, 76–79.

Jobling MA (2001) In the name of the father: surnames and genetics. *Trends Genet.* **17**, 353–357.

Jobling MA, Pandya A, Tyler-Smith C (1997) The Y chromosome in forensic analysis and paternity testing. *Int. J. Legal Med.,* **110**, 118–124.

Kayser M, Brauer S, Willuweit S et al. (2002) Online Y-chromosomal short tandem repeat haplotype reference database (YHRD) for US populations. *J. Forens. Sci.* **47**, 513–519.

Krawczak M (2001) Forensic evaluation of Y-STR haplotype matches: a comment. *Forens. Sci. Int.* **118**, 114–115.

Lander ES (1989) DNA fingerprinting on trial. *Nature* **339**, 501–505.

Lander ES, Budowle B (1994) DNA fingerprinting dispute laid to rest. *Nature* **371**, 735–738.

Lowe AL, Urquhart A, Foreman LA, Evett IW (2001) Inferring ethnic origin by means of an STR profile. *Forens.Sci.Int.* **119**, 17–22.

Lucassen A, Parker M (2001) Revealing false paternity: some ethical considerations. *Lancet* **357**, 1033–1035.

Macintyre S, Sooman A (1991) Non-paternity and prenatal genetic screening. *Lancet* **338**, 869–871.

Monson KL, Miller KWP, Wilson MR, DiZinno JA, Budowle B (2002) The mtDNA population database: an integrated software and database resource of forensic comparison. *Forens. Sci. Comm.* **4**, http://www.fbi.gov/hq/lab/backissu/april2002/miller1.htm.

Neel JV (1973) "Private" genetic variants and the frequency of mutation among South American Indians. *Proc. Natl Acad. Sci. USA* **70**, 3311–3315.

Olaisen B, Stenersen M, Mevåg B (1997) Identification by DNA analysis of the victims of the August 1996 Spitzbergen civil aircraft disaster. *Nature Genet.* **15**, 402–405.

Palmer JS, Duffy DL, Box NF et al. (2000) Melanocortin-1 receptor polymorphisms and risk of melanoma: is the association explained solely by pigmentation phenotype? *Am. J. Hum. Genet.* **66**, 176–186.

Parra EJ, Marcini A, Akey L et al. (1998) Estimating African American admixture proportions by use of population-specific alleles. *Am. J. Hum. Genet.* **63**, 1839–1851.

Parra EJ, Kittles RA, Argyropoulos G et al. (2001) Ancestral proportions and admixture dynamics in geographically defined African Americans living in South Carolina. *Am. J. Phys. Anthropol.* **114**, 18–29.

Pena SDJ, Chakraborty R (1994) Paternity testing in the DNA era. *Trends Genet.* **10**, 204–209.

Redd AJ, Agellon AB, Kearney VA et al. (2002) Forensic value of 14 novel STRs on the human Y chromosome. *Forens. Sci. Int.* **130**, 97–111.

Roewer L, Krawczak M, Willuweit S et al. (2001) Online reference database of Y-chromosomal short tandem repeat (STR) haplotypes. *Forens. Sci. Int.* **118**, 103–111.

Rolf B, Keil W, Brinkmann B, Roewer L, Fimmers R (2001) Paternity testing using Y-STR haplotypes: assigning a probability for paternity in cases of mutations. *Int. J. Legal. Med.* **115**, 12–15.

Rosenberg NA, Pritchard JK, Weber JL, Cann HM, Kidd KK, Zhivotovsky LA, Feldman MW (2002) Genetic structure of human populations. *Science* **298**, 2381–2385.

Saillard J, Forster P, Lynnerup N, Bandelt H-J, Nørby S (2000) mtDNA variation among Greenland Eskimos: the edge of the Beringian expansion. *Am. J. Hum. Genet.* **67**, 718–726.

Santos FR, Pandya A, Tyler-Smith C (1998) Reliability of DNA-based sex tests. *Nature Genet.* **18**, 103.

Sasse G, Muller H, Chakraborty R, Ott J (1994) Estimating the frequency of non-paternity in Switzerland. *Hum. Hered.* **44**, 337–343.

Shriver MD, Smith MW, Jin L, Marcini A, Akey JM, Deka R, Ferrell RE (1997) Ethnic-affiliation estimation by use of population-specific DNA markers. *Am. J. Hum. Genet.* **60**, 957–964.

Sibille I, Duverneuil C, de la Grandmaison GL, Guerrouache K, Teissiere F, Durigon M, de Mazancourt P (2002) Y-STR DNA amplification as biological evidence in sexually assaulted female victims with no cytological detection of spermatozoa. *Forens. Sci. Int.* **125**, 212–216.

Steinlechner M, Berger B, Niederstatter H, Parson W (2002) Rare failures in the amelogenin sex test. *Int. J. Legal. Med.* **116**, 117–120.

Sullivan KM, Mannucci A, Kimpton CP, Gill P (1993) A rapid and quantitative DNA sex test – fluorescence-based PCR analysis of X-Y homologous gene amelogenin. *Biotechniques* **15**, 636.

Sunderland E (1956) Hair-colour variation in the United Kingdom. *Ann. Hum. Genet.* **20**, 312–330.

Sykes B, Leiboff A, Low-Beer J, Tetzner S, Richards M (1995) The origins of the Polynesians: an interpretation from mitochondrial lineage analysis. *Am. J. Hum. Genet.* **57**, 1463–1475.

Sykes B, Irven C (2000) Surnames and the Y chromosome. *Am. J. Hum. Genet.* **66**, 1417–1419.

Thangaraj K, Reddy AG, Singh L (2002) Is the amelogenin gene reliable for gender identification in forensic casework and prenatal diagnosis? *Int. J. Legal. Med.* **116**, 121–123.

Thomas MG, Skorecki K, Ben-Ami H, Parfitt T, Bradman N, Goldstein DB (1998) Origins of Old Testament priests. *Nature* **394**, 138–140.

Thomas MG, Weale ME, Jones AL *et al.* (2002) Founding mothers of Jewish communities: Geographically separated Jewish groups were independently founded by very few female ancestors. *Am. J. Hum. Genet.* **70**, 1411–1420.

van Oorschot RAH, Jones MJ (1997) DNA fingerprints from fingerprints. *Nature* **387**, 767.

Werrett DJ (1997) The National DNA Database. *Forens. Sci. Int.* 88, 33–42.

Wilson JF, Erlandsson R (1998) Sexing of human and other primate DNA. *Biol. Chem.* **379**, 1287–1288.

Zerjal T, Xue Y, Bertorelle G *et al.* (2003) The genetic legacy of the Mongols. *Am. J. Hum. Genet.* **72**, 717–721.

α-thalassemia: A blood cell disorder resulting from a deficiency of the α-globin protein

3′ end: The end of a DNA or RNA molecule bearing a free hydroxyl (–OH) group (unattached to another nucleotide) on the 3′ carbon

5′ end: The end of a DNA or RNA molecule bearing a free hydroxyl (–OH) group (unattached to another nucleotide) on the 5′ carbon

Acclimatization: Physiological mechanisms producing short-term and sometimes rapid reversible responses to environmental change

Acculturation: The spread of cultural practices by learning from neighboring groups, rather than by the spread of people themselves; a possible mode of spread of agriculture in Europe, for example

Acheulean: One of the stone tool traditions of the Lower Paleolithic, including *bifaces* and cleavers; named after the French site of St. Acheul

Acrocentric: Of a chromosome – having the *centromere* close to one end of the chromosome

Adaptation: A character that has been favored by natural selection for its effectiveness in a particular role; an adaptation is not necessarily *adaptive*

Adaptive: Describing a character that functions currently to increase reproductive success; see *adaptation*

Adaptive immunity: The system of immune response that identifies a foreign agent, adapts to best recognize it and remembers for future reference, and thus is able to mount a more effective response on future exposure to the same agent

Additive tree: A *phylogeny* with branch lengths proportional to the amount of evolutionary change

Adenine: A *purine* base found within DNA – pairs with *thymine*

Admixture: The formation of a hybrid population through the mixing of two ancestral populations

Advantageous allele: An allele that increases the fitness of its carrier

Agonist: A drug that triggers an action, e.g., from a cell; contrasted with an antagonist, which blocks or nullifies an action

Allele: One of two or more alternative forms of a gene or DNA sequence at a specific chromosomal location

Allele frequency: The frequency of an allele within a population

Allelic architecture: The number and frequency of susceptibility alleles at a locus

Allen's rule: Shorter appendages are favored in colder climates because they have relatively high ratios of mass to surface area, and therefore retain heat

Alpha factor: The ratio of mutations that occur in the male germ-line compared to those that occur in the female germ-line, see *male-driven evolution*

Alphoid repeat: A repeat unit of ~ 170 bp found at the *centromeres* of chromosomes and organized into higher order repeat structures

Altaian: A person from the mountainous Altai region of southwest Siberia

Alu element: A member of the *SINE* class of interspersed repeated sequences. The element is about 290 bp long, and has a copy number of about 1 million

Alu insertion polymorphism: A polymorphism defined by the insertion of an *Alu element* into one allele

Amelogenin: A gene on the X and Y chromosomes commonly used in PCR-based sex testing

Amerind: A language superfamily proposed by Joseph Greenberg that includes all Native American languages not belonging to the *Eskimo-Aleut* and *Na-Dene* language families; regarded by some as controversial

Amino acid: One of the building blocks of proteins; 20 different amino acids are commonly found in proteins

Amplicon: A DNA sequence amplified in a specific PCR reaction

Amplified fragment length polymorphism, AFLP: Fragments amplified by PCR that are polymorphic in length due to variation in the position of restriction sites in the original source DNA

Anagenesis: Speciation through the accumulation of changes in a single species; contrasted with *cladogenesis*, speciation through branching

Analogous: Similarity not due to common ancestry

Anatomically modern humans: *Homo sapiens* showing modern head and body structure; contrasted with archaic forms of humans such as *Neanderthals*; anatomically modern humans do not necessarily show other modern characteristics such as modern behavior

Ancestral haplotype: The *haplotype* on which a new variant arose

Ancestry informative marker, AIM: A marker having alleles that are significantly more common in one group of populations than in other groups. Useful in admixture studies

Aneuploidy: Departure from the normal diploid number of 46 human chromosomes

Annealing: The stage of a PCR reaction in which *oligonucleotide primers* are allowed to bind to their complementary sites in the single-stranded template; typically carried out at 50–60°C

Anthropocentric: A view of the world centered on how it relates specifically to humans

Anthropogenic: Of an environmental phenomenon – caused by human activity

Antibody: A protein produced by white blood cells in response to *antigen*; antigen–antibody reactions are very specific

Antigen: A substance that elicits an immune response

Aotearoa: The Maori name for New Zealand

Apomorphic: A derived characteristic, one that has been altered from its ancestral state

Archipelago: A cluster of closely positioned islands

Arrays: Closely spaced DNA or protein molecules that allow the simultaneous investigation of many macromolecules

Ascertainment bias: Distortions in a dataset caused by the way that markers or samples are collected

Association study: see *Disease association study*

Assortative mating: Using phenotypic similarity to guide mate choice

Assortment: see *Lineage assortment* or *Independent assortment*

Aurignacian: One of the earliest *Upper Paleolithic* cultures, defined by its tools and associated with *anatomically modern humans* in Europe. Named after the first site where it was discovered, Aurignac in the Pyrenees

Aurochs: *Bos primigenius*, the progenitor species to modern cattle; an unusual singular word with the plural 'aurochsen'

Australoid: One of the human morphological types identified by early anthropologists, typified by Australian Aborigines

Austronesian: A large language family spoken throughout Island Southeast Asia and the Pacific

Autocorrelation: A correlation calculated between all items in a series

Autoradiography: The process of recording bands on a Southern blot that are hybridizing to radio-labeled *probes* by exposure to X-ray film

Autosome: One of the 22 biparentally inherited chromosomes, each present in two copies in males and females, and numbered 1 to 22

Back mutation: A mutation from the derived state back to the ancestral state

Background selection: The reduction in diversity at linked polymorphic markers as a result of nearby *negative selection*

Bacterial artificial chromosome, BAC: A DNA vector in which inserts of up to 300 kb long can be propagated in bacterial cells

Balancing selection: A selective regime that favors more than one allele and thus prevents the fixation of any allele

Bantu expansion: A complex series of population movements from western Africa into central and southern Africa, in which the Bantu languages, iron-working and farming may have been co-dispersed

Base: The informational part of a nucleotide molecule, the building block of DNA; the bases in DNA are *adenine* (A), *guanine* (G), *thymine* (T) and *cytosine* (C)

Base composition: The proportion of the four bases in a given DNA sequence

Base-pairing: The interaction of one base with another through hydrogen-bond formation – A pairs with T, and G with C. This interaction holds one strand of DNA together with its complementary strand

Base substitution: A mutation in which one base of DNA is exchanged for another

Bathymetry: Measurement of sea depth

Bayesian: Statistical methods based on Bayes' theorem that allows inferences to be drawn from both the data and any prior information

Bergmann's rule: Body size increases as climate becomes colder, because as mass increases, surface area does not increase proportionately

Beringian land bridge, Beringia: Land that joins the North American and Eurasian continents across the Bering straits at times of low sea level

Biallelic: Of a polymorphism – possessing two alleles only (see *diallelic, binary marker*)

Biface: Teardrop-shaped stone tool worked on both sides and most of the margin, sometimes called a *handaxe*; biface is the preferred term because it does not imply that the tool was used as an axe

Bifurcation: The splitting of an ancestral lineage into two daughter lineages

Binary marker: A polymorphism possessing two alleles only (see *biallelic, diallelic*)

Bioinformatics: The computational analysis of biological data including sequence analysis and expression studies

Biological species concept: One of a number of species definitions, based on an ability to interbreed in the wild

Bipedalism: Walking upright, on two feet, like modern humans

Blades: Flakes, commonly made from flint or similar material, that are long and narrow

Block: See *Haplotype block*

Bonobo: The pygmy chimpanzee, *Pan paniscus*

Bootstrap: A method for assessing how well supported by the data are individual *clades* within a *phylogeny*

Bottleneck: The reduction of genetic diversity that results from a dramatic reduction in population size

Brachycephalic: 'Broad-headed', with a *cephalic index* over 80 (see *dolichocephalic, mesocephalic*)

Broad sense heritability: The proportion of genetic variance that can be attributed to additive, dominant, and *epistatic* genetic effects

Buccal cells: Cells taken from inside the cheek with a brush or swab – a convenient source of small quantities of human genomic DNA

Candidate gene: A gene selected for possible involvement in a phenotype because of what is known about the physiological or biochemical basis of the phenotype, and the function of the gene product

Cardial ware: A distinctive type of pottery associated with the spread of farming in the northern Mediterranean region; impressed with the serrated edge of the cardium (cockle) shell

Carrier: A person who is *heterozygous* for a recessive mutation. They carry the mutation, but do not manifest the disease, although they can pass it on to a child if their partner has a mutation in the same gene

Caucasian: An old racial definition based on a skull from the Caucasus mountains, and including the peoples of western Europe. Commonly used to mean European, or of European descent

Cell line: Cultured cells established from primary sampled tissues such as skin fibroblasts or white blood cells

centiMorgan: A measure of recombination frequency – two markers separated by 1% recombination frequency are 1 centiMorgan (cM) apart

Centromere: The primary constriction of a chromosome, separating the short arm from the long arm, and the point at which *spindle fibers* attach to enable the chromosomes to move apart at cell division

Cephalic index: The ratio of the breadth to the length of the skull multiplied by 100

Châtelperonian: One of the first *Upper Paleolithic* cultures, defined by the tools used and associated with *Neanderthals*

Chromosomal segregation: The separation of chromosomes into daughter cells when a cell divides

Chromosome: The DNA–protein structure that contains part of the nuclear genome; diploid human cells have 46

Chromosome painting: The hybridization of labeled DNA from a single chromosome to a metaphase chromosome spread; useful in detecting chromosomal rearrangements within and between species

Clade: An evolutionary branch

Cladistics: The science of reconstructing evolutionary relationships by identifying common ancestors through the sharing among taxa of *derived* characteristics, rather than the sharing of *ancestral* characteristics

Cladogenesis: Speciation through the splitting of a single species into two noninterbreeding species

Cladogram: An evolutionary tree that encapsulates the relative ancestries of different taxa

Classical polymorphism: A polymorphism assayed by serological or protein electrophoretic means, used before DNA polymorphisms became available; e.g., blood groups

Cline: A gradient of gene frequencies from one region to another

Clovis point: A diagnostic stone tool fashioned by early settlers of the Americas

Coalescence time: The time taken (going backwards in time) for two or more lineages to coalesce (join) into their ancestral lineage

Coalescent theory: A branch of population genetics that achieves considerable computational gains by considering the ancestry of lineages backwards in time

Codominant selection: A selective regime in which both alleles contribute to fitness, such that the fitness of the heterozygote is distinguishable from those of the two homozygotes

Codon: A series of three adjacent bases in messenger RNA (and, by extension, in DNA) that encodes a specific *amino acid* in a protein

Colonization: The settlement by a group of previously uninhabited lands; this term can also be applied when the territory was previously occupied by another population, e.g. European colonization of the Americas

Commensal: A species that is not domesticated but has adopted a niche created by human activities

Common disease/common variant (CD/CV) hypothesis: The proposal, based largely on theoretical considerations, that genetic susceptibility to common diseases is expected to be due to common alleles at susceptibility loci

Comparative phylogeography: Comparing the geographical distribution of lineages among a number of species to identify the evolutionary processes common to those species

Complementary: A strand of DNA that will base pair with another strand

Complex disorder: A disorder including genetic and nongenetic components, in which the mode of inheritance is not *Mendelian*

Conchoidal fracture pattern: A fracture pattern showing smooth curved surfaces like the interior of a shell (conch)

Conservative substitution: A mutation causing a *codon* to be replaced by another codon encoding an *amino acid* whose physical properties are similar to those of the original amino acid. Often has no effect, or a comparatively mild effect

Constitutive: Always present (e.g., of gene expression), not subject to change over time

Control region: The segment of mtDNA containing the origin of replication and two *hypervariable segments*

Convergent evolution: The evolution of similar forms by two lineages that do not share a common ancestor with that form. For example, the evolution of wings by birds and bats

Contig: a set of contiguous clone (e.g., *BAC*) sequences, forming part of the genome sequence

Continental shelf: The submerged land of shallow gradient that lies between the sea shore and deep ocean bed

Coprolite: Fossilized feces; a source for ancient DNA analysis

Correlogram: A graphical display of how autocorrelation values vary depending on the distance between populations

Cosmid: A circular DNA vector in which inserts of typically ~ 40-kb long can be propagated in bacterial cells

Covalent bonds: Strong chemical bonds between atoms in a molecule requiring considerable energy to be broken

CpG dinucleotide: The sequence 5′–CG–3′ within a longer DNA molecule. The site of specfic *DNA methylation*, important in the control of gene expression

CpG island: A region of DNA, often associated with the 5′ region of a gene, in which unmethylated *CpG dinucleotides* are more plentiful than elsewhere in the genome

Cranium: The bones of the skull except the *mandible* (lower jaw)

Creole language: A fully fledged hybrid language formed by the mixing of two parental languages

Crossing-over, or **Crossover**: The exchange of DNA between chromosomes at *meiosis*

Cultural diffusion: see *Acculturation*

Cytoplasm: The material within a cell excluding the *nucleus*

Cytoplasmic segregation: The unequal segregation of different mitochondrial DNA types within a cell into daughter cells at cell division

Cytosine: A *pyrimidine* base found within DNA – pairs with *guanine*

D/L ratio: The proportions of *stereoisomeric* L- and D-forms of an *amino acid* in a sample

Degenerate oligonucleotide-primed PCR, DOP-PCR: A method for *whole-genome amplification* using short primers with random sequences (*degenerate primers*)

Degenerate primer: Short primer with a random sequence

Deleterious allele: An allele that decreases the fitness of an individual carrying it

Deme: see *Subpopulation*

Demic diffusion: An advancing wave of a particular population, driven by demographic growth

Denaturation: The separation of the strands of DNA, often by heating or treatment with alkali; in PCR, the first stage of a cycle, in which the temperature is typically 94°C

Denaturing high performance liquid chromatography, DHPLC: Method for mutation detection in which *heteroduplexes* are detected by their altered retention time on chromatography columns under near-denaturing conditions

Deoxyribonucleic acid, DNA: The informational macromolecule that encodes genetic information

Deoxyribonucleotide: The monomeric molecular component of the polymer DNA; often abbreviated as nucleotide

Deoxyribose: The sugar part of a deoxyribonucleotide, the building block of DNA

Developmental plasticity: The variation in developmental outcome as a result of environmental influences

Diabetes mellitus: A condition in which the amount of glucose in the blood is elevated, causing long-term damage to eyes, kidneys, nerves, heart and major arteries. Some forms of the disease have a genetic component. Often abbreviated to 'diabetes'

Diallelic: Of a polymorphism – possessing two alleles only (see *biallelic*, *binary marker*)

Diaspora: A multi-directional dispersal from a central homeland, often applied to Jewish populations

Diploid: Having two copies of the genome – e.g., a cell, or an organism

Disassortative mating: Using phenotypic disparity to guide mate choice

Disease association study: A study in which the frequencies of alleles in individuals with a disease are compared with those in normal control individuals in the hope of identifying a region of the genome associated with the disease

Disease heritage: The characteristic pattern of genetic diseases found within a population as a result of its evolutionary history

Distance matrix: A table whose elements are distances between the categories (often *haplotypes* or populations) arrayed in rows and columns

Diversifying selection: A selective regime that favors greater diversity than that expected under neutral evolution

Dizygotic, DZ: Of twins, formed from two independent *zygotes*, and therefore no more closely related than any pair of sibs. Also called nonidentical, or fraternal twins. As opposed to *monozygotic*

DNA fingerprint: A pattern of bands produced after Southern blotting and hybridization that is highly individual-specific and can be used in individual identification

DNA polymerase: An enzyme responsible for the synthesis of a DNA strand using the complementary DNA strand as a template

DNA repair: The set of processes that maintain DNA integrity and minimize the number of heritable mutations

Dolichocephalic: 'Long-headed', with a *cephalic index* below 75 (see *brachycephalic*, *mesocephalic*)

Domestication: The selective breeding of a plant or animal species to make it more useful to humans

Dominant mutation: A mutation that shows its complete phenotypic effect when present in only one copy (*heterozygous*)

Double-stranded DNA: DNA in its double-helical form, with two complementary polynucleotide strands attached to each other by *base pairing*

Drift: see *Genetic drift*

Dyad symmetry: Property of a segment of DNA whose sequence reads identically 5′ to 3′ on one strand, and 5′ to 3′ on the opposite strand; e.g., the sequence 5′–CG–3′ is also 5′–CG–3′ on the opposite strand

Dynamic mutation: A mutation, usually at a trinucleotide repeat locus, involving a large expansion of the repeat array; associated with trinucleotide repeat expansion disorders such as myotonic dystrophy and Huntington disease

Ectopic: Of a recombination event, between nonallelic sequences ('in the wrong place')

Effective population size (N_e): A quotient invented by Sewall Wright to compare the genetic drift experienced by different populations; contrasts with census population size

Einkorn wheat: The first wheat to be successfully domesticated – *Triticum monococcum*

Electropherogram: A trace showing the intensity of fluorescence as a function of molecular weight; peaks in the trace at a particular wavelength (color) correspond to a specifically labeled molecule of a particular size. Seen as the output of many gel and capillary-based electrophoresis devices in DNA sequencing, SNP typing and microsatellite analysis

Electrophoresis: Separating macromolecules by using a voltage gradient; usually carried out using a gel or capillary matrix

Emmer wheat: The most important plant domesticate of the Neolithic – *Triticum dicoccum*

Endemic: Of a disease, constantly prevalent in a certain region; of a species, naturally found in a certain restricted region

Endogamy: The practice of marrying within a social group

Endogenous mutagens: Molecules produced within the body that can cause changes in DNA sequence (mutations)

Endogenous retrovirus: Dispersed repetitive genomic element representing the remnant of a retroviral infection in our ancestors; often a substrate for recombination-mediated chromosomal rearrangements

Endonuclease: An enzyme that cleaves within a nucleic acid molecule (e.g., DNA)

Endosymbiotic: Describing an organism living within the cell of another in symbiosis; the state of the progenitor of *mitochondria*

Enhancer: A DNA sequence that regulates gene expression from a distance

Epicanthal fold: The internal skinfold of the eyelid, seen in many Asian populations

Epigenetic: Inherited without involving a change to the DNA sequence, for example a pattern of *DNA methylation* or protein binding

Episodic selection: Selection that is not constant over time

Epistasis: Interactions between genes influencing a complex trait

Epitope: A single antigenic determinant

Eskimo–Aleut: One of the three major families of Native American languages proposed by Joseph Greenberg, includes the languages spoken by the Inuit

Estimator: A quantity calculated from the sample data to estimate an unknown parameter in the population

Ethnonym: The name of a people or ethnic group

Euchromatin: The part of the genome containing transcriptionally active DNA, and which, unlike *heterochromatin*, is in a relatively extended conformation

Eurocentrism: A biased view that concentrates on European populations and contrasts other populations with them. An example is the description of many non-European populations as *lactose intolerant*

Eustatic: Factors that alter the ratio of global water present in ice and liquid form

Exaptation: A trait that arose for one purpose (or by chance) and is later used for another

Exclusion: In forensic genetics, the finding that a crime-scene sample has a different DNA profile to a suspect, or to another crime-scene sample thus demonstrating that they are derived from different individuals; term also used in *paternity testing*

Exon: One of the noncontiguous sections of a gene that together form the region present in the mature transcript; in many genes, exons comprise the coding region, but some genes have noncoding exons. See also *intron*

Exonuclease: An enzyme that degrades a DNA molecule from one of its ends

Extension: The stage of a PCR reaction in which *Taq* polymerase extends the new strands of DNA from the *oligonucleotide primers*

Falsifiability: The possibility of saying with certainty that a hypothesis is not correct

Fecundity: The biological capacity for reproduction

Fertility: The actual numbers of offspring produced

Fitness: The ability of an individual to survive and reproduce relative to the rest of the population

Fixation: The process by which one allele increases in a population until all other alleles go extinct and the locus becomes monomorphic

Fixation indices: Quotients devised by Sewall Wright to analyze the departure of genotype frequencies from *Hardy–Weinberg equilibrium*

Fluorescence *in situ* hybridization (**FISH**): Hybridization of a fluorescently labeled nucleic acid to a target DNA or RNA, usually immobilized on a microscope slide; a key technique in physical mapping of chromosomes

Forward simulations: Successive *in silico* modeling of population and/or molecular processes forward in time to generate multiple replicates of artificial genetic diversity datasets

Founder effect: Reduced genetic diversity in a population founded by a small number of individuals

Frameshift mutation: A mutation in a coding region that causes the three-base reading frame to be shifted by adding or subtracting a number of bases that is not a multiple of three

Frequency-dependent selection: A selective regime under which lower frequency variants are favored, which prevents their elimination and promotes diversity

F_{ST}: A family of measures of population subdivision devised by Sewall Wright, that can also be used as a *genetic distance* between populations

Gain-of-function mutation: A mutation that causes a gene product to possess a new function, often leading to a dominantly inherited phenotype

Gamete: An egg or sperm – these cells are haploid

Gametic phase: The way in which the alleles of two linked loci are associated together in a genotype

G-banding: A staining method that reveals dark- and light-staining bands in chromosomes, used to distinguish chromosomes from each other and recognize individual chromosomes

Gene: Part of a DNA molecule that encodes a protein or functional RNA molecule

Gene–culture coevolution: The interaction of genetic and cultural adaptations in human evolution

Gene conversion: A nonreciprocal exchange of sequence information between one DNA molecule and another

Gene diversity: A widely used measure of genetic diversity devised by Masatoshi Nei

Gene duplication: The generation of an identical copy of a segment of DNA, often located in tandem

Gene flow: The movement of genes resulting from the movement of people from one region to another, followed by successful reproduction and subsequent genetic contribution to the next generation

Gene genealogy: The ancestral relationships (genealogical history) among a number of sampled copies of a DNA segment

Gene therapy: The treatment of disease through the modification of a patient's genetic material

Gene tree: A phylogeny relating the ancestry of sequences at the same locus in different species or individuals, where *bifurcations* relate molecular divergences, and not speciation or population events

General reversible model: A model of sequence evolution in which each base substitution has its own rate of mutation, identical in either direction, and base composition is taken into account

Generation time: The average time between the birth of a parent and the birth of their offspring

Generation time hypothesis: The suggestion that differences in mutation rates between different evolutionary lineages can be accounted for by the different number of germ-line replications per unit of time

Genetic barrier: A geographical or cultural difference associated with elevated allele frequency change as a result of low gene flow

Genetic boundary: A geographical or cultural difference associated with elevated allele frequency change; may result from low gene flow or migration

Genetic code: The three-letter code that allows base sequences within messenger RNA molecules to be translated into specific *amino acids* within a protein

Genetic dating: A class of statistical methods used to estimate the time to the most recent common ancestor of a set of populations or molecules

Genetic differentiation: The process by which the genetic composition of two or more populations isolated from one another diverges over time

Genetic distance: A measure of the evolutionary relatedness of two populations or two molecules **OR**: Distance on a genetic map, defined by the frequency of recombination events, and measured in centiMorgans

Genetic drift: The random fluctuation of allele frequencies in a finite population due to chance variations in the contribution of each individual to the next generation

Genetic enhancement: The use of genetic techniques to engineer the introduction of desirable characteristics into an individual

Genetic isolation: see *Isolated population*

Genetic load: The burden on a population caused by the deaths required to eliminate deleterious mutations

Genetic map: An ordered sequence of genetic markers determined by analyzing recombination events. The distance between markers is determined by the probability of observing a recombination between them, as opposed to a *physical map*

Genome scan: An attempt to locate a gene of interest by linkage analysis carried out using a large number of anonymous polymorphic markers spread throughout the genome

Genomic disorder: A genetic disorder in which the underlying mutation is due to an underlying structural feature of the genome, such as deletion or duplication of a region through aberrant recombination

Genomics: The scientific discipline of mapping, sequencing and analyzing genomes

Genotype: The combination of allelic states of a set of polymorphic markers lying on a pair of chromosomes – the combination of two *haplotypes*

Germ line: The cell lineage culminating in eggs and sperm, the cells responsible for passing genetic information from one generation to the next

Glacial period: Period when the average temperature was colder than at present and parts of the Earth were covered by ice sheets

Glacial refugia: Regions in which populations took refuge after their retreat from the harsh environments of extreme latitude during recent *glacial periods*

Glottochronology: A statistical methods for calculating linguistic divergence times from the proportion of word cognates shared between different languages

Glumes: Leaf-like structures that protect the seeds in an ear of wheat

Goodness-of-fit: A class of statistical methods used to assess competing models on the basis of their fit to empirical data

Gracile: Of body form and skeletal remains, lightly built; as opposed to *Robust*

Grammatical: Pertaining to the phonology, rules of inflection and sentence structure of a language

Great circle distance: The distance between two points on the surface of a sphere (e.g., the world) calculated by using a trigonometric equation to take into account the curvature of the sphere

Group informed consent: *Informed consent* given not only by the participating individual, but also by family or wider group involved

Guanine: A *purine* base found within DNA – pairs with *cytosine*

Handaxe: see *Biface*

Haplogroup: Usually applied to a set of mtDNA or Y-chromosomal *haplotypes* that is defined by relatively slowly mutating markers, and that has more phylogenetic stability than other 'haplotypes'. For the Y chromosome, haplogroups are defined by binary markers such as SNPs, while 'haplotypes' are usually defined by *microsatellites*

Haploid: Having one copy of the genome; can describe a cell (e.g., a gamete), or an organism

Haploinsufficiency: A condition in which a gene product is only available from one allele of a diploid locus, and where this amount of product is not sufficient for normal function. One basis for a dominantly inherited phenotype

Haplosufficiency: A condition in which a gene product is only available from one allele of a diploid locus, but this amount of product is sufficient for normal function. This kind of mutation has a recessive mode of inheritance

Haplotype: The combination of allelic states of a set of polymorphic markers lying on the same DNA molecule, e.g., a chromosome or region of a chromosome

Haplotype block: The apparent haplotypic structure of recombining portions of the genome in which blocks of consecutive coinherited alleles are separated by short boundary regions

Haplotypic background: see *Ancestral haplotype*

Hardy–Weinberg equilibrium: The situation when the observed genotype frequencies within the population equate to the expected genotype frequencies calculated from known allele frequencies using the Hardy– Weinberg principle

Heinrich event: Brief spell of extreme cold during a *glacial period*

Hemizygous: Possessing only one allele as a result of normal chromosome constitution – e.g., males are hemizygous for all X chromosome-specific loci

Heritability: The proportion of variance in a trait that is explained by genetic factors

Heterochromatin: Highly condensed, transcriptionally inert segment of the genome often behaving abnormally in chromosomal banding; the part of the genome other than *euchromatin*

Heteroduplex: A double-stranded DNA molecule in which one strand contains a mutation with respect to the other. There is a base mismatch within such a heteroduplex, which can be detected by various mutation detection methods

Heteroplasmy: Possessing two or more different mitochondrial DNA sequences in the same cell, or individual; as opposed to *homoplasmy*

Heterozygosity: A measure of the diversity of a polymorphic locus; for a diploid locus, the average probability that the alleles carried by an individual are different from each other

Heterozygote advantage: A selective regime in which the heterozygote confers greater fitness than either homozygote

Heterozygous: Carrying two different alleles at a particular diploid locus

Heuristic: Of a model or representative device, aiding understanding or explanation

Hierarchical population structure: See *Population structure*

Histocompatibility: The degree to which tissue from one individual will be tolerated by the immune system of another

Histone: One of a family of small, basic (positively charged), highly conserved proteins associated with DNA in the chromosomes of all cells except sperm (which instead use protamines)

Hitchhiking: The increase in frequency of a neutral allele as a result of positive selection for a linked allele

Holliday junction: A four-stranded DNA structure occurring as an intermediate in homologous recombination

Holocene: The last ~ 11 KY, with an unusually warm and stable climate

Hominid/hominin: A species (extinct or extant) more closely related to humans than to our closest living relatives, chimpanzees and *bonobos*

Homologous: Phenotypic or genotypic characters that share a common ancestor

Homoplasmy: Possessing only one mitochondrial DNA sequence type in a cell, or individual; as opposed to *heteroplasmy*

Homoplasy: The generation of the same state by independent means (convergent evolution)

Homozygous: Carrying the same allele at each copy of a diploid locus

Horticulture: Nonintensive management of individual plants, rather than large populations of plants (agriculture)

Hotspot: see *Recombination hotspot*

Human leukocyte antigen (**HLA**): Different proteins, expressed on the surfaces of all nucleated cells, that can be recognized by the immune system; responsible for *histocompatibility*

Hunter–gatherer: A system of food supply based on the hunting of wild animals and gathering of wild plants

Hybridization: The process of annealing two homologous DNA molecules together

Hydantoin: Oxidized derivative of the base *cytosine* or *thymine*, blocking DNA polymerases – a problem in ancient DNA analysis

Hydrogen bond: Weak inter-atomic bond either within a molecule or between molecules; these are the bonds between bases in a base pair of DNA

Hypervariable region: see *Hypervariable segment*

Hypervariable segment: Part of the mitochondrial DNA molecule showing particularly high DNA sequence variability; two of them lie within the *control region* of mtDNA; also known as *hypervariable regions*

Iatrogenic: Of a disease or medical problem – caused by medical intervention

Iberian peninsula: The region of southwestern Europe that comprises Spain and Portugal

Ice age: see *Glacial period*

Ice sheet: A layer of ice covering a tract of land

Identity by descent: Property of alleles in an individual or in two people which are identical because they were inherited from a common ancestor; as opposed to *identity by state*

Identity by state: Property of alleles in an individual or in two people that are identical because of coincidental mutational processes, and not because they were inherited from a common ancestor (*identity by descent*)

Immunoreactivity: A method for assaying genetic diversity at the protein level by using the discriminating power of the immune system

in silico: Within a computer (in silicon)

in utero: Within the uterus (during gestation)

in vitro: Within laboratory equipment (in glass)

in vivo: Within a living organism (in life)

Inbreeding: Reproduction involving genetically closely related parental types

Inbreeding depression: The reduction in fitness observed in offspring resulting from inbreeding as a result of the elevated frequency of deleterious recessive homozygotes

Incidence: The proportion of people who develop a characteristic such as a disease over a particular period, for example a year or a lifetime

Indel: A mutation or polymorphism involving the insertion or deletion of DNA sequence

Independent assortment: The random segregation of each member of a chromosome pair to opposite poles of the cell during *meiosis*

Inferential methods: Statistical methods for making inferences about the processes that generated the data

Infinite alleles model: A population genetics model that assumes that each mutation generates a new allele not previously found within the population

Informative sites: A subset of *segregating sites* that are phylogenetically informative because neither allele is a *singleton*

Informed consent: Agreement to take part in an investigation that is given freely and based on an understanding of the work, including its risks and benefits

Interfertile: Of two species, capable of mating and producing offspring, though these are not necessarily themselves of normal fertility

Interglacial: Warm geological interval between *glacial periods*; we are currently living in an interglacial

Interpolation: The process of generating a value for a variable at a site intermediate between two sites at which the value for the variable is known empirically

Interstadial: Brief, relatively warm spell during a *glacial period*

Intra-allelic diversity: The diversity that has accrued among chromosomes containing a given allele

Intron: A transcribed but noncoding section of a gene that separates the *exons*

Ionizing radiation: Energetic electromagnetic radiation capable of directly breaking the DNA backbone, or producing reactive ions that modify DNA e.g., gamma- and X-radiation

Island model: A simple model of population structure in which two populations of identical size exchange alleles with equal probabilities

Isoform: One of a number of alternative three-dimensional structures that can be adopted by the same protein

Isolated population: A population that has experienced little gene flow from surrounding populations and is therefore differentiated from them

Isolation by distance: The decline of population similarity with geographical distance as a result of mating distances being less then the range of the species

Isostatic: Factors influencing local sea levels as a result of vertical movements of the earth's crust

Jukes–Cantor model: The simplest model of sequence evolution in which each base substitution has the same rate

Karyotype: The set of chromosomes in a cell or individual, revealed by cytogenetic analysis at certain stages of the cell cycle

Kilobase: A unit of a thousand bases, or (usually) a thousand base pairs

Kinetochore: The DNA–protein structure at the *centromere* of a chromosome to which *spindle fibers* attach

Kin-structured migration: The situation when migrating individuals are not a random sample from the source population but are members of the same family

Knuckle-walking: Walking on all fours, using the knuckles of the hands – like chimpanzees

L1 element: A member of the *LINE* class of interspersed repeated sequences. The full-length element is about 6.1 kb long, but many are truncated. The L1 copy number is about 0.5 million

Lactase persistence: The persistence of intestinal lactase into adulthood, maintaining the ability to digest lactose, the major sugar of milk

Lactose intolerance: The inability to digest lactose, the major sugar of milk, in adulthood; this is the ancestral condition, found in nearly all mammals

Language shift: The displacement of one language by another throughout an entire speech community

Lapita: A cultural complex, denoted in part by a distinctive pottery style, that is associated with the first settlement of *Remote Oceania*

Later Stone Age: The period following the *Middle Stone Age* in Africa, defined by its cultural assemblages such as complex tools; the term *Upper Paleolithic* originated in Europe and is not used in Africa, but the two are broadly contemporary

Levallois technique: A *Middle Paleolithic/Middle Stone Age* method of producing stone tools by first preparing a core and then removing the tool in a nearly finished state as one final flake

Lexical: Relating to items of vocabulary

Lexicostatistics: The statistical study of vocabulary differences between related languages

Likelihood: A statistical framework that considers which hypothesis out of a range of options best accounts for the observed outcome

Likelihood ratio test: A method for comparing the likelihood of different hypotheses

Lineage: A group of taxa sharing a common ancestor to the exclusion of other taxa

Lineage assortment: The process by which polymorphisms within a parental species are randomly distributed into daughter lineages during speciation. This can lead to discrepant gene trees

Lineage effects: Mutation rate differences among different evolutionary lineages

Linear pottery: A distinctive type of pottery associated with the early European Danubian farming culture of the *Neolithic*

Linearity: The property of a summary statistic that describes how closely its relationship with another variable of interest (often time) approximates to a straight line

LINE: Long interspersed nucleotide element – a dispersed long repeat sequence such as a *L1 element*

***Linienbandkeramik*, LBK**: see *Linear pottery*

Linkage analysis: The mapping, often using pedigrees, of a gene or trait on the basis of its tendency to be coinherited with polymorphic markers; see *LOD score*

Linkage disequilibrium (LD): Nonrandom association between alleles in a population due to their tendency to be coinherited because of reduced recombination between them

LOD score: The logarithm (in base 10) of the odds of linkage – the ratio of the likelihood that loci are linked to the likelihood that they are not linked

Long-PCR: A modified PCR protocol allowing long (up to 35 kb) fragments to be amplified

Low-copy repeat, LCR: Repeated sequences present on a chromosome in two or a few copies, between which *nonallelic homologous recombination* can occur. Usually a few kilobases in size

Macroevolution: The evolution of taxonomic groups above the species level

Major histocompatibility complex, MHC: The locus on human chromosome 6 that contains the major determinants of tissue compatibility between individuals

Male-driven evolution: The hypothesis that the majority of evolutionary mutations occur in the male germ-line as a result of the greater number of cell replications than in the female germ-line

Mandible: Lower jaw bone

Mass screen: In forensic genetics, the DNA profiling of a large number of volunteers within a geographical region in which a perpetrator is thought to live

Mastodon: An extinct elephant-like mammal

Match: In forensic genetics, the finding that a crime-scene sample has the same DNA profile as a suspect, or as another crime-scene sample; term also used in *paternity testing*

Mate choice: the situation whereby when one sex is of limited abundance and can exercise choice over their mates leading to sexual selection for attractive traits

Material culture: The physical remains of cultural processes

Matriline: All female-line relatives of a person (excluding the daughters of a man) – they will share a mtDNA haplotype

Matrilocality: A marital residence pattern whereby a husband moves into his wife's village (as opposed to *patrilocality*)

Maximum likelihood: A method for selecting the best hypothesis from a set of alternatives on the basis of which maximizes the likelihood of the outcome

Maximum parsimony: A method for selecting the best evolutionary tree from a set of alternatives on the basis of which contains the fewest evolutionary changes

Megabase: Unit of a million bases, or (usually) a million base pairs

Megafauna: Large-bodied vertebrates, now mostly extinct except in Africa

Megalithic: Built of large stones; including tombs, single standing stones, and alignments of stones. Characteristic of the Neolithic in the west of Europe

Meiosis: The series of events, involving two cell divisions, by which diploid cells produce haploid gametes

Melanesia: The group of Pacific islands that lies from New Guinea in the west to Fiji in the east. Now discredited as a useful unit of anthropological investigation as a result of the different settlement histories of the populations on these islands

Melanin: The most important pigment influencing skin and hair color

Melanocyte: The cell type in which *melanin* is synthesized; lies at the boundary between the dermis and epidermis

Melanosome: A vesicle within the *melanocyte* in which *melanin* is synthesized; transported into keratinocytes in the skin

Melting: The separation of the strands of DNA, often by heating or treatment with alkali; also called *denaturation*

Meltwater pulses: Rapid rises in sea levels that resulted from sudden ice sheet thaws during the warming at the end of the last *Ice Age*

Mendelian: Of a pedigree pattern, showing inheritance consistent with simple *recessive*, *dominant*, or sex-linked behavior

Mesocephalic: Having a *cephalic index* between 75 and 80 − neither long- (*dolichocephalic*) nor broad-headed (*brachycephalic*)

Messenger RNA, mRNA: An intermediate RNA molecule transcribed from a gene that is used as the template for the production of a protein by *translation*

Meta-analysis: The analysis of analyses; the statistical analysis of a large collection of results from individual studies for the purpose of integrating the findings

Metabolic rate hypothesis: The suggestion that different rates of sequence change among evolutionary lineages result from differences in the metabolic rates of those lineages

Metacentric: Of a chromosome − having the *centromere* close to the middle of the chromosome

Metaphase: A stage of cell division in which chromosomes are in a condensed state and aligned at the cell center prior to separation

Metaphase spread: The condensed chromosomes from a single cell at *metaphase* displayed on a slide

Meta-population: A group of populations connected by migration

Methylation: Modification of a cytosine base in DNA by the addition of a methyl group, to yield 5-methylcytosine; methylation of other bases occurs in other organisms

Microevolution: The processes of evolutionary change operating within species, at the population level

Micronesia: The group of small coral atolls in the Pacific that lie to the north of *Melanesia*

Microsatellite: A DNA sequence containing a number (usually ⩽50) of tandemly repeated short (2–6 bp) sequences, such as $(GAT)_n$. Often polymorphic, and also known as a *short tandem repeat (STR)*

Middle Paleolithic: The period between the *Lower Paleolithic* and *Upper Paleolithic* in Europe and Asia, defined by its artifacts such as flake tools. These were produced by both *Neanderthals* and modern humans

Middle Stone Age: The period between the *Acheulian* and *Later Stone Age* in Africa, defined by artifacts such as flake tools, some but not all made by the *Levallois technique*

Migration: The process of movement of a population (or individual) from one inhabited area to another

Migration drift equilibrium: A stable level of population subdivision that is reached when migration acting to homogenize subpopulations balances the genetic drift that differentiates them

Minimum efficient processing segment (MEPS): The shortest segment of DNA sequence identity between two DNA molecules that will allow efficient homologous recombination to occur; probably about 200 bp

Minisatellite: A DNA sequence containing a number (~ 10 to > 1000) of tandemly repeated sequences, each unit typically 10–100 bp in length. Sometimes hypervariable, and useful in DNA fingerprinting

Minisatellite variant repeat-PCR, MVR-PCR: A PCR-based method for assaying the positions of repeat units within minisatellite arrays that have variant DNA sequences

Mismatch: A site within a double-stranded DNA molecule where bases are noncomplementary. Formed *in vivo* in the mutation process, or *in vitro* as a *heteroduplex* in order to detect a sequence difference between DNA molecules

Mismatch distribution: The frequency distribution of pair-wise differences between a set of DNA sequences or *haplotypes*

Mismatch repair: A process that replaces a mispaired nucleotide in a DNA duplex with the correct base

Missense mutation: see *Nonsynonymous mutation*

Mitochondrion: A cellular organelle containing the molecular apparatus for several metabolic pathways, primarily concerned with energy generation; present in many (usually 1000s) of copies per cell, and contains its own circular genome (*mtDNA*). The organelle, and its genome, are maternally inherited

Mitosis: The series of events resulting in cell division; contrasted with *meiosis*

Molecular clock: The hypothesis that sequence evolution occurs at a sufficiently constant rate to allow divergence between two sequences to be accurately related to the time they split from a common ancestor

Molecular distance: A measure of evolutionary change between two DNA sequences or haplotypes

Molecular evolution: The study of changes to DNA through time

Molecular marker: A polymorphic DNA sequence deriving from a single locus, that can be used in linkage analysis or genetic diversity studies

Mongoloid: One of the morphological types of human identified by early anthropologists, typified by populations of East Asia

Monogenic: Of a disease, due to mutations in a single gene

Monophyletic: A group of lineages sharing a common ancestor to the exclusion of all others

Monozygotic, MZ: Of twins − formed from a single *zygote* that has split, and therefore genetically identical. As opposed to *dizygotic*

Monte-Carlo methods: see *Permutation test*

Mousterian: A Middle Paleolithic culture using tools made by the *Levallois technique*; mainly, but not exclusively, associated with *Neanderthals*

Movius Line: A boundary running across Asia, roughly from the Caucasus Mountains to the Bay of Bengal, distinguishing areas to the south and west where *bifaces* were common from those to the north and east where they are rare; first proposed by the anthropologist Movius

mtDNA: The circular genome carried by the *mitochondrion*

Muller's ratchet: The unidirectional accumulation of deleterious mutations by nonrecombining sequences as a result of the periodic elimination by genetic drift of the fittest *haplotype*

Multiallelic marker: A polymorphic marker possessing more than two alleles, e.g., a *microsatellite*

Multidimensional scaling: A type of multivariate analysis which allows multidimensional information to be displayed graphically (usually in two dimensions) with minimum loss of information

Multifactorial disorder: see *Complex disorder*

Multifurcation: see *Polytomy*

Multiple displacement amplification, MDA: A non-PCR-based method for *whole-genome amplification*, employing *degenerate primers* and a special polymerase from a bacterial virus

Multiplex: Of PCR, the simultaneous amplification of several loci using multiple primer pairs in a single reaction

Multivariate analyses: A class of statistical methods that extract information from multidimensional data

Mutagen: A chemical or physical cause of mutation

Mutation: Any change in a DNA sequence (usually with the exception of changes caused by *crossing over*)

Mutation-drift equilibrium: A stable level of genetic diversity reached when the rate at which new variants are introduced by mutation is balanced by their loss due to genetic drift

Mutation pressure: The decline in frequency of a *haplotype* as a result of mutations creating new haplotypes

Mutational bias: A pattern of mutation that results in an unequal accumulation of certain nucleotides

Na-Dene: One of the three groups of Native American languages proposed by Joseph Greenberg

Narrow sense heritability: Only that proportion of genetic variance that can be attributed to additive genetic effects

Natural selection: The differential contribution of individuals to the next generation on the basis of their ability to survive and reproduce

Neanderthal: A group of extinct humans who lived in Europe and West Asia between about 250 and 28 KYA. Their relationship to modern humans is debated. Also sometimes spelled 'Neandertal'

Near Oceania: An area within the Pacific Ocean that encompasses islands first settled over 25 KYA

Nearly neutral theory: A population genetic hypothesis that suggests patterns of genetic diversity are best explained if *nonsynonymous* mutations are slightly deleterious

Negative selection: A selective regime in which new mutations are almost exclusively deleterious and are removed from the population

Neighbor joining: A fast method for phylogenetic reconstruction

Neolithic: The New Stone Age; originally defined by the use of ground or polished stone implements and weapons; other characteristic features of the Neolithic are now taken to include the manufacture of pottery and the appearance of settlements

Neoteny: The retention of juvenile features in adults; a feature selected for in domesticated animals

Nested cladistic analysis: A statistical method that seeks to detect geographic patterns in genetic diversity by analyzing individually each clade of a phylogeny

Network: see *Phylogenetic network*

Neutral: Of a mutation or locus, having no effect on selective fitness

Neutral allele: An allele that does not affect the fitness of the carrier

Neutral theory of molecular evolution: The theory that the majority of mutations do not influence the fitness of their carriers

Neutrality test: A statistical method for exploring whether observed genetic diversity is compatible with neutral evolutionary processes

Node: The position of a *taxon* within a *phylogeny*, at the end of an evolutionary branch

Nomenclature: A naming system used to classify diversity for facilitating communication

Nonallelic homologous recombination, NAHR: Recombination occurring between sequences that are very similar in sequence, but are not allelic (*paralogs*)

Nonconservative mutation: A mutation causing a *codon* to be replaced by another codon encoding an *amino acid* whose physical properties are different from those of the original amino acid. Often has a serious effect on protein function

Nondisjunction: The failure of a chromosome pair to segregate correctly at meiosis: both chromosomes move to one pole of the cell, rather than one to each. Results in *aneuploidy*

Nonrecombining loci: Segments of the genome in which no recombination events have occurred since the most recent common ancestor of all extant copies

Nonrecombining portion of the Y: The part of the Y chromosome that escapes from meiotic recombination; sequences in this region are truly male-specific (sometimes abbreviated to NRY, or NRPY)

Nonsense mutation: A mutation that occurs within a *codon* and changes it into a *stop codon*

Nonsynonymous mutation: A base substitution in a gene that results in an *amino acid* change

Nuclear genome: The autosomes and sex chromosomes, residing in the *nucleus*; excludes mtDNA

Nucleotide: The monomeric molecular component of the polymers DNA or RNA; sometimes used as a shorthand for deoxyribonucleotide, the specific building block of DNA

Nucleotide diversity: A measure of genetic diversity based on the probability of two randomly chosen sequences from within the population having different bases at the same nucleotide position

Nucleotide excision repair: A versatile repair pathway, involved in the removal of a variety of bulky DNA lesions

Nucleus: The large body within the *cytoplasm* of a cell that contains the chromosomes

Numt: An insertion of mtDNA sequence within the nuclear genome (**nu**clear **mt**DNA insertion)

Obsidian: A volcanic rock resembling glass that has often been used for making stone tools

Oldowan: The oldest assemblage of archaeological remains recognized. *Lower Paleolithic*, dating back to 2.5 MYA, and named after the Olduvai Gorge in Tanzania, East Africa

Oligonucleotide primers: Short (typically 18–24 base), single-stranded, chemically synthesized DNA molecules, usually of specific sequence and used in opposing pairs to prime synthesis of a specific DNA target in *polymerase chain reactions* (PCR).

Optimality criteria: A set of rules that allow a single best option to be chosen from a number of possibilities

Oral histories: The collection of stories about the past handed down by word of mouth between generations

Orthologous: Homologous sequences that have diverged since splitting from their common ancestor as a result of a speciation event

Osteomalacia: An adult disease in which bones become softened, due to a deficiency of vitamin D (see *Rickets*)

Outbreeding: Reproduction involving distantly related parents

Outbreeding depression: A reduction in fitness of offspring resulting from the mating of genetically distinct parents

Outgroup: The *taxon* from a group of taxa that is known to diverge earliest

Overdominant selection: A selective regime under which the *heterozygote* is the fittest genotype, see *heterozygote advantage*

Pairwise: A comparison between two entities

Paleoclimatology: Study of climates in prehistory

Paleoecology: Study of prehistorical ecosystems

Paleoindians: Native American peoples prior to ~ 9 KYA

Paleolithic: The 'Old Stone Age', defined by archaeological remains such as stone tools; subdivided according to time and geographical region

Panmixis: Random mating throughout the entire range of a population

Papuan: The group of non-Austronesian languages spoken in and around the island of New Guinea, also used to refer to the inhabitants of Papua New Guinea in the eastern half of New Guinea

Paracentric: On one side of a *centromere*, for example, a paracentric inversion has both breakpoints on one chromosomal arm

Parallelism: see *Recurrent mutation*

Paralog: Highly similar nonallelic sequences resulting from a duplication event

Paralogous sequence variant, PSV: A difference in DNA sequence between *paralogs*

Parameter: A numerical characteristic of a population that is typically unknown and therefore estimated by taking a sample from the population. A quantity used in a model (e.g., mean) that can be calculated from data

Paraphyletic: A grouping that shares a common ancestor to the exclusion of many other lineages but does not include all descendants of that common ancestor

p arm: The short ('petit') arm of a chromosome

Parsimony principle: The principle that the best explanation is that which requires the least number of causal factors

Pastoralism: The herding of animals; often a nomadic practice, as animals are followed from one region to another as conditions dictate

Paternity testing: Determining whether or not a particular man is the father of a child, using DNA analysis

Paternity index: In *paternity testing*, the likelihood ratio of the hypothesis that the alleged father is indeed the true father, to the alternative hypothesis that the father is some unknown, unrelated man; the higher its value, the more likely is true paternity

Pathogenic: Relating to the causation of disease

Patriline: All male-line relatives of a man – they will share his Y chromosome haplotype

Patrilocality: The anthropologically defined practice whereby men tend to remain closer to their birthplace upon marriage than do women; opposite of *matrilocality*

Penetrance: The frequency with which a person carrying a particular genotype will manifest a disease

Pericentric: On both sides of the *centromere*, for example, a pericentric inversion has breakpoints on both chromosomal arms

Permutation test: A statistical test which assesses significance by randomizing the observed data many times, and comparing a test statistic calculated from these randomizations to the value of the test statistic calculated from the observed data

Phase: see *Gametic phase*

Phenocopy: An imitation of a phenotype through nongenetic effects

Phenotype: The observable characteristics of a cell or organism

Phylogenetic network: A graphical representation of evolutionary relationships that includes cycles or reticulations, and thus summarizes a collection of evolutionary trees

Phylogenetics: The study of genetic diversity through the construction of evolutionary trees

Phylogeny: A tree-like structure that represents evolutionary relationships among a set of taxa

Phylogeny reconstruction: The process of deducing the evolutionary tree underlying a set of data

Phylogeography: Analysis of the geographical distributions of different *clades* within a *phylogeny*

Physical anthropology: The study of humans through analysis of their physical, rather than social or cultural, characteristics

Physical map: A map in which the relative positions of markers are defined by the physical distance in base pairs between them, rather than by recombination frequencies (genetic linkage)

Plesiomorphic: A character state that has not undergone change since the ancestral state, as opposed to derived, or *apomorphic*, characters

Pleistocene: The geological epoch that lies between about 2 MYA and 10 KYA

Poisson distribution: A statistical distribution that has the property that the mean equals the variance

Polarity: A difference between one end of a DNA or RNA molecule and another, provided by the asymmetry of the sugar–phosphate backbone

polyA tail: A stretch of adenine ribonucleotides added after transcription to the 3′ end of most messenger RNA molecules

Polymerase chain reaction, PCR: The exponential amplification of a specific DNA sequence using specific *oligonucleotide primers*, a thermal cycling protocol and a thermally stable DNA polymerase

Polymorphism: The existence of two or more variants (of DNA sequences, proteins, chromosomes, phenotypes) at significant frequencies in the population. For DNA – any sequence variant, or sometimes any sequence variant at ≥1% frequency

Polynesia: A geographical area of the Pacific defined by a triangle with apices at *Aotearoa* (New Zealand), Hawaii and *Rapanui* (Easter Island), containing islands inhabited by speakers of Polynesian languages

Polynesian motif: A haplotype comprising four mutations in the HVS I segment of the mitochondrial genome found at highest frequencies in Polynesian populations

Polypeptide: A protein, *amino acids* joined by peptide bonds; sometimes implies a short protein

Polyphyletic: A grouping of evolutionary lineages that derive from many different ancestors and so do not share a common ancestor to the exclusion of any other lineages

Polyploidy: Having more than two copies (*diploidy*) of the genome per cell. Lethal in humans, but well tolerated in plant species

Polytomy: An evolutionary split in which more than two daughter lineages derive from a single ancestor

Population: A group of individuals that may be defined according to some shared characteristic which may be social or physical. Sometimes used in a theoretical sense to mean a group of individuals in which there is random mating

Population differentiation: The process by which allele frequencies in two or more populations diverge over time

Population genetics: The study of genetic diversity in populations and how it changes through time

Population mutation parameter (θ or 'theta'): A fundamental parameter of population genetics that encapsulates the expected level of genetic diversity in a randomly mating constant sized population not subject to selection when an equilibrium is reached between genetic drift and mutation

Population recombination parameter (ρ or 'rho'): A fundamental parameter of population genetics that encapsulates the expected level of *linkage disequilibrium* in a randomly mating constant sized population not subject to selection when an equilibrium is reached between genetic drift and recombination

Population structure: The absence of random mating within a population, often taken to mean that the population can be more accurately represented as being a *meta-population* comprising several *subpopulations*

Population subdivision: see *Population structure*

Population-specific allele, PSA: An allele that exhibits marked differences in frequencies between populations, far higher than the genome average, and is therefore useful for estimating admixture

Positional cloning: The isolation of a gene based on its position in the genome, as determined by *physical* and *genetic mapping* approaches

Positional candidate: A gene selected for possible involvement in a phenotype because of its perceived possible role in the phenotype of interest and because it maps within a previously identified candidate region

Positive selection: A selective regime that favors the fixation of an allele that increases the fitness of its carrier

pre-mRNA: The primary RNA product of transcription, before the splicing out of *introns* and other post-transcriptional modifications

Prevalence: The proportion of people who have a characteristic such as a disease at any one time

Primer: see *Oligonucleotide primer*

Principal Components Analysis: A type of *multivariate analysis* that allows multidimensional information to be displayed graphically with minimum loss of information

Private haplotype: A haplotype only found within a single population

Probe: A specific DNA sequence, typically 200 bp to a few kilobases long, that can be labeled and used to detect specific target sequences in hybridization, e.g., of a *Southern blot*; alternatively, a nucleotide sequence attached to a solid support as part of a DNA chip

Promoter: The regulatory region, located 5′ to a gene, containing sequences necessary for transcription to be initiated

Protein electrophoresis: The separation of protein variants by electrophoresis; used to define some classical polymorphisms

Proto-language: A reconstructed language ancestral to a group of extant languages

Pseudoautosomal: Of a region of the sex-chromosomes – displaying inheritance from both parents. These regions, at the tips of the X and Y chromosomes, are the only segments of the Y that undergo recombination

Pseudogene: A nonfunctional DNA sequence that shows a high degree of similarity to a nonallelic homologous gene

Pulsed-field gel electrophoresis, PFGE: A method for separating large (up to megabase-scale) molecules, by periodically alternating the direction of the electric field during electrophoresis

Punctate: Describing the heterogeneous distribution of recombination events along a chromosome, featuring *hotspots* and intervening 'cold regions'

Purifying selection: see *Negative selection*

Purine: A class of base containing two closed rings, including *adenine* and *guanine*

Pyrimidine: A class of base containing one closed ring, including *thymine* and *cytosine*

q-arm: The long ('queue') arm of a chromosome

Quantitative genetics: A discipline used in the identification of *quantitative trait loci* contributing to complex traits. It aims to subdivide the total variance in a phenotype into its genetic and environmental variance components

Quantitative trait locus: A polymorphic genetic locus identified through the statistical analysis of a continuously distributed trait (such as height). These traits are typically affected by more than one gene, as well as by the environment

Racemization: Structural transition of a molecule from one *stereoisomeric* form to the other

Rachis: The 'backbone' of an ear of seeds, in wheat, for example. Fragile in wild wheats when ripe, but toughened in domesticated wheats to facilitate harvesting

Racism: The mistaken belief that humans are divided into distinct categories, 'races', usually distinguished by a few phenotypic characteristics such as skin color; often associated with discriminatory behavior towards those considered to belong to other 'races'

Radiation hybrid distance: A measure of *genetic distance* based on the co-inheritance of markers in somatic cell hybrids containing subsets of the genome produced by fragmentation with radiation

Raggedness: A statistic to measure the smoothness of a distribution of a discrete variable

Random genetic drift: see *Genetic drift*

Randomly amplified polymorphic DNA, RAPD: Polymorphic markers generated by using short (8–12 bases long) primers to amplify random fragments of DNA

Rapanui: The Polynesian name for Easter Island, meaning 'big Rapa'

Recessive: A mutation that will not manifest in a phenotype unless both alleles at a diploid locus are mutated (*homozygous*)

Recombination: Exchange of DNA between members of a chromosomal pair, usually in *meiosis*

Recombination hotspot: A short (few kb) region of the genome in which recombination is significantly elevated over the genome average

Recombination/drift equilibrium: The balance reached when the rate at which drift removes haplotypes from the population is exactly matched by the rate at which recombination generates new haplotypes

Recurrent mutation: A mutation that independently generates a derived state previously observed within the population

Redundant: Of the genetic code – the property that some amino acids are encoded by more than one *codon*

Refugia: See *Glacial refugia*

Relative rates test: A simple statistical test to examine whether all lineages are evolving at the same rate

Relative risk: The risk (of developing a disorder, for example) for the individual in question compared with the risk for a randomly chosen individual

Remote Oceania: An area within the Pacific Ocean that encompasses islands first settled with the past 4000 years

Replication: The copying of a double-stranded DNA molecule to yield two double-stranded daughter molecules

Replication origin: Part of a DNA molecule in which DNA replication begins

Replication slippage: Errors in DNA replication in which the DNA polymerase 'slips' in a repeated tract, leading to gains or losses of repeat units. Thought to be a major mechanism of *microsatellite* mutation

Reproductive variance: The variation in number of offspring produced by a group of individuals

Resequencing: Taking a known sequence from an existing source, such as a database and determining it in several different individuals. A way to discover sequence polymorphism

Restriction enzyme: An *endonuclease*, usually isolated from a bacterium, that cleaves double-stranded DNA at a specific short sequence, typically 4–8 bp in length

Restriction fragment length polymorphism, RFLP: A polymorphism identified (typically in hybridization analysis after *Southern blotting*) by differences in the lengths of restriction fragments. Can be due to polymorphism in the restriction sites themselves, or variation in the length of a sequence between the sites

Reticulation: A closed loop observed within a phylogenetic network indicating the potential existence of *homoplasy*

Retrotransposon: A mobile DNA element that inserts into genomic location after transcription into RNA from an active genomic copy, then reverse transcription into DNA; examples are *Alu* and *L1* elements

Retrovirus: A virus with an RNA genome, and a *reverse transcriptase* function, allowing the genome to be copied into DNA prior to insertion into the chromosomes of a host cell; e.g., HIV

Reverse transcriptase: An enzyme that makes a DNA copy (a cDNA) of an RNA molecule; used by *retroviruses* prior to host genome integration, and encoded by some *L1 elements*

Reversion: see *Back mutation*

Ribonucleotide: The monomeric molecular component of the polymer RNA

Ribosome: A *cytoplasmic* multi-subunit protein/RNA complex that performs the function of *translation* of *messenger RNA* into protein

Rickets: A childhood disease in which bones become softened, due to a deficiency of vitamin D

Robust: Of body form and skeletal remains, heavily built; as opposed to *Gracile*

Rubicon: A point of no return, named after the river Rubicon in Italy, crossed at a crucial moment by Julius Caesar

Sahul: The land mass formed at lower sea levels by the joining of the islands of Australia, New Guinea and Tasmania

Satellite: A large tandem repeated DNA array spanning hundreds of kilobases to megabases, and composed of repeat units of a wide range of sizes that can display a higher-order structure Some satellites (e.g., at *centromeres*), are important functional components of chromosomes

Seafloor topography: The geographical patterns of variation in sea depth

Secondary product revolution: Applications of domestic animals other than for food; e.g., the use of hair and hides for clothing, and large animals for traction

Sedentism: Of a population – staying in one place; as opposed to nomadism or *pastoralism*

Segmental duplication: see *Paralog*

Segregating sites: The nucleotide sites that are polymorphic within a set of sequences

Segregation: see *Chromosomal segregation*

Selection coefficient: A quotient used to compare the fitness of different genotypes

Selective sweep: The rapid *fixation* of an advantageous allele and other alleles linked to it

Sequence alignment: The juxtaposition of a set of sequences such that nucleotide sites that derive from a common ancestor are aligned in a column

Sequence divergence: The number of evolutionary changes that distinguish two or more homologous sequences

Serological: Pertaining to the study of blood serum

Sex chromosomes: The X and the Y chromosomes, the constitution of which differs between the sexes

Sex-biased admixture: The generation of a hybrid population from two ancestral populations in which there are different contributions from males and females of either population

Sex-specific admixture: The generation of a hybrid population from two ancestral populations in which the contribution from one of the ancestral populations comes entirely from a single sex

Sexual dimorphism: Sex-specific differences in morphology

Sexual selection: A selective regime under which the characteristics that are selected for are those that enhance mate attractiveness or competitiveness

Shattering: Release of individual seeds from an ear of wheat, for example, on ripening

Short tandem repeat, STR: see *Microsatellite*

Silent-site substitution: see *Synonymous substitution*

Silver staining: A sensitive chemical means of detecting DNA or protein in electrophoretic gels

SINE: Short interspersed nucleotide element – a dispersed short repeated sequence of high copy number, such as an *Alu element*

Single nucleotide polymorphism, SNP: A polymorphism due to a base substitution or the insertion or deletion of a single base

Single-locus profiling: The use of single-locus probes, DNA probes specific for one genomic locus, to derive a DNA profile; as opposed to *multi-locus fingerprinting*

Single-strand conformation polymorphism: Method for mutation detection in which changed mobilities of single-stranded mutant molecules are detected on nondenaturing gels

Single-stranded DNA: A DNA molecule that comprises just a single polynucleotide chain

Singleton: An allele or haplotype that occurs only once within the population

Sister chromatids: The two copies of each chromosome, associated at their *centromeres* after DNA replication during *meiosis* or *mitosis*

Site frequency spectrum: The frequency distribution within the population of polymorphic nucleotide sites from a given sequence

Soma: The cells of the body other than the germ-line

Southern blotting: The technique of detecting specific restriction fragments of DNA. Digested and size-separated DNA is transferred from an agarose gel to a filter by blotting in a transfer solution; fragments are detected subsequently by *hybridization* using a labeled *probe*

Spatial autocorrelation: A statistical method that examines how the *autocorrelation* in gene frequencies between two populations depends on the distance between them

Species tree: A phylogeny that relates the evolutionary relationships among a set of species, where *bifurcations* represent speciation events

Sphenoid: A small bone at the base of the *cranium* whose development is thought to have a major role in determining cranial morphology

Spindle fibers: Fibers within the cell that attach to the *centromere* to move apart the replicated chromosomes at cell division

Splicing: The process of removal of intronic sequences from a pre-messenger RNA molecule, and the linking together of exonic sequences to form a transcript containing a single unbroken reading frame for translation

SRY: The testis-determining gene on the Y chromosome – *sex-determining region, Y*

Star (star-like) phylogeny: A special topology of a *phylogeny* in which each extant *taxon* is derived independently from the common ancestor of all taxa

Stepping-stone model: A model of population structure in which gene-flow can only take place between neighboring *subpopulations*

Stepwise mutation model, SMM: A model of *microsatellite* evolution in which the length of the microsatellite varies by single units at a fixed rate independent of repeat length and with the same probability of expansion and contraction

Stereoisomers: Chemically identical, but structurally different versions of the same molecule (e.g., an amino acid) – mirror-images of each other

Stochastic: The result of a random process, such that the outcome cannot be predicted precisely

Stop codon: A series of three adjacent bases in messenger RNA (and, by extension, in DNA) that instructs the protein translation machinery to stop translating – the end of the coding region of a gene (also termination codon)

Subpopulation: A randomly mating population that exchanges migrants with other populations to form a *meta-population*

Sugar-phosphate backbone: The structural component of DNA, excluding the information-bearing bases

Summary statistic: A statistic that reduces complex data to a single value

Sunda: The landmass formed at lower sea levels that encompassed many of the present-day islands of southeast Asia

Supernumerary: Of a chromosome, an additional (often small) chromosome in excess of the normal 46; may have no phenotypic effect

Surface: A three-dimensional representation of how a single variable varies in two dimensions

Synapomorphic: An *apomorphic* character state that is shared among at least two taxa

Synonymous substitution: A base substitution that replaces one *codon* with another that encodes the same *amino acid*

Synteny: The state of a set of genes being on the same chromosome

Synthetic map: A method for displaying geographically information on allele frequencies from several loci

tag SNPs: A minimal set of SNPs that define most of the haplotype diversity of a *haplotype block*

Taxon: An evolutionary unit of investigation

Telomere: A specialized DNA–protein structure at the tips of chromosomes, containing an array of short tandem hexanucleotide repeats. Protects the chromosome from degradation and from fusing with other chromosomes

Template: A DNA molecule forming the substrate for the synthesis of another, either in cellular DNA replication, or in *in vitro* DNA synthesis such as PCR

Teosinte: The wild progenitor of maize

Termination codon: see *Stop codon*

Thrifty genotype hypothesis: A suggestion that limiting food resources in the past have favored alleles that promote efficient bodily storage of energy reserves; in conditions of plentiful nutrition, the genotype predisposes to noninsulin dependent *diabetes mellitus*

Thrifty phenotype hypothesis: A suggestion that limiting maternal food resources cause physical and metabolic adaptations of the fetus that predispose to noninsulin dependent *diabetes mellitus* in later life

Thymine: A *pyrimidine* base found within DNA – pairs with *adenine*

Toponymy: The study of place names

Transcription: The production of an RNA molecule from a gene (DNA)

Transcriptome: The combined variability of all transcripts produced by the genome, including the products of alternative splicing

Transfer RNA, tRNA: An 'adaptor' molecule that bears a specific *amino acid* and recognizes the appropriate *codon* in messenger RNA, thus allowing that amino acid to be incorporated into a growing protein molecule

Transformation: An *in vitro* process whereby cells taken from a primary tissue (e.g., white blood cells) are 'immortalized', so that they continue to divide indefinitely; alternatively, the uptake and incorporation of exogenous DNA by a cell

Transition: A base substitution in which a *pyrimidine* base (C or T) is exchanged for another pyrimidine, or a *purine* base (A or G) is exchanged for another purine

Translation: The production of a protein from a *messenger RNA* molecule within the *ribosome*

Transnational isolate: An isolated population that has maintained its genetic distinctiveness despite being spread over a wide area

Transposition: The movement of a DNA sequence from one genomic location to another

Trans-species polymorphism: A polymorphism that has been maintained over a long period of time such that it is present in two species as a result of it being present in their common ancestor

Transversion: A base substitution in which a *pyrimidine* base (C or T) is exchanged for a *purine* base (A or G), or vice versa

Tree topology: The shape (branching pattern) of an evolutionary tree

Trichotomy: An internal node in a *phylogeny* from which three taxa descend

Trisomy 21: The state of having three copies of chromosome 21, which causes Down syndrome

Trypanosomiasis: A serious disease of cattle, caused by the trypanosome carried by the tsetse fly; also causes sleeping sickness in humans

Type 1 error: A false positive result; in forensic genetics, a false inclusion of a suspect

Type 2 error: A false negative result; in forensic genetics, a false exclusion of a suspect

Type specimen: The specimen used to delineate the defining characteristics of a type, e.g., of a hominid species

Ultrametric tree: A *phylogeny* in which the summed branch lengths of every taxon to their common ancestor is equal

Underdominant selection: A selective regime that favors either *homozygote* over the *heterozygote*. The heterozygote has the lowest fitness of all genotypes

Unique event polymorphism, UEP: A polymorphism representing a likely unique event in human history, such that all individuals sharing the derived allele share it by descent, rather than by state. All *Alu* insertion polymorphisms and many SNPs are examples of UEPs

Upper Paleolithic: The latest part of the Paleolithic in Europe and Asia, defined by its cultural assemblages and roughly equivalent to the Later Stone Age in Africa; mainly associated with modern humans, but at least one Upper Paleolithic culture (*Châtelperonian*) is associated with *Neanderthals*

Uracil: A *pyrimidine* base found within RNA – analogous to *thymine* in DNA, it pairs with *adenine*

Variable number of tandem repeat (VNTR) polymorphism: A polymorphism due to differing numbers of tandemly arranged repeat sequences; ranges from *satellites*, through *minisatellites*, to *microsatellites*. Sometimes used specifically to refer to minisatellites

Variance: A statistic that describes how widely spread are a number of estimates of the same *parameter*, by comparing the difference between each estimate and the mean of all estimates

Variant: A loosely used term for a rare polymorphism, usually without a phenotypic effect; sometimes used as a nonspecific term when the status of a mutation or polymorphism is unclear

Virtual heterozygosity: For a haploid system, the probability that two haplotypes drawn randomly from the population are different from each other; see *heterozygosity*

Wallace line: The line delineated by Alfred Wallace that divides two regions with distinct fauna in Island Southeast Asia

Wallacea: The biogeographically rich islands of southeast Asia that lay between the *Sahul* or *Sunda* landmasses

Wave of advance: see *Demic diffusion*

Whole genome amplification: Method for indiscriminate amplification of sequences from the entire genome

Wild-type: A nonmutant form of an allele, gene, cell or organism

Wright–Fisher population model: A simple idealized population model that forms the basis for much population genetic analysis. The population has the properties of constant size, equal sex ratio, nonoverlapping generations, random mating and each individual has the same probability of contributing to the next generation

X chromosome: One of the sex chromosomes, present in one copy in men, and two in women

Y chromosome: One of the sex chromosomes, present only in men, and male determining

Yeast artificial chromosomes, YAC: A DNA vector in which inserts megabases in length can be propagated in yeast cells

Younger Dryas event: A brief and rapid cooling during the general warming at the end of the *glacial period*; its end marks the beginning of the *Holocene*

Zoonosis: A disease of humans acquired from animals

Zygote: The fertilized egg, from which an individual will develop

Index